D1189397

THE NEW PUBLIC HEALTH

THE NEW PUBLIC HEALTH

AN INTRODUCTION FOR THE 21ST CENTURY

THEODORE H. TULCHINSKY
Ministry of Health, Jerusalem, Israel; and
School of Public Health, Hadassah-Hebrew University,
Jerusalem, Israel

ELENA A. VARAVIKOVA
First Moscow Medical Academy (Sechenov);
Central Research Institute of Public Health
Moscow, Russia

ACADEMIC PRESS
San Diego   New York   Boston   London   Sydney   Tokyo   Toronto

Academic Press
A Harcourt Science and Technology Company
525 B Street, Suite 1900, San Diego, California 92101-4495, U.S.A.
http://www.academicpress.com

Academic Press
Harcourt Place, 32 Jamestown Road, London NW1 7BY, UK
http://www.academicpress.com

Library of Congress Catalog Card Number: 99-66055

International Standard Book Number: 0-12-703350-5

PRINTED IN THE UNITED STATES OF AMERICA
00   01   02   03   04   05   MM   9   8   7   6   5   4   3   2   1

CONTENTS

1

A HISTORY OF PUBLIC HEALTH

2

EXPANDING THE CONCEPT OF PUBLIC HEALTH

3

MEASURING AND EVALUATING THE HEALTH OF A POPULATION

4

COMMUNICABLE DISEASES

5

NONCOMMUNICABLE CONDITIONS

6

FAMILY HEALTH

7

SPECIAL COMMUNITY HEALTH NEEDS

8

NUTRITION AND FOOD SAFETY

9

ENVIRONMENTAL AND OCCUPATIONAL HEALTH

10

ORGANIZATION OF PUBLIC HEALTH SYSTEMS

11

MEASURING COSTS: THE ECONOMICS OF HEALTH

12

PLANNING AND MANAGING HEALTH SYSTEMS

13

NATIONAL HEALTH SYSTEMS

14

HUMAN RESOURCES FOR HEALTH CARE

15

TECHNOLOGY, QUALITY, LAW, AND ETHICS

16

GLOBALIZATION OF HEALTH

ACKNOWLEDGMENTS

Our gratitude goes out to the many persons whose support and encouragement made this book possible. We especially thank Mrs. Joan Tulchinsky whose unstinting support, suggestions, and editing made a vital contribution to the content clarity and clarity of this book. We are grateful to the Soros Foundation (Mr. George Soros and Dr. Srdjan Matic) and the American Joint Distribution Committee (Mr. Steve Schwager and Dr. Martin Cherkasky) for their generous financial assistance for its preparation and printing in Russian. The help and advice of Dr. Richard Laster of the Amutah for Education and Health have been invaluable in the long process of preparation and its administration.

Dr. Joshua Cohen, retired Advisor on Health Policy to the Director General of the World Health Organization, Dr. Dianne Dayan-Coggan, former Chief Epidemiologist, Israel Defence Forces, Dr. Victoria Semenova, Medsoceconinform Public Health Institute, Moscow, and Dr. Milton Roemer, professor emeritus at the school of Public Health, University of California, Los Angeles, reviewed drafts of the text and made enormously valuable suggestions.

Dr. David Carpenter, Dean of the School of Public Health at the State University of New York (SUNY) at Albany, and Dr. Harvey Bernard of the New York State Department of Health provided encouragement and moral support from the earliest stages of preparation of the text, as did Dr. Michael Reich, of the Takemi Program at the Harvard School of Public Health, Mr. Yehoshua Matza, Minister of Health, and Profesor Michael Zilberman, Chief Scientist of the Ministry of Health of Isreal. The Hadassah-Hebrew University School of Public Health encouraged use of the textbook in its draft stages for teaching purposes in the International Master of Public Health course.

Valuable reviews, material, comments, and suggestions in the preparation of this book were provided by colleagues of the Israel Ministry of Health and of the School of Public Health at Hadassah-Hebrew University of Jerusalem, including Drs. Joe Abramson, Elliot Berry, Michael Davies, Gary Ginsberg, Charles Greenblatt, Hava Palti, Eric Peritz, Helen Pridan, Yehuda Neumark, and Elihu Richter. Students of the 1993, 1994, and 1995 classes of the International Master of Public Health program at the Hebrew University-Hadassah School of Public Health helped as commentators on the text as part of a seminar course. They reviewed the proposed contents and made suggestions for its further development from the student's point of view. Particular contributions were made by Dr. Emanuel Oyemakinde, Dr. Zhang He, Dr. Eduardo Villamour, and Mr. David Eshkol. Editorial assistants included Samar Abdelnour, Cilla Acker, Jeremy Alberga, Andra Bloom, Marla Clayman, Catherine Cogdell, Tina Eshaghpour, Shoshana Kahane, and Adrienne Williams.

Faculty members at the School of Public Health at the State University of New York (SUNY), Albany, including Drs. Susan Standfast, George DiFerdinando, and Guthrie Birkhead, Elizabeth Marshall, and Mr. David Momrow, reviewed portions of the draft and made helpful comments and suggestions. Colleagues at the School of Public Health, University of California, Los Angeles (UCLA), generously provided support, advice, content material, and editing of chapters in their areas of competence; we are especially grateful to Drs. Ruth and Milton Roemer and Dr. Lester Breslow. Other faculty who provided valuable reviews, comments, and suggestions included Drs. Ronald Andersen, John Froines, Lillian Gelberg, Gerald Kominski, Jeffrey Luck, Charlotte Neumann, Alfred Neumann, Stuart Schweitzer, and Barbara Vischer.

We also wish to thank the staff at Academic Press, especially Tari Paschall, Destiny Irons, Mark Sherry, and Paul Gottehrer for the hard work they put into this project. We could not have developed this book without the encouragement, support, and constructive input of our families, friends, students, and colleagues. We are grateful for their support and contributions to the international flavor of the book. The common goal is to improve public health knowledge and practice. The final responsibility is, of course, with the authors.

T. H. Tulchinsky
E. A. Varavikova

FOREWORD

The term "public health," both as a noun and as an adjective, has several meanings. They range from narrow to broad. The narrow pole defines certain organized preventive activities—especially with respect to environmental sanitation and the control of communicable diseases—intended to protect the health of populations.

The broad pole of public health defines a very wide scope of organized activities, concerned not only with the provision of all types of health services, preventive and therapeutic, but also with the many other components relevant to the operation of a national health system. These involve questions on health behavior and the environment as well as the production of resources (personnel and facilities), the organization of programs, the development of economic support, and the many strategies required to ensure equity and quality in the distribution of health services.

This book applies the broad definition of public health, although its emphasis is clearly on the side of health promotion. Accordingly, concern for hospitals is only tangential, with proper recognition of the large impact of hospital budgets on health system financing. The concern for the chronic, noncommunicable disorders—their measurement and prevention—is as diligent as that for communicable diseases, including the results of injury, trauma, and violence.

A full chapter is devoted to family health, including not only the customary maternal and child health services, but also the health of the elderly. Another chapter addresses issues surrounding mental health, mental retardation, and other special needs groups. Environmental health issues, which are seldom appreciated by medical professionals, are given full and proper attention.

Most important is the comprehensive examination of the many issues of health economics—the forces contributing to escalating costs and the strategies for controlling costs. These require elaboration of the planning and management of national health systems, along with background on the general theories of planning and management in modern society.

Analyses are offered of several classic models of health systems, as illustrated in Germany (where social security was born), Great Britain (and its National Health Service), Canada (where provinces took the initiative), Scandinavian countries, the former Soviet Union, and the freewheeling United States. An especially significant section is devoted to the recent changes in post-Communist Russia, where health status has declined markedly since the collapse of the former Soviet Union and where major reforms are now in progress.

All countries are engaged to some extent in reforming their national health systems. They recognize that organized community preventive services must be strengthened, along with personal health services (both preventive and therapeutic). Such advances contribute to overall knowledge and economic development.

In the United States, an increasing proportion of the population is being served by specific managed care organizations, either through health centers and hospitals such as in health maintenance organizations (HMOs), paid on a capitation basis, or in organized coverage through providers in independent practice. Such arrangements give incentives for maximizing primary health care and making prudent use of hospitalization and high-technology procedures.

Medical practitioners and medical students all too often interpret "public health" according to the narrow definition. Far away from this is clinical medicine focused on the individual patient. The overall national health system—of which both clinical patient care and organized community prevention are parts—tends to be overlooked entirely. It is the broad overview of national health systems that this book provides.

By applying this broad perspective, this book should help its readers to appreciate the place in the system that they personally hold. It should correct or counteract the myopia from which any dedicated specialist may suffer. By understanding, in proper perspective, his/her role in the overall health system, he/she should be able to play that role more effectively.

The concept of the "new public health" is an important contribution to the teaching of public health. It links the classic public health issues of environmental sanitation, hygiene, epidemiology, and other subjects developed over the past century to the newer issues of universal health care, economics, and management of health systems in a 21st century approach. It addresses the intellectual preparation of students to enable them to face the world of managed care and its many ramifications, as well as incorporating the targets of health promotion and community responsibility, evolving from the Alma-Ata strategy of primary health care as the path to attaining health for all.

The undergraduate medical student may well acquire, through education, a properly broad perspective, only to lose it with day-to-day absorption in private practice. The physician working in an organized framework is more likely to retain a broad perspective. It is reassuring, therefore, that the trends in health systems are increasingly toward organized service arrangements.

One cannot expect every medical practitioner to perform public health functions. There are graduate schools for training specialists in public health administration and health services management. Furthermore, not all health administrators need to be physicians. Many functions can be performed effectively and efficiently by specialists in management. Such health service managers, however, require adequate training in the many special characteristics and constraints of health systems.

The health of a population is not an incidental trait. It is central to the economic productivity of that population, not to mention the sense of personal well-being of its families and individuals. Application of resources to the health system, therefore, must be recognized not as expenditures or losses, but rather as investments for achievements of future benefits.

If the contents of this lucidly written book are absorbed by medical students and medical practitioners in a particular country, it should help to ensure the optimal application of that country's resources for the protection and advancement of human health.

> Milton I. Roemer*
> Emeritus Professor of Health Services,
> School of Public Health,
> University of California at Los Angeles
> (UCLA)

*Milton I. Roemer is an Emeritus Professor at the School of Public Health at the University of California, Los Angeles since 1962, previously having taught at Cornell University and Yale Medical School. He has served as a consultant to governments and international organizations on all continents. He is the author of 32 books, including a standard textbook of comparative health systems of the world, as well as over 400 articles on social aspects of medicine. Professor Roemer is widely known internationally and considered the leading scholar of this generation on the comparative study of national health systems throughout the world.

Introduction

The central theme for this book evolved from many years of teaching the principles of health organization to master of public health students from Africa, Latin America, the Caribbean, Asia, the United States, eastern Europe, and Russia. It also emerges from cumulative experience of the practice of public health in a variety of settings, including Canada, Israel, the West Bank and Gaza, the United States, Russia, and professional visits to Colombia, Azerbaijan, and Portugal.

While working together on a review of the health situation in Russia between 1992 and 1995 for an international organization, we saw a need for a new textbook of public health to bring current thinking and an international orientation in the broad field to new students and veteran practitioners in that country. The first edition of this book was published in the Russian language in November 1999 with distribution to medical and nursing schools, libraries, and a new School of Public Health in Moscow. It will hopefully help in the development of a sorely needed new idea of public health during the critical transition of the post-Soviet period, with a wide context of health systems elsewhere, including the United States and other countries.

The English language edition is based on the fact that health systems everywhere are undergoing reform towards a population-based approach as well as individual patient care. Both editions are intended to reach people entering the field of public health as undergraduate students or entry-level graduate students in their respective countries with an overview of the broad scope of the field. We hope this approach will also serve the needs of physicians, nurses, managers and policy makers, and many others in the health field to define a new approach, linking the fields of public health and clinical services.

The term "New" in no way deprecates the known and trusted elements of the

"Old" in public health. On the contrary, the New is a rediscovery and stands on the shoulders of the Old. The ancient Greek gods Aesculapius and Hygeia respectively represented medical therapy and prevention in health. The Mosaic law in health is based on two cardinal principles of Pikuah Nefesh, or sanctity of life, and Tikun Olam, literally repairing the world, in the sense of individual and collective responsibility to correct the faults in human society. These, together with the power of subsequent religious and social philosophies of health as an individual and community right, and the creativity of science, provide the basis for the practice and ethical underpinnings of the New Public Health.

Traditionally, public health has been defined as health of populations and communities. However, the New Public Health addresses the health of the individual, as any medical care provider, both directly with individuals and indirectly through communities and populations. The indirect approach is to reduce the risk factors in the environment, whether physical or social, such as in reducing exposure to contaminated water working or improving educational levels, while the direct approach works with the individual patient or client as does a doctor treating the patient or a program for vaccinating children.

The major components of public health, each separately and together, have records of great achievements and failures, but they are part of modern civilization and the desires of all other societies to emulate. While no one is exempt from death ultimately, no mother or father anywhere wants to lose a child, or a parent, especially from a preventable disease or condition. Yet, despite the experience that a combination of science and political commitment through well-planned interventions can greatly reduce the risk of that happening, we have failed to apply that knowledge as widely as we have implemented treatment services. The potential effectiveness of prevention and treatment services is diminished when organized and financed separately, sometimes because of conflicting economic incentives and ethical values.

The New Public Health is a synthesis of classical public health as experienced over the past several centuries, interacting with the biomedical, clinical and social sciences, economics and technology assessment, management, and experience of health systems as they have evolved and as they continue to develop. Improving the health of the individual requires both the direct and indirect approaches to illness and risk factors for disease. Many issues in this broader New Public Health deal with both the individual and the community, including prevention of waterborne disease, assurance of access to medical care through health insurance, organization of home and chronic care, managed care and district health systems, prevention of work and environmentally related disease, birth defects or genetic disorders, such as thalassemia, or development of new health professions, such as nurse practitioners and community health workers, and many others.

The New Public Health is not so much a concept as it is a philosophy which endeavors to broaden the older understanding of public health so that, for example, it includes the health of the individual in addition to the health of populations, and seeks to address such contemporary health issues as are concerned with equitable access to health services, the environment, political governance and social and economic development. It seeks to put health in the development framework to ensure that health is protected in public policy. Above all, the New Public Health is concerned with action. It is concerned with finding a blueprint to address many of the burning issues of our time, but also with identifying implementable strategies in the endeavor to solve these problems.*

The New Public Health includes all possible activities known to be useful and effective in promoting health and in the prevention, treatment, and rehabilitation of diseases for the individual, the community, and the population as a whole. It provides standards relevant to any country whether developed or developing, but application of the specifics depend on the particular health problems and economic status of each country, or region within a country.

The New Public Health incorporates a wide range of interventions in the physical and social environment, health behavior, and biomedical methods along with health care organization and financing. It links traditional hallmarks of public health, such as sanitation, communicable disease control, maternal and child health and epidemiology, with clinical services, health systems management, and health promotion. It recognizes that the health of the individual and the community are directly and indirectly affected by social and economic factors. An understanding of these concepts is essential to the design of effective health care interventions to prevent the occurrence of diseases or their complications.

The New Public Health takes into account the realities of resource allocation, economic factors, and priorities in health policy. It recognizes that resources for health care are limited even in the wealthiest societies, so that choices must be made as to the balance of programs and services provided, often made under the imperative of cost constraint and substitution of one type of service for another. Above all, it stresses that both society and the individual have rights and responsibilities in promoting and maintaining health through direct services and through healthy environmental and community health promotion.

Social advocacy is part of this challenging and rewarding field. Yet, public health is also the art of the possible. We cannot solve all problems of poverty and injustice, but we can improve survival and quality of life, step by step, one acre at a time, to achieve wondrous miracles. This requires the defining of measurable targets of improved health of the individual and the community.

The New Public Health addresses both the social and physical environment as well as the personal services that address individual health needs. It brings together those elements of public health that are community-oriented with personal care that is individual-oriented. One can no longer be separated from the other if we are to address the health needs of individuals and society in the 21st century.

---

*Ncayiyana D., Goldstein, G., Goon, E., and Yach, D. *In* "New Public Health and the WHO's Ninth General Program of Work: A Discussion Paper." Geneva, World Health Organization, 1995.

1

A History of Public Health

INTRODUCTION

History provides a background for the development of understanding and coping with health problems of communities. We can see through the eyes of history how societies conceptualized and dealt with disease. All societies had to face the realities of disease and death, and develop concepts and methods to manage them. These coping mechanisms form part of a world view associated with a set of religious or scientific beliefs, which in turn help to determine the curative and preventive approaches to health.

The history of public health is the story of the search for effective means of preventing disease in the population. Epidemic and endemic infectious disease stimulated thought and innovation in disease prevention on a pragmatic basis, often before the scientific basis of causation was established. The prevention of disease in the population revolves around defining diseases, measuring their occurrence, and seeking effective interventions.

Public health evolved with trial and error, and with expanding scientific medical knowledge, often stimulated by war and natural disasters. The need for organized health services grew as part of the development of community life, and, in particular, urban society. Religious and societal health beliefs influenced approaches to explaining and attempting to control communicable disease by sanitation, town planning, and provision of medical care. Where religious and social systems repressed scientific investigation and spread of knowledge, they were capable of inhibiting development of public health.

Modern society faces the ancient scourges of cholera and plague as well as the more prominent killers, cardiovascular diseases, trauma, and cancer. The advent of AIDS and new or newly resistant microorganisms affecting human society force us to seek new ways of preventing their potentially very serious consequences. The evolution of public health continues; pathogens change, as do the environment and the host. Facing the challenges ahead is aided by understanding of the past.

PREHISTORIC SOCIETIES

Earth is considered to be some 4.5 billion years old, with the earliest stone tools dated from 2.5 million years BC representing the presence of antecedents of man. *Homo erectus* lived from 1.5 million to 500,000 years ago and *Homo sapiens Neanderthalensis* about 110,000 BC. The Paleolithic Age was the food gathering stage of man's development, with people living in bands which survived by hunting and gathering food, with evidence of use of fire going back to some 230,000 years ago, and increasing sophistication of stone tools, jewelry, cave paintings, and religious symbols. Modern man evolved from *Homo sapiens sapiens,* probably originating in Africa and the Middle East about 90,000 years ago, and appeared in Europe during the Ice Age period 40,000–35,000 BC. During this period man spread over all the major land masses, following the retreating glaciers of the last Ice Age 11,000–8000 BC.

A Mesolithic Age or transitional phase of evolution from hunter-gatherer societies into the Neolithic Age of food-raising societies occurred at different periods in various parts of the world, first in the Middle East from 9000 to 8000 BC onward, reaching Europe about 3000 BC. The change from hunting, fishing, and gathering modes of survival to agriculture, was first by domestication of animals and then growing of wheat, barley, corn, root crops, and vegetables. Associated skills of storing and cooking of food, pottery, basket weaving, ovens, smelting, trade, and other skills led to improved survival techniques and population growth gradually spread to all parts of the world.

At each stage of human biological, technological, and social evolution man coexisted with the diseases associated with the environment and living patterns, finding herbal and mystical treatments for the diseases. Subject to famine and plagues, man called on the supernatural and magic to appease these forces to prevent calamities. Shamans or witch doctors acted to remove the harm by magical or religious practices along with herbal treatment acquired through trial and error. Adaptation to environmental hazards became essential with population growth and new environmental surroundings as well as the effects of human communal habitation.

Nutrition and exposure to communicable disease changed as mankind evolved. Social organization included tools and skills for hunting, clothing, shelter, fire for warmth and cooking of food for use and storage, burial of the dead, and removal of waste products from living areas. Adaptation of human society to the environment has been, and remains a central issue in health to the present time. This is a recurrent theme in the development of public health, facing daunting new challenges of adaptation and balance with the environment.

THE ANCIENT WORLD

Development of agriculture served growing populations unable to exist solely from hunting and stimulated the organization of more complex societies able to share in production and in irrigation systems. Trade, commerce, and government were associated with development of urban societies. Growth of population and communal living led to improved standards of living but also created new health hazards including the spread of diseases. These changes required community action to prevent disease and promote survival.

Eastern societies were the birthplace of world civilization. Empirical and religious traditions were mixed. Superstition and shamanism coexisted with practical knowledge of herbal medicines, midwifery, management of wounds or broken bones, and trepanation to remove "evil spirits" that resulted from blood clots inside the skull. All were part of communal life with variations in historical and cultural development. The invention of writing led to medical documentation. Requirements of medical conduct were spelled out as part of the general legal Code of Hammurabi in Mesopotamia (circa 1700 BC). This included regulation of physician fees and punishment for failure and set a legal base for the secular practice of medicine. Many of the main traditions of medicine were those based on empirical knowledge from magic or that derived from religion. Often medical practice was based on belief in the supernatural, and healers were believed to have a religious calling. Training of medical practitioners and regulating their practice and ethical standards evolved in a number of ancient societies.

Some ancient societies equated cleanliness with godliness and associated hygiene with religious beliefs and practices. Chinese, Egyptian, Hebrew, Indian, and Incan societies all provided sanitary amenities as part of the community religious belief system and took measures to provide water, sewage, and drainage systems, allowing for successful urban settlement. Personal hygiene was part of religious practice. Technical achievements in providing community hygiene slowly evolved as part of urban society.

Chinese practice in the twenty-first to eleventh centuries BC included digging of wells for drinking water; from the eleventh to the seventh centuries BC this included use of protective measures for drinking water and destruction of rats and rabid animals. In the second century BC, Chinese communities were using sewers and latrines. The basic concept of health was that of countervailing forces between the principles of yin (female) and yang (male), with emphasis on a balanced lifestyle. Medical care emphasized diet, herbal medicine, hygiene, massage, and acupuncture.

Ancient cities in India were planned with building codes, street paving, and covered sewer drains built of bricks and mortar. Indian medicine originated in herbal medicine and was associated with the mythical gods. Between 800 and 200 BC, Aurvedic medicine developed and with it, medical schools and public hospitals. Between 800 BC and AD 400, major texts of medicine and surgery were written. Primarily focused in the Indus Valley, the golden age of ancient Indian medicine began in approximately 800 BC. Personal hygiene, sanitation, and water supply engineering were emphasized in the laws of Manu. Pioneering physicians, supported by Buddhist kings, developed the use of drugs and surgery, and established schools of medicine and public hospitals as part of state medicine. Indian medicine played a leading role throughout Asia, as did Greek medicine in Europe and the Arab countries. With the Mogul invasion of AD 600, state support declined, and with it, Indian medicine.

Ancient Egyptian intensive agriculture and irrigation practices were associated with widespread parasitic disease. The cities had stone masonry gutters for drainage, and personal hygiene was highly emphasized. Egyptian medicine developed surgical skills and organization of medical care, including specialization, and training that greatly influenced the development of Greek medicine. The Eberus Papyrus, written 3400 years ago, gives an extensive description of Egyptian medical science, including isolation of infected surgical patients.

The Hebrew Mosaic Law of the five Books of Moses stressed prevention of disease through regulation of personal and community hygiene, reproductive and maternal health, isolation of lepers and other "unclean conditions," and family and personal sexual conduct as part of religious practice. It also laid a basis for medical and public health jurisprudence. Personal and community responsibility for health included a mandatory day of rest, limits on slavery and guarantees of the rights of slaves and workers, protection of water supplies, sanitation of communities and camps, waste disposal, and food protection, all codified in detailed religious obligations. Food regulation prevented use of diseased or unclean animals, and prescribed methods of slaughter improved the possibility of preservation of the meat. While there was an element of viewing illness as a punishment for sin, there was also an ethical and social stress on the value of human life with an obligation to seek and provide care. The concepts of sanctity of human life (*Pikuah Nefesh*) and improving the quality of life on Earth (*Tikun Olam*) were given overriding religious and social role in community life. In this tradition, the saving of a single human life was considered "as if one saved the whole world," with an ethical imperative to achieve a better earthly life for all. The Mosaic Law, which forms the basis for Judaism, Christianity, and Islam, codified health behaviors for the individual and for society, all of which have continued into the modern era as basic concepts in environmental and social hygiene.

In Cretan and Minoan societies, climate and environment were recognized as playing a role in disease causation. Malaria was related to swampy and lowland areas, and prevention involved planning the location of settlements. Ancient Greece placed high emphasis on healthful living habits in terms of personal hygiene, nutrition, physical fitness, and community sanitation. Hippocrates articulated the clinical methods of observation and documentation and a code of ethics of medical practice. He articulated the relationship between disease patterns and the natural environment (Air, Water, and Places) which dominated epidemiologic thinking until the nineteenth century AD. Preservation of health was seen as a balance of forces:

exercise and rest; nutrition and excretion; recognizing the importance of age and sex variables in health needs. Disease was seen as having natural causation, and medical care was valued, with the city-state providing free medical services for the poor and for slaves. City officials were appointed to look after public drains and water supply, providing organized sanitary and public health services. Hippocrates gave medicine both a scientific and ethical spirit lasting to the present time.

Ancient Rome adopted much of the Greek philosophy and experience concerning health matters, with high levels of achievement and new innovations in the development of public health. The Romans used skillful engineering of water supply and sewage and drainage systems, public baths and latrines, town planning, sanitation of military encampments, and medical care. Roman law also regulated businesses and medical practice. The influence of the Roman Empire resulted in the transfer of these ideas throughout much of Europe and the Middle East. Rome itself had access to clean water via 10 aqueducts supplying ample water for the citizens, as well as public drains. Marshlands were drained to reduce the malarial threat. Public baths were built to serve the poor, and fountains were built in private homes for the wealthy. Streets were paved, and organized garbage disposal served the cities.

Roman military medicine included well-designed sanitation systems, food supplies, and surgical services. Roman medicine, based on superstition and religious rites, with slaves as physicians, developed with Greek physicians who brought their skills and knowledge to Rome after the destruction of Corinth in 146 BC. Training as apprentices, Roman physicians achieved a highly respected role in society. Hospitals and municipal doctors were employed by Roman cities to provide free care to the poor and the slaves, but physicians also engaged in private practice, mostly on retainers to families. Occupational health was described with measures to reduce known risks such as lead exposure, particularly in mining. Weights and measures were standardized and supervised. Rome made important contributions to the public health tradition of sanitation, urban planning, and organized medical care. Galen, Rome's leading physician, perpetuated the fame of Hippocrates through his medical writings, basing medical assessment on the four humors of man (sanguine, phlegmatic, choleric, and melancholic). These ideas dominated European medical thought for nearly 1500 years until the advent of modern science.

THE EARLY MEDIEVAL PERIOD
(FIFTH TO TENTH CENTURIES AD)

The Roman Empire disappeared as an organized entity following the sacking of Rome in the fifth century AD. The eastern empire survived in Constantinople, with a highly centralized government. Later conquered by the Moslems, it provided continuity for Greek and Roman teachings in health. The western empire integrated Christian and Pagan cultures, looking at disease as punishment for sin. Possession by the devil and witchcraft were accepted as causes of disease. Prayer, penitence, and exorcising witches were accepted means of dealing with health problems. The ensuing period of history was dominated in health, as in all other spheres of human life, by the Christianity institutionalized in the Church. The secular political structure was dominated by feudalism and serfdom, associated with a strong military landowning class in Europe.

Church interpretation of disease was related to original or acquired sin. Man's destiny was to suffer on Earth and hope for a better life in heaven. The appropriate intervention in this philosophy was to provide comfort and care through the charity of church institutions. The idea of prevention was seen as interfering with the will of God. Monasteries with well-developed sanitary facilities were located on major travel routes and provided hospices for travelers. The monasteries were the sole centers of learning and for medical care. They emphasized the tradition of care of the sick and the poor as a charitable duty of the righteous and initiated hospitals. These institutions provided care and support for the poor, as well as efforts to cope with epidemic and endemic disease.

Most physicians were monks guided by Church doctrine and ethics. Medical scholarship was based primarily on the teachings of Galen. Women practicing herbal medicine were branded as witches. Education and knowledge were under clerical dominance. Scholasticism, or the study of what was already written, stultified development of descriptive or experimental science. The largely rural population of the European medieval world lived with poor nutrition, education, housing, and sanitary and hygienic conditions. Endemic and epidemic disease resulted in high infant, child, and adult mortality. Commonly, 75% of newborns died before the age of five. Maternal mortality was high. Leprosy, malaria, measles, and smallpox were established endemic diseases with many other less well-documented infectious diseases.

Between the seventh and tenth centuries, outside the area of Church domination, Moslem medicine flourished under Mohammedan rule primarily in Persia and later Baghdad and Cairo; Rhazes and Ibn Sinna (Avicenna) translated and adapted ancient Greek and Mosaic teachings, adding clinical skills developed in medical academies and hospitals. Piped water supplies were documented in Cairo in the ninth century. Great medical academies were established, including one in conquered Spain at Cordova. The Cordova medical academy was a principal center for medical knowledge and scholarship prior to the expulsion of the Moslems from Spain. The academy helped stimulate European medical thinking and the beginnings of western medical science in anatomy, physiology, and descriptive clinical medicine.

THE LATE MEDIEVAL PERIOD
(ELEVENTH TO FIFTEENTH CENTURIES)

In the later feudal period, ancient Hebraic and Greco-Roman concepts of health were preserved and flourished in the Moslem Empire. The twelfth century Jewish philosopher–physician Moses Maimonides, trained in Cordova and expelled to Cairo, helped synthesize Roman, Greek, and Arabic medicine with Mosaic concepts of isolation of infectious patients and sanitation.

Monastery hospitals were established between the eighth and twelfth centuries to provide charity and care to ease the suffering of the sick and dying. Monastery hospitals were described in the eleventh century in Russia. Monasteries provided centers of literacy, medical care, and the ethic of caring for the sick patient as an act of charity. The monastery hospitals were gradually supplanted by municipal, voluntary, and guild hospitals developed in the twelfth to sixteenth centuries. By the fifteenth century, Britain had 750 hospitals. Medical care insurance was provided by guilds to its members and their families. Hospitals employed doctors, and the wealthy had access to private doctors.

In the early middle ages, most physicians in Europe were monks, and the medical literature was compiled from ancient sources. In 1131 and 1215, Papal rulings increasingly restricted clerics from doing medical work, thus promoting secular medical practice. In 1224, the Emperor Frederick II of Sicily published decrees regulating medical practice, establishing licensing requirements: medical training (3 years of philosophy, 5 years of medicine), 1 year of supervised practice, then examination followed by licensure. Similar ordinances were published in Spain in 1238 and in Germany in 1347.

The Crusades (AD 1096–1270) exposed Europe to Arabic medical concepts, as well as leprosy. Hospitallers developed hospitals in Rhodes, Malta, and London to serve returning pilgrims and crusaders. The Moslem world had hospitals, such as Al Mansour in Cairo, available to all as a service provided by the government. Growing contact between the Crusaders and the Muslims through war, conquest, cohabitation, and trade introduced Arabic culture and diseases, and revised ancient knowledge of medicine and hygiene.

Leprosy became a widespread disease in Europe, particularly among the poor, during the early middle ages, but the problem was severely accentuated during and following the Crusades, reaching a peak during the thirteenth to fourteenth centuries. Isolation in leprosaria was common. In France alone there were 2000 leprosaria in the fourteenth century. The disease gradually disappeared during, and possibly as a result of, the bubonic plague in the mid fourteenth century.

As rural serfdom and feudalism were declining in western Europe, cities developed with crowded and unsanitary conditions. Towns and cities developed in Europe with royal charters for self-government, primarily located at the sites of former Roman settlements and at river crossings related to trade routes. The Church provided stability in society, but repressed new ideas and imposed its authority particularly via the Inquisition. Established by Pope Gregory in 1231, the Inquisition was renewed and intensified, especially in Spain in 1478 by Pope Sixtus, to root out heretics, Jews, and anyone seen as a challenge to the accepted Papal dogmas.

Universities established under royal charters, in Paris, Bologna, Padua, Naples, Oxford, Cambridge, and others, set the base for scholarship outside the realm of the Church. In the twelfth and thirteenth centuries there was a burst of creativity in Europe, with inventions including the compass, the mechanical clock, the waterwheel, the windmill, and the loom. Physical and intellectual exploration opened up with the travels of Marco Polo and the writings of Thomas Aquinas, Roger Bacon, and Dante. Trade, commerce, and travel flourished.

Medical schools were established in medieval universities, in Salerno, Italy, in the tenth century and in universities throughout Europe in the twelfth to fifteenth centuries; in Paris in 1110, Bologna 1158, Oxford 1167, Montpellier 1181, Cambridge 1209, Padua 1222, Toulouse 1233, Seville 1254, Prague 1348, Cracow 1364, Vienna 1365, Heidelberg 1386, Glasgow 1451, Basel 1460, and Copenhagen 1478. Physicians, recruited from the new middle class, were trained in scholastic traditions based on translations of Arabic literature and the ancient Roman and Greek texts, mainly Aristotle, Hippocrates, and Galen, but with some more current texts, mainly written by Arab and Jewish physicians.

Growth exacerbated public health problems in the newly walled commercial and industrial towns gradually forcing the definition of solutions. Crowding, poor nutrition and sanitation, lack of adequate water sources and drainage, unpaved streets, keeping of animals in towns, and lack of organized waste disposal created conditions for widespread infectious diseases. Municipalities developed protected water sites (cisterns, wells, and springs) and public fountains with municipal regulation and supervision. Piped community water supplies were developed in Dublin, Basel, and Bruges (Belgium) in the thirteenth century. Between the eleventh and fifteenth centuries, Novgorod in Russia used clay and wooden pipes for water supplies. Municipal bath houses were available, but European standards of personal hygiene were generally low in all social classes.

Medical care was still largely oriented to symptom relief with few resources to draw on. Traditional folk medicine survived especially in rural areas, but was suppressed by the Church as witchcraft. Physicians provided services for those able to pay, but medical knowledge was a mix of pragmatism, mysticism, and sheer lack of scientific knowledge. Conditions were ripe for vast epidemics of smallpox, cholera, measles, and other epidemic diseases fanned by the debased conditions of life and vastly destructive warfare raging throughout Europe.

The Black Death (pneumonia and bubonic plague), brought from the steppes of central Asia to Europe with the Mongol invasions, was transmitted via extensive trade routes throughout Europe by sea and overland. The Black Death was introduced into China with the Mongol invasions, bringing horrendous slaughter with halving of the population of China between AD 1200 and 1400. Between the eleventh and thirteenth centuries, during the Mongol–Tatar conquests, many widespread epidemics, including plague, were recorded in Rus (now Russia). The plague traveled rapidly with armies, caravan traders, and later by shipping as world trade expanded in the fourteenth to fifteenth centuries. The plague ravaged most of Europe between 1346 and 1350, killing between 24 and 50 million persons, or approximately one-third of the population, leaving vast areas of Europe underpopulated. Despite local efforts to prevent the disease by quarantine and isolation of the sick, the disease devastated whole communities.

BOX 1.1 "THIS IS THE END OF THE WORLD": THE BLACK DEATH

"Rumors of a terrible plague supposedly arising in China and spreading through Tartary (Central Asia) to India and Persia, Mesopotamia, Syria, Egypt, and all of Asia Minor had reached Europe in 1346. They told of a death toll so devastating that all of India was said to be depopulated, whole territories covered by dead bodies, other areas with no one left alive. As added up by Pope Clement VI at Avignon, the total of reported dead reached 23,840,000. In the absence of a concept of contagion, no serious alarm was felt in Europe until the trading ships brought their black burden of pestilence into Messina while other infected ships from the Levant carried it to Genoa and Venice.

"By January 1348 it penetrated France via Marseille, and North Africa via Tunis. Shipborne along coasts and navigable rivers, it spread westward from Marseille through the ports of Languedoc to Spain and northward up the Rhône to Avignon, where it arrived in March. It reached Narbonne, Montpellier, Carcassone, and Toulouse between February and May, and at the same time in Italy spread to Rome and Florence and their hinterlands. Between June and August it reached Bordeaux, Lyon, and Paris, spread to Burgundy and Normandy into southern England. From Italy during the summer it crossed the Alps into Switzerland and reached eastward to Hungary.

"In a given area the plague accomplished its kill within four to six months and then faded, except in the larger cities, where, rooting into the close-quartered population, it abated during the winter, only to appear in spring and rage for another six months."

Source: Tuchman BW. 1978. *A Distant Mirror: The Calamitous Fourteenth Century.* New York: Alfred A. Knopf Inc. (with permission).

Fear of a new and deadly disease, lack of knowledge, speculation, and rumor led to countermeasures which often exacerbated the spread of epidemics (as occurred in the late twentieth century with the AIDS epidemic). In western Europe, public and religious ceremonies and burials were promoted, which increased contact with infected persons. The misconception that cats were the cause of plague led to their slaughter, when they could have helped to stem the tide of disease borne by rats and by their lice to humans. Hygienic practices limited the spread of plague in Jewish ghettoes, leading to the blaming of the plague's spread on the Jews, with widespread massacres, especially in Germany and central Europe.

Seaport cities in the fourteenth century began to apply the biblical injunction to separate lepers by keeping ships coming from places with the plague waiting in remote parts of the harbor, initially for 30 days (treutina), then for 40 days (quarantina) (Ragusa in 1465 and Venice in 1485), establishing the public health act of quarantine, which on a pragmatic basis was found to reduce the chance of entry to the plague. Towns along major overland trading routes in Russia took measures to restrict movement in homes, streets, and whole towns during epidemics. All over Europe, municipal efforts to enforce isolation broke down as crowds gathered and were uncontrolled by inadequate police forces. In 1630, all officers of the Board of Health of Florence, Italy, were excommunicated because of efforts to prevent spread of the contagion by isolation of cases, thereby interfering with religious exercises to assuage God's wrath through appeals to divine providence.

The plague continued to strike with epidemics in London in 1665, in Marseille in 1720, Moscow in 1771, and Russia, India, and the Middle East through the nineteenth century. In sixteenth century Russia, Novgorod banned public funerals during plague epidemics, and in the seventeenth century, Czar Boris Godunov banned trade, prohibited religious and other ceremonies, and instituted quarantine type measures. Plague continued into the twentieth century (see *The Plague* by Albert Camus) with epidemics in Australia (1900), China (1911), Egypt (1940), and India (1995). The disease is endemic in rodents in many parts of the world, including the United States.

Guilds organized to protect economic interests of traders and skilled craftsmen developed mutual benefit funds to provide financial aid and other benefits for illness, death, widows and orphans, and medical care, as well as burial benefits for members and their families. The guilds wielded strong political powers during the late middle ages. These brotherhoods provided a tradition which was later expressed in the mutual benefit or Friendly Societies, sick funds, and insurance for health care based on employment groups.

The fourteenth century saw a devastation of the population of Europe by the plague, wars, and the breakdown of feudal society. It also set the stage for the agricultural revolution and later the industrial revolution. The period following the Black Death was innovative and dynamic. Lack of farm labor led to innovations in agriculture. Enclosures of common grazing land reduced spread of disease among animals, increased field crop productivity, and improved sheep farming, leading to development of the wool and textile industries and the search for energy sources, industrialization, and international markets.

THE RENAISSANCE (1500-1750)

Commerce, industry, trade, merchant fleets, and voyages of discovery to seek new markets led to development of a monied middle class and wealthy cities. Mines, foundries, and industrial plants flourished creating new goods and wealth. Partly as a result of the trade generated and the movement of goods and people, vast epidemics of syphilis, typhus, smallpox, measles, and the plague continued to spread across Europe. Malaria was still widespread throughout Europe. Rickets, scarlet fever, and scurvy, particularly among seamen, were rampant.

A virulent form of syphilis, allegedly brought back from America by the crews of Columbus, spread rapidly throughout Europe between 1495 and 1503, when it was described by Fracastorus. Control measures tried in various cities included examination and registration of prostitutes, closure of communal bathhouses, isolation in special hospitals, reporting of disease, and expulsion of sick prostitutes or strangers. The disease gradually decreased in virulence, but it remains a major public health problem to the present time.

In Russia, Czar Ivan IV (the Terrible) in the sixteenth century arranged to hire the court physician of Queen Elizabeth I, who brought with him to Moscow a group of physicians and pharmacists to serve the court. The Russian army had a tradition of regimental doctors. In the mid seventeenth century, the czarist administration developed pharmacies in major centers throughout the country for military and civilian needs, and established a State Pharmacy Department to control pharmacies and medications, education of doctors, military medicine, quarantine, forensic medicine, and medical libraries. Government revenues from manufacturing, sale, and encouragement of vodka provided for these services. Preparation of military doctors (*Lekars*) with 5–7 years of training was instituted in 1654. Hospitals were provided by monasteries, serving both civilian and military needs. In 1682, the first civic hospital was opened in Moscow, and in the same year, two hospitals were opened also in Moscow by the central government for care of patients and training of *Lekars*.

In European countries, growth of cities with industrialization and massive influx of the rural poor brought the focus of public health needs to the doorsteps of municipal governments. The breakdown of feudalism, the decline of the monasteries, and the land enclosures dispossessed the rural poor. Municipal and voluntary organizations increasingly developed hospitals, replacing those previously run by monastic orders. In 1601, the British Elizabethan Poor Laws defined the local parish government as being responsible for the health and social well-being of the poor. Municipal control of sanitation was weak. Each citizen was in theory held responsible for cleaning his part of the street, but hygienic standards were low with animal and human wastes freely accumulating.

During the Renaissance, the sciences of anatomy, physiology, chemistry, microscopy, and clinical medicine flourished. Medical schools in universities developed affiliations with hospitals, promoting clinical observation with increasing precision in description of disease. The contagion theory of disease, described in 1546 by Fracastorus and later Paracelsus, including the terms infection and disinfection, was contrary to the until-then sacrosanct miasma teachings of Galen.

From 1538, parish registers of christenings and burials were published in England as weekly and annual abstracts, known as the *Bills of Mortality*. Beginning in 1629, national annual Bills of Mortality included tabulation of death by cause. On the basis of the Bills of Mortality, novelist Daniel Defoe described the plague epidemic of London of 1665 one hundred years later.

In 1662, John Graunt in England published *Natural and Political Observations Upon the Bills of Mortality*. He compiled and interpreted mortality figures by inductive reasoning, demonstrating the regularity of certain social and vital phenomena. He showed statistical relationships between mortality and living conditions. Graunt's work was important because it was the first instance of statistical analysis of mortality data, providing a foundation for use of health statistics in the planning of health services. This established the sciences of demography and vital statistics and methods of analysis, providing basic measurements for health status evaluation with mortality rates by age, sex, and location. Also in 1662, William Petty took the first census in Ireland. In addition, he studied statistics on the supply of doctors and hospitals.

Microscopy, developed by Antony van Leeuwenhoek in 1676, provided a method of study of microorganisms. In the seventeenth century, the great medical centers were located in Leyden, Paris, and Montpelier. Bernardino Ramazzini published the first modern comprehensive treatise on occupational diseases in 1700.

In Russia, Peter the Great (1682–1725) initiated political, cultural, and health reforms. He sent young aristocrats to study sciences and technology in western Europe, including medicine. He established the first hospital-based medical school in Petersburg and then in other centers, mainly to train military doctors. He established the Anatomical Museum of the Imperial Academy of Sciences in St. Petersburg in 1717, and initiated a census of males for military service in 1722. In 1724, V. N. Tateshev carried out a survey by questionnaire of all regions of the Russian empire regarding epidemic disease and methods of treatment.

ENLIGHTENMENT, SCIENCE, AND REVOLUTION (1750–1830)

The Enlightenment, a dynamic period of social, economic, and political thought, provided great impetus for emancipation and rapid advancement of science and agriculture, technology, and industrial power. Changes in many spheres of life were exemplified by the American and French revolutions, along with the economic theory of Adam Smith (author of *The Wealth of Nations*), which developed political and economic rights of the individual. Improvements in agriculture created greater productivity and better nutrition. These were associated with higher birth rates and falling death rates, leading to rapid population growth. The agricultural revolution during the sixteenth and seventeenth centuries based on mechanization and in larger land units of production with less manpower, led to rural depopulation provided excess workers to staff the factories, mines, ships, homes, and shops of the industrial revolution, expanding commerce, and a growing middle class. Exploration and colonization provided expansion of markets that fueled the industrial revolution, growth of science, technology, and wealth.

An agricultural revolution of improved production of grain, milk, and meat through better land use, animal husbandry, and farm machinery resulted in greater agricultural productivity and food supply. Later introduction of new crops from the Americas, including the potato, the tomato, peppers, and maize, contributed to a general improvement in nutrition. This was supplemented by increasing availability of codfish from the Grand Banks, adding protein to the common diet.

Industrialized urban centers grew rapidly, and the crowded cities were ill-equipped to house and provide services for the new working class. Urban areas were characterized by poor housing, sanitation, and nutrition and harsh working conditions which produced appalling health conditions. During this period, documentation and statistical analysis developed in various forms, which became the basis for social sciences including demography and epidemiology. Intellectual movements of the eighteenth century defined the rights of man and gave rise to revolutionary movements to promote liberty and release from tyrannical rule, as in the American and French revolutions of 1775 and 1789. Following the final defeat of Napoleon at Waterloo in 1815, conservative governments were faced with strong middle-class movements for reform of social conditions, with important implications for health.

Eighteenth Century Reforms

The period of enlightenment and reason was characterized by the philosophers Locke, Diderot, Voltaire, Rousseau, and others. These men produced a new approach to science and knowledge derived from observations and systematic testing of ideas as opposed to instinctive or innate knowledge as the basis for human progress. The idea of the rights of man contributed to the American and French revolutions, but also to a widening belief that society was obliged to serve all rather than just the privileged. This had a profound impact on approaches to health and societal issues.

The late eighteenth century was a period of growth and development of clinical medicine, surgery, and therapeutics, as well as of the sciences of chemistry, physics, physiology, and anatomy. From the 1750s onward, voluntary hospitals were established in major urban centers in Britain, America, and on the continent. Medical–social reform involving hospitals, prisons, and lazarettos (leprosy hospitals) in Britain, led by John Howard (who published *On the State of Prisons* in 1777), produced substantive improvements in these institutions. During the French Revolution, Philippe Pinel removed the chains from patients at the Bicetre mental hospital near Paris and fostered reform of insane asylums. Reforms were further carried out in Britain by the Society of Friends (the Quakers), who built the York Retreat, providing humane care as an alternative to the inhuman conditions of the York Asylum.

Although Ramazzini's monumental work on occupational diseases was published in 1700, little progress was made in applying epidemiologic principles to this field. However, in the latter part of the century, interest in the health of sailors and soldiers led to important developments in military and naval medicine. Studies of diseases in various trades, such as metalworkers, bakers, shoemakers, and hatmakers, identified causative agents and methods of prevention. Observational studies of Percival Potts on scrotal cancer in chimney sweeps (1775), and Baker on the Devonshire colic (lead poisoning) in 1767, helped to lay the basis for development of investigative epidemiology.

Pioneers and supporting movements successfully agitated for reform in Britain through the parliamentary system. The antigin movement, aided by the popular newspapers (the "penny press") and the brilliant engravings of Hogarth, helped produce legal, social, and police reforms in English towns. Conditions for seamen were improved following the voyages of discovery of Captain James Cook, and the Spithead mutiny in the British fleet. The United States developed the Marine Hospitals Service for treatment and quarantine of seamen in 1798, which later became the U.S. Public Health Service. The antislavery movement led by protestant Christian churches goaded the British government to ban slavery in 1797 and the slave trade in 1807, using the Royal Navy to sweep the slave trade from the seas during the early part of the nineteenth century.

Applied Epidemiology

Scurvy (the Black Death of the Sea) was a major health problem among seamen during long sea voyages. In 1498, Vasco da Gama lost 55 men of his crew to scurvy during his voyages, and in 1535, Jacques Cartier's crew suffered severely from scurvy on his voyage of discovery to Canada. During the sixteenth century, Dutch seamen knew of the value of fresh vegetables and citrus fruit in preventing

---

BOX 1.2 JAMES LIND AND SCURVY, 1747

Captain James Lind, a physician serving Britain's Royal Navy, developed a hypothesis regarding the cause of scurvy based on clinical observations. In May 1747, on HMS *Salisbury,* Lind conducted the first controlled clinical epidemiologic trial by treating 12 sailors sick with scurvy with six different dietary regimens. The 2 sailors fed oranges and lemons became well and fit for duty within 6 days, while the others remained sick. He concluded that citrus fruits would treat and prevent scurvy. In 1757 he published his *Treatise on the Scurvy: An Inquiry on the Nature, Causes and Cure of That Disease.*

This discovery was adopted by progressive sea captains and aided Captain Cook in his famous voyages of discovery in the South Pacific in 1768–1771. By 1795, the Royal Navy adopted routine issuance of lime juice to sailors to prevent scurvy. Lind also instigated reforms in living conditions for sailors, thus contributing to improvement in their health and fitness and the functioning of the fleet. This doubled the time that a ship could remain at sea, providing an important advantage in the naval blockade of Napoleon-controlled Europe.

scurvy. In 1601, Purchas, and in 1617, a British naval doctor (John Woodall) recommended use of lemons and oranges in treatment of scurvy, but this was not widely practiced. During the seventeenth to eighteenth centuries, Russian military practice included antiscorbutic preparations, and use of sauerkraut for this purpose became common in European armies.

Scurvy was a major cause of sickness and death among sailors when supplies of fruit and vegetables ran out; it caused disease and death which seriously limited long voyages and contributed to frequent mutinies at sea. In 1747, James Lind carried out a pioneering epidemiologic investigation on scurvy among sailors on long voyages, leading to adoption of lemon or lime juice as a routine nutrition supplement for the British sailors. This study, its publication, and conclusions led to an organized nutritional intervention. Vitamins were not isolated until almost 150 years later, but the scientific technique of careful observation, hypothesis formulation and testing, followed by documentation established clinical epidemiologic investigation of nutrition in public health.

Jenner and Vaccination

Smallpox, a devastating and disfiguring epidemic disease, ravaged all parts of the world and was known since the third century BC. Described first by Rhazes in the tenth century, the disease was confused with measles and was widespread in Asia, the Middle East, and Europe during the Middle Ages. It was a designated cause of death in the Bills of Mortality in 1629 in London. Epidemics of smallpox occurred throughout the seventeenth to eighteenth and into the nineteenth centuries primarily as a disease of childhood, with mortality rates of 25–40% or more and disfiguring sequelae.

Smallpox was a key factor in the near elimination of the Aztec and other societies in Central and South America following the Spanish invasion. Traditions of prevention of this disease by inoculation or transmission of the disease to healthy persons to prevent them from a more virulent form during epidemics was reported in ancient China. This practice of variolation was brought to England in 1721 by Lady Mary Montagu, wife of the British ambassador to Constantinople, where it was common practice. It was widely adopted in England in the mid eighteenth century, when the disease affected millions of people in Europe alone. Catherine the Great in Russia had her son inoculated by variolation by a leading English practitioner.

Edward Jenner first used vaccination with cowpox to prevent smallpox in 1796. In 1800, vaccination was adopted by the British armed forces, and the practice spread to Europe, the Americas, and the British Empire. Denmark made vaccination mandatory in the early nineteenth century and soon eradicated smallpox locally. Despite some professional opposition, the practice spread rapidly from the upper classes and voluntary groups to the common people because of the fear of smallpox. Vaccination later became compulsory in many countries, with the ultimate public health achievement of global eradication in the late twentieth century.

BOX 1.3 JENNER AND SMALLPOX

In 1796, Edward Jenner (1749–1823), a country physician in Glouster-shire, England, investigated local folk lore that milkmaids were immune to smallpox because of their exposure to cowpox. He took matter from a cow-pox postule of a milkmaid, Sarah Nelmes, and applied it with scratches to the skin of a youngster named James Phipps, who was later inoculated with smallpox. He did not develop the disease. Jenner's 1798 publication, *An Enquiry into the Causes and Effects of the Variolae Vaccina,* described his wide scale use of *vaccination* and its successful protection against smallpox, and prophesied that "the annihilation of the smallpox, the most dreadful scourge of the human species, must be the result of this practice."

He promoted vaccination as a method to replace variolation, which was exposure of persons to the pustular matter of cases of smallpox, originally documented in ancient China in AD 320. Variolation was practiced widely in the eighteenth century and constituted a very lucrative medical business. Opposition to vaccination was intense, and Jenner's contribution was ignored by the scientific and medical establishment of the day, but rewarded by Parliament. Vaccination was adopted as a universal practice by the British military in 1800 and by Denmark in 1803. Vaccination became an increasingly wide practice during the nineteenth century. In 1977, the last case of smallpox was identified, and smallpox eradication was declared by the World Health Organization in 1980. Remaining stocks of the virus in the United States and Russia are to be destroyed in 2002.

FOUNDATIONS OF HEALTH STATISTICS AND EPIDEMIOLOGY

Registration of births and deaths form the basis of demography. Epidemiology as a discipline borrows from demography, sociology, and statistics. The basis of scientific reasoning in these fields emerged in the early seventeenth century with the inductive reasoning enunciated by Francis Bacon and applied by Robert Boyle in chemistry, Isaac Newton in physics, William Petty in economics, and John Graunt in demography. Bacon's writing inspired a whole generation of scientists in different fields and led to the founding of the Royal Society.

In 1722, Peter the Great began Russia's system of registration of births of male infants for military purposes. In 1755, M. V. Lomonosov led initiatives in establishing demography in Russia. He carried out surveys and studies of birth statistics, infant mortality, quality of medical care, alcoholism, and worker's health.

He brought the results of these studies to the attention of the government, which led to improved training of doctors and midwives, and epidemic control measures. Lomonosov also helped initiate the medical faculty of Moscow University (1765).

Daniel Bernoulli, a member of a European family of mathematicians, constructed life tables based on available data showing that variolation against smallpox conferred lifelong immunity and vaccination at birth increased life expectancy. Following the French Revolution, health statistics flourished in the mid nineteenth century in the work of Pierre Louis, who is considered the founder of modern epidemiology. Louis conducted several important observational studies, including one showing that blood letting, then a common form of therapy, was not efficacious, thereby contributing to a decline in this harmful practice. His students included Marc D'Epigne in France, William Farr in Britain, and others in the United States who became the pioneers in spreading "la methode numerique" in medicine.

Health statistics for social and public health reform took an important place in the work of Edwin Chadwick, Lemuel Shattuck, and Florence Nightingale. Recognizing the critical importance of accurate statistical information in health planning and disease prevention, Edwin Chadwick's work led to legislation establishing the Registrar-General's Office in Britain in 1836. William Farr became its Director-General and placed the focus of this office on public health. Farr's analysis of mortality in Liverpool, for example, showed that barely half of its native-born lived to their sixth birthday, whereas in England overall the median age at death was 45 years. As a result, Parliament passed the Liverpool Sanitary Act of 1846, creating a legislated sanitary code, a Medical Officer of Health position, and a local health authority.

The London Epidemiological Society, founded in 1850, became an active investigative and lobbying group for public health action. Its work on smallpox led to passage of the Vaccination Act of 1853, establishing compulsory vaccination in the United Kingdom. William Budd, a student of Louis and founding member of the London Epidemiological Society, investigated outbreaks of typhoid fever in his home village in 1839, establishing it as a contagious self-propagating disease spread by microorganisms, discrediting the miasma theory.

In 1842 in Boston, Massachusetts, Lemuel Shattuck initiated a statewide registration of vital statistics, which became a model elsewhere in the United States. His report was a landmark in the evolution of public health administration and planning. This provided a detailed account of data collection by age, sex, race, occupation, and uniform nomenclature for causes of diseases and death. He emphasized the importance of a routine system for exchanging data and information.

In the later part of the nineteenth century, Florence Nightingale highlighted the value of a hospital discharge information system. She promoted collection and use of statistics that could be derived from the records of patients treated in hospitals. Her work led to improved management and design of hospitals, military medicine and nursing as a profession.

SOCIAL REFORM AND THE SANITARY
MOVEMENT (1830–1875)

Following the English civil war in 1646, the army veterans of the Parliamentary forces called on Parliament to provide free schools and free medical care throughout the country as part of democratic reform. However, they failed to sustain interest or gain support for their revolutionary ideas in the postwar religious conflicts and restoration of the monarchy.

In Russia, the role of the state in health was promoted following initiatives of Peter the Great to introduce western medicine to the country. During the rule of Catherine the Great, under the supervision of Count Orlov, an epidemic of plague in Moscow (1771–1772) was suppressed by incentive payments to bring the sick for care. In 1784, a Russian physician, I. L. Danilevsky, defended a doctoral dissertation on "Government power—the best doctor." In the eighteenth and nineteenth centuries, reform movements promoted health initiatives by government. While these movements were suppressed (the Decembrists, 1825–1830) and liberal reform steps reversed, their ideas influenced later czarist reforms.

Following the revolution in France, the Constituent Assembly established a Health Commission. A national assistance program for indigents was established. Steps were taken to strengthen the *Bureaux de Sante* (Offices of Health) of municipalities which had dealt primarily with epidemics. In 1802, the Paris Bureau dealt with a wide range of sanitation, food control, health statistics, occupational health, first aid, and medical care issues. The other major cities of France followed with similar programs over the next 20 years, and in 1848 a central national health authority was established. Child welfare services were also developed in France in the mid part of the nineteenth century. The reporting of vital statistics became reliable in the German states and even more so in France, leading to the development of epidemiologic analysis of causes of death.

The government approach to public health was articulated by Johann Peter Franck for the Germanic states in his monumental series of books *A Complete System of Medical Police* (1779–1817). This text explained the government role in states with strong central governments and how to achieve health reform through administrative action. State regulations were to govern public health and personal health practices including marriage, procreation, and pregnancy. He promoted dental care, rest following delivery and maternity benefits, school health, food hygiene, housing standards, sanitation, sewage disposal, and clean water supplies. In this system, municipal authorities were responsible for keeping cities and towns clean and for monitoring vital statistics, military medicine, venereal diseases, hospitals, and communicable diseases.

This system emphasized a strong, even authoritarian role of the state in promoting public health including provision of prepaid medical care. This was a comprehensive and coherent approach to public health, elucidating the key role of municipal and higher levels of government. This work was influential in Russia where Franck spent the years 1805–1807 as Director of the St. Petersburg Medical Academy. However, because of its primary reliance on authoritarian governmental roles, this approach was resisted in most western countries, especially following the collapse of absolutist government ideas following the Napoleonic period.

Municipal (voluntary) boards of health were established in some British and American cities in the late eighteenth and early nineteenth centuries. A Central Board of Health was established in Britain in 1805, primarily to govern quarantine regulations to prevent entry of yellow fever and cholera into the country. Town life improved as sanitation, paving, lighting, sewers, iron water pipes, and filtration of water sources were introduced, although organization for development of such services was inadequate. Multiple agencies and private water companies provided unsupervised and overlapping services. London City Corporation had nearly 100 paving, lighting, and cleansing boards, 172 welfare boards, and numerous other health related authorities in 1830. These were later consolidated into the London Board of Works in 1855.

In Great Britain, early nineteenth century reform was stimulated by the Philosophic Radicals led by Jeremy Bentham, who advocated dealing with public problems in a rational and scientific way, initiating a reform movement including parliamentary, law, and education reform. Economic and social philosophers in Britain, including Adam Smith and Jeremy Bentham, argued for liberalism, rationalism, free trade, political rights, and social reform, all contributing to "the greatest good for the greatest number." Labor law reforms (the Mines and the Factory Acts) banning children and women from underground work in the mines and regulating reduction in the work day to 10 hours were adopted by the British Parliament in the 1830s to 1840s. The spread of railroads and steamships, the penny post (1840), and telegraphs (1846), combined with growing literacy and compulsory primary education introduced in Britain in 1876, dramatically altered local and world communication.

The British Poor Law Amendment Act of 1834 replaced the old Elizabethan Poor Laws, shifting responsibility for welfare of the poor from the local parish to the central government's Poor Law Commision. The parishes were unable to cope with the needs of the rural poor whose condition was deteriorating with agricultural innovations and enclosures. The old system was breaking down, and the new industrialization needed workers, miners, seamen, and soldiers. The new conditions forced the poor to move from rural areas to the growing industrial towns. The urban poor were forced into workhouses on the continent, greater resistance to reform led to more radical trends toward change through unsuccessful revolution, followed by deep conservatism.

Deteriorating housing, sanitation, and work conditions in Britain in the 1830s resulted in rising mortality rates recorded in the Bills of Mortality. Industrial cities like Manchester (1795) had established voluntary boards of health, but they lacked the authority to alter fundamental conditions to control epidemics and urban decay. The boards of health were unable to deal with sewage, garbage, animal control, crowded slum housing, privies, adulterated foods and medicines, industrial plants, or other social or physical pollution. Legislation in the 1830s in Britain and

Canada improved the ability of municipalities and boards of health to cope with sanitation of community water supplies and sanitation.

The work of Edwin Chadwick and the *Report on the Sanitary Conditions of the Labouring Population of Great Britain* (1842) initiated a further series of reforms through the Poor Law Commission. The British Parliament passed the Health of Towns Act and the Public Health Act of 1848. This established the General Board of Health, housing legislation, and other reforms, including municipal boards of health in the major cities and rural local authorities. Despite setbacks due to reaction to these developments, the basis was laid for the "Sanitary Revolution," dealing with urban sanitation and health conditions, as well as cholera, typhoid, and tuberculosis control.

In 1850, the Massachusetts Sanitary Commission, chaired by Lemuel Shattuck, was established to look into similar conditions in that state. Boards of health established earlier in the century became efficiently organized and effective in sanitary reform in the United States. The report of that committee has become a classic public health document. Reissued in the 1970s, it defines a comprehensive approach to public health that remains valuable to the present time.

The Chadwick (1842) and Shattuck (1850) reports developed the concept of municipal boards of health based on public health law with a public mandate to supervise and regulate community sanitation. This included urban planning, zoning, restriction of animals and industry in residential areas, regulation of working conditions, and other aspects of community infrastructure.

The interaction between sanitation and social hygiene was a theme promoted by Rudolph Virchow, the founder of cellular pathology and a social–medical philosopher. Virchow, the leading German physician in the mid nineteenth century, despite being an anticontagionist, promoted observation, hypothesis, and experimentation, helping to establish the scientific method and dispel philosophic approaches to medical issues. He was a social activist and linked health of the people to social and economic conditions, emphasizing the need for political solutions. Virchow played an important part in the 1848 revolutions in central and western Europe, the same year as the publication of the *Communist Manifesto* by Karl Marx. These all contributed to growing pressure on governments by workers' groups to promote better living, working, and health conditions in the 1870s.

In 1869, the Massachusetts State Board of Health was established and in the same year a Royal Sanitary Commission was appointed in the United Kingdom. The American Public Health Association (APHA), established in 1872, served as a professional educational and lobbying group to promote the interests of public health in the United States, often successfully prodding federal, state, and local governments to act in the public interests in this field. The APHA definition of appropriate services at each level of government set standards and guidelines for local health authorities to match themselves against. The organization of local, state, and national public health activities over the twentieth century in the United States owes much to the professional leadership and lobbying skills of this organization.

Max von Pettenkoffer in 1873 studied the high mortality rates in Munich, comparing them to rapidly declining rates in London. His public lectures on the value of health to a city led to sanitary reforms, as were being achieved in Berlin at the same time under Virchow's leadership. Pettenkoffer introduced laboratory analysis to public health practice and established the first professorial chair in hygiene and public health, emphasizing the scientific basis for public health. He is considered to be the first professor of experimental hygiene. Pettenkoffer promoted the concept of the value of a healthy city and stressed that health is the result of a number of factors and that public health is a community concern and measures taken to help those in need benefit all of the community.

In 1861, Russia freed the serfs and returned independence to universities. Departments of hygiene were established in the university medical schools in the 1860s and 1870s to train future hygienists, and to carry out studies of sanitary and health conditions in manufacturing industries. F. F. Erisman, a pioneer in sanitary research in Russia, promoted the connection between experimental science, social hygiene, and medicine, and he established a school of hygiene in 1896, later closed by the czarist government. In 1864, the government initiated the Zemstvos system of providing medical care in rural areas as a governmental program. These health reforms were implemented in 34 of 89 regions in Russia. Prior to these reforms, medical services in rural areas were practically nonexistent. Epidemics and the high mortality of the working population induced the nobility and new manufacturers in rural towns to promote Zemstvos public medical services. In rural areas previously served by doctors based in the towns traveling to the villages, local hospitals and delivery homes were established. The Russian medical profession largely supported free public medical care.

In 1883, Otto von Bismarck, Chancellor of Germany, introduced legislation providing mandatory insurance for injury and illness, and survivors benefits for workers in industrial plants. In the United Kingdom in 1911, Prime Minister Lloyd George established compulsory insurance for workers. This was followed by similar programs in Russia in 1912, and in virtually all central and western European countries by the 1930s. This chain of events led to the establishment of governmental responsibility for health of the population in virtually all developed countries by the 1960s (see Chapter 13).

The harsh conditions in the industrial and mining centers of Europe during the industrial revolution led to efforts in social reform preceding and contributing to sanitary reform even before the germ theory of causation of disease was proved and the science of microbiology established. Pioneering breakthroughs, based on trial and error challenged the established dogmas of the time, produced the sanitary revolution, still one of the important foundations of public health.

Snow on Cholera

The great cholera pandemics originating in India between 1825 and 1854 spread via increasingly rapid transportation to Europe and North America. Moscow lost some 33,000 persons in the cholera epidemic of 1829, which recurred in 1830–1831. In Paris, the 1832 cholera epidemic killed over 18,000 persons in 6 months or just over 2% of the population.

Between 1848 and 1854 a series of outbreaks of cholera occurred in London with large-scale loss of life. The highest rates were in areas of the city where two water companies supplied homes with overlapping water mains. One of these (the Lambeth Company) then moved its water intake to a less polluted part of the Thames River, while the Southwark and Vauxhall company left its intake in a part of the river heavily polluted with sewage. John Snow, a founding member of the London Epidemiological Society and anesthetist to Queen Victoria, investigated an outbreak of cholera in Soho from August to September 1854 in the area adjacent to Broad Street. He traced some 500 cholera deaths occurring in a 10 day period. Cases either lived close to or used the Broad Street pump for drinking water. He determined that the brewery workers and poorhouse residents in the area, using uncontaminated wells, escaped the epidemic. Snow concluded that the Broad Street pump was probably contaminated. He persuaded the authorities to remove the handle from the pump, and the already subsiding epidemic disappeared within a few days.

During September to October 1854, Snow investigated another outbreak, again suspecting water transmission. He identified cases of mortality from cholera by their place of residence and which water company supplied the home (Table 1.1). Snow calculated the cholera rates in a 4-week period in homes supplied by each of the two companies. Homes supplied by the Southwark and Vauxhall Water Company were affected by high cholera death rates while adjacent homes supplied by the Lambeth Company had rates lower than the rest of London. This provided overwhelming epidemiologic support for Snow's hypothesis that the cholera epidemic source was the contaminated water from the Thames River, distributed to homes in a large area of south London.

This investigation of a natural experiment, with a study and control group occurring in an actual disease outbreak, strengthened the germ theory supporters who were still opposed by powerful forces. It also led to legislation mandating filtration of water companies' supplies in 1857. *Vibrio cholerae* was not isolated until

TABLE 1.1   Deaths from Cholera Epidemic in Districts of London Supplied by Two Water Companies, Seven Weeks, 1854[a]

Water Supply Company	Number of houses	Deaths from cholera	Cholera Deaths per 10,000 houses
Southwark and Vauxhall	40,046	1,263	315
Lambeth	26,107	98	37
Rest of London	256,423	1,422	59

[a]Source: Snow J. On the mode of transmission of cholera. In: *Snow on Cholera: A Reprint of Two Papers.* New York: The Commonwealth Fund, 1936.

1883 during an investigation of waterborne cholera outbreaks in Egypt by Robert Koch. Snow's work on cholera has become one of the classic epidemiologic investigations, studied to this day for its scientific imagination and thoroughness, despite preceding discovery of the causative organism nearly 30 years earlier.

William Budd, physician at the Bristol Royal Infirmary, was a pioneer exponent of the germ theory of disease. He carried out a number of epidemiologic investigations of typhoid fever in the 1850s, finding waterborne episodes of the disease. He investigated an outbreak in 1853 in Cowbridge, a small Welsh village, where a ball attracted 140 participants from surrounding countries. Almost immediately afterward, many of those attending the ball became sick with typhoid fever. He found that a person with typhoid had been at the location some days before and that his excreta have been disposed of near the well from which water was drawn for the ball. Budd then concluded that water was the vehicle of transmission of the disease. He investigated other outbreaks and summarized his reports in *Typhoid Fever: Its Nature, Mode of Transmission and Prevention,* published in 1873, which is a classic work on waterborne transmission of enteric disease. These investigations contributed to the movement to disinfect public water systems on a preventive basis.

The brilliant epidemiologic studies of Snow and Budd set a new direction in epidemiology and public health practice, not only with waterborne disease. They established a standard for investigation of the distribution of disease in populations with the object of finding a way to interrupt the transmission of disease. Cholera and typhoid epidemics, however, continue to the present day.

Germ versus Miasma Theories

Until the early and mid part of the nineteenth century, there were debates as to the causation of disease. The miasma theory that disease was the result of environmental emanations or miasmas went back to Greek and Roman medicine, and Hippocrates (Air, Water, and Places). Miasmists believed that disease was caused by infectious mists or noxious vapors emanating from filth in the towns and that the method of prevention of infectious diseases was to clean the streets from garbage, sewage, animal carcasses, and wastes that were features of urban living. This provided the basis for the Sanitary Movement, with great benefit to improving health conditions. The miasma theory had strong proponents well into the later part of the nineteenth century.

The contagion or germ theory gained ground, despite the lack of scientific proof, on the basis of biblical and middle ages experience with isolation of lepers and quarantine of other infectious conditions. In 1546, Fracastorus published *De Contagione,* a treatise on microbiological organisms as the case of specific diseases. The germ theory was strengthened by the work of Antony van Leeuwenhoek, who invented the microscope in 1676. The invention of this apparatus is considered to be one of the few watersheds in the history of science. His research showing small microorganisms led to his recognition as a Fellow of the Royal Society of England in 1680. The germ theorists believed that microbes, such as those described by van Leeuwenhoek, were the cause of disease which could be transmitted from person to person or by contact with sewage or contaminated water.

Major contributions to resolving this issue came from the epidemiological studies of Snow and Budd in the 1850s showing the waterborne transmission of cholera and typhoid. The classic study of a measles epidemic in the remote Faroe Islands by Peter Panum in 1846 clearly showed the person-to-person transmission of this disease, its incubation period, and the lifelong natural immunity exposure gives (Box 1.4). The dispute continued, however, with contagionists and miasmists or sanitationists arguing with equal vehemence.

While the issue was debated heatedly until the end of the nineteenth century, practical application of sanitary reform was promoted by both theories. Increasing attention to sewage, water safety, and removal of waste products by organized municipal activities was adopted in European and North American cities. The sanitary revolution proceeded while the debates raged and as solid scientific proof of the germ theory accumulated, primarily in the 1880s. Fear of cholera stimulated New York City to establish a Board of Health in 1866. In the city of Hamburg, Germany, a Board of Health was established in 1892 only after a cholera epidemic at-

---

BOX 1.4 PANUM ON MEASLES IN THE FAROE ISLANDS, 1846

Peter Ludwig Panum, a 26-year-old newly graduated medical doctor from the University of Copenhagen, was sent to the Faroe Islands by the Danish government to investigate an outbreak of measles in 1846. There had been no measles since 1781 in the islands, located in the far reaches of the North Atlantic. During the 1846 epidemic, about 6,000 of the 7,782 islanders were stricken with measles, and 102 of them died of the disease or its sequelae. Panum visited all isolated corners of the islands, tracing the chain of transmission of the disease from location to location, and the immunity of those exposed during the 1781 epidemic. From his well-documented observations he concluded, contrary to prevailing opinion, that measles is a contagious disease spread from person to person, and that one attack gives lifelong immunity. His superb report clearly demonstrated the contagious nature of the disease, its incubation period, and the fact that it is not a disease of "spontaneous generation," nor generally dispersed in the atmosphere and spread as a "miasma," proving that isolation of cases was an effective intervention.

Source: Panum, P. L. Observations Made during the Epidemic of Measles on the Faroe Islands in the Year 1846. *In:* Roueche B (ed). 1963. *Curiosities of Medicine: An Assembly of Medical Diversions 1552–1962.* London: Victor Gollancz Ltd.

tacked the city, while neighboring Altona remained cholera-free because it had established a water filtration plant.

The specific causation of disease (the germ theory) has been a vital part of the development of public health. The bacteriologic revolution (see below), led by the work of Louis Pasteur and Robert Koch, provided enormous benefit to medicine and public health. But those who argued that disease is environmental in origin (the miasma theory) also contributed to public health because of their recognition of the importance of social or other environmental factors, such as poor sanitation and housing conditions or nutritional status, all of which increase susceptibility to specific agents of disease, or the severity of disease.

HOSPITAL REFORM

Hospitals developed by monasteries as charitable services were supplanted by voluntary or municipal hospitals mainly for the poor during and after the Renaissance. Reforms in hospital care evolved along with the sanitary revolution. Hospitals were dangerous pest houses in eighteenth century Europe. Reforms in hospitals in England were stimulated by the reports of John Howard in the late eighteenth century, becoming part of wider social reform in the early part of the nineteenth century. Professional reform in hospital organization and care started in the latter half of the nineteenth century under the influence of Florence Nightingale, Oliver Wendel Holmes, and Ignaz Semmelweiss. Clinical–epidemiological studies of "antiseptic principles" provided a new, scientific approach to improvement in health care.

In the 1840s puerperal fever was a major cause of death in childbirth, and was the subject of investigation by Oliver Wendel Holmes in the United States, who argued that this was due to contagion. In 1846, Semmelweiss, a Hungarian obstetrician at the Vienna Lying-In hospital, suspected that deaths from puerperal fever were the result of contamination on the hands of physicians transmitted from autopsy material to living patients. He showed that death rates among women attended by medical personnel were 2–5 times the rates among those attended by midwives. By requiring doctors and medical students to soak their hands in chlorinated lime after autopsies, he reduced the mortality rates among the medically attended women to the rate of the midwife-attended group.

Semmelweiss's work, although carefully documented, was slow to be widely accepted, taking some 40 years for general adoption. His pioneering investigation of childbed fever (streptococcal infection in childbirth) in Vienna contributed to improvement in obstetrics and a reduction in maternal mortality. In the 1850s, prevention of blindness in newborns by prophylactic use of silver nitrate eye drops, developed by Karl Crede in Leipzig, spread rapidly through the medical world.

Florence Nightingale's momentous work in nursing and hospital administration in the Crimean War (1854–1856) established the professions of nursing and mod-

---

BOX 1.5 CREDE AND PREVENTION OF GONOCOCCAL OPHTHALMIA NEONATORUM

Gonorrhea was common in all levels of society in nineteenth century Europe, and ophthalmic infection of newborns was a widespread cause of infection, scarring, and blindness. Karl Franz Crede, professor of obstetrics at the University of Leipzig, attempted to treat neonatal gonococcal ophthalmic infection with many medications. He discovered the use of silver nitrate as a treatment and introduced its use as a preventive measure between 1584–1860 with astonishing success. The prophylactic use of silver nitrate spread rapidly hospital by hospital, but it was decades before it was mandated widely. It was only in 1879 that the gonococcus organism was discovered by Neisser. Estimates of children saved from blindness by this procedure in Europe during the nineteenth century are as high as one million.

---

ern hospital administration. In the 1860s, she emphasized the importance of poor law and workhouse reform and training special district nurses for care of the sick poor at home. Nightingale's subsequent long and successful campaigns to raise standards of military medicine, hospital planning, supply services and management, hospital statistics, and community health nursing were outstanding contributions to development of modern, organized health care and antisepsis.

THE BACTERIOLOGIC REVOLUTION

In the third quarter of the nineteenth century the sanitary movement was rapidly spreading through the cities of Europe with demonstrable success in reducing disease in areas served by sewage drains, improved water supplies, street paving, and waste removal. At the same time innovations were occurring in hospitals, stressing hygiene and professionalization of nursing and administration. These were accompanied by giant breakthroughs in establishing the scientific and practical applications of bacteriology and immunology.

Pasteur, Cohn, Koch, and Lister

In the 1850s to 1870s, Louis Pasteur, a French professor of chemistry, brilliantly developed the basis for modern bacteriology as a cornerstone of public health. He established a scientific, experimental proof for the germ theory by his demonstration in 1854 of anaerobic microbial fermentation. Between 1856 and 1860 he showed how to prevent wine from spoilage due to contamination from foreign organisms by heating the wine to a certain temperature before bottling it to kill the undesired ferments. This led to the process of "pasteurization." Asked to investigate the threatened destruction of the French silk industry by epidemics destroying the silkworms, he discovered microorganisms (1865) causing the disease and devised growing conditions which eliminated the problem, raising scientific and industrial interest in the germ theory. This was followed by similar work in the beer industry (1871).

Pasteur went on to develop the science of immunology by working on vaccines. He produced vaccines from attenuation or weakening of an organism's strength by passing it successively through animals, recovering it and retransmitting it to other animals. In 1881, he inoculated hens with attenuated cultures of chicken cholera, then in an inspired experiment challenged them with virulent organisms and found them to be immune. In 1883 he produced a similar protective vaccine for swine erysipelas, and then in 1884–1885, a vaccine for rabies.

Rabies was widely feared as a disease transmitted to humans through bites of infected animals and was universally fatal. Pasteur reasoned that the disease affected the nervous system and was transmitted in saliva. He injected material from infected animals, attenuated to produce protective antibodies but not the disease. In 1885, a 14-year-old boy from Alsace was severely bitten by a rabid dog. Local physicians agreed that because death was certain, Pasteur, a nonphysician, be allowed to treat the boy with a course of immunization. The boy, Joseph Meister, survived, and similar cases were brought to Pasteur and successfully immunized. Pasteur was criticized in medical circles, but both the general public and scientific circles soon recognized his enormous contribution to public health.

Ferdinand Cohn (1828–1898), professor of Botany at Breslau University, developed and systematized the science of bacteriology using morphology, staining, and media characteristics of microorganisms, and trained a key generation of microbiological investigators. One students, Robert Koch (1843–1910), a German rural district medical officer, investigated anthrax by using mice inoculated with blood from sick cattle, transmitting the disease for more than 20 generations. He developed basic bacteriologic techniques including methods of culturing and staining bacteria. He demonstrated the organism causing anthrax, recovered it from sick animals, and passed it through several generations of animals, proving the transmission of specific disease by specific microorganisms.

In 1882, Koch demonstrated and cultured the tubercle bacillus. He then headed the German Cholera Commission visiting Egypt and India in 1883, isolating and identifying *Vibrio cholerae* (*Nobel prize,* 1905). He demonstrated the efficacy of filtration of water in preventing transmission of enteric disease including cholera. In 1883 Koch, adapting postulates on causation of disease from clinician–pathologist Jacob Henle (1809–1885), established criteria for attribution of causation of a disease to a particular parasite or agent (Box 1.6). These were fundamental to establishment of the science of bacteriology and the relationship of microorganisms to disease causation.

The Koch–Henle postulates in their pure form were too rigid, and would limit identification of causes of many diseases, but they were important to establish the germ theory and the scientific basis of bacteriology, dispelling the many other theories of disease which were still widespread in the late nineteenth century. These postulates were later adapted by Evans (1976) to include noninfectious disease-

BOX 1.6 THE KOCH–HENLE POSTULATES
ON MICROORGANISMS AS THE CAUSE
OF DISEASE

1. The organism (agent) must be shown to be present in every case of the disease by isolation in pure culture;
2. The agent should not be found in cases of any other disease;
3. Once isolated, the agent should be grown in a series of cultures, and then must be capable of reproducing the disease in experimental animals;
4. The agent must then be recovered from the disease produced in experimental animals.

Source: Last, J. M., 1995.

causing agents, such as cholesterol, in keeping with the changing emphasis in epidemiology to the noninfectious diseases.

In the mid-1860s, Joseph Lister in Edinburgh, under the influence of Pasteur's work and with students of Semmelweiss, developed a theory of "antisepsis." His 1865 publication *On the Antiseptic Principle in the Practice of Surgery* described the use of carbolic acid to spray operating theaters and to cleanse surgical wounds, applying the germ theory with great benefit to surgical outcomes. Lister's work on chemical disinfection for surgery in 1865 was a pragmatic development that was a major advance in surgical practice, and an important contribution to establishing the germ theory in nineteenth century medicine.

Vector-Borne Disease

Studies of disease transmission defined the importance of carriers (those who can transmit a disease without showing clinical symptoms) in transmission of diphtheria, typhoid, and meningitis. This promoted studies of diseases borne by intermediate hosts or vectors. Parasitic diseases of animals and man were investigated in many centers during the nineteenth century, including Guinea worm disease, tapeworms, filariasis, and veterinary parasitic diseases such as Texas cattle fever. David Bruce showed the transmission of nagana, a disease of cattle and horses in Zululand, South Africa, in 1894–1895, caused by the trypanosome parasite transmitted by the tsetse fly, leading to environmental methods of control of disease transmission. Alexandre Yersin and Shibasabro Kitasato discovered the plague bacillus in 1894, and in 1898 French epidemiologist P. L. Simmond demonstrated the plague was a disease of rats spread by fleas to humans.

Malarial parasites were identified by French army surgeon Alphonse Laveran (Nobel prize, 1907) in Algeria in 1880. Mosquitoes were suspected as the method of transmission by many nineteenth century investigators, and in 1897 Ronald

Ross (Nobel Prize 1902), a British army doctor in India, Patrick Manson in England, and Benvenuto Grassi in Rome demonstrated the transmission of malaria by the *Anopheles* mosquito. Yellow fever, probably imported with slaves from Africa, was endemic in the southern United States but spread to northern cities in the late eighteenth century. An outbreak in Philadelphia 1798 killed nearly 8% of the population. Outbreaks in New York killed 732 persons in 1795, 2,086 in 1798, and 606 in 1803. The Caribbean and Central America were endemic with both yellow fever and malaria.

The conquest of yellow fever also contributed to establishing the germ or contagion theory versus the miasma theory when the work of Cuban physician Carlos Finlay was confirmed by Walter Reed in 1901. His studies in Cuba proved the mosquito-borne nature of the disease, as a transmissible disease via an intermediate host (vector), but not contagious between man and man. William Gorgas applied this to vector control activities and protection of sick persons from contact with mosquitoes, resulting in an eradication of yellow fever in Havana within 8 months, and in the Panama Canal zone within 16 months.

This work showed the potential for control of vector-borne disease that has had important success in control of many tropical diseases, including yellow fever and,

---

BOX 1.7 HAVANA AND PANAMA:
CONTROL OF YELLOW FEVER AND
MALARIA, 1901–1906

The U.S. Army Commission on Yellow Fever led by Walter Reed, an Army doctor, worked with Cuban physicians Carlos Finlay and Jesse Lazear to experiment with yellow fever transmission in Cuba in 1901. Using volunteers, he demonstrated transmission of the disease from person to person by the specific mosquito, *Stegomyia fasciata*. The Commission accepted that "the mosquito acts as the intermediate host for the parasite of Yellow Fever."

Another U.S. Army doctor, William Gorgas, applied the new knowledge of transmission of yellow fever and the life cycle of the vector mosquito. He organized a campaign to control the transmission of yellow fever in Havana, isolating clinical cases from mosquitoes and eliminating the breeding places for the *Stegomyia* with Mosquito Brigades. Yellow Fever was eradicated in Havana within 8 months. This showed the potential for control of other mosquito-borne diseases, principally malaria with its specific vector, *Anopheles*. Gorgas then successfully applied mosquito control to prevent both yellow fever and malaria between 1904 and 1906, permitting construction of the Panama Canal. These diseases had defeated the French builder of the Suez Canal, Ferdinand de Lesseps, 20 years earlier in his attempt to build a Panama canal.

currently, Guinea worm disease and onchocerciasis. Malaria, although it has come under control in many parts of the world, has resurged in many tropical countries since the 1960s.

MICROBIOLOGY AND IMMUNOLOGY

Elie Metchnikoff's work in Russia in 1883 described phagocytosis, a process in which white cells in the blood surround and destroy bacteria, and his elaboration of the processes of inflammation and humoral and cellular response led to a joint Nobel Prize in 1908 with Paul Ehrlich. Other investigators searched for the bactericidal or immunological properties of blood that enabled cell-free blood or serum to destroy bacteria. This work greatly strengthened the scientific bases for bacteriologyand immunology.

Pasteur's co-workers, Emile Roux and Alexandre Yersin, isolated and grew the causative organism for diphtheria and suggested that the organism produced a poison or toxin which caused the lethal effects of the disease. In 1890, Karl Fraenkel in Berlin published his work showing that inoculating guinea pigs with attenuated organisms of diphtheria could produce immunity. At the same time Emile Behring in Germany with Japanese co-worker Shibasaburo Kitasato produced evidence of immunity production in rabbits and mice to tetanus bacilli. Behring also developed a protective immunization against diphtheria in humans with active immunization as well as an antitoxin for passive immunization of an already infected person (Nobel prize, 1901). By 1894, diphtheria antitoxin was ready for general use. The isolation and identification of new disease-causing organisms proceeded rapidly in the last decades of the nineteenth century. The diphtheria organism was discovered in 1884 by Edwin Klebs and Friedrich Loeffler (students of Koch), and a vaccine for it in 1912, leading to the control of this disease in many parts of the world. Between 1876 and 1898, many pathogenic organisms were identified, providing a basis for many advances in vaccine development.

During the last quarter of the nineteenth century, it was clear that inoculation of attenuated microorganisms could produce protection through active immunization of a host by generating antibodies to that organism, which would protect the individual when exposed to the virulent (wild) organism. Passive immunization could be achieved in an already infected person by injection of serum of animals injected with attenuated organisms. The serum from that animal helps to counter the effects of the toxins produced by the invading organism. Pasteur's vaccines were followed by those of Haffkine for cholera and plague, Richard Pfeiffer and Carroll Wright for typhoid, Albert Calmette and Alphonse Guerin for tuberculosis, and Arnold Theiler and Theobald Smith for yellow fever.

The twentieth century has seen the flowering of immunology in the prevention of important diseases in animals and in man based on the pioneering work of Jenner, Pasteur, Koch, and those who followed. Many major childhood infectious diseases have come under control by immunization in one of the outstanding achievements of twentieth century public health.

Poliomyelitis

Poliomyelitis was endemic in most parts of the world prior to World War II causing widespread crippling of infants and children, hence its common name of "infantile paralysis." The most famous polio patient was Franklin Delano Roosevelt, crippled by polio in his early 30s, who went on to become President of the United States. During the 1940s and 1950s poliomyelitis occurred in massive epidemics affecting thousands of North American children and young adults, with na-

---

BOX 1.8 ENDERS, SALK, SABIN, AND ERADICATING POLIOMYELITIS

In the early 1950s, John Enders and colleagues developed methods of growing polio virus in laboratory conditions, for which they were awarded a Nobel Prize in 1954. Jonas Salk (1914–1995) at the University of Pittsburgh developed the first inactivated (killed) vaccine (IPV) under sponsorship of a large voluntary organization which mobilized the resources to fight this dreaded disease. Salk conducted the largest field trial ever, involving 1.8 million children in 1954. The vaccine was rapidly licensed and quickly developed and distributed in North America and Europe, interrupting the epidemic cycle and rapidly reducing polio incidence to low levels.

Albert Sabin (1906–1994) at the University of Cincinnati developed a live, attenuated vaccine given orally (OPV), which was approved for use in 1961. This vaccine has many advantages: it is given easily, spreads its benefits to nonimmunized persons, and is inexpensive. It became the vaccine of choice and was used widely, reducing polio to a negligible level in most developed countries within a few years. Sabin later pioneered application of OPV through national immunization days (NIDs) in South America which contributed to control of polio there, and more recently, in many other countries such as China and India.

In 1987, the World Health Organization declared the target of eradication of poliomyelitis by the year 2000. With the help of international and national commitment, the Americas were declared polio-free in 1990, with a strong potential to achieve global eradication of polio early in the new century.

A combination of OPV and IPV, with their mutually complementary advantages, was used successfully to eradicate polio in endemic areas in the late 1970s and 1980s with continued importation of wild poliovirus, and it was adopted in the United States to eliminate vaccine-associated paralytic poliomyelitis in 1997.

tional hysteria of fear of this disease because of its crippling and killing power. In 1952, 52,000 cases of poliomyelitis were reported in the United States, bringing a national response and support for the "March of Dimes" Infantile Paralysis Association for research and field vaccine trials.

Based on the development of methods of isolating and growing the virus by Enders and colleagues, Jonas Salk developed an inactivated vaccine in 1955 and Albert Sabin a live attenuated vaccine in 1961. Salk's field trial proved the safety and efficacy of his vaccine in preventing poliomyelitis. Sabin's vaccine proved to be cheaper and easier to use on a mass basis and is the mainstay of polio eradication worldwide. Conquest of this dreaded, disabling disease has provided one of the most dramatic achievements of public health in the mid twentieth century with good prospects for elimination of poliomyelitis by 2000–2002.

Advances in Treatment of Infectious Diseases

Since World War II, advances in immunology as applied to public health led to the control and in some cases potential eradication of diphtheria, pertussis, tetanus, poliomyelitis, measles, mumps, rubella, and more recently hepatitis B and *Haemophilus influenzae* type b. The future in this field is promising and will play a central role in public health well into the twenty-first century.

Treatment of infectious diseases has also played a vital role in reducing the toll of disease and reducing its spread. The search for the "magic bullet" that would cure established infectious diseases has been long and fruitful. Since the discovery by Paul Ehrlich of an effective antimicrobial agent for syphilis (Salvarsan) in 1908, the sulfa drugs in the 1930s, and penicillin and streptomycin in the 1940s by Alexander Fleming and Selman Waksman (Nobel prizes, 1945 and 1952). These and later generations of antibiotics proved to be powerful tools in the treatment of infectious diseases.

Antibiotics and vaccines, along with improved nutrition, general health and social welfare, led to dramatic reductions in infectious disease morbidity and mortality. As a result, optimistic forecasts of the conquest of infectious disease led to widespread complacency in the medical and research communities. In the 1990s, organisms resistant to available antibiotics constituted a major problem for public health and health care systems. This has reached the stage where resistant organisms are developing faster than new generation antimicrobials can be developed, threatening a return of diseases thought to be controlled. The pandemic of AIDs and other newly emerging and reemerging infectious diseases will require new modalities of treatment and prevention including new vaccines, antibiotics, chemotherapeutic agents, and risk reduction through education.

MATERNAL AND CHILD HEALTH

Preventive care for the special health needs of women and children developed as public concerns in the late nineteenth century. Public concern regarding severe conditions of women and child labor grew to include the effects on health of poverty, poor living conditions and general hygiene, home deliveries, lack of prenatal care, and poor nutrition.

Preventive care as a service separate from curative medical services for women and children was initiated in the unhygienic urban slums of industrial cities in nineteenth century France in the form of milk stations (*gouttes de lait*). One village in France instituted an incentive payment to mothers whose babies lived to 1 year; this resulted in a decline in infant mortality from 300 per 1000 to 200 per 1000 within a few years. The plan was later expanded to a complete child welfare effort, especially promoting breast-feeding and a clean supply of milk to children, which had dramatic effects in reducing infant deaths.

The concept of child health spread to other parts of Europe and the United States with the development of pediatrics as a speciality with emphasis on child feeding. Henry Koplik in 1889 and Nathan Strauss in 1893 promoted centers to provide safe milk to pregnant women and children in the slums of New York City in order to combat summer diarrhea. The Henry Street Mission serving poor immigrant areas developed the model of visiting nurses and the milk station. The milk station concept combined with home visits was pioneered by Lillian Wald, who coined the term district or public health nurse. This became the basis for public prenatal, postnatal, and well-child care as well as school health supervision. Visiting Nursing Associations (VNAs) gradually developed throughout the United States to provide such services. Physicians' services in the United States were mainly provided on a fee-for-service basis for those able to pay, with charitable services in large city hospitals. The concept of direct provision of care to those in need by local authorities and by voluntary charitable associations, with separation between preventive and curative services, is still a model of health care in many countries up to the present time.

In Jerusalem from 1902, Shaarei Zedek Hospital kept cows to provide safe milk for infants and pregnant women. In 1911, two public health nurses came from New York to Jerusalem to establish milk stations (*Tipot Halav*, drop of milk stations) for poor pregnant women and children. This model became the standard method of Maternal and Child Health (MCH) provision throughout Israel, operating parallel to the Sick Funds which provided medical care. The separation between preventive and curative services persists to the present, and is sustained by the Israeli national government's obligation to assure basic preventive care to all regardless of insurance or ability to pay.

In the Soviet Union, institution of the state health plan in 1918 gave emphasis to maternal and child health, along with epidemic and communicable disease control. All services were provided free as a state responsibility through an expanding network of polyclinics and other services, and prenatal and child care centers, including preventive checkups, home visits, and vaccinations. Infant mortality declined rapidly even in the Asian republics with previously poor health conditions.

The emphasis placed on maternal and child health continues to be one of the major elements of public health up to the present time. Care of children and women in relation to fertility is the application of what later came to be called the "risk approach," where attention is focused on designing a health program for the most vulnerable groups in the population.

NUTRITION IN PUBLIC HEALTH

As infectious disease control and later maternal and child health became public health issues in the eighteenth to nineteenth centuries, nutrition gained recognition from the work of pioneers such as James Lind (see above). In 1882, Kanehiro Takaki, Surgeon-General of the Japanese navy, reduced incidence of beriberi among naval crews by adding meat and vegetables to their diet of rice. In 1900, Christian Eikman, a Dutch medical officer in the East Indies, found that inmates of prison camps who ate polished rice developed beriberi, while those eating whole rice did not (Nobel prize, 1929). He also produced beriberi experimentally in fowls on a diet of polished rice, thus establishing the etiology of the disease as a deficiency condition fulfilling a nutritional epidemiologic hypothesis.

In the United States, the pioneering Pure Food and Drug Act was passed in 1906, stimulated by journalistic exposures of conditions in the food industry and Upton Sinclair's famous 1906 novel *The Jungle.* The legislation established federal authority in food and labeling standards, originally for interstate commerce, but later for the entire country. This provided for a federal regulatory agency and regulations for food standards. The Food and Drug Administration (FDA) pioneered standards in this field used throughout the world.

In the early part of the twentieth century, the U.S. Department of Agriculture (USDA) established an extension service that promoted good nutrition in the poor farming areas of the country which, along with local women's organizations, created a mass movement involved in improving living standards by promoting better nutrition through education and community participation.

In 1911, the chemical nature of vitamin D was discovered, and a year later, Kasimir Funk coined the term vitamin ("vital amine"). In 1914, Joseph Goldberger of the U.S. Public Health Service established the dietary causes of pellagra and in 1928 he discovered the pellagra-preventing factor in yeast. In 1916, U.S. investigators defined fat-soluble vitamin A and water-soluble vitamin B, the latter later shown to be more than one factor. In 1922, Elmer McCollum identified vitamin D in cod-liver oil, which became a staple for child care for many decades. In the period 1931–1937, fluoride in drinking water was found to prevent tooth decay, and in 1932 vitamin C was isolated from lemon juice.

Iodization of salt to prevent iodine deficiency disorders (IDD) is one of the successes and failures of twentieth century public health. From studies in Zurich and in the United States, the effectiveness of iodine supplements in preventing goiter was demonstrated. Morton's iodized salt became a national standard in the United States. In Canada in 1979, iodized salt became mandatory along with other vitamin and mineral fortification of bread and milk (see Chapter 8). Rickets, still

BOX 1.9 GOLDBERGER ON PELLAGRA

"Mal de la rosa," first described in Spain by Casal in 1735, was common in northern Italy, when, in 1771, Frappolli described "pelle agra" or farmers' skin, common among poor farmers whose diet was mainly corn flour. In 1818, Hameau described a widespread skin disease among poor farmers in southern France. Roussel investigated and concluded that pellagra was endemic and due to poverty rather than a diet heavy in corn. His recommended reforms were implemented by the Department of Agriculture and raised standards of living among the poor farmers, including growing wheat and potatoes instead of corn, and the disease disappeared by the beginning of the twentieth century. Similar measures in Italy reduced growth of corn, and here too the disease disappeared.

The disorder was thought to be due to a toxin in raw corn or produced by digestion in the intestine. Lambrozo, in Verona in northern Italy, reported many cases of pellagra among mental patients, concluding that it was due to toxic material in corn. At the beginning of the twentieth century, the corn theory was less accepted and the common view was that pellagra was an infectious disease. British investigator L. V. Sambon, discoverer of the role of the tsetse fly in trypanosomiasis, in 1910 took the view that the disease was mosquito-borne.

Pellagra was first reported in the United States in 1906 as epidemic in a mental hospital in Alabama. In the first decades of the twentieth century, pellagra was considered the leading public health problem in the southern United States, where poverty was rampant. The medical community had no ideas as to the cause or prevention of this widespread disease, generally believed to be infectious in origin.

In 1913 Joseph Goldberger was appointed by the U.S. Public Health Service to investigate pellagra. He had previously worked on yellow fever, dengue, and typhus. He visited psychiatric hospitals and orphanages with endemic pellagra and was struck by the observation that the staff was not affected, suggesting that the disease was not infectious but may have been due to the diet. In one mental hospital, he eliminated pellagra by adding milk and eggs to the diet and concluded that the disease was due to a lack of vitamins and was preventable by a change in diet alone. Goldberger, trained in infectious disease, was able to recognize nontransmission from patients to staff. He went on to establish the nutritional base of the disease, and along with Lind, established nutritional epidemiology in public health.

Source: Rosen, G. A History of Public Health. Baltimore MD: Johns Hopkins University Press, 1993.

widely prevalent in industrialized countries before World War II, and still present in the 1950s, disappeared following fortification of milk with vitamin D. Prevention of IDD by iodization of salt has become an important goal in international health, and progress is being made toward universal iodization of salt in many countries where goiter, cretinism, and iodine deficiency are still endemic.

The international movement to promote proper nutrition has become a key factor in modern public health in developing countries to reduce the toll of the malnutrition–infection cycle, and in developed countries to prevent the noncommunicable diseases associated with overnutrition, including cardiovascular diseases, diabetes, and some cancers. Nutrition has become one of the key issues in the New Public Health, with international movements to eradicate vitamin and mineral (micronutrient) deficiency conditions, all of which are important, widespread, and preventable.

MILITARY MEDICINE

Warfare has been part of mankind's experience affecting not only combat personnel but also civilians. The casualties inflicted on civil populations by disease and injury during periods of conflict are not only historical events, but very much part of the twentieth century which may have seen the bloodiest warfare ever. Especially with the advent of weapons of mass destruction, and the spread of technology for long-range delivery systems increasingly in the hands of states with unresolved regional conflicts, the dangers are great for the twenty-first century.

Since organized conflict began, armies have had to deal with the health of soldiers as well as treatment of the wounded. Biblical injunctions on camp siting and sanitation were clear and adequate for the time. Roman armies excelled at construction of camps with care and concern for hygienic conditions, and food and medical services for the soldiers. Throughout history examples of defeat of armies by disease and lack of support services prove the need of serious attention to the health and care of the soldier. Studies of the casualties of war in each of the major conflicts contributes not only to military medicine but to knowledge of the care of civilian populations in natural or man-made disasters.

Nightingale's work at the British army's Scutari Hospital in the Crimean War provided an enormous contribution to knowledge and practice of hospital organization and management. On the opposing side of the same Crimean War, Nikolai Perogov, a military surgeon in the Russian czarist army, developed rectal anesthesia for field surgery, triage of wounded by degree of severity, and hygiene of wounds. Perogov also defined improved systems for management of the wounded in war theaters, which had applicability in civilian hospitals. The French army in World War I further developed the triage system of casualty clearance now used worldwide in military and disaster situations.

Nutrition of sailors on long sea voyages and the epidemiological study of scurvy, followed a century later by the work on beriberi, were important steps in identifying nutrition and its importance to public health. Bismarck's establishment of national health insurance and other benefits for workers was partly based on the need to improve the health of the general population in order to build mass armies of healthy conscripts (see Chapter 13). During conscription to the U.S. army in World War I, high rates of rejection of draftees as medically unfit for military service raised concerns for national health standards. Finding high rates of goiter in the draftees led to efforts to identify high risk areas and to reduce iodine deficiency in the civilian population by iodization of salt.

In the wake of World War I, a massive pandemic of influenza killed some 20 million persons. Epidemics of louse-borne typhus in Russia following the war and the Russian Revolution contributed to the chaos of the period. This prompted Lenin's statement "Either socialism will conquer the louse, or the louse will conquer socialism." In World War II, sulfa drugs, antimalarials, and antibiotics made enormous contributions to the Allied war effort and later to general health care and preserving the health of the population. Lessons learned in war for protection of soldiers from disease and treatment of burns, crash injuries, amputations, battle fatigue, and many other forms of trauma were brought back to civilian health systems. Much of modern medical technology was first developed for or tested by the military. As an example, sonar radio wave mechanisms developed to detect submarines were adapted after World War II as ultrasound, a common noninvasive diagnostic tool.

In the twentieth century the destructiveness of war has increased enormously with weapons of mass destruction, chemical, biological, and nuclear. The Nuremberg Trials addressed the Holocaust and unethical medical experimentation of the Nazi military on civilian and military prisoners. The brutalities of wars against civilian populations have tragically recurred even near the end of the twentieth century in genocidal warfare in Iraq, Rwanda, and the former Yugoslavia. Those tragedies produced massive casualties and numerous refugees, with resultant public health crises requiring intervention by local and international health agencies. International and national public health agencies have major responsibility for prevention of extension of some of the mass tragedies of the twentieth century recurring, perhaps on a bigger scale in the twenty-first century.

INTERNATIONALIZATION OF HEALTH

From the first international conference on cholera in 1851 in Cairo to the health organization of the League of Nations after World War I, the idea of international cooperation in health was organized. Following World War II, the establishment of the World Health Organization (WHO) in 1946, and the independence of many formerly colonized countries, international health began to promote widespread application of public health technology, such as immunization, to developing countries. The WHO charter defined health as "the complete state of physical, social and mental well-being, and not merely the absence of disease" produced a new idealistic target in health. The 1978 International Conference of Primary health Care in Alma-Ata, Kazakhstan, influenced many country health policies.

The tradition of international cooperation is continued by organizations such as the World Health Organization, the International Red Cross/Red Crescent (IRC), United Nations Children's Fund (UNICEF), and many others. Under the leadership of the WHO, eradication of smallpox by 1977 was achieved through united action, showing that major threats to health could be controlled by international cooperation. The potential for eradication of polio further demonstrates this principle.

Health has become globalized in many senses in the late twentieth century. Infectious diseases can spread rapidly, such as that seen with the AIDS epidemic since the 1980s, and, no less dangerous, patterns of chronic disease also spread by adoption of lifestyles from the industrialized countries to those countries that are developing middle-class communities with similar disease patterns. Many other factors affect health globally, including environmental degradation with global warming, accumulating ozone and toxic wastes, acid rain, nuclear accidents, loss of nature reserves such as the Amazon basin rain forests, and the human tragedy of chronic poverty of many developing countries. Global health issues are by their very nature beyond the capacity of individual or even groups of countries to solve. They require organized common efforts of governments, international agencies, and nongovernmental organizations to cooperate with each other, with industry, and with the media to bring about change and reduce the common hazards that abuse of the environment and social gaps cause.

Bringing health care to all the people is as great a challenge as feeding a rapidly growing global population. Successes in eradication of smallpox and control of many other diseases by public health measures show the potential for concerted international cooperation and action targeted to specific objectives that reduce disease and suffering.

THE EPIDEMIOLOGIC TRANSITION

As societies evolve, so do patterns of disease. These changes are partly the result of public health and medical care but just as surely are due to improved standards of living, nutrition, housing, and economic security, as well as changes in fertility and other family and social factors. As disease patterns change, so do appropriate strategies for intervention.

During the first half of the twentieth century infectious diseases predominated as causes of death even in the developed countries. Since World War II, a major shift in epidemiologic patterns has taken place in the industrialized countries, with the decline in infectious diseases and an increase in the noninfectious diseases as causes of death. An increase in longevity has occurred primarily from declining infant and child mortality, improved nutrition, vaccine-preventable diseases controlled and treatment of acute infectious disease with antibiotics was available. The rising incidence of cardiovascular diseases and cancer affect primarily older persons, leading to a growing emphasis in epidemiologic investigations on causative risk factors for these noninfectious diseases.

Studies of the distribution of noninfectious diseases in specific groups go back many centuries when the Romans reported excess death rates among specific occupation groups. These studies were updated by Ramazzini in the early eighteenth century. As noted earlier, in eighteenth century London, Percival Potts documented that cancer of the scrotum was more common among chimney sweeps than in the general population. Nutritional epidemiologic studies, from Lind on scurvy among sailors in 1747 to Goldberger on pellagra in the southern United States in 1914, focused on nutritional cases of noninfectious diseases in public health.

Observational epidemiological studies of "natural experiments" produced enormously important data in the early 1950s, when pioneering investigators in the United Kingdom, Richard Doll, Austin Bradford Hill, and James Peto, demonstrated a relationship between tobacco use and lung cancer. They followed the mortality patterns of British physicians from different causes, especially lung cancer. They found that mortality rates from lung cancer were 10 times higher in smokers than in nonsmokers. Epidemiologic studies pointing to the relationship of diet and hypertension with cardiovascular diseases also provided important material for public health policy, and raised public concern and consciousness in western countries of the impact of lifestyle on public health. These issues are discussed subsequently throughout this book.

While infectious diseases were coming under control, the disease of modern living, such as cardiovascular diseases, trauma, and cancer, in the 1950s and 1960s became the predominant causes of premature death. These are more complex than the infectious diseases, both in causation and the means of prevention. Public health in the latter part of the twentieth century has shown surprising success in combating this set of mortality patterns by interventions that come from a combination of improved medical care and a set of activities under the general title of "health promotion."

At the the end of the twentieth century, the need for linking public health with clinical medical care and organization of services is increasingly apparent. The decline in coronary heart disease mortality is accompanied by a slow increase in morbidity, and new epidemiologic evidence shows new risk factors that are not directly related to lifestyle but require medical care over long periods of time to prevent early recurrence and premature death.

THE EVOLUTION OF PUBLIC HEALTH
IN THE TWENTIETH CENTURY

The foundations of public health organization were laid in the second half of the nineteenth and first half of the twentieth centuries. Sanitation of water, waste removal, and food control developed by municipal and higher levels of government, establishment of organized local public health offices with state and federal grants, and the improving technology of vaccination all contributed to the control of communicable diseases. Organized public health services implemented the regulatory and service components of public health in the developed countries, with national standards of food and drugs, state licensing, and discipline of the health professions.

At the beginning of the twentieth century, there were few effective treatments for disease, but improved public health standards were resulting in reduced mortality and increasing longevity. As medical technology improved following World War II with antibiotics, antihypertensives, and antipsychotic therapeutic agents, the focus was on curative medical care, with a widening chasm between public health and medicine. Toward the end of the twentieth century a new interest in the commonality between the two is emerging as new methods of organizing and financing health care emerge to contain the rising costs of health care and as preventive methods showed increasing effectiveness.

National and state efforts to promote public health during the twentieth century widened in scope of activities and financing programs. This required linkage between governmental and nongovernmental activities for effective public health services. Dramatic scientific innovations brought vaccines and antibiotics which, coupled with improved nutrition and living standards, helped to control infectious disease as the major cause of death. In the developed countries, the advent of national or voluntary health insurance on a wide scale opened access to health care to high percentages of the population.

The modern era of public health from the 1970s to the end of the twentieth century has brought a new focus on noninfectious disease epidemiology and prevention. Studies of the impact of diet and smoking on cardiovascular diseases and smoking on lung cancer isolated preventable factors for chronic disease. As a result of these and similar studies of disease and injury related to the environment, modern public health has through health promotion and consumer advocacy played and continues to play significant roles in disease prevention and contributed to dramatic changes in mortality and morbidity patterns in a wide spectrum of diseases. For prevention of premature disease and death, more comprehensive approaches will be needed by public health and health care providers than have developed to date.

The utopian dream of international and national health agencies to achieve Health for All faces serious obstacles of inequities, lack of resources, distortions with overdevelopment of some services at the expense of others, and competing priorities. Managing health care to use resources more effectively is now a concern of every health professional. At the same time, public expectations are high for unlimited access to care, including the specialized and highly technical services that can overwhelm the budgetary and manpower resources available. The wealthiest and the poorest countries, and those in between, all face the problem of managing limited resources. How that will be achieved is part of the challenge we discuss as the New Public Health.

CREATING AND MANAGING HEALTH SYSTEMS

The provision of medical care to the entire population is one of the great challenges of public health. Governments of all political stripes are active in the field of health, as insurers, providers, or regulators of health care. As will be discussed in subsequent chapters, nations have many reasons to promote health for all, just as they require universal education and literacy. National interests in the late nineteenth and early twentieth centuries were defined to include having healthy populations, especially as workers and soldiers, and for national prestige. Responsibility for the health of the nation included measures for prevention of disease, but also financing and prepayment for medical and hospital care. National policies gradually took on measures to promote health, structures to evaluate health of the nation, and modification of policies in keeping with changing needs.

The health of a population requires access to medical and hospital services as well as preventive care and a healthy environment. Greek and Roman cities appointed doctors to provide free care for the poor and the slaves. Medieval guilds provided free medical services to their members. In 1883, Germany introduced compulsory national health insurance for workers because he wanted healthy workers and army recruits, which would give him a political advantage. In 1911, Britain's Chancellor of the Exchequer, Lloyd George, instituted the National Insurance Act, providing compulsory health insurance for workers and their families. In 1918, following the October revolution, the Soviet Union created a comprehensive state-operated health system with an emphasis on prevention, providing free comprehensive care in all parts of the country.

During the 1920s national health insurance was expanded in many countries in Europe. Following the Great Depression of the 1930s and hopes raised by the victory in World War II, important social and health legislation was enacted to provide health care to the populations of Britain, Canada, and the United States. In Britain, the welfare state, including the National Health Service (NHS), was developed by the Labour Government. In Canada, a more gradual development of improved social legislation took place in the period 1940–1970, including the establishment of national pensions and a national health insurance program. In the United States, social legislation was slow in coming because of the defeat of national health insurance legislation in Congress in 1946, and strong ideological opposition to "socialized medicine," but in 1965, universal coverage of the population over age 65 (Medicare) was instituted and coverage for the poor under Medicaid soon followed. In the latter part of the twentieth century virtually every country recognizes the importance of health for the social and economic well-being of its population.

The term health systems may imply a formalized structure or a network of functions that work together directly or indirectly to meet the health needs of the population through health insurance or health service systems for their populations. Private health insurance is still the predominant mode in the United States, but the elderly and the poor are covered by governmental health insurance; the country is seeking a way to achieve universal health coverage. Prepayment for health is through general tax revenues in many countries, and in others through payment by workers and employers to social security systems. Both developed and developing countries are involved in financing health care as well as research and training of health professionals.

The industrialized countries share increasing concerns of cost escalation, with rates of increase in health expenditures which prejudice economic growth. While health care is also a large-scale employer in all developed countries, the allocation of funds, reaching 13% of Gross National Product (GNP), is a major factor in stimulating health care reform. Many countries are also struggling to keep up with rising costs and competition for other social needs, such as education, employment, and social welfare, all important for national health and well-being. Some economic theories allocated no economic value to a person except as a producer. Liberal and social democratic political philosophies advocated an ethical concern and societal responsibility in health. Both approaches now concur that health has social and economic value. The very success of public health has produced a large increase in the percentage of the elderly in the population, raising many ethical and economic questions about health care consumption, the supply of services and social support systems.

For developing countries, providing health care for all the population is a distant dream. Limited resources and overspending on high technology facilities in the capital cities leaves little for primary care for the rural and urban poor. Despite this, there has been real progress in implementing fundamental primary care services such as immunization and prenatal care, but millions of preventable deaths still occur annually because of lack of basic primary care programs.

SUMMARY

The history of public health is directly related to the evolution of thinking about health. Ancient societies in one way or another realized the connection between sanitation and health and the role of personal hygiene, nutrition, and fitness. The sanctity of human life (*Pikuah Nefesh*) established an overriding human responsibility to save life derived from Mosaic Law from 1500 BC. The scientific and ethical basis of medicine was also based on the teachings of Hippocrates in the fourth century BC. Sanitation, hygiene, good nutrition, and physical fitness for health all had roots in many ancient societies including obligations of the society to provide care for the poor. These ethical foundations support efforts to preserve life even at the expense of other religious or civil ordinances.

Social and religious systems linked disease to sin and punishment by higher powers, viewed investigation or intervention by society, except for relief of pain and suffering, as interference with God's will. Childbirth was associated with pain, disease, and frequent death as a general concept of "in sorrow shall you bring forth children." Health care was seen as a religious charitable responsibility to ease suffering of the sinner.

The clear need and responsibility of society to protect itself by preventing entry or transmission of infectious diseases was driven home by the pandemics of leprosy, plague, syphilis, smallpox, measles, and other communicable diseases in the Middle Ages. The diseases themselves evolved, and pragmatic measures were gradually found which were meant to control their spread, such as isolation of lepers, quarantine of ships, and closure of public bathhouses. Epidemiologic investigations of cholera, typhoid, occupational diseases, and nutritional deficiency disorders in the eighteenth and nineteenth centuries began to show causal relationships and effective ways of intervention before scientific proof of causation was established. Even in the twentieth century, public health practice has continued to proceed on a pragmatic basis long before the full scientific basis of the causation of diseases has been worked out. Public health organizations to ensure basic community sanitation and other modalities of prevention evolved through the development of local health authorities, fostered, financed, and supervised by civic, state, or provincial and national health authorities as governments became more and more involved in health issues.

Freeing human thought from dogmas which restricted scientific exploration of health and disease led to the search for the natural causation of disease. This was of paramount importance in seeking interventions and preventive activities. This concept, first articulated in ancient Greek medicine, provided the basis for clinical and scientific observations leading to the successes of public health in the past two centuries. The epidemiologic method led to public health interventions before the biological basis of the condition was determined. Sanitation to prevent disease was accepted in many ancient societies, and codified in some as part of civil and religious obligations. Lind's investigation of scurvy, Jenner's discovery of vaccination to prevent smallpox, and Snow's investigation of cholera in London demonstrated in modern scientific epidemiologic terms the cause of disease and were accepted despite lack of contemporary biochemical and bacteriologic proof. They helped to formulate the methodology of public health.

Public health developed through pioneering epidemiologic studies, devising many forms of preventive medicine and community health promotion. Reforms pioneered in many areas, from abolition of slavery and serfdom to provision of state legislated health insurance, have all improved the health of the general population. In the last years of the twentieth century, the relationship between health and social and economic development has gained recognition internationally. The twentieth century has seen a dramatic expansion of the scientific basis for medicine and public health. Immunology, microbiology, pharmacology, toxicology, and epidemiology have provided powerful tools and results in improving the health status of the population. New medical knowledge and technology have come to be available to the general public in many countries in the industrialized world through the advent of health insurance. In this century, virtually all industrialized countries established systems of assuring access to care for all the population as essential for the health of the individual and the population.

Major historical concepts have had profound effects on the development of public health. Sickness as punishment for sin prevented attempts to control disease over many centuries. This mentality continues by "blaming the victim." AIDS patients are seen as deserving their fate because of their behavior, workers are believed to get injured because of their own negligence, and the obese person and the smoker are believed to deserve their illnesses because of the way they conduct their lives. Conversely, public health is associated with charity in which there is a religious responsibility for kindness and relief of suffering. Ethical controversies are still important in such diverse areas as managed care, reproductive health, cost–benefit analysis, euthanasia, and care of prisoners.

Acceptance of the right to health for all by the founders of the United Nations and the World Health Organization added a universal element to the mission of public health. This concept was embodied in the constitution of the WHO and given more concrete form in the Health for All concept of Alma-Ata which emphasizes the right of health care for everyone and the responsibility of governments to ensure that right. This concept also articulated the primary importance of prevention and primary care, which became a vital issue in the competition for resources between public health and hospital-oriented health care.

The lessons of history are important in public health. The basic issues of public health need to be relearned because new challenges for health appear while old ones reappear. The philosophical and ethical basis of modern public health is a belief in the inherent worth of the individual and his/her human right to a safe and healthful environment. The health and well-being of the individual and the community are interdependent. Investment in health, as in education, is a contributor to economic growth, as healthy and educated individuals contribute to a creative and economically productive society.

The New Public Health is derived from the experience of history. Organized activity to prevent disease and promote health had to be relearned from the ancient and post-industrial revolution worlds. As the twentieth century draws to a close, we need to learn from a wider framework how to use all health modalities, including clinical and prevention-oriented services, to effectively and economically preserve, protect, and promote the health of the individual and of society. The New Public Health, as public health did in the past, faces ethical issues which relate to health expenditures, priorities, and social philosophy. Throughout the course of this book, we discuss these issues and try to indicate a balanced approach toward the New Public Health.

HISTORICAL MARKERS

| 3000 BC | Dawn of Sumerian, Egyptian, and Minoan cultures—drains, flush toilets |
| 2000 | Indus Valley—urban society with sanitation facilities |

1700	The Code of Hammurabi—rules governing medical practice
2000	Indus Valley—urban society with sanitation facilities
1500	Mosaic Law—personal, food, and camp hygiene, segregating lepers, overriding duty of saving of life (*Pikuah Nefesh*) as religious imperatives
400	Greece—personal hygiene, fitness, nutrition, sanitation, municipal doctors, occupational health; Hippocrates—clinical and epidemic observation and environmental health
500 BC to AD 500	Rome—aqueducts, baths, sanitation, municipal planning, and sanitation services, public baths, municipal doctors, military, and occupational health
AD 170	Galen—physiology, anatomy, humors dominated western medicine until AD 1500
500–1000	Europe—destruction of Roman society and the rise of Christianity; sickness as punishment for sin, mortification of the flesh, prayer, fasting, and faith as therapy; poor nutrition and hygiene, pandemics; antiscience; care of the sick as religious duty
700–1200	Islam—preservation of ancient health knowledge, schools of medicine, Arab–Jewish medical advances (Ibn Sinna and Maimonides)
1000+	Universities and hospitals in Middle East and Europe
1000+	Rise of cities, trade, and commerce, craft guilds, municipal hospitals
1096–1272	Crusades—contact with Arabic medicine, hospital orders of knights, leprosy
1268	Roger Bacon publishes treatise on use of eyeglasses to improve vision
1348	Venice—Board of health and quarantine established
1348–1350	Black Death—origins in Asia, spread by armies of Genghis Khan, world pandemic kills 60 million in fourteenth century, 1/3 to 1/2 of the population of Europe
1300	Pandemics—bubonic plague, smallpox, leprosy, diphtheria, typhoid, measles, influenza, tuberculosis, anthrax, trachoma, scabies, and others until eighteenth century
1400–1600	Renaissance and enlightenment, decline of feudalism, rise of urban middle class, trade, commerce, exploration, new technology, arts, science, anatomy, microscopy, physiology, surgery, clinical medicine, hospitals (religious, municipal, voluntary)
1518	Royal College of Physicians founded in London
1532	Bills of Mortality published
1546	Girolamo Fracastorus publishes *De Contagione*—the germ theory
1562–1601	Elizabethan Poor Laws—responsibility for the poor on local government
1628	William Harvey publishes findings on circulation of the blood
1629	London Bills of Mortality specify causes of death
1639	Massachusetts law requires recording of births and deaths
1660s	Leyden University strengthens anatomical education
1661	John Graunt founds medical statistics
1661	Rene Descartes publishes first treatise on physiology
1662	Royal Society of London founded by Francis Bacon
1665	Great Plague of London
1673	Antony van Leeuwenhoek—microscope, observes sperm and bacteria
1667	Pandemics of smallpox in London; pandemic of malaria in Europe
1687	William Petty publishes *Essays in Political Arithmetic*
1700	Bernardino Rammazzini publishes compendium of occupational diseases
1701	London—75% of newborns die before fifth birthday
1701	Variolation against smallpox practiced in Constantinople, isolation practiced in Massachusetts
1710	English Quarantine Act
1720+	London—voluntary teaching in hospitals; Guy's, Westminster

1730+	Science and scientific medicine; Rights of Man, encyclopedias, agricultural and industrial revolutions, population growth—high birth rates, falling death rates
1733	Obstetrical forceps invented
1733	Stephen Hales measures blood pressure
1747	James Lind—case control study of scurvy in sailors
1750+	British naval hospitals established
1750+	John Hunter establishes modern surgical practice and teaching
1752	William Smellie publishes textbook of midwifery
1762	Jean Jacques Rousseau publishes *Social Contract*
1775	Percival Pott investigates scrotal cancer in chimney sweeps
1777	John Howard promotes prison and hospital reform in England
1779	Johann Frank promotes Medical police in Germany
1785	William Withering—foxglove (digitalis) treatment of dropsy
1788	British Legislation to protect boys employed as chimney sweepers
1796	Edward Jenner—first vaccination against smallpox
1797	Massachusetts legislation permitting local boards of health
1798	Philippe Pinel removes chains from insane in Bicetre Asylum
1800⎤	Britain and United States—Municipal Boards of Health
1830⎦	Sanitary and social reform, growth of science
1800	Vaccination adopted by British army and navy
1800+	Adam Smith, Jeremy Bentham—economic, social philosophers
1801	Vaccination mandatory in Denmark, local eradication of smallpox
1801	First national census, United Kingdom
1802	U.S. Marine hospitals founded; later U.S. Public Health Service
1804	Modern chemistry established—Humphrey Davey, John Dalton
1807	Abolition Act—mandates eradication of international slave trade by the Royal Navy
1827	Carl von Baer in St. Petersburg establishes embryology
1834	Poor Law Amendment Act documents harsh state of urban working class in U.K.
1837	Pierre Louis—French founder of modern epidemiology
1837	United Kingdom National Vaccination Board
1840s	Voluntary societies for reform, boards of health, mines and factory acts—improving work conditions
1842	Edwin Chadwick—Sanitary Commission links poverty and disease
1844	Horace Wells—anesthesia in dentistry then surgery
1848	U.K. Parliament passes Public Health Act establishing the General Board of Health
1850	Massachusetts—Shattuck Report of Sanitary Commission
1852	Adolph Chatin uses iodine for prophylaxis of goiter
1854	John Snow—waterborne cholera in London: the Broad Street pump
1854	Florence Nightingale, modern nursing and hospital reform—Crimean War
1855	London—mandatory filtration of water supplies and consolidation of sanitation authorities
1858	Louis Pasteur proves no spontaneous generation of life
1858	Rudolph Virchow publishes *Cellular Pathology;* pioneer in political–social health context
1858	Public Health and Local Government Act and Medical Act in United Kingdom—local health authorities and national licensing of physicians
1859	Charles Darwin publishes *On the Origin of Species*
1861	Emancipation of the serfs in Russia

1861	Ignaz Semmelweiss publishes *The Cause, Concept and Prophylaxis of Puerperal Fever*
1862	Louis Pasteur publishes findings on microbial causes of disease
1862	Florence Nightingale founds St. Thomas' Hospital School of Nursing
1862	Emancipation of slaves in United States
1864	Boston bans use of milk from diseased cows
1864	Russia—rural health as tax-supported local service through Zemstvos
1866	Gregor Johann Mandel, Czech monk, basic laws of heredity, basis of genetics
1867	Joseph Lister describes use of carbolic spray for antisepsis
1869	Dimitri Ivanovitch Mendeleev—periodic tables
1872	American Public Health Association founded and milk stations established in New York
1876	Robert Koch discovers anthrax bacillus
1879	Neisser discovers gonococcus organism
1879	U.S. National Board of Health established
1879	U.S. Food and Drug Administration established
1882	Robert Koch discovers the tuberculosis organism, tubercle bacillus
1880	Typhoid bacillus discovered (Laveran); leprosy organism (Hansen); malaria organism (Laveran)
1883	Otto von Bismarck—workmen's compensation, national health insurance for workers and their families in Germany
1883	Robert Koch discovers bacillus of cholera
1883	Louis Pasteur vaccinates against anthrax
1884	Diphtheria, staphylococcus, streptococcus, tetanus organisms identified
1885	Pasteur develops rabies vaccine; Escherich discovers coli
1886	Karl Fraenkel discovers pneumococcus organism
1887	Malta fever or brucellosis (Bruce) and chancroid (Ducrey) organisms identified
1887	U.S. National Institutes of Health founded
1890	Anti-tetanus serum (ATS)
1892	Gas gangrene organism discovered by Welch and Nuttal
1894	Plague organism discovered (Yersin, Kitasato); botulism organism (Van Ermengem)
1895	Wilhelm Roentgen—discovers electromagnetic waves (X rays) for diagnostic imaging
1897	London School of Hygiene founded
1897	Felix Hoffman—synthesizes acetylsalicylic acid (aspirin)
1904	Ivan Petrovitch wins Nobel Prize for work in conditioned reflexes, neurophysiology
1905	U.S. Pure Food Act
1905	Abraham Flexner—major report on medical education in U.S.
1905	Workman's Compensation Acts in Canada
1910	Paul Ehrlich—chemotherapy use of arsenical salvarsan for treatment of syphilis
1911	United Kingdom—compulsory health insurance for workers
1911	Casimir Funk investigates vitamins (vital amines)
1912	Emil Von Behring—diphtheria antitoxin
1912	Health insurance for industrial workers in Russia
1912	U.S. Children's Bureau established
1914	Joseph Goldberger—investigates cause and prevention of pellagra
1915	Tetanus prophylaxis and antitoxin for gas gangrene
1918	Pandemic of swine flu (influenza) kills some 20 million people
1918	Nikolai Semashko introduces U.S.S.R. national health plan
1921	Frederick Banting and Charles Best discover insulin in Toronto

1923	Health Organization of League of Nations
1924	David Cowie promotes widespread ionization of salt in the United States
1926	Pertussis vaccine developed
1928	Alexander Fleming discovers penicillin
1928	George Papanicolaou—Pap smear for detection of cancer of cervix
1929–1936	The Great Depression—widespread economic collapse, unemployment, poverty, and social distress in industrialized countries
1935	U.S. Social Security Act and the New Deal in the United States
1940	Charles Drew describes storage and use of blood plasma for transfusion
1941	Normal Gregg publishes on rubella as cause of congenital anomalies
1939–1945	World War II
	U.K. National Hospital Service—wartime nationalization of hospitals; The Beveridge Report in the United Kingdom—the "Welfare State" (1942)
	U.S. National Centers for Disease Control
	U.S. Emergency Maternity and Infant Care for families of servicemen
	U.S.S.R. wartime emergency medical structure
1946	World Health Organization founded
1946	National health insurance defeated in U.S. Congress
1946	U.S. Congress passes Hill–Burton Act assisting local hospital construction up to 4.5 beds/1000 population
1946	Tommy Douglas—Saskatchewan provincial hospital insurance plan
1948	United Kingdom establishes National Health Service
1953	James Watson and Francis Crick—structure of DNA
1954	Framingham study of heart disease risk factors
1954	Richard Doll reports on link of smoking and lung cancer
1954	Jonas Salk's inactivated poliomyelitis vaccine licensed
1956	Gregory Pincus reports on first successful trials of birth control pills
1960	Albert Sabin's live poliomyelitis vaccine licensed
1962	Francis Crick, Thomas Watson discover DNA, the genetic code
1963	Measles vaccine licensed
1965	U.S. passes Medicare for the elderly, Medicaid for the poor
1966	U.S. National Traffic and Motor Vehicle Safety Act
1966	U.S. Surgeon General's Report on Smoking
1967	Mumps vaccine licensed
1970	Rubella vaccine licensed
1971	Canada has universal health insurance in all provinces
1974	LaLonde Report—New Perspectives on the Health of Canadians
1977	WHO adopts Health for All by the Year 2000
1978	Alma-Ata Conference on Primary Health Care
1978	Hepatitis B vaccine licensed
1979	Canada adopts mandatory vitamin/mineral enrichment of foods
1979	WHO declares eradication of smallpox achieved
1981	First recognition of cases of acquired immune deficiency syndrome (AIDS)
1985	WHO European Region Health Targets
1985	Luc Montaignier publishes genetic sequence of HIV
1989	WHO targets eradication of polio by the year 2000
1989	Warren and Marshall show *Helicobacter pylori* as a treatable cause of gastritis and peptic ulcers
1989	International Convention on the Rights of the Child
1990	World Summit on Children, New York
1990	World Conference on Education for All, Jomtien
1990	W.F. Anderson performs first successful gene therapy

1990	Concern for newly emerging, reemerging diseases (HIV, Marburg, Ebola, cholera, mad cow disease, tuberculosis) and multidrug resistant organisms
1992	United Nations Conference on Environment and Development, Rio de Janiero
1992	International Conference on Nutrition
1993	World Conference on Human Rights, Vienna
1993	*World Development Report: Investing in Health* published by World Bank
1993	Russian Federation approves compulsory national health insurance
1994	International Conference on Population and Development, Cairo
1994	Clinton National Health Insurance plan defeated in U.S. Congress
1995	World Summit for Social Development, Copenhagen
1995	United Nations Fourth World Conference on Women, Beijing
1996	Second United Nations Conference on Human Settlement (Habitat II), Istanbul
1996	Explosive growth of managed care plan coverage in the United States
1997	Legal suits for damages against tobacco companies for costs of health effects of smoking, 33 states in the United States and other countries
1998	Clinton proposed legislation on patients rights in managed care
1998	FDA approves rotavirus vaccine
1998	WHO Health for All in the Twenty-first Century adopted
1998	U.S. National Academy of Sciences recommends routine vitamin supplements for adults
1999	U.S. Congress passes legislation regulating patients' rights in managed care

RECOMMENDED READINGS

Ashton, J. 1997. Chadwick Lecture—Is a healthy North West Region achievable in the 21st century? *Journal of Epidemiology and Community Health,* 53:370–382.

Baker, J. P. 1994. Women and the invention of well child care. *Pediatrics,* 94:527–531.

Markel, H. 1987. When it rains it pours: Endemic goiter, iodized salt, and David Murray Cowie, MD. *American Journal of Public Health,* 77:219–229.

Larson, E. 1989. Innovations in health care: Antisepsis as a case study. *American Journal of Public Health,* 79:92–99.

Monteiro, L. A. 1985. Florence Nightingale on public health nursing. *American Journal of Public Health,* 75:181–186.

Rosen, G. 1958. *A History of Public Health.* New York: MD Publications. Republished as Expanded Edition. Baltimore, MD: Johns Hopkins University Press, 1993.

Roueche, B. (ed). 1963. *Curiosities of Medicine: An Assembly of Medical Diversions 1552–1962.* London: Victor Gollancz Ltd.

BIBLIOGRAPHY

Barkan, I. D. 1985. Industry invites regulation: The passage of the Pure Food and Drug act of 1906. *American Journal of Public Health,* 75:18–26.

Buehler-Wilkerson, K. 1993. Bringing care to the people: Lillian Wald's legacy to public health nursing. *American Journal of Public Health,* 83:1778–1786.

Camus, A. 1947. *The Plague.* Middlesex, England: Penguin Modern Classics.

Carter, K. D. 1991. The development of Pasteur's concept of disease causation and the emergence of specific causes in nineteenth-century medicine. *Bulletin of the History of Medicine,* 65:528–548.

Diamond, J. 1997. *Guns, Germs and Steel: The Fates of Human Societies.* New York: W. W. Norton Co.

Garrison, F. H. 1929. *An Introduction to the History of Medicine,* Fourth Edition. Republished by WB Saunders Co., Philadelphia, 1966.

Grob, G. N. 1985. The origins of American psychiatric epidemiology. *American Journal of Public Health,* 75:229–236.

Hughes, J. G. 1993. Conception and creation of the American Academy of Pediatrics. *Pediatrics,* 92:469–470.

Light, D. W, Liebfried, S., Tennstedt, F. 1986. Social medicine vs professional dominance: The German experience. *American Journal of Public Health,* 76:78–83.

Mack, A. (ed). 1991. *The Time of the Plague: The History and Social Consequences of Lethal Epidemic Disease.* New York: New York University Press.

Marti-Ibanez, F. (ed). 1960. *Henry E. Sigerist on the History of Medicine.* New York: MD Publications.

Massachusetts Sanitary Commission. 1850. *Report of a General Plan for the Promotion of Public and Personal Health, Sanitary Survey of the State.* Reprinted by Arno Press & The New York Times, New York, 1972.

McCullough, D. 1977. *The Path Between the Seas: The Creation of the Panama Canal 1870–1914.* New York: Touchstone.

McNeill, W. H. 1989. *Plagues and Peoples.* New York: Anchor Books.

Monteiro, L. A. 1985. Florence Nightingale on public health nursing. *American Journal of Public Health,* 75:181–186.

Plotkin, S. L., Plotkin, S. A. 1994. A short history of vaccination. *In* Plotkin SA, Mortimer EA (eds) *Vaccines,* Second Edition. Philadelphia: WB Saunders.

Rather, L. J. (ed). 1958. *Disease, Life, and Man: Selected Essays by Rudolf Virchow.* Stanford: Stanford University Press.

Roberts, D. E., Heinrich, J. 1985. Public health nursing comes of age. *American Journal of Public Health,* 75:1162–1172.

Roemer, M. I. (ed). 1960. *Sigerist on the Sociology of Medicine.* New York: MD Publications.

Roemer, M. I. (1988). Resistance to innovation: The case of the community health center. *American Journal of Public Health,* 78:1234–1239.

Rosenberg, C. E. 1992. *Explaining Epidemics and Other Studies in the History of Medicine.* Cambridge, UK: Cambridge University Press.

Sinclair, U. 1906. The Jungle.

Slaughter, F. G. 1950. *Immortal Magyar: Semmelweiss, Conqueror of Childbed Fever.* New York: Henry Schuman.

Smith, I. S. 1990. *Patenting the Sun: Polio and the Salk Vaccine.* New York: Wm. Morrow & Co.

Snow, J. *Snow on Cholera: A Reprint of Two Papers.* New York: Commonwealth Fund, 1936.

Sorokina, T. S. 1994. *History of Medicine.* Moscow: PAIMS. [in Russian]

Starr, C. G. 1991. *A History of the Ancient World.* New York: Oxford University Press.

Tuchman, B. W. 1978. *A Distant Mirror: The Calamitous Fourteenth Century.* New York: Alfred A. Knopf Inc.

Wills, C. 1997. *Plagues: Their Origin, History, and Future.* London: Flamingo Press.

2

EXPANDING THE CONCEPT
OF PUBLIC HEALTH

INTRODUCTION

The evolution of public health from its ancient and recent roots, in the past two centuries especially, point to a continuing process, with revolutionary leaps forward. The study of history helps the student and the practitioner to monitor the process and to define where we came from, and contributes to understanding where we are going. This may seem academic, but it is vital to understand change to be able to deal with radical changes of direction that occur and will continue to develop with science, practice, and economics, which are major driving forces of change and public health development at the end of the twentieth century. For the coming generations, not just quality of life, but survival are at stake.

Originally, public health was seen as a discipline which studies and implements measures for control of communicable diseases, primarily by sanitation and vaccination. The sanitary revolution, which came before the development of modern bacteriology, made an enormous contribution to improved health, but improved nutrition, education, and housing were no less important. Maternal and child health and many other aspects of a growing public health network of activities played important roles, as have the physical and social environment and personal habits of living in determining health status.

Public health has evolved as a multidisciplinary field that includes the use of basic and applied science, education, the social sciences, economics, management, and communication skills to promote the welfare of the individual and the community. It is greater than the sum of its component elements and includes the art and politics of the coordination of diverse community and individual services. This holistic view, of balance and equilibrium, may be a renaissance of classical Greek and Mosaic traditions, applied with the broad knowledge and experience of public health and medical care at the end of the twentieth century as change continues to challenge our capacity to adapt.

BOX 2.1 ARISTOTLE ON HEALTH

"Health of body and mind is so fundamental to the good life that if we believe men have any personal rights at all as human beings, then they have
an absolute moral right to such measure of good health as society and society alone is able to give them."

The practitioner of public health is as much engaged in organization as involved
in direct caregiving. But, any health professional needs a thorough understanding
of the issues that are included in the New Public Health and how they evolved, interact, are put together in organizations, and are financed in various parts of the
world in order to understand changes going on before his/her eyes.

The New Public Health is comprehensive in scope. It relates to or encompasses all community and individual activities directed toward reducing factors that
contribute to the burden of disease and fosters those that relate directly to improved
health. Its programs range from immunization, health promotion, and child care to
food labeling and food fortification to the assurance of well-managed, accessible
health care services. The planning, management, and monitoring functions of a
health system are indispensable in a world of limited resources and high expectations. This requires well-developed information systems to provide the feedback
and control data needed for good management. It includes responsibilities and
coordination at all levels of government and by nongovernment organizations
(NGOs) and participation of a well-informed media and strong professional and
consumer organizations. No less important are clear designations of responsibilities of the individual for his/her own health, and of the provider of care for humane, high quality professional care.

EVOLUTION OF PUBLIC HEALTH

Many changes have signaled a need for transformation toward the New Public
Health. Religion, although still a major political and policy-making force in many
countries, is no longer the central organizing force in most societies. Societies have
evolved from large extended families and tribes to rural societies, cities, and national governments. With the growth of urban communities, and later multina-
tional economic systems, the health of individuals became more than just a local
problem. A person is not only a citizen of the village, city, or country in which he
or she lives, but of a "global village."

The agricultural revolutions of the fifteenth to seventeenth centuries which increased food supply, were followed only much later by knowledge of nutrition as

a public health issue. The scientific revolution of the seventeenth to nineteenth centuries provided means to describe and analyze the spread of disease and the poisonous effects of the industrial revolution, including crowded living conditions and pollution of the environment. In the mid twentieth century, a new agricultural "green revolution" had a great impact in reducing human deprivation internationally, yet the ecological bill has yet to be paid.

These changes and many other societal changes discussed in Chapter 1 have enabled public health to expand its potential and horizons, while developing its pragmatic and scientific base. Organized public health of the twentieth century has proved effective in reducing the burden of infectious diseases and has contributed to improved quality of life and longevity. In the last half-century, chronic diseases have become the primary causes of morbidity and mortality in the developed countries and increasingly in developing countries. Growing scientific and epidemiologic knowledge increases the capacity to deal with these diseases. Many aspects of public health can only be influenced by the behavior and risks to the health of individual persons. These require interventions that are more complex and relate to personal lifestyle as much as environmental, community, and societal factors.

Chronic diseases have come to the center stage in what has come to be called the epidemiological transition. At the same time, a revolution is in the making in the economics and management of health systems due to rapidly rising costs and a nearly universal recognition of the rights of people to have access to health care of reasonable quality. These have led public health professionals and institutions to become involved in the management and economics of health systems. Concepts such as objectives, targets, priorities, cost-effectiveness, and evaluation have become part of the New Public Health agenda. Understanding of how these concepts evolved helps the future health provider or manager to cope with the complexities of mixing science, humanity, and effective management of resources to achieve higher standards of health and to cope with new issues as they develop in the broad scope of the New Public Health for the twenty-first century.

HEALTH AND DISEASE

Health can be defined from many perspectives, ranging from factual statements of mortality and morbidity rates to idealized versions of human and societal perfection. An operational definition of health is as follows: a state of equilibrium of the person with the biological, physical, and social environment, with the object of maximum functional capability. Health is thus seen as a state characterized by anatomic, physiological and psychological integrity, and an optimal functional capability in the family, work, and societal roles (including coping with associated stresses), a feeling of well-being, and freedom from risk of disease and premature death.

There are many interrelated factors in disease and in their management. In 1878, Claude Bernard described the phenomenon of adaptation and adjustment of the internal milieu of the living organism to physiologic processes. This concept is fundamental to medicine. It is also central to public health because understanding the spectrum of events and factors between health and disease is basic to the identification of contributory factors affecting the balance, and to seeking the points of potential intervention to reverse the balance toward health.

From the time of Hippocrates and Galen, diseases were thought to be due to humors and miasmata or emanations from the environment. The miasma theory, while without basis in fact, was acted on in the early to mid nineteenth century with practical measures to improve sanitation, housing, and social conditions with successful results. The competing germ theory developed by pioneering epidemiologists (Pannum, Snow, and Budd), scientists (Pasteur, Cohn, and Koch), and practitioners (Lister and Semmelweiss) led to the science of bacteriology and a revolution in practical public health measures. The combined application of these two theories has been the basis of classic public health, with enormous benefits coming in the control of infectious disease.

Host–Agent–Environment Paradigm

In the host–agent–environment paradigm, a harmful agent comes through a sympathetic environment into contact with a susceptible host, causing a specific disease. This has dominated public health thinking until the mid twentieth century. The host is the person who has or is at-risk for a specific disease. The agent is the organism or direct cause of the disease. The environment includes the external factors which influence the host, his or her susceptibility to the agent, and the vector which transmits or carries the agent to the host from the environment. This explains the causation and transmission of many diseases. A vector is a mode of transmission by which the agent reaches the host from the environment. This paradigm, in effect (Fig. 2.1), joins together the contagion and miasma theories of disease causation. A specific agent, a method of transmission, and a susceptible host are involved in an interaction, which are central to the infectivity or severity of the disease. The environment can provide the carrier or vector of an infective (or toxic) agent, and it also contributes factors to host susceptibility e.g., poverty, low education level.

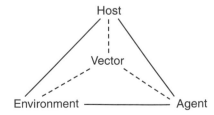

FIGURE 2.1   The host–agent–environment paradigm.

The Expanded Host–Agent–Environment Paradigm

The expanded host–agent–environment paradigm widens the definition of each of the three components (Fig. 2.2), in relation to both acute infectious and chronic noninfectious disease epidemiology. In the latter half of the twentieth century, this expanded host–agent–environment paradigm took on added importance in dealing with the complex of factors related to chronic diseases, now the leading causes of disease and premature mortality in the developed world, and increasingly in developing countries.

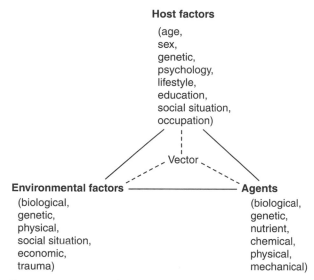

FIGURE 2.2   The expanded host–agent–environment paradigm.

Interventions to change host, environmental, or agent factors are the essence of public health. In infectious disease control the biological agent may be removed by pasteurization of food products or filtration and disinfection (chlorination) of water supplies to prevent transmission of waterborne disease. The host may be altered by immunization to provide immunity to the infective organism. The environment may be changed to prevent transmission by destroying the vector or its reservoir of the disease. A combination of these interventions can be used against a specific organism or disease.

Vaccine-preventable diseases may require both routine and special activities to boost herd immunity to protect the individual and the community. For other infectious diseases for which there is no vaccine, for example, malaria, control involves a broad range of activities including case finding and treatment to improve the individual's health and to reduce the reservoir of the disease in the population, as well as vector control to reduce the mosquito population. Tuberculosis control requires not only case finding and treatment, but understanding the contributing factors of social conditions, diseases with tuberculosis as a secondary condition (drug abuse and AIDS), agent resistance to treatment, and the inability of patients or carriers to complete treatment without supervision. Sexually transmitted diseases, not controllable by vaccines, require a combination of personal behavior change, health education, medical care, and skilled epidemiology.

With noninfectious diseases, intervention is even more complex, involving human behavior factors and a wide range of legal, administrative, and educational issues. There may be multiple risk factors, which have a compounding effect in disease causation, and they may be harder to alter than infectious diseases factors. For example, smoking in and of itself is a risk factor for lung cancer, but exposure to asbestos fibers has a compounding effect. Preventing exposure may be easier than smoking cessation. Reducing trauma morbidity and mortality is equally problematic.

The situation of a single, identifiable cause resulting in a disease is scientifically and practically of great value in modern public health, enabling such direct interventions as use of vaccines or antibiotics to protect or treat individuals from infection by a specific organism. The cumulative effects of several contributing or risk factors in disease causation are also of great significance in many disease entities, in relation to the infectious diseases, such as tuberculosis, or chronic diseases such as the cardiovascular group.

The health of an individual is affected by risk factors intrinsic to that person as well as by external factors. Intrinsic factors include the biological ones that the individual inherits and those life habits he or she acquires, such as smoking, overeating, or engaging in other high risk behaviors. External factors affecting individual health include the environment, the socioeconomic and psychological state of the person, the family, and the society in which he/she lives. Education, culture, and religion are also contributing factors to individual and community health.

There are factors that relate to health of the individual in which the society or the community can play a direct role. One of these is provision of medical care. Another is to ensure that the environment and community services include safety factors that reduce the chance of injury and disease, or include protective measures, for example, flouridation of a community water supply to improve dental health, and seat belt or helmet laws to reduce traffic injury and death. These modifying factors may affect the response of the individual or the spread of an epidemic (see Chapter 3). An epidemic may also include chronic disease, since common risk factors may cause an excess of cases in a susceptible population group, as compared to before the risk factor appeared, or in comparison to a group not exposed to the risk factor.

THE NATURAL HISTORY OF DISEASE

Disease is a dynamic process, not only of causation, but also of severity and the effects of intervention intended to modify outcome. Knowledge of the natural history of disease is fundamental to understanding where and with what means intervention can have the greatest chance for successful interruption or change in the disease process, for the patient, the family, or the community.

The natural history of a disease is the course of that disease from beginning to end. This includes the factors that relate to its initiation, its clinical course leading up to resolution, cure, continuation, or long-term sequelae (further stages or complications of a disease), and environmental or intrinsic (genetic or lifestyle) factors and their effects at all stages of the disease. The effects of intervention at any stage of the disease are part of the disease process (Fig. 2.3).

As discussed above, disease occurs in an individual when agent, host, and environment interact to create adverse conditions of health. The agent may be an infectious organism, a chemical exposure, a genetic defect, or a deficiency condition. A form of individual or social behavior may lead to injury or disease, such as reckless driving or risky sexual behavior. The host may be immune or susceptible, as a result of many contributing social and environmental factors. The environment includes the vector, which may be a malaria-bearing mosquito, a con-taminated needle shared by drug users, lead-contaminated paint, or an abusive family situation.

Assuming a natural state of "wellness," that is, optimal health or a sense of well-being, function, and absence of disease, a disease process may begin with a course of a disease, infectious or noninfectious, following a somewhat characteristic pattern described by clinicians and epidemiologists. Preclinical or predisposing events may be detected by a clinical history, with determination of risk including possible exposure or presence of other risk factors. Interventions, before and during the process, are intended to affect the later course of the disease.

The clinical course of a disease, or its laboratory or radiological findings, may be altered by medical or public health intervention, leading to the resolution or continuation of the disease with fewer or less severe secondary sequelae. Thus the intervention becomes part of the natural history of the disease. The natural history of an infectious disease in a population will be affected by the extent of prior vaccination or previous exposure in the community. Diseases particular to children are often so because the adult population is immune from previous exposure or vaccinations. Measles and diphtheria, primarily childhood diseases, now affect adults to a large

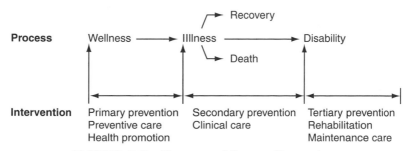

FIGURE 2.3   The process of disease and intervention.

extent because they are less protected by naturally acquired immunity or are vulnerable when their immunity wanes due to inadequate vaccination in childhood.

In chronic disease management, high costs to the patient and the health system accrue where preventive service or management are inadequate. The progress of a diabetic to severe complications is delayed by good management. The patient with advanced chronic obstructive pulmonary disease or congestive heart failure may be managed well and remain stable with careful management of medications, immunizations against influenza and pneumonia, and other care needs. If this fails, the patient might well require long and expensive care. Failure to provide adequate supportive care will show up in ways more costly to the health system and will prove more life threatening to the patient.

As in an individual, the phenomenon of a disease in a population may follow a course in which many factors interplay, and where interventions affect the natural course of the disease. The epidemiologic patterns of an infectious disease can be assessed in their population patterns, just as they can for individual cases. The classic mid-nineteenth century description of measles in the Faroe Islands by Panum showed transmission and the epidemic nature of the disease as well as the protective effect of acquired immunity.

SOCIETY AND HEALTH

The health of populations, like the health of individuals, depends on societal factors no less than on genetics, personal risk factors, and medical services. Social inequalities in health have been understood and documented in public health over the centuries. The Chadwick and Shattuck reports of the 1840–1850 period documented the relationship of poverty, bad sanitation, housing, and working conditions with high mortality, and ushered in the idea of social epidemiology. Political and social ideologies thought that the welfare state, including universal health care systems of one type or another, would eliminate social and geographic differences in health status.

From the introduction of compulsory health insurance in Germany in the 1880s to the failed attempt in the United States at national health insurance in 1995 (see Chapters 1 and 13), social reforms to deal with inequalities in health have focused on improving access to medical and hospital care. As almost all industrialized countries developed such systems, their contribution to improved health status was an important part of social progress, especially since World War II.

But even in societies with universal access to health care, persons of lower socioeconomic status suffer higher rates of morbidity and mortality from a wide variety of diseases. The findings of the Black Report (written by Douglas Black) in the United Kingdom in the early 1980s pointed out that the Class V population (unskilled laborers) had twice the total and specific mortality rates of the Class I population (professional and business) for virtually all disease categories, from infant mortality to death from cancer. The report was shocking because all Britons have had access to the NHS since its inception in 1948, with access to a complete range of free services, close relations to their general practitioners and good access to specialty services. These findings initiated reappraisals of the social factors that had previously been regarded as the academic interests of medical sociologists and anthropologists and marginal to medical care.

Although the epidemiology of cardiovascular disease is directly related to the classic risk factors of smoking, poor diet, and physical inactivity, differences in mortality from cardiovascular disease between different classes among British civil servants are not entirely explainable by these factors. The differences are also affected by social and economic issues that may relate to the psychological needs of the individual, such as the degree of control people have over their own lives. Blue collar workers have less control over their lives than their white collar counterparts, and have higher rates of coronary heart disease mortality than higher social class persons.

Social conditions affect disease distribution in all societies. In the United States and western Europe, tuberculosis has reemerged as a significant public health problem in urban areas partly because of high risk population groups, owing to poverty and alienation from society, as in the cases of homelessness, drug abuse, and HIV infection. In eastern Europe and the former Soviet Union, the recent rise in TB incidence has resulted from various social and economic factors in the early 1990s, including the large-scale release of prisoners. In both cases, diagnosis and prescription of medication are inadequate, and the community-at-large becomes at-risk because of the development of antibiotic-resistant strains of tubercle bacillus (readily spread by inadequately treated carriers, acting as human vectors.

Studies of socioeconomic status (SES) and health are applicable and valuable in many settings. In Alameda County, California, differences in mortality between black and white population groups in terms of survival from cancer became insignificant when controlled for social class. A 30-year follow-up study of the county population reported that low income families in California are more likely to have physical and mental problems that interfere with daily life, contributing to further impoverishment.

Studies in Finland and other locations showed that lower SES women use less preventive care such as smears for cervical cancer than women of higher SES, despite having greater risk of cervical cancer. Factors leading to SES inequalities relate to differences in risk behavior, socioemotional distress, occupational factors including exposures to toxic substances, a feeling of lack of control over one's own life, and inadequate family or community social support systems.

Health and disease are multifactorial and the risk factors associated with them require health care systems to take into account the social, physical, and psychological factors that limit the effect of even the best medical care. This applies to interventions by the medical care provider as well as by the wider health system, including prevention and public health. The paradigm of host–agent–environment takes on a new and wider frame of reference in which the sociopolitical environment and organized efforts of intervention affect the epidemiologic and individual clinical course of disease. The health system is meant to affect the occurrence or outcome of disease, either directly by primary prevention or treatment, or indirectly by reducing community or individual risk factors.

The effects of social conditions on health can be partly offset by interventions intended to promote healthful conditions, for example, improved sanitation, or through health services, used efficiently and made available to all. The approaches to preventing the disease or its complications may require physical changes in the environment, such as removal of the Broad Street pump handle to stop the cholera epidemic in London, or altering diets as in Goldberger's work on pellagra. The societal context in terms of employment, social security, female education, recreation, family income, cost of living, housing, and homelessness is relevant to the health status of a population. Income distribution in a wealthy country may leave a wide gap between the upper and lower socioeconomic groups, which affects health status. The New Public Health has a responsibility for advocacy of societal conditions supportive of good health.

MODES OF PREVENTION

When an objective has been defined in preventing disease, the next step is to identify suitable and feasible methods of achieving it, or a strategy with tactical objectives. This determines the method of operation and the resources needed to carry it out. The methods of public health are categorized as health promotion and primary, secondary, and tertiary prevention.

Health Promotion

Health promotion is a guiding concept involving activities intended to enhance individual and community health and well-being. It seeks to increase involvement and control of the individual and the community in their own health. It acts to improve health and social welfare, and to reduce specific determinants of diseases

---

BOX 2.2 MODES OF PREVENTION

Health promotion: Fostering individual and community standards of behavior conducive to good health, promoting legislative, social, or environmental conditions that reduce individual and community risk, and creating a healthful environment.

Primary prevention: Preventing a disease from occurring.

Secondary prevention: Making an early diagnosis and giving prompt and effective treatment to stop progress or shorten the duration and prevent complications from an already existing disease process.

Tertiary prevention: Preventing long-term impairments or disabilities as sequelae; restoring and maintaining optimal function once the disease process has stabilized, i.e., promoting functional rehabilitation.

and risk factors that adversely affect the health, well-being, and productive capacities of an individual or society, setting targets based on the size of the problem but also the feasibility of successful intervention, in a cost-effective way.

Health promotion is a key element of the New Public Health and is applicable in the community, the clinic or hospital, and in all other service settings. Some health promotion activities are government interventions such as mandating the use of seat belts in cars, requiring that children be immunized to come to school, declaring that certain basic foods must have essential mineral and vitamins added in order to prevent deficiency disorders in vulnerable population groups, and requiring that all newborns should be given prophylactic vitamin K to prevent hemorrhagic disease of the newborn. Setting food and drug standards, and raising taxes on cigarettes and alcohol to reduce their consumption, are also part of health promotion. Health promotion is practiced by persons with many professional backgrounds, working in many different organizational settings.

Raising awareness and informing people about health and lifestyle factors that might put them at risk requires teaching young people about the dangers of sexually transmitted diseases (STDs), smoking, and alcohol abuse to reduce risks associated with their social behavior. It might include disseminating information on healthy nutrition, for example, the need for folic acid supplements for women in the age of fertility and multiple vitamins for the elderly. Community and peer

---

BOX 2.3 THE ELEMENTS
OF HEALTH PROMOTION

1. Addresses the population as a whole in health-related issues, in everyday life as well as people at risk for specific diseases;
2. Directs action to risk factors or causes of illness or death;
3. Undertakes activist approach to seek out and remedy risk factors in the community that adversely affect health;
4. Promotes factors that contribute to a better condition of health of the population;
5. Initiates actions against health hazards, including communication, education, legislation, fiscal measures, organizational change, community development, and spontaneous local activities;
6. Involves public participation in defining problems, deciding on action;
7. Advocates relevant environmental, health, and social policy;
8. Encourages health professional participation in health education and health advocacy.

Source: Adapted from World Health Organization. *Ottawa Charter for Health Promotion.* Geneva: WHO, 1986.

group attitudes and standards affect individual behavior. Health promotion endeavors to create a climate of knowledge, attitudes, beliefs, and practices that are associated with better health outcomes.

Primary Prevention

Primary prevention refers to those activities that are undertaken to prevent the disease and injury from occurring. Primary prevention works with both the individual and the community. It may be directed at the host, to increase resistance to the agent (such as in immunization or cessation of smoking), or may be directed at environmental activities to reduce conditions favorable to the vector for a biological agent, such as mosquito vectors of malaria or dengue fever. Examples of such measures abound. Immunization of children prevents diseases such as tetanus, pertussis, and diphtheria. Chlorination of drinking water prevents transmission of waterborne gastroenteric diseases. Wearing seat belts in motor vehicles prevents much serious injury and death in road crashes. Reducing the availability of firearms reduces injury and death from intentional, accidental, or random violence.

Primary prevention also includes activities within the health system that can lead to better health. This may mean, for example, setting standards and ensuring that doctors not only are informed of appropriate immunization practices and modern prenatal care, but also are aware of their role in preventing cerebrovascular, coronary, and other diseases like cancer of the lung. In this role, the health-care provider serves as a teacher and guide, as well as a diagnostician and therapist. Like health promotion, primary prevention does not depend on doctors alone: both work to raise individual consciousness of self-care, mainly by raising awareness and information levels, empowering the individual and the community to improve self-care, to reduce risk factors, and to live healthier lifestyles.

Secondary Prevention

Secondary prevention is the early diagnosis and management to prevent complications from a disease. Public health interventions to prevent spread of disease include the identification of sources of the disease and the implementation of steps to stop it, as shown in Snow's closure of the Broad Street pump. Secondary prevention includes steps to isolate cases and treat or immunize contacts so as to prevent further cases of meningitis or measles in outbreaks. Needle exchange programs or distribution of condoms in schools or colleges help to prevent the spread of STDs and AIDS.

All health care providers have a role in secondary prevention, for example, in preventing strokes by early and adequate care of hypertension. The child who has an untreated streptococcal infection of the throat may develop complications which are serious and potentially life-threatening, including rheumatic fever, rheumatic valvular heart disease, and glomerulonephritis. When a patient is found to have elevated blood pressure, this can be managed by weight loss, physical fitness, smoking cessation, and the taking of medication, which together lower the risk of stroke. In the case of injury, competent emergency care, safe transportation, and good trauma care may reduce the chance of death and/or permanent handicap. Screening and high quality care in the community prevents complications of diabetes, including heart, kidney, eye, and peripheral vascular disease; they can also prevent hospitalizations, amputations, and strokes, thus lengthening and improving the quality of life. Health care systems need to be actively engaged in secondary prevention, not only as individual doctor's services, but also as organized systems of care.

Tertiary Prevention

Tertiary prevention involves activities directed at the host but also at the environment in order to promote rehabilitation, restoration, and maintenance of maximum function after the disease and its complications have stabilized. The person who has undergone a cerebrovascular accident or trauma will come to a stage where active rehabilitation can help to restore lost functions and prevent recurrence or further complications.

Treatment for a myocardial infarction or a fractured hip now includes early rehabilitation in order to promote maximum recovery and restoration of function. Providing a wheelchair, special toilet facilities, doors, ramps, and transportation services for paraplegics are often the most vital factors in rehabilitation. Public health agencies work with groups in the community with interest in promoting help for specific categories of disease or disability, to reduce discrimination, to eliminate physical or social barriers, to promote community awareness, and to finance special equipment or other needs of this group. Close follow-up and management of chronic disease, physical and mental, requires home care and assuring an appropriate medical regimen including drugs, diet, exercise, and support services. The follow-up of chronically ill persons to supervise the taking of medications, to monitor changes, and to support them in maximizing their independent capacity in activities of daily living are essential elements of the New Public Health.

DEMOGRAPHIC AND EPIDEMIOLOGIC TRANSITION

Public health uses a population approach to reach many of its objectives. This requires defining the population, including trends of change in the age–sex distribution of the population. The reduction of infectious disease as the major cause of mortality, coupled with declining fertility rates, resulted in changes in the age composition, or a demographic transition. Demographic changes, such as fertility and mortality patterns, are important factors in changing the age distribution of the population, resulting in a greater proportion of people surviving to older ages. Declining infant mortality, increasing educational levels of women, the availability of birth control, and other social and economic factors lead to changes in fertility patterns and the demographic transition—an aging of the population—with important effects on health services needs.

Age and sex distributions of a population affect and are affected by patterns of disease. Change in epidemiologic patterns, or an epidemiologic shift, is a change in predominant patterns of morbidity and mortality. The transition of infectious diseases becoming less prominent as causes of morbidity and mortality and being replaced by chronic and noninfectious diseases has occurred in both the developed and developing countries. The decline in mortality from chronic diseases, such as cardiovascular disease, represents a new stage of epidemiologic transition, creating an aging population with higher standards of health but also long-term community support and care needs. Monitoring and responding to these changes are fundamental responsibilities of public health, and readiness to react to new, local, or generalized changes in epidemiologic patterns is vital to the New Public Health.

Societies are not totally homogeneous in ethnic composition, levels of affluence, or other social markers. A society classified as a developing country may have substantial numbers of persons with incomes which promote overnutrition, so that disease patterns may include diseases of excesses, such as diabetes. On the other hand, affluent societies include population groups with disease patterns of poverty, including poor nutrition and low birth weight babies.

A further stage of epidemiologic transition has been occurring in the industrialized countries since the 1960s, with dramatic reductions in mortality from coronary heart disease, stroke, and, to a lesser extent, trauma. The interpretation of this epidemiologic transition is still not perfectly clear. How it occurred in the industrialized western countries but not in those of the former Soviet Union is a question whose answer is vital to the future of public health in Russia and eastern Europe. Developing countries must also prepare to cope with epidemics of noninfectious diseases, and all countries face renewed challenges from infectious diseases with antibiotic resistance or newly appearing infectious agents posing as major public health threats.

INTERDEPENDENCE OF HEALTH SERVICES

Each component of a health service may have developed with different historical emphases, operating independently as a separate service under different administrative auspices and funding systems, competing for limited health care resources. In this situation, preventive of community care receives less attention and resources than more costly treatment services. Figure 2.4 suggests a set of health services in an interactive relationship to serve a community or defined population, but the emphasis should be on the interdependence of these services one to the other and all to the comprehensive network in order to achieve effective use of resources and a balanced set of services for the patient, the client population, and the community.

Clinical medicine and public health each play major roles in primary, secondary, and tertiary prevention. Each may function separately in their roles in the

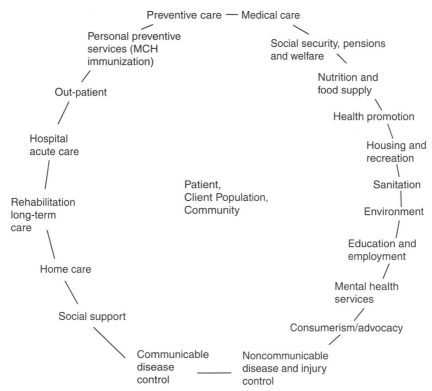

FIGURE 2.4   Community health as a network of services serving a defined population.

community, but optimal success lies in their integrated efforts. Allocation of resources should promote management and planning practices to promote this integration. There is a functional interdependence of all elements of health care serving a definable population. The components of health services to a population group interact with the patient or the client as the central figure. Effectiveness in use of resources means that the service most appropriate for the client's needs are those that should be applied. Long stays in an acute care hospital often occur when there are insufficient home care or community support services. This is wasteful and destructive to good patient care and costly to the health care economy. Linkage and a balance of services that meet client and community needs promote effective and efficient use of resources.

Separate organization and financing of services place barriers to appropriate provision of services for both the community and the individual patient. The interdependence of services is a challenge in health care organization for the future. Where there is competition for limited resources, pressures for tertiary services often receive priority over programs to prevent children from dying of preventable diseases. Public health must be seen in the context of all health care and must play an influential role in promoting prevention at all levels. Clinical services need public health in order to provide prevention and community health services that reduce the burden of disease, disability, and dependence on the institutional setting.

DEFINING PUBLIC HEALTH

The scope of public health has changed as expansion of medical sciences and public health experience have changed concepts of disease and its causes. In this chapter we examine how this, together with an expanding concept of health, has contributed to the development of the New Public Health. Health was traditionally thought of as a state of absence of disease, pain, or disability, but has gradually been expanded to include physical, mental, and societal well-being. Defining health and disease is basic to the search for methods of prevention.

In 1920, C. E. A. Winslow, professor of public health at Yale University, defined public health as follows:

> Public health is the Science and Art of (1) preventing disease, (2) prolonging life, and (3) promoting health and efficiency through organized community effort for
>
> (a) the sanitation of the environment,
> (b) the control of communicable infections,
> (c) the education of the individual in personal hygiene,
> (d) the organization of medical and nursing services for the early diagnosis and preventive treatment of disease, and
> (e) the development of social machinery to ensure everyone a standard of living adequate for the maintenance of health,
>
> so organizing these benefits as to enable every citizen to enjoy his birthright of health and longevity.
>
> Source: As quoted in Institute of Medicine. 1988. *The Future of Public Health*. Washington DC: National Academy Press.

Winslow's far-reaching definition remains a valid framework but unfulfilled where clinical medicine and public health have grown apart. Isolation from the financing and provision of medical and nursing care services left public health the task of meeting the health needs of the poor and underserved population groups, with inadequate resources and recognition.

Terms such as social hygiene, preventive medicine, community medicine, social medicine, and others have been used to denote public health over the past century. Preventive medicine is a combination of some elements of public health with clinical medicine. Public health deals with the individual just as the clinical health care provider does, as in the case of immunization programs. Clinical medicine also deals in the area of prevention in management of patients with hypertension or diabetes, and in doing so prevents the serious complications of these diseases. Preventive medicine focuses on a medical or clinical function, or what might be called personal preventive care, with stress on risk groups in the community and national efforts for health promotion.

Social Medicine and Community Health

Social medicine looks at illness in a social context, but it lacks the environmental and regulatory functions of public health. Community health implies a local form of health intervention, whereas public health more clearly implies a global approach, which includes action at the international, national, state, and local levels. There are issues in health that cannot be dealt with at the individual, family, or community levels, requiring global strategies and intervention programs.

Medical Ecology

In 1961, Kerr White and colleagues defined medical ecology as population-based research as the foundation for management of health care quality. This concept stresses a population approach, including those not attending and those using health services. This concept was based on previous work on quality of care, randomized clinical trials, medical audit, and structure–process–outcome research. It also addressed health care quality and management.

These themes influenced medical research by stressing the population from which clinical cases emerge as well as public health research with clinical outcome measures, themes that recur in development of health services research and, later, evidence-based medicine. This led to development of the Agency for Health Care Policy and Research and Development in the U.S. Department of Health and Human Services and evidence-based practice centers to synthesize fundamental knowledge for development of information for decision-making tools such as clinical guidelines, algorithms or pathways.

Community-Oriented Primary Care

Community-oriented primary care (COPC) is an approach to primary health care that links community epidemiology and appropriate primary care, using proactive responses to the priority needs identified. COPC, originally pioneered in South Africa and Israel by Sidney and Emily Kark and colleagues in the 1950s and 1960s, stresses that medical services in the community need to be molded to the needs of the population, as defined by epidemiologic analysis. COPC involves community outreach and education, as well as clinical preventive and treatment services.

COPC focuses on community epidemiology and an active problem-solving approach. This differs from national or larger scale planning that sometimes lose sight of the local nature of health problems or risk factors. COPC combines clinical and epidemiologic skills, defines needed interventions, and promotes community involvement and access to health care. It is based on linkages between the different elements of a comprehensive basket of services along with attention to the social and physical environment. A multidisciplinary team and outreach services are important for the program, and community development is part of the process.

In the United States, the COPC concept has influenced health care planning for poor areas, especially provision of federally funded community health centers in attempts to provide health care for the underserved since the 1960s. In more re-

---

BOX 2.4 FEATURES OF COMMUNITY-ORIENTED PRIMARY HEALTH CARE (COPC)

1. Essential features
   a. Clinical and epidemiologic skills
   b. A defined population
   c. Defined programs to address community health issues
   d. Community involvement
   e. Accessibility to health care—reducing geographic, fiscal, social, and cultural barriers
2. Desirable features
   a. Integration/coordination of curative, rehabilitative, preventive, and promotive care
   b. Comprehensive approach extending to behavioral, social, and environmental factors
   c. Multidisciplinary team
   d. Mobility and outreach
   e. Community development

Source: Tollman, S. 1991. Community oriented primary care: Origins, evolution, applications. *Social Science and Medicine,* 32:633–642.

---

cent years, COPC has gained wider acceptance in the United States, where it is associated with family physician training and community health planning based on the risk approach and "managed care" systems. Indeed, the three approaches are mutually complementary. As the new emphasis in health care reform in the late 1990s moves toward managed care, the principles of COPC will be important in promoting health and primary prevention in all its modalities, as well as tertiary prevention with follow-up and maintenance of the health of the chronically ill.

COPC stresses that all aspects of health care have moved toward prevention based on measurable health issues in the community. Through either formal or informal linkages between health services, the elements of COPC are part of the daily work of health care providers and community services systems. The U.S. Institute of Medicine issued *Report on Primary Care* in 1995, defining primary care as "the provision of integrated, accessible health care services by clinicians who are accountable for addressing the majority of personal health care needs, developing a sustained partnership with patients and practicing in the context of the family and the community." This formulation was criticized by the American Public Health Association as lacking a public health perspective and failing to take into account both the individual and the community health approaches. It is just this gap that COPC tries to bridge.

The community, whether local or national, is the site of action for many public health interventions. Moreover, understanding the characteristics of the community is vital to a successful community-oriented approach. By the 1980s, new patterns of public health began to emerge, including all measures used to improve the health of the community and at the same time working to protect and promote the health of the individual. The range of activities to achieve these general goals is very wide indeed, including individual patient care systems and the community-wide activities that affect the health and well-being of the individual. These include the financing and management of health systems, evaluation of the health status of the population, and steps to improve the quality of health care. They place reliance on health promotion activities to change environmental risk factors for disease and death. They promote integrative and multisectoral approaches and the international health teamwork required for global progress in health.

WORLD HEALTH ORGANIZATION'S DEFINITION OF HEALTH

The definition in the charter of the World Health Organization (WHO) of health as a complete state of physical, mental, and social well-being had the ring of utopianism and irrelevance to states struggling to provide even minimal care in adverse economic, social, and environmental conditions (Box 2.5). In 1977, a more modest goal was set for attainment of a level of health compatible with maximum feasible social and economic productivity. One needs to recognize that health and disease are on a dynamic spectrum that affects everyone; the mission for public health is to use a wide range of methods to prevent disease and premature death, and improve quality of life for the benefit of individuals and the community.

In the 1960s, most industrialized countries were concentrating energies and financing in health care on providing access to medical services through national insurance schemes. Developing countries were often spending scarce resources trying to emulate this trend. The World Health Organization was concentrating on categorical programs, such as eradication of smallpox and malaria, as well as the Expanded Program of Immunization and similar specific categorical programs. At the same time there was a growing concern that developing countries were placing much emphasis on curative services and not enough on prevention and primary care.

Alma-Ata: Health for All

The WHO and the United Nations Children's Education Fund (UNICEF) sponsored conference held in Alma-Ata, Kazakhstan, in 1978 was convened to refocus on primary care. The Alma-Ata Declaration stated that health is a basic human right, and that governments are responsible to assure that right for their citizens and to develop appropriate strategies to fulfill this promise. This proposition has come to be increasingly accepted in the international community. The conference stressed the right and duty of people to participate in the planning and implemen-

BOX 2.5 DEFINITIONS OF HEALTH AND
MISSION OF WORLD HEALTH
ORGANIZATION

The World Health Organization defines health as "a state of complete physical, mental and social well-being, not merely the absence of disease or infirmity" (WHO Constitution, 1948).

In 1978 at the Alma-Ata Conference on Primary Health Care, the WHO related health to "social and economic productivity in setting as a target the attainment by all the people of the world of a level of health that will permit them to lead a socially and economically productive life." Three general programs of work for the periods 1984–1989, 1990–1995, and 1996–2001 were formulated as the basis of national and international activity to promote health.

In 1995, the WHO, recognizing changing world conditions of demography, epidemiology, environment, political and economic status, addressed the unmet needs of developing countries and health management needs in the industrialized countries, calling for international commitment to "attain targets that will make significant progress towards improving equity and ensuring sustainable health development."

The 1999 object of the WHO is restated as "the attainment by all peoples of the highest possible level of health," as defined in the WHO constitution, by a wide range of functions in promoting technical cooperation, assisting governments, and providing technical assistance, international cooperation, and standards.

Source: *New Challenges for Public Health: Report of an International Meeting*. Geneva: World Health Organization, 1996, and 1999 website www://who.org/about who.

tation of their health care. It advocated the use of scientifically, socially, and economically sound technology. Joint action through intersectoral cooperation was also emphasized.

The Alma-Ata Declaration focused on primary health care as the appropriate method of assuming adequate access to health care for all. This approach was endorsed in the World Health Assembly (WHA) in 1977 under the banner "Health for All by the Year 2000" (HFA 2000). This was a landmark decision and has had important practical results. Many countries have gradually come to accept the notion of placing priority on primary care, resisting the temptation to spend high percentages of health care resources on high tech and costly medicine. Spreading these same resources into highly cost-effective primary care, such as immunization and nutrition programs, provides greater benefit to individuals and to society as a whole.

Alma-Ata provided a new sense of direction for health policy, applicable to developing countries and in a different way to the developed countries. During the 1980s this influenced national health policies in the developing countries and to a

BOX 2.6 DECLARATION OF ALMA ATA, 1978; A SUMMARY OF PRIMARY HEALTH CARE (PHC)

1. Reaffirms that health is a state of complete physical, mental, and social well-being, and not merely the absence of disease or infirmity, and is a fundamental human right.
2. Existing gross inequalities in the health status of the people, particularly between developed and developing countries as well as within countries, is of common concern to all countries.
3. Governments have a responsibility for the health of their people. The people have the right and duty to participate in planning and implementation of their health care.
4. A main social target is the attainment, by all peoples of the world by the year 2000, of a level of health that will permit them to lead a socially and economically productive life.
5. PHC is essential health care based on practical, scientifically sound, and socially acceptable methods and technology.
6. It is the first level of contact of individuals, the family, and the national health system bringing health care as close as possible to where people live and work, as the first element of a continuing health care process.
7. PHC evolves from the conditions and characteristics of the country and its communities, based on the application of social, biomedical, and health services research and public health experience.
8. PHC addresses the main health problems in the community, providing promotive, preventive, curative, and rehabilitative services accordingly.
9. PHC includes the following:
   a. Education concerning prevailing health problems and methods of preventing and controlling them;
   b. Promotion of food supply and proper nutrition;
   c. Adequate supply of safe water and basic sanitation;
   d. Maternal and child health care, including family planning;
   e. Immunization against the major infectious diseases;
   f. Prevention of locally endemic diseases;
   g. Appropriate treatment of common diseases and injuries;
   h. The provision of essential drugs;
   i. Relies on all health workers . . . to work as a health team.
10. All governments should formulate national health policies, strategies and plans, mobilize political will and resources, used rationally, to ensure PHC for all people.

Source: Website http://www.who.dk.policy/almaata.htm

lesser extent in the developed countries. For example, developing countries have accepted immunization and diarrheal disease control as high priority issues and achieved remarkable success in raising immunization coverage from some 10% to over 75% in just a decade.

Developed countries adopted these principles in a different way. In these countries, the concept of primary health care led directly to important conceptual developments in health, including HFA 2000, health objectives, and targets as part of national health planning. This systematic approach to individual and community health is part of the New Public Health.

The interactions among community public health, personal health services, and health-related behavior, including their management, is the essence of the New Public Health. How the health system is organized and managed affects the health of the individual and the population, as does the quality of providers. Health information systems with epidemiologic, economic, and sociodemographic analysis are vital to monitor health status and allow for changing priorities and management. Well-qualified personnel are essential to provide services, to manage the system, and to carry out relevant research and health policy analysis. Diffusion of data, health information, and responsibility helps to provide a responsive and comprehensive approach to meet the health needs of the individual and community. The physical, social, economic, and even the political environment are important determinants of health status of the population and the individual. Joint action (intersectoral cooperation) between public and nongovernment or community-based organizations is needed to achieve the well-being of the individual in a healthy society.

In the 1980s and 1990s, these ideas became part of an evolving New Public Health, spurred by epidemiologic changes, health economics, the development of managed care linking health system, and prepayment. Knowledge and self-care skills, as well as community action to reduce health risks, are no less important in this than the roles of medical practitioners and institutional care. All are parts of a coherent holistic approach to health.

SELECTIVE PRIMARY CARE

The concept of selective primary care, articulated in the 1960s by Walsh and Warren, addresses the needs of developing countries to select those interventions on a broad scale that would have the greatest positive impact on health, taking into account limited resources such as money, facilities, and manpower.

The term selective primary care is meant to define national priorities that are based not on the greatest causes of morbidity or mortality, but on common conditions of epidemiologic importance for which there are effective and simple preventive measures. Throughout health planning, there is an implicit or explicit selection of priorities for allocation of resources. Even in primary care, selection of targets is a part of the process of resource allocation. In modern public health, this process is more explicit. A country with limited resources and a high birth rate will emphasize maternal and child health before investing in geriatric care.

This concept has become part of the microeconomics of health care and technology assessment, discussed in Chapters 11 and 15, respectively, and is used widely in setting priorities and resource allocation. In primary care in developing countries, cost-effective interventions have been articulated by many international organizations, including iodization of salt, use of oral rehydration therapy (ORT) for diarrheal diseases, vitamin A supplementation for all children, expanded programs of immunization, and others that have the potential for saving hundreds of thousands of lives yearly at low cost. In developed countries, health promotion targeted to reduce accidents and risk factors for cardiovascular diseases are low cost public health interventions that save lives and reduce use of hospital care.

THE RISK APPROACH

The risk approach selects population groups on the basis of risk and helps to determine interventions priorities to reduce morbidity and mortality. The measure of health risk is taken as a proxy for need, so that the risk approach provides something for all, but more for those in need—in proportion to that need. In epidemiologic terms these are persons with higher relative risk or attributable risk.

Some groups in the general population are at higher risk than others for specific conditions. The Expanded Programme on Immunization (EPI), Control of Diarrheal Diseases (CD), and Acute Respiratory Disease (ARD) programs of the WHO are risk approaches to tackling fundamental public health problems of children in developing countries.

Public health places considerable emphasis on maternal and child health because these are vulnerable periods in life for specific health problems. Pregnancy care is based on a basic level of care for all, with continuous assessment of risk factors that require a higher intensity of follow-up. Prenatal care helps identify factors that increase the risk for the pregnant woman or her fetus/newborn. Efforts directed toward these special risk groups have the potential to reduce morbidity and mortality. High risk identification, assessment, and management are vital to a successful maternal care program.

Similarly, routine infant care is designed not only to promote the health of infants, but also to find the earliest possible indications of deviation and the need for further assessment and intervention to prevent a worsening of the condition. Low birth weight babies are at greater risk for many hazards and should be given special treatment. Screening of all babies is done on a routine basis for birth defects or congenital disorders such as hypothyroidism (CH) and phenylketonuria (PKU). Screening is followed by investigating and treating those found to have a clinical deficiency. This is an important element of infant care since infancy itself is a risk factor.

As will be discussed in Chapter 6 and others, epidemiology has come to focus on the risk approach with screening based on known genetic, social, nutritional, environmental, occupational, behavioral, or other factors contributing to the risk for disease. The risk approach has the advantage of specificity and is often used to initiate new programs directed at special categories of need. This approach can lead to narrow and somewhat rigid programs that may be difficult to integrate into a more general or comprehensive approach, but until universal programs can be achieved, selective targeted approaches are justifiable. Indeed, even when universal health coverage is established, it is still important to address health needs or issues of groups at special risk.

Working to achieve defined targets means making difficult choices. The supply and utilization of some services will limit availability for other services. There is an interaction, sometimes positive, sometimes negative, between competing needs and the health status of a population.

THE CASE-FOR-ACTION

Public health identifies needs by measuring and comparing the incidence or prevalence of the condition in a defined population with that in other comparable population groups and defines targets to reduce or eliminate the risk of disease. It determines ways of intervening in the natural epidemiology of the disease, and develops a program to reduce or even eliminate the disease.

Because of the interdependence of health services, as well as the total financial burden of health, it is essential to look at the costs of providing health care, and how resources should be allocated to achieve the best results possible. Health economics has become a fundamental methodology in policy determination. The costs of health care, the supply of services, the needs for health care or other health-promoting intervention, and effective means of using resources to meet goals are fundamental in the New Public Health. It is possible to err widely in health planning if one set of factors is over- or underemphasized. Excessive supply of one service diminishes availability of resources for other needed investments in health. If diseases are not prevented or their sequelae not well-managed, patients must use costly health care and are unable to perform their normal functions in society. Lack of investment in health promotion and primary prevention creates a larger reliance on institutional care, driving health costs upward. This also restrict flexibility in meeting patients' needs. The interaction of supply and demand for health services is an important determinant of the political economy of health care. Health and its place in national priorities is determined by the social and political philosophy of a government.

The case-for-action, or the justification for a public health intervention, is a complex of epidemiologic, economic, and public policy factors (Table 2.1). Each disease or group of diseases requires its own case-for-action. The justification for public health intervention requires sufficient evidence of the incidence/prevalence of the disease (see Chapter 3), the effectiveness and safety of an intervention, risk factors, safe means at hand to intervene, the human, social, and economic cost of the disease, political factors, and a policy decision as to the priority of the problem. This often depends on subjective factors, such as the guiding philosophy of the health system and the way it allocates resources.

TABLE 2.1   The Case-for-Action: Factors in Justifying Public Health Interventions

Ethics and potential	Issues
The right to health	Public expectation and social norms
Public advocacy	Concerned groups, the media, an individual
Need—epidemiologic and clinical	Morbidity, mortality, functional disability, physiologic indicators
Available technology	Documented effectiveness, safety, experience, acceptability, affordability
Precedent—"state of the art"	Good public health practice; what is done in leading centers of excellence, not necessarily consensus
Legal constraints and liability	Law and court decisions Providers, managed care and governments
Costs and benefits	Direct cost to health system; indirect cost to the individual, family, and society
Acceptability	Media and public opinion
Leadership	Political and professional
Quality of life	Optimizing human potential

Some interventions are so well established that no new justification is required to make the case, and the only question is how to do it most effectively. For example, infant vaccination is a cost-effective program for the protection of the individual child and the population as a whole. Whether provided as a public service or as a clinical preventive measure by a private medical practitioner, it is in the interest of public health that all children be immunized.

An outbreak of diarrheal disease in a kindergarten presents an obvious case-for-action, and a public health system must respond on an emergency basis, with selection of the most suitable mode of intervention. The considerations in developing a case-for-action, as outlined above. Need is based on clinical and epidemiologic evidence but also on the importance of an intervention in the eyes of the public. The technology available, its effectiveness and safety, and accumulated experience are important in the equation, as are its acceptability and affordability. The precedents for use of an intervention are also important. On epidemiologic evidence, if the preventive practice has been seen to provide reduction in risk for the individual and for the population, then there is good reason to implement it. The costs and benefits must be examined as part of the justification to help in the selection of health priorities.

New drugs, vaccines, and medical equipment are constantly becoming available, and each new addition needs to be examined among the national health priorities. Sometimes, due to cost, a country cannot afford to add a new vaccine to the routine. However, when there is good medical evidence for the vaccine, it can be applied for those at greatest risk. Although there are ethical issues involved, it may be necessary to advise parents or family members to independently purchase the vaccine. Clearly, recommending individual purchase of vaccine is counter to the principle of equity and solidarity. On the other hand, failure to advise parents of potential benefits to their children creates other ethical problems.

Mass screening programs involving complete physical examinations have not been found to be cost-effective or to significantly reduce disease. In the 1950s to 1960s routine general health examinations were promoted as an effective method of finding disease early. Since the late 1970s a selective and specific approach to screening has become widely accepted. This involves defining risk categories for specific diseases and bearing in mind the potential for remedial action. Early case finding of breast cancer by routine mammography has been found to be effective after age 35, and Papanicolaou smear testing to discover cancer of the cervix is timed according to risk category.

The factor of contribution to quality of life should be considered. A vaccine for varicella may be justified partly for the prevention of deaths or illness from chicken pox. A stronger argument is often based on the fact that this is a disease that causes children up to 2 weeks of moderate illness and may require parents to stay home with the child, resulting in economic loss to the parent and the society. The fact that this vaccination prevents the occurrence of herpes zoster or shingles later in life may also be a justification. Adoption of hepatitis B vaccine is justified on the grounds that it prevents cancer of the liver and hepatic failure in a small percentage of the population affected.

How many cases of a disease are enough to justify an intervention? One or several cases of some diseases, such as poliomyelitis, may be considered an epidemic in that each case constitutes or is an indicator of a wider threat. A single case of polio suggests that another 100 persons are infected but have not developed a recognized clinical condition. Such a case constitutes a public health emergency, and forceful organization to meet a crisis is needed. Current standards are such that measles epidemics indicate a failure of public health practice, even a few cases constitutes an outbreak of public health significance or an epidemic.

Assessing a public health intervention to prevent the disease or reduce its impact requires measurement of the disease in the population and its economic impact. There is no simple formula to justify a particular intervention, but the cost–benefit approach is now commonly required to make such a case-for-action. Sometimes public opinion and political leadership may oppose the views of the professional community, or may impose limitations of policy or funds that prevent its implementation. Conversely, professional groups may press for additional resources that compete for limited resources available to provide other needed health activities. Both the professionals of the health system and the general public need full access to health-related information to take part in such debates in a constructive way. To maintain progress a system must examine new technologies and justify their adoption or rejection (Chapter 15).

POLITICAL ECONOMY AND HEALTH

As the concept of public health has evolved, and the value of medical care has improved through scientific and technological advances, societies have identified health as a legitimate area of activity of collective bargaining and of government. With this process, the need for managing health care resources became more clearly defined as a public responsibility. In industrialized countries with very different political makeups, national responsibility for universal access to health became part of the social ethos. With that, the financing and managing of health services became part of a broad concept of public health, and economics, planning, and management came to be part of the New Public Health (discussed in Chapters 10–13).

Social, ethical, and political philosophies have profound effects on policy decisions including allocation of public moneys and resources. Investment in public health is an integral part of socioeconomic development. Governments are major suppliers of funds and leadership in health infrastructure development, provision of health services, and health payment systems. They also play a central role in the development of health promotion and the regulating environment, food, and drugs essential for community health.

In liberal social democracies, the individual is deemed to have a right to health care. The state accepts responsibility to assure availability, accessibility, and quality of care. In many developed countries, government has also taken responsibility to arrange funding and services that are equitably accessible and of high quality. Health care financing may involve taxation, allocation, or special mandatory requirements on employers to pay for health insurance. Services may be provided by a state-financed and regulated service or through non-government organization (NGOs) and/ or private service mechanisms. These systems allocate between 6 and 14% of gross national product (GNP) to health services, with some governments funding over 80% of health expenditures, for example, Canada and the United Kingdom.

In Marxist states, the state organizes all aspects of health care with the philosophy that every citizen is entitled to equity in access to health services. The state health system manages research, manpower training, and service delivery, even if operational aspects are decentralized to local health authorities. This model applied primarily to the Soviet model of health services, but ex-colonial health services also have many characteristics of this approach. These systems, except for Cuba, place financing of health low on the national priority, with funding under 4% of GNP. In the shift to market economies in the 1990s, these countries are struggling with declining health status and a difficult shift from a strongly centralized health system to a decentralized system with diffusion of powers and responsibilities. Promotion of market concepts in former Soviet countries may reduce access to care and create a new dilemma for their governments.

Former colonial countries, independent since the 1950s and 1960s, largely carried on the governmental health structures established in the colonial times. Most developing countries have given health a relatively low place in budgetary allotment, with expenditures under 3% of GNP. During the 1980s, there has been a trend in developing countries toward decentralization of health services and greater roles for NGOs, and the development of health insurance. Some, influenced by medical concepts of the mother countries, fostered development of specialty medicine in the major centers with little emphasis on the rural majority population. Soviet influence in many ex-colonial countries promoted state-operated systems. The World Health Organization promoted primary care, but the allocations favored city-based specialty care. Israel, as an ex-colony, used British ideas of public health together with central European sick funds and maternal and child health as major streams of development until the mid-1990s.

A growing new conservatism in the 1980s and 1990s in the industrialized countries is a restatement of old values in which market economics and individualistic social values are placed above the common good concepts of liberalism and socialism in its various forms. In the more extreme forms of this concept, the individual is responsible for his own health, including payment, and has a choice of health care providers that will respond with high quality personalized care.

Market forces, meaning competition in financing and provision of health services with rationing of services, based on fees or private insurance, willingness and ability to pay, have become part of the ideology of the new conservatism. This assumes that the patient (i.e., the consumer) will select the best service for his or her need, while the provider best able to meet consumer expectations will thrive. In its purest form, the state has no role in providing or financing of health services except those directly related to community protection and promotion of a healthful environment without interfering with individual choices. The state ensures that there are sufficient health care providers and allows market forces to determine prices and distribution of services with minimal regulation. The United States retains this policy in a highly modified form, with 85% of the population covered by some form of private or public insurance systems.

Modified market forces in health care are part of health reforms in many countries as they seek not only to ensure quality health care for all, but also to constrain costs. A free market in health care is costly and ultimately inefficient because it encourages inflation of provider incomes or budgets and increasing utilization of highly technological services. Further, even in the most free market societies, the economy of health care is highly influenced by many factors outside the control of the consumer and provider. The total national health expenditures in the United States rose rapidly until reaching 13.7% of GNP in 1996, the highest of any country, despite serious deficiencies for those with no or inadequate health insurance (approximately 30% of the population). This is compared to under 9% of GNP in Canada, which has universal health insurance under public administration. Following the 1994 defeat of President Clinton's national health program, the conservative Congress and the business community took steps to expand managed care in order to control costs resulting in a revolution in health care in the United States (Chapters 11 and 13).

Market reforms are being implemented even in many "socialized" health systems. These may be through incentives to promote achievement of performance indicators, such as full immunization coverage. Others are using control of supply, such as hospital beds or licensed physicians, as methods of reducing overutilization of services that generates increasing costs. Market mechanisms in health are aimed not only at the client but also at the provider. Incentive payment systems must work to protect the patient's legitimate needs, and conversely incentives that might reduce quality of care should be avoided. Fee for service promotes high rates of services such as surgery. Increasing private practice and user fees can adversely affect middle and low income groups, as well as employers, by raising costs of health insurance. Managed care systems, with restraint on fee-for-service medical practice, has emerged as a positive response to the market approach. Incentive systems in payments for services may be altered by government or insurance agencies in order to promote rational use of services, such as reduction of hospital stays. The free market approach is affecting planning of health insurance systems in previously highly centralized health systems in developing countries as well as redevelopment of health systems in former Soviet countries.

Despite political differences, reform of health systems has become a common factor in virtually all health systems in the 1990s, as each searches for cost-effectiveness, quality of care and universality of coverage. The new paradigm of health care reform sees the convergence of different systems to common principles. National responsibility for health goals and health promotion leads to national financing of health care with regional and managed care systems. Most developed countries have long since adopted national health insurance or service systems. Some governments may, as in the United States, insure only the highest risk groups such as the elderly and the poor, leaving the working and middle classes to private insurance. The nature and direction of health care reform affecting coverage of the population is of central importance in the New Public Health because of its effects on allocation of resources and on the health of the population.

HEALTH AND DEVELOPMENT

Individuals in good health are better able to study and learn, and be more productive in their work. Improvements in standard of living have long been known to contribute to improved public health; however, the converse has not always been recognized. Investment in health care was not considered high priority in many countries where economic considerations directed investment to the "productive" sectors such as manufacturing and large-scale infrastructure projects, such as hydroelectric dams.

Whether health is a contributor to economic development or a drain on societies' resources has been a fundamental debate between socially and market-oriented advocates. Classic economic theory, both free enterprise and Marxist, has tended to regard health as a drain on economies, distracting investment needed for economic growth. As a result, in many countries health has been given low priority in budgetary allocation, even when the major source of financing is governmental. This belief among economists and banking institutions prevented loans for health development on the grounds that such funds should focus on creating jobs and better incomes, before investing in health infrastructure. Consequently, development of health care has been hampered.

A socially oriented approach sees investment in health as necessary for the protection and development of "human capital," just as investment in education is needed for the long-term benefit of the economy of a country. In 1993, a landmark document issued by the World Bank, *World Development Report: Investing in Health,* articulated a new approach to economics in which health, along with education and social development, are considered essential contributors for economic development. While many in the health field have long recognized the importance of health for social and economic improvement, its adoption by leading international development banking may mark a turning point for investment in developing nations, so that health may be a contender for increased development loans.

The concept of an essential package of services discussed in that report establishes priorities in low and middle income countries for efficient use of resources based on the burden of disease and cost-effectiveness analysis of services. It includes both preventive and curative services targeted to specific health problems. It defines a minimum package of essential services, such as one district hospital bed per 1000 population, 1 physician and 2–4 nurses per 10,000 population, and a reorientation of government spending on health to establish equity in access for the poor and other neglected sectors of society for essential primary care services. Other services are termed "discretionary services," which could be financed by private or public sources. This is a valuable concept in public health.

HEALTH SYSTEMS: THE CASE FOR REFORM

As medical care has gradually become more involved in prevention, and as it has gradually moved into the era of managed care, the gap between public health and clinical medicine has narrowed. As noted above, many countries are engaged in reforms in their health care systems. The motivation is partly derived from the need for cost containment, or to extend health care coverage to underserved parts of the population. Countries without universal health are may still have serious inequities in distribution of or access to services, and may seek reform to reduce those inequities, perhaps under political pressures to improve the provision of services. The incentives, or case-for-reform, center on cost constraint, regional equity, and preserving or developing universal access and quality of care.

In some settings, a health system may fail to keep pace with developments in prevention and in clinical medicine. Some countries have overdeveloped medical and hospital care, neglecting important initiatives to reduce risk of disease. The process of reform requires setting standards to measure health status and the balance of services to optimize health. A health service set a target of immunizing 95% of infants with a national immunization schedule, but requires a system to monitor performance and incentives for changes.

A health system may have failed to adapt to changing needs of the population through lack or misuse of a health information program. As a result, the system may err seriously in its allocation of resources, with excessive emphasis on hospital care and insufficient attention to primary and preventive care. All health services should have mechanisms for correctly gathering and analyzing needed data for monitoring the incidence of disease and other health indicators, such as hospital utilization, ambulatory care, and preventive care patterns. For example, the United Kingdom's National Health Service periodically goes through a restructuring process of parts of the system to improve the efficiency of service. This involves organizational changes and decentralization by changing the means of regional allocation of resources.

A continuing evolution affect in demographic and epidemiologic patterns, health systems are also influenced by technology, financing, and organizational change. New problems continually emerge and priorities must be reassessed in order to reach effective methods of addressing them. Reforms may create unanticipated problems, such as professional or public dissatisfaction, which must be evaluated, monitored, and addressed as part of the evolution of public health.

ADVOCACY AND CONSUMERISM

Literacy, freedom of the press, and increasing public concern for social and health issues have contributed to the development of public health. The British medical community lobbied for restrictions on the sale of gin in the 1780s in order to reduce its damage to the working class. In the late eighteenth and the nineteenth centuries, reforms in society and sanitation were largely the result of strongly organized advocacy groups influencing public opinion through the press. Such pressure stimulated governments to act in regulating working conditions of mines and factories. Abolition of the slave trade and its suppression by the British navy in the early nineteenth century resulted from advocacy groups and their effects on public opinion through the press. Vaccination against smallpox was promoted by privately organized citizen groups, until later taken up by local and national government authorities.

Advocacy is the act of individuals or groups in publicly pleading for, supporting, espousing, or recommending a cause or course of action. The advocacy role of reform movements of the nineteenth century was the basis of the development of modern organized public health. This ranged from the reform of mental hospitals, nutrition for sailors, and labor laws to improve working conditions for women and children, to the promotion of universal education and improved living conditions for the working population. Reforms on these and other issues resulted from

---

BOX 2.7 THE PLIMSOLL LINE

Samuel Plimsoll, British member of Parliament elected for Derby in 1868, conducted a solo campaign for the safety of seamen. His book *Our Seamen* described ships sent to sea so heavily laden with coal and iron that their decks were awash. Seriously overloaded ships, deliberately sent to sea by unscrupulous owners, frequently capsized at sea, drowning many crew members, with the owners collecting inflated insurance fees. Overloading was the major cause of wrecks and thousands of deaths in the British shipping industry. Plimsoll pleaded for mandatory load-line markers to prevent any ships putting to sea when the marker was not clearly visible. Powerful shipping interests fought him every inch of the way, but he succeeded in having a Royal Commission established, leading to an act of Parliament mandating the "Plimsoll Line," the safe carrying capacity of cargo ships. This regulation was adopted by the U.S. Bureau of Shipping as the Load Line Act in 1929 and is now standard practice worldwide.

---

the stirring of the public consciousness by advocacy groups and the public media, all of which generated political decisions in parliaments. Such reforms were in large part motivated by fear of revolution throughout Europe in the mid nineteenth century and the early part of the twentieth century.

Trade unions, and before them medieval guilds, fought to improve hours and conditions of work as well as social and health benefits for their members. In the United States, collective bargaining through trade unions achieved widespread coverage of the working population under voluntary health insurance. Unions and some industries pioneered prepaid group practice, the predecessor of health maintenance organizations and managed care.

Through raising public consciousness on many issues, advocacy groups pressure governments to enact legislation to restrict smoking in public places, prohibit cigarette advertising, and mandate the use of bicycle helmets. Advocacy groups play an important role in advancing health based on disease groups, such as cancer, multiple sclerosis, and thalassemia, or advancing health issues, such as the LaLeche League promoting breast feeding. Some organizations finance services or facilities not usually provided within insured health programs. Such organizations, which can number in the hundreds in a country, advocate the importance of their special concern and play an important role in innovation and meeting community health needs. Advocacy groups, including trade unions, professional groups, women's groups, self-help groups, and so many others, focus on specific issues and have made major contributions to advancing the New Public Health.

Professional Advocacy and Resistance

The history of public health is replete with pioneers whose discoveries led to strong opposition and sometimes violent rejection by conservative elements and vested interests in medical, public, or political circles. Opposition to Jennerian vaccination, the rejection of Semmelweiss by colleagues in Vienna, and the opposition to the work of Pasteur, Florence Nightingale, and many others may deter other innovators. Opposition to Jenner's vaccination lasted well into the late nineteenth century in some areas, but its supporters gradually gained ascendancy, ultimately leading to global eradication of smallpox. These and other pioneers led the way to improved health, often after bitter controversy on topics later accepted and which, in retrospect, seem to be obvious.

Advocacy has sometimes had the support of the medical profession but slow response by public authorities. David Marine of the Cleveland Clinic and David Cowie, professor of pediatrics at the University of Michigan, proposed prevention of goiter by iodization of salt. Marine carried out a series of studies in fish, and then in a controlled clinical trial among schoolgirls in 1917–1919, with startlingly positive results in reducing the prevalence of goiter. Cowie campaigned for iodization of salt, with support from the medical profession. In 1924, he convinced a private manufacturer to produce Morton's iodized salt, which rapidly became popular throughout North America. Similarly, iodized salt came to be used in many parts of Europe. This was achieved mostly without governmental support or legislation, and iodine deficiency disorders (IDDs) remain a widespread condition estimated to have affected 1.6 billion people worldwide in 1995. The target of international eradication of IDDs by the year 2000 was set at the World Summit for Children in 1990, and WHO called for universal iodization of salt in 1994.

BOX 2.8 AN ENEMY OF THE PEOPLE

Advocacy is a function in public health that has been important in promoting advances in the field, and one that sometimes places the advocate in conflict with established patterns and organizations. One of the classic descriptions of this function is in Henrik Ibsen's play *An Enemy of the People,* in which the hero, a young doctor, discovers that the water in his community is contaminated. This knowledge is suppressed by the town's leadership, led by his brother the mayor, because it would adversely affect plans to develop a tourist industry in the small Norwegian town in the late nineteenth century. The young doctor is driven from the town, having been declared an "enemy of the people" and a potential risk. The term took on a far more sinister and dangerous meaning in George Orwell's novel *1984* and in totalitarian regimes of the 1930s to the present time.

Professional organizations have contributed to promoting causes such as child and women's health, and environmental and occupational health. The American Academy of Pediatrics has contributed to establishing and promoting high standards of care for infants and children in the United States, and to child health internationally. Hospital accreditation has been used for decades in the United States, Canada, and more recently in Australia and the United Kingdom It has helped to raise standards of hospital facilities and care by carrying out systematic peer review of hospitals, nursing homes, primary care facilities, mental hospitals, as well as ambulatory care centers and public health agencies.

Public health needs to be aware of negative advocacy, sometimes based on professional conservatism or economic self-interest. Professional organizations can also serve as advocates of the status quo in the face of change. Opposition by the American Medical Association (AMA) and the health insurance industry to national health insurance in the United States has been strong and succesful for many decades. In some cases, the vested interest of one profession may block the legitimate development of others, such as when ophthalmologists lobbied successfully against the development of optometry, now widely accepted as a legitimate profession.

Jenner's discovery of vaccination with cowpox to prevent smallpox was adopted rapidly and widely. However, intense opposition by organized groups of anti-vacinationists, often led by those opposed to government intervention in health issues and supported by doctors with lucrative variolation practices, delayed adoption of vaccination for many decades. Fluoridation of drinking water is the most effective public health measure for preventing dental caries, but it is still widely opposed, and in some places removed even after implementation, by well-organized antifluoridation campaigns. Opposition to legislated restrictions on private ownership of assault weapons and handguns is intense in the United States, led by well-organized, well-funded, and politically powerful lobby groups, despite the amount of morbidity and mortality due to gun-associated violent acts.

Progress may be blocked where all decisions are made in closed discussions, not subject to open scrutiny and debate. Public health personnel working in the civil service of organized systems of government may not be at liberty to promote public health causes. However, professional organizations may then serve as forums for the essential professional and public debate needed for progress in the field. Professional organizations such as the American Public Health Association (APHA) provide effective lobbying for the interests of public health programs and can make an important impact on public policy. In mid-1996, efforts by the Secretary of Health and Human Services (HHS) in the United States have brought together leaders of public health with representatives of the AMA and academic medical centers to try to find areas of common interest and willingness toward promoting the health of the population.

Public advocacy has played an especially important role in focusing attention on ecological issues. In 1995, Greenpeace, an international environment activist group, struggled to prevent dumping of an oil rig in the North Sea and forced a major oil company to find another solution that would be less damaging to the environment. It also carried on efforts to stop renewal of testing of atomic bombs by France in the South Pacific. International protests led to cessation of almost all testing of nuclear weapons. Such international revulsion led to cessation in the Canadian slaughter of seal pups, and decimation of the whale, and continues to promote cessation of destruction of the Amazonian rain forests.

International conferences help to create a worldwide climate of advocacy for health issues. International sanitary conferences in the nineteenth century were convened in response to the cholera epidemics. International conferences continue in the twentieth century to serve as venues for advocacy on a global scale, bringing forward issues in public health that are beyond the scope of individual nations. WHO, UNICEF, and other international organizations perform this role on a continuing basis (Chapter 16).

Consumerism

Consumerism is a movement that promotes the interests of the purchaser of goods or services. In the 1960s a new form of consumer advocacy emerged from the civil rights and antiwar movement in the United States. Concern was focused on the environment, occupational health, and the rights of the consumer. Rachel Carson stimulated concern by dramatizing the effects of DDT on wildlife and the environment. This period gave rise to environmental advocacy efforts worldwide, and even a political movement in western Europe, the Greens.

Ralph Nader showed the power of the advocate or "whistle-blower" who publicizes health hazards to stimulate active public debate on a host of issues related to the public well-being. Nader, a consumer advocate lawyer, developed a strategy for fighting against business and government activities and products which endangered public health and safety. His 1965 book *Unsafe at Any Speed* took issue with the U.S. automobile industry for emphasizing profit and style over safety. This led to the enactment of the National Traffic and Motor Safety Act of 1966, establishing safety standards for new cars. This was followed by a series of enactments including design and emission standards and seat belt regulations. Nader's work continues to promote consumer interests in a wide variety of fields, including the meat and poultry industries, coal mines, and promotes greater government regulatory powers regarding pesticide usage, food additives, consumer protection laws, rights to knowledge of contents, and safety standards.

Consumerism has become an integral part of free market economies, and the educated consumer does influence the quality, content, and price of products. Greater awareness of nutrition in health has influenced food manufacturers to improve packaging, content labeling, enrichment with vitamins and minerals, and advertisement to promote those values. Low fat diary products are available because of an increasingly sophisticated public concerned over dietary factors in cardiovascular diseases. The same process occurred in safe toys and clothing for children, automobile safety features such as car seats for infants, and other innovations that quickly became industry standards in the industrialized world.

Voluntarism and Community Involvement

Advocacy and voluntarism go hand in hand. Voluntarism takes many forms, including raising funds for the development of services or operating services needed in the community. It may be in the form of fund-raising to build clinics or hospitals in the community, or to provide medical equipment to the elderly or handicapped. It may take the form of retirees and teenagers working as hospital volunteers to provide services that are not available through paid staff, and to provide a sense of community caring for the sick in the best traditions of religious or municipal concerns for the sick. This can also be extended to prevention as in support for immunization programs, assistance for the handicapped and elderly in transportation, meals-on-wheels, and many other services that may not be provided by the "basket of services" provided by the state, health insurance, or public health services.

Community involvement can take many forms, and so can voluntarism. The pioneering role of women's organizations in promoting literacy and nutrition in North America during the latter part of the nineteenth and the early twentieth centuries profoundly affected the health of the population. The advocacy function is enhanced when an organization mobilizes voluntary activity and funds to promote changes or needed services, sometimes forcing official health agencies or insurance systems to revise their attitudes and programs to meet these needs.

THE HEALTH FIELD CONCEPT

By the early 1970s, Canada's system of federally supported provincial health insurance plans were in place coast to coast. The federal Minister of Health, Marc LaLonde, initiated a review of the national health situation, in view of concern over rapidly increasing costs of health care. This led to articulation of the health field concept in 1974, which defined health as a result of four major factors: human biology, environment, behavior, and health care organization. Lifestyle and environmental factors were seen as important contributors to the morbidity and mortality in modern societies. This concept gained wide acceptance, promoting new initiatives that placed stress on health promotion in response to environmental and lifestyle factors. Conversely, reliance primarily on medical care to solve all health problems could be counterproductive.

The health field concept came at a time when many epidemiologic studies were identifying risk factors for cardiovascular diseases and cancers that related to personal habits, such as diet, exercise, and smoking. The concept advocated that public policy needed to address individual lifestyle as part of the overall effort to improve health status. As a result, the Canadian federal government established health promotion as a new activity. This quickly spread to many other jurisdictions and gained wide acceptance in many industrialized countries.

Concern was expressed that this concept could become a justification for a "blame the victim" approach, in which those ill with a disease related to personal lifestyles, such as smokers or AIDS patients, are seen as having chosen to contract the disease. Such a patient might then be considered not entitled to all benefits of

BOX 2.9 THE HEALTH FIELD
CONCEPT—MARC LALONDE

1. Definition: Health is a result of factors associated with genetic inheritance, the environment, and personal lifestyle, and of medical care. Promotion of healthy lifestyles can improve health and reduce the need for medical care.
2. Elements:
   a. Genetic an biological factors;
   b. Behavioral and attitudinal factors (lifestyle);
   c. Environment, including economic, social, cultural, and physical factors;
   d. The organization of health care systems.

Source: LaLonde, M. 1974. *A New Perspective on the Health of Canadians: A Working Document.* Ottawa: Information Canada.

insurance or care that others may receive. The result may be a restrictive approach to care and treatment that would be unethical in the public health tradition and probably illegal in western jurisprudence. This concept was also used to justify withdrawal from federal commitments in cost sharing and escape from facing controversial health reform in the national health insurance program.

THE VALUE OF MEDICAL CARE IN PUBLIC HEALTH

During the 1960s, outspoken critics of health care systems, such as Ivan Illytch, questioned the value of medical care for the health of the public in the 1970s. This became a widely discussed nihilistic view toward medical care, which was influential in promoting skepticism regarding the value of the biomedical mode of health care, and antagonism toward the medical profession.

In 1976, Thomas McKeown presented a historical–epidemiological analysis showing that up to the 1950s medical care had only limited impact on mortality rates, although improvements in surgery and obstetrics were notable. He showed that crude death rates in England averaged about 30 per 1000 population from 1541 to 1750, then declined steeply to 22 per 1000 in 1851, 15 per 1000 in 1901, and 12 per 1000 in 1951 when medical care became truly effective. McKeown concluded that much of the improvement in health status over the past several centuries was due to reduced mortality from infectious diseases. This he related to limitation of family size, increased food supplies, improved nutrition and sanitation, specific preventive and therapeutic measures, and overall gains in quality of life for growing elements of the population. He cautioned against placing excessive reliance for health on medical care, much of which was of unproved effectiveness.

This skepticism of the biomedical model of health care was part of wider anti-establishment feelings of the 1960s and 1970s in North America. In 1984, Milton Roemer pointed out that the advent of vaccines, antibiotics, antihypertensives, and other medications contributed to great improvements in infant and child care, and in management of infectious diseases, hypertension, diabetes, and other conditions. Therapeutic gains continue to arrive from teaching centers around the world. Vaccine, pharmaceutical, and diagnostic equipment manufacturers continue to provide important innovations that have important benefits, but also raise the cost of health care. The latter issue is one which has stimulated the search for reforms and listing of priorities.

The value of medical care to public health, and vice versa, has not always been clear, neither to public health personnel nor to clinicians. The achievements of modern public health in controlling infectious diseases, and even more so in reducing the mortality and morbidity associated with chronic diseases such as stroke and coronary heart disease, were in reality a shared achievement between clinical medicine and public health (Chapter 5).

Preventive medicine has become part of all medical practice, with disease prevention through early diagnosis and health promotion through individual and community-focused activities. Risk factor evaluation determines appropriate screening and individual and community-based interventions. Medical care is crucial in controlling hypertension and in reducing the complications and mortality from coronary heart disease. New modalities of treatment are reducing death rates from first acute myocardial infarctions. Better management of diabetes prevents early onset of complications. At the same time, the contribution of public health to improving outcomes of medical care is equally important. Control of the vaccine-preventable diseases, improved nutrition, and preparation for motherhood contribute to improved maternal and infant outcomes. Promotions of reduced exposure to risk factors for chronic disease are a task shared by public health and clinical medical services. Both clinical medicine and public health contribute to improved health status and both are integral elements of the New Public Health.

HEALTH TARGETS

During the 1950s, many new management concepts emerged in the business community, such as management-by-objective, coined by Peter Drucker and developed at General Motors, with variants such as zero-based budgeting developed in the U.S. Department of Defense. They focused the activities of an organization and its budget on targets, rather than on previous allocation of resources. These concepts were applied in other spheres, but they influenced thinking in health, whose professionals were seeking new ways to approach health planning. The logical application was to define health targets and to promote the efficient use of resources to achieve those targets. This occurred in the United States and soon after in the WHO European region. In both cases, a wide-scale process of discussion and consensus building was used before reaching definitive targets. This process contributed to the adoption of the targets by many countries in Europe as well as by states and many professional and consumer organizations. The United States developed national health objectives in 1979 for the year 1990 and subsequently for the year 2000, with monitoring of progress in their achievement and development of further targets for 2010. Beginning in 1987, state health profiles are prepared by the Epidemiology Program Office of the Centers for Disease Control based on 18 health indicators recommended by a consensus panel representing public health associations and organizations.

The process requires defining the incidence of the disease or risk factor at a point in time (point A), which requires a competent surveillance system based on complete reporting of cases by physicians with laboratory confirmation, and epidemiologic monitoring. Defining the target of the program (point B) requires an effective professional consultative mechanism and informed discussion by knowledgeable people both inside and outside the operative agency. As in other spheres of life, the direct line between A and B may not be the shortest or most feasible direction; the achievement of the target may require indirect and multiple approaches. Nevertheless, defining a target is crucial to the process.

During the 1990s, a series of international conferences on health-related issues were held, focusing political attention on the plight of the world's children (1990), population and demography (Cairo 1994), world poverty (Stockholm 1995), and women's issues (Beijing 1995). The impact of such conferences and statements of intent are not easy to measure. Although it is easy to be skeptical and negative about their value, there are encouraging signs that national governments are influenced by the general movement to place greater emphasis in resource allocation and planning on primary care to achieve internationally agreed-upon goals and targets. The successful elimination of smallpox and rising immunization coverage in the developing countries, and increasing implementation of salt iodization, have shown that such goals are achievable.

United States Health Targets

While the United States has not succeeded in developing universal health care access, it has a strong tradition of public health and health advocacy. Federal, state, and local health authorities have worked out cooperative arrangements for financing and supervising public health and other services. With growing recognition in the 1970s that medical services alone would not achieve better health results, health policy leadership in the federal government formulated a new approach, in the form of developing specific health targets for the nation.

In 1979, the Surgeon General of the United States published the Report on Health Promotion and Disease Prevention (*Healthy People*). This document set five overall health goals for each of the major age groups for the year 1990, accompanied by 226 very specific health objectives. A review of progress made in the achievement of these objectives is listed in Table 2.2.

TABLE 2.2   United States Health Targets 1990 and 2000, with Actual Rates for 1990 and 1996

Objective	1987 Actual	1990 Goal	1990 Actual	2000 Goal	1996 Actual
Infant mortality rate (per 1000 live births)	10.1	9	9.2	7	7.3
Schoolchild mortality rate (1–14 years, per 100,000)	33.7	34	30.8	28	Not available
Adolescent mortality rate (15–24 years per 100,000)	97.8	94	115	85	89.6
Adult mortality rate (25–64 years per 100,000)	426.9	400	406.2	340	Not available

Source: *Health United States,* 1992 and 1998.

New targets for the year 2000 were developed in three broad areas: to increase healthy life spans, to reduce health disparities, and to achieve access to preventive health are for all Americans. These broad goals are supported by 297 specific targets in 22 health priority areas, each one divided into four major categories: health promotion, health protection, preventive services, and surveillance systems. This sets the public health agenda on the basis of measurable indicators that can be assessed year by year.

The process of working toward health targets in the United States has moved down from the federal level of government to the state and local levels. Professional organizations, nongovernmental organizations, as well as community and fraternal organizations are also involved. The states are encouraged to prepare their own targets and implementation plans as a condition for federal grants, and many states require county health departments to prepare local profiles and targets.

Diffusion of this approach encourages state and local initiatives to meet measurable program targets. It also sets a different agenda for local prestige in competitive terms, with less emphasis on the size of the local hospital or other agencies than on having the lowest infant mortality or the least infectious disease among neighboring local authorities.

European Health Targets

In 1985, the European Regional Office of WHO published its targets for HFA 2000 (Table 2.3). These were developed as a consensus among member nations. These 38 targets included measurable outcome and process indicators that serve member states as policy guidelines. The targets assume the following as prerequisites for health: freedom from fear of war; equal opportunity for all; satisfaction of basic needs, including food, education, clean water, sanitation, and decent housing; secure work and a useful social role; and political will and support.

European health targets have helped many countries to redefine health policies and priorities with a stronger role for prevention and health promotion. Recent re-

TABLE 2.3   European Region of the World Health Organization Health Targets
for the Year 2000

Objective	Targets	Specific targets
Health for All	Reduce inequalities between and within countries	
Overall reduction of disease	Eliminate specific diseases	Eliminate measles, poliomyelitis, neonatal tetanus, congenital rubella, diphtheria, congenital syphilis, indigeneous malaria
Infant mortality	No country with over 40 per 1000	Reduce to <15 per 1000
Maternal mortality	No country with over 25 per 1000	<25 per 100,000 to <10 per 100,000
Cardiovascular and cancer mortality	Reduce death rate for people under 65	Reduce by 15%
Road and occupational accident deaths	Reduce road, occupational, and home accidents	Road deaths <20 per 100,000; reduce occupational mortality by 50%
Lifestyles coducive to health	Environment conducive to health Strengthen individual capacity Improve knowledge Support systems Specific risk factors	Specific national and community programs Health promotion Improved nutrition Reduced smoking

Source: WHO Regional Office for Europe, 1985.

views carried out by the European office indicate an emphasis in many countries
on this approach in health planning, and real progress in achieving the targets. The
European region WHO has prepared a Health for All computer data set, also available on the Internet (www.who.dk), for monitoring of progress in selected health
indicators for all member countries.

United Kingdom's Health Targets

In 1990–1991, the United Kingdom's Ministry of Health initiated a consultative process that examined the nation's health strategy and targets (Table 2.4). The
strategy for health focused on improved life span, and the strategic objectives were
as follows:

1. To identify health problems;
2. To focus on promotion of good health and prevention of disease as much
   as on care and rehabilitation;
3. To recognize that health is determined by a whole range of factors (genetic
   inheritance, personal behavior, social circumstances, physical, and social
   environment);

TABLE 2.4   England's Progress to Health for All 2000

WHO European region HFA targets	England's progress
1. Reduce disparity between and within countries by 25%	Gap between manual and nonmanual classes persists, but marked reduction in infant mortality
2–4, 6. Improve ability to lead socially useful, fulfilling lives for general population and handicapped; 10% increase in disease- and disability-free years	Life expectancy above HFA target. Lack of disability data; little change in self-reported morbidity possibly due to changing expectations
5. Communicable disease control; no indigenous measles, polio, neonatal tetanus, diphtheris, pertussis, congential syphilis, or indigenous malaria	Making satisfactory progress; no polio from wild virus, declining measles adopted MMR vaccine, reducing cogenital rubella, congential syphilis; immunization coordinator in each district health authority.
7–8. Reducing infant mortality to <20 per 1000 and maternal mortality to <15 per 100,000 live births	Infant mortality rate 8.4 per 1000 in 1989 (and 7 in 1992); maternal mortality 6 per 100,000 in 1989.
9. Cardiovascular disease under age 65 reduced by 15%	Between 1980 and 1989 deaths from diseases of circulatory system fell from 121 to 85 per 100,000 population; ischemic heart disease mortality declined by 29% and from CVA by 32%
10. Mortality from cancer in people under age 65 should be reduced by at least 15%	All cancer deaths for those <65 fell by 8% in the 1980s; lung cancer (male) mortality (<65) fell by 29%
11–12. Accident deaths should be reduced by at least 25%; suicide rates should decline	Accident mortality is low in England and fell by 19% overall in the 1980s; suicide rates declined by 18%
13–17. Promote healthy lifestyles: family, knowledge, nutrition, limited use of alcohol, smoking reduction, reduced risk-taking behavior	Smoking declined from 39 to 32% of adults; diet contains less fat (fat consumption still high at 42% of total energy vs. recommended 35%); increased participation in sport/recreational activity; alcohol and illicit drug use increased
18–25. Protect the environment; community and international cooperation; monitor control measures for safe water supplies; reduce pollution of rivers, lakes, and seawater, air pollution, food contamination, and hazardous wastes; healthy housing and settlements	England met regional criteria in this area: Water Act of 1989 improved standards for water quality of lakes, rivers, and coastal waters; standards for smoke, sulfer dioxide, lead in air, and nitrogen dioxide are met.
26–31. Health systems based on primary, secondary, and tertiary care; health promotion; focus on	District health authorities, family health services authorities, and local health authorities coordinate services; medical audit systems and performance indicators are being devel-

(continues)

TABLE 2.4   *(continued)*

WHO European region HFA targets	England's progress
high risk, underserved persons and groups; coordinate at community level to ensure quality patient care	oped to promote efficient use of resources and high quality care
32– 38. Research and information to support HFA 2000 targets, policies, and strategies; planning, training, postgraduate education, technology assessment	Progress made in managerial processes for resource allocation, administration, records linkage; postgraduate medical, dental, and nursing education standards have been improved

Source: Secretary of State for Health, The Health of the Nation. A Consultative Document for Health in England. 1991, reprinted 1995.

4. To recognize that concerted action calls for greater cooperation between those involved at the national and local level, within and outside the NHS;
5. To secure a proper balance between central strategic direction and local and individual discretion, flexibility and initiative, and ensure that where responsibilities are devolved there is fair, but rigorous, scrutiny of performance and outcomes;
6. To secure the best possible set of resources, which will always be finite.

There are competing demands in society for expenditure by the government, so making the best use of resources—money and people—is therefore an important objective. Key subjects chosen for action were ischemic heart disease and stroke, cancer, mental illness, HIV and sexual health, and accidents.

International Health Targets

The World Summit for Children (WSC) held in 1990 was attended by 71 heads of state and reflected a broad commitment to national programs of action (NPAs) to achieve goals of improved health for children and mothers. Emphasis was placed on the need to mobilize the whole of society to achieve these goals, including neighborhood and civic associations, religious groups, voluntary agencies, professional associations, the business community, organized labor, and the universities. The NPAs are based on measurable and realistic goals and specific targets achievable by the year 2000.

The WSC reflected increasing support for national programs for child health. For example, basic immunization coverage rates had improved so that, in 1990, the developing countries achieved 80% immunization coverage resulting in the saving of 3 million lives each year, and protection of millions more from preventable or easily treatable diseases. Diarrheal disease death rates had been reduced by more than 1 million lives per year by the widespread use of ORT, one of

---

BOX 2.10 THE WORLD SUMMIT FOR CHILDREN 1990: PLAN OF ACTION— HEALTH GOALS

1. Eradication of polio by the year 2000.
2. Elimination of neonatal tetanus by 1995.
3. Reduction of measles cases by 90%, and 95% reduction of measles deaths.
4. Increase in immunization coverage to 90% of 1-year-old children and universal immunization of women in childbearing age for tetanus.
5. Decrease in child deaths by 50% from diarrheal diseases and reduction by 25% of incidence of diarrheal diseases.
6. Reduction by 33% of child deaths from acute respiratory infections.
7. Virtual elimination of vitamin A deficiency and iodine deficiency disorders.
8. Reduction of low birth weight rates to no more than 10%.
9. Reduction from 1990 levels by 33% in iron deficiency anemia rates in women.
10. Access of all women to prenatal care, trained birth attendants, referral for high risk pregnancies, and obstetric emergencies.

---

the great medical innovations in the past 40 years. The WSC set new objectives in these and other areas.

Involvement of the private sector in health achievements on a world scale is exemplified by the involvement of Rotary International in the program to eradicate poliomyelitis by the year 2000, a goal which is achievable. Already North and South America are free from domestic wild poliomyelitis cases (see Chapter 4). Many health targets depend on international as well as national policy, but implementation will require involvement of communities and the individual.

INDIVIDUAL AND COMMUNITY PARTICIPATION IN HEALTH

National policy in health ultimately relates to health of the individual. The various concepts outlined in the health field concept, community-oriented primary health care, health targets, and effective management of health systems can only be effective if the individual and his community are knowledgeable participants in seeking solutions. Involving the individual in his or her own health status requires raising levels of awareness, knowledge, and action. The methods used to achieve these goals include health counseling, health education, and health promotion (Fig. 2.5).

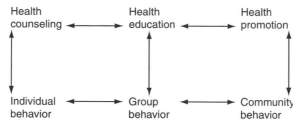

FIGURE 2.5   Health counseling–health education–health promotion paradigm.

Health counseling has always been a part of health care between the doctor or nurse and the patient. It raises levels of awareness of health issues of the individual patient. Health education has long been part of public health, dealing with promoting consciousness of health issues in selected target population groups. Health promotion incorporates the work of health education but takes health issues to the policy level of government and involves all levels of government and NGOs in a more comprehensive approach to healthier environment and personal lifestyles.

Health counseling, health education, and health promotion are among the most cost-effective interventions for improving health of the public. While costs of health care are rising rapidly, demands to control cost increases should lead to greater stress on prevention, and adoption of health education and promotion as an integral part of modern life. This should be carried out in schools, the workplace, the community, commercial locations (e.g., shopping centers), recreation, and in the political agenda.

Psychologist Abraham Maslow described a hierarchy of needs of human beings. Every human has basic requirements including physiological needs of safety, water, food, warmth, and shelter. Higher levels of need include recognition community, and self-fulfillment. These insights supported observations of efficiency studies such as those of Elton Mayo in the famous Hawthorne effect in the 1920s, showing that workers increased productivity when acknowledged by management in the objectives of the organization (see Chapter 11). In health terms, these translate into factors that motivate people to positive health activities when all barriers to health care are reduced.

Modern public health faces the problem of motivating people to change behavior; sometimes this requires legislation, enforcement, and penalties for failure to comply, such as in mandating car seat belt use. In others it requires sustained performance by the individual, such as the use of condoms to reduce the risk of STD and/or HIV transmission. Over time, this has been developed into a concept known as knowledge, attitudes, beliefs, and practices (KABP), a measurable complex that cumulatively affects health behavior (see Chapter 3). There is often a divergence between knowledge and practice, for example, the knowledge of the importance of safe driving practices, yet not putting this into practice. This concept is sometimes referred to as the KAP gap.

The health belief model has been a basis for health education programs, whereby a person's readiness to take action for health stems from a perceived threat of disease, a recognition of susceptibility to disease and its potential severity, and the value of health. Action by an individual may be triggered by concern and by knowledge. Barriers to appropriate action may be psychological, financial, or physical, including fear, time loss, and inconvenience. Spurring action to avoid risk to health is one of the fundamental goals in modern health care. The health belief model is important in defining any health intervention in that it addresses the emotional, intellectual, and other barriers to taking steps to prevent or treat disease.

Health awareness at the community and individual levels depends on basic education levels. Mothers in developing countries with primary or secondary school education are more successful in infant and child care than less educated women. Agricultural extension services reaching out to poor and uneducated farm families in North America in the 1920s were able to raise consciousness of good nutrition, and when this was supplemented by basic health education in the schools, generational differences could be seen in levels of awareness of the importance of balanced nutrition. Secondary prevention with diabetics and patients with coronary heart disease hinges on education and awareness of nutritional and physical activity patterns needed to prevent or delay a subsequent myocardial infarction.

Ottawa Charter for Health Promotion

The World Health Organization sponsored the First International Conference on Health Promotion, held in Ottawa, Canada, in 1986. The resulting Ottawa Charter defined health promotion and set out five key areas of action: building healthy public policy, creating supportive environments, strengthening community action, developing personal skills, and reorienting health services. The Ottawa Charter called on all countries to

> put health on the agenda of policy-makers in all sectors and at all levels, directing them to be aware of the health consequences of their decisions and to accept responsibility for health. Health promotion policy combines diverse but complementary approaches, including legislation, fiscal measures, taxation, and organizational change. It is a coordinated action that leads to health, income, and social policies that foster greater equity. Joint action contributes to ensuring safer and healthier goods and services, healthier public services, and cleaner, more enjoyable environments. Health promotion policies require the identification of obstacles to the adoption of healthy public policies in nonhealth sectors, and ways of removing them. The aim must be to make healthier choice the easier choice for policy makers as well. [Source: Health and Welfare Canada–World Health Organization, 1986.]

State and Community Models of Health Promotion

An effective approach to health promotion was developed in Australia where in the State of Victoria revenue from a cigarette tax has been set aside for health promotion purposes. This has the effect of discouraging smoking, and at the same time finances health promotion activities and provides a focus for health advocacy in terms of promoting cessation of cigarette advertising at sports events or on television. It also allows for assistance to community groups and local authorities to develop health promotion activities at the work place, in schools, and at places of recreation. Health activity in the workplace involves reduction of work hazards as well as promotion of healthy diet, physical fitness, and avoidance of risk factors such as smoking and alcohol abuse.

In the Australian model, health promotion is not the only persuasion of people to change their life habits; it also involves legislation and enforcement toward environmental changes that promote health. For example, this involves mandatory filtration, chlorination, and fluoridation for community water supplies to reduce waterborne disease and to promote dental health. It also involves vitamin and mineral enrichment of basic foods to prevent micronutrient deficiencies. These are at the level of national or state policy, and are vital to a health promotion program and local community action.

Community-based programs to reduce chronic disease by use of the concept of community-wide health promotion have developed in a wide variety of settings. Such a program to reduce risk factors for cardiovascular disease was pioneered in the North Karelia Project in Finland. This project was initiated as a result of pressures from the affected population of the province, which was aware of the high incidence of mortality from heart disease. Finland had the highest rates of coronary heart disease in the world and rural area of North Karelia was even higher than the national average. The project was a regional effort involving all levels of society, including official and voluntary organizations, to try to reduce risk factors for coronary heart disease. After 15 years of follow-up, there was a substantial decline in mortality with similar decline in a neighboring province taken for comparison, although the decline began earlier in North Karelia.

In many areas where health promotion has been attempted as a strategy, community-wide activity has developed with participation of NGOs or any valid community group as initiators or participants. Healthy Heart programs have developed widely with health fairs, sponsored by charitable or fraternal societies, schools, or church groups, to provide a focus for leadership in program development. A wider approach to addressing health problems in the community has developed into an international movement of "healthy cities."

Health Cities/Towns/Municipalities

Following deliberations of the Health of Towns Commission, chaired by Edwin Chadwick, the Health of Towns Association was founded in 1844 by Southwood Smith, a prominent reform leader of the Sanitary Movement, to advocate change to reduce the terrible living conditions of much of the population of cities in the United Kingdom. The Association established branches in many cities and promoted sanitary legislation and public awareness of the "Sanitary Idea" that overcrowding, inadequate sanitation, and absence of safe water and food created the conditions under which epidemic disease could thrive. In the 1980s, Ilona Kickbush, Trevor Hancock, and others promoted renewal of the idea that local authorities have a responsibility to build health issues into their planning and development processes.

Healthy Cities is an approach to health promotion that emerged in the 1980s, promoting urban community action on a broad front of health promotion issues

TABLE 2.5   The Qualities of Healthy Cities

1. A safe, clean physical environment	7. A diverse, vital, and innovative city
2. A stable ecosystem sustainable in the long term	8. Connection with the past—cultural,
3. A strong, supportive, nonexploitive community	biological heritage
4. A high level of participation/control by the public	9. A form compatible with these objectives
over decisions affecting their lives and well-being	10. An optimum level of public health and
5. Meeting basic needs (food, water, shelter, income,	medical services
saftey, and work) for all of the city's people	11. High health status; positve health, low
6. A wide variety of experiences and resources	levels of disease
for contact, interaction and commuication	

Source: WHO. Twenty Steps for Developing a Healthy Cities Project. Second Edition. Copenhagen: Regional Office for Europe, 1995.

(Table 2.5). Activities include environmental projects (such as recycling of waste products), improved recreational facilities for youth to reduce violence and drug abuse, health fairs to promote health awareness, and screening programs for hypertension, breast cancer, and others. It combines health promotion with consumerism and returns to the tradition of local public health action and advocacy.

The municipality, in conjunction with many NGOs, develops a consultative process and program development approach to improving the physical and social life of the urban environment and the health of the population. In 1995, the Healthy Cities movement involved 18 countries with 375 cities in Europe, Canada, the United States, the United Kingdom, South America, Israel, and Australia, an increase from 18 cities in 1986.

Montreal has had a Health Cities project since 1988, which is typical of healthy cities projects internationally (Table 2.6). It is a city with a population of 1.02 million in an area of 177 square kilometers. It is 61% French-speaking, and is the major urban center in Quebec. Average household income is $26,300 Canadian. The city is divided into neighborhoods, engaged in a wide range of activities fostered by the healthy cities project. The Healthy Cities project builds a coalition of municipal and voluntary groups working together in a continuing effort to improve quality of services, facilities, and the living environment.

TABLE 2.6   Montreal's Healthy City Program

1. Nuisance abatement in local quarries	7. Recreational facilites for youth
2. Gardens in poor neighborhoods	8. Restoration of neglected sites
3. Primary and secondary school improvements	9. Extension of public parks
4. Neighborhood profiles	10. Youth and community activities
5. Cooperative housing for low income families	11. Tree planting
6. Intercultural communication	12. Reduction of drug and crime environment

Source: Montreal Healthy Cities Project, 1994.

SOCIAL ECOLOGY AND HEALTH PROMOTION

Human ecology, a term introduced in 1921 by Park and Burgess, attempted to apply theory from plant and animal life to human communities. It evolved as a branch of demography and developed with applications in rural sociology and later medical sociology, which studied social and cultural contexts of disease, health risks, and health behavior.

Parallel to this, psychology developed subdisciplines of social, community, and environmental psychology. Health education developed as a discipline and function within public health systems in school health, rural nutrition, military medicine, occupational health, and many other aspects of preventive-oriented health care, discussed in later chapters. Health promotion as an idea evolved, in part, from the LaLonde health field concept and from growing realization in the 1970s that access to medical care was necessary but not sufficient to improve health of a population. The integration of the health behavior model, social ecological approach, environmental enhancement, or social engineering formed the basis of the social ecology approach to defining and addressing health issues (Table 2.7).

Individual behavior depends on many surrounding factors, while community health also relies on the individual; the two cannot be isolated from one another. The ecological perspective in health promotion works toward changing people's

TABLE 2.7   Health Promotion Approaches: Behavior Modification, Environment Enhancement, and Social Ecology

Health behavior model, change and lifestyle modification	Social ecological approaches	Environmental enhancement national, municipal, and community-based
Behavior modification	Cultural change models of health	Universal access to health care
Social learning theory	Biopsychosocial models of health	Environmental health
Health belief model	Stressful life events	Industrial hygiene
Theory of reasoned action	Ecology of human development	Social security
Theory of planned behavior	Public health psychology	Societal support
Risk perception theory	Medical sociology	Community organization
Fear arousal	Ethnography	Ergonomics/human factors
Protection-motivation theory	Social epidemiology	Health monitoring epidemiology
Health communicaitons	Social ecology of health	Urban planning, architecture
Mass media	Community health promotion	Regulation of housing zoning
	Public policy iniatives	Injury and disaster control
	Healthy communities	Food and drug control
		Nutrition and food fortification

Source: Modified from Stokols, D. 1996. Translating social ecological theory into guidelines for community health promotion. *American Journal of Health Promotion* 10: 282–298.

behavior to enhance health. It takes into account factors not related to individual behavior. These are determined by the political, social, and economic environment. It applies broad community, regional, or national approaches that are needed to address severe public health problems, such as control of HIV infection, tuberculosis, malnutrition, cardiovascular disorders, violence and trauma, and malignant disease.

DEFINING PUBLIC HEALTH STANDARDS

The American Public Health Association formulation of the public health role in 1995 entitled *The Future of Public Health in America* was presented at the annual meeting in 1996. The APHA periodically revises standards and guidelines for organized public health services provided by federal, state, and local governments (Table 2.8). These reflect the profession of public health as envisioned in the United States where access to medical care is limited for large numbers of the population because of lack of universal health insurance. Public health in the United States has been very innovative in determining risk groups in need of special care and finding direct and indirect methods of meeting those needs.

TABLE 2.8   Standards for Public Health Services, American Public Health Association, 1995

Vision	Healthy people in healthy communitites
Mission	Promote health and prevent disease
Goals	Prevent epidemics and spread of disease
	Protect against environmental hazards
	Prevent injuries
	Promote and encourage healthy behaviors
	Respond to disasters and assist communities in recovery
	Assure health status to identify community problems
Essential services	Monitor helath status to identify community problems
	Diagnose and investigate helath problems in the community
	Inform, educate, and empower people about health issues
	Develop policies and plans that support individual and community health efforts
	Enforce laws and regulations that protect helath and ensure safety
	Link people to needed personal health services and assure provision of health care when otherwise unavailable
	Assure an expert public health work force
	Evaluate effectiveness, accessibility, and quality of health services
	Research for new insights and inovative solutions to health problems

Source: American Public Health Association, The Nation's Health, March, 1995.

INTEGRATIVE APPROACHES
TO PUBLIC HEALTH

Public health involves both direct and indirect approaches. Direct measures in public health include immunization of children, modern birth control, hypertension, and diabetes case finding. Indirect methods used in public health protect the individual by community-wide means, such as raising standards of environmental safety, assurance of a safe water supply, sewage disposal, and improved nutrition.

In public health practice, the direct and indirect approaches are both relevant. To reduce morbidity and mortality from diarrheal diseases requires an adequate supply of safe water, and also education of the individual in hygiene and the mother in use of oral rehydration therapy (ORT). The targets of public health action therefore include the individual, the family, the community, the region, or the nation.

The targets for protection in infectious disease control are both the individual and the total group at risk. For vaccine-preventable diseases, immunization protects the individual but also has an indirect effect by reducing the risk even for unimmunized persons. In control of some diseases, individual case finding and management reduce risk of the disease in others and the community. For example, tuberculosis requires case finding and adequate care among high risk groups as a key to community control. In malaria control, case finding and treatment are essential together with environmental action to reduce the vector population, to prevent transmission of the organism by the mosquito to a new host.

Control of noncommunicable diseases, where there is no vaccine for mass application, depends on the knowledge, attitudes, beliefs, and practices of individuals at risk. In this case, the social context is of vital importance as is the quality of care to which the individual has access. Control and prevention of noninfectious diseases involve strategies using individual and population-based methods. Individual or clinical measures include professional advice to a client on how best to reduce the risk of the disease by early diagnosis and implementation of appropriate therapy. Population-based measures involve indirect measures with government action banning cigarette advertising, or direct taxation on cigarettes. Mandating food quality standards, such as limiting the fat content of meat, and requiring food labeling laws are part of control of cardiovascular diseases.

The way individuals act is central to the objective of reducing disease, since many noninfectious diseases are dependent on behavioral risk factors of the individual's choosing. Changing the behavior of the individual means addressing the way one sees his or her own needs. This can be influenced by the provision of information, but how a person sees his or her own needs is more complex than that. An individual may define needs differently than the society or the health system. Reducing smoking among women may be difficult to achieve if smoking is thought to reduce appetite and food intake, with the social message that "slim is beautiful." Reducing smoking among young people is similarly difficult if smoking is seen as fashionable and diseases such as lung cancer seem very remote. Recognizing how individuals define needs helps the health system design programs that influence behavior that is associated with disease.

Public health has become linked to wider issues as health care systems are reformed to take on both individual and population-based approaches. Public health and mainstream medicine found increasing common ground in addressing the issues of chronic disease, growing attention to health promotion and economics-driven health care reform. At the same time, the social ecology approaches showed success in slowing major causes of disease, including heart disease and AIDS, and the biomedical sciences provided major new technology for preventing major health problems including cancers, heart disease, genetic disorders, as well as infectious diseases.

Technological innovations unheard of just a few years ago are now commonplace, in some cases driving up costs of care and in others replacing older and less effective care. At the same time, resistance of important pathogenic microorganisms and to antibiotics and pesticides is producing new changes from diseases once thought to be under control, and "newly emerging infectious diseases" challenge the entire health community. New generations of antibiotics, antidepressants and antihypertensive medications, and other treatment methods are changing the way many conditions are treated. Research and development of the biomedical sciences are providing means of prevention and treatment that profoundly affect disease patterns where they are affectively applied.

The technological and organizational revolutions in health care are accompanied by many ethical, economic, and legal dilemmas. The choices in health care include heart transplantation, an expensive life-saving procedure, which may compete with provision of funds and manpower resources for immunizations for poor children or for health promotion to reduce smoking and other risk factors for chronic disease. New means of treating acute conditions such as myocardial infarction and peptic ulcers are reducing hospital stays, and improving long-term survival and quality of life. Less costly methods of intervention in some cases are producing improved outcomes. The choices in resource allocation can be difficult. These, in part, depend on the public's level of health information and legal protection, whether it be through individuals or advocacy or regulatory approaches for patients' rights. These are factors in a widening methodology of public health (Table 2.9).

THE NEW PUBLIC HEALTH

A WHO meeting in November 1995 on "new challenges for health" reported that the New Public Health was an extension, rather than a substitution of the traditional public health. It described organized efforts of society to develop healthy public policies: to promote health; to prevent disease; and to foster social equity within a framework of sustainable development. A new, revitalized public health

TABLE 2.9   Origins and Synthesis of the New Public Health

Classic public health	Social ecology	Biomedical care	Organization and financing
To end of nineteenth century			
Food and personal hygiene	Church and serfdom	Basic sciences	Private payment for the rich
Settlement health	Renaissance	Clinical sciences	Municipal doctors for poor
Quarantine	Agricultural revolution	Medical education	Charity, church, voluntary
Nutrition/fitness	Improved nutrition	Hospitals: church	hospital care
Vital statistics	Rise of cities	municipal, volun-	Guilds, mutual benefit, and
Epidemiology	Rights of man	tary, university	friendly societies for
Sanitation, miasma	Industrial revolution	Specialization	medical, pensions, and
theory	Labor laws	Therapeutics	burial benefits
Municipal organization	Universal education	Antisepsis	National health insurance
Bacteriology, germ	Social reform	Vaccines	for workers and families
theory	Political revolution	X ray	Sick funds and voluntary
Vaccines, immunology	Information revolution		health insurance
Control of infectious			
diseases			
Maternal and child health			
Health education			
To the 1980s			
Epidemiologic transition	Aging of populatin	Advancing medical	Collective bargaining
Declining mortality and	Rising expectations	sciences	health benefits
birth rates, aging of	Lifestyle and risk	Clinical specialization	Government resposibilty
population	factors	Diagnostics, imaging,	National health insurance
Demographic transistion	Social inequities	laboratory technology	or national health
Decreasing infectious	Social security	Therapeutics, antibiotics,	service
disease	The welfare state	antihypertensives,	Rising costs of health
Increase in non-	Governmental	cardiac, psychotropic	care
infectious disease	responsibilty of	drugs	Imbalance of hospital care
International health	health	Preventive medicine	and primary care
Eradication of smallpox	Advocacy	Home care	Health maintenance organi-
	Health promotion	Long-term care	zation
		Hospital vs community	Cost-benefit evaluation
		care	Regionalization
		Ambulatory surgery	Reforms
Toward 2000 and Beyond—The New Public Health			
Policy coordination	National health policy	University medical	National health targets
Evaluation of health status	Resource allocation	Postgraduate education	Decentralization/diffusion
Health promotion	Econmomic devel-	schools	of implementation
Regulation of food,	opment	Health management	District health systems
drugs, water, work	Social context	training	Managed care systems
site, toxic agents	Social security	Peer review systems	(HMOs)
trauma, environmental	Ecology and envi-	Accreditation	Modified market mecha-
risk factors	ment	Quality of care	nisms, regulation of sup-
Communicable disease	Nutrition and food	Targeted research	ply, incentives, fee
control	policy	Balance hospital/	control, competition,
Control chronic disease	Healthy public	community care,	managed care
Reduce risk factors	policy	long-term care,	Management accountability

(continues)

TABLE 2.9   *(continued)*

Classic Public Health	Social ecology	Biomedical care	Organization and financing
Special needs group	Healthy communities	home care, elder-	Economic assessment
Mental health	Intersectoral cooper-	ly housing, com-	Integrated health systems
Dental health	ation	munity services	
Health information	Advocacy	Integrated health sys-	
systems	Voluntarism	tems	
Epidemiologic systems	Community participa-	Managed care systems	
Planning and management	tion	Ethical issues	

must continues to fulfill the traditional functions of sanitation, protection and related regulatory activities, but in addition to its expanded functions'

> The New Public Health is not so much a concept as it is a philosophy which endeavors to broaden the older understanding of public health so that, for example, it includes the health of the individual in addition to the health of populations, and seeks to address such contemporary health issues as are concerned with equitable access to health services, the environment, political governance and social and economic development. It seeks to put health in the development framework to ensure that health is protected in public policy. Above all, the New Public Health is concerned with action. It is concerned with finding a blueprint to address many of the burning issues of our time, but also with identifying implementable strategies in the endeavor to solve these problems. [Source: Ncayiyan *et al.*, 1995.]

The New Public Health is therefore an evolving concept or approach drawing on many ideas and experiences in public health throughout the world. It is influenced by a growing recognition of social inequality in health, even in developed countries with universal health programs, and an acknowledgment of the failure of state-operated health services to cope with dramatic changes in disease patterns affecting their populations. The World Bank evaluation of cost-effective public health and medical interventions to reduce the burden of disease also contributed to the need to seek and apply new approaches to health. The New Public Health synthesizes traditional public health with management of personal services and community action for a holistic approach.

SUMMARY

The object of public health, like that of clinical medicine, is better health for the individual and for society. Public health works to achieve this through indirect methods, such as by improving the environment, or through direct means such as preventive care for mothers and infants. Clinical care focuses directly on the indi-

---

BOX 2.11 DEFINING A NEW
PUBLIC HEALTH

The New Public Health (NPH) is a comprehensive approach to protecting and promoting the health status of the individual and the society, based on a balance of sanitary, environmental, health promotion, personal, and community oriented preventive services, coordinated with a wide range of curative, rehabilitative, and long-term care services.

The NPH requires an organized context of national, regional, and local governmental and nongovernmental programs with the object of creating healthful social, nutritional, and physical environmental conditions. The content, quality, organization, and management of component services and programs are all vital to its successful implementation.

Whether managed in a diffused or centralized structure, the NPH requires a systems approach acting toward achievement of defined objectives and specified targets. The NPH works through many channels to promote better health; this includes all levels of government and parallel ministries; groups promoting advocacy, academic, professional, and consumer interests; private and public enterprises; insurance, pharmaceutical, and medical products industries; the farming and food industries; media, entertainment, and sports industries; legislative and law enforcement agencies, and others.

The NPH is based on responsibility and accountability for defined populations in which financial systems promote achievement of these targets through effective and efficient management, and cost-effective use of financial, human, and other resources. It requires continuous monitoring of epidemiologic, economic, and social aspects of health status as an integral part of the process of management, evaluation, and planning for improved health.

The NPH provides a framework for industrialized and developing countries, as well as countries in political–economic transition such as those of the former Soviet system. They are at different stages of economic, epidemiologic, and sociopolitical development, each attempting to assure adequate health for its population with limited resources.

---

vidual patient, mostly at the time of illness. But the health of the individual depends on the health and social programs of the society, just as the well-being of a society depends on the health of its citizens. The New Public Health consists of a wide range of programs and activities that link individual and societal health.

The old public health was concerned largely with the consequences of unhealthy settlements and with safety of food, air, and water. It also targeted the infectious, toxic, and traumatic causes of death, which predominated among young people and were associated with poverty. The new public health has emerged to meet a whole new set of conditions—those associated with increasing longevity and aging of the population, with the growing imortance of chronic diseases, with inequalities in health in affluent and developing societies, with environmental damage, and with ecological imbalance. Many of the underlying factors are believed to be amenable to prevention through social, environmental, or behavioral change and effective use of medical care.

The New Public Health idea evolved primarily since Alma-Ata, which articulated the concept of Health for All, followed by a trend in the late 1970s to establish health targets as a basis for health planning. During the late 1980s and early 1990s, the debate on the future of public health in the Americas intensified as health professionals looked for new models and approaches to public health research, training, and practice. This helped redefine traditional approaches of social, community, and preventive medicine. The search for the "new" in public health continued with a return to the Health for All concept, and a growing realization that health of the individual and of the society involves the management of personal care services and community prevention.

ELECTRONIC MEDIA WEBSITES

American Public Health Association http://www.apha.org/
Declaration of Alma Ata http://www.who.int/hpr.docs/almaata.html
World Health Organization http://www.who.org/

RECOMMENDED READINGS

Black, D. 1993. Deprivation and health. *British Medical Journal,* 307:1630–1631.
Centers for Disease Control. 1991. Consensus set of health status indicators for the general assessment of community health status. *Morbidity and Mortality Weekly Reports,* 40:449–451.
Editorial. 1991. What's new in public health. *Lancet,* 337:1381–1383.
Editorial. 1994. Population health looking upstream. *Lancet,* 343:429–430.
Green, L. W., Richard, L., Potvin, L. 1996. Ecological foundations of health promotion. *American Journal of Health Promotion,* 10:314–328.
Hancock, T. 1993. The evolution, impact and significance of Healthy Cities/Healthy Communities. *Journal of Public Health Policy,* 14:5–18.
Maiese, D. R. 1998. Data challenges and successes with Healthy People. *Healthy People 2000 Statistics and Surveillance,* Centers for Disease Control and Prevention, National Center for Health Statistics, 9:1–8.
Roemer, M. 1984. The value of medical care for health promotion. *American Journal of Public Health,* 74:243–248.
Schmidd, T. L., Pratt, M., Howze, E. 1995. Policy as intervention: Environmental and policy approaches to the prevention of cardiovascular diseases. *American Journal of Public Health,* 85:1207–1211.
Shea, S. (editorial). 1992. Community health, community risks, community action. *American Journal of Public Health,* 82:785–787.

Smith, G. D., Egger, M. 1992. Socioeconomic differences in mortality in Britain and the United States. *American Journal of Public Health,* 82:1079–1081.

Stokols, D. 1996. Translating social ecology theory into guidelines for community health promotion. *American Journal of Health Promotion,* 10:282–298.

Tollman, S. 1991. Community oriented primary care: Origins, evolution, applications. *Social Science and Medicine,* 32:633–642.

Walsh, J. A., Warren, K. S. 1979. Selective primary health care—An interim strategy for disease control in developing countries. *New England Journal of Medicine,* 301:967–974.

White, K., Williams, T. F., Greenberg, B. G. 1961. The ecology of medical care. *The New England Journal of Medicine,* 265:885–892.

BIBLIOGRAPHY

American Public Health Association. 1991. *Health Communities 2000: Model Standards for Community Attainment of the Year 2000 National Health Objectives,* Third Edition. Washington, DC: APHA.

American Public Health Association. 1995. Washington, DC: *The Nation's Health,* March 1995.

Ashton, J. (ed). 1992. *Health Cities.* Milton Keynes: Open University Press.

Ashton, J., Seymour, H. 1988. *The New Public Health: The Liverpool Experience.* Philadelphia: Open University Press.

Backett, E. M., Davies, A. M., Petros-Barvazian, A. 1984. *The Risk Approach in Health Care: With special Reference to Maternal and Child Health, Including Family Planning.* Public Health Papers Number 76. Geneva: World Health Organization.

Bootery, B., Kickbusch, I. (eds). 1991. *Health Promotion Research: Towards a New Social Epidemiology.* WHO Regional Publications, European Series, No. 37. Copenhagen: World Health Organization.

Downie, R. S., Fyfe, C., Tannahill, A. 1990. *Health Promotion: Models and Values.* Oxford: Oxford University Press.

Health and Welfare Canada–World Health Organization. 1986. *Ottawa Charter for Health Promotion: An International Conference on Health Promotion, Ottawa, Canada.*

Institute of Medicine. 1988. *The Future of Public Health.* Washington, DC: National Academy Press.

Kark, S. L. 1981. *Epidemiology and Community Medicine.* New York: Appleton-Century-Crofts.

LaLonde, M. 1974. *A New Perspective on the Health of Canadians: A Working Document.* Ottawa: Information Canada.

Lasker, R. D. (ed). 1997. *Medicine and Public Health: The Power of Collaboration.* New York: The New York Academy of Medicine.

McKeown, T. 1979. *The Role of Medicine.* Oxford: Blackwell.

Martin, C., McQueen, C. J. (eds). 1989. *Readings for a New Public Health.* Edinburgh: Edinburgh University Press.

Ncayiyana, D., Goldstein, G., Goon, E., Yach, D. 1995. *New Public Health and WHO's Ninth General Program of Work: A Discussion Paper.* Geneva: World Health Organization.

Nutting, P. A. (ed). 1990. *Community-Oriented Primary Care: From Principles to Practice.* Albuquerque: University of New Mexico Press.

Pan American Health Organization. 1992. *The Crisis in Public Health: Reflections for the Debate.* Washington, DC: PAHO.

Rose, G. 1993. *The Strategy of Preventive Medicine.* Oxford: Oxford University Press.

Secretary of State for Health. 1991. *The Health of the Nation: A Consultative Document for Health in England.* London: Her Majesty's Stationery Office. Reprinted 1995.

Siegel, P. Z., Frazier, E. L., Mariolis, P., Brackbill, R. M., Smith, C. 1993. Behavioral risk factor surveillance, 1991: Monitoring progress toward the nation's year 2000 health objectives. *Morbidity and Mortality Weekly Report* 42:1–21.

Smith, A., Jacobson, B. 1988. *The Nation's Health: A Strategy for the 1990s.* King Edward's Hospital Fund for London. London: Oxford University Press.

U.S. Public Health Service. *Health United States 1992.* Hyattsville, MD: U.S. Department of Health and Human Services, Public Health Service.

U.S. Public Health Service. *Health United States 1998.* Hyattsville, MD: U.S. Department of Health and Human Services, Public Health Service.

White, K. L. 1991. *Healing the Schism: Epidemiology, Medicine, and the Public's Health.* New York: Springer-Verlag.

World Bank. 1993. *World Development Report: Investing in Health.* New York: Oxford University Press.

World Health Organization. 1978. *Alma-Ata 1978. Primary Health Care.* Geneva: World Health Organization.

World Health Organization, Regional Office for Europe. 1985. *Targets for Health for All: Targets in Support of the European Strategy for Health for All.* Copenhagen: World Health Organization Regional Office for Europe.

World Health Organization. 1994. *Information Support for New Public Health Action at the District Level.* Report of a WHO Expert Committee. Technical Support Series Number 845. Geneva: World Health Organization.

World Health Organization. 1994. *Ninth General Programme of Work Covering the Period 1996–2001.* Geneva: World Health Organization.

World Health Organization, Regional Office for Europe. 1995. *Twenty Steps for Developing a Healthy Cities Project,* Second Edition. Copenhagen: World Health Organization, European Regional Office.

3

MEASURING AND EVALUATING THE HEALTH OF A POPULATION

INTRODUCTION

The question of what is health has been discussed in the previous chapters as an evolving concept. How we measure the health of populations is fundamental to improving their health status. Traditionally public health deals with the health of populations, while the New Public Health deals with the health of both individuals and population groups. The professional working with individual and community health needs to acquire the knowledge and skills necessary to measure and interpret the factors that relate to disease and health both in the individual person and in population groups. Demography and epidemiology are the basis of health information systems, but the social and basic medical sciences are also increasingly important in understanding public health, providing an expanding array of health status indicators and measures of the impact of interventions.

Demography deals, among other things, with the recording of the characteristics and trends of a population over time. Epidemiology involves the quantification of the extent of disease, its patterns of change, and the prevalence of risk factors associated with disease. Epidemiology deals with the causes, distribution, control, and outcome of disease in the population. Other disciplines provide information and insights needed for community and national health assessment. These include many social sciences including sociology, psychology, anthropology, and economics, as well as clinical fields, such as pediatrics and geriatrics, and basic sciences such as immunology and genetics.

This chapter is an introduction to epidemiology and health information systems intended to familiarize the student with basic terms, concepts, and methods. The scope of this text does not lend itself to detailed discussion of biostatistical and epidemiologic methods, but instead focuses on the basic ideas and their relevance to the New Public Health. This chapter should be used to fit epidemiology and health information systems into the context of the New Public Health, and not serve as an authoritative text on the subject. Specialized texts are listed at the end of this chapter.

EPIDEMIOLOGY

Health care providers are generally oriented toward individual patient assessment and care. However, even the specialized clinician must have a basic understanding that disease is not an event isolated to an individual, but affects population groups and communities alike.

Epidemiology is the study of health events in a population, used to understand disease process and outcome, to determine factors in causation, and to provide direction for medical or public health intervention. The distribution and determinants of health-related states or events in specified populations help to identify potential interventions and priorities to control of health problems. "Study" includes surveillance, observation, hypothesis testing, analytic research, and experiments. "Distribution" refers to analysis by time, place, and groups or classes of persons affected. "Determinants" are all the physical, biological, social, cultural, and behavioral factors that influence health. "Health-related states and events" include diseases, causes of death, behavior such as use of tobacco, compliance with preventive regimens, and provision and use of health services. "Specified populations" are those with common, identifiable characteristics that can be quantified. "Application to control . . ." emphasizes the ultimate purpose of epidemiology— to promote, protect, and restore health (from Last, J. M. (ed.). *A Dictionary of Epidemiology,* 1995).

Classically, the clinician examines a patient who presents himself or herself; the epidemiologist studies a population at risk. Both evaluate the effects of a preventive or treatment measure. But clinical care also involves understanding risk factors and the natural process of disease as affected by current methods of intervention. Epidemiology studies a population with a particular disease taking into account factors such as age, sex, ethnicity, exposure to known or suspected risk factors, and socioeconomic patterns, as well as the effect of various interventions. This is undertaken in order to comprehend the natural history of disease, its diagnostic criteria, appropriate methods of prevention or management, outcomes to be expected, and the costs and benefits of the different methods of control. Clinicians and epidemiologists depend on each other, and need the work of other fields, such as health economics and management, and the documented experience of interventions to improve care and efficient use of resources.

Epidemiologic Transition

An epidemiologic transition occurs when there are major changes in patterns of health and disease in a society. The determinants and consequences of the epidemiologic shift are profound for any society and especially important for the management of health services. The commonest causes of death until the mid

BOX 3.1 THE GOALS AND METHODS
OF EPIDEMIOLOGY

Goals

1. To eliminate or reduce health problems and their consequences;
2. To prevent their occurrence or recurrence.

Methods

1. Describe the distribution and size of disease problems in human populations;
2. Identify etiological (the cause of disease) factors in the pathogenesis of disease;
3. Provide data essential to the planning, implementation, and assessment of services for the prevention, control, and treatment of disease and to establish priorities among these services.

Source: The International Epidemiologic Association, 1996. Homepage, 1999: http://www.dundee.ac.uk/

twentieth century were the cumulative effects of infectious disease, poverty, poor nutrition, sanitation, hygiene, and housing, and the lack of effective preventive or curative health care. Mortality patterns, stable at very high rates for centuries, began declining in the industrialized countries from the mid eighteenth century, before effective preventive or medical care became available. This drop in mortality was largely due to improving nutrition, sanitation, education, and socioeconomic conditions (see Chapter 1).

In the nineteenth to twentieth centuries, a profound transition occurred in the industrialized countries as the diseases of "pestilence and famine" waned, and the chronic diseases became the leading causes of death. Many of these were associated with man-made environmental problems and personal lifestyle. This epidemiologic transition took place, in part, because of the cumulative effects of successful public health activities such as environmental sanitation, communicable disease control, and the success of vaccines and antibiotics in reducing the major diseases of childhood as well as improvements in living conditions.

Particularly during the 1950s to 1960s, rising standards of living were associated with increases in noncommunicable diseases, including cardiovascular diseases and specific kinds of trauma such as those associated with industrialization and the automobile. This transition is playing an important role in the disease patterns of developing countries as they urbanize and the middle class grows.

Since the 1960s, a new and equally profound epidemiologic transition has occurred with the decline of heart disease, stroke, and trauma as causes of death. These have contributed to increasing longevity. Greater health consciousness and self-care, improved social security for the elderly and handicapped, and advances in medical care have contributed to this phenomenon.

In the 1980s, a new epidemiologic challenge appeared with the advent of a pandemic of HIV infection and a return of diseases thought to have been under control. Potentially dangerous infectious diseases can be transmitted far from their original habitat in the rapid movement of populations throughout the globe. Other infectious diseases are becoming resistant to available treatments, with multidrug resistant (MDR) infectious diseases. "Newly emerging" diseases are a notable threat to the gains made in the health status of the industrialized world, and an even greater threat to the struggling health systems of developing countries (see Chapters 4 and 16). However, the chief threat to public health remains the massive deprivation in developing countries and poverty still present in the industrialized countries.

In the 1990s there were new findings in the epidemiology of infections causing chronic diseases of high prevalence, such as the finding of *Helicobacter pylori* bacterial infection in peptic ulcers with a strong relation to cancer of the stomach. The previously known relationship of hepatitis B to cancer of the liver, and chronic cirrhosis, took on new importance as an effective and inexpensive vaccine was available. Similarly a newly identified infectious agent was identified, the prion, transmitting Creutzfeld-Jakob Disease, a serious degenerative neurological disorder. New evidence suggest that infectious diseases may also be involved in causation of coronary heart disease and some forms of mental illness; and nutritional deficiencies are cofactors in a variety of diseases. These relationships increase the importance of combining epidemiologic and clinical investigations to confirm these relationships and to seek out preventive mechanisms.

SOCIAL EPIDEMIOLOGY

Epidemiology has evolved from its origins as a factor in sanitary statistics in the first half of the nineteenth century, as exemplified in the political arithmetic and vital statistics of Farr and the social statistics of Chadwick and Shattuck. It helped foster the sanitary movement with public health benefits of drains, sewage systems, and community sanitation. In the late nineteenth century through the first half of the twentieth century, epidemiology was associated with the germ theory of single agents relating to one specific disease, and public health activities focused on interruption of transmission or primary prevention through vaccinations. In the latter half of the twentieth century, chronic disease epidemiology showed associations between multiple risk factors and outcomes, without full understanding of the intervening factors or pathogenesis.

Chronic disease epidemiology led to application of risk control public health measures, affecting lifestyle (diet, exercise, smoking), agents (food, guns, cars), and environment (pollution, passive smoking). A new era of epidemiology is emerging at the end of the twentieth century in which organization, information, and application of biomedical technology are vital in population health. This involves a wider, multidisciplinary approach, where statisticians, economists, social scientists, health systems managers, and epidemiologists bring different skills to a more complex paradigm of public health.

Some landmarks of epidemiology are shown in Box 3.2. They are further discussed in Chapters 1, 4, 5, 8, and 13.

Recent studies from different countries show social inequalities in morbidity and mortality in regard to many different diseases. Such inequalities in France increased between 1970 and 1990 for coronary heart disease and have not been reduced for cerebrovascular disease. A study of late-stage diagnosis of colorectal cancer in New York State showed women and African-Americans as more likely to have late-stage cancer than men and whites. Individuals living in areas of low SES were significantly more likely to be diagnosed at late-stage than those living in higher SES areas. Similar patterns of socioeconomic disparity in mortality have been shown among men in the state of São Paulo, Brazil, with the poor having three times greater rates of mortality than the wealthy minority. In contrast, a study in Denmark of regional and social class variation in relative risk of death showed little social variation except for persons with no known address.

Social epidemiology, in some ways a return to nineteenth century traditions of Chadwick, Shattuck, and Farr (see Chapter 1), states that epidemiology is entangled with society, and that to understand causation of disease, it is essential to understand its historical and social context. This social epidemiology incorporates qualitative methodologies with ecological studies supplemented by methods of measuring associations between exposure and disease in individuals. Thus, epidemiology integrates itself into the New Public Health, rediscovering and linking both the population and social perspective.

EPIDEMIOLOGY IN BUILDING HEALTH POLICY

Epidemiology defined the relationship between environment and communicable diseases with statistical methods, leading to sanitary changes. This created an enormously successful application of public health in the first half of the nineteenth century. The golden period of infectious disease epidemiology in the late nineteenth and first half of the twentieth centuries established the basis for control of communicable disease, a revolution still in process. The mid twentieth century saw the development of chronic disease epidemiology, providing the basis for health promotion of lifestyle changes, contributing to reduced morbidity and mortality from cardiovascular disease and the potential for control of cancer, trauma, and other noncommunicable conditions.

The fundamentals of epidemiology are as vital for the student of health sciences as are the study off bacteriology, biochemistry, or surgery. It is equally important that health planners, economists, and others concerned with the macro aspects of health be conversant with epidemiology, so that they may adapt health services to the epidemiologic changes occurring in society.

Epidemiology is also essential to formulation of policy and operation of a health system. It is essential for the smooth functioning of a health system, as a method

BOX 3.2 SELECTED LANDMARKS IN EPIDEMIOLOGY

Social Epidemiology

1662   Graunt publishes *Natural and Political Observations Made upon the Bills of Mortality*

1836   Registrar General's Office established by U.K. Parliament

1842   Chadwick—*Report . . . Sanitary Condition of the Labouring Population of Great Britain*

1848   Virchow—"medicine is a social science"

1858   Simon maps mortality by district in relation to social and environmental conditions

1982   Black—social class differences in mortality in the U.K.

1995   Beijing Conference on Women, empowerment for health of women and children

Infectious Disease Epidemiology

1796   Jenner uses cowpox to vaccinate against smallpox

1854   Snow identifies and interrupts water transmission of cholera in London

1882   Koch demonstrates organisms of tuberculosis and cholera

1978   Eradication of smallpox achieved

1980s HIV and other newly emerging infectious diseases

2000+ Elimination of yaws, poliomyelitis, leprosy, dracunculiasis, and measles

Noninfectious Disease Epidemiology

1747   Lind demonstrates prevention of scurvy by citrus fruits

1775   Pott shows excess cancer of the scrotum in chimney sweeps

1914   Goldberger identifies nutritional cause of pellagra

1950   Doll and Hill relate cigarette smoking to lung cancer

1954   Framingham study reports on heart disease risk factors

1960s Reducing mortality from cardiovascular diseases, trauma

1990s Infections as causes of chronic diseases; MDR organisms

Health Policy Epidemiology

1883   Bismarck initiates workers compensation and national health insurance

1917   Semashko institutes Soviet state health system

1948   U.K. establishes National Health Service

1978   Declaration on Alma-Ata and Health for All 2000

1979   U.S. Surgeon General Healthy People and health targets

1990   Managed care revolution in U.S.

Source: Modified from Susser, M., and Susser, E. 1996. Choosing a future for epidemiology: eras and paradigms. *American Journal of Public Health,* 86:668–673.

of analysis, and as a monitoring tool. Assessment and monitoring of the health status of a population are, by their very nature, multifactorial. Preliminary and, perhaps, impressionistic reading of the situation, makes use of data available from routine sources, and serves to generate hypotheses for testing. Evaluation is a more formal and systematic approach to determining the quality of the health of a population as objectively as possible. All evaluations need to look at the input, process, and output of a system. The epidemiologic method is applied to measurements (indicators) of inputs (resources) of a health system, the process (manner) of their utilization, and outcomes of care (indicators of morbidity, mortality, or functional status of a population).

Epidemiology and demography are necessary, but not sufficient determinants of health policy. Other factors include the funds, manpower and facilities available, community attitudes, and political will. Epidemiology, health care financing, and resource allocation relate to supply and demand and ultimately policy. Analysis of this complex of factors provides the intelligence or feedback for managing the broad complexity of public health. The New Public Health integrates assessment, evaluation, and epidemiologic analysis with the organization, supply of health care, and other factors relating to health of the community as a whole.

RATES AND RATIOS

Measuring the extent of a disease (or a risk factor) in a population relates known cases to a population base, expressed as a rate or a ratio. Rates and ratios are used to compare occurence of the condition in various population groups.

Rates are measures of frequency of a phenomenon, such as occurrence of a health event (A), in a defined population (B), in a given time period. The components of a rate are the numerator (A), the denominator (B), the time in which the events occurred, and a multiplier to convert the fraction or ratio to a decimal number. This allows comparisons between the frequency of the event in different population groups, for instance, to compare mortality rates from a specific cause between the populations of countries, districts of a country, or population groups defined by social class, education level, ethnicity, or other characteristic.

The risk group may be the entire population defined by a geographic area, an occupational group, a school, a service system, a case load served by a hospital, or any other definable group of people. The population may also be persons who share a risk factor for disease, such as smokers, substance abusers, or persons who eat certain foods.

Designating the population-at-risk is a crucial aspect of any epidemiologic study and is subject to common errors. Defining cases of the disease or the risk factors being studied to provide the numerator is also fraught with difficulties. Not all cases of a disease may be reported, so that the true numerator may be much larger than the number of reported cases. This can occur in common infectious diseases (e.g., measles) or where many cases of disease are not clinically diagnosed and therefore go unreported (e.g., diabetes mellitus). There may be discrepancies such as in coronary heart disease where mortality has declined but morbidity patterns have remained high.

Crude rates are summary rates based on the actual number of events (e.g., births or deaths) reported in a total population in a given time period. Cause-specific rates measure defined conditions (e.g., TB) occurring in the total population or in a designated population group (e.g., age–sex groups) in a specified time period. The population used for annual rate calculations is usually estimated at July 1 of that year, i.e., the midyear population.

Ratios are proportions or fractions (A/B) where A and B are two separate and distinct quantities (neither being included in the other). A ratio may compare mortality rates between two distinct population groups, for example, comparing mortality rates from a specific disease in two population groups, one group exposed to a risk factor compared to the mortality rate in an unexposed population. This is called a risk ratio (RR), which is described later in this chapter.

DEMOGRAPHY

Demography is defined as "the study of populations, especially with reference to size and density, fertility, mortality, growth, age distribution, migration, and vital statistics and the integration of all these with social and economic conditions" (Last, 1995).

Vital statistics include births, deaths, population by age, sex, by location of residence, marital status, socioeconomic status, and migration. Birth data are derived from mandatory reporting of births, and mortality data from compulsory death certificates. Other sources of data are population registries including marriage, divorce, adoption, emigration, immigration as well as economic and labor force statistics compiled by governmental agencies, census data, and data from special household surveys. These form the basic data sets for demographers.

Demography measures trends over time of indices such as birth and death rates, rural–urban residential patterns, marriage and divorce rates and migrations, as well as social and economic conditions. Since public health deals with disease as it occurs in the population, the definition of populations and their characteristics is fundamental.

A census is an enumeration of the population recording the identity of all persons in every residence at a specified time. The census provides important information on all members of the household, including age, date of birth, sex, occupation, national origin, marital status, income, relation to head of the household, literacy, education levels, and health status (e.g., permanent handicapping conditions). Other information on the home and its facilities include type of building, number of rooms, electricity, major home appliances (e.g., stove, refrigerator), toilet and bathroom facilities (e.g., bathtub, shower), car ownership and home heating (stove or central), food purchases, and spending on clothing, entertainment, and other consumption items.

A census may assign persons according to their location at the time of the enumeration (*de facto*) or to the usual place of residence (*de jure*). A census tract is the smallest geographic area for which census data are aggregated and published. Data for larger geographic areas (metropolitan statistical areas) are also published. More extensive data may only be collected for a small percentage of the population. This is carried out over a period of years by a specialized national agency (e.g., Bureau of the Census in the United States, and the Central Bureau of Statistics, Office of Population, Censuses and Surveys in the United Kingdom).

Census data are published in multiple volume series with availability for research on computer disks, CD-ROMs, and on the internet. Usually intercensus surveys are carried out to determine trends in important economic or demographic data such as family incomes, nutrition, employment, and other social indicators. Accuracy of such a complex and costly process cannot be 100%, but great care is taken to assure maximum response and standardization in interview methods and processing to assure precision. Despite its limitations, the census is accepted as the basis of statistical definition of a population.

A demographic transition occurs when there is a substantive change in the age distribution of a population. Population growth is mainly affected by birth and death rates, although other factors such as migration, war, political chaos, famine, or natural disasters may affect population distribution. Changing population patterns also accompany economic development, a process known as demographic transition. This is often characterized by the following stages:

1. Traditional: high and balanced birth and death rates;
2. Transitional: falling death rates and sustained birth rates;
3. Low stationary: low and balanced birth and death rates;
4. Graying of the population: increased proportion of elderly as a result of decreasing birth and death rates, and increasing life expectancy;
5. Regression: migration or increasing death rates among young adults due to trauma, AIDS, early cardiovascular disease or war resulting in steady or declining longevity (demographic regression).

Fertility, mortality, disease patterns, and migration are the main influences on this transition within the population. The many factors that affect fertility decline and increasing longevity are outlined in Box 3.3. Education of women, urbaniza-

---

BOX 3.3 FACTORS IN FERTILITY DECLINE
AND INCREASING LONGEVITY

Factors in fertility decline

1. Education, especially of women;
2. Decreasing infant and child mortality, reduces pressure for more children to ensure survivors;

3. Economic development, improved standards of living, expectations, income levels;

4. Urbanization, changes family needs compared to rural society;

5. Birth control, supply, accessibility, and knowledge;

6. Government policy promoting fertility control as a health measure;

7. Mass media increases awareness of birth control, and aspiration to higher standards of living;

8. Health system development access to medical care;

9. Changing economic status, social role, and self-image of women;

10. Changing social, religious, and political–ideological values.

Factors in increasing longevity

1. Increasing family income and standards of living;

2. Improved nutrition, food supply, distribution, quality, and knowledge;

3. Control of infectious diseases;

4. Reduction in noninfectious disease mortality;

5. Safe water, sewage and garbage disposal, and adequate housing conditions;

6. Disease prevention, reducing risk factors, promoting healthy lifestyle;

7. Clinical care services with improved access and quality;

8. Health promotion and education activities of the society, community, and the individual;

9. Social security systems, e.g., child allowances, pensions, national health insurance;

10. Conditions of employment, recreation, economic, and social well-being.

---

tion, improved hygiene and preventive care, and economic improvement with better living conditions, and declining mortality of infants and children are the major factors. This is an important issue in developing countries where high fertility rates and declining mortality of children contribute to rapid population growth and poverty.

Population Pyramid

A population pyramid provides a graphic demonstration of the percentage of men and women in each age group in a total population (Fig. 3.1). A country or region with a wide population base has a high birth rate and a large percentage of its population under age 15, usually accompanied by limited resources and is a formula for continued poverty. A population pyramid with a narrow base (i.e., few young people) and a growing elderly population will have a smaller work force to

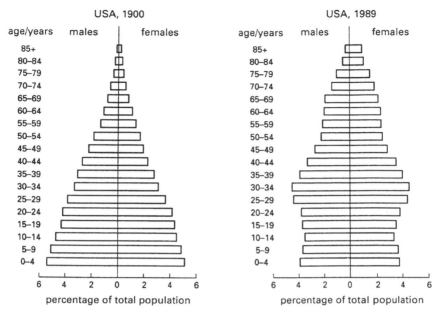

FIGURE 3.1.   Population pyramids for a developing country (Brazil, left) and the United States (right). [Source: Gray A (ed). 1993. *World Health and Disease*. Milton Keynes: Open University Press, with permission.]

provide for the "dependent age" population (i.e., both the young and the old). With a smaller working age population to support these costs, adverse economic consequences may prejudice costly pension and health services. Other factors may also effect the population pyramid, for example, the loss of a large number of people during wartime. This loss affects a particular age–sex group as well as fertility patterns both during and after the war, for example, a post-war "baby boom."

Measures of Fertility

Fertility is bearing of living children and is clearly determined by more than biological potential. Fertility is a complex issue, influenced by cultural, social, economic, religious, and even political factors. Economic prosperity may initially promote higher birth rates, but increases in education levels and economic prospects, as well as in survival of those born, promote restraint in population growth (Box 3.4).

Measures of Morbidity

Morbidity is a departure, subjective or objective, from a state of physiologic or psychological well-being or normal function. It can be measured as the number of persons who are ill, periods or spells of illness, or duration of illnesses (days, weeks, months). Morbidity is also described in terms of frequency or severity. Dis-

---

BOX 3.4 COMMONLY USED
FERTILITY RATES

1. Crude birth rate (CBR) = number of live births during the year (A) divided by the average midyear total population (B) × 1000. In 1991 there were 4,110,907 live births (A) in the United States, and the total resident population was 252,177,000 (B). The crude birth rate was therefore A/B × 1000 = 16.3 per 1000 population, a decline from 23.7 per 1000 in 1960 (−31%).

2. Total fertility rate (TFR) is the average number of children that would be born per woman if all women live to the end of their child bearing years and bore children according to age-specific fertility rates in that population. It tells how many children a woman has, on average, in a specific country or region.

Source: J. M. Last (ed). 1995. *A Dictionary of Epidemiology.* Third Edition. New York: Oxford University Press.

---

ability or incapacity rates measure the extent of long-term reduction of a person's capacity to function in society.

Morbidity data are derived from reported communicable diseases, or chronic, genetic and other conditions for which there are reporting systems and registries. They provide direction for etiological studies and for priorities and avenues for intervention to control the spread of disease. Morbidity is measured by incidence and prevalence rates, as well as severity and duration, though these are not usually available on routine reporting and may require special investigation. Incidence is the number of new cases of a disease in a population group in a given period of time. Incidence is more useful for acute conditions, whereas prevalence is more important in measuring chronic disease and assessing the long-term impact of a disease.

Latency is the time period between exposure to a disease-causing agent and the appearance of manifestation of the disease. For an infectious disease, it is called the incubation period. A disease may appear clinically days, weeks, months, or even years after exposure to the causative agent. whether microbiological, toxic, carcinogenic, or traumatic.

An attack rate is a specific incidence rate expressed as the percentage of the exposed population suffering from the disease. When the population is at risk for a limited period of time, such as during an epidemic, the study period can readily encompass the entire epidemic. The attack rate gives a measure of the extent of the epidemic and may provide information needed to control it. For example, if an epidemic of measles spreads from the initial, or index cases, with an increasing attack rate among the exposed population, a change in vaccination tactics may be needed in order to avoid rapid spread to other vulnerable groups.

BOX 3.5 MEASURES OF FREQUENCY OF DISEASE IN POPULATION GROUPS

Disease frequency is calculated with the following equation:

$$\text{Rate} = \frac{\text{number of cases in a given period}}{\text{population at risk in same period}} \times 10^{n}$$

where $10^{n} = 100, 1000, 10,000$, etc. The period is usually 1 year. If the period is a calendar year, the midyear estimated population on July 1 is generally used.

Incidence is the number of "new health-related events occuring in a population in a specified time period. The rate is calculated from the numerator of new events the denominator is the population at risk of experiencing the event during this period.

Prevalence is the total number of all individuals who have an attribute or a given disease or condition at a point in time or a designated time period. The prevalence rate is the prevalence divided by the population at risk, either at that point in time (point prevalence) or midway through the period (period prevalence).

Attack rate is the cumulative number of cases of a specified disease among the population known to be exposed to that disease over a defined period of time.

Prevalence is the number of cases of a disease or condition in a population group at a particular point in time (point prevalence) or during a specified time period (period prevalence), including both old and new cases. It allows comparison over time with the same or other population groups. Thus morbidity from a specific condition during 1 year can be compared to previous years, weeks, or months, and between countries or regions in a country.

The prevalence rate is calculated on the basis of the number of cases and the number of persons exposed, and may be compared to the nonexposed population. Estimation of case prevalence in an exposed population may be underreported if insufficient time has elapsed for a disease with a long latency period. An example of period prevalence is the number of cases of cancer among persons exposed to a carcinogenic agent in the past, for example, mesothelioma cases occurring in an ex-asbestos worker population over a 30-year latency period following exposure.

Measures of Mortality

Mortality data are based on the mandatory reporting of all deaths. A standard national death certificate provides basic information needed for demographic and epidemiologic purposes (Table 3.1). Modern epidemiology originated in studies

TABLE 3.1 Data Recorded on United States Death Certificates

1. Name	17. Parents' names
2. Sex	18. Parents' address
3. Date of death	19. Disposition of the body, method and place
4. Race, ethnicity	20. Name and address of certifying physician
5. Date of birth	21. Date and time of death
6. Age at last birthday	22. Causes of death[a]
7. City, town, or location of death	a. Immediate cause, e.g., pheumonia
8. Hospital or institution at death	b. Underlying cause, e.g., chronic ischemic
9. Citizenship	heart disease
10. Marital status	c. Other significant conditions, e.g., diabetes
11. Surviving spouse	23. Accident Y/N
12. ID or Social Security number	24. Date and hour of injury
13. Member of armed forces ever Y/N	25. Work injury, time and place
14. Occupation (usual)	26. Place of injury, home, farms, street, factory,
15. State and county of residence	office, other
16. City, town, street address	27. Suicide, homicide, pending investigation

[a]Underlying cause of death is used for official publication of cause-specific mortality.

of mortality derived from the Bills of Mortality (publication of deaths by location and cause) in the United Kingdom by John Graunt in 1662.

Death certificates are mandatory and must be signed by a licensed physician before the body can be buried or cremated and before insurance payment or inheritance can occur. The contents of the death certificate are important since the medically certified cause of death is the basis for mortality statistics. Personal data includes the age, sex, ethnicity, place of residence, and other variables such as occupation, and injury. Completeness of reporting, accuracy of diagnosis, and coding of causes of death may limit the conclusions that can be drawn from such data. In practice however, the data reported in large disease categories are an acceptable guide to actual events.

Causes of death recorded on the death certificate include the *immediate cause of death,* for example, cardiac arrest; the second and third lines includes *contributing conditions,* for example, acute myocardial infarction and congestive heart failure, with the fourth line being the *underlying cause,* for example, coronary heart disease. The death certificate is filed with a public registry office and forwarded to a vital records office where the causes of death are recorded by a registrar trained to federal standards to interpret and code medical diagnoses, according to the 10th International Classification of Disease (ICD-10), adopted by WHO in 1990 (see below).

Overall patterns of mortality are examined by age, sex, and ethnic group, by cause of death. Mortality trends will be discussed under communicable and noncommunicable disease in Chapters 4 and 5. National mortality trends give vital information on disease and changing epidemiologic patterns, allowing for regional as well as international comparisons and help to define health programs and targets.

BOX 3.6 COMMONLY USED MORTALITY
RATES AND RATIOS

1. *Crude death rate* (CDR) = number of deaths from all causes per 1000 population in a given year = $A/B \times 1000$ (total deaths/average population $\times$ 1000);

2. *Age-specific mortality rate* = number of deaths of persons in the specified age group per 1000 live population in that age group over a given period, usually 1 year;

3. *Cause-specific mortality rate* = number of deaths from a specific cause per 100,000 live population (estimated on July 1 of the given year); e.g., if the annual number of deaths from lung cancer in a given year is 400 in a population of 1 million, the cause-specific mortality rate is 400/1,000,000 = 40 lung cancer deaths per 100,000 population;

4. *Case fatality rate* (CFR) = number of deaths from a specified cause during a given period over the number of diagnosed cases of that disease during the same period $\times$ 100; e.g., 10 deaths from measles among 5000 cases gives a CFR of $10/5000 \times 100 = 0.2\%$;

5. *Proportional mortality rate* (PMR) for a specific cause = the number of deaths from that cause in a specified period over the total number of deaths in that population in the same time $\times$ 100; e.g., 25 deaths from road crashes/1000 total deaths from all causes $\times$ 100 = 2.5% (i.e., the denominator includes the numerator)

Source: Adapted from Last, J. M. (ed.). 1995. *A Dictionary of Epidemiology.* Third edition. New York: Oxford University Press with permission.

Mortality patterns can be studied in a particular year or over time. A cohort is usually a group of persons born in a particular year, but it can be any defined group being followed epidemiologically. Cohorts of persons born in particular years can be followed to observe and compare mortality patterns. With suitable age standardization, the mortality pattern of men born, for example, in 1900, 1920, 1940, and 1990 can be compared with each other.

Mortality statistics are fundamental to epidemiology and provide some of the most reliable data available. Epidemiologic analysis of mortality data depends on the registration of deaths with basic demographic data and causation of death as recorded by the physician certifying it. Total, age, and sex specific mortality are usually calculated on an annual basis, with the midyear population as the denominator. This provides crude, age-specific, cause-specific, and proportional mortality rates from which standardized mortality rates are calculated. Case fatality rates (CFRs) relate mortality from a cause to the incidence or prevalence of that disease.

Changes in mortality patterns may occur as a result of a number of factors affecting the outcome of a disease, such as changes in socioeconomic conditions, disease prevention, or methods of treatment. Diagnostic criteria or accuracy of death certificates may also change over time. Thus, a change in mortality may reflect a change in incidence of the disease or case fatality rates related to treatment methods and access to care, or changes in definition or classification of diseases (see Table 3.7).

Social Classification

The British Registrar General's Classification of Occupations was established in 1911 and is updated every 10 years. It is easy to use and provides an excellent demographic and epidemiologic tool that has been used in many studies of disease outcomes. It can help to illustrate the different health experiences of the various social classes, even within the universal National Health Service. It has become part of the data base of vital statistics and morbidity patterns in the United Kingdom.

The United States and most other western countries do not have social class data recorded on death certificates and therefore proxy measures of social classification must be used. Ethnicity, national origin, or religion are all such proxy measures. In the United States, race is recorded on death certificates and this mortality data can be analyzed by racial groups including American Indian or Alaskan Native, Asian or Pacific Islander, Black, Hispanic, and White. Education level and occupation are also recorded, but mortality data is generally presented by racial group, not social indicators.

Interrelationship between ethnicity and disease or mortality often masks other socioeconomic factors, such as higher levels of poverty or reduced access to medical care among African-American and Hispanic groups in the United States or immigrant groups in European countries. Because there are wide variations in socioeconomic and educational levels within ethnic or racial groups, and many confounding factors in ethnicity or race that might affect disease patterns, analysis of data classified this way should be interpreted carefully.

BOX 3.7 BRITISH REGISTRAR GENERAL'S CLASSIFICATION OF OCCUPATION OF HEAD OF HOUSEHOLD

Class I   Professional and business occupations, e.g., physicians, banker
Class II   Intermediate occupations, e.g., schoolteacher, storekeeper
Class IIIN   Nonmanual occupations, e.g., clerk
Class IIIM   Manual skilled occupations, e.g., foreman
Class IV   Partly skilled occupations, e.g., salesperson, factory worker
Class V   Unskilled occupations, e.g., porter, waiter

Social class is increasingly identified as a major variable in health status. It serves as a proxy measure for many health related issues, such as nutrition, access to care, and dependence on occupations with hazards or with little opportunity for personal development or lacking security. Social class variations in health status exist even where universal access health systems are well established, but social differences are less pronounced in the Nordic countries where social gaps are generally less than in states with less developed social welfare systems.

LIFE EXPECTANCY

Life expectancy is an important health status indicator based on average number of years a person at a given age may be expected to live given current mortality rates. Life expectancy can be measured at age 0, or any other specific age, representing expected survival time once a person has reached that specific age, for example, age 15, 60, or 75.

Life expectancy at birth from 1900–1990 in the United States (Table 3.2) increased dramatically in the first half of the century, reflecting mainly reduction in infectious diseases and adverse conditions of maternity and infancy. The second half of the century was characterized by an increase, then a decrease in cardiovascular diseases as a cause of mortality and an increase in cancer and trauma-related deaths, so that life expectancy increased, but at a lower rate than in the earlier period.

Specific life expectancy rates are used in chronic disease epidemiology to summarize patterns of mortality and survival in a population, such as a population with breast cancer. This is important in clinical epidemiology where studies of effectiveness of specific interventions are assessed.

Person-Time

Person-time is a concept which combines, by multiplication, the number of persons exposed and the time of their exposure to a disease-causing agent. Usually, this is in person-years of exposure, such as to an occupational carcinogen or smok-

TABLE 3.2   Life Expectancy in Years at Birth and Changes, United States, Selected Years, 1900–1996

Population	Year							Change (%)	
	1900	1950	1960	1970	1980	1990	1996	1900–1950	1950–1996
Male	46.3	65.6	66.6	67.1	70.0	71.8	73.1	+41.7	+11.4
Female	48.3	71.1	73.1	74.7	77.4	78.8	79.1	+47.2	+11.3
Total	47.3	68.2	69.7	70.8	73.7	75.4	76.1	+44.2	+11.6

Source: National Center for Health Statistics. 1998. *Health United States 1998,* Hyattsville, MD.

ing. Thus if a person is exposed to the agent (e.g., tobacco) for 10 years, and another person in the study group smokes for 12 years, then they contribute 10 and 12 person-years, respectively, to the total sum of the years (or other units of time) that the persons in the study population have been exposed to the agent under investigation. This may be modified by degree of exposure, such as the duration of smoking and the average number of cigarettes smoked over that period of time. The degree of exposure as well as duration can also be measured. Person-distance is used in relating exposure to risk of motor vehicle crashes (person-kilometers).

SENTINEL EVENTS

Sentinel events are taken as measures of problems in a health care process. They are events, such as avoidable deaths, which should not happen if all goes well with preventive and curative care of acceptable standards are in place, except rarely. Avoidable deaths will vary according to the state of health development of a country, and each country may define its own sentinel events for review. Sentinel events include maternal or surgical deaths, medication errors, or infections occurring in hospital that may jeopardize the health of a patient. Such events occur frequently enough to pose both a health risk and economic burden to the hospital, the insurer, and, of course, the patient.

In infectious disease epidemiology, the index case is the first or initial group of cases of a condition that come to attention, providing the first clues in an outbreak or epidemic. In noninfectious conditions, the sentinel event may be a death, where the investigation of the circumstances may help to understand the process of the disease or the care that was received. Any case of polio or several epidemiologically linked cases of measles in a country previously free of the disease may be considered sentinel events which should not happen and their investigation may show errors of omission or commission which explain the event and which point to needed remedial action.

Reporting and data systems should be arranged to indicate avoidable deaths from vital records or hospital discharge information systems. Comparison between areas might also include avoidable deaths as a health status indicator. There are selected conditions which are generally preventable or treatable, and therefore warrant investigation when they occur. Maternal deaths (i.e., deaths associated in time and related to pregnancy or the postpartum period), deaths within 24 hours of hospital admission, or deaths following surgery, are examples of sentinel events which are uncommon and should always be investigated. Deaths from appendicitis or appendectomy, tonsillectomy, hysterectomy, tubal ligation, or other surgical procedures that are elective, should be investigated as sentinel avoidable deaths until other explanatory factors are found. Nosocomial (hospital-acquired) infections are a major cause of mortality, increased length of hospital stay, and health care expenditures. This requires an active program of surveillance and prevention within the care setting.

BOX 3.8 MEASURES OF THE
BURDEN OF DISEASE

1. Years of potential life lost (YPLL): YPLL is a measure of the impact of a particular disease or condition on society by measuring the sum of years lost due to early deaths resulting from a particular cause had these persons lived to a normal life span, compared to the total persons dying from that cause.

2. Disability-adjusted life year (DALY): DALYs are units for measuring the global burden of disease and the effectiveness of health interventions and changes in living conditions. DALYs are calculated as the present value of future years of disability-free life that are lost as a result of premature death or disability occurring in a particular year.

3. Quality-adjusted life year (QALY): QALYs are an adjustment or reduction of life expectancy reflecting chronic conditions, disability, or handicap, derived from survey, hospital discharge, or other data. Numerical weighting of severity of disability is established on the basis of patient and health professional judgment.

Source: Adapted from Last, J. M. (ed). 1995. *A Dictionary of Epidemiology,* Third Edition. New York: Oxford University Press with permission.

THE BURDEN OF DISEASE

The burden of disease (BOD) is measured by combining morbidity and mortality data. The BOD concept is used in examining health issues internationally from the perspective of determining cost-effective interventions especially in developing countries. It helps place priority determination for resource allocation, and to focus international aid to assist recipient countries (Chapter 16). The BOD is increasingly used as an epidemiologic research instrument. This approach however may be too inclusive and mask social and other factors in disease which are multifactorial in origin.

Years of Potential Life Lost

Years of potential life lost (YPLL) are calculated based on age-specific rates of mortality, or disability. They provide a refinement in epidemiology which has added important new perspectives in the analysis of specific problems. The leading causes of death in the United States, as in most developed countries, are coronary heart disease, cancer, and stroke. However, when the data are examined from the point of view of years of potential life lost, trauma (unintentional injuries, homicides, and suicides) becomes the leading cause of death. Years of potential life lost is a better reflection of the impact of diseases on a society than other mortality rates and should be taken into account when determining national health pri-

TABLE 3.3   Years of Potential Lost Life before Age 75 for Selected Causes of Death,
All Races and Sexes, Age Adjusted, United States, Selected Years 1980–1996

Disease Group	1980	1985	1990	1995	1996
All causes	9814	8793	8518	8128	7748
Heart disease	1878	1664	1363	1259	1223
Stroke	303	251	221	212	210
Neoplasms	1815	1776	1714	1588	1554
Unintentional injuries	1688	1366	1263	1156	1137
HIV	na[a]	na	366	570	402
Suicide	403	405	406	406	388
Homicide	461	357	466	436	395
Diabetes mellitus	115	110	133	150	154
Chronic obstructive lung disease	141	156	157	161	161
Chronic liver disease, cirrhosis	259	196	167	150	146

Source: National Center for Health Statistics. 1998. *Health United States 1998,* Hyattsville, MD.
[a]na, Not available.

orities. Trends in YPLL for the years 1980–1996 are shown in Table 3.3, showing
a large drop in YPLL from total and most specific causes. There was also a substantial decline in YPLL for total and some categories from 1995 to 1996, especially in HIV, suicide, and homicide deaths.

Qualitative Measures of Morbidity and Mortality

Quality adjusted life years (QALYs) and disability adjusted life years (DALYs)
are calculations of morbidity introduced recently in international health literature.
Other terms used include disability-free life expectancy (DFLE), and health expectancy, both measures of mortality, morbidity, and impairment or disability.
BOD measures are used to assess cost-effectiveness of specific interventions (see
Glossary). The World Bank calculates the variation in BOD between demographic regions, varying from nearly 600 DALYs lost per 1000 population in sub-
Saharan African countries compared to approximately 120/1000 in the industrialized countries. These measures are being used in economic analyses of health status, helping to focus on outcome measures to justify resource allocation by comparing benefits in terms of reduced mortality and morbidity.

MEASUREMENTS

As has been seen, epidemiology and public health are dependent on quantitative observations to establish relationships and possible points of intervention.
Therefore, an appreciation of methods of handling statistics and their interpretation is fundamental. A complete presentation of this field is beyond the scope of
this text; however, some general concepts are important to establish.

Interpretation of statistical events requires a familiarity with methods of gathering the basic information and in processing it. Statistics is "the science and art of collecting, summarizing and analyzing data that are subject to random variation" (Last, 1995). Biostatistics is the application of statistics to biological problems.

A variable is any factor being studied which is considered to affect health status and which can be measured. It may be an attribute, phenomenon, or event that can have different values, such as age, sex, socioeconomic status, or exposure to a toxic or infectious agent. A dependent variable is the outcome being studied. An independent variable is the characteristic being observed or measured which is hypothesized to cause or contribute to the event or outcome being studied, but is not itself influenced by that event.

The null hypothesis is the assumption that one variable has no association with another variable, and that two or more populations being studied do not differ from one another. A statistical test is used to decide whether the null hypothesis comparing the distribution of a population to another population may be rejected or accepted, that is, whether any differences observed may be due to chance alone. Statistical testing thus provides the basis for inference or decisions regarding the results of a study as "statistically significant" and to what degree.

A confounding variable (confounder) is a factor other than the one being studied that is associated both with the disease (dependent variable) and with the factor being studied (independent variable). A confounding variable may distort or mask the effects of another variable on the disease in question. For example, a hypothesis that coffee drinkers have more heart disease than non-coffee drinkers may be influenced by another factor (Fig. 3.2). Coffee drinkers may smoke more ciga-

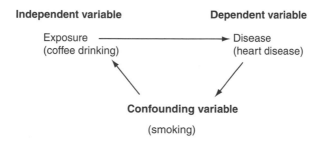

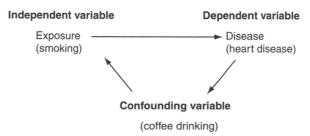

FIGURE 3.2.   Independent, dependent, and confounding variables in a study. [Source: Beaglehole R, Bonita R, Kjellstrom T. 1993. *Basic Epidemiology.* Geneva: World Health Organization.]

rettes than non-coffee drinkers, so smoking is a confounding variable, since smoking was not originally studied. The increase in heart disease may be due to the smoking and not the coffee.

We are often limited to observational studies for evidence of causal relations. Experimental studies may not be possible for many technical, ethical, or other reasons. The proper causal interpretation of the relations from carefully developed epidemiologic studies is vital to the development of effective measures of prevention.

NORMAL DISTRIBUTION

A normal distribution is a continuous, symmetrical, bell-shaped frequency distribution of observations (Fig. 3.3). A normal distribution has upper and lower values that may extend to infinity, but it has an arithmetic mean, mode, and median from a central point.

Mean, median, and mode are measures of central tendency in a group of numbers. The mean is the average value of the observations (i.e., the sum of values of the observations divided by the number of observations). The median value is the midpoint value where half of the observations are equal or less, and half are equal or greater. It is the middle observation when a set of observation numbers is arranged in increasing values. The mode is the most frequently occurring value in a set of observations. In a normal distribution, the mean, median, and mode are all equal to one another.

The standard deviation is the summary value of how widely dispersed the values are around the mean or median. The symmetrical bell-shaped (Gaussian) curve represents the normal distribution of biological characteristics, such as heart rate, height, weight, or blood pressure in a normal population group. In such a dis-

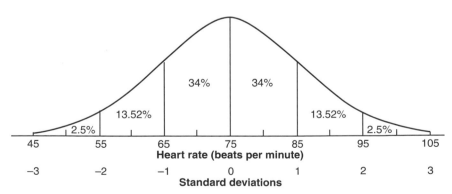

FIGURE 3.3.   Normal distribution. [Source: Last JM (ed). 1995. *A Dictionary of Epidemiology*, Third Edition. New York: Oxford University Press, p. 114.]

BOX 3.9 MEAN, MEDIAN, AND MODE

The following are the scores of 20 students on a math test, ranked in order:

38 39 58 69 78 79 84 84 84 87
89 89 90 90 92 93 98 95 96 99

Mean: 1626/20 = 81.3, or the sum of all values divided by the number of observations;
Median: 88, half the values lie above and below this value;
Mode: 84, the most frequently occurring value.

---

tribution, approximately two-thirds of the observations fall within one standard deviation and approximately 95% fall within two standard deviations of the mean.

Normality may be defined in several senses. It is a range of variation in a given population, within two standard deviations below and above the mean, or between specified percentiles, for example, the 10th and 90th of the distribution. Normally also refers to the limits of a range of a test or measurement and is an indication of the finding being conducive to good health.

Deciding when a group of observations is "normal" or "abnormal" requires defining cut-off points, both in clinical medicine and in epidemiology. In clinical medicine, deciding what is a normal blood pressure, cholesterol level, or growth of a child is based on norms determined from a large number of observations of what is assumed to be a "normal" population. For example, growth patterns of children used as an international standard are based on data derived from a white, middle-class American population (see Chapter 6).

STANDARDIZATION OF RATES

Standardization of rates is a set of techniques used to remove, as far as possible, the effects of different age and sex distributions when comparing mortality or disease incidence between two or more populations of countries or regions or other designated populations. A standard population in this procedure is one in which the age and sex composition is known and therefore is used as a benchmark in order to compare rates for a number of different population groups. For example, comparisons between different states in the United States or countries in Europe would use a U.S. or European population distribution. Standardization can be done by direct or indirect methods.

Direct Method of Standardization

The direct method of standardization is used when age-specific mortality rates are known for the populations being compared to a standard population. These

TABLE 3.4 Crude and Age-Standardized Mortality Rates (per 100,000 Population) for Diseases of the Circulatory System in Selected Countries, 1980

Country	Crude rate	Standardized rate, all ages	Age-specific rate 45–54 years	55–64 years
Finland	491	277	204	631
New Zealand	369	254	184	559
France	368	164	97	266
Japan	247	154	95	227
Egypt	192	299	301	790
Venezuela	115	219	177	497
Mexico	95	163	132	327

Source: Beaglehole *et al.*, 1993, p. 25; calculated from data in *World Health Statistics Annual, 1986.*

rates are then applied to the standard population to calculate the expected numbers of deaths for each age group in the population, as if its age composition were the same as in the standard population. They are then totaled and divided by the total standard population to give a summary age-adjusted rate. Standardized death rates can be calculated for specific diseases. For example, if one is comparing lung cancer death rates in a number of cities to see if there are differences that might be attributed to external factors such as air pollution patterns, the data for each city can be compared by using standardized cause-specific mortality rates.

The direct standardization of rates is an important method of comparing mortality patterns between cities, districts, regions, and countries. Table 3.4 shows the important differences seen in presenting data with age standardization, where a country such as Egypt, with a low crude death rate from circulatory diseases because of the youthfulness of the population, but may have very high age-specific rates.

Indirect Method of Standardization

The indirect method of standardization is used when age-specific mortality rates for the study population are not available, or if the numbers in some age groups are too small. This method uses age-specific rates from a standard population to calculate the expected number of the same health event for the population being studied (Fig. 3.4). The expected number of deaths or cases thus calculated is then compared to the actually observed number of deaths or cases. The ratio of observed/expected is then multiplied by 100 to give the standardized mortality ratio (SMR), which now shows the comparisons free of confounding factors such as different age distribution. The SMR thus allows for comparison of one national, regional, or other specified population group to a selected standard population for which specific rates are known. This same method is also used to calculate morbidity as standardized incidence ratios (SIRs) or other health related observations.

$$\text{SMR (SIR)} = \text{O/E} = \frac{\text{observed deaths (cases)}}{\text{expected deaths (cases)}} \times 100$$

FIGURE 3.4.   Formula for calculating standardized mortality rates (SMR) and standardized incidence rates (SIR).

Standardized mortality (incidence) ratios (SMRs or SIRs) are therefore the crude rate or the total number of deaths or cases occurring in the study group, compared to the expected number of deaths if that population had experienced the same specific death rates as the standard population. The standard population provides a strong base of comparison as it is larger in size, with less likelihood of random variation.

Standardized mortality ratios can be calculated for a specific population group at special risk and compared to a standard population to see if it is vulnerable to higher rates. A group of persons who have been employed in a specific industry and exposed to asbestos may after a long latency period develop mesothelioma. The SMR for a population of ex-asbestos workers in a 25-year follow-up study is seen in Table 3.5.

In the United Kingdom, the SMR is used as the adjustment factor for allocation of funds to district health authorities. Following a lengthy examination of many alternatives, the SMR was believed to incorporate many variables affecting health, including age, sex, socioeconomic, and environmental factors. Areas with higher than expected mortality may have more disease or higher case fatality rates resulting from a greater prevalence of risk factors (genetic, environmental, and/or socioeconomic). Excess mortality may also be due to less access to or poorer quality of health care. Extra resources are made available on this basis to deal with the poorer health status of the population. This is a practical method of addressing regional differences in health, providing a high degree of equity in resource allocation. It takes into account greater need in some areas than in others. The SMR applies epidemiologic methods to improve management practice in health.

TABLE 3.5   Mesothelioma Deaths among Former Asbestos Workers in Israel, 1950–1990

Study group (n)	4401
Number of mesothelioma deaths in study group	26
Expected deaths from national population rates[a]	0.12
SMR	26/0.12 = 216.7

Source: Tulchinsky et al., 1992.

[a]Expected deaths were derived from applying age-specific mesothelioma mortality rates of the total population of Israel to the study group.

SAMPLING

A sample is a selected subset of a population which may be random or nonrandom and representative or nonrepresentative of the general population. Sampling is the process of selecting a subset rather than all individuals in the population in order to study a variable or set of variables in the total population.

A cluster sample selects units such as a family or home, rather than an individual, using a nonrandom method of selection from a random starting point, such as every tenth house. It provides relatively easy access to a population sample for nutrition or immunizations surveys. It is more representative of the population than a sampling based on those who arrive at a health facility, which may be a nonrandom population although convenient and easy to get.

A nonrandom sample is one in which a form of bias is introduced into the sampling process. One example is convenience. A convenience sample is a group of persons who are readily accessible, such as volunteer blood donors or people who appear at a health fair for blood pressure examination. The bias in such samples is that there may well be a self-selection process, not representative of the total population. A selection of a group at special risk, for example, might entail choosing districts with known low immunization coverage in order to attempt to define the reasons for this. Such a study would then be applicable to those districts and, although not generalizable to the total population, could provide valuable information affecting the immunization program.

When all individuals in the population have an equal chance of being selected, the group is known as a random sample. This is often done through assigning each person in the group a number, then selecting the sample from a table of random numbers until the desired sample size is reached. A stratified random sample divides the population into subgroups, for example, by age, sex, socioeconomic status, or region of residence, followed by the selection of a random sample from each subgroup. For example, if 20% of the population are in the age group 40–59 and 20% of the sample comes from this age group, and similarly for other age groups, then all strata are fairly represented with regard to numbers of persons in the sample.

Conclusions based on sample results may be attributable to the population from which the sample is taken. Extrapolation to the total population or a different population is a judgment which is justified only if qualified by description of the sampling methods used and their potential biases. Despite these limitations, careful sampling is valuable in assessing a particular characteristic in a larger population and should give results which are reproducible by other investigators.

POTENTIAL ERRORS IN MEASUREMENT

Data must be assessed as to its validity, reliability, and consistency. It should also be considered for its biological plausibility. These all affect the degree

---

BOX 3.10 OBSERVATION MEASUREMENT ISSUES IN EPIDEMIOLOGY

1. Validity: degree to which a measure actually measures what it claims to measure.
2. Accuracy: extent to which a measure conforms to or agrees with the true value.
3. Precision: quality of being sharply defined.
4. Reliability, reproducibility: stability seen when a measure is repeated under similar conditions.
5. Instrumental error: includes all sources of variation inherent in the test itself.
6. Digit preference: consistent bias by observer rounding of numbers, e.g., to the nearest whole number.
7. Interobserver variation: differences in observation between different observers of the same phenomenon.
8. Individual observer variation: when the same observer records the same observation differently due to changes within the observer, not the observed.
9. Bias: an effect or inference that departs systematically from the true value.
10. Spurious: an apparent but not genuine epidemiologic relationship.

Source: Adapted from Last, J. M. (ed.) 1995. *A Dictionary of Epidemiology,* Third Edition. New York: Oxford University Press.

---

to which inferences can be made and generalizations drawn from the study sample.

Reproducibility or reliability is the degree of stability of the data when the measurement is repeated under similar conditions. If the findings of two persons carrying out the same test (such as the measurement of blood pressure) are very close, the observations show a high degree of interobserver reproducibility. However, it is common in medicine that standardization of even relatively objective measurements by different observers, such as radiologists reading the same X-rays or cardiologists reading the same cardiograms, show high degrees of variability. Instrument standardization, observer training in common standards, and standardization of recording observations are needed to assure acceptable standards of reliability in any data set. Measuring the same patient at different times can produce different results (as in measuring blood pressure or blood sugar), such that standardization of conditions of recording or timing the test are essential to assure comparable data. Standardization of test requires, as part of quality control, sending samples tested in one laboratory to a reference laboratory to see if the test results are the same. It is important to minimize sources of bias (Box 3.11).

BOX 3.11 SOURCES OF BIAS

1. Assumption bias: errors from faulty logic, premises, or assumptions on which the study is based.

2. Response bias: systematic error due to differences between those who choose or volunteer for a study as compared to those who do not.

3. Selection bias: error due to inclusion of those who appear and are included in a study, leaving out those who did not arrive because they had died, were cured without care, were not interested, etc.

4. Sampling bias: error when sampling methodology does not ensure that all members of the reference population have a known and equal chance of being selected for the sample.

5. Observer bias: error due to differences between observers; may be between observers (interobserver) or by the same observer on different occasions (intraobserver).

6. Detection bias: systematic error due to faulty methods of diagnosis or verification of cases in a survey.

7. Design bias: systematic bias due to faulty design of the study.

8. Information bias: flaws in measuring exposure or outcome resulting in data being not comparable.

9. Measuring instrument bias: faulty calibration, inaccurate measuring instruments, contaminated reagents, incorrect dilutions/mixing of reagents.

10. Interviewer bias: conscious or subconscious selection in gathering of data.

11. Reporting bias: self-report selective reporting, suppressing, or exaggerating of information, e.g., history of STDs.

12. Publication bias: editors prefer positive results so that a distorted perception of an issue may occur.

13. Bias due to withdrawals: loss of cases from the sample by withdrawal or nonappearance in follow-up.

14. Ascertainment bias: error due to the type of patients seen by the observer, or in the diagnostic process affected by the culture, customs, or idiosyncracies of the provider of care.

Source: From Last, J. M. (ed.) 1995. *A Dictionary of Epidemiology,* Third Edition. New York: Oxford University Press.

SCREENING FOR DISEASE

Screening for disease may be carried out in a mass basis of a whole population, as was commonly done in the past for tuberculosis. When done with a number of tests it is called multiphasic screening. Screening may target a group at special risk,

---

BOX 3.12 VALIDITY, SENSITIVITY, AND SPECIFICITY OF SCREENING TESTS

	Population Tested		
Test Result	With Disease	Without Disease	Total
Positive	Have disease and test positive (TP)	No disease but test positive (FP)	*TP + FP*
Negative	Have disease and test negative (FN)	No disease and test negative (TN)	*FN + TN*
Total	*TP + FN*	*TN + FP*	*TP + FP + TN + FN*

$$\text{Sensitivity} = \frac{TP}{TP + FN} \times 100 \quad \text{Specificity} = \frac{TN}{(B + D)} \times 100$$

Source: Gorlis, L. 1996. *Epidemiology.* Philadelphia: W. B. Saunders Company, p. 60.

---

such as blood lead screening among workers exposed to lead at their place of work or on children living in the vicinity of a plant using lead. Screening may be part of patient care when the caregiver routinely tests, for example, for blood sugar or cholesterol. Accuracy of a test is usually measured in terms of sensitivity and specificity. Targetted screening may be required by law as in the case of newborn screening for PKU, hypothyroidism, and other congenital disorders. The value of the screening test is defined as to its degree of sensitivity and specificity, as well as its costs and benefits for screening or not screening.

Sensitivity is the proportion of truly diseased persons in the screened population who are identified as such by a screening test. This is sometimes called the true positive rate. Specificity is the proportion of truly nondiseased persons who are identified as not having the disease, that is, it measures the probability of correctly identifying a nondiseased person with a screening test, or the true negative rate. A test which produces too many false positives or false negatives is not valid (Box 3.12).

False negatives occur when a negative laboratory result appears in a person who has the condition for which the test is being conducted. The condition is present but does not show up on the initial screening test or data set. If screening for phenylketonuria is done too soon after birth, some cases may be missed and will only appear later. False negatives reduce the effectiveness of the screening program.

False positive results are those cases in which a positive laboratory result occurs in a person without the condition for which the test is being conducted. Not everyone with an isolated elevated reading of blood pressure has true hypertension. False positive results must be checked because they cannot be excluded without confirmation by more specific testing, such as repeated blood pressure readings. Precision is the quality of sharp definition of the test. If a laboratory test for environmental contamination is accurate to parts per billion as compared to parts per million, then the precision is enhanced.

Screening for disease and risk factors is a common and necessary part of public health. In order to be valuable, screening requires a valid test and a significant condition with a high prevalence in the population. Screening for breast cancer, carcinoma of the cervix, and many other conditions are part of the armamentarium of public health and contribute to lowering mortality and improving survival rates for these diseases. Screening is justified for a condition which is serious and treatable, but uncommon as for phenylketonuria (PKU).

EPIDEMIOLOGIC STUDIES

Epidemiological methods of study are important, not only to define disease in the population, but also to look for specific risk factors for the disease to the population. Epidemiologic studies permit analysis of a risk factor, variable, or of an intervention on a population having or at-risk for a condition, as compared to a similar population not subject to the risk or intervention. This permits testing of new hypotheses and innovations in medicine and public health.

Epidemiologic studies are classified as observational or experimental. The observational study allows nature to take its course, with no intervention. Experimental studies involve activities whose purpose is to reduce the risk of disease occurring and may be controlled in a field or community trial. Retrospective and prospective studies refer to the time of the event being studied, that is, did the disease occur before the time of the study or is it a study of future events?

Observational Studies

Observational studies are those where the population is studied, but nature is allowed to take its course. They may be descriptive or analytical. Descriptive studies are limited to describing the occurrence of a disease in a population, which is often the first step in investigation, as it may provide clues for more specific investigation. Analytical studies go further by looking for specific variables which may be causally associated with the disease. Analytical studies include ecological cross-sectional, case–control, and cohort studies. Each type of study has potential for biases, discussed above.

Descriptive Epidemiology. Descriptive epidemiology uses observational studies of the distribution of disease in terms of person, place, and time. The study describes the distribution of a set of variables, without regard to causal or other hypotheses. Personal factors include age, sex, SES, educational level, ethnicity, and occupation. The place of occurrence can be defined by natural or political boundaries, and can also include such variables as location of residence, work, school, or recreation. Time factors include the relationship between exposure and occurrence of the disease, as well as time trends, which are generally divided into three types:

1. Secular trends: long-term variations;
2. Cyclic changes: periodic fluctuations on an annual or other basis;
3. Short-term fluctuations: as is seen in epidemic disease outbreaks.

Time trends contribute to understanding the natural history of epidemics of acute infectious diseases, such as measles, waterborne disease, or of noncommunicable diseases, such as stroke, or certain cancers, such as among selected occupational groups. Epidemiology also examines the frequency of physiological or pathophysiological events, such as hypertension or growth retardation, and health-related events such as smoking and other risk factors in human activities or events. Descriptive epidemiology includes analytic, ecological, cross-sectional (prevalence), case control, and cohort studies, which are described below.

Analytical Studies.   Analytical studies are concerned with establishing causes or contributory risk factors to disease, including social, economic, psychological, or political conditions that impinge on health. This helps to define programs to intervene in order to reduce the burden of disease in the population. Analytical epidemiology has made vital contributions to modern medicine through identification of key risk factors, such as higher rates of cancer of the lung among smokers and higher rates of stroke among people with hypertension. Analytical studies may include cross-sectional (or prevalence studies), as well as retrospective and prospective studies.

1. ECOLOGICAL STUDIES.   Ecological studies are those in which the analysis is of populations or groups of people, not individuals. An example is a study of the association between income levels, social class, or employment status and cancer or cardiovascular disease mortality. There is a danger in drawing hard conclusions from such studies. An ecological fallacy is a bias in which the association between aggregated variables based on group characteristics does not necessarily represent the association at the individual level. It is therefore an error in inference to conclude that there is a causal relationship.

Studies showing an apparent correlation between quality of drinking water and mortality rates from heart disease have not been substantiated as indicating a "cause–effect" relationship. It would be an inappropriate conclusion (ecological fallacy) to infer from this finding alone that exposure to water of a particular level of water hardness necessarily influences an individual's chances of developing or dying of heart disease. Spurious relationships may be difficult to disprove and may mislead the investigator, when there are unrecognized confounding variables that affect the apparent relationship.

Ecological studies can be valuable in generating hypotheses for further investigation and intervention. For example, comparison of SMRs for disease categories from routine mortality sets can identify regions with high rates of a specific disease, such as lung cancer or diabetes-related conditions, or motor vehicle accidents (MVAs), which require follow-up, investigation, and possibly intervention even before more complete epidemiologic studies can be carried out. Studies have shown higher rates of cardiovascular disease mortality for blacks compared to whites in the United States. However, further analysis shows that there are gradients for cardiovascular mortality for both whites and blacks according to median family income, such that SES emerges as a more important factor than race.

2. CROSS-SECTIONAL STUDIES OR PREVALENCE STUDIES.   These examine the relationship between specific diseases and health-related factors as they exist in individuals in a population at a particular time. The population may then be divided into subgroups, with and without the disease, and the characteristics of each member of each group analyzed for different variables, for example, age, sex, region of residence, occupation, and social class. Comparisons of these variables may indicate a higher risk for disease in one population group as compared to an otherwise similar comparable population.

3. CASE–CONTROL STUDIES.   Case–control studies are observational studies of persons with the disease or other outcome variable of interest and a suitable control (comparison, reference), group of persons without the disease. These studies are retrospective, taking a case and looking backward in time for potential causes of the disease. They compare two similar population groups, one with the disease or condition and the other without. The two groups, cases and controls, are then compared for exposure to the possible risk factor. An example is the study of the occurrence of children with a limb defect born in Germany in the late 1950s, which showed that of those born with this defect, 41 out of 46 mothers had taken the medication thalidomide, an antinausea pill promoted for use during pregnancy, whereas none of the 300 control mothers with normal children had done so. This study led first to preventing this drug from approval for use in the United States, and later to its banning in countries where it was already in use. The odds ratio is commonly used to summarize findings of case control studies. It is a ratio of the odds of exposure among cases to the odds of exposure among controls. Case–control studies may be vital to define the differences between the sick and the control groups in an epidemic or outbreak situation.

4. COHORT STUDIES.   Cohort studies are also referred to as prospective, longitudinal or follow-up studies. They examine a population which is initially free of the disease, dividing the population into subgroups according to exposure to a potential risk factor. The relative risk is a ratio of risk of disease in the two groups, that is, exposed and unexposed. Such studies can yield incidence rates as well as the magnitude of risk of disease for the population. In addition, cohort studies permit the observation of many outcomes. Important morbidity data are sometimes not readily available in the general population reporting systems so that special

study populations are required. The Framingham Study of risk factors for heart disease which followed the population of Framingham, Massachusetts, continuously since 1949 is an example (see Chapter 2). Observational studies of particular population groups have provided important public health advances over the past 50 years.

A natural experiment is a situation in which naturally occurring circumstances result in two similar population groups, one exposed to a supposed causal factor and one not exposed as a study or control group. This term is derived from John Snow's 1850 study of Londoners exposed to drinking water supplied by two different water companies, one group having high rates of cholera and the other low rates. This term is currently used in investigating epidemiological events, regarding each event as a unique situation for which relevant factors need to be defined and to the extent possible, linked to the disease.

Experimental Epidemiology

Experimental studies are studies of conditions under the direct control of the investigator, conducted as closely as possible to a laboratory experiment. Experimental epidemiology involves changing a variable and measuring the effect in one or more population groups. Clinical epidemiology applies experimental epidemiologic research methods to clinical problems and practice. It includes promoting the use of epidemiologic knowledge in the clinical care of individual patients. Clinical epidemiology also contributes knowledge to the planning and operation of health care systems.

1. CONTROLLED TRIALS.   Controlled trials are epidemiologic experiments designed to study an intervention (preventive or therapeutic). It requires a random method of allocating the cases to the experimental or the control group, then both are observed for change over time in relation to the condition being studied. If the people in both the test and control groups do not know which group they are in, the study is called blind. If in addition, the people judging the outcome are also not aware whether the person tested is in the test or control group, the trial is called double blind. Further, if those analyzing the data also do not know who was in each group, the study may be called triple blind. This helps to avoid various biases which limit the value of a study. If the difference in outcomes is statistically significant for the control group and the treatment group, then the treatment is deemed to have been effective. Assignment to the treatment or the control group is by random selection.

2. FIELD TRIALS.   Field trials follow people who are disease-free in two groups, one with and one without a specific intervention, to determine if it affects the risk of developing the disease. This is often used to test a new vaccine in a susceptible population. The field trial conducted by Jonas Salk of inactivated poliomyelitis vaccine in 1956 demonstrated its protective effect and safety in some

1.5 million American children, and was subsequently adopted throughout the world. Field trials are part of the process of approval for new vaccines and medications.

3. COMMUNITY TRIALS.   Community trials are conducted on whole communities measuring the effect of a risk factor or intervention. This cannot easily be randomized since the entire community is selected, and it may be difficult to isolate the community from changes going on in the general population. Community-based heart disease prevention programs such as North Karelia and others (Minnesota Heart Health Project, Pawtucket Heart Health Project in Rhode Island, and many others) are difficult to evaluate, with a conflict between experimental design and community realities. Regional programs for prevention of heart disease cannot be isolated from time trends in the surrounding communities, limiting the interpretation of outcomes measured.

Nevertheless, community trials are necessary in evaluating health interventions to reduce risks or adverse health outcomes. They often rely on performance or utilization indicators as proxies. For example, a village health worker program may increase earlier and more frequent use of prenatal care or immunization coverage, but measurement of outcome variables may be difficult in field conditions.

ESTABLISHING CAUSAL RELATIONSHIPS

Classically, the search for causation in medicine and in public health is for the agent–host–vector relationship, with the agent being a specific causative organism. In infectious disease epidemiology, this has provided the scientific basis for immunology and control of vaccine preventable diseases, and for sanitation to prevent transmission of food and waterborne diseases. Criteria for causation include strength of the association, biological plausibility, consistency with other investigations, and dose–response relationship. Biological plausibility is a test of the plausibility of a causal association based on existing biological or medical knowledge. Consistency with other investigations means that the findings are similar to those of other studies. The dose–response relationship is that in which a change in amount, intensity, or duration of exposure is associated with a change (increase or decrease) in a specified outcome.

Even in infectious disease control, the public health reality is often more complex than the single causation model. Tuberculosis deaths fell during the nineteenth century, presumably due to improved nutrition and living conditions, and were further reduced in the early part of the twentieth century before the antibiotic era by a combination of improved nutrition and symptomatic treatment. Mortality from measles dropped dramatically despite its endemicity (the continuing presence of a disease in a given geographic area) prior to the successful vaccine introduced in the 1960s. This can be attributed to rising standards of living and improved means of treatment of complications. Even today, the mortality rate from measles is seen to be affected by improving the nutrition of children and by vitamin A supplementation.

For noncommunicable diseases, causation is even more clearly multifactorial, and a risk factor for one disease may also be a contributor to increased risk for another disease. Diet has been established as a major risk factor for coronary heart disease, as well as diabetes, and hypertension. Diabetes is a major risk factor for coronary heart disease, stroke, renal, eye, and peripheral vascular disease. Nutrition is an important contributor to certain cancers, so that the multiple-factor causation of disease cannot be ignored.

Risk factors for disease are those aspects of personal behavior or lifestyle, occupational or environmental exposure, social and economic conditions, or inborn or inherited characteristics which, on the basis of epidemiologic evidence, are known to be associated with health-related conditions considered important to prevent. Noninfectious diseases are often multifactorial and exacerbated by "risk factors," and measurement of prevalence of risk factors, or intervening variables, is important to epidemiologic assessment of the future risk of such diseases. Prevalence of smoking serves as an indicator of future potential of lung cancer and cardiovascular diseases. Body fat index, blood pressure, and serum cholesterol levels measured in the community serve as indicators of risk for coronary heart disease. These measurements indicate individual and community risk, and to measure effectiveness of health promotion programs.

NOTIFIABLE DISEASES

Morbidity data are reported by doctors, usually based on compulsory reporting of specific infectious and noninfectious diseases. Some diseases such as plague, cholera, yellow fever, louse-borne typhus, and louse-borne relapsing fever are notifiable by international convention. Locally endemic diseases are notifiable under national public health laws in order to monitor their prevalence and the impact of public health measures (see Chapter 4). Additional diseases reported routinely in other countries include; water and food-borne disease, chemical poisonings, botulism, leishmaniasis, septicemia, chlamydia trachomatitis (genital), gonococcal ophthalmia, and listeriosis. Other diseases or health events may be added to routine reporting (or to special surveys) according to endemic environmental conditions. Reporting of infectious diseases is one of the most important foundations of public health practice.

SPECIAL REGISTRIES AND REPORTING SYSTEMS

Special registries are important to establish a basis for the epidemiologic study of clinical states and vital health events pertinent to the population (Table 3.6). They provide both clinical follow-up and epidemiologic information. Priorities

TABLE 3.6   Health Event Reporting and Registries

Health-related events	Public health registries—mandatory or voluntary
1. Birth	1. Vital statistics
2. Death	2. Notifiable infectious diseases
3. Marriage and divorce	3. STDs and HIV
4. Communicable diseases	4. Tuberculosis
5. Chronic diseases	5. Cancer
6. Birth defects	6. Birth defects
7. Abortions and other pregnancy events	7. Congenital screening for PKU, hypothyroidism
8. Birth weights	8. Birth weight
9. Growth and development indicators	9. Neurological disorders
10. Nutritional status indicators	10. Psychiatric/mental health
11. Behavioral risk factors	11. Thalassemia
12. Health behaviors	12. Sickle cell disease
13. Air and water quality	13. Hospital discharge information systems
14. Environmental hazards	14. Battered children
15. Occupational safety and health hazards	15. Blind and partially sighted persons
16. Mental illness	16. Deaf and hearing impaired
17. Animal reservoirs and health	17. Disability
18. Vaccination and drug reactions	18. At-risk workers groups
19. Nosocomial infections	19. Coronary heart disease
20. Injuries, poisonings, trauma	20. Diabetes

Source: Modified from Declich and Carter, 1994.

may vary from country to country, but the basic registry needs in health care include a range of conditions, including infectious diseases, cancer, birth defects, and hospital discharge information systems. Data from cancer, birth defect, and low birth weight registries can give valuable clues for environmental exposures of public health importance.

Disease registries should be coordinated into unified health information systems. Individual identification numbers for each member of the population enables the use of data from related special registries. However, protective measures must be in place to ensure privacy and to prevent the misuse of this data for unethical purposes, but safeguard mechanisms can be built into data systems to protect the privacy of the individual.

Linkages between data sets allow important epidemiologic correlations to be studied. For example, linking data sets for cancer registries, vital records, and hospital discharge information systems may enhance investigation of specific medical conditions, such as monitoring longevity and hospital use for childhood cancer. It may also be used to compare morbidity and mortality patterns for specific conditions by comparing hospitalizations with mortality patterns.

Studies from vital statistics registries may raise epidemiologic questions or hypotheses which need further investigation. The role of the special survey becomes important as the follow-up to initial findings. Intervention can then be planned on the basis of these investigations. As an example, a review of vital statistics in New York State (1987) showed that some infant deaths were reported during the 1980s as a result of hemorrhagic disease of the newborn (HDN), a disease preventable by prophylactic vitamin K injections of newborns. A follow-up study of the State Hospital Discharge Information system showed a substantial number of hospital discharges with the diagnosis of HDN (first to fourth diagnosis) during the same time period. A case record review of infant deaths with HDN as a diagnosis (first to fourth diagnosis) showed that some two-thirds of the cases did not receive vitamin K at all, or not until after bleeding had already begun. As a result, the State Department of Health adopted mandatory vitamin K prophylaxis for newborns. Record linkage between hospitalization data and the individual cases would have made such a study more readily achievable.

The importance of records linkage might also be demonstrated by the following epidemiologic question. Mortality from cardiovascular disease has fallen dramatically in industrialized countries since its peak in the early 1960s. This can be attributed to changes in nutrition, smoking, and other risk factors. Others suggest that much of this change in mortality is due to improved treatment and not changes in the prevalence of the basic disease process. Which one is it? Studies linking hospitalization patterns with mortality data for cardiovascular diseases may help answer this question.

DISEASE CLASSIFICATION

Because comparative statistics are vital in monitoring the health status of a population, it has been essential to develop internationally accepted standard nomenclature and a coding system in order to minimize differences in classification. The Bills of Mortality used in the seventeenth century defined 17 categories. Classification of disease by anatomic sites or body system was initiated by William Farr at the Second International Statistical Congress in Paris, 1855.

After World War I, the League of Nations supervised revisions of the international classification of diseases (ICD), and since the 1948 sixth revision, the ICD has been updated at about 10-year intervals by the WHO. The tenth revision of the International classification of diseases (ICD-10) came into general use in 1993. The classification is broken down into many subcategories with coding to indicate precise disease and procedure groups (Table 3.7). Similarly, mental health classification of disorders have been developed (Chapter 7).

HOSPITAL DISCHARGE INFORMATION

Admission to a hospital is a major medical event, not less important from an epidemiologic point of view than the reporting of a death or an infectious disease. A hospital discharge data system is an informational, planning, budgeting, epi-

TABLE 3.7   International Classification of Diseases (ICD-10)

1. Certain infectious and parasitic diseases	A00–B99
2. Neoplasms	C00–D84
3. Diseases of the blood and blood-forming organs and certain disorders involving the immune mechanism	D50–D89
4. Endocrine, nutritional, and metabolic diseaes	E00–E90
5. Mental and behavioral disorders	F00–F99
6. Diseases of the nervous system	G00–G99
7. Diseases of the eye and adnexa	H00–H59
8. Diseases of the ear and mastoid process	H60–H95
9. Diseases of the circulatory system	I00–I99
10. Diseases of the respiratory system	J00–J99
11. Diseases of the digestive system	K00–K93
12. Diseases of skin and subcutaneous tissue	L00–L99
13. Diseases of musculoskeletal system, connective tissue	M00–M99
14. Diseases of the genito-urinary system	N00–N99
15. Pregnancy, childbirth, and puerperium	O00–O99
16. Certain conditions originating in perinatal period	P99–P95
17. Congenital malformations, chromosonal abnormalities	Q00–Q99
18. Symptoms, signs, and abnormal clinical, or laboratory findings not classified elsewhere	R00–R99
19. Injury, poisoning, and some other external causes	S00–T98
20. External causes of morbidity and mortality	V01–Y98
21. Factors influencing health and contact with health services	Z00–Z99

Source: Website http://www.who.int/whosis/icd10/descript.htm

demiologic and quality control tool in modern health care. It involves gathering a basic data set on all hospital discharges, input of data into a central file on a regular basis, and processing the data for administrative and epidemiologic purposes. This requires a basic data retrieval form for all hospitalized patients and a system of reporting and analysis with computerized data retrieval preferred.

Hospital statistics were originally promoted by Florence Nightingale. The Uniform Hospital Discharge Information System (UHDIS) evolved due to the importance of hospital utilization in the economics of health care. Introduced in the 1960s by the U.S. National Center for Health Statistics (NCHS), it provided the basis for development of diagnosis related groups (DRGs), which have become the major mode of payment for hospitals in United States and in some other countries since the 1980s. Use of the international classification of diseases allows for comparisons between data sets, regions, and countries.

A central governmental professional unit is needed at the state level to plan, train, and supervise data retrieval and to process and interpret the output data. Data provided by all hospitals provide a complete picture of the entire population using all hospital services, rather than just those services provided for by an individual hospital in the region. This is necessary as people residing in a hospital catchment area may be hospitalized in another region by referral or for emergency care.

BOX 3.13 THE UNIFORM HOSPITAL DISCHARGE INFORMATION SYSTEM (UHDIS)

1. Planning: Organizing based on admission and surgical rates, utilization by age and sex, diagnosis, length of stay, and "small area analysis" which compares practice patterns and use or excess and waste of resources; search for new methods to promote patient flow to alternative care facilities, e.g., minimal supervisory residential care, ambulatory, or home care.

2. "Case-mix" analysis: Makeup of the hospital case load, looking for common diagnoses or rare events which might be of epidemiological significance, or may have administrative and quality control importance.

3. Budgeting: Planning within the hospital and in relation to referral sources based on utilization patterns by diagnosis and department.

4. Quality of care monitoring: Determination of aberrant practice, complications or outcomes, e.g., excess surgical rates, infections, mortality.

5. Epidemiology: Tracing and mapping epidemics of communicable diseases and identifying localizations and sources; using "tracer conditions" to pick out medically and epidemiologically significant events such as strokes or diabetes mellitus; supplementing national or regional mortality data.

6. Research: Through case finding of particular clinical events which may then be analyzed for related variables, e.g., incidence of coronary heart disease to compare with mortality patterns, intracranial hemorrhages, and administration of prophylactic vitamin K to newborns, or follow-up of patients with coronary artery bypass procedures.

7. Linkage with other registries: Linkage with death records, cancer, or other special disease registries; relating hospitalization events to special disease registries, such as birth defects, cystic fibrosis, asbestosis, and mesothelioma; supplementing a cancer registry.

8. Economic analysis: This is an essential aspect of modern health care and the use of hospital care and its alternatives is central to health economics; linked data from various registries and hospitalization data can provide data for important cost-effectiveness and other economic planning models.

Developing countries should avoid the mistakes of compiling vast amounts of undigestible data on ambulatory care utilization. Instead, scarce financial and personnel resources should be focused on more significant and higher quality data associated with hospitalizations. Fewer centers are involved in hospital care than in ambulatory care, so that data retrieval is easier to control. Most importantly, the less common event of hospitalization is medically and epidemiologically more significant because it consumes 40–75% of health care finances. A UHDIS may be seen as a priority information system after the reporting of infectious diseases, mortality cancer, and birth defects.

The three primary users of information flow in a hospital information system are clinical medicine, epidemiology, and managerial services. However, much of the development of information systems in recent years has been for managerial purposes. Good data should be easy to interpret for managers and clinicians alike. This requires informatics staff (knowledgeable of modern technology) to tailor the data reporting method so that the manager and others can analyze the data for their needs. The data should be given in a manageable format and training provided for its users.

Hospital discharge provides a basis for epidemiologic monitoring and control of diseases and simple research information. Analysis of hospital discharge data, especially mortality, surgical complications, and excessive length of stay, reflects the quality of care. Interregional variations in hospital utilizations provided a clear premise for designing and implementing policies. Hospital discharge data studies permit case-mix studies, show trends in care patterns, and provide a basis for peer review within a hospital and between hospitals. They provide material for analysis and policy formation at the clinical level, as well as for hospital management and planning, for example, in development of ambulatory care, reducing admissions, and length of stay for services better provided on an outpatient basis.

Hospitalization rates are few fewer in number, varying by age group, ranging from 3 to 25 admissions per 1000 population. Limitations of the data include factors such as lack of standardization of diagnostic criteria. Some patients do not reach a hospital for economic or other reasons; they may have transportation problems, or may have died prior to admission. Others may be unaware of the existence of some health services, or are simply afraid of it. Moreover, the denominator for rates is missing because the hospitals may not have a defined catchment population. Nevertheless, hospital discharge information is an important tool for planning, monitoring, and evaluation of health services.

Ambulatory care utilization is generally vast in numbers and too large a data set for effective monitoring. The number of ambulatory care visits may range from 4 to 10 per person per year, depending on the country. Ambulatory care data are of poorer quality because they are usually in broad categories of diagnosis, such as musculoskeletal and respiratory complaints, which comprise the bulk of visits. Ambulatory care can be monitored selectively through sampling or monitoring of representative sentinel centers to provide examples for wider replication. Specific components of ambulatory care should be monitored, such as infants and school age children receiving immunizations, attendance for prenatal care, birth control services, screening for hypertension and diabetes, or breast cancer screening, as particular health goals. With increasing trends for ambulatory care surgery and medical care, linkage of such data with inpatient care is needed in order to ensure continuity of comparisons with previous patterns of care.

HEALTH INFORMATION SYSTEMS
(INFORMATICS)

Information is needed for the management of any health system. It is vital to establishing objectives, developing programs, and managing the use of resources. Modern information technology, or informatics, provides the tools for analysis and policy formation to adjust the service. This is as much a part of health care as the cardiograph or ultrasound machines. It provides the feedback, "imaging," or cybernetics potential for management.

Dissemination of information is no less vital than its collection, or interpretation in central offices. Reporting of vital data is meaningless unless the data are processed and fed back to the service system in a regular, timely, and usable fashion or in current computer terminology in a user-friendly manner. Modern health information monitors the operation of a health care system. This includes component parts such as objects (hospital buildings), persons (health personnel), services, policy (equity), finance, organization, administration, regulation, quality assurance, and health promotion. The component parts interact to support the system as a whole. Interaction is made possible through information and communications technology and driven by financing and organizational imperatives.

Health care services are a source of increasing expense to governments and individuals. As a result, governments throughout the world are recognizing the importance of health information for effective health services management and plan-

---

BOX 3.14 FUNCTIONS OF HEALTH
INFORMATION SYSTEMS

1. Comparisons: Using historical, regional, national, or international patterns and standards.

2. Assessment: An overview of health status of a population based on available data, the professional literature, field visits, and interviews with key health personnel and community representatives.

3. Evaluation: Monitoring use of resources, performance, and outcomes of programs as part of total quality management.

4. Prediction: Using current data to predict trends in disease and utilization patterns, costs, potential outcomes, program planning, policy formulation, and priority setting.

5. Explanation: Data to understand disease patterns, risk factors, and service utilization of population of a district and determine causal relations, or need for intervention.

ning. The requirement for public accountability has led to the design of policies to ensure appropriate quantity, quality, and effectiveness of care with the best use of resources. This has created substantial requirements for information. While each country must develop its own health information system and uniform health information systems, such as that developed by WHO European Region, provides for comparative analysis. That system provides a timely (current or real time) spectrum of vital statistics, demography, and key outcome measures, as well as data on health care resources and utilization. Each country should provide local, district, community or municipal, and regional health profiles. This information should be widely distributed and available for analysis and discussion to the media, the public, and health professionals. Data are of little value if locked away and unavailable for regular circulation and dissemination to a wide audience, who require this information in order to make an informed contribution to policy analysis and formation.

Precision is limited by the quality of the data, but even limited data are extremely important in epidemiology and for health planning. Some infectious diseases are reported less stringently than others, partly because of lesser concern by physicians, but also because the clinical presentation may be atypical, or some cases may be entirely subclinical. A clinical case of polio may represent a hundred subclinical cases. Many infectious diseases of public health importance (e.g., measles, rubella) are underreported because nonimmunized, vulnerable children may not be brought to medical care despite mandatory reporting requirements, while some reported cases are not confirmed by laboratory evidence. Nevertheless, reported cases are the basis for monitoring and policy formation. Awareness of the direction and magnitude of errors will enable the user to determine the validity of the data.

Making health information data available on a routine basis to providers and managers of services helps promote an awareness of the overall operation of the health system in which they are involved. Information provides the basis for accountability, which implies that the provider of care or the manager of a health system is responsible for and must report on the results of his or her work, including unintended outcomes. Any system of service requires a system of accountability in order to maintain standards and to provide the consumer with an assurance of quality care.

In a centrally managed system, reporting of services provided is part of the chain of command. In a decentralized system, such data may be derived from billing patterns from hospitals or physician payments. It is then transferred to the higher levels of the health service administration and used for decision-making and planning. Those who provide the data should be informed of the outcome, including resultant operational decisions.

The World Health Organization Technical Committee on Information Systems emphasizes that the more active and innovative a health policy is, the greater the need for information. Data collection and processing requires planning and training. While massive data banks are not helpful, well-selected and widely available information systems targeted to vital events in the health process can promote flexibility and relevance in the planning of health services.

SURVEILLANCE, REPORTING, AND PUBLICATION

Publication and wide distribution of weekly summaries of specified reportable diseases is essential to maintain the viability of reporting and promote meaningful use of the data. The Centers for Disease Control (CDC) of the U.S. Public Health Service publishes and widely distributes the *Morbidity and Mortality Weekly Report* (*MMWR*) reporting on national and international epidemiologic events through surveys and special reports. The weekly report is supplemented by indepth special reviews of important public health topics.

BOX 3.15 FACTORS AFFECTING THE VALUE OF DATA

1. Relevancy: Are we getting the right data? Are some data collected no longer useful?

2. Coverage: Does the data help identify high risk groups?

3. Quality: How good do data need to be to be useful? Limitations of data are a factor in decision making.

4. Acceptability: Are the data collected acceptable in terms of design, cost, and ethical standards?

5. Timeliness: How recent are the data? How long a time series is needed to show temporal patterns?

6. Accessibility: Are the data available to those who need them? Are the data suitable for publication? Are published and distributed on the Internet and in hard copy?

7. Usability: Are the data in usable format? Are they presented in a user-friendly (i.e., easy to access and use for the nonspecialist)? Can you generate summaries, graphs, and tabulations?

8. Cost: What does it cost to collect and process the data?

9. Sensitivity: Does the information reveal what you are looking for?

10. Specificity: To what extent do the data relate to the issue of concern? Does the information also relate to other societal issues?

11. Vertical or horizontal reporting/data aggregation: Is data reported by disease or service category and by population at risk; is data aggregated by region of residence?

12. Biological plausibility: The criterion that an observed or presumed

---

causal association is compatible with existing biological and medical knowledge. Can it be explained from a biological perspective?

13. Equity: Does the data show inter-regional and social-class variation
and inequity?

14. Dissemination: Information obtained, collated, and analyzed must
be organized and available to those who report the raw data, to professionals who need data to monitor health status and to plan health services and
health promotion needs of the population.

WHO publishes the *Weekly Epidemiologic Record* (*WER*) which reports country epidemiologic events and offers a summary of infectious diseases internationally. Other countries publish their own weekly or monthly epidemiologic
summaries of reportable diseases and related laboratory findings, such as the
Canada Communicable Disease Report, as well as *Chronic Diseases in Canada*
for noninfectious diseases and related laboratory findings.

Reporting systems and publication of the data are both vital to epidemiologic
monitoring of infectious and noninfectious disease trends. Regular circulation to
field personnel increases the sense of awareness and participation in epidemiologic
monitoring and the idea that the reporting is put to good use. Awareness of the reported data helps local health providers and managers in managing their services
more effectively. Providing current data as the events unfold promotes a sense of
involvement and challenge for achievement of goals, such as high coverage of immunization and rapid control of disease outbreaks.

The Internet has become a vital tool for public health, for reporting and obtaining data, and for access to the world literature. Many resources such as the
MMWR, WER, and major journals are available on-line. Newsgroups enable convenient and immediate discussions by professionals on particular topics such as
Promed for current infectious disease reporting from around the world (Chapter
4). Similarly, the internet permits literature searches and access to interest groups
on virtually any topic in health. This allows people to be in contact with and to obtain support from many others in their field. The WHO home page (www.who.org)
provides access to its component departments and regional offices.

ASSESSING THE HEALTH OF THE INDIVIDUAL

Physicians and other health professionals are trained to assess the health of the
individual patient seeking care. This involves more than dealing with the chief
complaint, requiring a history of the present illness as well as a wider review of
body functions, family and occupational history, physical examination, and labo-

---

BOX 3.16 ASSESSING THE HEALTH
STATUS OF THE INDIVIDUAL

1. Current chief complaint;
2. Personal data—age, sex, ethnicity, education, marital status, children, living situation;
3. Occupational history;
4. Family history;
5. Personal history;
6. Functional inquiry—systems review;
7. Summary of risk factors—family history, hypertension, diabetes, smoking, sedentary lifestyle, high fat diet, occupation, alcohol use, stress, other;
8. History of the present illness;
9. Physical examination;
10. Differential diagnosis;
11. Other medical problems;
12. Investigation: laboratory, cardiographic, imaging, other;
13. Presumptive or working diagnosis;
14. Treatment and its effects;
15. Definitive diagnosis;
16. Management of other medical problems;
17. Follow-up management and monitoring;
18. Long-term health needs.

---

ratory and imaging tests. A differential diagnosis and treatment for the presumptive diagnosis define follow-up to observe the course of the disease, the outcomes of diagnostic tests, and the effects of intervention. Caregivers must take into account the effects of the process on the patient, the family, and the community. Providers must also be concerned about costs of care, alternative methods of looking after the patient to meet changing needs, and promote early and maximum recovery. Continuous monitoring and reevaluation is a key part of the process. There are many parallels in care of the individual and care of the population.

ASSESSMENT OF POPULATION HEALTH

Health service administration is being increasingly decentralized in many countries, while the concept of healthy cities/municipalities is becoming more widespread. These developments have increased the need and value of health profiles

BOX 3.17 ANNUAL HEALTH PROFILE OF CALI, COLOMBIA

The Secretary of Health of the Municipality of Cali, Colombia, publishes an annual epidemiologic review which is part of the Healthy Municipality profile for the city and its subdivisions. This report includes the following data for the 2 million residents of this city:

1. General description of the city
2. Demographic data
3. Population distribution by age and sex
4. Population pyramid
5. Age distribution curve
6. Population by socioeconomic status
7. Subdivision of the city
8. Health indicators—fertility, life expectancy
9. Age-specific fertility
10. Mortality rates—general, infant, maternal, trauma
11. Mortality by leading causes
12. Age-specific mortality rates
13. Infant mortality by age and leading causes
14. Infant mortality: 1970–present
15. Infant and child mortality for diarrheal diseases: 1980–present
16. Maternal mortality: 1985–present
17. Maternal mortality by cause
18. Death rates from homicide: 1982–present
19. Age-specific life expectancy by sex
20. Years of potential life lost (YPLL) by cause
21. Age-specific YPLL by sex
22. Principal causes of morbidity in the social security health service
23. Reportable diseases
24. Immunization coverage
25. Institutions operated
26. Annual budget (in current and constant currency and US$)
27. Staff of the municipal health service: 1986–present year
28. Budget sources—municipal and national

Source: Secretaria de Salud Municipal. Perfil Epidemiologico de Santiago de Cali. Cali, Colombia: Municipio de Santiago de Cali, and PAHO–Healthy Cities Website.

at the community, county, and district levels. This type of health profile provides management with regular monitoring of the health situation, including resources, utilization, morbidity, and mortality. This is the application of modern health informatics at a community level and does not require advanced computer capacity or skills. Annual reports in a standard format using all existing data sources can be brought together in a user-friendly manner to provide valuable health status monitoring.

District or community health information systems increase the potential for local health authorities and communities to have greater power in determining local health policy. National health authorities need to provide the guidance of health targets and resources that may be used flexibly to meet local needs. But, supervision and regulation of national health authorities are essential to assure that resources are well-used, that targets are being met, and to reduce inequalities between regions.

European Region of the World Health Organization has developed a user-friendly computer program for one thousand health indicators, including sociodemographic, mortality, morbidity, health resources, utilization, and lifestyle indicators. These can readily be produced in tabular or graphic form with time trends and mapping capability. This is accessible free of charge to everyone with a personal computer, internet access, and modest computer skills via website http://www.who.dk.

As with individual health assessment, evaluation of the health status of a population is based on the accumulation of a portfolio of observations and data from a variety of sources and their interpretation, with comparisons to international, national, or regional patterns or standards. Community health assessment (CHA) begins with identification of the main health problems or chief complaints as understood by key health professionals and the community, or identified from regular community health profiles.

Information should be derived about the community's socioeconomic status (SES), the resources available for health care, how they are distributed and how they are distributed and how services are utilized, as well as morbidity, mortality, and other "outcome" measures which help to describe or compare health status (Table 3.8). Health measures include how care is provided, how it governs or monitors itself, and how the system is accountable for its component services. The knowledge, attitudes, beliefs, and practices (KABP) of the people and health providers and the way in which society addresses risk factors for ill-health may also be important determinants of health status.

Gathering the data necessary to monitor itself should be part of the standard functions of a health system. This provides for accountability in use of public resources and maintains a self-correcting feature of the system. CHAs help point out health risk factors at the population level, and if carried out in a timely and regular fashion, changes can be made without inordinately long waiting periods and without any unnecessary increase in morbidity or mortality.

The CHA is part of the health planning process; it may be designed to monitor the impact of an intervention program meant to deal with a particular health problem, such as coronary heart disease, or a set of risk factors for disease, such as smoking. The CHA is also part of program evaluation, especially in community trials, with an evaluation protocol based on a multiphasic approach and data from many sources.

TABLE 3.8   Evaluation of Population Health of a Community, District, State, or Country

Factor	Topics	Example indicators
Geography	Climate, topography, density, settlement, economy	Tropical, temperate, mountains, desert Distances from medical facilities
Demography	Vital statistics	Population size, age/sex, and urban/rural
Socioeconomic	Ethnic, cultural, religious practices Community and family economic status	Per capita and family income, education, literacy (women), employment, religious affiliation, social attitudes, occupations
Nutrition	Supply and use of major food groups Food safety and quality	Under- and overnutrition Risk group identification Monitoring child growth patterns, anemia Food fortification
Environment and occupational	Water, air, waste, and sewage disposal, toxic wastes, radiological hazards Industrial or agricultural toxic materials	Ambient air pollutants, bacteriologic and chemical qualities of community and recreational water, radiation and radon levels, heavy metal levels in soil, water
Health care system	Organization Prepid coverage Finance total and internal allocation	Decentralized administration and finances Integration of local services and finances Total resources, % GNP and per capita (US$), spent on health care Percentage of population with full, partial, or no health benefit insurance
Health resources	Expenditures per capita Hospital beds per capita Long-term care facilities Clinics Personnel, doctors, nurses per capita	Expenditure by type of service, preventive, curative, hospital Acute care beds per 1000 Special hospital beds per 1000 Long-term care facilities per 1000 Doctors, nurses per 10,000
Health care utilization	Hospitals, general, chronic, and mental Ambulatory care Preventive services	Admissions and days of care per 1000 population Physician visits per person per year Immunization coverage at age 2 years Ambulatory surgery, home care measures
Health outcomes	Morbidity Mortality Functional/physiologic status "Tracer conditions"	Infectious and chronic disease incidence/prevalence Infant, child, maternal, age-sex specific mortality rates by cause, cardiovascular disease, trauma Anemia of infancy, pregnancy, blood lead levels Lower limb ampuation rates
Process (quality) of care	Professional care standards Accreditation by external agency Peer review Records review Mortality case review	Criteria for hysterectomy, including second opinion Immunization and child health monitoring rates Correction of deficiencies from accreditation Departmental reviews of cesarian, infection rates Maternal and infant mortality case-by-case reviews
Cost and benefits	Examine specific diseases, procedures, services, or health promotion	Cost benefit of second dose of measles vaccine, bicycle helmets, air bags in cars, antismoking campaign among high school girls
Knowledge, attitude, beliefs, and practices (KABP)	General population Risk groups Patients Staff	Diet, smoking, eating, alcohol use, and exercise habits Birth control, rights of women AIDS/STDs related issues, condom use

Defining the Population

The population served by a health system must be defined in terms of its age and sex distribution. This is one of the key factors in the planning of health care services, as different age groups have different needs. Women, children, and the elderly utilize more health services and institutional care. The health status of the elderly is affected by the major chronic diseases and their associated disability and mortality patterns. While increasing longevity is associated with a healthier elderly population, the demand for care still grows with age. The elderly, and increasingly the very elderly (those over 85) are high users of health services, including institutional care in hospitals and long-term facilities.

Socioeconomic Status

Health is affected by standards of living and therefore analysis of income and its distribution is a component of the process of assessing the health status of a population. The national average income is often simplified to the GNP or GDP per capita, representing the average of the total production of goods and services of a nation. Real income may vary by state or district, ethnic group, educational levels, sex, or family size. These and many other factors may affect the distribution of wealth in the population.

Living conditions, as reflected in housing standards, density of housing, and crowding (persons per room or per square meter), are dependent on family income. Services, such as electricity, running water, indoor toilet and bathing facilities, as well as other service facilities in the home (e.g., refrigerators, toilets, baths, stoves, and central heating) are also important measures of health-related socioeconomic conditions. Adverse economic conditions prejudice health status in measurable ways. In developing countries, the poverty–disease–malnutrition cycle affects children, women, and the elderly predominately, but reduces potential for economic growth. Even in industrialized countries, there is unevenness in the patterns of income and of health status, and health status of the upper social class is much better than that of the unskilled workers for many health indicators. Where there are large gaps between the rich and the poor, such as in the United States, there is poorer health status than in countries with smaller social gaps such as Japan and the Scandinavian countries.

Educational level of parents is an important factor in family health. In the case of the father in a family, education level is often a direct determinant of income. In the case of the mother, education relates to income, but even more strongly to successful health care of infants and children. Mothers with higher levels of education, as measured by years of school attendance, are more likely to absorb new knowledge regarding self care in pregnancy and care of the infant in areas such as nutrition, immunization, and routine baby care. Better educated women tend to have fewer pregnancies, not only because of knowledge of need and methods of birth control, but also because of greater self-awareness and different life goals. Ethnic, cultural, political, and religious beliefs and practices have important implications for health, in such areas as the status of women, mental health, family structure, nutrition, substance use and abuse, as well as birth control and abortion. These beliefs and practices can affect attitudes toward issues such as national health insurance and the funding of health care.

Nutrition

Appropriate nutrition, overnutrition, and undernutrition are fundamental determinants of the health of a population. Overnutrition places a heavy burden of morbidity and mortality on the health system, with such diseases as diabetes, coronary heart disease, hypertension, and stroke and their complications. Undernutrition in the form of gross malnutrition is rare in the industrialized countries, but extremely common in many developing nations. In all societies there are groups at risk for overt or subclinical malnutrition, such as anemia, iodine deficiency, osteoporosis, and others. Society's actions to prevent malnutrition in vulnerable groups is an indicator of the well-being of that society. Public health and economic measures to promote good quality of food and its accessibility to the population, fortification of basic foods, school lunch programs, and meal services for the elderly and chronically ill are health promotion programs that show the level of organized community responsibility for its members (Chapter 8).

Special surveys, such as low birth weight or nutritional status conditions are needed to provide nutrition status data. Periodic large-scale national surveys, such as the National Health and Nutrition Examination Surveys (NHANES), initiated in 1971 in the United States, provide meaningful information on nutrition status in the country. Within the United States, they provide vital information for adjusting recommended dietary allowances and national, state, or local nutrition programs. This is of great importance for the food industry which is obliged to follow federal government standards of labeling and content of packaged and processed foods.

Environment and Occupation

Safety of community water, management of solid and toxic wastes, air and noise pollution, and ambient air standards, are all factors in health of the community. Organized public health has traditionally focused on these issues, but they remain public policy issues in virtually all countries and internationally. Healthy societies are dealing with these issues with a very high degree of public awareness, sometimes overcoming strong economic interest groups to force improved attention to the environment by governments, communities, and businesses. Environment includes housing, recreation, schools, parks, urban and rural planning, and many other aspects of community life that are addressed in "healthy community" initiatives. Employment of children and work in hazardous industries are health issues. Societies that tolerate toxic and dangerous work settings create health hazards that are preventable, but costly to treat. Unemployment, job insecurity, loss of health insurance with change of employer, and how income levels for many workers all contribute to poor health (Chapter 9).

HEALTH CARE FINANCING
AND ORGANIZATION

How a nation finances and organizes health care is an important aspect of health status evaluation. Where there is universal coverage of the population, either through health insurance or through a state-operated health care program, the population in principle has equity in access to care. Financial access, however, does not guarantee actual access because the distribution and supply of services are important variables in utilization. Financing of health services and their organization are related issues, discussed in Chapters 10–13, that must be recognized as part of the process of assessing the health status of the population of a country or region. Assurance of access to medical and hospital care does not assure that appropriate or effective services are provided.

How services link facilities of different levels of intensity of care and costs is a basic issue in health reform in many countries. How preventive care is provided to special groups in the population (infants, children, adults, the elderly, the chronically ill) and how these fit together as a holistic entity, interacting to serve the community are important in determining the status of health and health costs of a community or a country.

Health Care Resources

While overall financing for health is an important determinant of the level of health care available, no less important is how the resources are spent, that is, what is the internal allocation of total health care financing. The major resources for health care are primary care services, hospitals, and long-term care facilities. All countries have finite health budgets. One aspect of health services can only grow at the expense of another.

Hospitals are the largest segment of the health care system in terms of expenditures and may consume more than 50% of total expenditures. The supply of hospital beds is, therefore, a central factor in the health care economy. The number of hospital beds per 1000 population is a key indicator for health economics. The hospital bed-to-population ratio varies widely, from 2.5–16 care beds per 1000 in Organization of Economic Cooperation and Development (OECD) countries, with most countries reducing hospital bed supplies rapidly since the 1980s.

Age distribution of the population affects morbidity and therefore hospitalizations; countries with a high percentage of elderly may need more hospital facilities, as well as alternative care services, such as home and long-term institutional care services. Innovations in health care organization are influencing health planning in with many developed countries reducing acute care hospital admissions and length of stay by a variety of incentive and management systems (see Chapters 11–13). Health planning requires facing up to political and other pressures to increase hospital bed supplies beyond real need, also at the expense of other services.

The ratio of medical doctors per 10,000 population also varies widely. A high ratio may indicate an overpopulation of specialists and a lack of primary care services, while a low ratio may indicate a need for more doctor training. Countries in eastern Europe have high doctor-to-population ratios and low ratings on health status indicators (such as SMRs for trauma) than in countries with fewer doctors. Nurse-to-population ratios are also equally variable, but typically, many countries which have high levels of physician manpower have relatively low numbers of nursing personnel. The number of nurses registered to practice often overstates the actual supply since many nurses never practice following graduation, work only part-time, or stay in the profession only for a short period.

Excessive supply of medical doctors, inequitable distribution, relative shortages of nurses, inefficient development of community health programs, and inefficient use of community health workers are important issues in many countries (see Chapter 15). These all have economic and health outcome implications, requiring continuous review and reassessment in each country, and application of lessons learned from other countries.

Utilization of Services

Rapid cost increases have fostered a search for efficient ways of organizing and financing health services. In the United States, the development of the diagnostic related group method of payment for hospital services has reduced hospital length of stay. Health maintenance organizations (HMOs) have been successful in providing comprehensive care with less hospitalization and fewer hospital beds than traditional fee-for-service practice. This has led policy makers and the business community to focus on "managed care" systems to meet the need to extend insurance coverage and to control costs.

While supply of services is important, actual utilization patterns are also a valuable part of the overall evaluation program. Hospital care is a key issue because of its dominance in the economics of health care. Monitoring hospital performance indicators can play an important role in determining the effective functioning of the health care system.

Surgical and other procedure rates are continuing issues in health systems management. Hysterectomy rates vary widely among Canadian provinces, from 639 per 100,000 in Newfoundland to 426 per 100,000 in Alberta, and vary by a factor of 18 within Ontario on a county-to-county basis. A study conducted in Saskatchewan showed that the introduction of mandatory second opinions resulted in dramatic reductions in hysterectomy rates. Appendectomy rates in Germany are as high as three times the rates in other countries, with no epidemiologic explanation.

Studies abound in the United States showing differential utilization of health services by black and white populations for coronary heart bypass procedures, for localized as compared to radical surgery for lumps in the breast, and for mammography and other services currently considered beneficial to the patient. These differences generally are primarily due to differences in health insurance coverage, but other socioeconomic or ethnic variables may also be responsible. Excess surgical procedures, for example, cesarean sections, are a widespread problem in countries where fee-for-service is the method of payment, but the amount of surgery is also related to the number of surgeons and fee-for-service payments.

Health Care Outcomes

While it is clear that health status is affected by many social and economic factors, the general state of the country's health is often described by epidemiologic indicators, such as mortality, morbidity indicators of health status. Epidemiologic information on communicable and noncommunicable diseases helps to determine a potential for intervention and alteration of the natural history of the disease.

BOX 3.18 MORTALITY AND MORBIDITY "OUTCOME INDICATORS" OF HEALTH STATUS OF A POPULATION

1. Mortality related indicators
   a. Infant and child mortality rates (IMRs);
   b. Maternal mortality rates (MMRs);
   c. Crude mortality rates (CMRs);
   d. Age-specific mortality rates;
   e. Cause-specific mortality rates—infectious, noninfectious;
   f. Life expectancy (LE) at age 0, 1, 65, and other ages;
   g. Standardized mortality rates (SMRs)—total specific;
   h. Years of potential life lost (YPLL);
   i. Quality adjusted life years (QALYs);
   j. Disability adjusted life years (DALYs).
2. Morbidity outcome indicators
   a. Incidence of vaccine-preventable disease;
   b. Incidence of waterborne disease;
   c. Incidence of food-borne disease;
   d. Incidence/prevalence of tuberculosis;
   e. Incidence/prevalence of STDs/AIDS;
   f. Incidence of malaria, other tropical diseases;
   g. Prevalence of noninfectious diseases—cardiovascular diseases, diabetes, cancer, trauma;
   h. Prevalence of disabling conditions;
   i. Prevalence of risk factors.
3. Physiological indicators
   a. Risk facotrs—smoking; alcohol and drug use; unsafe sexual practices; high risk behavior regarding motor vehicles, violence, drug usage, suicide;

   b. Nutritional status—growth patterns of infants and children; body
      mass index of adults; dietary patterns; biochemical indicators (blood
      sugar, cholesterol, lipids, vitamins A, B, C, D); anemia among in-
      fants, children, and women; iodine status, environment.
4. Functional indicators
   a. Work and school absence;
   b. Psychomotor function;
   c. Work capacity;
   d. School performance;
   e. Fitness test performance;
   f. Activities of daily living (ADL).

Outcome indicators (Box 3.18) include a variety of measures from routing data sources and special surveys. DALYs and QALYs (described earlier) attempt to quantify mortality and quality of life measures for comparisons and for analysis of specific interventions. In addition, physiologic or functional indicators such as activities of daily living measure patient perfomance. Special surveys for clinical signs of undernutrition or assessment of anthropometric measures (growth and body size) should be supplemented by biochemical level and hematological surveys to establish patterns of undernutrition. Special surveys of nutrition status and disability, school performance, and other indicators of functional status are important aspects of health status evaluation.

Quality of Care

The quality of care will be reviewed extensively in Chapter 11, but is mentioned here as part of evaluation of health in any population. How available funds are spent to address the health problems specific to that population is part of the community health assessment. The findings of such evaluations are meant to affect resource allocation to address unmet needs.

Costs and Benefits

Analysis of costs and benefits, reviewed in more detail in Chapter 11 on economics and health policy, will be mentioned here only briefly. Evaluation of the health status of a population requires examination of the choices made in resource allocation in a particular geographical area. This is of concern, not only to the planner, but also to the provider of health care and to the public. If priorities in resource allocation promote highly technological medicine, then primary care may lag behind in resources, and the health status of the population may be compromised. Cost–benefit analyses can contribute to establishing priorities within a health care system.

SUMMARY—FROM INFORMATION TO KNOWLEDGE TO POLICY

Information is the basis for planning, organizing, managing, and providing high quality care. The process begins with basic vital statistics and the epidemiology of infectious and noninfectious diseases in order to identify the health needs of the population. It extends into health information systems in order to manage and monitor the functioning of the health care system. Surveillance of health events at national, regional, and community levels depends on building information systems and linking data to provide community health profiles. This is fundamental to monitoring and managing health systems. It requires clear policy to ensure that information systems do not exist to serve only those who process the data at national levels, but are returned to the community level and linked with other data sources in readily usable formats.

Information is widely available in health statistics and published data of all kinds. Health policy formulation requires seeking the appropriate information and making intelligent use of it. Educating health workers in using information coordination and streamlining of data helps them understand the relevance and impact of their actions. Information systems and the properly organized and disseminated flow of data are vital for management. It is as important to the functioning of the system as an intelligence service is to a military operation. The vast mechanism of a health service operates in the dark without such a service.

Throughout the world, health care systems are under critical scrutiny due to concerns over costs, accessibility, appropriateness, quality, and outcomes of care. The effectiveness of a health system is frequently on the political agenda. Quality assurance and accountability are critical in the operation of any health system. Health expenditures must be increasingly justified in terms of their need and cost-effectiveness, policy formulation, strategies, and priorities, taking into account economic, sociological, and political factors.

Curbing the soaring costs of health care is not a matter of choice for governments and individuals if the WHO policy of Health for All is to be achieved. One means of reaching the goals and objectives of this policy is to develop an efficient health information system. Knowing the population, the epidemiologic patterns of its diseases, and its health care services and utilization, are all part of monitoring and feedback systems essential to allow the health system to evaluate health status, and keep pace with changes. They are therefore essential elements of the New Public Health.

ELECTRONIC SOURCES

Canada
   Statistics Canada http://www.statcan.ca
   Canadian Institute for Health information http://www.cihi/stats/canhe.htm

Israel Center for Disease Control http://www.health.gov/icdc

Organization for Economic Development (OECD)
OECD in Figures http://www.oecd.org/publications/figures

United Kingdom
Government Statistical Service http://www.statistics.gov.uk/stats/health.htm
Deparment of health http://www.doh.gov.uk/dhome.htm

United States
Centers for Disease Control and Prevention (CDC) http://www.cdc.gov/
Census Bureau-Statistical Abstract of the Untied States
http://www.census.gov./prod/3/98pubs/98stab/cc98stab.htm
Dartmouth Atlas of Health Care in the United States 1998
http://dartmouth.edu%7Eatlas/toc98.html
cecs.atlas@dartmouth.edu
National Center for Health Statistics, Health, United States 1999
http://www.cdc.gov/nchswww/products/pubs/pubd/hus/99husdes.htm
National Information Center on Health Service Research and Health
Technology. Health Statistics Sources
http://www.nih.gov/nichsr/stats/contents/contents.html

World Health Organization (WHO) http://www.who.int/whosis/
Burden of Disease http://www.hsph.harvard.edu/organizations/bdu/gbdmain.htm
European Region. Health for All Data Set. WHO Copenhagen
http://www.who.dk/
Pan American Health Organization
PAHO Country Health Profiles http://www.paho/english/country.htm
World Health Report, 1998 http://www.who.int/whr/1998/whr-en.htm

RECOMMENDED READINGS

Black, D. 1993. Deprivation and health. *British Medical Journal,* 307:1630–1631.
Declich, S., Carter, A. O. 1994. Public health surveillance: Historical origins, methods and evaluation. *Bulletin of the World Health Organization,* 72:285–304.
Feinleib, M. [editorial]. 1993. From information to knowledge: Assimilating public health data. *American Journal of Public Health,* 83:1205–1207.
Fletcher, R. H. 1992. Clinical medicine meets modern epidemiology—and both profit. *Annals of Epidemiology,* 2:325–333.
Pearce, N. 1996. Traditional epidemiology, modern epidemiology, and public health. *American Journal of Public Health,* 86:678–683.
Susser, M., Susser, E. 1996. Choosing a future for epidemiology: I. Eras and paradigms; and II. From black box to Chinese boxes and eco-epidemiology. *American Journal of Public Health,* 86:668–673.

BIBLIOGRAPHY

Beaglehole, R., Bonita, R., Kjellstrom, T. 1993. *Basic Epidemiology.* Geneva: World Health Organization.

Bennet, S., Woods, T., Liyanage, W. M., Smith, D. L. 1991. A simplified general method for cluster-sample surveys of health in developing countries. *World Health Statistics Quarterly,* 44:98–106.

Centers for Disease Control. 1992. Proceedings of the 1992 International Symposium on Public Health Surveillance, December 1992. *Morbidity and Mortality Weekly Report, Supplement,* 41:1–218.

Dean, A. G., Dean, J. A., Burton, A. H., Dicker, R. C. 1990. *EPI Info, Version 5: A Word Processing Database and Statistics System for Epidemiology on Microcomputers.* Atlanta, GA: Centers for Disease Control.

Duncan, B. D., Rumel, D., Zedlmanwiocz, A., Mengue, S. S., Dos Santos, S., Dalmaz , A. 1995. Social inequality in mortality in São Paulo State, Brazil. *International Journal of Epidemiology,* 24:359–365.

Elliott, P., Cuzick, J., English, D., Stern, R. [eds]. 1992. *Geographical and Environmental Epidemiology: Methods for Small Area Studies.* World Health Organization, Regional Office for Europe, Copenhagen: Oxford University Press.

Gordis, L. 1996. *Epidemiology.* Philadelphia: WB Saunders Co.

Gray, A. (ed). 1993. *World Health and Disease.* Milton Keynes: Open University Press.

Ibrahim, M. A. 1985. *Epidemiology and Health Policy.* Rockville, MD: Aspen.

International Epidemiologic Association. 1996.

Kuller, L. H. [editorial]. 1995. The use of existing databases in morbidity and mortality studies. *American Journal of Public Health,* 85:1198–1199.

Lang, T., Duceimetiere P. 1995. Premature cardiovascular mortality in France: Divergent evolution between social categories from 1970–1990. *International Journal of Epidemiology,* 24:331–339.

Last J. M. (ed). 1995. *A Dictionary of Epidemiology,* Third Edition. New York: Oxford University Press.

Lilienfeld, D. E., Stolley, P. D. 1994. *Foundations of Epidemiology,* Third Edition. New York: Oxford University Press.

National Center for Health Statistics. 1998. *Health United States, 1998.* Hyattsville, Maryland.

Poikolainen, K., Eskola, J. 1995. Regional and social class variation in the relative risk of death from amenable causes in the city of Helsinki, 1980–1986. *International Journal of Epidemiology,* 24:114–118.

Secretaria de Salud Municipal. 1992. *Perfil Epidemiologio de Santiago de Cali.* Cali, Colombia: Municipio de Santiago de Cali.

Smith, G. D., Egger, M. [editorial]. 1992. Socioeconomic differences in mortality in Britain and the United States. *American Journal of Public Health,* 82:1079–1081.

Tulchinsky, T. H. 1982. Evaluation of personal health services as a basis for health planning: A review with applications for Israel. *Israel Journal of Medical Sciences,* 18:197–209.

Tulchinsky, T. H., Ginsberg, G. M., *et al.* 1992. Mesothelioma mortality among former asbestos-cement workers in Israel, 1953–1990. *Israel Journal of Medical Sciences,* 28:542–547.

United States Public Health Service. *Health United States 1998.* Hyattsville, MD: U.S. Department of Health and Human Services.

Vaughan, J. P., Morrow, R. H. 1989. *Manual of Epidemiology for District Health Management.* Geneva: World Health Organization.

World Health Organization. 1994. *Information Support for New Public Health Action at the District Level.* Report of a WHO Expert Committee. Technical Support Series 845. Geneva: World Health Organization.

4

COMMUNICABLE DISEASES

INTRODUCTION

At the end of the twentieth century, communicable disease control is still a central task of public health. This chapter describes communicable diseases and programs for their prevention, control, elimination, and eradication. Control of communicable disease requires a systems approach using available resources effectively, mobilizing environmental measures, immunization, as well as clinical and health systems. Rapid transportation and communication make a virus outbreak in any part of the world an international concern. An outbreak of disease will often be covered by international media within a matter of hours or days. A basic understanding of infectious diseases is therefore an expectation of any student, just as a general knowledge of family health, chronic disease, nutrition, and economics are part of the modern public health culture.

The material presented in this chapter is intended to give an overview for the student or review for the public health practitioner, with an emphasis on the applied aspects of communicable disease control. We have relied for the content of this chapter on several standard references, especially Benenson's *Control of Communicable Diseases Manual,* Sixteenth Edition, published by the American Public Health Association (1995), and *Jawetz, Melnick and Adelberg's Medical Microbiology,* Twenty-first Edition (Brooks *et al.,* 1998), along with *Morbidity and Mortality Weekly Report* of the Centers for Disease Control and Prevention (CDC) as well as electronic sources such as Promed, the American Academy of Pediatrics, and WHO websites. The references listed will augment the limited discussion possible in this text.

PUBLIC HEALTH AND THE CONTROL OF COMMUNICABLE DISEASE

Organized public health grew out of the sanitation movement of the mid nineteenth century which sought to reduce the environmental and social factors in com-

---

BOX 4.1 COMMUNICABLE DISEASE

A communicable disease "is an illness due to a specific infectious agent or its toxic products that arises through transmission of that agent or its products from an infected person, animal, or inanimate reservoir to a susceptible host." Transmission may be direct from person to person, or indirect through an intermediate plant or animal host, vector, or the inanimate environment.

Source: Benenson, A. S. (ed). 1995. *Control of Communicable Diseases Manual,* Sixteenth Edition. Washington, DC: American Public Health Association, p. 533.

---

municable disease. Traditionally, the prevention and control of communicable diseases has been accomplished by sanitation, safe water and food supply, isolation, and by immunization.

The potential for infectious disease to disturb or destroy human life still exists and may increase as infectious diseases evolve and escape current man-made control mechanisms. The spread of bubonic plague throughout Europe and Asia in the fourteenth century and subsequent pandemics of smallpox, tuberculosis, syphilis, measles, cholera, and influenza show the explosive potential and epidemic nature of infectious diseases. The spread of AIDS since the 1980s, the cholera epidemic in South America, and diphtheria in the former Soviet Union in the 1990s, remind us why communicable disease control is still one of the major responsibilities of public health.

Both the miasma (environment–host) and bacteriologic (agent–host) theories contributed to great achievements in the control of communicable disease in the first half of the twentieth century. The emergence of the germ theory in the late nineteenth century led to the sciences of bacteriology and immunology, growing out of the work of Jenner, Pasteur, Koch, Lister, and many others (see Chapter 1). The control of the vaccine-preventable diseases has been a boon to mankind, saving countless millions of lives and providing a cornerstone for public health. Despite this, millions of children still die annually from preventable diseases. Infectious diseases of childhood are still tragically undercontrolled internationally. Infectious diseases also undermine health of other vulnerable groups in the population, such as the elderly and the chronically ill, thereby playing a major role in the economics of health care.

Great strides have been made in the control of communicable diseases through vaccination and environmental sanitation, but the field of infectious disease continues to be dynamic. New infectious disease threats are providing great challenges to public health. Increasing resistance to therapeutic agents augment the need for new strategies and coordination between public health and clinical services. Together, these make up what are termed the emerging infectious diseases. Understanding the principles and methodologies of communicable disease control, and eradication, is important for all health providers and public health personnel so as to be able to cope with the scale of these problems and to absorb new technologies as they emerge from scientific advances and experience, and their successful application.

THE NATURE OF COMMUNICABLE DISEASE

An infectious disease may or may not be clinically manifest so that a person may carry the disease agent without having the full illness. Acute infectious diseases are brief, intense or short-term, self-limiting illness, but may have long-term sequelae of great public health importance, such as streptococcal infection and glomerulonephritis or rheumatic heart disease. Other infectious diseases are chronic with their own long-term effects, such as HIV infection or peptic ulcers. Others have both short-term and long-term effects, such as hepatitis infections. The stages of the infectious disease include

1. Exposure and infection;
2. Presymptomatic stage;
3. Nonmanifest or subclinical disease;
4. Clinically manifest disease and its progression;
5. Resolution, recovery, remission, relapse, suprainfection, or death; and
6. Long-term sequelae.

Each disease has its own characteristic organism and natural history from onset to resolution. Many infectious diseases may remain at a presymptomatic or subclinical stage without progressing to clinical symptoms and signs. Even a subclinical disease may cause an immunologic effect producing immunity. The drama of infectious disease is exemplified in the tragic event of the plague in the fourteenth century and its periodic recurrence as in the epidemic of 1665 in London, described by Daniel Defoe (Box 4.2).

BOX 4.2 DANIEL DEFOE—A JOURNAL OF THE PLAGUE YEAR, LONDON, 1665

"It was about the beginning of September 1664, that I, among the rest of my neighbors, heard, in ordinary discourse that the plague had returned again in Holland; for it had been very violent there, and particularly in Amsterdam and Rotterdam, in the year 1663, whither they say, it was brought, some said from Italy, others from the Levant, among some goods, which were brought home by their Turkey fleet; others said it was brought from Candia; others from Cyprus. It mattered not from whence it came; but all agreed it was come into Holland again.

"It was now mid-July and the plague, which had chiefly raged at the other end of town . . . began to now come eastwards toward the part where I lived. It was to be observed, indeed, that it did not come straight on toward us; for the city, that is to say within the walls, was indifferently healthy still; nor was it got over the water into Southwark; for though there died that week 1,268 of all distempers, whereof it might be supposed above 900 died of the plague, yet there was but 28 in Southwark, Lambeth parish included; whereas in the parishes of St. Giles and St. Martin-in-the-Fields alone there died 421."

Source: Defoe, D. 1723. *A Journal of the Plague Year.* Winnipeg: Meridian Classic, 1984, reprint.

HOST–AGENT–ENVIRONMENT TRIAD

The agent–host–environment triad, discussed in Chapter 2, is fundamental to the success of understanding transmission of infectious diseases and their control, including those well known, those changing their patterns, and those newly emerging or escaping current methods of control. Infection occurs when the organism successfully invades the host body, where it multiplies and produces an illness.

A host is a person or other living animal, including birds and arthropods, who provides a place for growth and sustenance to an infectious agent under natural, as opposed to experimental, conditions. Some organisms, such as protozoa or helminths, may pass successive stages of their life cycle in different hosts, but the primary or definitive host is the one in which the organism passes its sexual stage. The secondary or intermediate host is where the parasite passes the larval or asexual stage. A transport host is a carrier in which the organism remains alive, but does not develop.

An agent of an infectious disease is necessary, but not always sufficient to cause a disease or disorder. The infective dose is the quantity of the organism needed to cause clinical disease. A disease may have a single agent as a cause, or it may occur as a result of the agent in company with contributory factors, whose presence is also essential for the development of the disease. A disease may be present in an infected person in a dormant form such as tuberculosis, or a subclinical form, such as poliomyelitis or HIV. The virulence or pathogenicity of an infective agent is the capacity of an infectious agent to enter the host, replicate, damage tissue, and cause disease in an exposed and susceptible host. Virulence is indicated by the severity of clinical disease and case fatality rates.

The environment provides a reservoir for the organism, and the mode of transmission, by which the organism reaches a new host. The reservoir is the natural habitat where an infectious agent lives and multiplies, from which it can be transmitted directly or indirectly to a new host. The reservoir refers to the natural habitat of the organism, which may be in people, animals, arthropods, plants, soil, or substances in which an organism normally lives and multiplies, and on which it depends for survival or in which it survives in a dormant form.

Contacts are persons or animals who have been in association with an infected person, animal, or contaminated inanimate object, or environment that might provide an opportunity for acquiring the infective agent. Persons or animals that harbor a specific infectious agent, often in the absence of discernible clinical disease, and who serve as a source of infection or contamination of food, water, or other materials, are carriers. A carrier may have an inapparent infection (a healthy carrier) or may be in the incubation or convalescent stage of the infection.

CLASSIFICATIONS OF COMMUNICABLE DISEASES

Communicable diseases may be classified by a variety of methods: by organism, by mode of transmission, by methods of prevention (e.g., vaccine preventable, vector controllable), or by major organism classification, that is, viral, bacterial, and parasitic disease.

A virus is a nucleic acid molecule (RNA or DNA) encapsulated in a protein coat or capsid. The virus is not a complete cell and can only replicate inside a complete cell. The capsid may have a protective envelope of a lipid containing membrane. The capsid and membrane facilitate attachment and penetration of a host cell. Inside the host cell, the nucleic molecule may cause the cell's chromosomes to be changed in its own genetic material or so that there is cellular manufacture and virus replication. Viroids are smaller RNA structures without capsids which can cause plant disease. Prions are recently discovered (Stanley Prusiner, Nobel prize, 1997) variants of viruses or viroids which are the infective agents cause of scrapie in sheep, and similar degenerative central nervous system diseases in cattle and in man (mad cow disease or Creutzfeld-Jakob disease in humans).

Bacteria are unicellular organisms that reproduce sexually or asexually, grow on cell-free media, and can exist in an environment with oxygen (aerobic) or in one lacking oxygen (anaerobic). Some may enter a dormant state and form spores where they are protected from the environment and may remain viable for years. Bacteria include a nucleus of chromosomal DNA material within a membrane surrounded by cytoplasm, itself enclosed by the cellular membrane. Bacteria are often characterized by their coloration under Gram's stain, as gram-negative or gram-positive, as well as by their microscopic morphology, colony patterns on growth media, by the diseases they may cause, as well as by antibody and molecular (DNA) marking techniques. Bacteria include both indigenous flora (normal resident) bacteria and pathogenic (disease causing) bacteria. Pathogenic bacteria cause disease by invading, overcoming natural or acquired resistance, and multiplying in the body. Bacteria may produce a toxin or poison that can affect a body site distant from where the bacterial replication occurs, such as in tetanus. Bacteria may also initiate an excessive immune response, producing damage to other body tissues away from the site of infection, e.g., acute rheumatic fever and glomerulonephritis.

Parasitology studies protozoa, helminths, and arthropods that live within, on, or at the expense of a host. These include oxygen-producing, flagellate, unicellular organisms such as *Giardia* and *Trichomonas,* and amoebas such as *Entamoeba* important in enteric and gynecologic disorders. Sporozoa are parasites with complex life cycles in different hosts, such as cryptosporidium or malarial parasites. Parasitic disease usually refers to infestation, with fungi, molds, and yeasts that can affect humans. Helminths are worms that infest humans especially in poor sanitation and tropical areas.

MODES OF TRANSMISSION OF DISEASE

Transmission of diseases is by the spread of an infectious agent from a source or reservoir to a person (Table 4.1). Direct transmission from one host to another occurs during touching, biting, kissing, sexual intercourse, and projection via droplets, as in sneezing, coughing, or spitting, or by entry through the skin. Indirect transmission includes via aerosols of long-lasting suspended particles in air, fecal-oral transmission such as food and waterborne as well as by poor hygenic conditions with inanimate materials, such as soiled clothes, handkerchiefs, toys, or other objects.

Vector-borne diseases are transmitted via crawling or flying insects, in some cases with multiplication, and development of the organism in the vector, as in

TABLE 4.1　Classification of Infectious Diseases by Principal Modes of Transmission

Mode	Method	Examples
Direct	Physical contact	Leprosy, impetigo, scabies, anthrax
Direct	Sexual contact	HIV, syphilis, gonorrhea, herpes genitalis, hepatitis B, chlamydia, human papillovirus
Direct/ indirect	Airborne droplets and aerosols	Viral exanthems (measles), streptococcal diseases, various upper and lower respiratory tract diseases, tuberculosis, Legionnaire's disease, influenza
Indirect	Blood and blood products	HIV, hepatitis B, hepatitis C
Indirect	Oral–Fecal Hygiene Foodborne Waterborne	Cholera, shigella, salmonella, typhoid, botulism, campylobacter, staph aureus, cryptosporidium, listeria, worms, giardia, hepatitis A, rotavirus, enteroviruses, poliovirus, adenoviruses, entameba histolytica
Indirect	Trans cutaneous	Vector-borne via insects (arthropod): malaria, viral hemorrhagic fevers, schistosomiasis, plague Animal bite (zoonoses): rabies Health care (iatrogenic): hospital infections, HIV, hepatitis B Self-injected (illicit drug users): HIV, hepatitis B
Vertical	Congenital	Congenital rubella syndrome, congenital syphilis, gonorrheal ophthalmia, cytomegalovirus (CMV)
	Maternal–fetal	HIV, rubella, syphilis, hepatitis B, gonorrhea, chlamydia

malaria. The subsequent transmission to humans is by injection of salivary gland fluid during biting, e.g., congenital syphilis, or by deposition of feces, urine or other material capable of penetrating the skin through a bite wound or other trauma. Transmission may occur with insects as a transport mechanism, as in salmonella on the legs of a housefly.

Airborne transmission occurs inderectly via infective organisms in small aerosols that may remain suspended for long periods of time and which easily enter the respiratory tract. Small particles of dust may spread organisms from soil, clothing, or bedding.

Vertical transmission occurs from one generation to another, or from one stage of the insect life cycle to another stage. Maternal–infant transmission occurs during pregnancy (transplacental), delivery, as in gonorrhoea, breast-feeding, e.g., HIV, with transfer of infectious agents from mother to fetus or newborn.

IMMUNITY

Resistance to infectious diseases is related to many host and environmental factors, including age, sex, pregnancy, nutrition, trauma, fatigue, living and socio-economic conditions, and emotional status. Good nutritional status has a protective effect against the results of an infection. Vitamin A supplements reduce complication rates of measles and enteric infections. Tuberculosis may be present in an individual whose resistance is sufficient to prevent clinical disease, but the infected person is a carrier of an organism which can be transmitted to another or cause clinical disease if the person's susceptibility is reduced.

Immunity is resistance to infection resulting from presence of antibodies or cells that specifically act on the microorganism associated with a specific disease or toxin. Immunity to a specific organism can be acquired by having the disease, that is, natural immunity, or by immunization, active or passive, or by protection

---

BOX 4.3 VACCINES AND PREVENTION

"The Greeks had two gods of health, Aesculapius and Hygeia, therapy and prevention, respectively. Medicine in the twentieth century retains those two concepts, and vaccination is a powerful means of prevention. What follows is information on the vaccines that together with sanitation, make modern society possible, and that if wisely used will continue to bestow on mankind the gift of prevention, which according to proverb is worth far more than cure."

Source: Plotkin, S. A., Mortimer, E. A. 1994. *Vaccines.* Second edition. Philadelphia: WB Saunders (with permission).

BOX 4.4 BASIC TERMS IN IMMUNOLOGY OF INFECTIOUS DISEASES

Infectious agent: a pathogenic organism (e.g., virus, bacteria, rickettsia, fungus, protozoa, or helminth) capable of producing infection or an infectious disease.

Infection: the process of entry, development, and proliferation of an infectious agent in the body tissue of a living organism (human, animal, or plant) overcoming body defense mechanisms, resulting in an inapparent or clinically manifest disease.

Antigen: a substance (e.g., protein, polysaccharide) capable of inducing specific response mechanisms in the body. An antigen may be introduced into the body by invasion of an infectious agent, by immunization, inhalation, ingestion, or through the skin, wounds, or via transplantation.

Antibody: a protein molecule formed by the body in response to a foreign substance (an antigen) or acquired by passive transfer. Antibodies bind to the specific antigen that elicits its production, causing the infective agent to be susceptible to immune defense mechanisms against infections e.g., humoral and cellular.

Immunoglobulins: antibodies that meet different types of antigenic challenges. They are present in blood or other body fluids, and can cross from a mother to fetus *in utero,* providing protection during part of the first year of life. There are five major classes (IgG, IgM, IgA, IgD, and IgE) and subclasses based on molecular weight.

Antisera or antitoxin: materials prepared in animals for use in passive immunization against infection or toxins.

Source: Jawetz, Melrick, and Adelberg, *Medical Microbiology,* 1998.

---

through elimination of circulation of the organism in the community. Immunity may be by antibodies produced by the host body or transferred from externally produced antibodies. The body also reacts to infective antigens by cellular responses, including those that directly defend against invading organisms and other cells which produce antibodies.

The immune response is the resistance of a body to specific infectious organisms or their toxins provided by a complex interaction of antibodies and cells including

a. B Cells (bone marrow and spleen) produce antibodies which circulate in the blood, i.e., humoral immunity;
b. T Cell-mediated immunity is provided by sensitization of lymphocytes of thymus origin to mature into cytotoxic cells capable of destroying virus-infected or foreign cells;

c. Complement, a humoral response which causes lysis of foreign cells;
d. Phagocytosis, a cellular mechanism which ingests foreign microorganisms (macrophages and leukocytes).

SURVEILLANCE

Surveillance of disease is the continuous scrutiny of all aspects of occurrence and spread of disease pertinent to effective control of that disease. Maintaining ongoing surveillance is one of the basic duties of a public health system, and is vital to the control of communicable disease, providing the essential data for tracking of disease, planning interventions, and responding to future disease challenges. Surveillance of infectious disease incidence relies on reports of notifiable diseases by physicians, supplemented by individual and summary reports of public health laboratories. Such a system must concern itself with the completeness and quality of reporting and potential errors and artifacts. Quality is maintained by seeking clinical and laboratory support to confirm first reports. Completeness, rapidity, and quality of reporting by physicians and laboratories should be emphasized in undergraduate and postgraduate medical education. Enforcement of legal sanctions may be needed where standards are not met. Surveillance of infectious diseases includes the following:

1. Morbidity reports from clinics to public health offices;
2. Mortality reports from attending doctors to vital records;
3. Reports from selected sentinel centers;
4. Special field investigations of epidemics or individual cases;
5. Laboratory monitoring of infectious agents in population samples;
6. Data on supply, use, and side effects of vaccines, toxoids, immune globulins;
7. Data on vector control activities such as insecticides use;
8. Immunity levels in samples of the population at risk;
9. Review of current literature on the disease;
10. Epidemiologic and clinical reports from other jurisdictions.

Epidemiologic monitoring based on individual and aggregated reports of infectious diseases provide data vital to planning interventions at the community level or for the individually exposed patient and his contacts, along with other information sources such as hospital discharge data and monitoring of sentinel centers. These may be specific medical or community sites that are representative of the population and are able to provide good levels of reporting to monitor an area or population group. A sentinel center can be a pediatric practice site, a hospital emergency room, or other location which will provide a "finger on the pulse" to assess the degree and kind of morbidity occurring in the community. It can also include monitoring in a location previously known for disease transmission, such as Hong Kong in relation to influenza.

Epidemiologic analysis provided by government public health agencies should be published weekly, monthly, and annually and distributed to a wide audience of public health and health-related professionals throughout the country. Feedback to those in the field on whose initial reports the data are based is vital in order to promote involvement and improved quality of data, as well as to allow evaluation of the local situation in comparison to other areas. In a federal system of government, national agencies report regularly on all state or provincial health patterns. State or provincial health authorities provide data to the counties and cities in their jurisdictions. Such data should also be readily available to researchers in other government agencies, universities, and other academic settings for further research and analysis both on internet and hard-copy publications.

Notifiable diseases are those which a physician is legally required to report to state or local public health officials, by reason of their contagiousness, severity, frequency, or other public health importance (Table 4.2). Public health laboratory services provide validation of clinical and epidemiologic reports. They also pro-

TABLE 4.2   Notifiable Infectious Diseases in the United States

AIDS/HIV	Mumps
Anthrax	Pertussis (whooping cough)
Botulism	Plague
Brucellosis (undulant fever)	Poliomyelitis, paralytic
Chancroid	Psittacosis
Chlamydia trachomatis, genital infection	Rabies (animal and human)
Cholera	Rocky Mountain spotted fever
Coccidiomycosis	Rubella
Cryptosporidiosis	Rubella congenital syndrome
Diphtheria	Salmonellosis
Encephalitis (California, eastern and western	Shigellosis
equine, St. Louis)	Streptococcal disease, invasive group A
Escherichia coli 0157:H7	Streptococcal pneumonia, drug-resistant
Gonorrhea	invasive
Haemophilus influenza	Streptococcal toxic shock syndrome
Hansen disease (Leprosy)	Syphilis (primary, secondary, total of all stages)
Hemolytic uremic syndrome (post diarrhoea)	Tetanus
Hepatitis (A, B, Cnon-A/non-B)	Toxic shock syndrome
Legionnellosis	Trichinosis
Lyme disease	Tuberculosis
Malaria	Tularemia
Measles	Typhoid fever (cases/carriers)
Meningococcal disease	Yellow fever

Source: Centers for Disease Control. 1997. Case definitions for infectious conditions under public health surveillance. *Morbidity and Mortality Weekly Review*. 46(RR-10): 1–55; also available at CDC website www.cdc.gov/epo/mmwr/preview/mmwrhtml/0047449.htm

Note: Other diseases for which individual state monitoring may be required include: amebiasis, meningitis (aseptic and other bacterial), campylobacteriosis, cyclosporiosis, dengue fever, ehrlichiosis, genital herpes, genital warts, giardiasis, granuloma inguinale, leptospirosis, listeriosis, *lymphogranuloma venereum,* mucopurulent cervicitis, nongonococcal urethritis, lymphogranuloma venereum, pelvic inflammatory disease, rheumatic fever, tularemia, varicella and others.

vide day-to-day supervision of public health conditions, and can monitor communicable disease and vaccine efficacy and coverage. In addition, they support standards of clinical laboratories in biochemistry, microbiology, and genetic screening.

Nosocomial Infections

Nosocomial or hospital-acquired infections constitute a major health hazard associated with care in institutions. In the United States, they occur in 5–10% of hospital admissions and are the cause of lengthening of hospital stay and an estimated 30,000 deaths per year. In developing countries, nosocomial infection rates may occur in up to 65% of hospitalizations. This category of infectious disease most commonly includes infections of the urinary tract, surgical wounds, lower respiratory tract (pneumonias), and blood poisoning or septicemias. In the United States, up to 60% of hospital-acquired infections are caused by multidrug resistant organisms. Staphylococcus infections resistant to many current antibiotics, for example, methicillin and vancomycin, are a notable cause of prolongation of hospitalization or even death. The increasing number of immunodeficient patients has increased the importance of prevention of nosocomial infections.

Where standards of infection control are lacking, in both developed and developing countries, hospital staff are vulnerable to serious infection. In developing countries, deadly new viruses, such as Ebola and Marburg viruses mainly affect nursing, medical, and other staff as secondary cases. Surveillance and control measures are important elements of hospital management. Hospital epidemiologists and infection control staff are part of modern hospital staffing.

The cost to the health system of nosocomial infections is a major consideration in planning health budgets. Reducing the risk of acquiring such infections in hospital justifies substantial expenditures for hospital epidemiology and infection control activities. With diagnostic related group payment for hospital care (by diagnosis rather than by days of stay) the good manager has a major incentive to ensure that the risk of nosocomial infections is minimized, since they can greatly prolong hospital stays, raising patient dissatisfaction and health care costs.

ENDEMIC AND EPIDEMIC DISEASE

An endemic disease is the constant usual presence of a disease or infectious agent in a given geographic area or population group. Hyperendemic is a state of persistence of high levels of incidence of the disease. Holoendemic means that the disease appears early in life and affects most of the population, as in malaria or hepatitis A and B in some regions.

An epidemic is the occurrence in a community or region of a number of cases of an illness in excess of the usual or expected number of cases. The number of cases constituting an epidemic varies with the disease, and factors such as previous epidemiological patterns of the disease, time and place of the occurrence, and the population involved must be taken into account. A single case of a disease long absent from an area, such as polio, constitutes an epidemic, and therefore a public health emergency because a clinical case may represent a hundred carriers with nonparalytic or subclinical poliomyelitis. In the 1990s, two to three or more cases of measles linked in time and place may be considered sufficient evidence of transmission and presumed to be an epidemic. A pandemic is occurrence of a disease over a very wide area, crossing international boundaries, affecting a large proportion of the population.

Epidemic Investigation

Each epidemic should be regarded as a unique natural experiment. The investigation of an epidemic requires preparation and field investigation in conjunction with local health and other relevant authorities. Verification of cases and the scope of the epidemic will require case definition and laboratory confirmation. Tabulation of known cases according to time, place, and person are important for immediate control measures and formulation of the hypothesis as to the nature of the epidemic. An epidemic curve is a graphic plotting of the distribution of cases by the time of onset or reporting, which gives a picture of the timing, spread, and extent of the disease from the time of the initial index cases and the secondary spread.

Epidemic investigation requires a series of steps. This starts with confirmation of the initial report and preliminary investigation, defining who is affected, determining the nature of the illness and confirming the clinical diagnosis, and recording when and where the first (index) and follow-up (secondary) cases occurred, and how the disease was transmitted. Samples are taken from index case patients (e.g., blood, feces, throat swabs) as well as from possible vectors (e.g., food, water, sewage, environment). A working hypothesis is established based on the first findings, taking into account all plausible explanations. The epidemic pattern is studied, establishing common source or risk factors, such as food, water, contact, environment, and drawing a time line of cases to define the epidemic curve.

How many are ill (the numerator) and what is the population at risk (the denominator) establish the attack rate, namely, the percentage of sick among those exposed to the common factor. What is a reasonable explanation of the occurrence; is there a previous pattern, with the present episode a recurrence or new event? Consultation with colleagues and the literature helps to establish both a biological and epidemiologic plausibility. What steps are needed to prevent spread and recurrence of the disease? Coordination with relevant health and other officials and providers is required to establish surveillance and control systems, document and distribute reports, and respond to the public's right to know.

The first reports of excess cases may come from a medical clinic or hospital. The initial (sentinel or index) cases provide the first clues that may point to a common source. Investigation of an epidemic is designed to quickly elucidate the cause and points of potential intervention to stop its continuation. This requires skilled investigation and interpretation. Epidemiologic investigations have defined many public health problems. Rubella syndrome, Legionnaire's disease, AIDS, and Lyme and hantavirus diseases were first identified clinically when unusually large numbers of cases appeared with common features. The suspicions that were raised led to a search for causes and the identification of control methods.

A working hypothesis of the nature of an epidemic is developed based on the initial assessment, the type of presentation, the condition involved, and previous local, regional, national, and international experience. The hypothesis provides the basis for further investigation, control measures, and planning additional clinical and laboratory studies. Surveillance will then monitor the effectiveness of control measures. Communication of findings to local, regional, national, and international health reporting systems is important for sharing the knowledge with other potential support groups or other areas where similar epidemics may occur.

The Centers for Disease Control and Prevention (CDC), originally organized in 1942 as the Office for Malaria Control in War Areas, is part of the U.S. Public Health Service. As of 1993, the CDC had a budget of $1.5 billion, and 7300 employees include epidemiologists, microbiologists, and many other professionals. The CDC includes national centers for environmental health and injury control, chronic disease prevention and health promotion, infectious diseases, prevention services, health statistics, occupational safety and health, and international health.

The Epidemic Intelligence Service (EIS) of the CDC in the United States is an excellent model for the organization of the national control of communicable diseases. Young clinicians are trained to carry out epidemiologic investigations as part of training to become public health professionals. EIS officers are assigned to state health departments, other public health units, and research centers as part of their training, carrying out epidemic investigation and special tasks in disease control.

The CDC, in cooperation with the WHO, has developed and offers free of charge, a personal computer program to support field epidemiology, including epidemic investigations (EPI-INFO), which can be accessed and down-loaded from the worldwide web. This program should be adopted widely in order to improve field investigations, to encourage reporting in real time, and to develop high standards in this discipline.[1]

CDC's *Morbidity and Mortality Weekly Report* (*MMWR*) is a weekly publication of the CDC's epidemiologic data, also available free on the internet. It includes special summaries of reportable infectious diseases as well as noncommunicable diseases of epidemiologic interest. *MMWR* publishes periodic special reports of important infectious and noninfectious diseases with comprehensive reviews of the literature and recent investigative work by the CDC and other organizations.

---

[1]Epidemiologic investigation may be arranged at the CDC or WHO by contacting the Epidemiology Program Office, Mailstop G34, Centers for Disease Control, Atlanta, Georgia 30333, or by telephone 404-639-2709 or fax 404-639-3296; or the World Health Organization, 1211 Geneva 27, Switzerland, or by telephone 41-22-791-2111 or fax 41-22-791-0746.

TABLE 4.3 Methods of Prevention or Control of Infectious Diseases by Type of Organism

Control of major infectious diseases	Viruses	Bacteria	Parasites
Vaccination: preexposure to protect individuals and the community (herd immunity); postexposure for individual protection (e.g., for rabies following animal bite, or contact after exposure to measles cases); or immunization to prevent infected meat or milk transfer of disease to humans (e.g., brucellosis)	Rabies, polio, measles, rubella, mumps, hepatitis B, influenza, varicella, and hepatitis A	Diphtheria, pertussis, tetanus, tuberculosis, anthrax, brucella, pneumococcal pneumonia, *Haemophilus influenz* B	Malaria vaccines being tested
Environmental measures: water, sewage, and vector control (e.g., chlorination of water, fly, tick, and mosquito control)	Hepatitis A, rotaviruses, polio, arboviruses, tick- and mosquito-borne viruses (yellow fever)	Salmonella, shigella, cholera, Legionnaire's disease, *E. coli*, Lyme disease	Malaria, onchocerciasis, dracunculiasis, schistosomiasis, worms, cryptosporidiosis, giardiasis
Educational/social/behavior measures: to promote self-care and self-protection to reduce risk (e.g., safe sexual practices to prevent STDs and HIV)	HIV, hepatitis B	Diarrheal diseases, syphilis, gonorrhea, chancroid	Malaria, scabies. onchocerciasis, dracunculiasis
Animal and food control: to reduce transmission (e.g., pasteurization of milk, radiation of food)	Rabies	Brucellosis, salmonellosis, coliforms	Tapeworms
Case finding and treatment: to cure, prevent transmission and reduce the carrier population (e.g., blood, sputum screening)	Rabies, herpes, CMV, HIV	Tuberculosis, STDs. rheumatic fever	Malaria, worms, dracunculiasis, onchocerciasis, schistosomiasis, tapeworms
Occupational measures: to protect persons exposed at place of work (e.g., immunization of food handlers, health and child care workers)	HIV, hepatitis, measles, rubella, arboviruses	Brucellosis, tuberculosis, anthrax	Hydatid cyst, trichinosis

CONTROL OF COMMUNICABLE DISEASES

Although an infectious disease is an event affecting an individual, it is communicable to others, and therefore its control requires both individual and community measures of protection. Control of the disease is a reduction in its incidence, prevalence, morbidity, and mortality. Elimination of a disease in a specified geographic area may be achieved as a result of intervention programs such as individual protection against tetanus; elimination of infections such as measles requires stoppage of circulation of the organism. Eradication is success in reduction to zero of incidence of the disease and presence in nature of the organism, such as with smallpox. Extinction means that a specific organism no longer exists in nature or in laboratories.

Public health applies a wide variety of tools for the prevention of infectious diseases and their transmission. It includes activities ranging from filtration and disinfection of community drinking water to environmental vector control, pasteurization of milk, and immunization programs (see Table 4.3). No less important are organized programs to promote self protection, case finding, and effective treatment of infections to stop their spread to other susceptible persons (e.g., HIV, sexually transmitted diseases, tuberculosis, malaria). Planning measures to control and eradicate specific communicable diseases is one of the principal activities of public health and remains so for the twenty-first century.

Treatment

Treating an infection once it has occurred is vital to the control of a communicable disease. Each person infected may become a vector and continue the chain of transmission. Successful treatment of the infected person reduces the potential for an uninfected contact person to acquire the infection. Bacteriostatic agents or drugs such as sulfonamides inhibit growth or stop replication of the organism, allowing normal body defenses to overcome the organism. Bacteriocidal drugs such as penicillin act to kill pathogenic organisms.

Traditional medical emphasis on single antibiotics has changed to use of multiple drug combinations for tuberculosis and more recently for hospital-acquired infections. Antibiotics have made enormous contributions to clinical medicine and public health. However, pathogenic organisms are able to adapt or mutate and develop resistance to antibiotics, resulting in drug resistance. Wide-scale use of antibiotics has led to increasing incidence of resistant organisms. Multidrug resistance constitutes one of the major public health challenges at the end of the twentieth century. Antiviral agents (e.g., ribovarin) are important additions to medical treatment potential, as are "cocktails" of antiviral agents for management of HIV infection. Antibiotic use is a health problem requiring attention of clinicians and their teachers as well as the public health community and health care managers, representing the interaction of health issues across the entire spectrum of services.

Methods of Prevention

Organized public health services are responsible for advocating legislation and for regulating and monitoring programs to prevent infectious disease occurrence and/or spread. They function to educate the population in measures to reduce or prevent the spread of disease.

Health promotion is one of the most essential instruments of infectious disease control. It promotes compliance and community support of preventive measures. These include personal hygiene and safe handling of water, milk, and food supplies. In sexually transmitted diseases, health education is the major method of prevention.

Each of the infectious diseases or groups of infectious diseases have one or more preventive or control approaches (Table 4.3). These may involve the coordinated intervention of different disciplines and modalities, including epidemiologic monitoring, laboratory confirmation, environmental measures, immunization, and health education. This requires teamwork and organized collaboration.

Very great progress has been made in infectious disease control by clinical, public health, and societal means since 1900 in the industrialized countries and since the 1970s in the developing world. This is attributable to a variety of factors, including organized public health services; the rapid development and wide use of new and improved vaccines and antibiotics; better access to health care; and improved sanitation, living conditions, and nutrition. Triumphs have been achieved in the eradication of smallpox and in the increasing control of other vaccine-preventable diseases. However, there remain serious problems with TB, STDs, malaria, and new infections such as HIV, and an increase in multiple drug-resistant organisms.

VACCINE-PREVENTABLE DISEASES

Vaccines are one of the most important tools of public health in the control of infectious diseases, especially for child health. Vaccine-preventable diseases

TABLE 4.4   Annual Incidence of Selected Vaccine-Preventable Infectious Diseases in Rates per 100,000 Population Selected Years, United States, 1950–1996

Disease	1950	1960	1970	1980	1985	1990	1996
Diphtheria	3.8	0.5	0.2	0	0	0	0
Pertussis	79.8	8.2	2.1	0.8	1.5	1.8	2.9
Poliomyelitis	22.0	1.8	0	0	0	0	0
Measles	211.0	245.4	23.2	6.0	1.2	11.2	0.2
Mumps	na[a]	na	55.6	3.9	1.3	2.2	0.3
Rubella	na	na	27.8	1.7	0.3	0.5	0.1
Hepatitis A	na	na	27.8	12.8	10.0	12.6	11.7
Hepatitis B	na	na	4.1	8.4	11.5	8.5	4.0

Source: *Health United States,* 1990, 1998.

(VPDs) are those diseases preventable by currently available vaccines (Table 4.4). The word vaccine comes from use of cowpox (vaccinia virus) to stimulate immunity to smallpox, first demonstrated by Jenner in 1796. The term is now generally used for all immunizing agents.

The body responds to invasion of disease-causing organisms by antigen–antibody reactions and cellular responses. Together, these act to restrain or destroy the disease-causing potential. Strengthening this defense mechanism through im-

---

BOX 4.5 DEFINITIONS OF IMMUNIZING AGENTS AND PROCESSES

Vaccines: a suspension of live or killed microorganisms or antigenic portion of those agents presented to a potential host to induce immunity to prevent the specific disease caused by that organism. Preparation of vaccines may be from:

a. Live attenuated organisms which have been passed repeatedly in tissue culture or chick embryos so that they have lost their capacity to cause disease but retain an ability to induce antibody response, such as polio-Sabin, measles, rubella, mumps, yellow fever, BCG, typhoid, and plague.

b. Inactivated or killed organisms which have been killed by heat or chemicals but retain an ability to induce antibody response; they are generally safe but less efficacious than live vaccines and require multiple doses, such as polio-Salk, influenza, rabies, and Japanese encephalitis.

c. Cellular fractions usually of a polysaccharide fraction of the cell wall of a disease-causing organisms, such as pneumococcal pneumonia or meningococcal meningitis.

d. Recombinant vaccines produced by recombinant DNA methods in which specific DNA sequences are inserted by molecular engineering techniques, such as DNA sequences spliced to vaccinia virus grown in cell culture to produce influenza and hepatitis B vaccines.

Toxoids or antisera: modified toxins are made nontoxic to stimulate formation of an antitoxin, such as tetanus, diphtheria, botulism, gas gangrene, and snake and scorpion venom.

Immune globulin: an antibody-containing solution derived from immunized animals or human blood plasma, used primarily for short-term passive immunization, e.g., rabies, for immunocompromised persons.

Antitoxin: an antibody derived from serum of animals after stimulation with specific antigens and used to provide passive immunity, e.g., tetanus.

Source: Brooks, G. E., Butel, J. S., Morse, S. A. 1998. *Jawetz, Melnick and Adelberg's Medical Microbiology,* Twenty-first edition. Stamford, CT: Appleton & Lange; *Harrison's Textbook of Internal Medicine* (1998).

munization is one of the outstanding achievements of public health, as treatment of infectious diseases by antimicrobials is a major element of clinical medicine.

Immunization (vaccination) is a process used to increase host resistance to specific microorganisms to prevent them from causing disease. It induces primary and secondary responses in the human or animal body:

a. Primary response occurs on first exposure to an antigen. After a lag or latent period of 3–14 days (depending on the antigen) specific antibodies appear in the blood. Antibody production ceases after several weeks but memory cells that can recognize the antigen and respond to it remain ready to respond to a further challenge by the same antigen.

b. Secondary (booster) response is the response to a second and subsequent exposure to an antigen. The lag period is shorter than the primary response, the peak is higher and lasts longer. The antibodies produced have a higher affinity for the antigen, and a much smaller dose of the antigen is required to initiate a response.

c. Immunologic memory exists even when circulating antibodies are insufficient to protect against the antigen. When the body is exposed to the same antigen again, it responds by rapidly producing high levels of antibody to destroy the antigen before it can replicate and cause disease.

Immunization protects susceptible individuals from communicable disease by administration of a living modified agent, or subunit of the agent, a suspension of killed organisms or an inactivated toxin (see Table 4.5) to stimulate development of antibodies to that agent. In disease control, individual immunity may also protect another individual.

Herd immunity occurs when sufficient persons are protected (naturally or by immunization) against a specific infectious disease reducing circulation of the organism, thereby lowering the chance of an unprotected person to become infected. Each pathogen has different characteristics of infectivity, and therefore different levels of herd immunity are required to protect the nonimmune individual.

Immunization Coverage

The critical proportion of a population that must be immunized in order to interrupt local circulation of the organism varies from disease to disease. Eradication of smallpox was achieved with approximately 80% world coverage, followed by concentration on new case findings and immunization of contacts and surrounding communities. For highly infectious diseases, such as measles, immunization coverage of over 95% is needed to achieve local eradication.

Immunization coverage in a community must be monitored in order to gauge the extent of protection and need for program modification to achieve targets of disease control. Immunization coverage is expressed as a proportion in which the numerator is the number of persons in the target group immunized at a specific age, and the denominator is the number of persons in the target cohort who should have been immunized according to the accepted standard:

$$\text{Vaccine coverage} = \frac{\text{no. persons immunized in specific age group}}{\text{no. persons in the age group during that year}} \times 100$$

Immunization coverage in the United States is regularly monitered by the National Immunization Survey by a household survey in all 50 states, as well as selected urban areas considered to be at high risk for undervaccination. An initial telephone survey is followed by confirmation, where possible, from documentation from the parents or health care providers. The survey for July 1994–June 1995 examined children born between August 1991 and November 1993 (i.e., aged 19–35 months, median age 27 months). The results show improving coverage, with 95% having received three or more doses of DPT (diphtheria, pertussis, and tetanus), 88% with three or more doses of OPV (oral polio vaccine), 92% with three or more doses of *Haemophilus influenzae,* type b (Hib), but only 62% with three or more doses of hepatitis B. However, only 75% had received all recommended vaccines at the recommended ages.

Present technology allows for control or eradication of important infectious dis-

TABLE 4.5 Development of Vaccines by Period of Development and Type of Vaccine

Period/ century	Live attenuated	Killed, whole organism	Purified protein or polysaccharide	Genetically engineered
Eighteenth century	Smallpox (1798)	na[a]	na	na
Nineteenth century	Rabies (1885)	Hog cholera (1886) Typhoid (1896) Cholera (1896) Plague (1897)	Diphtheria antitoxin (1888)	na
Early twentieth century	BCG tuberculosis (1927) Yellow fever (1935)	Pertussis (1926) Influenza (1936) Rickettsia (1936)	Diphtheria (1923) Tetanus toxoid (1927) Influenza A (1936)	na
Post-World War II	Yellow fever (1953) Polio, Sabin (1963) Measles (1963) Mumps (1967) Rubella (1970) MMR (1971)	Influenza (1945) Tetanus toxoid (1949) Typhoid (1952) Polio, Salk (1955) Anthrax (1970)	Diphtheria toxoid (1949) Pneumococcis (1976, 1983) Meningococcus (1962) Tick-borne encephalitis	na
1980–1999	Adenovirus (1980) Typhoid (1992, 1995) (salmonella Type21a, Vi) Varicella (1995) Lyme disease (1998) Rotavirus (1998)	Rabies (1980, human diploid cell) Japanese encephalitis (1993) Hepatitis A (1995)	*Hemophilus influenz* b (1985) Hepatitis B (1981, plasma) Pertussis, acellular (1993)	Hepatitis B (1987) recombinant (yeast or mammalian cell derived)
2000–2010 Anticipated	New vaccines for pneumococcal, meningococcal disease, influenza, parainfluenza, respiratory syncytial virus (RSV), H. pylori, human papillomavirus (HPV), streptococcus, HIV, hepatitis C, adenoviruses			

Note: Years developed or licensed in the United States.

Source: Modified from Plotkin SA, Mortimer EA. 1994. *Vaccines,* Second Edition. Philadelphia: Saunders; and Centers for Disease Control, 1999. Vaccines universally recommended for children—United States, 1990–98. *Morbidity and Mortality Weekly Report,* 48:243–248.

[a]na, Not available.

eases that still cause millions of deaths globally each year. Other important infectious diseases are still not subject to vaccine control because of difficulties in their development. In some cases, a microorganism can mutate with changes. Viruses can undergo antigenic shifts in the molecular structure in the organism, producing completely new subtypes of the organism. Hosts previously exposed to other strains may have little or no immunity to the new strains.

Antigenic drift refers to relatively minor antigenic changes which occur in viruses. This is responsible for frequent epidemics. Antigenic shift is believed to explain the occurrence of new strains of influenza virus necessitating, for example, annual reformulation of the influenza vaccine associated with large scale epidemics and pandemics. New variants of poliovirus strains are similar enough to the three main types so that immunity to one strain is carried over to the new strain. Molecular epidemiology is a powerful new technique used to specify the geographic origin of organisms such as poliomyelitis and measles viruses, permiting tracking of the source of the virus and epidemic.

Combinations of more than one vaccine is now common practice with a trend to enlarging the cocktail of vaccines in order to minimize the number of injections, and visits required. This reduces the number of visits to carry out routine immunization saving staff time and costs, as well as increasing compliance. There are virtually no contraindications to use of multiple antigens simultaneously. Examples of vaccine cocktails include DPT (diphtheria, pertussis, and tetanus) in combination with *Haemophilus influenzae* b, poliomyelitis, and varicella, or MMR (measles, mumps, and rubella) vaccines.

Interventions in the form of effective vaccines save millions of lives each year and contribute to improved health of countless children and adults throughout the world. Vaccination is now accepted as one of the most cost-effective health interventions currently available. Continuous policy review is needed regarding allocation of adequate resources, logistical organization, and continued scientific effort to seek effective, safe, and inexpensive vaccines for other important diseases such as malaria and HIV. New technology of recombinant vaccines, such as that of hepatitis B, holds promise for important vaccine breakthroughs in the decades ahead.

Internationally, much progress was made in the 1980s in the control of vaccine-preventable diseases. At the end of the 1970s, fewer than 10% of the world's children were being immunized. WHO, UNICEF, and other international organizations mobilized to promote an Expanded Programme on Immunization (EPI) with a target of reaching 80% coverage by 1990. Immunization coverage increased in the developing countries, preventing some 3 million child deaths annually. Bacillus Calmette-Guérin (BCG) coverage rose from 31 to 85%; poliomyelitis with OPV (three doses) from 24 to 80%, and tetanus toxoid for pregnant women from 14 to 54%. Since 1991, there has been a decline in coverage in some parts of the world, mainly in sub-Saharan Africa.

The challenge remains to achieve control or eradication of vaccine-preventable diseases, thus saving millions of more lives. Part of the HFA stresses the EPI approach, which includes immunization against diphtheria, pertussis, tetanus, poliomyelitis, measles, and tuberculosis. An extended form of this is the EPI PLUS program which combines EPI with immunization against hepatitis B and yellow fever and, where appropriate, supplementation with vitamin A and iodine. The success in international eradication of smallpox is now being followed by a campaign to eradicate poliomyelitis and other important infectious diseases.

Vaccine-Preventable Diseases

Diphtheria. Diphtheria is an acute bacterial disease of the tonsils, nasopharynx, and larynx caused by the organism *Corynebacterium diphtheriae.* It occurs in colder months in temperate climates where the organism is present in human hosts and is spread by contact with patients or carriers. It has an incubation period of 2–5 days. In the past, this was primarily an infection of children and was a major contributor to child mortality in the prevaccine and preantibiotic eras. Diphtheria has been virtually eliminated in countries with well-established immunization programs.

In the 1980s, an outbreak of diphtheria occurred in the countries of the former Soviet Union among people over age 15. It reached epidemic proportions in the 1990s, with 140,000 cases (1991–1995) with 1100 deaths in 1994 in Russia alone. This indicates a failure of the vaccination program in several respects: it used only three doses of DPT in infancy; no boosters were given at school age or subsequently; the efficacy of diphtheria vaccine may have been low, and coverage was below 80%.

Efforts to control the present epidemic include mass vaccination campaigns for persons over 3 years of age with a single dose of dT (diphtheria and tetanus) and increasing coverage of routine DPT vaccines to four doses by age 2 years. The epidemic and its control measures have led to improved coverage with dT for those over 18 years, and 93% coverage among children aged 12–23 months.

WHO recommends three doses of DPT in the first year of life and a booster at school entry. This is considered by many to be insufficient to produce long-lasting immunity. The United States and other industrialized countries use a four-dose schedule and recommend periodic boosters for adults with dT.

Pertussis. Pertussis is an acute bacterial disease of the respiratory tract caused by the bacillus *Bordetella pertussis.* After an initial coldlike (catarrhal) stage, the patient develops a severe cough which comes in spasms (paroxysms). The disease can last 1–2 months. The paroxysms can become violent and may be followed by a characteristic crowing or high pitched inspiratory whooping sound, followed by expulsion of a tenacious clear sputum, often followed by vomiting. In poorly immunized populations and those with malnutrition, pneumonia often follows and death is common.

Pertussis declined dramatically in the industrialized countries as a result of widespread coverage with DPT. However, because the pertussis component of the vaccine caused some reactions, many physicians avoided its use, using DT alone. During the 1970s in the United Kingdom, many physicians recommended against vaccination with DPT. As a result, pertussis incidence increased with substantial mortality rates. This led to a reappraisal of the immunization program, with institution of incentive payments to general practitioners for completion of vaccination schedules. As a result of these measures, vaccination coverage, with resulting pertussis control, improved dramatically in the United Kingdom.

Pertussis continues to be a public health threat and recurs wherever there is inadequate immunization in infancy. A new acellular vaccine is ready for widespread use and will be safer with fewer and less severe reactions in infants, increasing the potential for improved confidence and support for routine vaccination. Use of the new vaccine is spreading in the United States and forms part of the U.S. recommended vaccination schedule.

Tetanus. Tetanus is an acute disease caused by an exotoxin of the tetanus bacillus (*Clostridium tetani*) which grows anaerobically at the site of an injury. The bacillus is universally present in the environment and enters the human body via penetrating injuries. Following an incubation period of 3–21 days, it causes an acute condition of painful muscular contractions. Unless there is modern medical care available, patients are at risk of high case fatality rates of 30–90% (highest in infants and the elderly).

Antitetanus serum (ATS) was discovered in 1890 and during World War I, ATS contributed to saving the lives of many thousands of wounded soldiers. Tetanus toxoid was developed in 1993. The organism, because of its universal presence in the environment, cannot be eradicated. However, the disease can be controlled by effective immunization of every child during infancy and school age. Adults should receive routine boosters of tetanus toxoid once very decade.

Newborns are infected by tetanus spores (tetanus neonatorum) where unsanitary conditions or practices are present. It can occur when traditional birth attendants at home deliveries use unclean instruments to sever the umbilical cord, or dress the severed cord with contaminated material. Tetanus neonatorum remains a serious public health problem in developing countries. Immunization of pregnant women and women of childbearing age is reducing the problem by conferring passive immunity to the newborn. The training of traditional birth attendants in hygienic practice and the use of medically supervised birth centers for delivery also decreases the incidence of tetanus neonastorum.

Elimination of tetanus neonatorum by the year 2000 was made a health target by the World Summit of Children in 1990. In that year, the number of deaths from neonatal tetanus was reported by UNICEF as 700,000 infants worldwide, declining to 600,000 in 1993. Immunization of pregnant women increased from under 20% in 1984 to 52% in 1995–1997. Despite progress, coverage is still too low to achieve the target of elimination.

Poliomyelitis. Polio virus infection may be asymptomatic or cause an acute nonspecific febrile illness. It may reach more severe forms of aseptic meningitis and acute flaccid paralysis with long-term residual paralysis or death during the acute phase. Poliomyelitis is transmitted mainly by direct person-to-person contact, but also via sewage contamination. Large-scale epidemics of disease, with attendant paralysis and death, occurred in industrialized countries in the 1940s and

1950s, engendering widespread fear and panic and thousands of clinical cases of "infantile paralysis".

Growth of the poliovirus by John Enders and colleagues in tissue culture in 1949 led to development of the first inactivated polio vaccine by Jonas Salk in the mid-1950s and gave hope and considerable success in the control of the disease. The development of the live attenuated oral poliomyelitis vaccine by Albert Sabin, licensed in 1960, added a new dimension to its control because of the effectiveness, low cost, and ease of administration of the vaccine. The two vaccines in their more modern forms, enhanced strength inactivated polio vaccine (eIPV), and triple oral polio vaccine (TOPV), have been used in different settings with great success.

Oral polio vaccine (OPV) induces both humoral and cellular, including intestinal, immunity. The presence of OPV in the environment by contact with immunized infants and via excreta of immunized persons in the sewage gives a booster effect in the community. Immunization using OPV, in both routine and National Immunization Days (NIDs) has proven effective in dramatically reducing poliomyelitis and circulation of the wild virus in many parts of the world. Use of the enhanced strength IPV (eIPV) produces early and high levels of circulating antibodies, as well as protecting against the vaccine-associated disease.

In rare cases OPV can cause vaccine-associated paralytic poliomyelitis (VAPP), with a risk of 1 case per 520,000 with initial doses, and 1 case per over 12 million with subsequent doses. Approximately eight to ten cases of VAPP occur annually in the United States, with clinical, ethical, and legal implications. Use of IPV as initial protection eliminates this problem. Experience in Gaza and the West Bank in the 1970s and 1980s, and later in Israel, showed that a combination of IPV and OPV is effective in overcoming endemic and imported poliovirus. OPV requires multiple doses to achieve protective antibody levels. Where there are many enteroviruses in the environment, as is the case in most developing countries, interference in the uptake of OPV may result in cases of paralytic poliomeylitis among persons who have received 3 or even 4 doses of adequate OPV.

Controversy as to the relative advantages of each vaccine continues. The OPV program of mass repeated vaccination in control of poliomyelitis in the Americas established the primacy of OPV in practical public health, and the momentum to eradicate poliomyelitis is building. A combined schedule of IPV and OPV would eliminate the wild virus and protect against vaccine-associated disease. The sequential use of IPV and OPV was adopted as part of the routine infant immunization program in the United States in 1997, but IPV alone was adopted in 1999.

There are concerns that exclusive use of either vaccine alone will not lead to the desired goal of eradication of polyomyelitis. Progress in global eradication of polio has been impressive. Global coverage of infants with three doses of OPV reached 81% in 1996 as compared to 83% in 1995. The African region of WHO had an increase in OPV coverage from 58% in 1995 to 60% in 1996. National immunization days (NIDs) were conducted in 62 countries in 1995 and 82 in 1996, covering 419 million children in 1996. Mopping up operations to reinforce coverage of children in still endemic areas is proceeding along with increased emphasis on acute flaccid paralysis (AFP) monitoring. Confirmed polio cases reported continued at 5–6,000 per year in 1997–1998. With continued national and international emphasis, and support of WHO, Rotary International, UNICEF, donor countries, and others, there is a real prospect of a world without polio, if not by the year 2000, then or shortly thereafter.

Measles. Measles is an acute disease caused by a virus of the paramyxovirus family. It is highly infectious with a very high ratio of clinical to subclinical case ratio (99/1). Measles has a characteristic clinical presentation with fever, white spots (Koplik spots) on the membranes of the mouth, and a red blotchy rash appearing on the 3rd–7th day lasting 4–7 days. Mortality rates are high in young children with compromised nutritional status, especially vitamin A deficiency.

The measles virus evolved from a virus disease of cattle (rinderpest) some 3000–5000 years ago, becoming an important disease of humans with high mortality rates in debilitated, poorly nourished children, and significant mortality and morbidity even in industrialized countries. In the prevaccine era, measles was endemic worldwide, and even in the late 1990s it remains one of the major childhood infectious diseases. It is one of the commonest causes of death for school age children worldwide. Despite earlier predictions that measles deaths would be halved to 500,000 by 1996, WHO reported 1.1 million measles deaths in that year and over 1 million in 1997. Eradication in the first decade of the next century is a feasible goal, provided that there is an adequate international effort. Measles immunization increased from under 40% worldwide in 1985 to 79% in 1995–1996, but 56% in sub-Saharan Africa.

Single-dose immunization failed to meet control or eradication requirements even in the most developed parts of the world. A live vaccine, licensed in 1963, was later replace by a more effective and heat stable vaccine, but still with a primary vaccination failure rate (i.e., fails to produce protective antibodies) of 4–8%, and secondary failure rate (i.e., produces antibodies but protection is lost over time) of 4%. A two-dose policy incorporates a booster dose, usually at school-age, in addition to maximum feasible infant coverage of children in the 9–15 month period (timing varies in different countries). Catch-up campaigns among school-age children should be carried out until the routine two-dose policy has time to take full effect. Nearly universal primary education in developing countries, offers an opportunity for mass coverage of school age children with a second dose of measles and a resulting increase of herd immunity to reduce the transmission of the virus. The two-dose policy adopted in many countries, should be supplemented with catch-up campaigns in schools to provide the booster effect for those previously immunized and to cover those previously unimmunized, especially in developing countries.

The CDC considers that domestic transmission in the United States has been interrupted and that most localized outbreaks were traceable to imported cases. South America and the Caribbean countries are now considered free of indigenous measles, based on their successful use of NIDs, although a large epidemic occurred in 1999 in Brazil. It now appears that eradication has become a feasible target during the early part of the next century, with a strategy of levels of coverage in infancy with a two-dose policy, supplemented by catch-up campaigns to older children and young adults, and outbreak control.

Mumps. Mumps is an acute viral disease characterized by fever, swelling, and tenderness usually of the parotid glands, but also other glands. The incubation period ranges between 12 and 25 days. Orchitis, or inflammation of the testicles, occurs in 20–30% of postpubertal males and oophoritis, or inflammation of the ovaries, in 5% of postpubertal females. Sterility is an extremely rare result of mumps. Central nervous system involvement can occur in the form of aseptic meningitis, almost always without sequelae. Encephalitis is reported in 1–2 per 10,000 cases with an overall case fatality rate of 0.01%. Pancreatitis, neuritis, nerve deafness, mastitis, nephritis, thyroiditis, and pericarditis, although rare, may occur. Most persons born before 1957 are immune to the disease, because of the nearly universal exposure to the disease before that time.

The live attenuated vaccine introduced in the United States in 1967 is available as a single vaccine or in combination with measles and rubella as the measles-mumps-rubella (MMR) vaccine. It provides long-lasting immunity in 95% of cases. Mumps vaccine is now recommended in a two-dose policy with the first dose of MMR given between 12 and 15 months of age and a second dose given either at school entry or in early adolescence. MMR in two doses is now standard policy in the United Sates, Sweden, Canada, Israel, the United Kingdom, and other countries. The incidence of mumps has consequently declined rapidly. Local eradication of this disease is worthwhile and should be part of a basic international immunization program.

Rubella. Rubella (German measles) is generally a mild viral disease with lymphadenopathy and a diffuse, raised red rash. Low grade fever, malaise, coryza, and lymphadenopathy characterize the prodromal period. The incubation period is usually 16–18 days. Differentiation from scarlet fever, measles, or other febrile diseases with rash may require laboratory testing and recovery of the virus from nasopharyngeal, blood, stool, and urine specimens.

In 1942, Norman Gregg, an Australian ophthalmologist, noted an epidemic of cases of congenital cataract in newborns associated with a history of rubella in the mother during the first trimester. Subsequent investigation demonstrated that intrauterine death, spontaneous abortion, and congenital anomalies occur commonly when rubella occurs early in pregnancy.

Congenital rubella syndrome (CRS) occurs with single or multiple congenital anomalies including deafness, cataracts, microophthalmia, congenital glaucoma, microcephaly, meningoencephalitis, congenital heart defects, and others. Moderate and severe cases are recognizable at birth, but mild cases may not be detected for months or years after birth. Insulin-dependent diabetes is suspected as a late sequela of congenital rubella. Each case of CRS is estimated to cost some $250,000 in health care costs during the patient's lifetime.

Prior to availability of the attenuated live rubella vaccine in 1969, the disease was universally endemic, with epidemics or peak incidence every 6–9 years. In unvaccinated populations, rubella is primarily a disease of childhood. In areas where children are well vaccinated, adolescent and young adult infection is more apparent, with epidemics in institutions, colleges, and among military personnel.

A sharp reduction of rubella cases was seen in the United States following introduction of the vaccine in 1970, but increased in 1978, following rubella epidemics in 1976–1978. A further reduction in cases was followed by a sharp upswing of rubella and CRS in 1988–1990. An outbreak of rubella among the Amish in the United States, who refuse immunization on religious grounds, resulted in 7 cases of CRS in 1991. It is now thought that vaccination of sufficient numbers in the United States reduced circulation of the virus and protected most vulnerable groups in the population.

In the past, immunization policy in some countries was to vaccinate school girls aged 12 to protect them for the period of fertility. The current approach is to give a routine dose of MMR in early childhood, followed by a second dose in early school age to reduce the pool of susceptible persons. Women of reproductive age should be tested to confirm immunity before pregnancy and immunized if not already immune. Should a woman become infected during pregnancy, termination of pregnancy previously recommended is now managed with hyperimmune globulin.

The infection of pregnant women during their first trimester of pregnancy is the primary public health implication of rubella. The emotional and financial burden of CRS, including the cost of treatment of its congenital defects, makes this vaccination program cost-effective. Its inclusion in a modern immunization program is fully justified. Elimination of CRS syndrome should be one of the primary goals of a program for prevention of vaccine-preventable disease in developed and developing countries. Adoption of MMR and the two-dose policy will gradually lead to eradication of rubella and rubella syndrome.

Viral Hepatitis.   Viral hepatitis is a group of diseases of increasing public health importance due to their large scale worldwide prevalence, their serious consequences, and our increasing ability to take preventive action. Viral hepatic infectious diseases each have specific etiologic, clinical, epidemiologic, serologic, and pathologic characteristics. They have important short- and long-term sequelae. Vaccine development is of high priority for control and ultimate eradication.

HEPATITIS A.   Hepatitis A (HAV) was previously known as infectious hepatitis or epidemic jaundice. HAV is mainly transmitted by the fecal–oral route. Clinical severity varies from a mild illness of 1–2 weeks to a debilitating illness lasting several months. The norm is complete recovery within 9 weeks, but a fulminating or even fatal hepatitis can occur. Severity of the disease worsens with increasing age. HAV is sporadic/endemic worldwide. Improving sanitation raises the age of exposure, with accompanying complications. It now occurs particularly in persons from industrialized countries when exposed to situations of poor hygiene, or among young adults when traveling to areas where the disease is endemic. Common source outbreaks occur in school-aged children and young adults from case contact, or from food contaminated by infected handlers. Hepatitis A may be a serious public health problem in a disaster situation.

Prevention involves improving personal and community hygiene, with safe chlorinated water and proper food handling. Hepatitis A vaccine has been recently licensed for use in the United States, and will probably soon be recommended for routine vaccination programs, as well as for persons traveling to endemic areas.

HEPATITIS B.   Hepatitis B (HBV) once called serum jaundice, was thought to be transmitted only by injections of blood or blood products. It is now known to be present in all body fluids and easily transmissible by household and sexual contact, perinatal spread from mother to newborn, and between toddlers. However, it is not spread by the oral–fecal route.

Hepatitis B virus is endemic worldwide and is especially prevalent in developing countries. Carrier status with persistent viremia varies from <1% of adults in North America to 20% in some parts of the world. Carriers have detectable levels of HBsAg, the surface antigen (i.e., Australian antigen), in their blood. High risk groups in developed countries include intravenous drug users, homosexual men, persons with high numbers of sexual partners, those receiving tattoos, body piercing or acupuncture treatments, and residents or staff of institutions such as group homes and prisons. Immunocompromised and hemodialysis patients are commonly carriers of HBV. HBV may also be spread in a health system by use of inadequately sterilized reusable syringes, as in China and the former Soviet Union. Transmission is reduced by screening blood and blood products for HBsAg and strict technique for handling blood and body fluids in health settings.

HBV is clinically recognizable in less than 10% of infected children but is apparent in 30–50% of infected adults. Clinically HBV has an insidious onset with anorexia, abdominal discomfort, nausea, vomiting, and jaundice. The disease can vary in severity from subclinical, very mild to fulminating liver necrosis, and death. It is a major cause of primary liver cancer, chronic liver disease, and liver failure, all devastating to health and expensive to treat.

Hepatitis B virus is considered to be the cause of 60% of primary cancer of the liver in the world and the most common carcinogen after cigarette smoking. The WHO estimates that more than 2 billion people alive today have been infected with HBV. It is also estimated that 350 million persons are chronic carriers of HBV, with an estimated 1–1.5 million deaths per year from cirrhosis or primary liver cancer. This makes hepatitis B control a vital issue in the revision of health priorities in many countries.

Strict discipline in blood banks and testing of all blood donations for HBV, as well as HIV, and hepatitis C, is mandatory, with destruction of those with positive tests. Contacts should be immunized following exposure with HBV immunoglobulin and HBV vaccine. The inexpensive recombinant HBV vaccine should be adopted by all countries and included in routine vaccination of infants. Catch-up immunization for older children is also desirable. Immunization programs should include those exposed at work, such as health, prison, or sex workers and adults in group settings. HBV immunization has been included in WHO's EPI-PLUS expanded program of immunization.

HEPATITIS C.   First identified in 1989, and previously known as non-A, non-B hepatitis, hepatitis C (HCV) has an insidious onset with jaundice, fatigue, abdominal pain, nausea, and vomiting. It may cause mild to moderate illness, but chronicity is common going on to cirrhosis and liver failure. The CDC estimates that 4 million Americans are chronically infected with HCV, with 8000–10,000 resulting deaths per annum, and the main cause of liver transplants. HCV is transmitted most commonly in blood products, but also among injecting drug users (90% of intravenous drug users were HCV positive in a Vancouver study in 1998), and is also a risk for health workers. The disease may also occur in dialysis centers and other medical situations. Person-to-person spread is unclear. Prevention of transmission includes routine testing of blood donations, antiviral treatment of blood products, needle exchange programs, and hygiene. The WHO in 1998 has declared hepatitis prevention as a major public health crisis, with an estimated 170 million persons infected worldwide (1996), stressing that this "silent epidemic" is being neglected and that screening of blood products is vital to reduce transmission of this disease as for HIV. HCV is a major cause of chronic cirrhosis and liver cancer. No vaccine is available at present, but an experimental vaccine is undergoing field trials. Interferon and ribavirin treatment is reportedly effective in 40% of cases.

HEPATITIS D.   Hepatitis D virus (HDV) also known as Delta hepatitis, may be self-limiting or progress to chronic hepatitis. It is caused by a viruslike particle which infects cells along with HBV as a coinfection or in chronic carriers of HBV. HDV occurs worldwide in the same groups at risk for HBV. It also occurs in epidemics and is endemic in South America, Africa, and among drug users. Prevention is by measures similar to those for HBV. Management for HDV is by passive immunity with immunoglobulin for contacts and high risk groups, and should include HBV vaccination as the diseases often coincide. There is currently no vaccine for HDV.

HEPATITIS E.   Hepatitis E virus has an epidemiological and clinical course similar to that of HAV. There is no evidence of a chronic form of HEV. One striking characteristic of HEV is its high mortality rate among pregnant women. The disease is caused by a viruslike particle with an incubation period of 15–64 days and is most common in young adults. Sporadic cases as well as epidemics have been identified in India, Pakistan, Burma, China, Russia, Mexico, and North Africa. HEV results from waterborne epidemics or as sporadic cases in areas with poor hygiene, spread via the oral–fecal route. It is a hazard in disaster situations with crowding and poor sanitary conditions. Prevention is by safe management of water supplies and sanitation. Disease management is supportive care; passive immunization is not helpful and no vaccine is currently available.

Haemophilus influenzae type b.   *Haemophilus influenzae* type b (Hib) is a bacteria which causes meningitis and other serious infections in children under 18 months of age. Before the introduction of effective vaccines, as many as 1 in 200 children developed invasive Hib infection. Two-thirds of these had Hib meningitis, with a case fatality rate of 2–5%. Long-term sequelae such as hearing impairment and neurological deficits occurred in 15–30% of survivors.

The first Hib vaccine was licensed in 1985, based on capsular material from the bacteria. Extensive clinical trials in Finland demonstrated a high degree of efficacy, but less impressive results were in seen in postmarketing efficacy studies. By 1989, a conjugate vaccine based on an additional protein cell capsular factor capable of enhancing the immunologic response was introduced. Several conjugate vaccines are now available.

The conjugate vaccines are now combined with DPT as their schedule is simultaneous with that of the DPT. Although the Hib vaccine has been found to be cost-effective, despite initially being as costly as all the basic vaccines combined (i.e., DPT, OPV, MMR, and HBV). For this reason, its use thus far has been limited to industrialized countries. The vaccine is a valuable addition to the immunologic armamentarium. It showed dramatic results in local eradication of this serious early childhood infection in a number of European countries and a sharp reduction in the United States.

Impressive field trials in the Gambia showed a sharp reduction in mortality from invasive streptococcal diseases. The price of the vaccine has also fallen dramatically since the mid 1990s. As a result, in 1997, the World Health Organization recommended inclusion of Hib vaccine in routine immunization programs in developing countries.

Influenza.   Influenza is an acute viral respiratory illness characterized by fever, headache, myalgia, prostration, and cough. Transmission is rapid by close contact with infected individuals and by airborne particles with an incubation period of 1–5 days. It is generally mild and self-limited with recovery in 2–7 days. However, in certain population groups, such as the elderly and chronically ill, infection can lead to severe sequelae. Gastrointestinal symptoms commonly occur in children. During epidemics, mortality rates from respiratory diseases increase because of the large numbers of persons affected, although the case fatality rates are generally low.

Over the past century, influenza pandemics have occurred in 1889, 1918, 1957, and 1968, while epidemics are annual events. The influenza pandemic of 1918 caused millions of deaths among young adults, by some estimates killing more than had died in World War I. It was the fear of recurrence of this pandemic which led the CDC to launch a massive immunization program in the United States in 1976 to prevent swine flu (the virus was a strain antigenically similar to that of the 1918 pandemic influenza) from spreading from an isolated outbreak in an army camp. The effort was stopped after millions of persons were immunized with an urgently produced vaccine when serious reactions occurred (Guillain-Barre Syndrome, (i.e., a type of paralysis), and when no further cases of swine flu were seen. This demonstrated the difficulty of extrapolating scenarios from a historical experience.

BOX 4.6 HIGH RISK GROUPS
RECOMMENDED FOR ANNUAL
INFLUENZA VACCINATION

1. Adults and children with chronic cardiovascular and respiratory conditions under medical supervision;
2. Residents of long-term care facilities, such as nursing homes;
3. Adults over 65 years of age;
4. Patients on long-term aspirin therapy who are at risk for Reye's syndrome following influenza infection;
5. Persons with HIV infection or immunosuppression;
6. Medical personnel;
7. Employees of nursing homes and long-term care facilities;
8. Home care staff and contacts of high risk individuals.

Source: From Cassens, B. 1992. *Preventive Medicine and Public Health,* Second Edition. Malvern, PA: Harwal Co., p. 99.

Each year, epidemiologic services of the WHO and collaborating centers such as the CDC recommend which strains should be used in vaccine preparation for use among susceptible population groups. These vaccines are prepared with the current anticipated epidemic strains. The three main types of influenza (A, B, and C) have different epidemiological characteristics. Type A and its subtypes, which are subject to antigenic shift, are associated with widespread epidemics and pandemics. Type B undergoes antigenic drift and is associated with less widespread epidemics. Influenza Type C is even more localized.

Active immunization against the prevailing wild strain of influenza virus produces a 70–80% level of protection in high risk groups. The benefits of annual immunization outweigh the costs, and it has proven to be effective in reducing cases of influenza and its secondary complications such as pneumonia and death from respiratory complications in high-risk groups.

Pneumococcal Disease.   Pneumococcal diseases, which are caused by *Streptococcus pneumoniae,* include pneumonia, meningitis, and otitis media. The 23 capsular types of pneumococci selected out of 83 known types of the organism for the vaccine are those responsible for 88% of pneumococcal pneumonia cases and 10–25% of all pneumonia cases in the United States, and are responsible for some 40,000 deaths per year.

This vaccine has been found to be cost-effective for high risk groups, including persons with chronic disease, HIV carriers, patients whose spleens were removed, the elderly, and those with immunosuppressive conditions. It should be included in preventive-oriented health programs, especially for long-term care of the

---

**BOX 4.7 RISK GROUPS RECOMMENDED
FOR PNEUMOCOCCAL VACCINE**

Given once to the following categories of persons at high risk:

1. People over 65 years of age;
2. The chronically ill, e.g., with cardiovascular, respiratory, liver, renal disease, diabetes mellitus, renal disease, cancer, sickle cell disease, or cirrhosis;
3. Asplenic patients;
4. Adult immunocompromised patients, including HIV positive persons;
5. Children 2 years or older who are chronically ill, or immunocompromised;
6. Persons traveling abroad.

Source: Cassens, B. 1992. *Preventive Medicine and Public Health*, Second Edition. Malvern, PA: Harwal Co., p. 95.

---

chronically ill. Because pneumococci cause bacterial meningitis, pneumococcal vaccine may be a future candidate for use in routine immunization programs for children (over age 2).

Varicella (Chicken Pox, Shingles, Herpes Zoster). Varicella is an acute, generalized virus disease caused by the varicella zoster virus (VZV). Despite its reputation as an innocuous disease of childhood, varicella patients can be quite ill. A mild fever and characteristic generalized red rash lasts for a few hours, followed by vesicles occurring in successive crops over various areas of the body. Affected areas may include the membranes of the eyes, mouth, and respiratory tract. The disease may be so mild as to escape observation or may be quite severe, especially in adults. Death can occur from viral pneumonia in adults and sepsis or encephalitis in children. Neonates whose mothers develop the disease within 2 days of delivery are at increased risk with a case fatality rate of up to 30%.

Long-term sequelae include herpes zoster or shingles with a severely painful, vesicular rash along the distribution of sensory nerves, which can last for months. Its occurrence increases with age and it is primarily seen in the elderly. It can, however, occur in immunocompromised children (especially those on cancer chemotherapy), AIDS patients, and others. Some 15% of a population will experience herpes zoster during their lifetimes. Reye's syndrome is an increasingly rare but serious complication from varicella or influenza B. It occurs in children and affects the liver and central nervous system. Congenital varicella syndrome with birth defects similar to congenital rubella syndrome has been identified recently. Varicella vaccine is now recommended for routine immunization at age 12–18 months in the United States, with catch-up for children up to age 13 years and for occupationally exposed persons in health or child care settings. Varicella vaccine is also recommended for nonpregnant women of child bearing years. Cost–benefit studies indicate a 1:5 ratio if both direct and indirect costs are included (see Chapter 11). Varicella vaccine is likely to be added to a "cocktail vaccine" containing DPT, polio (IPV), and Hib.

Meningococcal Meningitis. Meningococcal meningitis, caused by the bacterium *Neisseria meningitides,* is characterized by headache, fever, neck stiffness, delirium, coma, and/or convulsions. The incubation period is 2–10 days. It has a case fatality rate of 5–15% if treated early and adequately, but rises up to 50% in the absence of treatment. There are several important strains (A, B, C, X, Y, and Z). Serogroups A and C are the main causes of epidemics, with B causing sporadic cases and local outbreaks. Transmission is by direct contact and droplet spread.

Meningitis (group A) is common in sub-Saharan African countries, but epidemics have occurred worldwide. During epidemics, children, teenagers, and young adults are the most severely affected. In developed countries, outbreaks occur most frequently in military and student populations. In 1997, meningococcal meningitis spread widely in the "meningitis belt" in Central Africa.

Epidemic control is achieved by mass chemoprophylaxis with antibiotics (e.g., rifampin or sulfa drugs) among case contacts, although the emergence of resistant strains is a concern. Vaccines against serotypes A and C (bivalent) or A, C, W, and Y are available. Their use is effective in epidemic control and prevention institutions and military recruits, especially for A and C serogroups.

ESSENTIALS OF AN IMMUNIZATION PROGRAM

Vaccination is one of the key modalities of primary prevention. Immunization is cost-effective and prevents wide-scale disease and death, with high levels of safety. Despite the general consensus in public health regarding the central role of vaccination, there are many areas of controversy and unfulfilled expectations.

A vaccination program should aim at 95% coverage at appropriate times, including infants, school children, and adults. Immunization policy should be adapted from current international standards applying the best available program to national circumstances and financial capacities (Table 4.6). Public health personnel with expertise in vaccine-preventable disease control are needed to advise ministries of health and the practicing pediatric community on current issues in vaccination and to monitor implementation and evolution of control programs. Controversies and changing views are common to immunization policy, so that discussions must be conducted on a continuing basis. Policy should be under continuing review by a ministerially appointed national immunization advisory committee, including professionals from public health, academia, immunology, laboratory sciences, economics, and relevant clinical fields.

TABLE 4.6   Recommended Childhood Immunization Schedule,[a] United States, 1999

Vaccine	Birth	1 mo	2 mo	4 mo	6 mo	12 mo	15–18 mo	4–6 yrs	11–12 yrs	14–16 yrs
						Age (months)				
Hepatitis B	Hep B1									
			Hep B2			Hep B3			Hep B#	TD
DPT/DTaP			DPT1	DPT2	DPT3	DPT4		DPT5		
Hib			Hib1	Hib2	Hib3	Hib4				
Poliovirus[b]			IPV1	IPV2		OPV1		OPV2		
MMR						MMR1		MMR2 (4–6 or 11–12)		
Varicella						Var1		Var#		
Rotavirus			Rv1	Rv2	Rv3					

Source: Center for Disease Control, 1999. Recommended childhood immunization schedule—United States, 1999. *Morbidity and Mortality Weekly Report,* 48:12–16.

[a]OPV, Oral polio vaccine; IPV, inactivated polio vaccine; DPT or DPaT, diphtheria, pertussis, tetanus; preferably the acellular preparation (DTaP) and tetanus toxoid (DPT4 can be given at age 12 months if 6 months elapsed since previous DPT); Td, Diphtheria and tetanus; MMR, measles, mumps, rebella; Hep B, hepatitis B; Hib, Haemophilus influenzae type b; Var, varicella zoster virus; RV, rotavirus; #, for those who are not immunized in infancy.

[b]During 1999, the recommendation for polio virus was changed to 3 doses of IPV in infancy.

Vaccine supply should be adequate and continuous. Supplies should be ordered from known manufacturers meeting international standards of good manufacturing practice. All batches should be tested for safety and efficacy prior to release for use. There should be an adequate and continuously monitored cold chain to protect against high temperatures for heat labile vaccines, sera, and other active biological preparations. The cold chain should include all stages of storage, transport, and maintenance at the site of usage. Only disposable syringes should be used in vaccination programs to prevent any possible transmission of blood-borne infection.

A vaccination program depends on a readily available service with no barriers or unnecessary prerequisites, free to parents or with a minimum fee, to administer vaccines in disposable syringes by properly trained individuals using patient-oriented and community-oriented approaches. Ongoing education and training on current immunization practices are needed. Incentive payments by insuring agency or managed care systems promote complete, on-time coverage. All clinical encounters should be used to screen, immunize, and educate parents/guardians.

Contraindications to vaccination are very few; vaccines may be given even during mild illness with or without fever, during antibiotic therapy, during convalescence from illness, following recent exposure to an infectious disease, and to persons having a history of mild/moderate local reactions, convulsions, or family history of sudden infant death syndrome (SIDS). Simultaneous administration of vaccines and vaccine "cocktails" reduces the number of visits and thereby improves coverage; there are no known interferences between vaccine antigens.

Accurate and complete recording with computerization of records with automatic reminders helps promote compliance, as does co-scheduling of immunization appointments with other services. Adverse events should be reported promptly, accurately, and completely. A tracking system should operate with reminders of upcoming or overdue immunizations; use mail, telephone, and home visits, especially for high risk families, with semiannual audits to assess coverage and review patient records in the population served to determine the percentage of children covered by second birthday. Tracking should identify children needing completion of the immunization schedule and assess the quality of documentation. It is important to maintain up-to-date, easily retrievable medical protocols where vaccines are administered, noting vaccine dosage, contraindications, and management of adverse events.

All health care providers and managers should be trained in education, promotion, and management of immunization policy. Health education should target parents as well as the general public. Monitoring of vaccines used and children immunized, individually and by category of vaccination can be facilitated by computerization of immunization records, or regular manual review of child care records. Where immunization is done by physicians in private practice, as in the United States, determination of coverage is by periodic surveys.

Regulation of Vaccines

Inspection of vaccines for safety, purity, potency, and standards is part of the regulatory function. Vaccines are defined as biological products and are therefore subject to regulation by national health authorities. In the United States, this comes under the legislative authority of the Public Health Service Act, as well as the Food, Drug and Cosmetics Act, with applicable regulations in the Code of Federal Regulations. The federal agency empowered to carry out this regulatory function is the Center for Drugs and Biologics of the Federal Food and Drug Administration.

Litigation regarding adverse effects of vaccines led to inflation of legal costs and efforts to limit court settlements. The U.S. federal government enacted the Child Vaccine Injury Act of 1988. This legislation requires providers to document vaccines given and to report on complications or reactions. It was intended to pay benefits to persons injured by vaccines faster and by means of a less expensive procedure than a civil suit for resolving claims. Using this no-fault system, petitioners do not need to prove that manufacturers or vaccine givers were at fault. They must only prove that the vaccine is related to the injury in order to receive compensation. The vaccines covered by this legislation include DTP, MMR, OPV, and IPV.

Vaccine Development

Development of vaccines from Jenner in eighteenth century to the advent of recombinant hepatitis B vaccine in 1987, and of vaccines for acellular pertussis, varicella, hepatitis A, and rotavirus in the 1990s, has provided one of the pillars of public health and led to enormous savings of human life. Vaccines for viral infections in humans for HIV, respiratory syncytial virus, papilloma, Epstein-Barr virus, dengue fever, and hantavirus are under intense research with genetic approaches using recombinant techniques. The potential for the future of vaccines will be greatly influenced by scientific advances in genetic engineering, with potential for development of vaccines attached to bacteria or protein in plants, which may be given in combination for an increasing range of organisms or their harmful products.

Recombinant DNA technology has revolutionized basic and biomedical research since the 1970s. The industry of biotechnology has produced important diagnostic tests, such as for HIV, with great potential for vaccine development. Traditional whole organism vaccines, alive or killed, may contain toxic products that may cause mild to severe reactions. Subunit vaccines are prepared from components of a whole organism. This avoids the use of live organisms that can cause the disease or create toxic products which cause reactions. Subunit vaccines traditionally prepared by inactivation of partially purified toxins are costly, difficult to prepare, and weakly immunogenic. Recombinant techniques are an important development for production of new whole cell or subunit vaccines that are safe, inexpensive, and more productive of antibodies than other approaches. Their potential contribution to the future of immunology is enormous.

Molecular biology and genetic engineering have made it feasible to create new, improved, and less costly vaccines. New vaccines should be inexpensive, easily administered, capable of being stored and transported without refrigeration, and given orally. The search for inexpensive and effective vaccines for groups of viruses causing diarrheal diseases led to development of the rotavirus vaccine. Some "edible" research focuses on the genetic programming of plants to produce vaccines and DNA. Vaccine manufacturers, who spend huge sums of money and years of research on new products, tend to work on those which will bring great financial rewards for the company and are critical to the local health care community. This has led to less effort being made in developing vaccines for diseases such as malaria. Yet industry plays a crucial role for continued progress in the field.

CONTROL/ERADICATION
OF INFECTIOUS DISEASES

Since the eradication of smallpox, much attention has focused on the possibility of similarly eradicating other diseases, and a list of potential candidates has emerged. Some of these have been abandoned because of practical difficulties with current technology. Diseases that have been under discussion for eradication have included measles, TB, and some tropical diseases, such as malaria and dracunculiasis. Eradication is defined as the achievement of a situation whereby no further cases of a disease occur anywhere and continued control measures are unnecessary. Reducing epidemics of infectious diseases, through control and eradication in selected areas or target groups, can in certain instances achieve eradication of the disease. Local eradication can be achieved where domestic circulation of an organism is interrupted with cases occurring from importation only. This requires a strong, sustained immunization program with adaptation to meet needs of importation of carriers and changing epidemiologic patterns.

Smallpox

Smallpox was one of the major pandemic diseases of the Middle Ages and its recorded history goes back to antiquity. Prevention of smallpox was discussed in ancient China by Ho Kung (circe AD 320), and inoculation against the disease was practiced there from the eleventh century AD. Prevention was carried out by nasal inhalation of powdered dried smallpox scabs. Exposure of children to smallpox when the mortality rate was lowest assumed a weakened form of the disease, and it was observed that a person could only have smallpox once in a lifetime. Isolation and quarantine were widely practiced in Europe during the sixteenth and seventeenth centuries.

Variolation was the practice of inoculating youngsters with material from scabs of pustules from mild cases of smallpox in the hope that they would develop a mild form of the disease. Although this practice was associated with substantial mortality, it was widely adopted because mortality from variolation was well below that of smallpox acquired during epidemics. Introduced into England in 1721 (see Chapter 1) it was commonly practiced as a lucrative medical specialty during the eighteenth century. In the 1720s, variolation was also introduced into the American colonies, Russia, and subsequently into Sweden and Denmark.

Despite all efforts, in the early eighteenth century smallpox was a leading cause of death in all age groups. Toward the end of the eighteenth century an estimated 400,000 persons died annually from smallpox in Europe. Vaccination, or the use of cowpox vaccinia virus to protect against smallpox, was initiated late in the eighteenth century. In 1774, a cattle breeder in Yorkshire, England, inoculated his wife and two children with cowpox to protect them during a smallpox epidemic. In 1796, Edward Jenner, an English country general practitioner, experimented with inoculation from a milkmaid's cowpox pustule to a healthy youngster, who subsequently proved resistant to smallpox by variolation (see Chapter 1). Vaccination, the deliberate inoculation of cowpox material, was slow to be adopted universally, but by 1801, over 100,000 persons in England were vaccinated. Vaccination gathered support in the nineteenth century in military establishments, and in some countries which adopted it universally.

Opposition to vaccination remained strong for nearly a century based on religious grounds, observed failures of vaccination to give lifelong immunity, and because it was seen as an infringement of the state on the rights of the individual. Often the protest was led by medical variolationists whose medical practice and large incomes were threatened by the mass movement to vaccination. Resistance was also offered by "sanitarians" who opposed the germ theory and thought cleanliness was the best method of prevention. Universal vaccination was increasingly adopted in Europe and America in the early nineteenth century and eradication of smallpox in developed countries was achieved by the mid twentieth century.

In 1958, the Soviet Union proposed to the World Health Assembly a program to eradicate smallpox internationally and subsequently donated 140 million doses of vaccine per year as part of the 250 million needed to promote vaccination of at least 80% of the world population. In 1967, WHO adopted a target for the eradication of smallpox. A program was developed which included a massive increase in coverage to reduce the circulation of the virus through person-to-person contact. Where smallpox was endemic, with a substantial number of unvaccinated persons, the aim of the mass vaccination phase was 80% coverage.

Increasing vaccination coverage in developing countries reduced the disease to periodic and increasingly localized outbreaks. In 1967, 33 countries were considered endemic for smallpox, and another 11 experienced importation of cases. By 1970, the number of endemic countries was down to 17, and by 1973 only 6 countries were still endemic, including India, Pakistan, Bangladesh, and Nepal. In these countries, a new strategy was needed, based on a search for cases and vaccination of all contacts, working with a case incidence below 5 per 100,000. The program then moved into the consolidation phase, with emphasis on vaccination of newborns and new arrivals. Surveillance and case detection were improved with case contact or risk group vaccination. The maintenance phase began when surveillance and reporting were switched to the national or regional health service with intensive follow-up of any suspect case. The mass epidemic era had been controlled by mass vaccination, reducing the total burden of the disease, but eradication required the isolation of individual cases with vaccination of potential contacts.

Technical innovations greatly eased the problems associated with mass vaccination worldwide. During the 1920s, there was wide variation in sources of smallpox vaccine. In the 1930s, efforts to standardize and further attenuate the strains used reduced complication rates from vaccinations. The development of lyophilization (freeze-drying) of the vaccine in England in the 1950s made a heat-stable vaccine that could be effective in tropical field conditions in developing countries. The invention of the bifurcated needle (Bernard Rubin 1961) allowed for easier and more widespread vaccination by lesser trained personnel in remote areas. The net result of these innovations was increased world coverage and a reduction in the spread of the disease. Smallpox became more and more confined by increasing herd immunity, thus allowing transition to the phase of monitoring and isolation of individual cases.

In 1977 the last case of smallpox was identified in Somalia, and in 1980 the WHO declared the disease eradicated. No subsequent cases have been found except for several associated with a laboratory accident in the United Kingdom in 1978. The WHO recommends that the last stores of smallpox virus should be destroyed in 1999. The cost of the eradication program was $112 million or $8 million per year. Worldwide savings are estimated at $1 billion annually. This monumental public health achievement set the precedent for eradication of other infectious diseases. The World Health Assembly decided to destroy the last two remaining stocks of the smallpox virus in Atlanta and Moscow in 1999. Destruction of the remaining stock was delayed in 1999 to 2002 because of concern that illegal stocks may be held by some states or potential bioterrorists for potential use in weapons of mass destruction, concern regarding the appearance of monkeypox and a wish to use the virus for further research.

Eradication of Poliomyelitis

In 1988, the WHO established a target of eradication of poliomyelitis by the year 2000. Global immunization coverage with three doses of OPV increased from some 45% in 1984 to over 80% in 1990, with a slight decline in the period 1991–1993. Support from member countries and international agencies such as UNICEF and Rotary International has led to widescale increases in immunization coverage throughout many parts of the world. The World Health Organization promotes use of OPV only as part of routine infant immunization or National Immunization Days (NIDs). This strategy has been successful in the Americas and in China, but India and the Middle East remain problematic. Eradication of wild poliomyelitis by the year 2000 will require flexibility in vaccination strategies and may require the combined approach, using OPV and IPV, as adopted in the United States in 1997 to prevent vaccine-associated clinical cases. The combination of OPV and IPV may be needed where enteric disease is common and leads to interference in OPV uptake, especially in tropical areas where endemic poliovirus and diarrheal diseases are still found. The World Bank estimated that achievement of global eradication would save $300 million annually in the United States alone.

Other Candidates for Eradication

Since the eradication of smallpox, discussion has focused on the possibility of similarly eradicating other diseases, and a list of potential candidates has emerged. Some of these have been abandoned because of practical difficulties with current technology. Diseases that have been under discussion for eradication have included measles, TB, and tropical diseases such as malaria and dracunculiasis.

Eradication of malaria was thought to be possible in the 1950s when major gains were seen in malaria control by aggressive case environmental control, case finding, and management. However, lack of sustained vector control and an effective vaccine has prevented global eradication. Malaria control suffered serious setbacks because of failure in political resolve and capacity to continue support needed for necessary programs. In the 1960s and 1970s, control efforts were not sustained in many countries, and a dreadful comeback of the disease occurred in Africa and Asia in the 1980s. The emergence of mosquitoes resistant to insecticides, and malarial strains resistant to antimalarial drugs, have made malaria control even more difficult and expensive.

Renewed effort in malaria control may require new approaches. Use of community health workers (CHWs) in small villages in highly endemic regions of Colombia resulted in a major drop in malaria mortality during the 1990s. The CHWs investigate suspect cases by taking clinical histories and blood smears.

BOX 4.8 CRITERIA FOR ASSESSING ERADICABILITY OF DISEASES, INTERNATIONAL TASK FORCE FOR DISEASE ERADICATION (ITFDE)

1. Scientific feasibility
    a. Epidemiologic vulnerability; lack of nonhuman reservoir, ease of spread, no natural immunity, relapse potential;
    b. Effective practical intervention available; vaccine or other primary preventive or curative treatment, or vectoricide that is safe, inexpensive, long lasting, and easily used in the field;
    c. Demonstrated feasibility of elimination in specific locations, such as an island or other geographic unit.
2. Political will/popular support
    a. Perceived burden of the disease; morbidity, mortality, disability, and costs of care in developed and developing countries;
    b. Expected cost of eradication;
    c. Synergy of implementation with other programs;
    d. Reasons for eradication versus control.

Source: Morbidity and Mortality Weekly Review 1992. *Update International Task Force for Disease Eradication 1991.* 41:40–42.

They examine smears for malaria parasites and a diagnosis is made. Therapy is instituted and the patient is followed. Quality control monitoring shows high levels of accuracy in reading of slides compared to professional laboratories.

In the late 1970s, there was widespread discussion in the literature of the potential for eradication of measles and TB. Measles eradication was set back as breakthrough epidemics occurred in the United States, Canada, and many other countries during the 1980s and early 1990s, but regional eradication was achieved combining the two-dose policy with catch-up campaigns for older children or in National Immunization Days, as in the Caribbean countries.

Tuberculosis has also increased in the United States and several European countries for the first time in many decades. Unrealistic expectations can lead to inappropriate assessments and policy when confounding factors alter the epidemiologic course of events. Such is the case with TB, where control and eradication have receded from the picture. This deadly disease has returned to developed countries, partly in association with the HIV infection and multiple-drug-resistant strains, as well as homelessness, rising prison populations, poverty, and other deleterious social conditions. Directly observed therapy is an important recent breakthrough, more effective in use of available technology and will play a major role in TB control in the twenty-first century.

Future Candidates for Eradication

A decade after the eradication of smallpox was achieved, the International Task Force for Disease Eradication (ITFDE) was established to systematically evaluate the potential for global eradicability of candidate diseases. Its goals were to identify specific barriers to the eradication of these diseases that might be surmountable and to promote eradication efforts.

The subject of eradication versus control of infectious diseases if of central public health importance as technology expands the armamentarium of immunization and vector control into the twenty-first century. The control of epidemics, followed by interruption of transmission and ultimately eradication, will save countless lives and prevent serious damage to children throughout the world. The smallpox achievement, momentous in itself, points to the potential for the eradication of other deadly diseases. The skillful use of existing and new technology is an important priority in the New Public Health. Flexibility and adaptability are as vital as resources and personnel.

TABLE 4.7   Potential Disease Candidates for Control and for Eradication, 1998

Organism	Control—Elimination as a Public Health Problem	Eradicable—Regional/Global
Bacterial diseases	Pertussis	Diphtheria
	Neonatal tetanus	Haemophilus influenzae b
	Congenital syphilis	
	Trachoma	
	Tuberculosis	
	Leprosy	
Viral disease	Hepatitis B	Poliomyelitis
	Hepatitis A	Measles
	Yellow fever	Rubella
	Rabies	Mumps
	Japanese encephalitis	
Parasitic disease	Onchocerciasis	Dracunculiasis
	Malaria	Chagas' disease
	Helminthic infestation	Filariasis
	Schistosomiasis	Echinococcus
	Leichmaniasis, visceral	Taeniasis
Noninfectious disease	Lead poisoning	
	Silicosis	
	Protein energy malnutrition	
	Micronutrient malnutrition	
	Iodine deficiency	
	Vitamin A deficiency	
	Folic acid deficiency	
	Iron deficiency	

Source: Goodman RA, Foster KL, Trowbridge FL, Figuero JP (eds). 1998. Global Disease Elimination and Eradication as Public Health Strategies: Proceedings of a Conference Held in Atlanta, Georgia, USA, 23–25 February 1998. *Bulletin of the World Health Organization,* 76 (Supplement 2):1–161.

Selecting diseases for eradication is not purely a professional issue of resources such as vaccines and manpower, organization and financing. It is also a matter of political will and perception of the burden of disease. There will be many controversies. The selection of polio for eradication while deferring measles when polio kills few and measles kills many may be questioned. The CDC published criteria for selection of disease for eradication are shown in Box 4.8.

The WHO, in a 1998 review of health targets in the field of infectious disease control for the twenty-first century, selected the following targets: eradication of Chagas' disease by 2010; eradication of neonatal tetanus by 2010; eradication of leprosy by 2010; eradication of measles by 2020; eradication of trachoma by 2020; reversing the current trend of increasing tuberculosis and HIV/AIDS.

In 1998, a conference in Atlanta, Georgia, reviewed the subject, which is still very much in a state of flux. Table 4.7 summarizes the selection of diseases which are presently seen as controllable and those considered to be potentially eradicable. The subject will be under review in the years ahead.

TUBERCULOSIS

Tuberculosis (TB) is caused by a group of organisms including *Mycobacterium tuberculosis* in humans and *M. bovis* in cattle. The disease is primarily found in humans, but it is also a disease of cattle and occasionally other primates in certain regions of the world. It is transmitted via airborne droplet nuclei from persons with pulmonary or laryngeal TB during coughing, sneezing, talking, or singing. The initial infection may go unnoticed, but tuberculin sensitivity appears within a few weeks. About 95% of those infected enter a latent phase with a lifelong risk of reactivation. Approximately 5% go from initial infection to pulmonary TB. Less commonly, the infection develops as extrapulmonary TB, involving meninges, lymph nodes, pleura, pericardium, bones, kidneys, or other organs.

Untreated, about half of the patients with active TB will die of the disease within 2 years, but modern chemotherapy almost always results in a cure. Pulmonary TB symptoms include cough and weight loss, with clinical findings on chest examination and confirmation by findings of tubercle bacilli in stained smears of sputum and, if possible, growth of the organism on culture media, and changes in the chest X-ray. Tuberculosis affects people in their adult working years, with 80–90% of cases in persons between the ages of 15 and 49. Its devastating effects on the work force and economic development contribute to a high cost-effectiveness for TB control.

The tubercle bacillus infects approximately 1.7 billion people in the world today, causing over 7 million cases and nearly 3 million deaths in 1997. During 1995, new cases of TB included 2.8 million (40%) in southeast Asia and the western Pacific regions of WHO, with 2.3 million cases in India, and 0.5 million in Indonesia. By 2005, the incidence of TB may increase to 11.9 million new cases per year, a 58% increase over 1990. Between 1990 and 1999, WHO estimates there were

88 million new cases of TB, of which 8 million cases were in association with HIV infection. During the 1990s, an estimated 30 million persons died of TB, including 2.9 million with HIV infection.

A new and dangerous period for TB resurgence has resulted from parallel epidemiologic events: first, the advent of HIV infection and second, the occurrence of multiple drug resistant TB (MDRTB), that is, organisms resistant at least to both Isoniazid (INH) and rifampicin, two mainstays of TB treatment. MDRTB can have a case fatality rate as high as 70%. HIV reduces cellular immunity so that people with latent TB have a high risk of activation of the disease. It is estimated that HIV negative persons have a 5–10% lifetime risk of TB; HIV positive people have a risk of 10% per year of developing clinical tuberculosis.

Drug resistance, the long period of treatment, and the socioeconomic profile of most TB patients combine to require a new approach to therapy. Directly observed treatment, short-course (DOTS), has shown itself to be highly effective with patients in poor self-care settings, such as the homeless, drug users, and those with AIDS. The strategy of DOTS uses community health workers to visit the patient and observes him or her taking the various medications, providing both incentive, support, and moral coercion to complete the needed 6 to 8 month therapy. DOTS has been shown to cure up to 95% of cases, at a cost of as little as $11 per patient. It is one of the few hopes of containing the TB pandemic.

In 1994, WHO released a new strategy for control of tuberculosis over the next decade. The plan calls for new guidelines for control, new aid funds for developing countries, and enlistment of NGOs to assist in the fight. The new guidelines stress short-term chemotherapy in well-managed programs of DOTS, stressing strict compliance with therapy for infectious cases with a goal of an 85% cure rate. Even under adverse conditions, DOTS produces excellent results. It is one of the most cost-effective health interventions combining public health and clinical medical approaches.

Tuberculosis incidence in the United States decreased steadily until 1985, increased in 1990, and has declined again since. From 1986 to 1992, there was an excess of 51,600 cases over the expected rate if the previous decline in case incidence had continued. This rise was largely due to the HIV/AIDS epidemic and the emergence of MDRTB, but also greater incidence among immigrants from areas of higher TB incidence, drug abusers, the homeless, and those with limited access to health care. This is particularly true in New York City, where MDRTB has appeared in outbreaks among prison inmates and hospital staff.

From 1992 to 1997, TB incidence in the United States declined by 26% and in some states, including New York, by 50% or more. This turnaround was due to stronger TB control programs that promptly identified persons with TB and initiated and ensured completion of appropriate therapy. Aggressive staff training, outreach, and case management approaches were vital to this success. Concern over rising rates among recent immigrants and the continued challenge of HIV/AIDS and coincidental transmission of hepatitis A, B, and C among drug users and marginal population groups show that continued support for TB control is needed.

Bacillus Calmette-Guérin (BCG) is an attenuated strain of the tubercle bacillus used widely as a vaccination to prevent TB, especially in high incidence areas. It induces tuberculin sensitivity or an antigen–antibody reaction in which antibodies produced may be somewhat protective against the tubercle bacillus in 90% of vaccinees. Although the support for its general use is contradictory, there is evidence from case–control and contact studies of positive protection against TB meningitis and disseminated TB in children under the age of 5. In some developed, low-incidence countries, it is not used routinely but selectively. It may also be used in asymptomatic HIV-positive persons or other high risk groups.

The BCG vaccine for tuberculosis remains controversial. While used widely internationally, in the United States and other industrialized countries, it is thought to hinder rather than help in the fight against TB. This concern is based on the usefulness of tuberculin testing for diagnosis of the disease. Where BCG has been administered, the diagnostic value of tuberculin testing is reduced, especially in the period soon after the BCG is used. Studies showing equivocal benefit of BCG in preventing tuberculosis have added to the controversy. While those in the field in the United States continue to oppose the use of BCG, internationally it is still felt to be of benefit in preventing TB, primarily in children.

A 1994 metaanalysis of the literature of BCG carried out by the Technology Assessment Group at Harvard School of Public Health concluded:

> On average, BCG vaccine significantly reduces the risk of TB by 50%. Protection is observed across many populations, study designs, and forms of TB. Age at vaccination did not enhance predictiveness of BCG efficacy. Protection against tuberculous death, meningitis, and disseminated disease is higher than for total TB cases, although this result may reflect reduced error in disease classification rather than greater BCG efficacy. [Colditz *et al.*, *JAMA*, 1994.]

BOX 4.9 CONTROL OF TUBERCULOSIS

1. Identifying persons with clinically active TB;
2. Diagnostic methods—clinical suspicion, sputum smear for bacteriologic examination, tuberculin skin testing, chest radiograph;
3. Case finding and investigation programs in high risk groups;
4. Contact investigation;
5. Isolation techniques during initial therapy;
6. Treatment, mainly ambulatory, of persons with clinically active TB;
7. Treatment of contacts;
8. Directly observed treatment, short-course (DOTS), where compliance suspect;
9. Environmental control in treatment settings to reduce droplet infection;
10. Educate health care providers on suspicion of TB and investigation of suspects.

Currently, the WHO recommends use of BCG as close to birth as possible as part of the Expanded Programme of Immunization (EPI).

Tuberculosis control remains feasible with current medical and public health methods. Deterioration in its control should not lead to despair and passivity. The recent trend to successful control by DOTS despite the growing problem of MDRTB suggest that control and gradual reduction can be achieved by an activist, community outreach approach. The WHO in 1999 made TB control one of its major priorities, expressing grave concern that the MDR organism, now widely spread in countries of Asia, eastern Europe, and the former Soviet Union, may spread the disease much more widely. The disease constitutes one of the great challenges to public health at the start of the new century.

STREPTOCOCCAL DISEASES

Acute infectious diseases caused by Group A streptococci include streptococcal sore throat, scarlet fever, puerperal fever, septicemia, ersypelas, cellulitis, mastoiditis, otitis media, pneumonia, peritonsillitis (quinsy), wound infections, toxic shock syndrome, and fasciitis, the "flesh eating bacteria." *Streptococcus pyogenes* group A include some 80 serologically distinct types which vary in geographic location and clinical significance. Transmission is by droplet, person-to-person direct contact, or by food infected by carriers. Important complications from a public health point of view include acute rheumatic fever and acute glomerulonephritis, but also skin infections and pneumonia.

Acute rheumatic fever is a complication of strep A infection that has virtually disappeared from industrialized countries as a result of improved standards of living and antibiotic therapy. However, outbreaks were recorded in the United States in 1985, and an increasing number of cases have been seen since 1990. In developing countries, rheumatic fever remains a serious public health problem affecting school age children, particularly those in crowded living arrangements. Long-term sequelae include disease of the mitral and aortic heart valves, which require cardiac care and surgery for repair or replacement with artificial valves.

Acute glomerulonephritis is a reaction to toxins of the streptococcal infection in the kidney tissue. This can result in long-term kidney failure and the need for dialysis or kidney transplantation. This disease has become far less common in the industrialized countries, but remains a public health problem in developing countries.

The streptococcal diseases are controllable by early diagnosis and treatment with antibiotics. This is a major function of primary care systems. Recent increases in rheumatic fever may herald a return of the problem, perhaps due to inadequate access to primary care in the United States for large sectors of the population, along with increased social hygiene problems.

Where access to primary care services is limited, infections with streptococci can result in a heavy burden of chronic heart and kidney disease with substantial health, emotional, and financial tolls. Measures to improve access to care and public information are needed to assure rapid and effective care to prevent chronic and costly conditions.

ZOONOSES

Zoonoses are infectious diseases transmissible from vetebrate animals to humans. Common examples of zoonoses of public health importance in nonindustrialized countries include brucellosis and rabies. In industrialized countries, salmonellosis, "mad cow disease" and influenza have reinforced the importance of relationships of animal and human health. Strong cooperation between public health and veterinary public health authorities are required to monitor and to prevent such diseases.

Brucellosis

Brucellosis is a disease occurring in cattle (*Brucella abortus*), in dogs (*Br. canis*), in goats and sheep (*Br. melitensis*), and in pigs (*Br. suis*). Humans are affected mainly through ingestion of contaminated milk products, by contact, or inhalation. Brucellosis (also known as relapsing, undulant, Malta, or Mediterranean fever) is a systemic bacterial disease of acute or insidious onset characterized by fever, headache, weakness, sweating, chills, arthralgia, depression, weight loss, and generalized malaise. Spread is by contact with tissues, blood, urine, vaginal discharges, but mainly by ingestion of raw milk and dairy products from infected animals. The disease may last from a few days to a year or more. Complications include osteoarthritis and relapses. Case fatality is under 2%, but disability is common and can be pronounced.

The disease is primarily seen in Mediterranean countries, the Middle East, India, central Asia, and in Central and South America. Brucellosis occurs primarily as an occupational disease of persons working with and in contact with tissues, blood, and urine of infected animals, especially goats and sheep. It is an occupational hazard for veterinarians, packinghouse workers, butchers, tanners, and laboratory workers. It is also transmitted to consumers of unpasteurized milk from infected animals. Animal vectors include wild animals, so that eradication is virtually impossible. Diagnosis is confirmed by laboratory findings of the organism in blood or other tissue samples, or with rising antibody titers in the blood, with confirmation by blood cultures.

Clinical cases are treated with antibiotics. Epidemiologic investigation may help track down contaminated animal flocks. Routine immunization of animals, monitoring of animals in high risk areas, quarantining sick animals, destroying infected animals, and pasteurizing milk and milk products prevents spread of the disease. Control measures include educating farmers and the public not to use unpasteurized milk. Individuals who work with animals (cattle, swine, goats, sheep, dogs, coyotes) should take special precautions when handling animal carcasses and materials. Testing animals, destroying carriers, and enforcing mandatory pasteurization will restrict the spread of the disease. This is an economic as well as public health problem, requiring full cooperation between ministries of health and of agriculture.

Rabies

Rabies is primarily a disease of animals, with a variety of wild animals serving as a reservoir for this disease, including foxes, wolves, bats, skunks, and raccoons, who may infect domestic animals such as dogs, cats, and farm animals. Animal bites break the skin or mucous membrane, allowing entry of the virus from the infected saliva into the bloodstream. The incubation period of the virus is 2–8 weeks; it can be as long as several years or as short as 5 days, so that postexposure preventive treatment is a public health emergency.

The clinical disease often begins with a feeling of apprehension, headache, pyrexia, followed by muscle spasms, acute encephalitis, and death. Fear of water ("hydrophobia") or fear of swallowing is a characteristic of the disease. Rabies is almost always fatal within a week of onset of symptoms. The disease is estimated to cause 30,000 deaths annually, primarily in developing countries. It is uncommon in developed countries.

Rabies control focuses on prevention in humans, domestic animals, and wildlife. Prevention in humans is based on preexposure prophylaxis for groups at risk (e.g., veterinarians, zoo workers) and postexposure immunization for persons bitten by potentially rabid animals. Because reducing exposure of pets to wild animals is difficult, immunization of domestic animals is one of the most important preventive measures. Prevention in domestic animals is by mandatory immunization of household pets. All domestic animals should be immunized at age 3 months and revaccinated according to veterinary instructions.

Prevention in wild animals to reduce the reservoir is successful in achieving local eradication in settings where reentry from neighboring settings is limited. Since 1978, the use of oral rabies immunization has been successful in reducing the population of wild animals infected by the rabies virus. Rabies eradication efforts, using aerial distribution of baits containing fox rabies vaccine in affected areas of Belgium, France, Germany, Italy, and Luxembourg, have been underway since 1989. The number of rabies cases in these affected areas has declined by some 70%. Switzerland is now virtually rabies-free because of this vaccination program. The potential exists for focal eradication, especially on islands or in partially restricted areas with limited possibilities of wild animal entry. Livestock need not be routinely immunized against rabies, except in high risk areas. Where bats are major reservoirs of the disease, as in the United States, eradication is not presently feasible.

Salmonella

Salmonella, discussed later in this chapter under diarrheal diseases, is one of the commonest of all infectious diseases among animals and is easily spread to humans via poultry, meat, eggs, and dairy products. Specific antigenic types are associated with food-borne transmission to humans, causing generalized illness and gastroenteritis. Severity of the disease varies widely, but the diseases can be devastating among vulnerable population groups, such as young children, the elderly, and the immunocompromised. Epidemiologic investigation of common food source outbreaks may uncover hazardous food handling practices. Laboratory confirmation or serotypes helps in monitoring the disease. Prevention is by maintaining high standards of food hygiene in processing, inspection and regulation, food handling practices, and hygiene education.

Anthrax

Bacillus anthracis causes a bacterial infection in herbivore animals. Its spores contaminate soil, worldwide. It affects humans exposed in occupational settings. Transmission is cutaneous by contact, gastrointestinal by ingestion, or respiratory by inhalation. It has gained recent attention (Iraq, 1997) as a highly potent agent for germ warfare or terrorism. Limited supplies of vaccine are available.

Creutzfeld-Jakob Disease

Creutzfeld-Jakob disease is a degenerative disease of the central nervous system linked to consumption of beef from cattle infected with bovine spongiform encephalopathy. It is transmitted by prions in animal feed prepared from contaminated animal material and in transplanted organs. This disease was identified in the United Kingdom linked to infected cattle leading to a 1997 ban on British beef in many parts of the world and slaughter of large numbers of potentially contaminated animals.

Other Major Zoonotic Diseases

The tapeworm causing diphyllobothriasis (*Diphyllobothrium latum*) is widespread in North American freshwater fish, passing from crustacean to fish to humans by eating raw freshwater fish. It is especially common among Inuit peoples and may be asymptomatic or cause severe general and abdominal disorder. Food hygiene (freezing and cooking of meat) is recommended; treatment is by anthelminthics.

Leptospiroses are a group of zoonotic bacterial diseases found worldwide in rats, raccoons, and domestic animals. It affects farmers, sewer workers, dairy and abattoir workers, veterinarians, military personnel, and miners with transmission by exposure to or ingestion of urine-contaminated water or tissues of infected animals. It is often asymptomatic or mild, but may cause generalized illness like influenza, meningitis, or encephalitis. Prevention requires education of the public in self protection and immunization of workers in hazardous occupations, along with immunization and segregation of domestic animals and control of wild animals.

VECTOR-BORNE DISEASES

Vector-borne diseases are a group of diseases in which the infectious agent is transmitted to humans by crawling or flying insects. The vector is the intermediary between the reservoir and the host. Both the vector and the host may be affected by climatic condition; mosquitoes thrive in warm, wet weather and are suppressed by cold weather; humans may wear less protective clothing in warm weather.

Malaria

The only important reservoir of malaria is humans. Its mode of transmission is from person to person via the bite of an infected female anopheles mosquito (Ronald Ross, Nobel prize, 1902). The causative organism is a single cell parasite with four species: *Plasmodium vivax, P. malariae, P. falciparum,* and *P. ovale.* Clinical symptoms are produced by the parasite invading and destroying red blood cells. The incubation period of approximately 12–30 days, depending on the specific plasmodium involved. Some strains of *P. vivax* may have a protracted incubation period of 8–10 months and even longer for *P. ovale.* The disease can also be transmitted through infected blood transfusions. Confirmation of diagnosis is by demonstrating malaria parasites on blood smears.

Falciparum malaria, the most serious form, presents with fever, chills, sweats, and headache. It may progress to jaundice, bleeding disorders, shock, renal or liver failure, encephalopathy, coma, and death. Prompt treatment is essential. Case fatality rates in untreated children and adults are above 10%. An untreated attack may last 18 months. Other forms of malaria may present as a nonspecific fever. Relapse of the *P. ovale* may occur up to 5 years after initial infection; malaria may persist in chronic form for up to 50 years.

Malaria control advanced during the 1940s–1960s through improved chlovaquine treatment and use of DDT for vector control with optimism for eradication of the disease. However, control regressed in many developing countries as allocations for environmental control and case findings/treatment were reduced. There has also been an increase in drug resistance, so that this disease is now an extremely serious public health problem in many parts of the world. The need for a vaccine for malaria control is now more apparent than ever.

The World Health Organization estimated that, in 1997, sub-Saharan Africa (SSA) had 270 million new malaria cases, with 5% of children up to age 5. Over 1 million deaths occur annually from malaria more than two-thirds of them in SSA. Large areas, particularly in forest or savannah regions with high rainfall, are holoendemic. In higher altitudes, endemicity is lower, but epidemics do occur. Chloroquine-resistant *P. falciparum* has spread throughout Africa, accompanied by an increasing incidence of severe clinical forms of the disease. The World Bank estimates that 11% of all disability-adjusted life years (DALYs) lost per year in SSA are from malaria, which places a heavy economic burden on the health systems.

In the Americas, the number of cases detected has risen every year since 1974, and the WHO estimates there to have been 2.2–2.5 million cases in 1991. The nine most endemic countries in the Americas achieved a 60% reduction in malaria mortality between 1994 and 1997.

Southeast Asian region reports some 3.4 million cases of malaria in 1996 and 8000 deaths from TB. This accounts for more than one-third of all non-African malaria cases. There is an increase in resistant strains to the major available drugs and of the mosquitoes to insecticides in use.

Vector control, case finding, and treatment remain the mainstay of control. Use of insecticide-impregnated bed nets and curtains, and residual house spraying, and strengthened vector control activities are important, as are early diagnosis and carefully monitored treatment with monitoring for resistance. Control of malaria will ultimately depend on a safe, effective, and inexpensive vaccine. Attempts to develop a malaria vaccine have been unsuccessful to date due to the large number of genetic types of *P. falciparum* even in localized areas. A Colombian-developed vaccine is being field-tested with partial effectiveness. Research in vaccines for malaria has also been hampered by the fact that it is a relatively low priority for vaccine manufacturers because of the minimal potential for financial benefit. Research on malaria concentrates on the pharmacological aspects of the disease because of increasing drug resistance.

In 1998, WHO has initiated a new campaign to "Roll Back Malaria" and maintain the dream of eradication in the future. Effective low technology interventions include community-based case finding, early treatment of good quality, insecticide use, and vector control. The use of community health workers in endemic areas, has shown promising results. Local control and even eradication can be achieved with currently available technology. This requires an integration of public health and clinical approaches with strong political commitment.

Rickettsial Infections

The rickettsia are obligate parasites, i.e., they can only replicate in living cells, but otherwise they have characteristics of bacteria. This is a group of clinically similar diseases, usually characterized by severe headache, fever, myalgia, rash, and capillary bleeding causing damage to brain, lungs, kidneys, and heart. Identification is by serological testing for antibodies, but the organisms can also be cultured in laboratory animals, embryonic eggs, or in cell cultures. The organisms are transmitted by arthropod vectors such as lice, fleas, ticks, and mites. The diseases caused millions of deaths during war and famine periods prior to the advent of antibiotics.

These diseases appear in nature in ways that make them impossible to eradicate, but clinical diagnosis, host protection, and vector control can help reduce the burden of disease and deal with outbreaks that may occur. Public education regarding self-protection, appropriate clothing, tick removal, and localized control measures such as spraying and habitat modification are useful.

Epidemic typhus, first identified in 1836, is due to *Rickettsia prowazekii*. Spread primarily by the body louse, typhus was the cause of an estimated 3 million deaths, i.e., during war and famine, in Poland and the Soviet Union from 1915–1922. Untreated, the fatality rate is 5–40%. Typhus responds well to antibiotics. It is currently largely confined to endemic foci in central Africa, central Asia, eastern Europe, and South America. It is preventable by hygiene and pediculicides such as DDT and lindane. A vaccine is available for exposed laboratory personnel.

Murine typhus is a mild form of typhus due to *Rickettsia typhi,* which is found worldwide and spread in rodent reservoirs. Scrub typhus, also known as Tsutsugamushi or Japanese river fever, is located throughout the Far East and the Pacific islands, and was a serious health problem for U.S. armed forces in the Pacific during World War II. It is spread by the *Rickettsia tsutsugamushi* and has a wide variation in case fatality according to region, organism, and age of patient.

Rocky Mountain spotted fever is a well-known and severe form of tick-borne typhus due to *Rickettsia rickettsii,* occurring in western North America, Europe, and Asia. Q. fever is a tick-borne disease caused by *Coxiella burnetii* and is worldwide in distribution, usually associated with farm workers, in both acute and chronic forms. Regular anti-tick spraying of sheep, cows, and goats helps protect exposed workers. Protective clothing and regular removal of body ticks help protect exposed persons.

Arboviruses (Arthropod-Borne Viral Diseases)

Arthropod-borne viral diseases are caused by a diverse group of viruses which are transmitted between vertebrate animals (often farm animals or small rodents) and people by the bite of blood-feeding vectors such as mosquitoes, ticks, and sandflies and by direct contact with infected animal carcasses. Usually the viruses have the capacity to multiply in the salivary glands of the vector, but some are carried mechanically in their mouthparts.

These viruses cause acute central nervous system infections (meningo-encephalitis), myocarditis, or undifferentiated viral illnesses with polyarthritis and rashes, or severe hemorrhagic febrile illnesses. Arbovirus diseases are often asymptomatic in vertebrates but may be severe in humans. Over 250 antigenetically distinct arboviruses are associated with disease in humans, varying from benign fevers of short duration to severe hemmorhagic fevers. Each has a specific geographic location, vector, clinical, and virologic characteristics. They are of international public health importance because of the potential for spread via natural phenomena and modern rapid transportation of vectors and persons incubating the disease or ill with it, with potential for further spreading at the point of destination.

Encephalitides

Arboviruses are responsible for a large number of encephalitic diseases characterized by mode of transmission and geographic area. Mosquito-borne arboviruses causing encephalitis include Eastern and Western equine, Venezuelan, Japanese, and Murray Hill encephalitides. Japanese encephalitis is caused by a mosquito-borne arbovirus found in Asia and is associated with rice-growing areas. It is characterized by headache, fever, convulsions, and paralysis, with fatality rates in severe cases as high as 60%. A currently available vaccine is used routinely in endemic areas (Japan, Korea, Thailand, India, and Taiwan) and for persons traveling to infected areas. Tick-borne arboviruses causing encephalitis include the Powassan virus, which occurs sporadically in the United States and Canada. Tick-borne encephalitis is endemic in eastern Europe, Scandinavia, and the former Soviet Union. An epidemic of mosquito-borne encephalitis in New York City in 1999 included 54 cases and 6 deaths, due to the West Nile Fever virus, never before found in the United States.

Rift Valley Fever. Rift Valley fever (RVF) is a virus spread by mosquitoes and other insect vectors. It affects animals and humans who are in direct contact with the meat or blood of affected animals. The virus causes a generalized illness in humans with encephalitis, hemorrhages, retinitis and retinal hemorrhage leading to partial or total blindness, and death (1–2%). It also causes universal abortion in ewes and a high percentage of death in lambs.

The normal habitat is in the Rift Valley of eastern Africa (the Great Syrian–African Rift), often spreading to southern Africa, depending on climactic conditions. The primary reservoir and vector is the *Aedes* mosquito, and affected animals serve to multiply the virus which is transmitted by other vectors and direct contact with animal fluids to humans.

An unusual spread of RVF northward to the Sudan and along the Aswan Dam reservoir to Egypt in 1977–1978 caused hundreds of thousands of animal deaths, with 18,000 human cases and 598 deaths. RVF appeared again in Egypt in 1993. This disease is suspected to be one of the ten plagues of Egypt leading to the exodus of the Children of Israel from Egypt during pharaonic-biblical times.

In 1997, an outbreak of RVF in Kenya, initially thought to be anthrax, with hundreds of cases and dozens of deaths, was related to abnormal rainy season and vector conditions. Satellite monitoring of rainfall and vegetation is being used to predict epidemics in Kenya and surrounding countries. Animal immunization, monitoring, vector control, and reduced contact with infected animals can limit the spread of this disease.

Hemorrhagic Fevers

Arboviruses can also cause hemorrhagic fevers. These are acute febrile illnesses, with extensive hemorrhagic phenomena (internal and external), liver damage, shock, and often high mortality rates. The potential for international transmission is high.

Yellow Fever. Yellow fever is an acute viral disease of short duration and varying severity with jaundice. It can progress to liver disease and severe intestinal bleeding. The case fatality rate is <5% in endemic areas, but may be as high as 50% in nonendemic areas and in epidemics. It caused major epidemics in the Americas in the past, but was controlled by elimination of the vector, *Aedes aegypti.* A live attenuated vaccine is used in routine immunization endemic areas and recommended for travelers to infected areas. Determining the mode of transmission and vector control of yellow fever played a major role in the development of public health (see Chapter 1). In 1997, the WHO reported 200,000 cases and 30,000 deaths from yellow fever globally.

Dengue Hemorrhagic Fever. Dengue hemorrhagic fever is an acute sudden onset viral disease, with 3–5 days of fever, intense headache, myalgia, arthralgia,

> ## BOX 4.10 DENGUE FEVER AND DENGUE HEMORRHAGIC FEVER, 1996–1997
>
> Dengue fever, a severe influenza-like illness, and dengue hemorrhagic fever are closely related conditions caused by four distinct viruses transmitted by *Aedes aegypti* mosquitos. Dengue is the world's most important mosquito-borne virus disease. A total of 2,500 million people worldwide are at risk of infection. An estimated 20 million cases occur each year, of whom 500,000 need to be hospitalized. This is a spreading problem, especially in cities in tropical and subtropical areas. Major outbreaks were reported in Colombia, Cuba, and many other locations in 1997.
>
> Source: World Health Organization. 1998. *World Health Report 1998*. Geneva: WHO.

gastrointestinal disturbance, and rash. Hemorrhagic phenomena can cause case fatality rates of up to 50%. Epidemics can be explosive, but adequate treatment can greatly reduce the number of deaths. Dengue occurs in Southeast Asia, the Pacific islands, Australia, West Africa, the Caribbean, and Central and South America. An epidemic in Cuba in 1981 included more than 500,000 cases, and 158 deaths. Vector control of the *A. aegypti* mosquito resulted in control of the disease during the 1950s–1970s, but reinfestation of mosquitoes led to incresased transmission and epidemics in the Pacific Islands, Caribbean, Central and South America in the 1980s and 1990s.

Outbreaks in Vietnam included 370,000 cases in 1987, another 116,000 cases in 1990, and a similar sized outbreak in 1997. Indonesia had over 13,000 cases in 1997 with 240 deaths, and in 1998 over 19,000 cases (January–May) with at least 531 deaths. In 1998, epidemics of dengue were reported in Fiji, the Cook Islands, New Caledonia, and northern Australia.

The WHO estimates 140,000 deaths and 3.1 million cases worldwide in 1997. Monkeys are the main reservoir, and the vector is the *A. aegypti* mosquito. No vaccine is currently available, and management is by vector control.

Other Hemmorhagic Fevers

Lassa Fever. Lassa fever was first isolated in Lassa, Nigeria, in 1969 and is widely distributed in West Africa, with 200,000–400,000 cases and 5000 deaths annually. It is spread by direct contact with blood, urine, or secretions of infected rodents and by direct person-to-person contact in hospital settings. The disease is characterized by a persistent or spiking fever for 2–4 weeks, and may include severe hypotension, shock, and hemorrhaging. The case fatality rate is 15%.

Marburg Disease. Marburg disease is a viral disease with sudden onset of generalized illness, malaise, fever, myalgia, headache, diarrhea, vomiting, rash, and hemorrhages. It was first seen in Marburg, Germany, in 1967, following exposure to green monkeys. Person-to-person spread occurs via blood, secretions, organs, and semen. Case fatality rates can be over 50%.

Ebola Fever.   Ebola fever is a viral disease with sudden onset of generalized illness, malaise, fever, myalgia, headache, diarrhea, vomiting, rash, and hemorrhages. It was first found in Zaire and Sudan in 1976 in outbreaks which killed more than 400 persons. It is spread from person to person by the blood, vomitus, urine, stools, and other secretions of sick patients, with a short incubation period. The disease has case fatality rates of up to 90%. An outbreak of Ebola among laboratory monkeys in a medical laboratory near Washington, D.C., was contained with no human cases. The reservoir for the virus is thought to be rodents.

An outbreak of Ebola in May 1995 in the town of Kikwit, Zaire, killed 245 persons out of 316 cases (78% case fatality rate). This outbreak caused international concern that the disease could spread, but it remained localized. Another outbreak of Ebola virus occurred in Gabon in early 1996, with 37 cases, 21 of whom had direct exposure to an infected monkey, the remainder by human-to-human contact, or not established; 21 of the cases died (57%). This disease is considered highly dangerous unless outbreaks are effectively controlled. In Zaire, lack of basic sanitary supplies, such as surgical gloves for hospitals, almost ensures that this disease will spread when it recurs.

Lyme Disease

Lyme disease is characterized by the presence of a rash, musculoskeletal, neurologic, and cardiovascular symptoms. Confirmation is by laboratory investigation. It is the most common vector-borne disease in the United States, with 33,000 cases reported between 1993 and 1995. It primarily affects children in the 5–14 age group and adults aged 30–49. Lyme disease is preventable by avoiding contact with ticks, by applying insect repellant, wearing long pants and long sleeves in infected areas, and by the early removal of attached ticks. Several U.S. manufacturers produced vaccines which are approved for animal and human use.

BOX 4.11 LYME DISEASE

In the mid 1970s, a mother of two young boys who were recently diagnosed with arthritis in the town of Lyme, Connecticut, conducted a private investigation among other town residents. She mapped each of the six arthritis cases in the town, cases which had occurred in a short time span among boys living in close proximity. This suggested that this syndrome of "juvenile rheumatoid arthritis" was perhaps connected with the boys playing in the woods. She presented her data to the head of Rheumatology at Yale Medical School in New Haven, who investigated this "cluster of a new disease entity." Some parents reported that their sons had experienced tick bites and a rash before onset of the arthritis. A tick-borne, spiral shaped bacterium, a spirochete, *Borrelia burgdorferi,* was identified as the organism, and ticks shown to be the vector. Cases repond well to antibiotic therapy.

In 1996 over 16,000 cases (6.2 per 100,000) were reported from 45 states, an increase from 11,000 in 1965 and 13,000 in 1994. Cases were mainly located in the northeast, north central, and mid-Atlantic regions. The disease accounts for over 90% of vector-borne disease in the United States and was the ninth leading reported infection in 1995. Lyme disease has been identified in many parts of North America, Europe, the former Soviet Union, China, and Japan. A newly licensed vaccine is effective for people exposed to ticks but not general usage. Personal hygiene for protection from ticks and environmental modification are important to limit spread of the disease.

Source: CDC, 1996, *MMWR,* 45:481–484; and CDC, 1997, *MMWR,* 46, no. 23. Lyme disease website http://www.cdc.gov/ncidad/disease/lyme/lyme.htm

PARASITIC DISEASES

Medically important parasites are animals that live, take nourishment, and thrive in the body of a host, which may or may not harm the host, but never brings benefit. They include those caused by unicellular organisms such as protozoa, which include amoebas (malaria, schistosomiasis, amebiasis, and cryptosporidium), and helminths (worms), which are categorized as nematodes, cestodes, and trematodes.

Public health continues to face the problems of parasitic diseases in the developing world. Increasingly, parasitic diseases are being recognized in industrialized countries. Giardiasis and cryptosporidium infections in waterborne and other outbreaks have occurred in the United States. Parasitic diseases are among the most common causes of illness and death in the world, e.g., malaria. Milder illnesses such as giardiasis and trichomoniasis cause widespread morbidity. Intestinal infestations with worms may cause of severe complications, although they commonly cause chronic low-grade symptomatology and iron deficiency anemia.

Echinococcosis

Echinococcosis (hydatid cyst disease) is infection with *Echinococcus granulosus,* a small dog tapeworm. The tapeworm forms unilocular (single, noncompartmental) cysts in the host, primarily in the liver and lungs, but they can also grow in the kidney, spleen, central nervous system, or in bones. Cysts, which may grow up to 10 cm in size, may be asymptomatic or, if untreated, may cause severe symptoms and even death. This parasite is common where dogs are used with herd grazing animals and also have intimate contact with humans.

The Middle East, Greece, Sardinia, North Africa, and South America are endemic areas, as are a few areas in the United States and Canada. The human disease has been eliminated in Cyprus and Australia. While the dog is the major host, intermediate hosts include sheep, cattle, pigs, horses, moose, and wolves. Preventive measures include education in food and animal contact hygiene, destroying wild and stray dogs, and keeping dogs from the viscera of slaughtered animals.

A similar, but multilocular, cystic hydatid disease is widely found in wild animal hosts in areas of the northern hemisphere, including central Europe, the former Soviet Union, Japan, Alaska, Canada, and the north-central United States. Another echinococcal disease (*Echinococcus vogeli*) is found in South America, where its natural host is the bush dog and its intermediate host is the rat. The domestic dog also serves as a source of human infection.

Surgical resection is not always successful, and long-term medical treatment may be required. Control is through awareness and hygiene as well as the control of wild animals that come in contact with humans and domestic animals. Control may require cooperation between neighboring countries.

Tapeworm

Tapeworm infestation (taeniasis) is common in tropical countries where hygienic standards are low. Beef (*Taenia saginata*) and pork (*T. solium*) tapeworms are common where animals are fed with water or food exposed to human feces. Freezing or cooking meat will destroy the tapeworm. Fish tapeworm (*Diphyllobothrium latum*) is common in populations living primarily on uncooked fish, such as Inuit people. These tapeworms are usually associated with northern climates.

Toddlers are especially susceptible to dog tapeworm (*Dipylidium caninum*), which is present worldwide, and domestic pets are often the source of oral–fecal transmission of the eggs. The disease is usually asymptomatic. Similarly, dwarf tapeworm (*Hymenolepis nana*) is transmitted through oral–fecal contamination from person to person, or via contaminated food or water. Rat tapeworm (*Hymenolepis diminuta*) also mostly affects young children.

Onchocerciasis

Onchocerciasis (river blindness) is a disease caused by a parasitic worm, which produces millions of larvae that move through the body causing intense itching, debilitation, and eventually blindness. The disease is spread by a blackfly that transmits the larva from infected to uninfected people. It is primarily located in sub-Saharan Africa and in Latin America, with over 120 million persons at risk. Control is by a combination of activities including environmental control by larvicidal sprays to reduce the vector population, protection of potential hosts by protective clothing and insect repellents, and case treatment.

A WHO-initiated program for onchocerciasis control started in 1974 is sponsored by four international agencies: the Food and Agriculture Organization (FAO), the United Nations Development Program (UNDP), the World Bank, and WHO. It covers 11 countries in sub-Saharan Africa, focusing on control of the blackfly by destoying its larvae, mainly via insecticides sprayed from the air. Prevalence in 1997 was reported by WHO as over 17 million persons.

The program has been successful in protecting some 30 million persons and helping 1.5 million infected persons to recover from this disease. WHO estimates that the program will have prevented 500,000 cases of blindness by the year 2000 and has freed 25 million hectares of land for resettlement and cultivation. The program cost $570 million. This investment is considered by the World Bank to have a return of 16–28% in terms of large scale land reuse and improved output of the population. A WHO program, the African Program for Onchocerciasis Control (APOC), started in 1996, uses a new drug (Ivermectin) and selective vector control efforts by spraying. This involves 30 countries in Africa, and 6 in a similar program in south America. See website http://www/who.int/ocp and is financed by many donor countries, internation organizations, Merck & Company, and NGOs.

Dracunculiasis

Dracunculiasis (Guinea worm disease) is a parasitic disease of great public health importance in India, Pakistan, and Central and West Africa. It is an infection of the subcutaneous and deeper tissues caused by a large (60 cm) nematode, usually affecting the lower extremities and causing pain and disability. The nematode causes a burning blister on the skin when it is ready to release its eggs. After the blister ruptures, the worm discharges larvae whenever the extremity is in water. The eggs are ingested in contaminated water and the larva released migrate through the viscera to locate as adults in the subcutaneous tissue of the leg. Incubation is about 12 months. The larva released in water are ingested by minute crustaceans and remain infective for as long as a month.

Prevention is based on improving the safety of water supplies and by preventing contamination by infected persons. Education of persons in endemic areas to stay out of water sources and to filter drinking water reduces transmission. Insecticides remove the crustaceans. Chlorine also kills the larvae and the crustaceans which prologue larval infectivity. There is no vaccine. Treatment is helpful, but not definitive.

Dracunculiasis was traditionally endemic in a belt from West Africa through the Middle East to India and central Asia. It was successfully eliminated from central Asia and Iran and has disappeared from the Middle East and from some African countries (Gambia and Guinea).

The World Health Organization has promoted the eradication of dracunculiasis. Major progress has been made in this direction. Worldwide prevalence is reported to have been reduced from 12 million cases in 1980 to 3 million in 1990, 152,814 in 1996, and 77,863 cases in 1997. Eradication was anticipated for the year 2000, and in 1995 the WHO established a commission to monitor and certify eradication in formerly endemic areas. India's reported cases fell from 17,000 in 1987 to 900 in 1992, and the country was free of transmission in 1997. In 1997, formerly high prevalence countries such as Kenya reported no cases in 1997, while Chad, Senegal, Cameroons, Yemen, and the Central African Republic less than 30 cases each. Eradication of this disease appears to be imminent.

The WHO eradication program was developed successfully as an independent program with its own direction and field staff, but further progress will require the integration of this program with other basic primary care programs in order to be self-sustaining as an integral part of community health. Community-based surveillance systems for this disease are being converted to work for monitoring of other health conditions in the community.

Schistosomiasis

Schistosomiasis (snail fever or bilharziasis) is a parasitic infection caused by the trematode (blood fluke) and transmitted from person to person via an intermediate host, the snail. It is endemic in 74 countries in Africa, South America, the Caribbean, and Asia. There are an estimated 200 million persons infected worldwide and more than 600 million at risk for the disease. The clinical symptoms include fever, nausea, vomiting, abdominal pain, diarrhea, and hematuria. The organisms *Schistosoma mansoni* and *S. japonicum* cause intestinal and hepatic symptoms, including diarrhea and abdominal pain. *Schistosoma haematobium* affects the genitourinary tract, causing chronic cystitis, pyelonephritis, with high risk for bladder cancer the ninth most common cause of cancer deaths globally. Infection is acquired by skin contact with freshwater containing contaminated snails. The cercariae of the organism penetrate the skin, and in the human host it matures into an adult worm that mates and produces eggs. The eggs are disseminated to other parts of the body from the worm's location in the veins surrounding the bladder or the intestines, and may result in neurological symptoms.

Eggs may be detected under microscopic examination of urine and stools. Sensitive serologic tests are also available. Treatment is effective against all three major species of schistosomiasis. Eradication of the disease can be achieved with the use of irrigation canals, prevention of contamination of water sources by urine and feces of infected persons, treatment of infected persons, destruction of snails, and health education in affected areas. Persons exposed to freshwater lakes, streams, and rivers in endemic areas should be warned of the danger of infection. Mass chemotherapy in communities at risk and improved water and sanitation facilities are resulting in improved control of this disease.

Leishmaniasis

Leishmaniasis causes both cutaneous and visceral disease. The cutaneous form is a chronic ulcer of the skin, called by various names, e.g., rose of Jericho, oriental sore, and Aleppo boil. It is caused by *Leishmania tropica, L. brasiliensis, L. mexicana,* or the *L. donovani* complex. This chronic ulcer may last from weeks to more than a year. Diagnosis is by biopsy, culture, and serologic tests. The organism multiplies in the gut of sandflies (*Phlebotomus* and *Lutzomi*) and is transmitted to humans, dogs, and rodents through bites. The parasites may remain in the untreated lesion for 5–24 months, and the lesion does not heal until the parasites are eliminated.

Prevention is through limiting exposure to the phlebotomines and reducing the sandfly population by environmental control measures. Insecticide use near breeding places and homes has been successful in destroying the vector sandflies in their breeding places. Case detection and treatment reduce the incidence of new cases. There is no vaccine, and treatment is with specific antimonials and antibiotics.

Visceral Leishmaniasis

Visceral leishmaniasis (kala azar) is a chronic systemic disease in which the parasite multiplies in the cells of the host's visceral organs. The disease is characterized by fever, the enlargement of the liver and spleen, lymphadenopathy, anemia, leukopenia, and progressive weakness and emaciation. Diagnosis is by culture of the organism from biopsy or aspirated material, or by demonstration of intracellular (Leishman–Donovan) bodies in stained smears from bone marrow, spleen, liver, or blood.

Kala azar is a rural disease occurring in the Indian subcontinent, China, the southern republics of the former U.S.S.R., the Middle East, Latin America, and sub-Saharan Africa. It usually occurs as scattered cases among infants, children, and adolescents. Transmission is by the bite of the infected sandfly with an incubation period of 2–4 months. There is no vaccine, but specific treatment is effective and environmental control measures reduce the disease prevalence. This includes the use of antimalarial insecticides. In localities where the dog population has been reduced, the disease is less prevalent.

Trypanosomiasis

Sleeping Sickness. Sleeping sickness a disease caused by *Trypanosoma brucei*, transmitted but the tsetse fly, primarily in the African savannahs, affecting cattle and humans. Some 55 million persons are at risk in sub-Saharan Africa. WHO reported 200,000 new cases, a total prevalence of 300,000 cases, and 150,000 deaths from this disease in 1996. Prevention depends on vector control, and effective treatment of human cases.

Chagas Disease (American trypanosomiasis)

Chagas disease is a chronic and incurable vector and blood transfusion borne parasitic disease (*Trypanosoma cruzi*) which causes disability and death. It affects some 17 million persons mainly in Latin America, with some 300,000 new cases and 45,000 deaths occurring annually. About 30% of affected persons develop severe heart disease. Brazil, which accounts for 40% of the cases prevalent in Latin America, achieved elimination of transmission in 1998, after Uruguay (1996) and Venezuela (1997) and followed by Argentina (1999). Elimination of transmission is projected by WHO by the year 2010.

Control is difficult, but control measures include reducing the animal host and vector insect population in its habitat by ecological and insectiside measures, education of the population in prevention by clothing, bednets, and repellents, and with chemotherapy for case management.

Other Parasitic Diseases

Amebiasis. Amebiasis is an infection with a protozoan parasite (*Entamoeba histolytica*) which exists as an infective cyst. Infestation may be asymptomatic or cause acute, severe diarrhea with blood and mucus, alternating with constipation.

Amebic colitis can be confused with ulcerative colitis. Diagnosis is by microscopic examination of fresh fecal specimens showing trophozoites or cysts. Transmission is generally via ingestion of fecal-contaminated food or water containing cysts, or by oral–anal sexual practices. Amebiasis is found worldwide. Sand filtration of community water supplies removes nearly all cysts. Suspect water should be boiled. Education regarding hygienic practices with safe food and water handling and disposal of human feces are the basis for control.

Ascariasis. Ascariasis is infestation of the small intestine with the roundworm *Ascaris lumbricoides,* which may appear in the stool, occasionally the nose or mouth, or may be coughed up from lung infestation. The roundworm is very common in tropical countries, where infestation may reach or exceed 50% of the population. Children aged 3–8 years are especially susceptible. Infestation can cause pulmonary symptoms and frequently contributes to malnutrition, especially iron deficiency anemia. Transmission is by ingestion of infective eggs, common among children playing in contaminated areas, or via the ingestion of uncooked products of infected soil. Eggs may remain viable in the soil for years. Vermox and other treatments are effective. Prevention is through education, adequate sanitary facilities for excretion, and improved hygienic practices, especially with food. Use of human feces for fertilizer, even after partial treatment, may spread the infestation. Mass treatment is indicated in high prevalence communities.

Pinworm Disease or Enterobiasis. Pinworm disease (oxyuriasis) is common worldwide in all socioeconomic classes; however, it is more widespread when crowded and unsanitary living conditions exist. The *Enterobius vermicularis* infestation of the intestine may be symptomless or may cause severe perianal itching or vulvovaginitis. It primarily affects schoolchildren and preschoolers. More severe complications may occur. Adult worms may be seen visually or identified by microscopic examination of stool specimens or perianal swabs. Transmission is by the oral–fecal ingestion of eggs. The larvae grow in the small intestine and upper colon. Prevention is by educating the public regarding hygiene and adequate sanitary facilities, as well as by treating cases and investigating contacts. Treatment is the same as for ascariasis. Mass treatment is indicated in high prevalence communities.

Ectoparasites. Ectoparasites include scabies (*Sarcoptes scabiei*), the common bed bug (*Cimex lectularius*), fleas, and lice, including the body louse (*Pediculus humanis*), pubic louse (*Phthirius pubis*), and the head louse (*Pediculus humanus capitis*). Their severity ranges from nuisance value to serious public health hazard. Head lice are common in schoolchildren worldwide and are mainly a distressing nuisance. The body louse serves as a vector for epidemic typhus, trench fever, and louse-borne relapsing fever. In disaster situations, disinfection and hygienic practices may be essential to prevent epidemic typhus. The flea plays an important role in the spread of the plague by transmitting the organism from the rat to humans. Control of rats has reduced the flea population, but during war and disasters, rat and flea populations may thrive. Scabies, which is caused by a mite, is common worldwide and is transmitted from person to person. The mite burrows under the skin and causes intense itching. All of these ectoparasites are preventable by proper hygiene and the treatment of cases. The spread of these diseases is rapid and therefore warrants attention in school health and public health policy.

LEGIONNAIRE'S DISEASE

Legionnaire's disease (Legionnellosis) is an acute bacterial disease caused by *Legionnelae*, a gram-negative group of bacilli, with 35 species and many serogroups. The first documented case was reported in the United States in 1947, and the first disease outbreak was reported in the United States in 1976 among participants of a war veterans convention. General malaise, anorexia, myalgia, and headache are followed by fever, cough, abdominal pain, and diarrhea. Pneumonia followed by respiratory failure may follow. The case fatality rate can be as high as 40% of hospitalized cases. A milder, nonpneumonic form of the disease (Pontiac fever) is associated with virtually no mortality.

The organism is found in water reservoirs and is transmitted through heating, cooling, and air conditioning systems, as well as from tap water, showers, saunas, and jaccuzzi baths. The disease has been reported in Australia, Canada, South America, Europe, Israel, and on cruise ships. Prevention requires the cleaning of water towers and cooling systems, including whirlpool spas. Hyperchlorination of water systems and the replacement of filters is required where cases and/or organisms have been identified. Antibiotic treatment with erythromycin is effective.

LEPROSY

Leprosy (Hansen's disease) was widely prevalent in Europe and Mediterranean countries for many centuries, with some 19,000 leprosaria in the year 1300. Leprosy was largely wiped out during the Black Death in the fourteenth century, but continued in endemic form until the twentieth century. Leprosy is a chronic bacterial infection of the skin, peripheral nerves, and upper airway. In the lepromatous form, there is diffuse infiltration of the skin nodules and macules, usually bilateral and extensive. The tuberculoid form of the disease is characterized by clearly demarcated skin lesions with peripheral nerve involvement. Diagnosis is based on clinical examination of the skin and signs of peripheral nerve damage, skin scrapings, and skin biopsy.

Transmission of the *Mycobacterium leprae* organism is by close contact from person to person, with incubation periods of between 9 months and 20 years (average of 4–8 years). Rifampicin and other medications make the patient noninfectious in a short time, so that ambulatory treatment is possible. Multidrug therapy (MDT) has been shown to be highly effective in combating the disease, with a very low relapse rate. Treatment with MDT ensures that the bacillus does not develop drug resistance. MDT is covering 91% of known cases in 1996, according to WHO reports, as compared to only 55% in 1994. The increase has been associated with improved case finding. BCG may be useful in reducing tuberculoid leprosy among contacts. Investigation of contacts over 5 years is recommended.

The disease is still highly endemic primarily in five countries, India, Brazil, Indonesia, Myanmar, and Bangladesh, and is still present in some 80 countries in Southeast Asia, including the Philippines and Burma, sub-Saharan Africa, the Middle East (Sudan, Egypt, Iran), and in some parts of Latin America (Mexico, Colombia) with isolated cases in the United States. World prevalence has declined from 10.5 million cases in 1980, 5.5 million in 1990, to less than 1 million cases in 1995. The World Health Organization expects to eliminate leprosy as a public health problem by the year 2000, defined as prevalence of less than 1 per 10,000 population, or less than 300,000 cases.

TRACHOMA

Trachoma is currently responsible for 6 million blind persons or 15% of total blindness in the world. The causative organism, *Chlamydia trachomatis,* is a bacteria which can survive only within a cell. It is spread through contact with eye discharges, usually by flies, or household items (e.g., handkerchiefs, washcloths). Trachoma is common in poor rural areas of Central America, Brazil, Africa, parts of Asia, and some countries in the eastern Mediterranean. The resulting infection leads to conjunctival scarring and if untreated, to blindness. WHO estimates there are 148 million cases of active disease in 46 endemic countries. Hygiene, vector control, and treatment with antibiotic eye ointments or simple surgery for scarring of eyelids and inturned eyelashes prevent the blindness. A new drug, azithromycin, is effective in curing the disease. The WHO is promoting a program for the global elimination of trachoma using azithromycin and hygiene education in endemic areas.

Chlamydia (*Chlamydia pneumonia*) is suspected of playing a role in coronary artery disease by intraarterial infection, with plaque formation and occlusion of the artery by thrombi consisting mainly of platelets. If borne out, this will provide potential for low cost intervention to reduce the burden of the leading worldwide cause of death.

SEXUALLY TRANSMITTED DISEASES

Sexually transmitted diseases (STDs) are widespread internationally with an estimated 330 million new cases per year, with 5.8 million new cases, over 30 million total cases, and 2.3 million deaths (1997), AIDS has captured world attention over the past decade. The global burden of STDs is enormous (Table 4.8), and the public health and social consequences are devastating in many countries.

TABLE 4.8   Estimated Worldwide Incidence of Major Sexually Transmitted Diseases, 1997

Disease	Organism responsible	New cases
Trichomoniasis	*Trichomonas vaginalis*	170 million
Chlamydia, genital	*Chlamydia trachomatis*	89 million
Gonorrhea	*Neisseria gonorrhoea*	62 million
Genital papilloma	Human papilloma virus	30 million
Anogenital herpes	Herpes simplex virus	20 million
Syphilis	*Treponema pallidum*	12 million
HIV	*Human immunodeficiency virus* (HIV)	5.8 million
Chancroid	Haemophilus ducreyi	2 million

Source: WHO, 1998, *World Health Report 1998.*

Sexually transmitted diseases, especially in women, may be asymptomatic, so that severe sequelae may occur before patients seek care. Infection by one STD increases risk of infection by other diseases in this group.

Syphilis

Syphilis is caused by the spirochete *Treponema pallidum.* After an incubation period of 10–90 days (mean = 21), primary syphilis develops as a painless ulcer or chancre on the penis, cervix, nose, mouth, or anus, lasting 4–6 weeks. The patient may first present with secondary syphilis 6–8 weeks (up to 12 weeks) after infection with a general rash and malaise, fever, hair loss, arthritis, and jaundice. These symptoms spontaneously disappear within weeks or up to 12 months later. Tertiary syphilis may appear 5–20 years after initial infection. Complications of tertiary syphilis include catastrophic cardiovascular and central nervous system conditions. Early antibiotic treatment is highly effective when given in a large initial dose, but longer term therapy may be needed if treatment is delayed.

Gonorrhea

Gonorrhea (GC) is caused by the bacterium *Neisseria gonorrhoeae.* The incubation period is 1–14 days. Gonorrhea is often associated with concurrent chlamydia infection. In women, GC may be asymptomatic or it may cause vaginal discharge, pain on urination, bleeding on intercourse, or lower abdominal pain. Untreated, it can lead to sterility. In men, GC causes urethral discharge and painful urination. Treatment with antibiotics ends infectivity, but untreated cases can be infectious for months. Drug resistance to penicillin and tetracycline has increased in many countries so that more expensive and often unavailable drugs are necessary for treatment. Prevention of gonococcal eye infection in newborns is based on routine use of antibiotic ointments in the eyes of newborns.

Other Sexually Transmitted Diseases

Chancroid.   Chancroid is caused by *Haemophilus ducreyi.* In women chancroids may cause a painful, irregular ulcer near the vagina, resulting in pain on intercourse, urination, and defection, but it may be asymptomatic. In men it causes a painful, irregular ulcer on the penis. The incubation period is usually 3–5 days, but may be up to 14 days. An individual is infectious as long as there are ulcers, usually 1–3 months. Treatment is by erythromycin or azithromycin.

Herpes Simplex.   Herpes simplex is caused by herpes simplex virus types 1 and 2 and has an incubation period of 2–12 days. Genital herpes causes painful blisters around the mouth, vagina, penis, or anus. The genital lesions are infectious for 7–12 days. Herpes may lead to central nervous system meningoencephalitis infection. It can be transmitted to newborns during vaginal delivery, causing infection, encephalitis, and death. Cesarian delivery is therefore necessary when a mother is infected. Anti-viral drugs are used in treatment, orally, topically, or intravenously.

Chlamydia.   Chlamydia is caused by *Chlamydia trachomatis.* In women, it is usually asymptomatic but may cause vaginal discharge, spotting, pain on urination, lower abdominal pain, and pelvic inflammatory disease (PID). In newborns, chlamydia may cause eye and respiratory infections. In men, chlamydia causes urethral discharge and pain on urination. The incubation period is 7–21 days and the infectious period is unknown. Treatment for chlamydia is doxycycline, azithromycin, or erythromycin. Chlamydia infection, not necessarily venereal in transmission, may be transmitted to newborns of infected mothers. *Chlamydia pneumoniae,* presently under investigation as a possible cause or contributor to coronary heart disease, and is widespread in poor hygenic conditions.

Trichomoniasis.   Trichomoniasis is caused by *Trichomonas vaginalis.* The incubation period is 4–20 days (mean = 7). In women, trichomoniasis may be asymptomatic or may cause a frothy vaginal discharge with foul odor, and painful urination and intercourse. In men, the disease is usually mild, causing pain on urination. Treatment is by metronidazole taken orally. Without treatment, the disease may persist and remain infectious for years.

Condyloma.   Condyloma or viral wart is caused by human papilloma virus (HPV). It is a sporadic disease which may be associated with cervical neoplasia and cancer of the cervix. HPV includes many types associated with a variety of conditons. The search for a HPV vaccine to prevent cancer of the cervix looks promising.

Control of Sexually Transmitted Infections

In areas where a full range of diagnostic services is lacking, a "syndromic approach" is recommended for the control of STDs. The diagnosis is based on a group of symptoms and treatment on a protocol addressing all the diseases that could possibly cause those symptoms, without expensive laboratory tests and repeated visits. Early treatment without laboratory confirmation helps to cure persons who might not return for follow-up, or may place them in a noninfective stage so that even without follow-up they will not transmit the disease. STD incidence between 1950 and 1996 is shown in Table 4.9, with decline overall except around 1990, with subsequent further fall in incidence.

Screening in prenatal and family planning clinics, prison medical services, and

TABLE 4.9   Annual Rates (Cases per 100,000 Population) of Reportable Sexually Transmitted Disease,[a] United States, Selected Years 1950–1996

Disease	1950	1960	1970	1980	1985	1990	1996
Syphilis (all stages)	146	69	45	31	29	54	20
Gonorrhea	192	145	297	445	384	278	124
Chancroid	3.3	0.9	0.7	0.4	0.9	1.7	0.1

Source: *Health United States, 1998.*

[a]The increase in syphilis in 1985–1990 and subsequent decline by more than 50% in reported cases includes all three stages of the disease as well as congential syphilis. Rates are cases per 100,000 population, rounded.

in clinics serving prostitutes, homosexuals, or other potential risk groups will detect subclinical cases of various STDs. Treatment can be carried out cheaply and immediately. For instance, the screening test for syphilis costs $0.10 and the treatment with benzathine penicillin injection costs about $0.40 in 1998. Partner notification is a controversial issue, but may be needed to identify contacts who may be the source of transmission to others.

Control of STDs through a syndrome approaach based on primary care providers is being promoted by WHO. Health education directed at high risk target groups is essential. Providing easy and cost-free access to acceptable, nonthreatening treatment is vital in promoting the early treatment of cases and thereby reducing the risk of transmission.

Promoting prevention through the use of condoms and/or monogamy requires long-term educational efforts that are now fostered by the HIV/AIDS pandemic. Increased use of condoms for HIV prevention is associated with reduced risk of other STDs. Training medical care providers in STD awareness should be stressed in undergraduate and continuing educational efforts including personal protection as care givers.

HIV/AIDS

Human immunodeficiency virus (HIV) is a retrovirus that infects various cells of the immune system, and also affects the central nervous system. Two types have been identified: HIV1, worldwide in distribution, and the less pathogenic HIV2, found mainly in West Africa. HIV is transmitted by sexual contact, exposure to blood and blood products, perinatally, and via breast milk. The period of communicability is unknown, but studies indicate that infectiousness is high, both during the initial period after infection and later in the disease. Antibodies to HIV usually appear within 1–3 months.

Within several weeks to months of the infection, many persons develop an acute self-limited flulike syndrome. They may then be free of any signs or symptoms for months to more than 10 years. Onset of illness is usually insidious with nonspecific symptoms, including sweats, diarrhea, weight loss, and fatigue. AIDS represents the later clinical stage of HIV infection. According to the revised CDC case definition (1993), AIDS involves any one or more of the following: low CD_4 count, severe systematic symptoms, opportunistic infections such as pneumocystis pneumonia or TB, aggressive cancers such as Kaposi's sarcoma or lymphoma, and/or neurological manifestations, including dementia and neuropathy. The WHO case definition is more clinically oriented, relying less on often unavailable laboratory diagnoses for indicator diseases.

AIDS was first recognized clinically in 1981 in Los Angeles and New York. By mid-1982 it was considered an epidemic in those and other U.S. cities. It was primarily seen among homosexual men and recipients of blood products. After initial errors, testing of blood and blood products became standard and has subsequently closed off this method of transmission. Transmission has changed markedly since the initial onslaught of the disease, with needle sharing among intravenous drug users, heterosexual, and maternal–fetal transmission becoming major factors. Comorbidity with other STDs apparently increases HIV infectivity and may have helped to convert the epidemiology to a greater degree of heterosexual transmission.

The disease grew exponentially in the United States (Table 4.10), but incidence of new cases nas declined since 1993. AIDS has become a major public health problem in most developed and developing countries, reaching catastrophic proportions in some sub-Saharan African countries affecting up to 30% of the population.

HIV-related deaths were the eighth leading cause of all deaths in 1993 in the U.S., the leading cause among men aged 25–44 years of age, and the fourth leading cause for women in this age group. By 1996, AIDS had been diagnosed in 548,000 persons and 343,000 had died. It is estimated that up to 1 million persons are HIV infected in the United States.

Globally, deaths from AIDS totalled 2.3 million in 1997, with an estimated 11.7 million person having died from this pandemic up to 1997. In 1998, an estimated 3.1 million person were HIV infected with 5.8 million new infection in 1997.

The declining incidence of new cases in the industrialized countries may be the result of greater awareness of the disease and methods of prevention of transmission. Improving early diagnosis and access to care, especially the combined therapy programs that are very effective in delaying onset of symptoms, are important parts of public health management of the AIDS crisis. Until an effective vaccine is available, preventive reliance will continue to be on behavior risk-reduction and other prevention strategies such as needle and condom distribution among high risk population groups.

Throughout the world, HIV continues to spread rapidly, especially in poor countries in Africa, Asia, and South and Central America. The United Nations reports that 21 million persons are living with HIV/AIDS, 90% of them in developing countries, where transmission is 85% by heterosexual contact. Every day, more than 8500 persons are infected, including 1000 children. In Thailand, 1 person in 50 is now infected. In sub-Saharan Africa 1 person in 40 is infected, and in some cities as many as 1 person in 3 carries the virus. Estimations of new infections per

TABLE 4.10   Acquired Immunodeficiency Syndrome (AIDS) Cases (000s), by Age/Sex Group,[a] United States, 1985–1997

| | Cases (000s) in year | | | | | | | | | | | | |
Group	1985	1986	1987	1988	1989	1990	1991	1992	1993	1994	1995	1996	1997[b]
Total	8.2	13.1	21.1	30.7	33.6	41.7	43.7	45.8	102.4	77.4	71.3	66.7	30.2
Males ≥ 13	7.5	12.0	19.1	27.1	29.6	36.3	37.6	39.1	85.4	62.9	57.2	52.8	23.3
Women ≥ 13	0.5	1.0	1.7	3.0	3.4	4.5	5.3	6.0	16.0	13.3	13.1	13.2	6.6
Children < 13[c]	0.1	0.2	0.3	0.6	0.6	0.7	0.7	0.7	0.9	1.0	0.7	0.7	0.3

Source:*Health United States*, 1998.
[a]The CDC expanded its criteria for AIDS case definition in 1985, 1987, and 1993. This accounts for some of the increases in cases (especially among women in 1993).
[b]The year 1997 includes cases reported up to June 30, 1997.
[c]Children includes both sexes up to 13 years of age. Numbers are rounded.

year in sub-Saharan Africa range from 1 to 2 million persons, while in Asia the range is from 1.2 to 3.5 million new infected persons per year.

Lessons are still being learned from the AIDS pandemic. The explosive spread of this infection, from an estimated 100,000 people in 1980 to an anticipated 40 million persons HIV infected, shows that the world is still vulnerable to pandemics of "new" infectious diseases. Enormous movements of tourists, business people, truck drivers, migrants, soldiers, and refugees promote the spread of such diseases. Widespread sexual exchange, traffic in blood products, and illicit drug use all promote the international potential for pandemics. War and massive refugee situations promote rape and prostitution, worsening the AIDS situation in some settings in Africa.

HIV has arrived in almost every country. However, there is the somewhat hopeful indication that the rate of increase, has slowed in the United States. This may be an indication either of higher levels of self-protective behavior, or that the most susceptible population groups have already been affected and the spread into the general population is at a slower rate. It is also possible that this may yet prove to be only a lull in the storm, as heterosexual contact becomes a more important mode of transmission.

The Eleventh International Conference on AIDS, held in Vancouver, Canada, in July 1996, reported signs that combinations of several drugs from among a number of antiretroviral medications are showing promise to suppress the AIDS virus in infected people. At a current annual price of $10,000–15,000 per patient, these sums well beyond the capacity of most developing countries. Development of methods of measuring the HIV viral load have allowed for better evaluation of potential therapies and monitoring of patients receiving therapy. In developed countries, transmission by blood products has been largely controlled by screening tests; transmission among homosexuals has been reduced by safe sex practices; transmission to newborns has been reduced by recent therapeutic advances. Safe sex practices and condom use may have helped in reducing heterosexual transmission. Further advances in therapy and prevention with a vaccine are expected over the next decade.

The HIV/AIDS pandemic is one of the great challenges to public health for the 21st century due to its complexity, its international spread, its sexual and other modes of transmission, its devastating and costly clinical effects, and its impact on parallel diseases such as tuberculosis, respiratory infections, and cancer. The cost of care for the AIDS patient can be very high. Needed programs include home care and community health workers to improve nutrition and self-care, and mutual help among HIV carriers and AIDS patients. The ethical issues associated with AIDS are also complex regarding screening of pregnant women, newborns, partner notification, reporting, and contact tracing, as well as financing the cost of care.

DIARRHEAL DISEASES

Diarrheal diseases are caused by a wide variety of bacteria, parasites, and viruses (Table 4.11) infecting the intestinal tract and causing secretion of fluids and dis-

TABLE 4.11   Major Organism Associated with Diarrheal Diseases

Classification	Organism
Bacteria	*Salmonella, Shigella, Escherichia coli, Vibrio cholerae, Bacillus cereus, Campylobacter jejuni*
Virus	Enteroviruses, rotaviruses, astroviruses, calciviruses, coronaviruses, small round virus group, Norwalk group
Parasites/protozoa	Schistosoma, Giardia lamblia, Cryptosporidium, Entamoeba histolytica

solved salts into the gut with mild to severe or fatal complications. In developing countries, diarrheal diseases account for half of all morbidity and a quarter of all mortality. Diarrhea itself does not cause death, but the dehydration resulting from fluid and electrolyte loss is one of the most common causes of death in children worldwide. Deaths from dehydration can be prevented by use of oral rehydration therapy (ORT), an inexpensive and simple method of intervention easily used by a nonmedical primary care worker and by the mother of the child as a home intervention. In 1983, diarrheal diseases were the cause of almost 4 million child deaths, but by 1996 this had declined to 2.4 million, largely under the impact of increased use of ORT.

Diarrheal diseases are transmitted by water, food, and directly from person to person via oral–fecal contamination. Diarrheal diseases occur in epidemics in situations of food poisoning or contaminated water sources, but can also be present at high levels when common source contamination is not found. Contamination of drinking water by sewage and poor management of water supplies are also major causes of diarrheal disease. The use of sewage for the irrigation of vegetables is a common cause of diarrheal disease in many areas.

Salmonella

Salmonella are a group of bacterial organisms causing acute gastroenteritis, associated with generalized illness including headache, fever, abdominal pains, and dehydration. There are over 2000 serotypes of salmonella, many of which are pathogenic in humans, the most common of which are *Salmonella typhimurium, S. enteritidis,* and *S. typhi.* Transmission is by ingestion of the organisms in food, derived from fecal material from animal or human contamination. Common sources include raw or uncooked eggs, raw milk, meat, poultry and its products, as well as pet turtles or chicks. Fecal–oral transmission from person to person is common. Prevention is in safe animal and food handling, refrigeration, sanitary preparation and storage, protection against rodent and insect contamination, and the use of sterile techniques during patient care. Antibiotics may not eliminate the carrier state and may produce resistant strains.

Shigella

Shigella are a group of bacteria that are pathogenic in man, with four groups: Type A = *Shigella dysenteriae,* Type B = *S. flexneri,* Type C = *S. boydii,* and Type D = *S. sonnei.* Types A, B, and C are each further divided into a total of 40 serotypes. Shigella are transmitted by direct or indirect fecal–oral methods from a patient or carrier, and illness follows ingestion of even a few organisms. Water and milk transmission occurs as a result of contamination. Flies can transmit the organism, and in nonrefrigerated foods the organism may multiply to an infectious dose. Control is in hygienic practices and in the safe handling of water and food.

Escheria coli

E. coli are common fecal contaminants of inadequately prepared and cooked food. Particularly virulent strains such as O157:H17 can cause explosive outbreaks of severe (enterohemmorhagic) diarrhoeal disease with a hemolytic-uremic syndrome and death, as occurred in Japan in 1998 with cases and deaths due to a food-borne epidemic. Other milder strains cause travellers diarrhoea and nursery infections. Inadequately cooked hamburger, unpasturized milk, and other food vectors are discussed under food safety in Chapter 8.

Cholera

Cholera is an acute bacterial enteric disease caused by *Vibrio cholerae,* with sudden onset, profuse painless watery stools, occasional vomiting, and, if untreated, rapid dehydration, and circulatory collapse, and death. Asymptomatic infection or carrier status, and mild cases are common. In severe, untreated cases, mortality is over 50%, but with adequate treatment, mortality is under 1%. Diagnosis is based on clinical signs, epidemiologic, serologic and bacteriologic confirmation by culture. The two types of cholera are the classic and el Tor (with Inaba and Ogawa serotypes).

In 1991, a large scale epidemic of cholera spread through much of South America. It was imported via a Chinese freighter, whose sewage contaminated shellfish in Lima harbor in Peru (Box 4.12). The South American cholera epidemic has caused hundreds of thousands of cases and thousands of deaths since 1991.

Prevention requires sanitation, particularly the chlorination of drinking water, prohibiting the use of raw sewage for the irrigation of vegetable crops, and high standards of community, food, and personal hygiene. Treatment is prompt fluid therapy with electrolytes in large volume to replace all fluid loss. Oral rehydration should be accomplished using standard ORT. Tetracycline shortens the duration of the disease, and chemoprophylaxis for contacts following stool samples may help in reducing its spread. A vaccine is available but is of no value in the prevention of outbreaks.

Viral Gastroenteritis

Viral gastroenteritis can occur in sporadic or epidemic forms, in infants, children, or adults. Some viruses, such as the rotaviruses and enteric adenoviruses, af-

---

BOX 4.12 THE CHOLERA PANDEMIC IN SOUTH AMERICA, 1991–1998

In the 1980s, Peruvian officials stopped the chlorination of community water supplies because of concern over possible carcinogenic effects of trihalomethanes, a view encouraged by officials of the U.S. Environmental Protection Agency (EPA) and the U.S. Public Health Service. In January 1991, a Chinese freighter arrived in Lima, Peru, and dumped bilge (sewage) in the harbor, apparently contaminating local shellfish. Consumption of raw shellfish is a popular local delicacy (ceviche) and associated with cases of cholera seen in local hospitals.

Contamination of local water supplies from sewage resulted in the geometric increase in cases, and by the end of 1992 the Pan American Health Organization (PAHO) reported an epidemic of 391,000 cases and 4002 deaths. The epidemic spread to 21 countries, and in 1992 there were a further 339,000 cases and 2321 deaths spreading over much of South America, continuing in 1999.

In the United States, 102 cases of cholera were reported in 1992; of these, 75 cases and 1 death were among passengers of an airplane flying from South America to Los Angeles in which contaminated seafood was served. In 1993, 91 cases of cholera were reported in the United States which were unrelated to international travel. These occurred mostly among persons consuming shellfish from the Gulf coast with a strain of cholera similar to the South American strain, also possibly introduced in ship ballast. Cholera organisms are reported in harbor waters in other parts of the United States (Promed, 1999).

Sources: Anderson, C. 1991. Cholera epidemic traced to risk miscalculation. *Nature,* 354:255; CDC. 1993. Update cholera—Western hemisphere, 1992. *MMWR,* 42:89–91; CDC. 1993. Isolation of *Vibrio cholerae* O1 from Oysters—Mobile Bay, 1991–1992. *MMWR,* 42:91–93; Promed, 1999.

---

fect mainly infants and young children, and may be severe enough to cause hospitalization for dehydration. Others such as Norwalk and Norwalk-like viruses affect older children and adults in self-limited acute gastroenteritis in family, institution, or community outbreaks.

Rotaviruses. Rotaviruses cause acute gastroenteritis in infants and young children, with fever and vomiting, followed by watery diarrhea and occasionally severe dehydration and death if not adequately treated. Diagnosis is by examination of stool or rectal swabs with commercial immunologic kits. In both developed and developing countries, rotavirus is the cause of about one-third of all hospitalized cases for diarrheal diseases in infants and children up to age 5. Most children in developing countries experience this disease by the age of 4 years, with the majority of cases between 6 and 24 months. In developing countries, rotaviruses are estimated to cause over 800,000 deaths per year. The virus is found in temperate climates in the cooler months and in tropical countries throughout the year. Breast-feeding does not prevent the disease but may reduce its severity. Oral rehydration therapy is the key treatment. A live attenuated vaccine was approved by the FDA in 1998 and adopted in the 1999 U.S. recommended routine vaccination programs for infants.

Adenoviruses. Adenoviruses, Norwalk, and a variety of other viruses (including astrovirus, calcivirus, and other groups) cause sporadic acute gastroenteritis worldwide, mostly in outbreaks. Spread is by the oral–fecal route, often in hospital or other communal settings, with secondary spread among family contacts. Food-borne and waterborne transmission are both likely. These can be a serious problem in disaster situations. No vaccines are available. Management is with fluid replacement and hygienic measures to prevent secondary spread.

Parasitic Gastroenteritis

Giardiasis. Giardiasis (caused by *Giardia lamblia*) is a protozoan parasitic infection of the upper small intestine, usually asymptomatic, but sometimes associated with chronic diarrhea, abdominal cramps, bloating, frequent loose greasy stools, fatigue, and weight loss. Malabsorption of fats and vitamins may lead to malnutrition. Diagnosis is by the presence of cysts or other forms of the organism in stools, duodenal fluid, or in intestinal mucosa from a biopsy. This disease is prevalent worldwide and affects mostly children. It is spread in areas of poor sanitation and in preschool settings and swimming pools, and is of increasing importance as a secondary infection among immunocompromised patients, especially those with AIDS.

Waterborne giardia was recognized as a serious problem in the United States in the 1980s and 1990s, since the protozoa is not readily inactivated by chlorine, but requires adequate filtration before chlorination. Person-to-person transmission in day-care centers is common, as is transmission by unfiltered stream or lake water where contamination by human or animal feces is to be expected. An asymptomatic carrier state is common. Prevention relies on careful hygiene in settings such as day-care centers, filtration of public water supplies and the boiling of water in emergency situations.

Cryptosporidium. Cryptosporidium parvum is a parasitic infection of the gastrointestinal tract in man, small and large mammals and vertebrates. Infection may be asymptomatic or cause a profuse, watery diarrhea, abdominal cramps, general malaise, fever, anorexia, nausea, and vomiting. In immunosuppressed patients, such as persons with AIDS, it can be a serious problem. The disease is most common in children under 2 years of age and those in close contact with them, as well as in homosexual men. Diagnosis is by identification of the cryptosporidium organism cysts in stools. The disease is present worldwide. In Europe and the United States, the organism has been found in <1 to 4.5% of individuals sampled. Spread is common by person-to-person contact by fecal–oral contamination, especially in such settings as day-care centers. Raw milk and waterborne outbreaks have also been identified in recent years. A large waterborne disease outbreak due to cryptosporidium occurred in Milwaukee in 1986 described in Chapter 9. Management is by rehydration and prevention is by careful hygiene in food and water safety.

Helicobacter pylori. *Helicobacter pylori,* first identified in 1986, is a bacterium causally linked to duodenal ulcers and gastritis, contributing to high rates of gastric cancer (Chapter 5). It is an important example of the link between infection and chronic disease. This has enormous implications for prevention of cancer of the stomach, chronic peptic ulcers and large-scale use of hospitals and other medical resources (see Chapter 5).

A Program Approach to Diarrhoel Disease Control

The control of diarrheal diseases requires a comprehensive program involving a wide range of activities, including good management of food and water supplies, education in hygiene, and, particularly where morbidity and mortality are high, education in the use of Oral Rehydration Therapy (ORT).

Oral rehydration therapy (ORT) is considered by UNICEF and WHO to have resulted in the saving of 1 million lives each year in the 1990s. Proper management of an episode of diarrhea by ORT (Table 4.12), along with continued feeding, not only saves the child from dehydration and immediate death, but also contributes to early restoration of nutritional adequacy, sparing the child the prolonged effects of malnutrition.

The World Summit for Children (WSC) in 1990 called for a reduction in child deaths from diarrheal diseases by one-third and malnutrition by one-half, with em-

TABLE 4.12   WHO Formula for Oral Rehydration Therapy (ORT)

Ingredients	Amount (g/liter)	Concentration Ion	(mmol/liter)
Sodium chloride (NaCl)	3.5	Sodium	90
Trisodium citrate, dihydrate, or sodium bicarbonate (NaHCO$_3$)	2.9 (or 2.5)	Citrate[a]	20 citrate[b]
Potassium chloride (KCl)	1.5	Potassium	10 of potassium 80 of chloride
Glucose (anhydrous)	20.0	Glucose	111

Source: World Health Organization, 1992, *Readings on Diarrhoea: Student Manual;* Benenson, 1995, *Control of Communicable Diseases Manual.*

[a]Or 2.5 g sodium bicarbonate.

[b]Or 30 mmol bicarbonate.

phasis on the widest possible availability, education for, and use of ORT. This requires a programmatic approach. Public health leadership must train primary care doctors, pediatricians, pharmacists, drug manufacturers, and primary care health workers of all kinds in ORT principles and usage. They must be backed by the widest possible publicity to raise awareness among parents.

Oral rehydration therapy is an important public health modality in developed countries as well as in developing countries. Diarrhoeal disease may not cause death as frequently in developed countries, but it is still a significant factor in infant and child health and, even under the most optimal conditions, can cause setbacks in the nutritional state and physical development of a child. Use of ORT does not prevent the disease (i.e., it is not a primary prevention), but it is excellent in secondary prevention, by preventing complications from diarrhoea, and should be available in every home for symptomatic treatment of diarrheal diseases.

An adaptation of ORT has found its place in popular culture in the United States. A form of ORT, marketed as "sports drinks," is used in sports where athletes lose large quantities of water and salts in sweat and insensible loss from the respiratory tract. The wider application of the principles of ORT for use in adults in dry hot climates and in adults under severe physical exertion with inadequate fluid/salt intake situations requires further exploration.

Management of diarrheal diseases should be part of a wider approach to child nutrition. The child who goes through an episode of diarrheal disease may have a faltering in growth and development. Supportive measures may be needed following the episode as well as during it. This involves providing primary care services that are attuned to monitoring individual infant and child growth. Growth monitoring surveillance is important to assess the health status of the individual child and the child population. Supplementation of infant feeding with vitamins A and D, and iron to prevent anemia are important for routine infant and child care, and more so for conditions affecting total nutrition such as a diarrheal disease.

ACUTE RESPIRATORY INFECTIONS

In the developing world, respiratory infections account for over one-quarter of all deaths and illnesses in children. As diarrheal disease deaths are reduced, the major cause of death among infants in developing countries is becoming acute respiratory infections (ARIs). In industrialized countries, ARIs are important for their potentially devastating effects on the elderly and chronically ill. They are also the major cause of morbidity in infants in developed countries, causing much anxiety to parents even in areas with good living conditions. Cigarette smoking, chronic bronchitis, poorly controlled diabetes or congestive heart failure, and chronic liver and kidney disease increase susceptibility to ARIs. ARIs place a heavy burden on health care systems and individual families. Improved methods of management of such chronic diseases are needed to reduce the associated toll of morbidity, mortality, and the considerable expenses of health care.

Acute respiratory infections are due to a broad range of viral and, to a lesser extent, bacterial infections. It is the latter which can progress to pneumonia with mortality rates of 10–20%. Acute viral respiratory diseases include those affecting the upper respiratory tract, such as acute viral rhinitis, pharyngitis, and laryngitis, as well as those affecting the lower respiratory tract, tracheobronchitis, bronchitis, bronchiolitis, and pneumonia. ARIs are frequently associated with vaccine-preventable diseases, including measles, varicella, and influenza. They are caused by a large number of viruses, producing a wide spectrum of acute respiratory illness. Some organisms affect any part of the respiratory tract, while others affect specific parts and all predispose to bacterial secondary infection. While children and the elderly are especially susceptible to morbidity and mortality from acute respiratory disease, the vast numbers of respiratory illnesses among adults cause large-scale economic loss from work absence.

Bacterial agents causing upper respiratory tract infection include group A streptococcus, mycoplasma pneumonia, pertussis, and parapertussis. Pneumonia or acute bacterial infection of the lower respiratory tract and lung tissue may be due to pneumococcal infection with *Streptococcus pneumoniae*. There are 83 known types of this organism, distinguished by capsule characteristics; 23 account for 88% of pneumococcal infections in the United States. An excellent polyvalent vaccine based on these types is available for high risk groups such as the elderly, immunodeficient patients, and persons with chronic heart, lung, liver, blood disorders, or diabetes.

Opportunistic infections attack the chronically ill, especially those with compromised immune suystems, often with life-threatening ARIs. Mycoplasma (primary atypical pneumonia) is a lower respiratory tract infection which sometimes progresses to pneumonia. TB and *Pneumonocytis carynia* are especially problematic for AIDS patients. Other organisms causing pneumonias include *Chlamydia pneumoniae, H. influenza, klebsiella pneumonia, Escherichia coli,* Staphylococcus, rickettsia (Q fever), and *Legionella.* Parasitic infestation of lungs may occur with nematodes (e.g., ascariasis). Fungal infections of the lung may be caused by aspergillosis, histoplasmosis, and coccidiomycosis, often as a complication of antibiotic therapy.

Access to primary care and early institution of treatment are vital to control excess mortality from ARIs. In developed countries, ARIs as contributors to infant deaths are largely a problem in minority and deprived population groups. Because these groups contribute disproportionately to childhood mortality, infant mortality reduction has been slower in countries such as the United States and Russia than in other industrialized countries. The continuing gap in mortality rates between white and black children in the United States can, to a large extent, be attributed to ARIs and less access to organized primary care. Children are brought to emergency rooms for care when the disease process is already advanced and more dangerous than had it been attended to professionally earlier in the process. Many field trials of ARI prevention programs have been proved successful involving parent education and training of primary care workers in early assessment and, if necessary, initiation of treatment. This needs field testing in multiple settings.

Reliance on vaccines to prevent respiratory infectious diseases is not currently feasible. ARIs are caused by a very wide spectrum of viruses, and the development of vaccines in this field has been slow and limited. The vaccine for pneumococcal pneumonia has been an important breakthrough, but it is still inadequately utilized by the chronically ill because of its limitations, costs, and lack of sufficient awareness, and it is too expensive for developing countries. Improvements in bacterial and viral vaccine development will potentially help to reduce the burden of ARIs. A programmatic approach with clinical guidelines and education of family and care givers is currently the only feasible way to reduce the still enormous morbidity and mortality from ARIs on the young and the elderly.

COMMUNICABLE DISEASE CONTROL IN THE NEW PUBLIC HEALTH

The success of sanitation vaccines and antibiotics led many to assume that all infectious diseases would sooner or later succumb to public health and medical technology. Unfortunately, this is a premature and even dangerous assumption. Despite the longstanding availability of an effective and inexpensive vaccine, the persistence of measles as a major killer of 1 million children per year represents a failure in effective use of both the vaccine and the health system. The resurgence of TB and malaria have led to new strategies, such as managed or directly observed care, with community health workers to assure compliance needed to render the patient noninfectious to others and to reduce the pool of carriers of the disease.

Current successes in reducing poliomyelitis, dracunculiasis, onchocerciasis, and other diseases to the point of eradication has raised hopes for similar success in other fields. But there are many infectious diseases of importance in developed and developing countries where existing technologies are not fully utilized. Oral rehydration therapy (ORT) is one of the most cost-effective methods of preventing excess mortality from ordinary diarrheal diseases, and yet is not used on sufficient scale.

Biases in the financing and management of medical insurance programs can result in underutilization of available effective vaccines. Hospital-based infections cause large-scale increases in lengths of stay and expenditures, although application of epidemiologic investigation and improved quality in hospital practices could reduce this burden. Control of the spread of AIDS using combined medical therapies is not financially or logistically possible in many countries, but education for "safe sex" is effective. Community health worker programs can greatly enhance tuberculosis, malaria, and STD control, or in AIDS care, promote prevention and appropriate treatment.

In the industrialized and mid-level developing countries, epidemiologic and demographic shifts have created new challenges in infectious disease control. Prevention and early treatment of infectious disease among the chronically ill and the elderly is not only a medical issue, it is also an economic one. Patients with chronic obstructive lung disease (COPD), chronic liver or kidney disease, or congestive heart failure are at high risk of developing an infectious disease followed by prolonged hospitalization.

SUMMARY

Public health has addressed, and will continue to stress the issues of communicable disease as one of its key issues in protecting individual and population health. Methods of intervention include classic public health through sanitation, immunization, and well beyond that into nutrition, education, case finding, and treatment, and changing human behavior. The knowledge, attitudes, beliefs, and practices of policy makers, health care providers, and parents is as important in the success of communicable disease control as are the technology available and methods of financing health systems. Together, these encompass the broad programmatic approach of the New Public Health to control of communicable diseases.

In a world of rapid international transport and contact between populations, systems are needed to monitor the potential explosive spread of pathogens that may be transferred from their normal habitat. The potential for the international spread of new or reinvigorated infectious diseases constitute threat to mankind akin to ecological and other man-made disasters.

The eradication of smallpox paved the way for the eradication of poliomyelitis, and perhaps measles, in the foreseeable future. New vaccines are showing the capacity to reduce important morbidity from rubella syndrome, mumps, meningitis, and hepatitis. Other new vaccines on the horizon will continue the immunologic revolution into the twenty-first century.

As the triumphs of control or elimination of infectious diseases of children continue, the scourge of HIV infection continues with distressingly slow progess an effective vaccine or cure for the disease it engenders. Partly as a result of the HIV/AIDS, TB staged a comeback in many countries where it was thought to be merely a residual problem. At the same time an old/new method of intervention using directly observed short-term therapy has shown great success in controlling the TB epidemic. The resurgence of TB is more dangerous in that MDRTB has become a widespread problem. This issue highlights the difficulty of keeping ahead of drug resistance in the search for new generations of antibiotics, posing a difficult challenge for the pharmaceutical industry, basic scientists as well as public health workers.

The burden of infectious diseases has receded as the predominant public health problem in the developed countries but remains large in the developing countries. With increases in longevity and increased importance of chronic disease in the health status of the industrial and mid-level developing nations, the effects of infectious disease on the care of the elderly and chronically ill is of great importance in the New Public Health. Long-term management of chronic disease needs to address the care of vulnerable groups, promoting the use of existing vaccines and antibiotics. Most important is the development of health systems that provide close monitoring of groups at special risk for infectious disease, especially patients with chronic diseases, the immunocompromised, and the elderly. The combination of traditional public health with direct medical care needed for effective control and eradication of communicable diseases is an essential element of the New Public Health. The challenge is to apply a comprehensive approach and management of resources to define and reach achievable targets in communicable disease control.

ELECTRONIC MEDIA

Access to e-mail and the Internet are vital to current practice of public health and nowhere is this more important than in communicable diseases. There are many such information sites and these will undoubtedly expand in the coming years. Several sites are given as examples. The Internet has great practical implications for keeping up to date with rapidly occurring events in this field.

Eurosurveillance Weekly is available at eurowkly@eurosurv.org or at website http://www.eurosurv.org

Gideon, outstanding encyclopedia database on infectious diseases (available via mdcassoc@ix.netcom.com at reduced price for Promed users, and free to sub-Saharan African sites); website http://www.cyinfo.com

Infectious disease early warning system via wilsonml@biology.lsa.umich.edu or web server http://eotest2.gsfc.nasa.gov/IDP/form.html (NB: Capitalized letters must be capitalized)

Morbidity and Mortality Weekly Reports is available on the Internet via the CDC home page and may be downloaded; consult CDC's home page at www.cdc.gov

Promed is an excellent, free report on current events in communicable diseases internationally; join via owner-promed@usa.healthnet.org

Weekly Epidemiologic Bulletin of the WHO is available on the World Wide Web via http://www.who.ch/programmes/emc/news.htm

World Health Organization, Diseases and Vaccines website, http://www.who.int/gpv-dvacc and http://www.who.int/gpv-surv/country/

RECOMMENDED READINGS

Centers for Disease Control. 1992. Update: International Task Force for Disease Eradication, 1990 and 1991. *Morbidity and Mortality Weekly Report,* 41:40–42.

Centers for Disease Control. 1994. Addressing emerging infectious disease threats: A prevention strategy for the United States. Executive summary. *Morbidity and Mortality Weekly Report,* 43(RR-5):1–18.

Centers for Disease Control. 1997. Update: trends in AIDS incidence—United States, 1996. *Morbidity and Mortality Weekly Report,* 46:861–867.

Centers for Disease Control. 1998. One thousand days until the target date for global poliomyelitis eradication. *Morbidity and Mortality Weekly Report,* 47:234–239.

Centers for Disease Control. 1998. Tuberculosis morbidity—United States, 1997. *Morbidity and Mortality Weekly Report,* 47:253–257.

Centers for Disease Control. 1998. Measles—United States, 1997. *Morbidity and Mortality Weekly Report,* 47:273–277.

Centers for Disease Control. 1998. National adult immunization awareness week—October 11–17,

1998; and influenza and pneumococcal vaccination levels among adults aged ≥65 years. *Morbidity and Mortality Weekly Report,* 47:797–803.

Centers for Disease Control. 1998. Impact of the sequential IPV/OPV schedule on vaccination coverage—United States, 1997. *Morbidity and Mortality Weekly Report,* 47:1017–1019.

Centers for Disease Control. 1998. Advances in global measles control and elimination: Summary of the 1997 international meeting. *Morbidity and Mortality Weekly Report,* 47(RR-11):1–23.

Centers for Disease Control. 1999. Recommended childhood immunization schedule—United States, 1999. *Morbidity and Mortality Weekly Report,* 48:12–16.

Centers for Disease Control. 1999. Impact of vaccines universally recommended for children—United States, 1990–1998. *Morbidity and Mortality Weekly Report,* 48:243–248.

Centers for Disease Control. 1999. Progress toward global poliomyelitis eradication. *Morbidity and Mortality Weekly Report,* 48:893–897.

Goodman, R. A., Foster, K. L., Trowbridge, F. L., Figuero, J. P. (eds). 1998. Global Disease Elimination and Eradication as Public Health Strategies: Proceedings of a Conference Held in Atlanta, Georgia, USA, 23–25 February 1998. *Bulletin of the World Health Organization,* 76(Supplement 2):1–161.

Peter, G. 1992. Childhood immunizations. *The New England Journal of Medicine,* 327:1794–1800.

Weekly Epidemiologic Record. 1999. Rotavirus vaccines: WHO position paper. *Weekly Epidemiologic Record,* 74:33–38.

BIBLIOGRAPHY

Anderson, R. 1992. *Infectious Diseases of Humans: Dynamic and Control.* Oxford: Oxford University Press. 1992.

Basch, P. 1994. *Vaccines and World Health: Science, Policy, and Practice.* New York: Oxford University Press.

Benenson, A. S. (ed). 1995. *Control of Communicable Diseases Manual,* Sixteenth Edition. Washington, DC: American Public Health Association.

Brooks, G. E., Butel, J. S., Morse, S. A. 1998. *Jawetz, Melnick and Adelberg's Medical Microbiology,* Twenty-first Edition. Stamford, CT: Appleton & Lange.

Cassens, B. 1992. *Preventive Medicine and Public Health,* Second Edition. Malvern, PA: Harwal Co.

Colditz, G. A., Brewer, T. F., Berkey, C. S., Wilson, M. E., Burdick, E., Fineberg, H. V., and Mosteller, F. 1994. Efficacy of BCG vaccine in the prevention of tuberculosis. Meta-analysis of the published literature. *Journal of the American Medical Association,* 271:698–702. (See also Brewer *et al.,* 1996. Evaluation of tuberculosis control policies using computer simulation. *JAMA,* 246:1898–1903 and related articles.)

Cook, G. C. 1996. *Manson's Tropical Diseases,* Twentieth edition. London: Saunders.

Cutts, F. T., Smith, P. G. (eds). 1994. *Vaccination and World Health.* Chichester: Wiley and Sons.

Mandel, G. L. 1994. *Principles and Practice of Infectious Diseases.* Edinburgh: Churchill, Livingstone.

Plotkin, S. A., Mortimer, E. A. 1994. *Vaccines,* Second Edition. Philadelphia: WB Saunders.

VACCINE-PREVENTABLE DISEASE

American Academy of Pediatrics. 1997. Immunization of adolescents: recommendations of the advisory Committee on Immunization Practices, the American Academy of Pediatrics, the American Academy of Family Physicians and the American Medical Association. *Pediatrics,* 99:479–488. (http://www.aap.org/policy/re9711.html)

American Academy of Pediatrics. 1999. Combination vaccines for childhood immunization: Recommendations of the Advisory Committee on Immunization Practices, the American Academy of

Pediatrics, the American Academy of Family Physicians and the American Medical Association. *Pediatrics,* 103:1064–1077.

American Academy of Pediatrics, Committee of Infectious Diseases. 1999. Poliomyelitis prevention: Revised recommendations for use of inactivated and live oral poliovirus vaccines. *Pediatrics,* 103:171–172.

Centers for Disease Control. 1993. Diphtheria outbreak—Russian Federation, 1990–1993. *Morbidity and Mortality Weekly Report,* 42:840–841, 847.

Centers for Disease Control. 1993. Resurgence of pertussis—United States, 1993. *Morbidity and Mortality Weekly Report,* 42:952–953, 959–960.

Centers for Disease Control. 1994. Rubella and congenital rubella syndrome—United States, January 1, 1991–May 7, 1994. *Morbidity and Mortality Weekly Report,* 43:397–401.

Centers for Disease Control. 1996. Compendium of animal rabies control, 1996: National Association of State Public Health Veterinarians. *Morbidity and Mortality Weekly Report,* 45(RR-3):1–9.

Centers for Disease Control. 1998. Progress toward elimination of *Haemophilus influenzae* type b disease among infants and children in the United States, 1987–1997. *Morbidity and Mortality Weekly Report,* 47:993–998.

Centers for Disease Control. 1997. Tetanus surveillance—United States, 1991–1994. *Morbidity and Mortality Weekly Report,* 46(SS-2):15–25.

Centers for Disease Control. 1998. Recommendations and reports—Vaccine use and strategies for elimination of measles, rubella, and congenital rubella syndrome and control of measles: Recommendations of the Advisory Committee on Immunization Practices. *Morbidity and Mortality Weekly Report,* 47(RR-8):1–59.

Centers for Disease Control. 1998. National, state and urban area vaccination coverage levels among children aged 19–35 months—United Sates, July, 1996–June, 1997. *Morbidity and Mortality Weekly Report,* 47:108–116.

Centers for Disease Control. 1998. Varicella related deaths among children—United States, 1997. *Morbidity and Mortality Weekly Report,* 47:365–368.

Centers for Disease Control. 1999. Progress toward global poliomyelitis eradication. *Morbidity and Mortality Weekly Report,* 48:416–421.

Centers for Disease Control. 1999. Ten great public health achievements—United States, 1900–1999. *Morbidity and Mortality Weekly Report,* 48:241–243, 243–248.

Tulchinsky, T. H., Abed, Y., Shaheen, S., Toubassi, N., Sever, Y., Schoenbaum, M., Handsher, R. 1989. A ten-year experience in control of poliomyelitis through a combination of live and killed vaccines in two developing areas. *American Journal of Public Health,* 79:1648–1652.

Tulchinsky, T. H., Ginsberg, G. M., Abed, Y., Angeles, M. T., Akukwe, C., Bonn, J. 1993. Measles control in developing and developed countries: the case for a two-dose policy. *Bulletin of the World Health Organization,* 71:93–103.

World Health Organization. 1999. Integration of vitamin A supplementation with immunization. *Weekly Epidemiological Record,* 74:1–6.

OTHER COMMUNICABLE DISEASES

Anderson, C. 1991. Cholera epidemic traced to risk miscalculation. *Nature,* 354:255.

Centers for Disease Control. 1993. Update cholera—Western hemisphere, 1992. *Morbidity and Mortality Weekly Report,* 42:89–91.

Centers for Disease Control. 1993. Isolation of *Vibrio cholerae* O1 from oysters—Mobile Bay, 1991–1992. *Morbidity and Mortality Weekly Report,* 42:91–93.

Centers for Disease Control. 1993. Estimates of future global tuberculosis morbidity and mortality. *Morbidity and Mortality Weekly Report,* 42:961–964.

Centers for Disease Control. 1994. Arbovirus disease—United States, 1993. *Morbidity and Mortality Weekly Report,* 43:385–387.

Centers for Disease Control. 1994. Update: outbreak of Legionnaire's disease associated with a cruise ship, 1994. *Morbidity and Mortality Weekly Report,* 43:574–575.

Centers for Disease Control. 1994. Rift Valley Fever—Egypt 1993. *Morbidity and Mortality Weekly Report,* 43:693, 699–700.

Centers for Disease Control. 1996. The role of BCG vaccine in the prevention and control of tuberculosis in the United States: A joint statement by the Advisory Council for the Elimination of Tuberculosis and the Advisory Committee on Immunization Practices. *Morbidity and Mortality Weekly Report,* 45(RR-4);1–18.

Centers for Disease Control. 1997. Update: Trends in AIDS incidence—United States, 1964. *Morbidity and Mortality Weekly Report,* 46:861–867.

Centers for Disease Control. 1997, Case definition for infectious conditions under public health surveillance. *Morbidity and Mortality Weekly Report,* 46(RR-10):1–55.

Centers for Disease Control. 1998. 1998 Guidelines for treatment of sexually transmitted diseases. *Morbidity and Mortality Weekly Report,* 47(RR-1):1–118.

Centers for Disease Control. 1998. Primary and secondary syphilis—United States, 1997. *Morbidity and Mortality Weekly Report,* 47:493–497.

Dolin, P. J., Raviglione, M. C., Kochi, A. 1994. Global tuberculosis incidence and mortality during 1990–2000. *Bulletin of the World Health Organization,* 72:213–220.

Halstead, S. B. 1992. The 20th century pandemic: Need for surveillance and research. *World Health Statistics Quarterly,* 45:292–298.

Slutsker, L., Ries, A. A., Greene, K. D., Wells, J. G., Hutwagner, L., Griffin, P. M. 1997. *Escherichia coli* O157:H7 diarrhoea in the United States: clinical and epidemiologic features. *Annals of Internal Medicine,* 126:505–513.

UNICEF. (1995–1998). *The State of the World's Children 1995, 1996, 1997, and 1998.* New York: United Nations Children's Fund, Oxford Press.

World Health Organization. 1990. *The Rational Use of Drugs in the Management of Acute Diarrhoea in Children.* Geneva: WHO.

World Health Organization. 1994. The malaria situation in 1991. *Bulletin of the World Health Organization,* 72:160–164.

World Health Organization. 1994. *AIDS: Images of the Epidemic.* Geneva: WHO.

World Health Organization. 1996. Dracunculiasis: Global surveillance summary. *Weekly Epidemiologic Record,* 71:141–148.

World Health Organization. 1996. Progress toward the elimination of leprosy as a public health problem. *Weekly Epidemiologic Record,* 71:149–156.

World Health Organization. 1996. *The World Health Report 1996: Fighting Disease, Fostering Development.* Geneva: WHO.

World Health Organization. 1997. *The World Health Report 1997: Conquering Suffering, Enriching Humanity.* Geneva: WHO.

World Health Organization. 1998. *Health for All in the Twenty-first Century.* EB101/8. Geneva: WHO.

World Health Organization. 1998. *The World Health Report 1998: Life in the Twenty-first Century: A Vision for All.* Geneva: WHO.

World Health Organization. 1999. *The World Health Report 1999: Making a Difference.* Geneva: WHO.

5

NONCOMMUNICABLE CONDITIONS

INTRODUCTION

The "diseases of modern life" or noninfectious conditions have become the leading causes of morbidity and mortality in developed countries. This epidemiologic transition is taking place in many developing countries as well. Causation in noninfectious (chronic) disease is complex, and prevention must take into account multiple contributory or risk factors. Despite the complexity, and often for reasons not well understood, dramatic success has been achieved in reducing stroke and heart disease death rates in many countries over the past 20 years. Cancer and trauma death rates, key elements of noninfectious disease patterns, have proved more difficult to reduce.

Chronic conditions place great demands on health systems. At the same time they complicate the potential for primary, secondary, and tertiary prevention to reduce mortality and disability as well as the burden of disease and disability. The New Public Health stresses all aspects of prevention and care and is therefore increasingly required as a broad approach to this wide group of conditions and the needs associated with them.

THE RISE OF CHRONIC DISEASE

The change from the predominance of infectious diseases to noninfectious diseases as the leading causes of premature mortality has occurred primarily since the end of World War II. Antibiotics and vaccines, along with improved living standards, sanitation, nutrition, and safe water, brought about a reduction in mortality rates from infectious diseases and an increase in life expectancy. Infectious diseases, while still important, are no longer the primary concerns in public health in the developed countries, and similar trends are appearing in the developing countries.

TABLE 5.1   Leading Causes of Death (Rates per 100,000 Population), United States, 1900 and 1995

1900	Rate	%	1990	Rate	%
All causes	1719	100.0	All causes	880	100.0
Pneumonia, influenza	202	11.8	Heart diseases	281	32.0
Tuberculosis	194	11.3	Malignant neoplasms	204	23.2
Gastritis, enteritis, colitis	143	8.3	Cerebrovascular diseases	60	6.8
Heart diseases	137	8.0	Chronic obstructive lung diseases	40	4.5
Symptoms, senility, ill-defined conditions	118	6.8	Accidents	34	3.9
Vascular lesions central nervous system	107	6.2	Pneumonia, influenza	32	3.6
Chronic nephritis, sclerosis	81	4.7	Diabetes mellitus	23	2.6
Unintentional injuries	72	4.2	HIV infection	16	1.8
Malignant neoplasms	64	3.7	Suicide	12	1.3
Diphtheria	40	3.3	Cirrhosis, chronic liver diseases	10	1.1
All other causes	561	32.6	All other causes	168	19.2

Source: Data compiled from National Center for Health Statistics, in *Chronic Disease Epidemiology and Control*. Second Edition. 1998. Washington, D.C.: American Public Health Association with permission.

The dramatic shift in causes in death from a predominance of infectious to the noninfectious diseases is seen in Table 5.1, which lists the leading cases of death in the United States in 1900 and 1990, reflecting changes in all the developed countries over the century.

Chronic diseases as the leading cause of morbidity and mortality are associated with a number of demographic and epidemiologic factors. First, the decline in infectious disease mortality resulted in greater longevity, increasing the numbers of persons surviving to ages when cancer and heart disease are common. Second, changes in lifestyle increased risk factors such as smoking, lack of exercise, diets rich in unhealthy fats and sugars, and risk taking behavior, so that cardiovascular diseases and cancer became the leading causes of disease, disability, and death. Third, trauma and chronic diseases are major contributors to rising costs of health care and the economics of health care. Fourth, public health experience and new scientific knowledge are leading to new forms of prevention and medical treatment that are reducing the burden of disease and disability from chronic conditions.

Identification of many risk factors and methods of early detection of these diseases has increased the potential for interventions to lower the prevalence of these diseases and their complications. As a result of these changes, the scope of public health has broadened. In this chapter we examine the major chronic diseases and their effects on the health of the population. We also look at the risk factors that contribute to them and at the interventions needed to reduce their prevalence in the population.

Causes of important chronic conditions are being identified with interventions and treatment that will further alter disease patterns in the years ahead. In addition, prevention methods, including the secondary and tertiary prevention carried out by health providers, are playing an important and measurable role in reducing the burden of disease, as is health promotion.

TABLE 5.2 United States Health Expenditures by Diagnostic Group and by Type of Service

Diagnostic group	Expenditures 1987 (in 1993 US$, billions)[a]	% of Total	Category of service	Expenditures 1997 (in 1997 US$, billions)[b]	% of Total
Cardiovascular	79.6	13.9	Hospital care	371.1	34.0
Injury	69.1	12.1	Physician services	217.6	19.9
Neoplasm	49.6	8.7	Dental services	50.6	4.6
Genitourinary	49.3	8.7	Other professional services	61.9	5.7
Pregnancy/birth	39.7	6.9	Home health care	32.3	3.0
Respiratory	38.3	6.7	Drugs, other medical supplies	108.9	10.0
Digestive	35.9	6.3	Vision, other medical durables	13.9	1.3
Musculoskeletal	27.7	4.8	Nursing home care	82.8	7.6
Other circulatory	20.2	3.5	Other personal care	29.9	2.7
Mental health	19.3	3.4	Research	18.0	1.6
Well care	17.4	3.0	Construction	16.9	1.5
Congenital anomalies	8.7	1.5	Public health (government)	38.5	3.5
Medical misadventure	6.9	1.2	Administration	50.0	4.6
Miscellaneous	110.6	19.3	Total expenditures (billions)	1,092.4	100
Total	572.3	100	Per capita expenditures (dollars)	3,925.0	—

Sources: Centers for Disease Control 1994. Medical care spending—United States. *MMWR*, 43:581–586; and Levit K, Cowan C, Braden B, Stiller J, Senseni A, Lazenby H. 1998. National health expenditures in 1997; more slow growth. *Health Affairs*, November/December: 99–110.

[a]Total health expenditures from public and private sources, excluding nursing homes, dental, nonmedical mental health services, and insurance processing costs. The 1987 National Medical Expenditure Survey (NMES-2) was a population-based, longitudinal survey for the civilian noninstitutionalized population for the year 1987, translated into 1993 dollars. Subsequent surveys have included source of payment and type of service, but not diagnostic category.

[b]Includes total health expenditures, public and private, including nursing homes.

TABLE 5.3   Activities of Daily Living as Indicators of Independent Functional Capacity

Personal care ADLs	Household management ADLs	Socialization ADLs
Bathing	Using telephone	Contacts—family, friends, neighbors
Dressing	Shopping	Visits
Toileting	Food preparation	Social activites
Eating	Light housekeeping	Sex
Walking, movement	Light maintenance	Regular exercise, walking 1/4 mile
Transfer in and out of bed, chairs, bath	Laundry	Using bus, car
Continence	Managing money	Religious activity
	Managing own medications	Hobbies, activities

THE BURDEN OF CHRONIC CONDITIONS

Chronic conditions place a heavy burden on the individual, the family, and society as a whole in terms of morbidity and mortality. Measurement of the burden of disease (discussed in Chapter 3) is a fundamental responsibility of public health agencies. The economic burden for health care of the population, whether funded by national responsibility as in most industrialized countries (see Chapters 10–13) or a mix of public and private expenditures, is seen in Table 5.2.

What is more difficult to quantify is the burden of disease on the individual, the family, and the community. Traditional measures of morbidity and mortality are supplemented by QALYs and DALYs, as seen in Chapter 3, but the physical and emotional burden of caring for someone is almost impossible to quantify. The burden of chronic disease on the individual is reflected on his or her ability to function in the normal activities of daily life. The level of function of a person with a chronic condition is measured by ability to perform activities of daily living (ADL), as seen in Table 5.3.

Chronic conditions may result in disabilities which impede optimal function in normal daily functions or activities. Activities of daily living measure independent capacity regarding personal care, household management, and socializing. They help determine the level and amount of home care needed, or the type of facility the patient may need. ADL measure function of the patient, but not the emotional, physical, and financial stress on the caregiver in a family.

RISK FACTORS AND CAUSATION
OF CHRONIC CONDITIONS

The criteria for causation in infectious disease, discussed in Chapter 1, are known as the Koch–Henle postulates. The criteria for causation in chronic disease, called the Evans criteria, are outlined in box 5.1. They articulate the relationship of predisposing or risk factors to the causation of chronic disease, and are impor-

BOX 5.1 CRITERIA FOR CAUSATION—
THE EVANS CRITERIA

1. Distribution of the hypothesized cause in the population should be similar to the distribution of the disease in the population;
2. Incidence of the disease should be significantly higher in those exposed to the hypothesized cause than in those not so exposed;
3. Exposure to the hypothesized cause should be more frequent among those with the disease than in controls without the disease, when all other risk factors are held constant;
4. The disease should temporally follow exposure to the hypothesized cause;
5. The greater the dose or length of exposure, the greater the likelihood of occurrence of the disease;
6. For some diseases, a spectrum or biological gradient of host responses (from mild to severe) should follow exposure to the hypothesized cause, in relation to the degree of exposure:
7. The association between the hypothesized causes and the disease should be found in various populations when different methods of study are used;
8. Other explanations for the association should be ruled out;
9. Elimination or modification of the hypothesized cause should decrease the incidence of the disease;
10. Modification of the host's response on exposure to the hypothesized cause should decrease or eliminate the disease;
11. In experimental settings, the exposed population should have the disease more frequently than the nonexposed;
12. All relationships and findings should have biological and epidemiologic plausibility.

tant in analyzing the relative importance of factors contributing to disease in a population.

For noninfectious disease, the causation may be complex, with many factors contributing to causation. Rarely is a single necessary factor sufficient by itself to produce the disease. Risk factors for the chronic diseases are summarized in Table 5.4. This shows the multiple risk factors of many disease groups, with low socioeconomic status a factor in nearly all the disease groups examined.

Risk factors in noncommunicable diseases have been identified by important epidemiologic studies over the past four decades. The classic work of Jeremy Morris showed the difference in risk for coronary heart disease between London bus drivers as compared to bus conductors. The conductors who are both physically required to climb bus stairs constantly and experience less stress as part

TABLE 5.4 Chronic Disease Risk Factors

Risk factor	Cardiovascular disease	Cancer	Chronic lung disease	Diabetes	Cirrhosis	Musculo-skeletal disease	Neurological disorders
Tobacco use	+	+	+	0	0	+	?
Alcohol abuse	+	+	0	0	+	+	+
High cholesterol	+	0	0	0	0	0	0
Hypertension	+	0	0	0	0	0	+
Diet	+	+	?	+	0	+	+
Physical inactivity	+	+	0	+	0	+	0
Obesity	+	+	0	+	0	+	+
Stress	?	?	0	0	0	0	0
Environmental tobacco smoke	?	+	+	0	0	0	0
Occupation	+	+	+	0	?	+	?
Air pollution	+	+	+	0	0	0	0
Low socioeconomic status	+	+	+	+	+	+	0

Source: Modified from Brownson *et al.*, 1998, p. 4.

of their job had lower rates of CHD mortality than the sedentary high tension drivers.

Smoking was identified as a risk factor in U.S. studies by Ernst Wynder and others in the 1940s and 1950s. In a longitudinal study of British physicians by Richard Doll and Bradford Hill in the 1950s, 35-year-old male cigarette smokers were shown to have less chance of surviving to age 65 (73%) as compared to non-smokers (85%) and ex-smokers (81%). The U.S. Surgeon General's Report on Smoking and Health of 1964 summarized the hundreds of published studies up to that time and concluded that cigarette smoking was a major health hazard and a cause for lung cancer, coronary heart disease, chronic pulmonary lung disease, and stroke. Subsequent Surgeon General's Reports in 1983 and 1984 attributed 30% of coronary heart disease deaths, and 80–90% of chronic obstructive lung disease deaths, to smoking. Smoking reduction has become one of the pillars of modern public health, with a decline in current cigarette smoking rates from 33.5% in 1979 to 24.7% in 1997, but is increasing among high-school students (whites>blacks).

The famous longitudinal heart study in Framingham, Massachusetts (1948), provided important epidemiologic data showing that hypertension, smoking, and elevated cholesterol are associated with increased risk of cardiovascular diseases. The Framingham study pioneered the epidemiologic approach to gain insight into causes of cardiovascular diseases. This was a prospective cohort study to quantify risks for these diseases, both in terms of absolute and relative risk. Its observations have led to causal inferences, e.g., elevated blood pressure with increased risk of stroke.

Subsequent studies have elaborated on the Framingham findings. High blood pressure (hypertension) refers to elevated levels of systolic and/or diastolic blood pressure and is associated with increased risk of morbidity and mortality from miocardial infarction, stroke, and renal disease. Concepts of normality have been replaced with guidelines for "optimal" values of blood lipid and blood pressure for long-term freedom from cardiovascular disease. Atherogenic potential for serum total cholesterol was shown to be derived from the low density lipoprotein cholesterol (LDL-cholesterol) fraction which is positively related to CHD incidence. High density lipoprotein cholesterol (HDL-cholesterol) is inversely related to CHD, since it removes cholesterol from tissues. Risk of CHD is independently related to each of these lipoprotein fractions. Therefore, the ratio of total to HDL-cholesterol is an efficient lipid risk profile.

The Alameda County study of Lester Breslow and colleagues followed a cohort (nearly 7000 people from the 1 million population) from 1965 to the present and identified seven health practices that were associated with reduced mortality and disability patterns. The health practices studied were: excessive alcohol intake cigarette smoking, obesity, sleeping less or more than 7–8 hours/day, physical inactivity, eating between meals, and not eating breakfast. Adjusting for age, sex, health status, and social networks, the occurrence of disability was half as great for those with good health practices as compared to poor practices. These patterns were prevalent in the cohort followed in the 1960s and 1970s. The sixth follow-up assessment will occur in 1999. The reduction in disability as well as mortality showed the effectiveness of healthy lifestyles on maintaining health, and not only in prolonging life.

CHRONIC MANIFESTATIONS
OF INFECTIOUS DISEASES

Infections as causes of chronic diseases are of very great importance for public health because such associations can lead to new treatments or preventive measures. Some of these associations are well established; others are reported but as yet are insufficiently confirmed. Once confirmed, the search for vaccines could replicate the success of immunization in control of the acute infectious diseases of childhood. This will be a central issue in public health in the new century.

The number of established and proposed relationships between certain organisms and chronic diseases is growing. The relationship of hepatitis B with chronic hepatitis, cirrhosis, and hepatic carcinoma provide the justification for wide-scale immunization to protect individuals, especially those in developing countries who are especially at risk for hepatitis B infection. Specific human papillomaviruses (HPVs) are associated with cervical carcinoma. Screening for cancer of the cervix is an important public health modality, but education in hygienic practices can help to reduce the spread of these organisms. A human papillomavirus vaccine (HPV) based on genetic engineering technology is expected to be available for prevention of cancer of the cervix early in the twenty-first century. Varicella virus is associated with herpes zoster and postherpetic neuralgia. Varicella vaccine now recommended for routine childhood immunization will eliminate this problem when the vaccine is used in routine childhood vaccination for a number of years.

Bacterial infections can also have long-term sequelae. Examples include peptic ulcer disease and chronic gastritis from the bacterim *Helicobacter pylori* (Box 5.2). In 1991, four case–control studies quantified and confirmed that *H. pylori* is associated with stomach cancer. This included demonstration of antibodies and growth of the organism before and after diagnosis of cancer. One study showed the relation of severity of infection and risk of stomach cancer. About 60% of patients with cancer of the stomach had previous infection, with the odds ratio of 1.6 to 17.7, suggesting a strong association between *H. pylori* infection and stomach cancer.

Standard treatment of gastric and duodenal ulcers now focuses on gastroscopy and biopsy, tests for *H. pylori,* and treatment with antihistamine to reduce acid production (H_2 inhibitors) and inexpensive antibiotics, with follow-up by breath tests for the organism. This will further reduce hospitalizations and surgery for chronic peptic ulcer complications and expensive long-term drug treatment with hydrogen ion antagonists or inhibitors. *H. pylori* is accepted as the major cause of peptic ulcers, now a treatable condition giving new life to the issue of infection as a cause of chronic disease.

Long-term sequelae of gonorrhoea and other STDs include sterility, and untreated syphilis can lead to long-term neurological deterioration and valvular heart disease. These and other examples of established relations and others that are still unproven hypotheses are shown in Table 5.5. The number of examples of infectious diseases causing chronic disease will increase with development of the bio-

BOX 5.2 HELICOBACTER PYLORI, PEPTIC ULCER, AND STOMACH CANCER

Mortality in the United States attributed to duodenal ulcer declined from 4 in 1950 in just over 1 per 100,000 population in 1980, while gastric ulcer mortality remained constant at 1 per 100,000. Hospitalizations for duodenal ulcer for males declined from 28 to 8 per 10,000 population from 1965 to 1982, but still some 5000 deaths occurred per year (2 per 100,000) from bleeding, perforation, or obstruction from peptic ulcers. In 1975, costs associated with this group of diseases was estimated at $1.3–2.6 billion. The introduction of powerful hydrogen ion antagonists in 1977 contributed to further decline in hospitalization rates from peptic ulcers. Many factors were considered important in causation for this disease, including aspirin, alcohol, smoking, coffee, stress, occupation, and genetic tendencies.

Helicobacter (Campylobacter) pylori, first reported in Lancet in 1983 as a curved bacterium associated with gastritis, is present in more than half the world's population. Once acquired, it persists and can cause disease decades after infection, acting as a "microbial parasite." The organism was shown in the pylorus of the gastric aspirates of 58 of 100 patients with peptic ulcer by an Australian physician Barry Marshall with J. R. Warren, in 1984. Marshall experimentally ingested the organism himself and developed nausea, stomach pain, and foul breath. They suggested that the organism might be the cause of peptic ulcer and stomach cancer. Treatment of peptic ulcer, after gastroscopy, biopsy, and demonstration of the organism, followed by a short course of antibiotics is now standard treatment.

An ecological study in the United States showed high correlation between areas with high prevalence of the organism and high prevalence of cancer of the stomach. In 1998, *H. pylori* was reported to have been found in surface water, suggesting a major reservoir and method of transmission of this organism. Chlorination apparently kills the organism.

Source: National Institute of Health Consensus Statement. 1994. *Helicobacter pylori* in peptic ulcer disease. NIH *Consensus Statement February* 7–9, Washington D.C. 12:1–23. Website http://www.opd.nih.gov/consensus/cons/094/094_intro.htm

logical sciences, as reflected by identification of prions as transmitters of such diseases as scrapie in sheep, bovine spongiform encephalitis (BSE) in cattle, and Creutzfeld-Jacob Disease (CJD) in humans.

During the 1990s, increasing evidence of *Chlamydia pneumoniae* infection as a cause of coronary heart disease has provided fresh impetus to the search for new approaches to prevention of the leading cause of death in industrialized and many developing countries. Initial reports from Finland, the United Kingdom, and Italy were followed by supportive studies in the United States. The association has not

TABLE 5.5   Infectious Disease Causing Chronic Diseases, Established and Hypothesized
Relationships

Established relationships	Hypothesized relationships
STDs with sterility	
Syphilis with neurological and cardiac diseases	
Streptococcal infection with chronic rheumatic heart disease and glomerulonephritis	Chlamydia infection with coronary heart disease and stroke
Hepatitis B and C chronic infection with cirrhosis of liver and cancer of liver (82%)	Fungal infection with polycystic kidneys
Human papillomavirus (HPV) with cancer of the cervix (83%)	Borna disease virus (BDV) and schizophrenia, transmitted from birds and mammals to humans
Heliobacter (Campylobacter) pylori with peptic ulcer and stomach cancer (55%)	Measles and Crohn's disease
Epstein-Barr virus with Burkitt's lymphoma in Africa	Hantavirus and hypertensive renal disease
Schistosomiasis with bladder cancer in developing countries (4%)	
AIDS with Kaposi's sarcoma and non-Hodgkin's lymphoma	
Bovine spongiform encephalitis (BSE) with Creutzfeld–Jakob disease (CJD)	

Source: World Health Organization, 1996; Morse, 1993; and *Journal of Emerging Diseases,* 1998.

yet satisfied the current criteria for causation of the Koch–Henle postulates, but the weight of evidence is mounting. More basic science and clinical trials are needed, and in 1999, at least two large prospective clinical–epidemiologic trials with antibiotic treatment as a preventive measure for CHD are underway. The consensus of cardiologists' opinion is that it is still premature to consider antibiotic treatment for this purpose. However, the search for a chlamydia vaccine may soon become an objective of the first order of importance for public health.

CARDIOVASCULAR DISEASES

Cardiovascular disease refers to a group of diseases of the heart and blood vessels, including coronary or ischemic heart disease, hypertension, and cerebrovascular disease (stroke). These are associated with atherosclerosis, excess fats in the diet and lipids in the body, and often with impairment of endocrine functions related to glucose metabolism and diabetes mellitus. Coronary heart disease (CHD) is a disease of the lining and blockage of the arteries that supply the muscles of the heart. The coronary arteries may become blocked with plaques and thromboses, thus cutting off the blood and oxygen supply (ischemia, i.e., ischemic heart disease) and leading to death (necrosis) of heart muscle, an acute myocardial infarction (AMI), or a heart attack. Other diseases of the heart include rheumatic diseases with damage to valves of the heart and other diseases of the heart muscle. Depending on the extent of heart muscle damage, a patient may go into congestive heart failure (CHF) due to weakened function of the heart as a pump and resulting congestion of fluids in the lungs and other tissues.

Cerebrovascular disease (CVD) is blockage or bleeding from blood vessels supplying the brain, causing death of areas of brain tissue, otherwise known as a cerebrovascular accident (CVA) or stroke. If this is the dominant side of the brain, that is, the left side for right-handed people, then depending on the area of permanent damage after recovery there are varying amounts of motor and mental limitations. Partial occlusion may cause transient ischemic attacks (TIAs) resulting in brief loss of motor and mental function. In both AMIs and CVAs the area of dead tissue will be surrounded by tissue that is inflamed and damaged. Treatment is intended to restore blood flow and minimize the inflammation or swelling (edema). Immediate care is vital to minimize damage and has a direct effect on maximizing recovery.

Cardiovascular diseases are the leading causes of death in the developed and in many developing countries. They have been called the diseases of "modern living" because etiologically they are due to factors such as smoking, lack of physical exercise, and a diet rich in animal fats. These are collectively known as "lifestyle" risk factors and are amenable to change by the individual at-risk and through community interventions. The most important lessons since the 1950s in public health is that these risk factors and diseases can be reduced dramatically by suitable public health interventions.

Precise measurement of the prevalence of these diseases through epidemiologically sound community-based prevalence studies is difficult or impossible at the present time because of lack of standard diagnostic criteria and terminology. Measures such as mortality rates, and hospitalization data may not give true prevalence rates, but they do provide important time trends and interarea comparisons that are valid for planning public health interventions. Cardiovascular diseases have common pathophysiological features with diabetes, that relate to nutrition, exercise, and other lifestyle factors. Cerebrovascular and coronary heart disease have important clinical differences, risk groups, as well as in screening, measurement, and interventions required.

Trends in Cardiovascular Disease Mortality

Mortality from coronary heart disease and from cerebrovascular disease increased in most western countries from the 1920s, reaching their peak levels in the 1950s. Mortality began to decline some 10 years later than cerebrovascular disease and continues to decline in the mid-1990s. Cerebrovascular disease, declining since the turn of the twentieth century, began a steep decline in the 1960s which is continuing in the late 1990s.

Table 5.6 shows a dramatic decline for deaths from cardiovascular diseases and all causes in the United States from 1950 to 1995. Age adjusted death rates for ischemic heart disease declined in the United States by 55% from 1950 to 1995, and stroke mortality by 70%, while mortality from all causes fell by 40%. The rates of

TABLE 5.6   Age-Adjusted Death Rates per 100,000 Population for Total and Selected Causes of Death, United States, 1950–1996

| Cause | Year | | | | | | | % change 1950–1996 |
	1950	1960	1970	1980	1985	1990	1996	
All causes	842	761	714	586	549	520	492	−42
Disease of the heart	307	286	254	202	181	152	135	−56
Cerebrovascular disease	89	80	66	41	33	28	26	−77
Cancer	125	126	130	133	134	135	128	+2

Source: *Health, United States,* 1998.

decline were not uniform in all regions or population groups. Average annual decline over this period was 3% for women and 3.8% for men. In the United States, rates of heart disease and cerebrovascular disease mortality for men and women, blacks and whites, have all declined, but at varying rates.

Despite the decline in coronary heart disease in most industrialized countries, this is still the largest cause of death. In the United States, there are some 1.5 million heart attacks each year causing nearly half a million deaths. Regional variation by state is wide. New York State has had higher mortality from ischemic heart disease than the overall U.S. pattern for many decades, declining for men over age 35 by 28% from 1979 to 1992 while national rates declined by 33% (respectively from 738 to 530 per 100,000, and 583 to 389 per 100,000). Within the state, with roughly half the population in upstate New York and half in New York City, there are major differences in ethnic mix, economic factors, and lifestyle. Ischemic heart disease is higher in New York City's population, while mortality from stroke is higher in the upstate population. IHD mortality is higher in New York City than upstate for both white and nonwhite population groups. This suggests that management of hypertension is better for the people of New York City versus the upstate population, and in the white populations compared to the nonwhite populations.

Mortality rates from cardiovascular diseases vary widely between countries, as between regions of a country, for men and women, or for different time periods in the same country (Table 5.7). Some countries experienced the peak of mortality from CVD in the early to mid-1950s, followed by a dramatic and sustained decline. In others, the peak was reached in the mid- to late 1960s or early 1970s followed by a more moderate rate of decline. In some countries of eastern Europe the mortality rates for coronary heart disease and strokes continue to increase in the 1990s, but may have peaked in 1994–1995.

Intercountry comparisons can be helpful in evaluation of public health needs and priorities. The experience of one country or region is not necessarily applicable directly to another, but the trends in cardiovascular disease mortality are now well established with declines of 40–50% in many countries. All countries or re-

TABLE 5.7 Trends in Mortality from Cardiovascular Disease in Selected Industrialized Countries, 1952–1967 and 1970–1985

Country	% Change male, 1952–1967	% Change male, 1970–1985	% Change female, 1952–1967	% Change female, 1970–1985
Japan	−8.4	−51.7	−31.7	−57.5
United States	−7.9	−38.8	−25.5	−38.9
Canada	−6.4	−38.8	−33.3	−42.8
New Zealand	+15.7	−31.9	−19.0	−33.6
Sweden	+2.0	−6.4	−38.0	−30.3
United Kingdom	+5.2	−20.9	−22.0	−26.7
Poland	na	+35.7	na	+12.4
Hungary	+13.7	+29.5	−21.4	+4.9

Source: Uemura and Pisa, 1988.

gions within a country that have persistent high rates should review their program priorities in public health. Figure 5.1 shows standardized death rates for a number of countries in the European Region of the World Health Organization. Finland has been a country with very high rates of cardiovascular mortality, but since the 1970s it experienced a sharp decline, as has Israel and, to a lesser degree, Sweden. Denmark and the United Kingdom did not show reductions until the late 1970s,

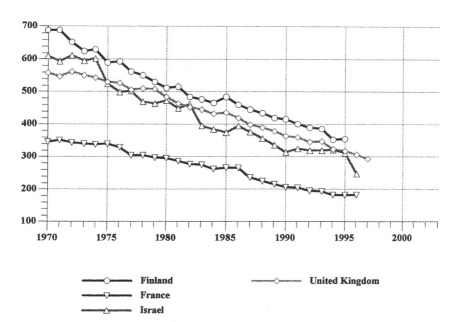

FIGURE 5.1 Standardized mortality rates (per 100,000 population) from cardiovascular heart diseases, selected countries, 1970–1997.

possibly due to later adoption of new innovations treatment of AMIs and in emphasis on dietary and smoking risk reduction.

Many studies of CVD morality have shown racial, sexual, and regional differences, but there may be many other contributing factors to these differences, which are important to identify so as to be able to plan suitable intervention programs. Such factors include education, diet, access to health care, knowledge, attitudes, and practices.

Hypertension, labile or fixed, systolic or diastolic, mild or severe, for any age or sex group, is an independent contributor for CHD. In addition, glucose intolerance or diabetes are atherogenic, especially for women. Familial history of CHD also confers excess risk, as does smoking, physical inactivity, and fatty diet. Multivariate analysis has established risk factors for intervention programs.

Early and good quality medical care during and after an AMI can reduce case fatality rates. Secondary prevention after a first AMI can reduce the risk or delay repeat myocardial infarctions and increase long-term survival. Primary prevention to reduce risk factors is also an important aspect of reducing the burden of CVD. A review of the literature and computer modeling of the experience in the United States, published in 1997, attributed less than one-third of the reduction of mortality rates between 1980 and 1990 to primary prevention, while improved treatment accounted for half the reduction, with secondary prevention, such as routine use of aspirin and beta blockers following AMI, accounting for the rest.

Diffusion of medical interventions, discussed in Chapter 15, is sometimes seen as too rapid, but use of simple, low–cost medical technology, such as aspirin and beta blockers, both proven to be highly effective in reducing risk of second AMIs, were not adopted by a majority of practitioners in a late 1990s survey in the United States. However, using a key informant approach of respected local physicians, the percentage of local doctors using these secondary prevention medications rose sharply.

Stroke risk factors include cardiac disease, atrial fibrillation, systolic hypertension, left ventricular hypertrophy, diabetes, excessive fat and sugar in diet, cigarette smoking, family history of early strokes, and low socio-economic status, as well as previous stroke or transient ischemic episodes. Reduction of stroke deaths is dependent on detection and management of hypertension and its control by changes in lifestyle and by supportive medication with long-term management and follow-up. Where stroke mortality is high, public health and medical services need to cooperate in developing education, screening, and management programs to reduce risk factors.

With declining mortality, the long-term problem of congestive heart failure (CHF) resulting from myocardial infarction and hypertension is increasing. An estimated 4.8 million Americans had CHF in 1998. Doctors office visits for CHF increased from 1.7 million in 1980 to 2.9 million in 1993. Hospitalization rates for persons aged 45–64 increased from some 140 per 100,000 to 200 per 100,000 in the same years. The U.S. National Heart, Lung and Blood Institute (NHBI of the NIH) in 1996 estimated that $17.8 billion was spent in 1993 on care of CHF pa-

---

BOX 5.3 HYPERTENSION—SECONDARY
PREVENTION

1. Hypertension case finding and control contributes greatly to the decline in cardiovascular mortality, especially due to stroke and heart failure. A 2 mm decrease in the diastolic blood pressure in a population results in a decline of more than 5% in a population's risk of developing CHD.
2. Screening for hypertension by examining blood pressure is recommended:
   a. Throughout pregnancy;
   b. For all adults up to age 65 at least every 5 years for adult men and women aged 16–64 years (Canadian Task Force on Screening, 1979), or either yearly or every other year (U.S. Preventive Services Task Force, 1989) based on the diastolic blood pressure;
   c. For all adults over age 65 at least every 2 years.
3. Management of hypertension is a combination of diet, smoking cessation, stress management, weight loss, salt restriction, and if necessary diuretics and other antihypertensive medications.

Source: Adapted from Cassens BJ. 1992. *Preventive Medicine and Public Health,* Second Edition.

---

tients for hospitalization, doctor's office care, home care, nursing homes, and medications. The NHBI expects an increasing burden of CHF care needs as more cardiac patients survive longer with their disease and as the population ages.

Prevention of Cardiovascular Diseases

The decline in cardiovascular disease mortality common in the industrialized countries over the past 25 years has been attributed to many factors, without clear evidence of the relative importance of each factor. This decline in mortality include a possible decline in prevalence or severity (case-fatality rates) of the disease resulting from reduced prevalence of risk factors. Higher standards of living, leisure and recreation, greater awareness of healthful nutrition and availability of appropriate foods at reasonable cost, and wider community and individual awareness of risk factors have played a role in this reduction. Improved quality and increased access to medical care, improved medications and other interventions, improved transportation, cardiopulmonary resuscitation, and emergency care helps in reducing case fatality rates. Their overall impact on mortality is still not easily separable from the changes in lifestyle and other risk factors.

The public health implications of these alternative explanations are substantial. Current data do not allow for clear distinctions of the contribution of each to the decline in mortality. Prudent public health would continue to place stress on all of these and attempt to strengthen the trend, particularly in state or local areas where higher than average rates prevail.

Primary prevention of cardiovascular diseases involves a spectrum of health promotion activities related to reduction of specific risk factors. This includes reduction of smoking and of obesity, better dietary habits, and healthful foods, with increased physical activity. Prevention requires a public health approach based on creating population-based and individual changes in the factors that contribute to heart attack or stroke, along with good medical supervision.

Implementation of primary and secondary preventive activities to reduce the burden of cardiovascular disease involves building institutional support, education of the public, community-based risk factor reduction activities, a healthy working environment, adequate information systems to monitor morbidity and risk factors, and a well-informed medical community. Medical interventions include identification and aggressive treatment of hypertension and diabetes, counseling to promote healthful dietary habits, a tobacco-free lifestyle, regular physical activity, and a supportive psychosocial environment. These are both individual and population-wide issues. The medical care provider is involved in screening, treatment, and advising patients of the importance of reducing risk factors both before and after the onset of symptoms. The client has to take personal responsibility for many aspects of prevention as a vital factor in screening for and sustained management of high blood pressure, reduction in elevated blood lipids, reduction of obesity, screening and management of diabetes, as well as in smoking cessation, and healthier exercise and dietary regimes.

Community-based programs have been designed to change life habits as a method of reducing the risks of CHD have become accepted parts of public health. Studies of community intervention programs such as the Minnesota Five Cities and the Stanford projects showed a small direct response when the target groups were broadly defined, but other studies focusing on specific target groups, such as prevention of teenage smoking, were more effective. Social behavior studies, economic status indicators, educational levels, ethnicity, and other factors all relate to risk behavior and help to define high risk target groups and methods of approach, the effect of social pressures for risk behavior as in dangerous driving, smoking, and binge drinking.

Governments and manufacturers may directly or indirectly promote smoking or alcohol consumption. Where governments permit advertising of alcohol and cigarettes in the media or fail to restrict cigarette use in public places there is a tacit approval of the practice. Manufacturers promote sales of their products among the young and the poor, and in developing countries where defense mechanisms such as awareness of health risks are lower than in a middle-class population. When governments profit from cigarette and alcohol consumption, there is a conflict of interests between health issues and government revenues.

A 1992 conference on heart health in Victoria, British Columbia, called on all community agencies to join forces in eliminating this modern epidemic by adopting new policies, making regulatory changes, and implementing health promotion and disease prevention programs directed at entire populations. Health personnel were encouraged to work on health promotion with the community, including the media, the education system, the social sciences, professional associations, government agencies, the private sector, international and private voluntary organizations, and community health coalitions.

A programmatic approach from the Victoria conference on prevention of cardiovascular disease includes the following:

1. Education: educate the public, health providers, community groups, governments in risk factor reduction.

2. Food policy: reduce fat content of milk and meat products and reduced salt content of processed foods, working with ministries of agriculture, industry, and commerce, as well as with dairies, meat producers, and food manufacturers.

3. Reduce smoking: increase cost of cigarettes through taxation, ban advertising, ban smoking in work and public places, and devote some revenue from cigarette taxes to health promotion and public education against smoking.

4. Promote physical exercise: promote personal and community attitudes and facilities encouraging participation in regular physical activity.

5. Reduce obesity: encourage individual and community-based health promotion.

6. Community-based initiatives: promote healthy lifestyle, including smoking cessation and fitness promotion; raise consciousness of health self-care issues; teach cardiopulmonary resuscitation (CPR).

7. Medical care: promote primary prevention techniques including screening for risk factors, management of hypertension, and patient counseling for risk factor reduction and stress management.

8. Screening: screen for risk factors such as diabetes, elevated blood lipids, and hypertension, and counsel as to findings and implications.

9. Emergency and hospital care: reduce case fatality rates, perform CPR, transport rapidly to designated medical centers with intensive care with current standards of antithrombotic agents (aspirin, streptokinase, or others) at district hospitals and ballooning, stents, or coronary artery bypass procedures at referral hospitals.

10. Rehabilitation: promote maximum recovery and function at work and in personal life; adopt preventive approaches to stop the pathological process.

Countries with very high rates of cardiovascular disease should develop aggressive intersectoral intervention programs at the national and community levels. Concerted national efforts of the Ministry of Health, other relevant ministries, the media, NGOs, food manufacturers, and local communities are needed in order to reverse the high rates of CVD in some U.S. states and northern and eastern Europe. Legislation and litigation against cigarette manufacturers may affect the sale and pattern of smoking, but the problem is a major public health issue.

This topic should also be placed high on the national agenda in the developing countries. It is particularly relevant to those at the midlevel of development where an epidemiological transition is underway, and the cardiovascular diseases are emerging as the leading cause of death. In developing countries, the problem of cardiovascular disease is hidden beneath more acute issues of high morbidity and mortality from infectious diseases, and those associated with maternal and child health. But there is a pattern of increasing cardiovascular disease morbidity and mortality affecting the growing middle class, with changes in diet and smoking. Age-specific mortality rates from cardiovascular diseases in Egypt are very much higher than in most European countries (see Chapter 3), and cardiovascular disease mortality is now the principal cause of death in many developing countries. Smoking is promoted for commercial reasons in developing countries, without countervailing health hazard warning labels, limits on advertising, and other restrictive legislation. This warrants attention by health authorities in developing countries and application of lessons learned in the industrialized countries over the past 30 years.

CHRONIC LUNG DISEASE

Chronic lung disease (CLD) is an important, diverse, and mostly preventable group of diseases which cause extensive morbidity and mortality. In 1995, CLD was the fifth leading cause of death in the United States accounting for 5% of all deaths, and measurable impairment of lung function in 4–6% of the U.S. population, up to 13% in some population groups. CLD can largely be prevented with good primary care and education for self-care.

When associated with acute respiratory infection, such as influenza, bronchitis, or pneumonia, CLD can result in lengthy hospitalization and premature death. Cough, shortness of breath, restricted exercise tolerance (e.g., climbing stairs), and difficulty sleeping are frequent symptoms, with impaired clearance of sputum and reduced lung capacity.

Asthma

Asthma is an intermittent, reversible condition of airway obstruction in response to various stimuli, resulting in wheezing and shortness of breath due to variable airflow. Usually appearing first in children up to age 5 years, it affects an estimated 14–15 million persons in the United States, including 4.8 million of those (6.9%) aged 0–18 years. It is the most common chronic disease among children.

During the 1980s, asthma prevalence and mortality increased in the United States and other countries. In the United States between 1982 and 1991, age-adjusted death rates from asthma per 1 million persons aged 5–34 years of age increased by 40% overall, but more for females. Annual mortality rates were more than fivefold higher in the black population than among whites. In 1994–1995, asthma was the cause of death for 11,274 persons (2.1 per 100,000 population).

Although the specific etiology of asthma is unknown, it is associated with familial, infectious, allergic, environmental, and psychosocial factors. Risk factors include animal allergens (usually from pets), household dusts and mites, primary and secondary cigarette smoke, outdoor allergens, and pollutants. Air pollution may be a factor in the increase in asthma; 63% of cases live in areas where pollution exceeds recommended levels. Ozone pollution, the result of hydrocarbons and nitrogen oxide emissions from motor vehicles or other sources mixed in the presence of sunlight, causes increased wheezing, coughing, and chest tightness, especially among susceptible children who play outdoors in polluted environments. The higher rates of asthma among blacks in the United States may be related to lower socioeconomic conditions and residence in inner city communities. Conversely, keeping children indoors for safety reasons in poorly ventilated older homes may contribute to increasing asthma.

Identification and removal of antigens, the use of bronchodilators and corticosteroids, and treatment of infections as they occur are the major methods of management. Education plays a key role in asthma control, for patients, their families, school personnel, and for the general public. Mortality may be increased by medication overuse, substance abuse, and cigarette smoking. Health care providers need continuing education regarding this condition and its management, especially during pregnancy, and regarding the possibilities and problems associated with medication usage.

Chronic Obstructive Pulmonary Disease

The term chronic obstructive pulmonary disease (COPD) represents advanced stages of chronic respiratory disease with airflow impairment due to chronic bronchitis affecting the smaller airways. This includes a variety of conditions resulting from damage to lung tissue, chronic narrowing of the respiratory tract, and obstruction of airflow. The term includes chronic bronchitis, emphysema, and other causes of COPD. Chronic bronchitis affects 20% of the adult male population of the United States. Chronic lung disease was the cause of 100,000 deaths in 1994–1995.

Chronic bronchitis and asthma can lead to emphysema, with fixed expansion and rigidity of lung tissue and reduced oxygen exchange capacity. Increasing shortness of breath, cough, loss of exercise ability, sleeplessness, repeated infections, hospitalizations, and death are common. The triad produces respiratory cripples who may function with limitations until a respiratory infection causes hospitalization and ultimately death.

Tobacco use is the greatest cause of COPD. Reduction of smoking is crucial to prevention, as is reduction of indoor pollution in the workplace and home, and outdoor pollution (see Chapter 9). Secondary and tertiary prevention should include annual influenza vaccination and pneumococcal vaccine usage. Careful monitoring of the patient at home for changing symptoms and early signs of infection would prevent long and costly hospitalization.

Occupational Lung Diseases

Occupational lung diseases are a group of conditions associated with workplace exposures to dusts and vapors, which act as irritants, carcinogens, or immunologic agents. Primary prevention by reducing exposure levels and secondary prevention by close medical follow-up of exposed workers and ex-workers for many years after exposure when the associated diseases become apparent are an important part of occupational health. The risk of such exposures causing serious disease is accentuated by cigarette smoking and environmental pollution (see Chapter 9).

Coal Workers Pneumoconiosis.   Frequently called black lung disease, coal workers pneumoconiosis (CWP) is due to prolonged exposure to coal dust for 10 years or more, with diagnosis by chest X-ray or by biopsy.

Silicosis.   Silicosis is due to chronic inhalation of crystalline silica. The disease progresses over 20–40 years from cough and sputum production to crippling COPD due to massive fibrosis of lung tissue. Some 2000 cases are reported in the United States annually. Persons suffering from silicosis are at increased risk for tuberculosis and should undergo routine testing.

Asbestosis.   Asbestosis is caused by asbestos exposure. It is a fibrous deterioration of lung tissue associated with increased rates of cancer of the lung (especially among smokers) and mesothelioma, a rapidly lethal form of cancer specific to occupational or community exposure to asbestos products.

Byssinosis.   Byssinosis is both an acute and chronic disease caused by exposure to cotton dust, flax, or hemp resulting in shortness of breath, chest tightness, and chronic cough. Exposure over periods of 10 years or more causes reduced pulmonary function and COPD.

Occupational Asthma.   Occupational asthma is bronchial restriction following exposure to agents to which the individual has become sensitized. Asthma with airflow obstructive symptoms may be severe, and may be chronic. The variety of occupations at risk ranges from electronic workers to hairdressers, involving people exposed to a wide variety of dusts, chemical agents, and animal antigens.

DIABETES MELLITUS

Diabetes mellitus is a common chronic condition with disturbed carbohydrate metabolism resulting from deficiency in production of insulin in the pancreas or impaired function of the insulin produced, resulting in increased levels of blood glucose. Diabetes is a major disease in its own right and a risk factor for cardiovascular diseases, including coronary heart disease, stroke, and peripheral vascular disease, as well as damage to nerves, kidneys, eyes, and other organs. There are two major types of diabetes: Type 1 or insulin-dependent diabetes mellitus (IDDM) and Type 2 or non-insulin-dependent diabetes mellitus (NIDDM).

Diabetes prevalence increased in the United States by 69% during the period

1959–1966, a further 41% from 1966 through 1973, by 21% from 1973 through 1980, and by 4% from 1980 through 1989. In 1996, diabetes affected 15.7 million Americans, one-third of whom were undiagnosed. Over 700,000 new cases of diabetes are diagnosed annually, including some 12,000 among schoolchildren. Prevalence rates were approximately twice as high among blacks than among whites in the United States. Reported diabetes rates increased by 28% among black males during 1980–1989. Native Americans and Hispanic peoples also have high rates of diabetes and its complications. Diabetes, which was ranked among the 10 leading causes of death since the 1930s, in 1996 was the seventh leading cause, with over 61,000 deaths directly attributed to the disease. Diabetes is also a major contributor to cardiovascular disease, the leading cause of blindness and end stage renal disease, and lower limb amputations.

The discovery of insulin and its use by Frederick Banting and Charles Best in Toronto in 1921 (Nobel prize, 1923, to Banting) gave patients with IDDM (then called juvenile diabetes and later Type 1 diabetes) the opportunity to live full lives provided they were carefully managed. The more recent development of simpler

---

BOX 5.4 DIABETES IN THE UNITED
STATES—PREVALENCE,
COMPLICATIONS, COST, AND
PREVENTABILITY

Diabetes is the seventh leading cause of death in the United States. Diabetes affects some 16 million Americans, with half being unaware they have the disease. They are at very high risk for ischemic heart disease, stroke, kidney, eye, and peripheral vascular disease as well as direct effects of diabetes. Controlling high blood pressure could reduce stroke death rates in people with diabetes by 75–90%, and coronary heart disease by 25–50%. Diabetic eye disease is the commonest cause of blindness, with 25,000 new cases annually, 90% of which could be prevented by early diagnosis and treatment. Diabetic lower extremity disease causes 57,000 leg, foot, and toe amputations per year; more than half of these could be prevented, with cost savings of $600 million annually. Direct and indirect costs of diabetes in the United States amount to more than $92 billion annually, and account for 27% of Medicare costs or $30 billion annually. Community-based intervention programs include improved prevention, education, and service designed to reduce complications of diabetes. Such programs should be directed at populations with high prevalence rates of diabetes, and epidemiologic monitoring should attempt to identify high prevalence groups.

Source: United States Senate Appropriations Subcommittee on Labor, Health, and Human Services (Chairman, Senator Arlen Specter), *Action Alert, The Nation's Health,* August, 1996.

monitoring, improved insulin preparations, and insulin pump devices have made management easier and more effective for the Type 1 diabetic. Insulin preparations that may be inhaled are under clinical trial. Oral hypoglycemic medications provide effective help for Type 2 diabetics in combination with weight reduction, physical fitness, and careful dietary management.

Using national mortality and hospitalization databases, it is possible to compare regional patterns of diabetes and its complications including coronary heart disease, cerebrovascular disease, eye disease, and reduced blood supply to lower limbs. Identifying population groups with higher incidence/prevalence of these conditions permits development of targeted community-based intervention programs to improve early diagnosis and management of diabetes and related conditions.

Prevention of Diabetes and Its Complications

Prevention is feasible for diabetes at the primary, secondary, and tertiary levels.

1. Health promotion involves education of the public to a high level of awareness of diabetes, its risk factors, and its complications.

2. Primary prevention identifies obesity as a risk factor for diabetes, and promotes its prevention through good nutritional practices to reduce the risk of Type 2 diabetes.

3. Secondary prevention to prevent compilations of diabetes is aided by early case finding and management. Routine screening for diabetes will uncover early cases for whom treatment can reduce the severity of complications. Screening is recommended for:

   a. persons with family history of diabetes;
   b. patients with cardiovascular, renal, and eye diseases;
   c. during pregnancy;
   d. follow-up for women with glucose intolerance in pregnancy, or history of infants weighing over 4000 g;
   e. obese persons.

For persons with abnormal blood glucose levels, follow-up management includes instruction and monitoring to prevent serious complications. Diet, exercise, regular urine and blood sugar testing, personal hygiene, and foot care should be followed closely in ambulatory and home care. In addition, medication, usually oral hypoglycemic agents, are sometimes needed to control blood sugar levels in Type 2 diabetes. Insulin is always needed for control for Type 1 diabetes. Social and medical support may be needed for persons unable to care adequately for themselves, especially elderly, poor, and poorly educated people.

4. Tertiary prevention aims to restore function and prevent further deterioration. For those with established peripheral vascular disease and potential gangrene, foot care can delay amputations; for amputees, rehabilitation care is essential to prevent total dysfunction. Close follow-up of diabetics by ophthalmologists promotes prevention of diabetic retinopathy by early treatment by photocoagulation (Box 5.5).

The St. Vincent Declaration, developed at a World Health Organization sponsored conference in 1990, set targets for reducing the complications of diabetes mellitus for the European region of the World Health Organization member countries (Box 5.6). The European region of the WHO has promoted development of national diabetes programs (NDPs) in countries from western Europe to those in central Europe and the central Asian republics of Kyrgyzstan and Azerbaijan. Each is created by broad-based working groups in each country with representatives of ministries of health, diabetes nursing and patient's groups, as well as medical experts in the field. Similar initiatives launched in 1997 in the United States are promoting the National Diabetes Education Program (sponsored by the CDC and NIH) with participation of the American Diabetes Association and other NGOs. Such programs are intended to promote public and health care provider awareness of the seriousness of the problem and promote integrated approaches to care and improved access to diabetes care.

END STAGE RENAL DISEASE

End stage renal disease (ESRD) is defined as reduced renal function (to <10% of normal capacity) requiring dialysis or kidney transplantation for survival. ESRD follows severe kidney damage from infection, glomerulonephritis, hypertension, drug reactions, or diabetes. About 40 persons in 100,000 in the United

---

BOX 5.5 PREVENTION OF BLINDNESS FROM DIABETES: STOCKHOLM REGION, 1989–1996

A program of early prevention to prevent blindness in diabetics in Stockholm County (population 1.7 million) in Sweden was initiated in 1989, with mailings to 10,000 known diabetics registered in 100 primary care centers. The program involved 500 general practitioners and 2 hospital departments of internal medicine and pediatrics, creating a diabetic registry. Biannual examinations were carried out by mobile teams using a standard protocol (the London protocol) including history, test of visual acuity, high quality fundus photography, and tonometry. Photographs were graded and externally validated at a second center (the Hammersmith Hospital, London). Cases registered were followed and referred for early photocoagulation. Annual incidence of all cases of blindness per 100,000 residents of the region decreased by 46% (from 0.63 to 0.34 per 100,000).

Source: Karolinska Institute, St. Eric's Eye Hospital and the Department of Social Medicine, Uppsala University, as reported in EuroDiabcare, *St. Vincent's Newsletter,* Summer, 1996.

BOX 5.6 THE ST. VINCENT DECLARATION ON DIABETES MELLITUS

Goal:

Sustained improvement in health experience and a life approaching normal expectation in quality and quantity: prevention and cure of diabetes and its complications by intensifying research efforts.

Objectives:

1. Better diabetes care through constructive use of information;
2. Better diabetes care through demonstration and education;
3. Better diabetes care through partnerships.

Methods:

1. Develop comprehensive programs for detection and control of diabetes, with community support;
2. Raise awareness of the population, and health care professionals;
3. Organize training and teaching in diabetes care;
4. Ensure care of diabetic children by specialized teams;
5. Reinforce existing centers of excellence in diabetes;
6. Promote independence, equity, and self-sufficiency for diabetics;
7. Remove hindrances to integration of diabetics into society;
8. Ensure quality assurance for laboratories and technical procedures for diabetics;
9. Promote international cooperation in diabetes research and development;
10. Take urgent action to initiation and accelerate cooperative activities to implement these recommendations.

Targets:

1. Reduce new cases of blindness due to diabetes by $\frac{1}{3}$ or more;
2. Reduce new cases reaching end stage renal disease by $\frac{1}{3}$ or more;
3. Reduce cases of limb amputations for diabetic gangrene by $\frac{1}{2}$;
4. Reduce morbidity/mortality from coronary heart disease in diabetics by programs of risk factor reduction;
5. Improve pregnancy outcome in diabetics to that of nondiabetic women.

Source: World Health Organization, Regional Office, Europe and International Diabetes Federation (Europe), 1997. Website: http://www.who.dk/cpa/pr99/pr9928e.htm

States have ESRD; almost 100,000 persons are on chronic dialysis, with more than 20,000 persons having had renal transplantation. Almost half of these are due to diabetes mellitus and most follow a long period of chronic renal failure. Prevalence varies widely between different countries and among ethnic groups in the United States. High rates occur among some ethnic groups such as Mexican and Hispanic Americans, African-Americans, North American Indians, Maoris, and Australian aborigines, probably due to high rates of diabetes and hypertension in these groups.

Prevention efforts to reduce the prevalence of ESRD should include the following:

1. Identification and effective treatment of streptococcal throat infections;
2. Careful use of medications with potential for renal damage;
3. Prompt treatment of urinary tract infection;
4. Screening for the early detection of diabetes mellitus and hypertension;
5. Proper monitoring and control of diabetes mellitus and hypertension.

End stage renal disease is an important public health issue because it is partly preventable and it is a large consumer of health care resources. ESRD accounted for $5.4 billion in medical costs in the United States in 1988. Kidney transplantation is the most cost-effective method of treatment, about one-third the cost of long-term hemodialysis, but is seriously restricted by lack of donors. Continuous ambulatory peritoneal dialysis is a cost-effective option, especially where donors are limited. Usually done by the patient or his family at home, peritoneal dialysis allows the patient to engage in normal daily activities, and eliminates costly hospitalization and hemodialysis equipment needs.

The United Kingdom has adopted a national health target to reduce ESRD by one-third, based on recommendations of diabetologists and the St. Vincent Declaration. The United States health targets call for reducing ESRD prevalence by the year 2000 to 13 per 100,000 population, from nearly 14 in 1987, and for the black population to 30 per 100,000.

Ethical issues in ESRD occur in both developed and developing countries. Some developing countries with specialty medical centers use donor kidneys purchased from poor persons, providing kidneys for medical tourists. In developed countries ESRD prevalence is increasing as the population ages as this is an age-related condition. Hemodialysis is a lifesaving procedure, and as costs of health care increase, important ethical issues arise as use of hemodialysis in very debilitated patients is increasing.

CANCER

Cancer is the second leading cause of death in the industrialized countries, and is rapidly emerging as a major factor in the epidemiology of developing countries. In the United States, the lifetime probability of developing cancer is 1 in 3. From 1930–

1990, there was an increase in age-adjusted cancer mortality rates, but they have been declining since (Table 5.8). The rise in lung cancer rates account for the increase, while incidence of new cases for some, mainly stomach, uterine, and colo-rectal cancers have declined. Breast cancer mortality has remained relatively constant.

Cancer is responsible for much suffering and heavy economic demands on society in terms of health services, loss of work, and premature mortality. The potential for prevention of cancer is so important that any public health program must include it as part of its duties.

Many advances have been made in clinical management and in understanding of the causes of some cancers. There is strong and increasing evidence of specific risk factors which are amenable to public health intervention. The most striking example is smoking, directly related to lung cancer, and a factor in bladder and cervical cancers. Exposure to chemical carcinogens, such as asbestos, is of major public health interest, as is prevention of liver cancer by immunization against hepatitis B. Recent research indicates that dietary factors are important contributors to colorectal, breast, and possibly lung and cervical cancers as well. As discussed earlier, recent evidence linking *H. pylori* infection as the cause of gastritis and stomach cancer, and incidence of both is falling internationally.

In the United States, rates of all cancers among blacks are higher than whites (relative risk of 1.1), while cancer mortality is still higher (1.3:1). Cancer rates also vary widely among other ethnic groups, with Japanese, Filipino, American, Indian, and Mexican Americans having low age adjusted rates. Japanese-Americans have much higher cancer rates than Japanese of Japan, strongly suggesting that lifestyle differences, such as diet and smoking, are the causation or trigger events.

There are important social class difference in cancer mortality in the United States. In the United States, the relative risk (RR) for black men to die of lung cancer was 0.8 compared to whites in 1950, but by 1990 the RR had increased to 1.5. These differences, like those of higher rates of CHD mortality among African-American males, were thought to be race-related, but the difference are largely explained by socioeconomic differences, perhaps related to different smoking, oc-

TABLE 5.8　Age-Adjusted Death Rates from Malignant Neoplasm of the Lung,[a] by Sex and Race, United States, Selected Years, 1950–1996

Group	1950	1960	1970	1980	1990	1996
Total population	12.8	19.2	28.4	36.4	41.4	39.3
White male	21.6	34.6	49.9	58.0	59.0	52.6
Black male	16.9	36.6	60.8	82.0	91.0	78.5
White female	4.6	5.1	10.1	18.2	26.5	28.0
Black female	4.1	5.5	10.9	19.5	27.5	27.6

Source: *Health, United States*, 1998.

[a]Rates per 100,000 population, standardized to the 1970 U.S. population. Lung includes trachea and bronchus.

cupation, dietary patterns and lesser access to medical care. This is borne out by studies by mortality rate differences by social class seen in the United Kingdom and the Scandinavian countries.

Lung cancer is the leading cancer cause of death in the United States with 149,000 deaths in 1993 out of 500,000 total cancer deaths. Lung cancer is also the most common cancer worldwide, 2.3 times more common in men than women, but has surpassed the breast as the leading cancer site in women. Survival rates are low (13% 5-year survival in the United States), so that screening for lung cancer is of dubious value. Primary prevention to reduce the number of new smokers and to encourage cessation of smoking are the most crucial public health activities for lung cancer prevention.

Cancer patterns for women in the United States, seen in Table 5.9, show a slight decrease in total cancer mortality rates between 1950 and 1996, while lung cancer increased dramatically. Over the same period, breast cancer among these women remained almost constant. In 1987, lung cancer surpassed breast cancer as a cause of death among women; however, breast cancer remains a greater cause of potential years of life lost, since it affects younger women than lung cancer.

Cancer mortality rates for 1970–1994 for men and women in Canada have remained stable for total cancer. There has been a fourfold increase in lung cancer mortality for women, but declining mortality from colorectal, uterus, cervix, and stomach cancer. Mortality from cancer of the breast and ovary remained essentially the same.

Cancer incidence rates vary widely between countries and within countries. A comparison of total cancer mortality between some countries in the WHO European Region show very marked differences, as seen in Figure 5.2, with declining rates since the mid-1970s in some countries (e.g., Finland), and since the mid-1980s in others (e.g., United Kingdom), and rising rates (e.g., Denmark and Spain). Even neighboring Scandinavian countries show quite marked differences in total, lung, and breast cancer. The explanations of such differences are not at all clear, but there may be differences both in risk factors, for example, environment, diet, smoking, as well as in treatment. Such international comparisons can generate hypotheses for further investigation.

TABLE 5.9   Age-Adjusted Cancer Death Rates for Females by Selected Cancers,[a] United States, Selected Years, 1950–1996

Cancer	1950	1960	1970	1980	1990	1996
Respiratory tract	4.6	5.2	10.1	18.3	26.2	27.5
Breast	22.2	22.3	23.1	22.7	23.1	20.2
Total (all cancer)	120.8	111.2	108.8	109.2	112.7	108.8
Total mortality (all causes)	688.4	590.6	532.5	432.6	390.6	381.0

Source: *Health, United States*, 1998.
[a]Rates per 100,000 population, standardized to the 1970 U.S. population.

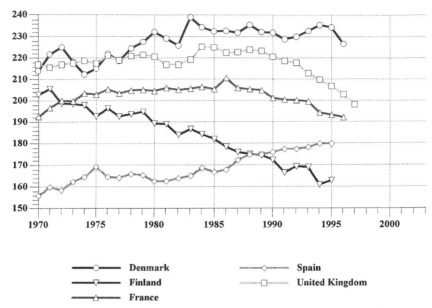

	Denmark		Spain
	Finland		United Kingdom
	France		

FIGURE 5.2   Standardized mortality rates (per 100,000 population) from all malignant neoplasms, selected European countries, 1970–1997. WHO European Region. Health for All Data Set, 1999.

Lung cancer mortality declined in countries such as Finland by 28% from 1981 to 1995, while in the United Kingdom, this rate declined by some 17% from its peak in the mid-1980s. In the United Kingdom, lung cancer mortality declined between 1969 and 1989 for men in the age group 55–64 by 37% and for men age 45–54 by 52%. During the same period lung cancer mortality rates for U.K. women increased by 71% for the age group 55–64. The relative risk of men to women dying from lung cancer in the age group 55–64 declined from 6:1 to 2.7:1 over this time period.

Prevention and Cancer

Primary prevention in cancer requires a reduction of major risk factors: smoking (lung and bladder), fatty diet (breast and colorectal), carcinogenic chemical exposures (mesothelioma, lung, leukemia, lymphoma), exposure to infectious carcinogenesis agents (hepatic), multiple sexual partners (cervical cancer), and excess sunlight exposure (melanoma). Lung and skin cancer, occupational related cancers, and some gastrointestinal cancers are most amenable to primary prevention, by reduced smoking and exposure to sunlight, asbestos, radon, and other carcinogens. Immunization against hepatitis B is protective against most liver cancer.

Secondary prevention in clinical services or in screening programs for high risk groups focuses on early case finding. The medical care system has a major role to play, with a variety of activities, such as screening for breast, prostate, cervical, and colorectal cancer. Occupational, environmental, and social factors all play important roles in cancer. As a result, intersectoral preventive activities are needed for cancer prevention.

Cancer prevention involves reaching the total population to increase awareness of risk factors and promoting change in them. Health care providers have a vital role to educate the public, to decide when and how to screen for disease, and to cross socioeconomic barriers in reaching patients.

More research will bring important breakthroughs in this field. Identification of genes that increase risk or cause cancer, infective agents associated with carcinogenesis, medications which stop or delay progress of cancer and other research appear very promising. The evidence for a causal relationship between nutrition and cancer is increasingly convincing.

BOX 5.7 CANCER PREVENTION

1. Primary prevention
    a. Smoking cessation.
    b. High vegetable, fruit and grain, low fat diet.
    c. Eliminate exposure to asbestos products.
    d. Limit exposure to high intensity sunlight.
    e. Reduce exposure to chemical carcinogens, e.g., pesticides.
    f. Hepatitis B vaccination.
    g. Test homes for radon and reduce levels where high.
    h. Reduce sexual risk behavior of multiple partners.
2. Secondary prevention—early diagnosis
    a. Lung cancer: Screening by X-ray and or pulmonary cytology is not currently cost-effective nor does it contribute to improved outcome of lung cancer.
    b. Breast examination: monthly self-examination, and mammography every 2–3 years for women aged 40–50 and annually or more frequently for those over 50. High risk populations are white women of higher socioeconomic status over age 40, low or zero parity, women with family history of breast cancer, benign breast lesions, or previous breast cancer.
    c. Colo-rectal cancer: rectal examination annually for men and women over age 45, including stools for occult blood (three samples), and sigmoidoscopy for high risk persons over age 50. High risk groups include those over age 45 and persons with colitis, familial polyposis, or familial cancer of the colon.
    d. Cervical cancer: screening by Pap smear to detect neoplasia is

cost-effective especially for high risk groups. Recommended for screening
by the 1996 U.S. Task Force every 1–3 years depending on the following
risk factors:

     i. Sexually active women;
    ii. Low socioeconomic status;
   iii. Prison inmates or prostitutes;
   iv. History of STDs;
    v. Early onset of sexual intercourse;
   vi. Multiple partners;
  vii. Previous induced abortions;
 viii. Previous squamous cell dysplasia;
   ix. Unmarried mothers.

Source: Adapted from Canadian Task Force on Screening (1994) and U.S. Task Force on
Screening (1996).

CHRONIC LIVER DISEASE

Chronic liver disease and cirrhosis together were the seventh leading cause of
death in the age groups 25–64 in the United State in 1996. Cirrhosis is also a major contributing cause of morbidity, causing 1% of all hospital discharges. Patients
with liver failure are high users of hospital care mainly because of the serious complications, such as gastrointestinal bleeding from esophageal varices. This is a
group of diseases related to chronic alcohol consumption and chronic viral hepatitis infection (mainly hepatitis B and C). The risk of cirrhosis is high among long-
term heavy users of alcohol and is related to amounts consumed daily. Other nutritional factors, such as vitamin B deficiency, may be secondary contributing factors.

Hepatitis B infection, transmitted in blood products, in body fluids, and in
household contacts, is common in many countries. An estimated 2 billion persons
are infected, and along with 350 million carriers they are at high risk for cirrhosis
of the liver and primary liver cancer. Hepatitis C prevalence worldwide is reported by the WHO as 170 million cases, with infection rates varying from under 1%
in Canada, Australia, and parts of western Europe, to 1–2.4% in the United States
and much of Europe, India, and most of South America, and to rates of up to 10%
of the population in China, much of Africa, and South America. Aflatoxin exposure in foods is a major contributor to chronic liver disease in developing countries.

Liver cancer and liver cirrhosis are major public health problems, with hepatitis B being the cause of 60–80% of primary liver cancer, especially in developing
countries of sub-Saharan Africa, east and southeast Asia, and the Pacific basin. The
WHO estimates there were 505,000 primary liver cancer deaths globally in 1997.
As there are over 2 billion persons infected and 350 million carriers, the WHO recommends prevention by inclusion of hepatitis B vaccine in routine infant vacci-

nation programs and catch-up immunization of other age groups. Hepatitis C, discovered only in 1988, is estimated to affect up to 3% of the world's population, and chronic carriers are also at risk for liver cirrhosis and primary liver cancer. Screening of blood and blood products for hepatitis C should be standard practice worldwide; however, the virus is commonly spread through unsanitary intravenous drug usage. There is still no vaccine for hepatitis C. Prevention of cirrhosis focuses on reducing daily consumption of alcohol and promoting universal immunization against hepatitis B. Needle exchange programs reduce transmission of hepatitis among intravenous drug users.

ARTHRITIS AND MUSCULOSKELETAL DISORDERS

Arthritis and musculoskeletal conditions are among the most common causes of physical disability, visits to doctors, and hospitalizations. Arthritis is the leading cause of disability in the United States, affecting as much as half of the population over age 65. It was associated with total direct and indirect costs of nearly $65 billion in the last National Health Interview Survey (1992). Arthritis and other rheumatic conditions (bursitis, lupus, fibromyalgia) are among the most common chronic conditions, affecting an estimated 43 million persons in the United States; by 2020 rheumatoid arthritis (RA) is expected to affect 60 million persons. It constitutes the leading cause of disability and is two to three times more common in women than men, also occurring in children. In 1997, arthritis and musculoskeletal disorders accounted for 2.4% of hospital admissions (744,000 admissions, 61% women) and days of care (approximately 4 million days, 66% women) in the United States.

Osteoporosis.   Osteoporosis is a major cause of fractures, disability, and death, as well as costly institutional care. The U.S. National Health Examination Survey study of 1996 estimated that between 26 and 38 million Americans have osteoporosis and are at risk for hip fractures, related fractures occurring annually, primarily among persons 65 years of age and over. The annual costs of medical care for fractures in 1986–1996 among this group increased from $7 to $14 billion. Postmenopausal women are the primary group at risk, due to low bone mass. The lifetime risk of a woman to suffer an osteoporosis-related fracture is about 40%, and the chance of developing a hip fracture equals her combined risk of developing cancer of the breast, uterus, or ovary combined (U.S. National Osteoporosis Foundation, website 1999 http://www.nof.org/). Screening of those at risk by X-ray assessment of bone density is now being replaced by a relatively inexpensive, portable ultrasound instrument for use in ambulatory clinics. Postmenopausal women presenting with hip fractures show vitamin D deficiency.

Osteoporosis is amenable to primary prevention and reducing risk factors. Primary prevention is directed toward adolescent, young adult, and perimenopausal women to assure adequate physical activity, avoidance of smoking, adequate dietary intake of calcium and vitamin D, reduction of excess alcohol consumption, and prevention of falls. For postmenopausal women, prevention should also include home safety measures, bone density screening, and hormone replacement therapy to inhibit bone resorption.

Degenerative Osteoarthritis. Osteoarthritis is a degenerative disorder, increasing in prevalence as a population ages. It is especially common in knees in women and in hips for men. It is strongly correlated with both obesity and increasing age. Reduced capacity in performing one or more activities of daily living may affect as many as 21% of people 18 years of age and older and 43% of persons over 55 years of age. This results in a heavy burden on the economy from medical costs and reduced productivity.

Reduction of obesity is the major preventive modality, requiring personal counseling and a climate of community attitudes that promotes healthy and sensible weight reduction. Treatment focuses on relief of symptoms and increasing mobility. Surgical replacement of hip or knee followed by physiotherapy and other rehabilitation measures are well-established procedures to prevent deterioration and dependency and improve quality of life.

Rheumatoid Arthritis and Gout. Rheumatoid arthritis (RA) is an autoimmune disease causing chronic inflammation of joints with stiffness, pain, deformity, and limitations of activities of daily living, affecting as many as 1% of adults. It is two to three times more common in women than men, and can also occur in children. RA is associated with premature mortality from infectious diseases of the respiratory tract. Supportive medical and other care should include immunization (e.g., influenza and pneumococcal pneumonia) as well as treatment of the primary condition mainly with anti-inflammatory agents.

Gout is a metabolic disorder, causing deposition of uric acid crystals in and around joints, especially those in the foot. Gout is also associated with high lead exposure in certain occupational groups, such as painting, plumbing, and ship building. National health survey data in the United States report a prevalence rate of 6 and 13 per 1000 for women and men, respectively. Reduction of lead exposure and early diagnosis and management of gout improves outcome and quality of life. Follow-up care is essential in management.

Low Back Syndromes. Low back pain from muscle injury, abnormalities of the vertebrae, disks, or joints, with or without compression of the lumbar nerves (sciatica), cause direct and indirect costs to the U.S. economy of $15–25 billion annually. It is especially common among industrial workers, equally for men and women and mainly in the 20–40 year age group. Recovery is usual (90%), but chronic states often lead to expensive and mostly inappropriate surgical interventions. Prevention involves industrial engineering and education to reduce back strain, especially in heavy lifting work.

NEUROLOGICAL DISORDERS

Neurological disorders are an important burden on the affected individual and society in terms of disability, loss of productivity, premature mortality, and in health costs. The direct annual cost of chronic neurological disorders exceeds $140 billion in the United States, and the loss of 400 million working days. This group of diseases includes multiple sclerosis, Parkinson's disease, neurodegenerative disorders, and Alzheimer's disease. Causes of neurological disorders, include acute and chronic trauma to the brain and peripheral nerves, infection, and chemical poisoning.

Alzheimer's Disease.   Alzheimer's disease is a brain disorder occurring later in life, possibly related to a genetic disorder. It is the leading cause (50–60%) of dementia among adults. It usually occurs after the age of 50 years, and more commonly in women than in men (1.6:1). There is no primary prevention as yet, but there is some evidence of benefit from hormone replacement and vitamin supplement therapy. Case management requires support for the family caregivers from community health resources. Other dementias, or organic brain syndromes, are due to cerebrovascular disease, Parkinson's disease, Creutzfeld-Jakob disease, and AIDS.

Parkinson's Disease.   Parkinson's disease is common after age 50, with characteristic tremor, stiff walking gait, slowness of movement, and muscular rigidity. Rates vary from 108 to 347 per 100,000 in northern Europe to 44–81 per 100,000 in Asia, Africa, and the Mediterranean. Genetic susceptibility is suspected. Treatment aimed at improving functional status is important along with support for caregivers, again to promote maximum independent function and avoid institutionalization. As the muscular rigidity affects chest muscles, respiratory infections are common, so that immunizations for influenza and pneumonia are valuable.

Multiple Sclerosis.   Multiple sclerosis (MS) is a disorder of the myelin sheath of neurons, leading to impairment of vision, weakness, tremor, incoordination, and loss of sensation and bladder and bowel control. It most commonly occurs between the ages of 20 and 50, and primarily in areas distant from the equator, for example, Canada, northern Europe, and Australasia. There is supportive evidence that this disease may be genetic, but other evidence points to an infectious agent origin. Prevention in MS focuses on good case management to reduce complications and promote rehabilitation. New medications are thought to reduce relapses and increase the length of remissions in this disease. Immunizations for influenza and pneumonia are recommended.

Epilepsy or Seizures.   Epilepsy is characterized by uncontrollable convulsions starting abruptly, with or without warning symptoms, and with or without loss of consciousness. These are due to disturbances of cerebral function due to abnormal electrical activity in the brain. Isolated seizures can occur in anyone with brain hypoxia, hypoglycemia, and in children with fevers.

The World Health Organization estimates that epilepsy affects one in every 130 persons, or 40 million persons worldwide, 80% of whom are in developing countries, with some 2 million new cases per year. Genetic factors, infections, and brain injuries are among the major causes of epilepsy. These include infection and toxicity in the prenatal period (e.g., maternal cocaine use), asphyxia and trauma during birth, postnatal infections causing febrile convulsions, infections of the central nervous system (i.e., meningitis and encephalitis), parasitic disease (e.g., malaria, schistosomiasis), and brain damage by alcohol, trauma, and toxic substances (e.g., lead, pesticides). Prevention of epilepsy is an important reason for good quality prenatal care, safe delivery, complete infant and childhood immunization, control of fever in children, control of infectious and parasitic diseases, reduction of brain injury (e.g., home and automobile accidents), and genetic counseling.

Brain and Spinal Cord Injury. Traumatic brain injury (TBI) is associated with 52,000 deaths and accounts for one-third of all injury deaths in the United States. Approximately 80,000 persons who survive TBIs incur some loss of function, residual disability, and increased medical care needs. Head injury causes prolonged hospitalization and often irreversible brain damage. The main causes are automobile injuries (40–50%), falls (25–30%), and sports or recreation injuries (5–10%).

Spinal cord injuries are extremely common results of automobile crashes (50% of cases), recreational injuries (15%), and falls (15%). If the injury is at the cervical spine level it causes quadriplegia and if lower in the spine, paraplegia. This condition is most common among young adults aged 15–24. The estimated number of cases annually in the United States is 110,000–200,000, requiring extensive hospital and rehabilitation services.

Brain and cord injuries are prevented by safer motor vehicle, work, and home environments. Helmets for bicycle and motorcycle drivers, football and hockey players, and other sports with records of head injury can reduce the incidence and seriousness of head trauma. Safe transportation of the injured and management in emergency rooms of hospitals may help reduce the extent of brain damage from the trauma.

VISUAL DISORDERS

Blindness is defined as visual impairment sufficient to prevent the person from performing work for which sight is essential. The WHO estimates global blindness at 45 million and 180 million visually impaired, with the vast majority in developing countries where blindness prevalence rates commonly exceed 1 to 2% of the population. Globally, the major causes of blindness are as follows (in millions): cataract (19), glaucoma (6.4), trachoma (5.6), childhood blindness (1.5), onchocerciasis (0.3), and other causes (10).

The major causes of blindness in the United Kingdom, for persons over 65 years of age, include senile macular degeneration (47%), cataract (20%), glaucoma (12%), diabetic retinopathy (2%), and others (19%). A study in England showed male rates increasing from 3 per 100,000 population in the 15–29 year age group to over 400 in the 75 and over group, and from 2 to 475 per 100,000 among females in these age groups.

Vitamin A deficiency is a common cause of visual impairment in children under age 5 in developing countries, causing blindness in half a million children per year and visual impairment in millions more. International efforts are being made to prevent this by vitamin A supplements. Onchocerciasis (river blindness), responsible for 1 million blind inhabitants of Africa and a smaller number in Latin America, is responding to improved control measures, with important economic as well as health benefits in many countries in West Africa.

Untreated gonorrhea, syphilis, measles, cataracts, and glaucoma are also important causes of blindness in developing countries. Trachoma, which when untreated leads to marked conjunctivitis and eyelid deformities causing conjunctival abrasions, can lead to blindness. This is widespread in the Middle East, Africa, and in some parts of Latin America. It is a disease of poverty, crowded living conditions, and lack of sanitation. A high percentage of cases of blindness are totally preventable by basic public health measures. Simple, inexpensive treatment are cost-effective and readily applied on a wide scale where there are planned governmental or NGO programs.

Prevention of blindness requires careful treatment of diabetes, screening of and treatment for glaucoma, cataract removal, and care of eyes using sunglasses in high sunlight areas. Public health measures also include use of safety glasses, shatterproof windows, seat belts and air bags on cars, prevention of infectious causes (STDs, measles, rubella), early treatment of eye diseases, and proper control of oxygen in incubators to prevent congenital retinal atrophy.

HEARING DISORDERS

Hearing loss is an important handicapping condition. Those who are deaf without speech can learn to communicate by hand and finger signs or writing. Those with minimal hearing may learn to lip-read and to speak. Hard of hearing persons have some useful hearing but require supplemental lipreading. The psychological stress of deafness on the individual, family, and the community should be considered in developing prevention programs. Detection and correction of hearing loss can have a profound effect on an individual's well-being.

Hearing loss is probably the most common handicapping medical condition, affecting large percentages of the population. The WHO estimates that 121 million people worldwide have disabling hearing impairment. In the United States, 10 million people have noise-induced hearing loss; another 20 million are exposed to hazardous noise levels in their place of employment. Prevention programs can reduce the burden of this problem.

Hearing loss may be conductive or sensory due to a neurological defect. Conductive loss is due to obstruction in the ear canal, or in the middle ear. These cases can be treated mechanically or surgically. Neurosensory hearing loss is due to damage of specialized hearing cells in the inner ear due to aging, noise trauma, infection (e.g., measles or mumps), birth defects, metabolic, autoimmune disorders, side effects of medications, or unknown causes. Some causes of hearing loss can be prevented by adequate vaccination of children and limiting the use of medications that can cause hearing loss to situations when there are no valid alternatives, with monitoring of blood levels.

Noise control, especially in the workplace, is important in the prevention of hearing loss. Noise measurement in occupational settings includes decibel levels, frequency of sound waves (cycles per second or hertz), loudness as perceived by the listener, and time or duration of exposure. Community noise levels of aircraft near airports, motor vehicle traffic, gardening equipment, and rock music for teenagers are difficult to control, but should be considered in urban planning requirements. Preventive programs include modifying machinery, erecting sound barriers, and using ear protective devices. Public health programs should be implemented in schools with the use of mobile hearing units as well as education in methods of reducing ear damage from excess noise. Infants should be screened for hearing ability before the age of 3 months, with appropriate follow-up management of treatment and education.

TRAUMA, VIOLENCE, AND INJURY

Trauma, or external injury, is a broad category that includes accidents, poisonings, suicide, homicide, and violence. In many countries, trauma is the leading cause of death because of its greater frequency among the young and the middle aged. It is often the leading cause of years of potential life lost (YPLL) in most developed countries and has become a major focus of intervention in modern public health program development. The possible interventions are many, with high potential for beneficial effects on longevity (Table 5.10).

Trauma morbidity can be reduced by public health measures including primary, secondary, and tertiary prevention. Primary prevention reduces risk factors that are associated with trauma by such measures as enforcement of measures against alcohol abuse with driving, motorcycle helmet, car seat belt, and speed limitation laws. Secondary prevention involves early and adequate medical care at the scene of an accident and rapid transportation to a hospital trauma center. Prevention of consequences of the trauma by such intervention as cardiopulmonary resuscitation, maintaining an airway, stopping bleeding, and treatment of shock at the accident site can reduce case fatality rates. Tertiary prevention involves effective and early rehabilitation by which the degree of disability and long-term management are made more effective, as in cases of head injury.

An accident is a sudden, unintended event that may be associated with human injury. The term does not imply that the event is not potentially anticipated, since

TABLE 5.10   Classification of Injuries and Primary Prevention Measures

Form of Injury	Regulatory control measures	Prevention education
Motor vehicle accidents	Mandatory seat belt, child safety seats and speeding laws; police enforcement of speeding and drunk driving laws lowering permissible blood alcohol levels; raising age for driving permits; restriction of driving privileges in high risk groups; car safety examination and enforcement, air bags, structure standards; pedestrian-safe crosswalks	Driver education: alcohol and drug awareness campaigns; pedestrian safety education
Falls	Safety devices for children and the elderly: nonskid carpets, rails in homes of elderly, bathroom rails; monitoring of medication use	Education and awareness campaigns
Burns	Regulators on home heating systems, regulation for manufacturers of appliances; flame-resistant toys, children's clothing, bedding products; building code standards for electrical wiring, smoke detectors, doors opening outwardly	Promote fire prevention awareness and occupational safety guidelines; education for care of thermal injuries
Poisonings	Manufacturers' labeling; childproof lids on medication and household chemicals	Education regarding dangers of household medications, chemicals, safe storage, labeling, and closure
Domestic violence	Attention to child and women injuries that may originate in domestic violence and abuse; police and medical alertness; shelters for abused persons; imprisonment and/or therapy for abusers	Education in reporting of domestic violence
Occupational injury	Enforcement of safety standards, employer criminal liability for unsafe conditions and injuries; monitoring small workshops and large industries, building sites, fisheries, lumbering sites	Worker and employer education
Sports injury	Mandatory use of helmets for motorcycle, bicycle riders, in sports, professional and amateur; Consumer Product Safety Commision monitoring of recreational sports/play equipment	Teaching safe sports practices in gymnastics and contact sports; good coaching and adjudication
Suicide	Raising awareness of the general population, especially health workers, teachers, social workers	Hot lines
Drowning	Water safety commission enforces guidelines for supervision of recreational swimming places, fencing of swimming poops; lifegaurds trained in rescue and CPR at public beaches and swimming pools	Swimming and boating safety awareness; swimming education in schools, summer camps, CPR

there may be neglect of fundamental preventive procedures and therefore increased risk of the event occurring. If a driver is drunk and exceeding the speed limit, the crash and deaths constitute a criminal liability. If a plant operator allows or requires workmen to do work in dangerous conditions without adequate safety measures, then the event is not accidental but one which could have been anticipated and probably prevented, and also constitutes criminal liability.

Some injuries and deaths inflicted intentionally include homicide, rape, assault, battery, child abuse, and suicide. Public health is concerned with intentional injury as well as unintentional injury and attempts to minimize their effects through preventive oriented programs, planning, organization, public education, training, and rescue operations.

Intervention to prevent violence is designed to break specific cycles. Violence prevention involves increased awareness by teachers, police, social workers, health professionals, and the public at-large in spotting potential and actual signs of violence, especially of abused children or women. Other forms of violence prevention include gun control, preventing weapons from entering schools, "hot lines" for victims to telephone to seek help, shelters for potential and actual victims, self-defense training, rapid response of police, enforcing drinking restrictions, and promotion of supervised teenage recreational activities. The patterns of mortality from various causes of trauma in the United States is seen in Table 5.11.

Motor Vehicle Accidents

The commonest cause of loss of life from trauma is motor vehicle accidents. Motor vehicle crashes are associated with death, injury, and medical care costs. In the United States, annual death tolls from automotive crashes are 45,000 persons per year, with 60% under age 35 years. Transportation-related injuries cause some 50% of brain and spinal cord injuries. The U.S. health targets for the year 2000 include reduction in road crash mortality from 19 per 100,000 population in 1987 to 17 per 100,000 in the year 2000.

Motor vehicle injuries are a problem not only in developed countries, but also in developing countries where road crash rates are extremely high. Alcohol use may be less restricted than in developed countries, road safety less advanced, cars poorly maintained, and drivers less experienced.

Minimizing the effects of trauma from car crashes involves a complex of legislation, enforcement, education, and technical development initiatives. Passenger restraints (seat belts, safety child car seats, crash air bags) and helmets for motorcycle and bicycle riders have demonstrated great potential for reducing death and injury from transport-related crashes. Prevention of crashes caused by speeding, alcohol abuse, fatigue, or negligence involves dealing with the human factor, based on strict enforcement and education. Laws and enforcement should ensure that

TABLE 5.11   Age-Adjusted Mortality from Selected Causes of Trauma, Total Population, Rates per 100,000 Population, United States, Selected Years, 1950–1996

Case	1950	1960	1970	1980	1990	1996
Motor vehicle	23.3	22.5	27.4	22.9	18.5	16.2
Homicide	5.4	5.2	9.1	10.8	10.2	8.5
Suicide	11.0	10.6	11.8	11.4	11.5	10.8

Source: *Health, United States,* 1998.

vehicles are in proper working order, especially lights and brakes. Governments must ensure adequate road lighting, elimination of barriers to visibility, and well-maintained roads.

Strict driver testing especially for persons under age 25 and those over 75 should be enforced. Statistically, women are safer driers than men and should be encouraged to drive buses, trucks, and military transport. Maintaining speed limits to less than 55 miles per hour (90 kilometers per hour; 50 mph or 80 kph for trucks) reduces accident fatality rates, as do nonrigid central and side road barriers.

Law enforcement is a major factor in reducing road crash deaths and injuries. Compulsory seat belt laws are major contributors to reduced severity of injuries, and enforcement increases compliance. Enforcement of speed limits and the incarceration of drunk drivers have a strong deterrent effect and reduce death and severe injury rates. Strong police enforcement, heavy fines, criminal liability, and loss of driving permits for offenders should be permitted.

The physical relationship of speed with extent of injury is based on laws of mechanical energy or momentum. A person traveling at 55 miles per hour (90 km/hour) has four times the kinetic energy of a person traveling at 30 miles per hour (45 km/hour). Injury is more severe at higher rates of speed. Permitting speed limits over 55 miles per hour (90 km/hour) or nonenforcement of the speed limits results in more serious injuries. This has been documented in studies of mortality patterns, which rose when speed limits were raised and declined when speed limits were lowered. Countries, especially those with limited resources, should reduce speed limits and invest in public transport and improved police enforcement and driver education rather than in costly highway systems.

BOX 5.8 INTERVENTIONS TO PREVENT OR MITIGATE MOTOR VEHICLE INJURIES

1. Mandatory seat belt legislation and enforcement;
2. Testing and enforcement of alcohol standards for drivers;
3. Administrative suspension of driving licenses;
4. Motorcycle and bicycle helmets mandatory and enforced;
5. Enforcement of speed limits of 55 miles/hour (90 km/hour) on intercity roads (50 mph or 80 kph for trucks);
6. Enforcement of minimum age drinking laws;
7. Mandatory child care seats;
8. Driver and passenger air bags;
9. Vehicle and road design standards;
10. Education and public policy commitment.

Source: Centers for Disease Control. 1992. Setting the national agenda for injury control in the 1990s. *MMWR*, 41:1–4.

Emergency care has made important strides in the past several decades. This has undoubtedly contributed to declining mortality rates from trauma. Central trauma units serving larger populations are more likely to have a broad multiservice potential and get better results from a larger volume of cases and experience. Organized ambulance services with well-trained and supervised paramedics taking patients to central trauma centers have been shown to be effective in reducing mortality and collateral damage rates. These programs are of vital importance in achieving goals for injury control and reducing deaths and serious complications of trauma.

Domestic Violence

Family or domestic violence is more readily identified and brought to public attention than in previous generations, so that apparent increases may be due to better reporting. Some factors that determine these intentional injuries include socioeconomic status, alcohol use, and family history of members who may have been abused themselves. Data are not readily available for incidence of child abuse, sexual abuse, or spousal abuse. Data are not readily available for incidence of child abuse, sexual abuse, or spousal abuse. Increased public awareness in recent years has led to increased reporting. Prevention requires strong public concern, police and court intervention with enforced therapy, and/or imprisonment for repeat offenders.

Suicide and Suicide Attempts

Internationally, there are wide variations in suicide rates, with the majority of cases occurring among adolescent men and the elderly. From the 1950s to the 1990s, suicide rates in Canada among young adults (15–24 years) rose by 317% for men and 257% for women. Male suicide rates in Central and South American countries are below 10 per 100,000 population, but in the United States and Canada they are above 20 per 100,000. In Europe, overall suicide rates are lower in the Mediterranean countries (Italy, Spain, Greece) and above 40 per 100,000 in France, Switzerland, Sweden, Denmark, Finland, Austria, and Belgium. Suicide rates in Hungary are above 100 per 100,000 population for men and 30 per 100,000 for women.

Suicide is the eighth leading cause of death in the United States. Among persons aged 15–19 mortality from suicide remained stable, from 11.0 per 100,000 in 1950 to 10.8 in 1996, but this masks an age/sex difference with a major decline in all age groups over age 45, and a nearly threefold increase in the 15–24 year age group. For each ten suicide attempts there is one successful suicide, with men more common in the latter group. It is estimated that 30% of suicides are the result of mental disorders, with the remainder due to decisions regarding life circumstances, low self-esteem, binge drinking, and situational depression.

Restriction of access to methods of suicide, such as firearms, is believed to reduce successful suicide, and restriction of publicity about suicides is thought to reduce suicide rates. Threats of suicide should be taken seriously; health care providers, teachers, counselors, and religious leaders should be instructed in suicide prevention and how to assist people through periods of depression. Mental health and supportive counseling must be part of any health care system because the suicidal individual requires immediate attention and care. Telephone counseling by volunteers on hot lines have proved to be valuable services in suicide prevention.

Homicide

Homicide has become one of the major causes of death in some countries, such as in Colombia. In young adult males between the ages of 15 to 24 years in the United States, homicide is the fourth leading cause of death. The epidemiologic analysis of murders shows a relation to drug traffic, both involving rich drug competitors and street-level violence for control of the street traffic. Random violence among school children is a frequent event, as are drive-by or "road-rage" shootings, often resulting in child deaths. Murders associated with rival gangs and random violent crime with murder are now common in many former Soviet countries. Gang violence in U.S. cities is matched by concern in rural areas where murderous rampages by preteen youths with access to and training in the use of deadly weapons occurs with increasing frequency. Gun control legislation has made some minor gains, but weapons remain accessible to large sectors of the U.S. population.

Sources of data for epidemiologic analysis are available from national and local police, but hospital emergency room records may be more useful since more than 50% of violent crime goes unreported to police. Murder rates have declined in the United States during the 1990s, with the decline often attributed to improved economic conditions, low unemployment, stricter punishment laws, and successful community policing, but the United States remains well above other industrialized countries in homicide rates.

Prevention of Violence

In many countries, violence is one of the leading causes of death, especially among teenagers and young adult males. Domestic violence leading to homicide is one of the most common causes. Prevention of violence and violence-related injuries is a major public health concern because of the large scale of loss of life and personal injury as well as the long-term damage to society. Interventions involve the whole of society, not the health system alone. Nevertheless public health has an advocacy role to play.

Violence inspired by religious, nationalistic, or other political motives is a fact of life in both developed and developing countries, sometimes occurring with shocking ferocity. These events, whether bombs in buses, subways, aircraft, or buildings or "ethnic cleansing" warfare, cause enormous physical and psychological trauma that must concern health care providers and public health personnel.

Dramatic events involving school violence and mass homicide by teenagers have occurred in the late 1990s in small communities all over the world, which are for the most part low crime areas, most recently in Jonesboro, Arkansas, and Littleton, Colorado, as well as in Australia, Canada, and Scotland. A 1997 CDC re-

BOX 5.9 PREVENTION OF VIOLENCE AND INJURIES DUE TO VIOLENCE

1. Improve recognition of victims of family violence, and encourage punitive measures against perpetrators;

2. Alert medical facilities to warning signs and obligations to report domestic violence;

3. Develop community specific programs for violence reduction through education and enforcement;

4. Restrain availability of firearms by legal restriction and enforcement;

5. Monitor chronic abusers of alcohol and other drugs with violent tendencies for treatment or incarceration;

6. Train children in self-protective and preventive measures;

7. Promote public and school education against violent behavior;

8. Develop mental health services with prevention of violence and suicide as major goals;

9. Reduce binge drinking by means of taxes, police enforcement, and education;

10. Promote crisis intervention and hostel services for rape and abuse victims;

11. Restrict violence on television during child watching hours;

12. Provide hot lines and shelters for abuse victims;

13. Train health care providers in violence and abuse situation identification and counseling;

14. Promote awareness of domestic violence using the public media.

Source: U.S. Public Health Service Task Force. 1996. *Guide to Clinical Preventive Services.* Second Edition. Baltimore: Williams and Wilkins.

port of a survey taken nationwide showed that 8.5% of U.S. students have carried weapons to school and 7.4% of students reported being threatened with weapons during the 12 months preceding the survey. Many people attribute these staggering statistics to a preponderance of violence in the media, whether through films, television, or the Internet. The Internet has become a forum for politically extreme and neo-Nazi ideals, as well as an avenue for illegal weapons dealing. The consensus is that this gives troubled youth the opportunity to act out fantasies of murder and mayhem with the goal of either satisfying a need for revenge, acting on an ideological hatred, or achieving notoriety. Within the context of high rates of risk-taking behavior in adolescent males, greater vigilance on the part of parents, health care providers, the educational system, and the community in seeking out warning signs, along with significant restrictions on gun sales, would help to rectify this serious problem.

CHRONIC CONDITIONS AND
THE NEW PUBLIC HEALTH

The burden of chronic conditions is an important factor in the use of health services. Primary, secondary, and tertiary prevention are parts of good clinical practice and good public health practice in the wider context of the New Public Health. The linkage between clinical care and public health requires strengthening with incentives and administrative changes that will foster wider implementation of existing techniques of care that can sustain the chronically ill in their own homes in the community. The economics of health care are an inevitable element of health policy, and the search for cost-effective ways to care for patients and prevent disease and disability are part of health care.

Chronic conditions are often associated with an acute crisis of a disease resulting from many factors developing over time. Improved care can prevent many crises exacerbating a long-standing disease process. The health care system has a responsibility to prevent these acute events, so that hospital intensive care units will not be filled by chronically ill patients whose balance was upset by a medical crisis not addressed early enough. The onus of their prevention and management falls on all components of a health system.

It may be more difficult and complex to prevent an accident or a diabetic foot amputation than to treat its results, but the benefit to the individual and the community is infinitely greater. The New Public Health includes health promotion, care of the ill, in a context of limited resources and advancing medical technology, preserving individual dignity and rights, and ethical concerns.

SUMMARY

Chronic conditions are major public health problems in most industrialized countries, and are rapidly becoming so in developing countries. Cardiovascular disease, cancer, and trauma are the major causes of death in most western countries, but the leading cause of years of potential life lost is trauma. Increasing longevity, improved nutrition, social support, and medical care are creating an increasingly elderly population living longer and healthier than previous generations. The public health challenge is to promote healthy middle-aged and elderly populations by reducing risk factors through health promotion and effective medical care.

The New Public Health involves working partnerships between clinical services to prevent and control chronic conditions, and to prevent or delay onset of their complications. Dramatic lowering of mortality and morbidity from cerebrovascular and coronary heart diseases has been accomplished by this approach. The potential for prevention for increasing the well-being of those affected by these conditions should be a central element of national health policy.

ELECTRONIC MEDIA

American Diabetes Association. http://www.diabetes.org/

Arthritis Foundation. http://www.arthritis.org/

National Cancer Institute. http:/www.nci.nih.gov/

National Eye Institute. http://www.nei.nih.gov/

National Heart, Lung and Blood Institute, National Institutes of Health, Congestive heart failure in the United States: A new epidemic (1996). http://www.nhbi.nhi.gov/nhlbi/cardio/other/gp/CHF.htm

National Institute of Arthritis and Skin Diseases. http://www.nih.gov/niams

National Institute of Deafness and other Communication Disorders. http://www. nih.gov/nided

National Institute of Diabetes and Digestive and Diseases of the Kidney. (NIDDK). http://www.niddk.nih.gov/

National Institutes of Health. http://www.nih.gov/icd/

National Institute of Health Consensus Program. http://www.opd.nih.gov/consensus/cons

National Osteoporosis Foundation. http://www.nof.org/

World Health Organization, European Region, Health for all data base (1999). See website for instructions to download HFA database. http://www.who.dk/ country/readwin1.htm

RECOMMENDED READING

Andrews, H. F., Kerner, J. F., Zauber, A. G., Mandelblatt, J., Pittman, J., Struening, E. 1994. Using census and mortality data to target small areas for breast, colorectal, and cervical cancer screening. *American Journal of Public Health,* 84:56–61.

Braunwald, E. 1997. Shattuck lecture—Cardiovascular medicine at the turn of the millennium: Triumphs, concerns, and opportunities. *New England Journal of Medicine,* 337:1360–1369.

Breslow, L., Breslow, N. 1993. Health practices and disability: Some evidence from Alameda County. *Preventive Medicine,* 22:86–95.

Campbell, L. A., Kuo, C.-C., Grayston, J. T. 1998. *Chlamydia pneumoniae* and cardiovascular disease. *Emerging Infectious Diseases.* 4:571–579.

Centers for Disease Control. 1992. A framework for assessing the effectiveness of disease and injury prevention. *Morbidity and Mortality Weekly Report,* 48:649–656.

Centers for Disease Control. 1997. National diabetes awareness month. *Morbidity and Mortality Weekly Report,* 46:1013–1027.

Centers for Disease Control. 1999. Achievements in public health, 1900–1999. Decline in deaths from heart disease and stroke, 1900–1999. *Morbidity and Mortality Weekly Report,* 48:649–656.

Elders, M. J., Perry, C. L., Erikson, M. P., Giovano, G. A. 1994. The Report of the Surgeon General: Preventing tobacco use among young people. *American Journal of Public Health,* 84:543–547.

Heise, L. 1993. Violence against women: The hidden health burden. *World Health Statistics Quarterly,* 46:78–85.

Hoffman, C., Rice, D., Sung, H-Y. 1996. Persons with chronic conditions: Their prevalence and costs. *Journal of the American Medical Association,* 276:1473–1479.

Jeanneret, O., Sand, E. A. 1993. Intentional violence among adolescents and young adults: An epidemiological perspective. *World Health Statistics Quarterly,* 46:34–51.

Kannel, W. B., Wolf, P. A. [editorial]. 1992. Inferences from secular trend analysis of hypertension control. *American Journal of Public Health,* 82:1593–1595.

LeBoff, M. S., Kohlmeister, L., Hurwitz, S., Frankin, J., Wright, J., Glowacki, J. 1999. Occult vitamin D deficiency in postmenopausal U.S. women with acute hip fracture. *Journal of the American Medical Association,* 281:1505–1511.

Mosterd, A., D'Agostino, R. B., Silbershatz, H., Sytkowski, P. A., Kannel, W. B., Grobbee, D. E., Levy, D. 1999. Trends in the prevalence of hypertension, antihypertensive therapy, and left ventricular hypertrophy from 1950 to 1989. *New England Journal of Medicine,* 340:1221–1238.

Sytkowski, P. A., D'Agostino, R. B., Belanger, A., Kannel, W. B. 1996. Sex and time trends in cardiovascular disease incidence and mortality: The Framingham Heart Study, 1950–1989. *American Journal of Epidemiology,* 143:338–350.

Winkleby, M. A., Taylor, C. B., Jatulis, D., Fortmenn, S. P. 1996. The long-term effects of a cardiovascular disease prevention: The Stanford Five-City Project. *American Journal of Public Health,* 86:1773–1779.

BIBLIOGRAPHY

Advisory Board, International Heart Health Conference. 1992. *The Victoria Declaration on Heart Health.* Ottawa: Health and Welfare, Canada.

Blaser, M. 1997. *Heliobacter pylori* persistence and injury of the human stomach. *In* Brown, F., Burton, D., Doherty, P., Breslow, L., Beck, J. L., Morgenstern, H., Fielding, J. E., Moore, A., Carmel, M., Higa, J. (eds). Development of a health risk appraisal for the elderly (HRA-E). *American Journal of Health Promotion,* 11:337–343.

Brownson, R. C., Remington, P. L., Davis, J. L. (ed.). 1998. Second Edition. *Chronic Disease Epidemiology and Control.* Washington, DC: American Public Health Association.

Canadian Task Force on Periodic Health Examination. 1994. *The Canadian Guide to Clinical Preventive Health Care.* Ottawa: Health Canada.

Centers for Disease Control. 1993. Public health focus: Prevention of blindness associated with diabetic retinopathy. *Morbidity and Mortality Weekly Report,* 42:191–195.

Centers for Disease Control. 1993. Public health focus: Physical activity and the prevention of coronary heart disease. *Morbidity and Mortality Weekly Report,* 42:669–672.

Centers for Disease Control. 1993. Mortality trends for selected smoking related cancers and breast cancer—United States, 1950–1990. *Morbidity and Mortality Weekly Report,* 42:857–866.

Centers for Disease Control. 1993. Surveillance for diabetes mellitus, United States, 1980–1989. *Morbidity and Mortality Weekly Report,* 42:SS-2:1–20.

Centers for Disease Control. 1994. Deaths resulting from firearm and motor-vehicle-related injuries—United States, 1968–1991. *Morbidity and Mortality Weekly Report,* 43:37–42.

Centers for Disease Control. 1994. Surveillance for smoking-attributable mortality and years of potential life lost, by state—United States, 1990. *Morbidity and Mortality Weekly Report,* 43:SS-1: 1–8.

Centers for Disease Control. 1995. Asthma—United States, 1982–1992. *Morbidity and Mortality Weekly Report,* 44:952–955.

Centers for Disease Control. 1996. Asthma mortality and hospitalization among children and young adults—United States. *Morbidity and Mortality Weekly Report,* 45:350–353.

Centers for Disease Control. 1999. Achievements in Public Health, 1900–1999. Motor vehicle safety: a 20th century public health achievement. *Morbity and Mortality Weekly Report,* 48:369–374.

Centers for Disease Control. 1999. Tobacco use—United States, 1900–1999; and cigarette smoking among adults—United States, 1997. *Morbidity and Mortality Weekly Report,* 48:986–993, 993–996.

Cole, P., Amoateng-Adjepong, Y. [editorial]. 1994. Cancer prevention: Accomplishments and prospects. *American Journal of Public Health,* 84:8–10.

Diekstra, R. F., Gulbinat, W. 1993. The epidemiology of suicidal behavior: A review of three continents. *World Health Statistics Quarterly,* 46:52–68.

Dobrossy, L. (ed). 1994. *Prevention in Primary Care: Recommendations for Promoting Good Practice.* CINDI 2000. Copenhagen: WHO Regional Office for Europe.

Doll, R., Peto, R. 1976. Mortality in relation to smoking: 20 years observation on male British doctors. *British Medical Journal,* 2:1525–1536.

Guide to Clinical Preventive Services. 1996. *Report of the U.S. Preventive Services Task Force.* Baltimore: Williams & Wilkins.

Hunink, M. G. M., Goldman, L., Tosteson, A. N. A., Mittleman, M. A., Goldman, P. A., Williams, L. W., Tsevat, J., Weinstein, M. C. 1997. The recent decline in mortality from coronary heart disease, 1980–1990: The effect of secular trends in risk factors and treatment. *Journal of the American Medical Association,* 277:535–542.

Jousilahti, P., Tuomilehto, J., Korhonen, H. J., Vartianinen, E., Puska, P., Nissinen, A. 1994. Trends in cardiovascular disease risk factor clustering in eastern Finland: Results of 15-year follow-up of the North Karelia Project. *Preventive Medicine,* 23:6–14.

Manson, J. E., Tosteson, H., Ridker, P. M., Satterfield, S., Hebert, P., O'Connor, G. T., Buring, J. E., Hennekens, C. H. 1992. The primary prevention of myocardial infarction. *The New England Journal of Medicine,* 326:1406–1416.

Marmot, M., Elliott, P. (eds). 1994. *Coronary Heart Disease Epidemiology: From Aetiology to Public Health.* New York: Oxford Medical Publications.

Morse, S. S. 1993. *Emerging Infectious Diseases.* New York: Oxford University Press.

National Institute of Health Consensus Statement. 1994. *Helicobacter pylori* in peptic ulcer disease. *NIH Consensus Statement,* Feb. 7–9, 12:1–23.

Nichols, E. S., Peruga, A., Restrepo, H. E. 1993. Cardiovascular disease mortality in the Americas. *World Health Statistics Quarterly,* 46:134–150.

Perry, C. L., Kelder, S. H., Murray, D. M., Klepp, D. I. 1992. Community wide smoking prevention: Long-term outcomes of the Minnesota Heart Health Program and the Class of 1989 Study. *American Journal of Public Health,* 82:1210–1216.

Uemura, K., Pisa, Z. 1988. The trends in cardiovascular disease mortality in industrialized countries since 1950. *World Health Quarterly,* 41:155–178.

United States Public Health Service. 1964. The Surgeon General's Report on Smoking and Health 1964. Washington, DC: U.S. Department of Health Education and Welfare.

United States Senate Appropriations Subcommittee on Labor, Health, and Human Services. 1996. *Action Alert, The Nation's Health.* Washington, D.C.: American Public Health Association.

Waller, J. A. 1994. Reflections on a half century of injury control. *American Journal of Public Health,* 84:664–670.

World Health Organization. 1992. Epidemiology and public health aspects of diabetes. *World Health Statistics Quarterly,* 45:314–381.

World Health Organization MONICA Project. 1988. Geographical variation in the major risk factors of coronary heart disease in men and women aged 35–64 years. *World Health Statistics Quarterly,* 41:115–140.

World Health Organization. 1994. *Prevention of Diabetes Mellitus.* Technical Report Series 844. Geneva: WHO.

World Health Organization. 1996. *The World Health Report 1996: Fighting Disease, Fostering Development.* Geneva: World Health Organization.

World Health Organization. 1997. *The World Health Report 1997: Conquering Suffering and Enriching Humanity.* Geneva: World Health Organization.

World Health Organization. 1998. *The World Health Report 1998: Life in the 21st Century: A Vision for All.* Geneva: World Health Organization.

World Health Organization, Regional Office for Europe. 1990. *Is the Law Fair to the Disabled?* European Series No. 29. Copenhagen: Regional Office for Europe.

6

FAMILY HEALTH

INTRODUCTION

The family structure provides an important foundation for physical and emotional health of the individual and the community. Marital and family status and interactions among family members affect each person's health and the well-being of the community and nation. This is in the context of cultural, economic, and social patterns unique to each society, with important commonalities of the role of the family in health. However, each person is a unique individual who passed through life stages with changing health needs and support systems not only in the family, but also in peer groups and society more widely.

Family health issues relate to specific periods involving fertility, infancy, childhood, adolescence, adulthood, and the elderly as well as the relationships among family members. Each age has specific health problems in which prevention and other health services play an important role. This chapter addresses the health needs of family members at different stages of life. No one population group is in isolation from another. Poor pregnancy outcome affects mother, child, family, and the community. The family of the person who is chronically ill, injured, or killed at work or on the road suffers economically, emotionally, and socially. The public health and medical care systems must be sensitive to the special needs of the family by providing appropriate health promotion, disease prevention, medical care, and support programs for each member of the family and the family as a whole.

The New Public Health includes traditional elements of public health, such as maternal and child health, but within a larger context of total family needs. In the United States, single teenage pregnancies entail a multitude of family and societal problems. In societies where women's rights are restricted or repressed, the health of women and children suffer measurably. The role of public health is then to advocate societal change and provide direct answers to the health needs generated by specific societal patterns, whether this relates to high rates of morbidity for women and children due, for example, to lack of primary care services, or whether the problem is lack of health insurance coverage for children.

THE FAMILY UNIT

The family is the basic social support unit in virtually all human societies, providing the basis for childbearing and child rearing. It has important roles in stability of basic physiological and psychological needs as well as the economic basis of the members of the unit. The family provides the key environment for the emotional needs, socialization, mutual help, and nurturing needed by adults as well as children.

The family is usually a group of two or more people related by marriage, common agreement, birth, or adoption who reside together in the same household, and may consist of one, two, or more generations. The family unit includes people living together, engaged in sexual relations, in fertility, and in rearing of children through their many stages of development before the reach independent adulthood. It also includes caring for elderly parents and relatives, as well as maintaining close contact with adult siblings and children, themselves in the process of childbearing and child rearing.

The nuclear family typically includes a male and female couple related by marriage, or living together by common consent, with or without children. Increasingly, family units include single or divorced mothers, or fathers, living alone with children, or with a parent or other person related or unrelated by marriage. The definition is being widened to include couples of the same sex, with or without children from previous heterosexual relationships, from artificial insemination, or through adoption.

The extended family is multigenerational and consists of the nuclear family and relatives of both parties, whether or not living in close geographic proximity. The extended family provides a broader basis of mutual support. It still exists in western societies, but frequently in an altered form. Multigenerational families consisting of single, divorced, or widowed women with children are now common, in association with high divorce rates, single parenthood, and increasing longevity, especially for women. Same-sex households can leave children of one gender without role models.

In developing countries, families often consist of large numbers of children born to poorly educated parents living in poverty. The father or less commonly the mother may be absent for long periods while working in a distant place. This can create serious health hazards for all family members. In societies where death of adults occurs from civil wars, famine, or infectious diseases such as AIDS, raising of children by single parents, neighbors, or older siblings is common. Abandonment of children is also common in such situations.

Divorce and single parenthood are often associated with relative poverty, creating additional stresses in the functioning of the family unit. The absence of one parent, usually the male, also strains the family. Abusive parenting, including psychological and sexual abuse and other violent behaviors, creates a heavy burden on children in their most vulnerable years, with long-term psychological damage.

In the United States, there has been a decline in the percentage of children in two-parent family settings. In 1980, 86% of all households in the white population were defined as two-parent families, declining to 81% in 1997. Among the African-American population, 56% of all families were in two-parent family homes in 1980, declining to 46% in 1997, while female-headed black households increased from 40% to 47%. For family groups with children under age 18 years, the percentage of one-parent family groups in black homes was 64% in 1997, as compared to 26% in white homes and 36% in Hispanic homes. This represents a high percentage of black homes, and children, at high risk for poverty, insecurity, and insufficient health insurance. This results in lack of access to family doctors and regular care for some 25% of American children aged 6 or below, with poor utilization of immunization and frequent use of hospital emergency rooms. In Canada, single-parent families make up 13% of all families, but all are included in universal health care. In the United Kingdom, just over one-quarter of all births occur to unwed mothers, including cohabiting parents, but all are included in the National Health Service so that society ensures children access to care despite their family circumstances.

Changes in family structure affect the health status of adults as well as children. Increases in longevity of women, more than men, together with high divorce rates, produce an excess of elderly women in relatively good health who live alone. The capacity of women to adapt to this condition is positively affected by the strong social support systems in the industrialized countries where universal health insurance, national pensions, pensions from prior work of spouses, and legal protection of economic rights in divorce have produced a relatively well-protected elderly population group.

Diseases of aging from middle age onward include cardiovascular and other degenerative conditions affecting men at an age nearly 10 years younger than women. Cancers are the second leading cause of death. The epidemiological and demographic transitions affect the family in many ways. With aging of the population, care of the elderly has become an increasing factor in family life, not only in provision of health care, but in the economics of health, for the family and the society. Where health insurance fails to fully cover long-term care, serious financial burdens are placed on middle-aged children of elderly parents. With aging of the population, it is not uncommon for middle-aged or elderly children to care for very elderly parents. Where society does bear the cost, the importance of prevention and control of chronic disease becomes essential to avert financial breakdown of the health system.

MATERNAL HEALTH

Women's health issues relate to their many roles: as individuals, workers, wives, mothers, and daughters. These demand lifelong responsibilities for knowledge, self-care, and family leadership in health related issues, such as nutrition, hygiene, education, exercise, safety, fertility, child care, and care of the elderly. Changes in social roles of women create extra demands and risks in health.

Fertility

Childbirth and care of infants and children have been traditional concerns of public health because of the related vulnerability and historically large-scale loss of life. Improved nutrition and living conditions as well as proper care in pregnancy and at delivery have combined to reduce maternal mortality to a rare event in developed countries. In least developed countries, however, maternal mortality can be as high as 1800 per 100,000 live births, whereas in the industrialized countries it is under 15 per 100,000. The differences between these populations may be found in socioeconomic conditions and the general health status of women, including education, age at fertility, previous number of births, the time or space between pregnancies, and the general care given by the health system.

Proper nutrition, communicable disease control, and education ensure that females reach the age of fertility physically and intellectually prepared for childbirth and child rearing. Literacy for women contributes to improved infant survival rates. The literate mother can refer to written materials instead of depending solely on community traditions and is better able to cope with the complexities of a health care system. Greater education is also likely to lead to a better chance of employability and greater family income.

Fertility-related health issues include preparation for and timing of pregnancy, and professional care during and after pregnancy. Using health education and prevention services, public health provides the necessary resources through which these issues may be addressed. Cessation of smoking, alcohol, drug use, and risk taking behavior at least for the duration of the pregnancy is an important part of protection of the fetus from harm.

Fertility patterns are affected by social, cultural, religious, and other factors, including the technology available for birth control. With the availability of safe and effective birth control measures and increasing education and job opportunities, fertility patterns have changed in the industrialized countries. Ethnic differences can be substantial within a country. For U.S. African-American women, the crude birth rate fell from 21.9 per 1000 in 1991 to 17.8 in 1996, compared to 15.4 and 14.1 per 1000, respectively, for white women.

Fertility rates in many developed countries have declined considerably since the peak "baby boom" that occurred in the years following World War II. This has generated concern that the birth rate is lower than that needed to sustain the population even at current levels. Developing countries have the opposite problem in that high birth rates with declining morbidity are placing a heavy economic, social, and medical burden on the community while the cultural and religious ethos produce little support for family planning.

Globally, despite declining birth rates, the world's population grew from 1 billion in 1804 to 2 billion in 1927, 3 billion in 1960, and 6 billion in 1999. It is anticipated to increase to 7 billion in 2013 and 8 billion in 2028 (United Nations Population Division, 1999).

Family planning and spacing of pregnancy is a vital issue in developing countries where the burden of frequent pregnancies contributes to high maternal and infant mortality rates. The traditional means of fertility limitation in these countries was through prolonged breast-feeding, which has declined with the promotion of commercial baby formulas. Improvements in socio-economic conditions, with female education, infant and child survival, and the strengthening of family planning services are likely to provide the needed impetus for increased contraceptive use and fertility decline in developing countries.

Public Health Concerns of Fertility

Provision of birth control along with prenatal, delivery, and postpregnancy care are among the central roles of any health care service. High levels of maternal mortality of the past in the industrialized countries are still present in developing countries. Traditions of high fertility rates, unattended deliveries, and unsafe practices of traditional birth attendants (TBAs) and female genital mutilation greatly contribute to the poor health status of women (in developing countries).

Safe prenatal care and birth practices, along with better nutrition and general health status of women, reduced maternal morbidity and mortality dramatically in the twentieth century, but the health gap remains enormous between the industrialized and least developed countries. Fertility rates have been falling in most parts of the world as individual education levels improve, as economic incentives in rural areas for more children are replaced by incentives for fewer children, and as urbanization and a search for better standards of health and living conditions occurs (Table 6.1).

As the technology of birth control becomes widely available, better educated women have the power to control their own fertility. There is an encouraging pattern of fertility decline in all parts of the world, including sub-Saharan Africa which had increasing fertility rates well into the 1980s. Public health promotes spacing of pregnancy to improve health outcomes for both the mother and the newborn, with wider implications for family health.

TABLE 6.1   Total Fertility Rates by World Demographic Regions,[a] Selected Years, 1960–1997

Region/group	1960	1970	1980	1990	1997
Sub-Saharan Africa	6.6	na	6.7	6.3	5.9
India	5.9	5.6	4.9	3.7	3.1
China	5.7	5.3	2.8	2.2	1.8
South Asia	6.1	na	5.2	4.0	3.4
Latin America and Caribbean	6.0	5.1	4.2	3.2	2.7
Middle East and North Africa	7.1	6.2	5.9	5.1	4.4
CIS/CEE and Baltic states	3.0	2.2	2.2	2.3	1.8
Industrialized countries	2.8	2.5	1.9	1.7	1.7
World	5.0	na	na	3.2	2.8

Source: World Development Report, 1993; UNICEF, 1999.
[a]Total fertility rates (see Chapter 3 and Glossary). CIS/CEE refers to the former Soviet Union and countries of eastern Europe. na, Not available.

In many societies, infertility is a problem associated with important personal distress and social stigma. Sexually transmitted diseases (STDs) are a major cause of infertility, so that prevention and management of STDs are important aspects of managing infertility. Treatment of infertility is associated with high cost, not only in financial expenditures, but also in emotional trauma. Modern services to treat infertility, including stimulation of ovulation, *in vitro* fertilization, and surrogate parenting, raise many ethical and financial issues. Despite these and other problems of multiple births and high rates of very low birth weight infants with associated perinatal and developmental problems, infertility treatment is very much a part of modern health care systems.

Family Planning

Family planning enables a woman to determine the time, spacing, and frequency of pregnancy, as well as adoption of children. It includes a range of methods for preventing or expelling a conception while maintaining a normal sex life. The technology for birth spacing has been revolutionized, so that there are now safe and effective contraceptive methods widely available at reasonable cost in industrialized and developing countries (Table 6.2). Birth control is a public health issue and responsibility. Lack of knowledge or access to birth control results in high rates of birth or resorting to abortion for birth control, which can complicate subsequent pregnancies. High fertility rates or pregnancies late in life pose extra hazards both to the mother and the newborn and are a burden to the family unit.

Modern birth control methods include pharmacological and chemical prevention of conception (hormone pills, implants or shots, spermicides) and physical methods [male and female condoms and intrauterine devices (IUDs)]. Traditional methods of prevention, such as breast-feeding, the rhythm method, and coitus interruptus, are less reliable than current methods, mainly the pill, condoms, and the IUD.

Spacing of pregnancies by modern methods of birth control is a fundamental right of women. In many areas of the world, birth control remains a religious and political issue as well as a power struggle between the sexes. The Roman Catholic Church is strongly opposed to birth control, as are many political regimes for dif-

TABLE 6.2   Percentage of Married Women Using Some Form of Contraception, World Regions, 1960–1965 to 1990–1998

Region	1960–1965	1990–1998
Sub-Saharan Africa	5	16
Middle East and North Africa	2	50
South Asia	15	39
Latin America and Caribbean	11	65
East Asia and Pacific	17	75

Source: UNICEF, State of the World's Children, 1994, 1999.

ferent reasons. However, in many Catholic countries the birth rate has fallen and in some is below the population replacement level. Use of birth control has increased dramatically over the past several decades in Latin America and in east Asia. Family planning has not been well understood or available in the former Soviet republics. They lack a supply and widespread awareness of modern contraception, relying on frequent abortion instead.

Despite the wide availability of contraception, teenage pregnancies account for a relatively large percentage of total births in the United States. This is a serious public health problem, being more prevalent in lower socioeconomic groups with problems of single mothers with welfare dependency and is associated with poor health outcomes. Teenage unmarried pregnancies are a complex of social, educational, and labor force problems not easily addressed. It is a growing problem in other industrialized countries as well. In developing countries, teenage, and even child, marriage produces a wide range of maternal health problems of physical and emotionally immature mothers trapped in a life role with no chance to become educated or employed.

Maternal Mortality and Morbidity

A maternal death is defined by the WHO as a death of a woman while pregnant or within 42 days of termination of pregnancy, from any cause related to or aggravated by the pregnancy or its management, but not from accidental or incidental causes. Maternal deaths are subdivided into two groups:

1. Direct—obstetrical deaths resulting from obstetric complications of the pregnant state;
2. Indirect—obstetric deaths resulting from preexisting disease or conditions not directly due to obstetric causes.

In past centuries, childbirth was life threatening. In 1664, a report of Hotel Dieu Hospital in Paris indicated a maternal death rate of 33% among hospitalized parturient women. In England between 1660 and 1680, 1 woman in 44 was reported to have died in childbirth. Semmelweiss in Vienna in the 1850s reported death rates of 2% in women during or following delivery and rates of 10% in medically supervised wards, due to lack of precautions against transmission of puerperal fever through unhygienic practices of the medical staff. His work led to the control of streptococcal septicemia, a major cause of maternal death (Chapter 1).

In the United States, maternal mortality declined from 670 per 100,000 live births in 1930, levels similar to those of least developed countries in the 1980s–1990s, to just under 80 in 1950 and 6 in 1996. However, maternal mortality rates for blacks remaining triple those of the white population. This dramatic decline in maternal mortality, representative of the trend in most industrialized countries, can be attributed to many factors including better standards of living and nutrition, improved medical care, and advances in obstetrical knowledge and training. The transfer of deliveries into maternity wards of general hospitals contributed to the improved safety of maternity. During this period, fertility rates fell as birth control

TABLE 6.3 Maternal Mortality Rates,[a] by Selected Population Groups and Years,
United States, 1930–1996

Group	1930	1940	1950	1960	1970	1980	1990[b]	1996
White	na	na	53.1	22.4	14.4	6.7	5.1	4.1
Black	na	na	na	92.0	65.5	24.9	21.7	19.9
Total	670	380	73.7	32.1	21.5	9.4	7.6	6.4

Source: *Health, United States,* 1991, 1998.
[a]Rates per 100,000 live births. na, Not available.
[b]Health target for 1990, not more than 5 per 100,000 overall and for all local and ethnic group rates.

became universally available, permitting the spacing of pregnancies and the occurrence of pregnancy by choice, an important factor contributing to lower maternal morbidity and mortality rates. Low rates of maternal mortality also reflect
greater access to medical care as more Americans became insured or could pay
privately, and growing, quality of care.

Despite overall improvement, the gap in maternal mortality rates between black
and white populations in the United States remains high. The relative risk of maternal mortality in blacks and whites has remained over 4 to 1 since 1960, despite
a decline of over 75% for both black and white women. There remains a wide gap
related to age and pregnancy preparation as well as adequacy of care, which continues to be a serious public health problem, especially for poor black teenage
pregnant women.

Internationally, an estimated 585,000 women died in 1997 from complications
of pregnancy and childbirth, 99% of whom are in developing countries. Maternal
mortality rates range from over 1000 per 100,000 (i.e., 1% for all births) in least
developed countries between 1990 and 1997 to 100–500 in mid-level developing
countries, 50 per 100,000 in the former Soviet countries, and less than 10 per
100,000 in industrialized countries, with some at 5 per 100,000 (e.g., Israel, Sweden, and Switzerland).

The World Bank estimates that the extension of prenatal, delivery, and postpartum care to 80% of the world's population would reduce by 40% the burden of
disease associated with unsafe childbirth. They suggest that an appropriate intervention program should include information, education, communication, transportation, community-based obstetrics, and district hospital facilities. High maternal mortality rates are associated with high birth rates and delivery by untrained
personnel (Table 6.4).

Maternal mortality reviews are a common method of investigating maternal
deaths in order to ascertain preventable causes. Such reviews can help to identify
factors leading to maternal deaths and provide important lessons for medical, obstetrical, and community health staff. They are requirements for hospital accreditation in Canada and the United States (see Chapters 13 and 16).

TABLE 6.4   Women's Health Indicators by Level of Country Development

Country's Level of development	Crude birth rate		% Births attended by trained personnel 1990–1997	Maternal mortality rate 1990
	1970	1997		
Least developed	48	39	28	1100
Developing	38	25	55	470
CIS/CEE[a]	20	14	93	85
Industrialized	17	12	99	13

Source: UNICEF, State of the World's Children, 1999.
[a]CIS/CEE refers to the former Soviet Union and countries of eastern Europe.

Pregnancy Care

The goals of prenatal, delivery, neonatal, and infancy care are to provide the mother and child with the optimal conditions and supervision to ensure the best possible outcome of the pregnancy, preserving the health and well-being of the mother and providing the newborn with the greatest chance of survival and optimal development. Since pregnancy is fraught with potential problems for the mother and the fetus, professional prenatal care and equally important self-care by the pregnant woman are necessary. Prepregnancy and prenatal care should be early and complete for mother and child.

Public health programs have the responsibility to assure prenatal care for the entire population either through direct provision of care or by obstetrical services in managed care or private practice settings. Pregnancy is often a planned event, so that preparation for pregnancy is feasible. Preparation for pregnancy includes a general examination, inquiry as to possible genetic problems, STD and HIV testing, nutrition status assessment and counseling, folic acid and iron supplements, smoking, alcohol and drug use cessation, and other counseling as required. The parents should be provided with reading material on pregnancy and parenting. Good prenatal care presupposes a basic program of visits with extra care for those at special risk.

Prenatal care in the community includes early presentation, high risk assessment and referral, and continuous care throughout the pregnancy. Early diagnosis of pregnancy is important in permitting the woman to attend prenatal care as early as possible, hopefully in the first trimester. Early presentation for prenatal care provides the opportunity to assess the health of the mother and to advise her on appropriate nutrition and self-care. In addition, it establishes a working relationship between the mother and the caregiver. Both the mother and the father should become involved in prenatal preparation including prenatal classes and exercises. Early detection of potential complications offers better outcomes of care and genetic disorders can be identified early.

Prenatal care should be built around three principles, as suggested by the American College of Obstetrics and Gynecology (ACOG) and a Delphi Group Panel,

U.S. Department of Health and Human Services, in the late 1980s. These principles are

1. Early and continuous risk assessment;
2. Health promotion;
3. Medical and psychological intervention as needed.

The ACOG also recommends prepregnancy consultation to discuss pregnancy and infant care issues. The ACOG suggests that prenatal care for a normal pregnancy should include a total of 13–15 doctor's visits, prenatal classes on physical care, and preparation for delivery, breast-feeding, and infant care. The mother's routine checkups should include monitoring of her nutritional state, weight gain, blood pressure, and potential complications. The frequency of visits and the content of normal prenatal care for healthy women vary widely among countries and are sometimes considered "excessive." However, good care should not be taken for granted. The achievement of low maternal mortality and morbidity along with low rates of risk for the newborn from low birth weight and associated developmental problems should reinforce the importance of close supervision throughout a pregnancy.

In some countries, such as Israel and France, prenatal care is separated from regular primary care and is offered in maternal and child health centers (MCH) or women's clinics. In others, such as Holland and the United Kingdom, family practitioners provide prenatal and obstetrical services. In still others, such as the United States and Canada, both specialist obstetricians and general practitioners provide prenatal and delivery services. In Israel public health nurses are the major providers of prenatal care and midwives do most hospital deliveries. Whether prenatal care is part of primary care services or a separate public health service, the goals must include universal coverage starting before or very early in pregnancy, risk assessment with referral and care, and ready access to specialized care. Health education provided in prenatal classes for the pregnant woman and her partner offers additional counseling and peer group support.

In developing countries, improved access to prenatal care and increased use of mid level health workers in "baby friendly" birth centers, coupled with training and supervision of traditional birth attendants, spacing of pregnancies, risk assessment, and referral, are all needed to lower present high rates of maternal and perinatal mortality.

The pregnant woman should be provided with a complete medical record of her prenatal care during scheduled visits so that when she arrives in the maternity unit the provider of care will have adequate information. Even if she has not been seen by that provider previously, the record grants a measure of continuity that is to the benefit of both. The patient can feel that her record of previous care is available when needed, and the provider has the benefit of all previous documentation on the patient. A public health system should ensure adequate records systems for maternity care in order to reduce unnecessary complications and mortality in the birth process.

High Risk Pregnancy

Low risk pregnancies are those in healthy women between the ages 18 and 34, who present at least once in the first trimester, who have had no more than three previous normal live births, no previous stillbirths or obstetrical complications, such as gestational diabetes or preeclampsia, no history of drug or alcohol abuse, and no major medical conditions such as hypertension or kidney disease. Such women should be followed in a routine prenatal care program.

High risk pregnancies (HRP) should be identified as early as possible so that these patients can be given special care for her benefit and especially for the well being of the fetus–newborn. Identification and the management of high risk factors initially and throughout pregnancy improve pregnancy outcomes for the mother and the newborn. Some predictors of HRP are age, previous obstetric difficulties, malnourishment, poverty, many previous pregnancies, early or late maternal age, women who attend STD clinics, and women who smoke cigarettes, drink alcohol, or take drugs. Risk factors may include social and economic factors such as adverse housing, finances, and working conditions.

The medical and obstetrical history provides evidence of previous risks such as frequent abortion, complications in pregnancy, or medical conditions that could affect the mother during the pregnancy or at the time of delivery. Pregnancy under age 16 or 17 or over age 35 should automatically define the pregnancy as being at higher than normal risk. Grand multiparity (i.e., more than five previous births) or a first pregnancy (primigravida) should also be considered as an extra risk for the mother but more so for the newborn.

A scoring system provides a set of standards or guidelines to risk assessment to assist the primary care provider in early detection and referral of patients on the basis of a reasonably objective set of criteria for high risk factors. Detailed guidelines are needed to implement this kind of standard for high risk pregnancy and monitoring are of the value to improved pregnancy care. The form and guidelines developed should take into account local risk factors, such as high consanguinity rates in some societies or chronic malnutrition in the population. Scoring systems produce a cumulative risk assessment by adding those factors which of themselves would not mean the pregnancy is high risk, but taken together indicate potential problems. The actual cutoff points for age, parity, and education levels can be adjusted to conditions in each country, but the principle of national standards is important.

The HRP assessment and referral form outlined in Table 6.5, developed and used in a rural primary care setting, was adopted throughout a governmental health system in a developing area (West Bank and Gaza). In this case A = low risk, B = medium risk, and C = high risk. Clear definitions and staff instruction are required. In some cases one C requires referral (e.g., hypertension), or several Cs or Bs means mandatory referral. The format and scoring system can vary, such as scoring each topic from 1 to 10 with systematic compilation of the score. Referral to HRP clinics needs not only a clear indication or reasons for referral, but also thorough assessment by specialists and feedback to the referring physician or oth-

TABLE 6.5   High Risk Pregnancy Scoring and Referral Form

I. Personal Data	II. Social/Personal	III. Obstetrical History
Name	Age (<17, >35)	Gravida (1 or >5)
Identity Number	Education (<6 years)	Para (0 or >5)
	Marital status	Abortions (>2)
Date of Birth	Economic status	Miscarriages (1+)
Adress	Consanguinity	Fetal deaths (1)
	Smoking	Bleeding in T3
City/town/village	Alcohol	Stillbirths (1)
	Drug use	Previous cesarian (1)
Clinic	Home conditions	Preterm deliveries
Date last menstrual period	Summary _____	Birth weights
Date of first visit	_____	Infant deaths (1)
	A B C	Toxemia (1)
		Birth defects (1)
		Summary _____
		_____ A B C
IV. Medical History	**V. Present Pregnancy**	**VI. Summary of Risk Factors**
Diabetes	Number of present pregnancy	**I. Personal** A B C
Hypertension	Time first visit T1 T2 T3	
Renal disease	Weight before pregnancy (<50 kg)	**II. Social** A B C
Heart disease	Height <145 cm	
Chronic chest disease	Bleeding	**III. Obstetrical History** A B C
Blood disorder	Preclampia	
Endocrine disease	Rh	**IV. Medical History** A B C
Phlebitis	Multiple pregnancy	
Other	Abnormal presentation	**V. Present Pregnancy** A B C
Summary _____	Summary _____ A B C	
_____ A B C		**VI. Total score** A B C

Summary of Reasons for Referral: _____

Date: _____   Signature: _____
                                   Position: _____
Report of High-risk Clinic: _____
– –
– –
Date: _____   Signature: _____
                                   Position: _____

er primary care provider. The HRP clinic should send the patient back to the referring center with a report of findings and a clear recommended plan of action.
The HRP clinic may continue to follow the patient because of the risk factors, but
the referring provider should have this information and help in its implementation.

A well-developed HRP assessment–referral–follow-up system contributes to
improved outcomes for both mothers and newborns, preventing costly long-term
consequences of maternal and infant morbidity and mortality. Its role in prevent-

ing complications during and following delivery is well justified on medical, public health, and economic grounds.

The issues may be different in developing and developed countries, but the principles are similar. In the United States there are substantial population groups that cannot get prenatal care for financial or bureaucratic reasons, including those who may be at highest risk. In Russia, the HRP approach is essential if the country is to bring down its high maternal mortality rates. As previously noted, countries with both universal access and well-developed obstetrical care and risk assessment have low rates of maternal mortality (Australia, Belgium, Denmark, Israel, Norway). Maternal mortality does respond to improved access to medical care, as seen in its rapid decline in Italy and Portugal during the 1970s following introduction of their national health services.

LABOR AND DELIVERY

Delivery is a highly personal service but with important public health aspects. Assurance of safe delivery affects both mother and newborn. Over the past century the place of delivery has evolved from the private home to maternity or lying-in hospitals to maternity units in general hospitals, and to "mother and baby friendly" birth centers (in hospital settings). Delivery in maternity units in hospitals in developed countries has made an important contribution to reduced maternal and neonatal mortality and should be considered a prime goal in public health and primary care programs.

Hospital delivery is encouraged in many countries by giving maternity grants paid by national insurance or social security to the mother for covering costs of delivery in a hospital. This was done previously in France and Israel before the advent of national health insurance in order to promote hospital delivery. France continues this tradition by providing grants to women who come to prenatal care early and frequently during pregnancy. In developing countries the place of delivery needs to be addressed, making use of existing resources with a goal of reducing the current extremely high rates of maternal morbidity and mortality.

Dangerous emergencies can occur at any stage of labor for which cesarian section, blood transfusions, general anesthesia, and fetal resuscitation or treatment of respiratory distress syndrome may be necessary, so that delivery in a general hospital obstetric department is safer than at home or in a free-standing maternity home. Electronic fetal monitoring during labor does not necessarily reduce fetal intracranial damage, retardation, and cerebral palsy, as compared to clinical fetal auscultation, but provides reassurance to the anxious mother and assists the harried staff of a busy obstetrical department.

Intrapartum care should be made as homelike as possible and include the husband or other person desired by the pregnant woman. Care during labor and delivery requires information from the prenatal period, by transmission of a pregnancy care record, carried by the woman to the delivery room. This record summarizes her medical, obstetrical, and prenatal care, especially parity, date of last menstrual period, risk factors, and prenatal care findings. A mother-carried record is preferable, since access to prenatal clinic records may be limited and may not include care rendered by other providers, such as emergency rooms or private physicians.

As noted above, some countries, including some with very low maternal mortality rates, use qualified midwives to perform routine deliveries while obstetricians are responsible for supervision and complex deliveries. Licensing of qualified midwives may be necessary in the United States to meet current obstetrical needs, especially where there is a shortage of obstetricians. Qualification for midwives is usually an additional year of study for registered professional or academic level nurses (Chapter 15).

Fetal monitoring during labor (intrapartum) by auscultation has been standard practice since the work of Evory Kennedy in Dublin in the 1830s. Since the late 1950s electronic fetal monitoring has become widespread and reassuring to parents and of help to staff on a busy obstetric ward. Hospital maternity wards should be "mother, family, and baby friendly," promoting a homelike atmosphere with both midwives and medical support.

Attendance by trained anesthetists and pediatricians should be considered for routine cases and mandatory for high risk cases. The presence of a pediatrician or pediatric nurse practitioner at the time of birth gives greater assurance of appropriate early care for the newborn, especially important in cases of low birth weight, respiratory distress, or other complications at birth. In countries with low overall infant mortality, early neonatal deaths are now a major element in that mortality. Further reduction in infant mortality rates requires greater attention to this risk period when resuscitation and other immediate lifesaving procedures may be needed.

In the United States in recent years, there has been a small trend to return to home deliveries with highly trained midwives and guaranteed backup services. Home deliveries are declining in the United Kingdom with an increase in specialist obstetric departments in general hospitals. Home births may be an adequate alternative to delivery in the hospital for low risk pregnancies, but only when carried out by well-trained professional midwives with the backup of emergency medical teams able to deal within minutes with the complications that can occur with any delivery. It is a trend that could lead to increases in maternal and infant death as an unfortunate side effect. Delivery in maternity homes or free-standing birth centers is an alternative to maternity units in general hospitals, but should be avoided in developed countries. The homes or centers lack full facilities in the event of complications unanticipated before the onset of labor. The place and methods of delivery may vary but should take into account not only the wishes of the pregnant woman but the safety of the mother and the newborn as well. High risk mothers, including all primiparas, should always be delivered in a hospital setting.

Safe Motherhood Initiatives

Safe motherhood is an initiative of the United Nations launched in 1987 with the goal of ensuring that women go through pregnancy and childbirth safely, and give birth to healthy children. Every year hundreds of thousands of women die or suffer serious complications from pregnancy and childbirth. Safe motherhood begins before conception with proper nutrition and a healthy lifestyle. It continues with planned pregnancy, appropriate prenatal care, prevention of complications when possible, and early and effective treatment of complications when they occur. It proceeds with labor at term with adequate care but without unnecessary interventions, followed by delivery of a healthy infant and provision of a healthy postpartum environment that supports the physical and emotional needs of the woman, infant, and family. This requires the support of many diverse sectors of the nation, the family, educators, employers, health care providers, community groups, and others. It is an investment in the physical, emotional, social, and economic well-being of women, their children, and their families and, by extension, the nation.

Home births are still common in developing countries. They are unattended or attended to by traditional birth attendants (TBAs). This may be the only choice the mother has, in which case the public health authorities should train and supervise the TBAs. Special attention should be given to hygiene, prevention of tetanus, and identification and referral of high risk cases. The World Health Organization promotes guidelines for integrating TBAs into the health care system in developing countries. Setting standards and supervising of TBAs is vital to reduce the current large-scale loss of mothers and newborns in developing countries. Primiparas should be referred to a district hospital for delivery, as should other high risk cases including grand multiparas, women over age 35, multiple pregnancies, late presentation for care, those with previous complicated deliveries, malnutrition, or chronic illness, or those with previous infant deaths.

Public health agencies should ensure hospital delivery wherever possible, and ensure that hospitals are staffed with well-trained personnel and that early and consistent prenatal care is utilized by all pregnant women. District hospital facilities should be equipped for obstetrical and newborn emergencies, including surgical, anesthesia, blood bank, and resuscitation staff and facilities for normal and cesarian deliveries. All care by community and district care centers and by TBAs should be documented in appropriate record systems developed by the public health authorities.

District- and community-based obstetric service centers should be developed to provide prenatal risk assessment and care, working with TBAs both in a supportive and supervisory capacity. Community-based delivery facilities should be equipped and staffed to perform normal deliveries and safe abortions and handle emergencies such as preeclampsia, hemorrhage, and manual removal of placenta. TBAs should be trained to arrange evacuation of high risk mothers and babies to district hospitals. Professional midwives or nurses should be appointed to supervise TBAs including visits to the villages and direct one-on-one training. TBAs performing deliveries on high risk patients except in absolute emergencies or nonpreventable situations should be fined by the public health authority.

Supervision of TBAs involves a process of regulation, education, and incentives to meet higher standards of practice. The health authority is empowered to license and regulate all health providers and to discipline offenders who practice without a license or do not meet current standards of practice. TBAs can be motivated to participate in training programs with incentives including provision of essential equipment and supplies. TBAs should be required to keep a log of deliveries with basic details and outcomes recorded in a state issued log book. If they are illiterate, they can be required to engage a literate person to keep such records, for regular inspection by the TBA supervisor. Training should include prenatal and delivery danger signs, hygiene in delivery, neonatal resuscitation and infancy care, and high risk situations. TBAs can also be trained in regard to STDs, HIV, anemia, iodine deficiency, and obstetrical or medical risks.

Care of the Newborn

The newborn is totally dependent and vulnerable to many factors that may inhibit optimal growth and development. This dependency begins before conception and continues during the prenatal period, the delivery, and up to the end of the first year of life. Care should be given to assure warmth and suction to reduce the chance of aspiration and assure breathing immediately following delivery. Many of the risk situations are preventable, and the drastic decline in infant mortality seen over the past century indicates this potential.

Survival of newborns falls gradually as birth weight declines. Newborns weighing less than 2500 grams (5 pounds 8 ounces) are considered low birth weight, with normal birth weight between 2500 and 4000 grams (5 pounds 8 ounces to 8 pounds 13 ounces). In developed countries the low birth weight rate is usually below 7% and in developing countries often over 10%. Babies born weighing less than 1500 grams (3 pounds 5 ounces, very low birth weight) and under 1000 grams (2 pounds 3 ounces, ultra low birth weight) have lower rates of survival and are often left with permanent handicapping conditions. A major purpose of good prenatal care is to ensure continuance of the pregnancy beyond 36 weeks and as close to full term as possible in order to promote full maturity and development of the fetus *in utero* and ensure greater chance of survival of the newborn. To reduce the factor of multiple births, the term low birth weight is sometimes used for single births only.

The newborn deserves the best care available. This requires immediate assessment and immediate response to resuscitation needs or other necessary interventions. The Apgar score, developed by Virginia Apgar in the early 1950s, provides an important standard method of evaluation of the newborn and its need for investigation and intensive care (Table 6.6). Scores of less than 6 at birth may indicate a need for resuscitation and specialized attention, and low scores after 2 minutes indicate an infant at risk.

Where home births take place, the potential for transport of the newborn with difficulties to a neonatal care center should be part of the preparation. Early detection and intervention reduces unnecessary morbidity and mortality of the newborn.

All newborns should receive well-trained professional care including resuscitation if needed and routine eye care to prevent ophthalmic infection with gono-

TABLE 6.6 Apgar Score

Sign	Score		
	0	1	2+
Heart rate (beats/minute)	Absent	Slow (<100)	100+
Respirations/ hypoventilation	Absent	Weak cry	Strong cry
Muscle tone/motion	Limp	Some reflection	Active
Reflex irritability	No response	Grimace	Cough or sneeze
Body color	Blue or pale	Body pink extremities blue	Completely pink

Source: Kendig, 1992.

coccus. Vitamin K should be given routinely either by injection or orally, in order to prevent hermorrhagic disease of the newborn (HDN). As many as 20% of newborns have low levels of prothrombin, a blood clotting factor, and are therefore subject to potentially serious and even fatal bleeding, which is preventable by injected or oral vitamin K.

Screening for HIV infection if the mother's HIV status is unknown and for signs of drug or alcohol effects in newborns is important in high risk settings. While these

---

BOX 6.1 RECOMMENDED STANDARDS FOR ROUTINE CARE OF THE NEWBORN: THE AMERICAN ACADEMY OF PEDIATRICS

1. Resuscitation by trained personnel should be available at all high risk pregnancy deliveries; transfer to neonatologist if necessary;
2. Suction mouth and nose and establish airway;
3. Clean, warm, and cover newborn; give to mother;
4. Stimulate the newborn by rubbing back and soles of feet;
5. Oxygen via Ambu bag with face mask, if needed: examine chest for adequate ventilation;
6. Apgar score at 1 and 2 minutes following delivery;
7. Apgar score at 5 minutes following delivery;
8. If prolonged resuscitation is needed further Apgar scores should be taken and recorded;
9. Establish bonding and breast-feeding by giving newborn to mother as soon as possible;
10. Vitamin K is routinely given to prevent hemorrhagic disease of the newborn;

11. Eye care is given with antibiotic ointment to prevent chlamydia or gonococcal eye infection;
12. Cord blood is taken for serology;
13. Umbilical cord is cut and tied applying antiseptic or antibiotic;
14. Hepatitis B vaccine is given, the first of three doses;
15. Newborn should be loosely wrapped, not swaddled;
16. Newborn should preferably stay with mother ("rooming in");
17. Father's visits and participation in care are to be encouraged;
18. Breast-feeding is strongly encouraged;
19. Complete examination by a pediatrician or primary care physician;
20. Family and sibling visits are desirable.

Source: Kendig, 1992; and American Academy of Pediatrics/American College of Obstetrics and Gynecology. 1992. Guidelines for Perinatal Care. Third Edition. Washington, DC.: ACOG, and Elk Grove Village, IL: AAP.

may be seen as part of clinical practice, it is up to modern public health agencies to establish the norms and training systems needed to foster high standards of care.

Complete examination by a pediatrician searches for birth defects and normal development and lays the basis for continued care in early infancy. The newborn should be examined for length, head circumference, birth weight, gestational assessment, and jaundice, with results recorded. Examination also includes assessment of the heart and vascular and neurological systems. Screening for congenital disease or abnormalities such as congenital cardiac defects, Down's syndrome, congenital dislocated hips (CDH), club feet, undescended testes, phenylketonuria (PKU), hypothyroidism, hemolytic anemias (thalassemia and sickle-cell disease), galactosemia, and other inborn errors of metabolism should occur in this period with follow-up of any suspicious or positive findings. Early diagnosis of abnormal conditions allows the child to benefit from suitable therapy.

Establishment of breast-feeding is a vital event for mother and newborn and should be encouraged by all health personnel. This is important for bonding between mother and infant and for the optimal development of the infant. It should continue for at least the first half year of life especially and longer if adequately supplemented by solid foods. Breast-feeding is discussed under nutrition (Chapter 8) and in globalization of health (Chapter 16).

Care in the Puerperium

The puerperium is the 6- to 8-week period from the time of delivery of the placenta up to the resumption of normal ovulation. In the nursing mother, the puerperium may be prolonged because of the effects of lactation on hormonal balance. This is an important time for the new mother and newborn to recover from the birth process and for their physiological and psychological adjustment. Adequate follow-up of the mother and infant soon after delivery helps to prevent complications and promotes optimal health for both.

Immediate postpartum concerns focus on assurance of complete expulsion of the placenta, contraction of the uterus, absence of bleeding from any site, or phlebitis. Basic examinations to protect the new mother's health include a breast examination, determination of the presence of any cervical or perineal tears or infections, and a hemoglobin test. Management of any residual problems should be part of the care process. The mother is provided with information on a variety of different topics including breast-feeding, nutrition, spacing for the next pregnancy, and birth control. She is also given any needed support and counseling to address the potential for postpartum depression which can become a debilitating disorder.

The mother's recovery from the pregnancy and birth process should be aimed at restoring her strength and normal family life. Postpartum depression is common and may need to be addressed by the care provider. Concern to lose weight rapidly may contradict the mother's nutritional demands, which are increased if she is breast-feeding. Iron supplements should continue in the puerperium and while breast-feeding to restore the iron lost to her during pregnancy, delivery, and breast-feeding. If necessary, Rh immune globulin to prevent future maternal–fetal blood immune reaction should be handled by obstetricians. Also if necessary, rubella immunization should be given to the postpartum mother to prevent possible future congenital rubella syndrome (Chapter 4).

Postnatal care should include a home visit by a public health nurse to assess and assist the adaptation of mother and infant and the home setting. A medical examination toward the end of the puerperium is an important element of pregnancy care because it is the first opportunity for complete assessment of the newborn after discharge from the hospital. The key concerns for the totally dependent newborn in this period are breast-feeding and bonding, hygiene and cleanliness, weight gain, vitamin supplements, beginning of immunization, and acceptance into the family. First immunizations for hepatitis B, BCG (where used), and poliomyelitis begin as close to birth as possible and the routine immunization program in the first 6 weeks (see Chapter 4). For home deliveries, postnatal home visits should assess and weigh the newborn, administer vitamin K, initiate the vaccination program and routine screening tests, as well as counsel on breast-feeding and infant care.

Counseling by the health care provider is very important to the new mother whether this is her first pregnancy or one of many. Family planning to promote spacing of pregnancy should be discussed as part of the health provider's attention and educational efforts. The need for support and advice is high in this period to reduce anxiety and to address many issues relating to maternal and infant health. Training of health providers working with delivery and newborn care should stress this role along with the professional skills of managing the delivery and immediate neonatal care.

GENETIC AND BIRTH DISORDERS

An estimated 8000 infants die and 150,000 are born annually with birth defects, with 8 billion dollars spent on their care. Prevention and management of birth defects and genetic disorders are therefore of public health concern, not only because of their high prevalence in many specific population groups, but also because of the financial and social cost to the family and community. Genetic disorders are increasingly both treatable and preventable. Screening, community education, and individual patient genetic counseling are likely to become even more important with the rapid development of gene therapy.

Prevention of birth defects requires screening for genetic diseases, congenital diseases, and Rh incompatibility, education, folic acid supplementation before and during pregnancy, reduced exposure to teratogenic chemicals at home, at work, or in the community, and prevention of some infections in pregnancy. Rh incompatibility prevention depends on screening of pregnant women and management where incompatibility between the maternal and fetal blood can develop. Screening for congenital diseases, such as phenylketonuria (PKU) and hypothyroidism, must be followed up for confirmation of suspected cases and managed appropriately with suitable standards of care. Other birth defects, such as congenital hip dislocation, which may occur in as many as 50 per 1000 births, and congenital cataracts are sought on clinical examination of the newborn, with hearing tested within 3 months. Findings should be referred for follow-up care in pediatric services.

The range of birth defects is wide and includes chromosomal and developmental abnormalities as well as metabolic, hematologic, musculo-skeletal, car-

---

BOX 6.2 BIRTH DEFECT PREVENTION

1. Community wide prevention
   a. Iodine fortification of salt to prevent iodine deficiency brain damage;
   b. Folic acid fortification of flour to prevent neural tube defects;
   c. Rubella immunization before pregnancy to prevent congenital rubella syndrome;
   d. Treatment of STDs to prevent congenital syphilis, gonococcal, CMV, and HIV infections;
   e. Hepatitis B immunization to prevent maternal–fetal transmission;
2. Prenatal prevention
   a. Cessation of smoking, alcohol, and drug intake to prevent fetal damage;
   b. Prepregnancy and pregnancy nutrition counseling to promote optimal fetal development;
   c. Genetic counseling for parents at risk (e.g., consanguineous marriages);
   d. Avoidance of unnecessary medication;
   e. Iron, folic acid, and multivitamin supplements (and iodine if salt is not fortified);

f. Risk assessment and referral to a high risk care program;

g. Prenatal care protocol starting in first trimester;

h. Fasting blood sugar to screen for gestational diabetes;

i. Good management of chronic diseases (e.g., diabetes and cardiac disease) before and during pregnancy;

j. Prevention of exposure to teratogenic agents before and during pregnancy (e.g., anesthetic agents);

k. Prevention of exposure to lead and mercury ingestion;

l. Hemoglobin, Rh testing, Coombs test, urinalysis;

m. Adequate weight gain (11–14 kg);

n. Sonography for suspected multiple gestation, structural, or placental abnormalities;

o. Fetal monitoring, fetal size, position, and heart rate;

p. Safe delivery with minimal sedation and early resuscitation to prevent brain damage at birth.

Source: U.S. Preventive Services Task Force, 1996.

diovascular, and neurological disorders. Genetic disorders are transmitted by abnormal chromosomes from one or both parents. Dominant genes are those which cause the genetic disorder in a heterozygote, that is, a person who has one copy of the abnormal and one copy of the normal gene. If only one parent is affected, but there is a 50% probability of offspring being affected. In recessive or homozygous conditions, both parents are carriers of the abnormal gene. Each child has a 25% chance of being affected, a 50% chance of being a carrier, and a 25% chance of being unaffected. If only one parent is a carrier of the homozygous gene, the disorder is not passed on although the child may also be a carrier.

Rhesus Hemolytic Disease of the Newborn.   Rhesus hemolytic disease of the newborn is a serious condition of breakdown of red cells in the newborn due to an incompatibility between fetal and maternal blood. It occurs when the mother is Rh negative and the fetus Rh positive (inherited from the father). Fetal red blood cells can mingle with the maternal blood during delivery causing the mother to produce antibodies to the fetal blood. This may affect subsequent pregnancies, causing hemolysis of fetal blood, producing anemia, jaundice, brain damage, or death. This was the cause of many stillbirths and neonatal deaths, until the development of exchange transfusions in newborns reduced the death rate. In the 1970s, anti-D immunoglobulin was introduced. This is given to Rh negative women following birth of an Rh positive baby to prevent the mother from developing antibodies which would affect the next pregnancy. Treatment with anti-D immunoglobulin should eliminate this disease, a major victory of preventive care.

Neural Tube Defects. Neural tube defects (NTDs) are defects of the neural tube, the precursor of the brain and spinal cord. This group includes a range of abnormalities, from anencephaly, a markedly defective development of the brain which usually results in death within a few hours of birth, to spina bifida, a defective closure of the vertebral column. NTDs vary in degree of severity but many require extensive surgical and medical care. NTDs occur in 2 per 1000 births in North America, but the incidence is declining as a result of screening in pregnancy and primary prevention through folic acid supplementation during prepregnancy and in early pregnancy. Screening for neural tube defects became possible in the early 1970s with amniocentesis and testing of amniotic fluid and later in blood tests for alpha-fetoproteins (AFP). Ultrasound can also detect NTDs.

During the 1980s, British investigators showed that folic acid given before pregnancy greatly reduces the chance of this abnormality developing. Prepregnancy care is not common, and compliance with recommended folic acid supplementation is under 30%, so that the addition of folic acid to bread was adopted by the United States Food and Drug Administration (FDA) in 1996 and will likely be adopted in Canada, the United Kingdom, and elsewhere for primary prevention of this disorder. In April 1998, the FDA and CDC recommended prepregnancy and pregnancy supplementation of folic acid, 400 µg/day, to augment the folic acid in bread in order to assure 100% RDA intake (see Chapter 8). CDC expects fortification of bread and supplements for women in the age of fertility to substantially reduce the current 3000 NTD cases born annually in the U.S.

Cerebral Palsy. Cerebral palsy (CP) is a group of neurological disorders occurring in about 1 in 400 births, causing motor disabilities. It may be associated with mental retardation, seizure disorders, motor spasticity, or sensory problems. CP is related to low birth weight (<2500 grams), and especially very low birth weight (<1500 grams), as well as intracranial hemorrhage, Rh incompatibility, intrauterine and birth trauma, maternal exposure to heavy metals such as mercury, and other unidentified factors. About 20% of CP cases are due to intrauterine fetal hypoxia. Preventive measures include reducing low birth weight births by improving maternal nutrition, smoking cessation during pregnancy, improving prenatal care and care in labor and delivery, giving vitamin K at birth, and reducing infant trauma. Use of professionally trained midwives reduces risk of CP. Prevention is limited by lack of identification of many causative factors.

Mental Retardation. Severe mental retardation (intelligence quotient or IQ of less than 50) occurs in between 3 and 5 per 1000 newborns. Retardation may be related to prenatal factors and is often associated with CP and seizures. More than one-third of cases are attributed to chromosomal abnormalities. Prenatal diagnosis, widespread use of amniocentesis in pregnant women over age 35, and termination of affected pregnancies have contributed to reducing retardation. Phenylketonuria, congenital rubella, congenital hypothyroidism, and maternal infections with toxoplasmosis and cytomegaloviruses can all cause severe retardation. Pregnancy complications such as toxemia, urinary tract infections, and anemia increase the risk of retardation. Mild retardation (IQ of 50–70) may relate to both perinatal and postnatal factors, including low birth weight or asphyxia at birth. Prevention requires well-organized prenatal, perinatal, and postnatal care.

Down's Syndrome.   Down's syndrome is a relatively common genetic disorder, occurring in 1 per 650 live births in the United States. The risk increases rapidly with increasing maternal age; for mothers over age 40 it occurs in 1 per 40 births. Although this condition occurs more commonly in older mothers, the majority of cases are in younger women. It is the most common cause of mental retardation in industrialized countries. In cases at risk, prenatal screening is done by amniocentesis, chorion villus sampling, and fetal blood sampling, allowing for the parental choice of termination of pregnancy should the chromosomal abnormality be detected. Chorion villus sampling can be performed in the first trimester, while amniocentesis can be performed in the second trimester. Chorionic villus sampling has itself been reported to be associated with minor birth defects and is currently under review. Pregnant women over age 35 should be screened as early as possible in pregnancy. Biochemical markers will help to increase screening with less invasive procedures. The total incidence of new births of Down's cases fell by 50% between 1960 and 1978 in the United States, probably because of reduced fertility among older women, widespread use of birth control, and prenatal screening.

Down's syndrome babies are mentally retarded, often with congenital cardiac defects and gastrointestinal obstruction. These patients now survive well into their 30s, and half to their 50s. Institutional care for a lifetime is estimated to cost $300,000, although emphasis has moved to self-care in the community. Education and training for Down's syndrome children has been emphasized in recent years. With vocational and life skills training, many are able to live independently or in groups in the community rather than in institutions. Those who are integrated into the community often live longer and have a better quality of life as compared to those raised in institutions.

Cystic Fibrosis.   Cystic fibrosis (CF) is the most common lethal genetic disease in the United States white population, occurring in 1 per 3500 for North Americans of Northern European descent, but 1:14,000 for black and 1:11,500 for Hispanic newborns. The disease is a recessive genetic defect. The gene is present in 5% of the white population and 2% of blacks in the United States. The defect in homozygotes causes production of abnormally thick mucus in the lungs, intestines, and other glands.

The clinical disturbance is chronic obstructive lung disease, repeated infections, and destruction of lung tissue. This leads to frequent hospitalization and death, even with the best of care. Survival rates have improved dramatically, from 20 years in 1970 to 30 years in 1994. Prenatal screening can be done with chorionic villus sampling or amniocentesis. Screening of pregnancies with a previous CF birth can prevent a second child with this condition, but general population screening is not yet available. Early diagnosis and support and education of parents can significantly improve the duration and quality of life of the affected person. Treatment is complex and costly, involving a multidisciplinary approach. Gene therapy techniques may be available in the next few years to improve case management.

Sickle-cell Disease.   Sickle-cell disease is an amino acid defect of red blood cells which affects 1 in every 400 of black newborns in the United States and is even more common in Africa, where it may have a protective effect against malarial parasites. The disease is caused by a recessive gene affecting hemoglobin structure, usually with a benign course in the carrier state, but with episodic pain due to small blood vessel closure in the homozygous state. Clinical symptoms appear in the second half of the first year of life. The patient develops moderately severe anemia, with an increased susceptibility to severe bacterial infections (meningitis, pneumonia, septicemia) and growth retardation. Screening of newborns has been recommended since 1972 in the United States; in practice, however, screening is done selectively with 20 states screening 57% of black infants. The U.S. target is to achieve 95% coverage by the year 2000. Identification of cases and carriers is important in order to assure prompt care in crises, for preventive use of penicillin which reduces infection rates, and to ensure they receive influenza and pneumococcal vaccines. Carriers should receive genetic counseling related to marriage and pregnancy.

Thalassemia.   Beta thalassemia is a recessive genetic disorder of hemoglobin structure. Beta thalassemia minor is usually without clinical significance. Beta thalassemia major, the homozygous state, is characterized by hemolytic anemia (i.e., early breakdown of red blood cells). Beta thalassemia major, also known as Cooley's anemia or Mediterranean anemia, is widespread throughout the Middle East, southern Europe, and across southern India and southeast Asia. It is ultimately fatal for those afflicted, but with current standards of treatment, including blood infusions and chelating agents (i.e., iron binding for excretion) to reduce iron overload and hemochromatosis, patients survive into their thirties.

The WHO estimates that 240,000 deaths, 290,000 new cases, and a total of 2.32 million cases of thalassemia and sickle-cell disorder occurred in 1995. The disorders of hemoglobin are widespread, originally localized in the Mediterranean areas, spreading to southeast Asia for the thalassemias and sub-Saharan Africa for sickle-cell disease. Because of migration, these diseases are now spread worldwide, with 10% of the population at risk in the United States.

Prevention of thalassemia is one of the success stories of genetic public health. Preventive approaches to this disease since the 1970s produced dramatic results in reducing the number and rate of new cases of beta thalassemia major in Sardinia, Cyprus, Greece, the United Kingdom, Canada, and other locations where this disease has been endemic among peoples of Mediterranean origin. Premarital screening, health education in schools, and access to prenatal screening are all part of a program to reduce new cases of this disease. The World Health Organization stresses the public health importance of this disease and recommends adoption

BOX 6.3 ERADICATION OF THALASSEMIA
MAJOR IN CYPRUS AND SARDINIA

Virtual eradication of thalassemia major has been achieved in formerly endemic areas. In Cyprus, 14% of the population were of carrier status for beta thalassemia and 1% of the Cyprus population (1 in 158 births) had the disease. Due to an intervention program initiated in the 1970s, only rarely are new cases of beta thalassemia major born in Cyprus and other Mediterranean locations such as Sardinia and Greece. This has been achieved by a long-term preventive program consisting of public education, screening for carriers, and counseling. Marriage between carriers is reduced by a community education program, and, when marriage does occur, careful screening of all pregnancies and termination of affected pregnancies reduces the number of thalassemia major births. The success of the Cyprus approach provides a model of control of a genetic disorder via a combination of health education, screening, and community support.

Source: Angastiniotis *et al.* 1986. How thalassemis was controlled in Cyprus. *World Health Forum*, 7:291–297.

of demonstrably successful preventive approaches to member states having this problem.

Screening for congenital anemias should take place at birth and at school age, because clinical cases may not appear until several years after birth. Gene carriers need to know at an age when they can understand the limitations this may place on them. In each pregnancy when both parents are carriers, chorionic villus sampling or amniocentesis should be carried out to determine if the fetus is affected. Abortion is currently recommended if the fetus is affected with thalassemia major, but bone marrow transplantation is showing promise in the treatment of new cases. The success of primary prevention in reducing new cases of thalassemia provides a model for application to disorders affecting other population groups, such as sickle-cell anemia.

Phenylketonuria. Phenylketonuria (PKU) is an inborn error of metabolism transmitted by a recessive gene. It occurs in approximately 1 per 12,000 births in North America, but this includes a wide range from 1 per 5000 in Scottish Irish parents to 1 per 300,000 in blacks. It involves a mutation in DNA which causes inadequate production of an enzyme needed to metabolize phenylalanine, an amino acid. The fetus is unaffected because the maternal metabolism handles the excess phenylalanine. However, if the newborn lacks the enzyme needed, phenylalanine accumulates in the blood, leading to brain and neurological disorders and severe retardation. If discovered early, PKU can be managed with a special low-phenylalanine diet with no significant retardation.

Screening of all newborns for this condition is widespread in developed countries and reduces the number of cases requiring long-term institutional care. Where screening is not universal, pregnant women with previous PKU children should be tested during pregnancy and the newborn tested at birth.

Congenital Hypothyroidism.   Congenital hypothyroidism (CH) is a relatively common congenital disorder occurring in about 1 per 3500–4000 live births in a Caucasian population with a range of 1–20 per 10,000 live births in different population groups, depending on genetic makeup and consanguinity rates. This genetic disorder causes an inefficient development of the thyroid and may be confused with iodine deficiency disorders common in areas with deficient iodine in water and soil. Therapy with thyroid replacement in CH cases prevents the severe intellectual impairment that otherwise occurs. Screening within 48 hours of birth should uncover cases for long-term follow-up and management.

Fetal Alcohol Syndrome.   Fetal alcohol syndrome is caused by the toxic influence of alcohol on the fetus via the placenta. Alcohol consumption prior to and during pregnancy increases the risk of behavioral and cognitive disorders. Exposure of the fetus to maternal alcohol consumption, especially binge drinking (i.e., more than five drinks on one occasion), is a cause of fetal growth retardation and anomalies of the central nervous system. Fetal alcohol syndrome is completely preventable, as it only occurs when the mother drinks alcohol.

Tay-Sachs Disease.   Tay-Sachs disease is an inborn error of metabolism associated with progressive mental deterioration and loss of vision by 4–8 months of age and death by 3–4 years. It affects 1 in 2500 births among Jews of eastern European origin. Tay-Sachs disease can be prevented by screening these risk groups before marriage, at the time of marriage, or early in pregnancy. Prenatal screening by amniocentesis or chorionic villus sampling will confirm whether the fetus is affected. In such cases termination of pregnancy is recommended. These screening programs have reduced the incidence of new cases.

G6PD.   Glucose-6-phosphate dehydrogenase (G6PD) deficiency is a genetic disorder common in Mediterranean populations, including Jews of North African or Middle Eastern origin (Sephardic Jews), Greeks, southern Italians, southeast Asians, and southern Chinese. It results in episodes of hemolytic anemia due to an infection, reactions to certain foods (e.g., fava beans), or reactions to oxidant drugs such as sulfonamides, antipyretics, and antimalarials. The degree of hemolysis varies with the agent and the degree of the enzyme deficiency. Identifying the condition helps the patient avoid exposure to hemolytic inducing agents and start prompt treatment in crises.

Familial Mediterranean Fever.   Familial Mediterranean fever is a recessive hereditary condition found in Arabs, Armenians, and Sephardic Jews with periodic fevers and pains in the chest, abdomen, and joints. Control is by genetic counseling.

INFANT AND CHILD HEALTH

Public health has long played a major leadership role in improving the health of children by provision of care and regulation of conditions to prevent disease, provide early and adequate care of illness, and promote health. Pediatrics developed as a clinical specialty under the leadership of Abraham Jacobi who opened the first pediatric clinic in New York City in 1860 based on German models. The first children's hospital in the United States was opened in 1865 in Philadelphia and continues to operate to the present time.

The American Medical Association, recognizing that women and children had health needs apart from those of the general population, established a section to address those needs in 1879, leading to the founding of the American Paediatric Society in 1888. Pediatrics emerged as a separate specialty from general medicine with emphasis on the treatment of children's diseases and birth disorders, the prevention of infectious diseases, and infant nutrition. Well-child care was pioneered in the United States based on milk stations, adapted from those in France ("Gouttes de Lait"), in the poor immigrant neighborhoods in New York City, leading to the development of public health nursing and the visiting nurse function.

The first textbook of pediatrics was written in 1869 by J. L. Smith, professor of Children's Diseases at Bellevue Hospital Medical College in New York. It was followed by Holt's *Diseases of Infancy and Childhood* in 1896. The latter is now in its nineteenth edition, and its current version is known as *Rudolph's Pediatrics*. The U.S. federal government established the Children's Bureau in 1913 to collect data on maternal and infant mortality. This later became the Bureau of Maternal and Child Health, which was subsequently empowered by federal legislation to provide grants to states for maternal and child health services. The American Academy of Pediatrics (AAP) has come to the forefront of advocacy for improved child health standards, pioneering in development of clinical guidelines and in promoting professional standards in a wide range of child health topics, from breast-feeding to cystic fibrosis screening, in parallel to the American College of Obstetrics and Gynecology (ACOG).

In the wider context of health internationally, maternal and child health are among the major priorities with special focus on primary health care. Primary health care, promoted since the Alma-Ata Conference, has placed emphasis on infant immunization, diarrheal disease control, breast-feeding and nutrition practices, and the prevention of deficiency disorders. The WHO, UNICEF, and many NGOs have provided leadership in development of primary health care in the developing world. This return to basics in health care has been of benefit to the industrialized world as well, as health care costs soared during the 1970s and 1980s. In the United States, public health had to cope additionally with the lack of health insurance for a substantial portion of the population, and even at the end of the twentieth century, federal initiatives in health are focused on providing health benefits to about 15% of the population who lack health insurance of the total U.S. child population of 75 million.

Fetal and Infant Mortality

A live birth is defined by the World Health Organization and the United States National Center for Health Statistics as a completed expulsion or extraction from its mother of a product of conception, irrespective of the duration of the pregnancy, which after separation, breathes or shows any other evidence of life such as heartbeat, umbilical cord pulsation, or definite movement of voluntary muscles, whether or not the umbilical cord has been cut or the placenta is attached. Each product of such a birth is considered live born. The delivery can be described as spontaneous vaginal, forceps, vacuum extraction, cesarian section, or vaginal delivery after previous cesarian.

A fetal death or stillbirth is a death prior to the complete expulsion or extraction from its mother of the product of conception, irrespective of the duration of pregnancy. The death is indicated by the fact that after such separation, the fetus does not breathe or show any other evidence of life, such as beating of the heart, pulsation of the umbilical cord, or definite movement of voluntary muscles. For statistical purposes, tabulations are shown for fetal deaths with stated or presumed gestation of 20 weeks or more and of 28 weeks or more; the latter are known as late fetal deaths. Other rates of infant mortality are seen in Box 6.4.

The infant mortality rate (IMR) is a generally accepted indicator of the health status of a population for internal regional and international comparisons, because it represents the cumulative effect of many socioeconomic, environmental, and health service factors. The industrialized countries have IMRs under 9 per 1000 live births and many 5 or even 4 per 1000 (Table 6.7). In these countries, most infant deaths are a result of congenital anomalies and perinatal conditions associated with prematurity and the neonatal period, usually occurring during the first week of life (early neonatal deaths).

In developing countries, most infant deaths are due to acute respiratory infections, diarrheal diseases, and prematurity (LBW), with measles and tetanus remaining large-scale causes of infant wastage. Developing countries generally have IMRs over 30 per 1000, with high neonatal and postneonatal mortality. Preventive health measures, such as immunization, breast-feeding and nutritional supplementation, good management of acute respiratory infections, and use of oral rehydration for diarrheal diseases, are effective mainly in the postneonatal period. Neonatal mortality rates can be reduced by reducing the number of low birth weight births, providing good maternal nutrition and prenatal care, minimizing birth injury, and giving good care immediately following birth. Together, these constitute the child survival package of services that are one of the main thrusts of public health.

The United States emphasizes targeted public health programming with attempts to provide care to people without access to prepaid medical care. Federal health officials set a goal to reduce the country's infant mortality rate and then expanded access to prenatal and infant care in order to reach this goal. From 1950 to 1995, the IMR of the United States fell from 29.2 to 7.6 per 1000 live births, a decline of 79%. The U.S. health target established in 1979, when the IMR was 15

BOX 6.4 RATES IN INFANT MORTALITY

1. Stillbirth rate $= \dfrac{\text{no. of stillbirths}}{\text{total live births} + \text{still births (per annum)}} \times 1000$

2. Perinatal mortality rate $= \dfrac{\text{no. stillbirths} + \text{no. deaths in the first week of life}}{\text{total live births (per annum)}} \times 1000$

3. Early neonatal mortality rate $= \dfrac{\text{no. of deaths in first week of life}}{\text{total live births (per annum)}} \times 1000$

4. Late neonatal mortality rate $= \dfrac{\text{no. of deaths between 7 and 28 days of life}}{\text{total live births (per annum)}} \times 1000$

5. Neonatal mortality $= \dfrac{\text{no. of deaths in first 28 days of life}}{\text{total live births (per annum)}} \times 1000$

6. Postneonatal mortality $= \dfrac{\text{no. of deaths between 28 and 364 days of life}}{\text{total live births (per annum)}} \times 1000$

7. Infant mortality rate $= \dfrac{\text{no. of deaths from birth to 364 days}}{\text{total live births (per annum)}} \times 1000$

per 1000, was to reduce the IMR to no more than 9 per 1000 live births by 1990, and this was nearly achieved. However, the rates are substantially different for white and black infants. Although mortality rates have declined for all ethnic groups the black infant mortality rates remain more than double the rates for white

TABLE 6.7 Mortality Rates for Infants,[a] by Race, United States, and Percentage Change, 1950–1995

Group	1950	1960	1970	1980	1990	1995	% change 1950–1995
Neonatal white	19.4	17.2	13.8	7.5	4.8	4.1	−79
Neonatal black	27.8	27.8	22.8	14.6	11.6	9.6	−65
Postneonatal white	7.4	5.7	4.0	3.5	2.8	2.2	−70
Postneonatal black	16.1	16.5	9.9	7.3	6.4	5.0	−69
Total infant mortality	29.2	26.0	20.0	12.6	9.2	7.6	−74

Source: *Health, United States,* 1998
[a]Rates per 1000 live births.

BOX 6.5 LEADING CAUSES
OF INFANT MORTALITY

Developed countries—IMR $< 20/1000$

1. Congenital anomalies
2. Sudden infant death syndrome
3. Prematurity/low birth weight
4. Respiratory distress syndrome
5. Newborn effects of maternal complications

Developing countries—IMR $> 30/1000$

1. Acute respiratory infections
2. Diarrheal diseases
3. Prematurity/low birth weight
4. Measles
5. Newborn effects of maternal complications
6. Neonatal tetanus

---

infants, reflecting socioeconomic differences and lesser access to medical care for the black population (Table 6.7).

Sudden infant death syndrome (SIDS) occurs in late infancy and into the second year of life. In the United Kingdom it constituted 45% of postneonatal deaths in 1990. SIDS may be due to a variety of causes, with maternal smoking during pregnancy as a risk factor. SIDS also occurs more often in disadvantaged homes.

INFANCY CARE AND FEEDING

The new infant is dependent on a healthy caring mother with support of the family and health providers. Physical and emotional warmth, cleanliness, and bonding with the mother are important in care and feeding (see Chapter 8). Healthful feeding in infancy is a vital issue in infant care and for primary care services, and is as important in infant survival and well-being as are infectious disease control and immunization. Therefore, it has an important place in maternal education and in health provider orientation.

Breast-feeding should be established early and encouraged as the sole feeding for 4–6 months, because of the ideal composition of breast milk nutritionally and for immunologic protection of the infant. Duration of breast-feeding should ideally continue to 1 year or more. Vitamins A and D should be given daily from

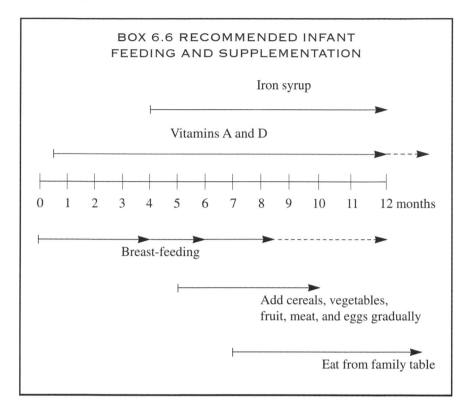

BOX 6.6 RECOMMENDED INFANT
FEEDING AND SUPPLEMENTATION

about 2 weeks after birth to prevent clinical and subclinical deficiency conditions for both breast- and formula-fed infants.

Prevention of iron deficiency anemia (IDA) is important because of its effects on the infant between 6 and 24 months of age when rapid brain growth and psychomotor development are at their peak. Iron supplementation should be given to all infants (from 4 to 12 months for normal birth weight babies and from 2 months for those with low birth weight) as iron-fortified formulas, supplemental iron, and/ or iron-enriched foods. If the child is not receiving iron-enriched formula, iron syrup, preferably with vitamin C, is given daily from 4 to 12 months at a level of 7–15 mg/day according to weight. Secondary prevention is by screening all infants at age 9–12 months to determine if hemoglobin is normal or if further care or investigation is needed.

Following cessation of breast-feeding, infant formula enriched with iron should be given preferably up to 1 year. Whole cow's milk should not be used until the infant is close to 1 year of age. Cereals fortified with iron and vitamins are also important in infant nutrition along with fruit, vegetables, meat, and eggs for optimal growth and development and prevention of vitamin and mineral deficiencies. Later in the first year, the infant should be eating at the family table.

The physical and emotional love of the mother, father, and family provide the security and stimulus needed by the infant to develop physically, intellectually, and emotionally. Stimulus by playing with and speaking to the baby help promote the normal development of the infant.

Immunization is discussed in detail in Chapter 4, and the current U.S. pediatric immunization schedule is presented. While there are many vaccination schedules and much controversy on this topic, the central importance of coverage of the infant with vaccination on schedule is of paramount importance for a successful child health program, including both infancy and school age follow-up.

ANTICIPATORY COUNSELING

An important element of the role of the pediatrician, family physician, or other health provider is counseling parents on what is to be expected at each stage of the child's development. This includes warnings of potential difficulties that might be experienced and how to cope with them. The American Academy of Pediatrics publishes detailed guideline questions for the practitioner addressing these issues from early infancy to the teenage years (see website http://www.aap.org/).

DOCUMENTATION, RECORDS, AND MONITORING

Standardized, user-friendly child health records for the primary care clinic help to promote national standards of child care practice, especially for preventive health care (Table 6.8). They help guide the busy practitioners, improve legal protection and professional peer group standards. Design of records systems and training of staff in their use should be given careful attention in any health care system, including private medical practice.

Well-constructed child health records set standards of care expected of the health provider. They furnish continuity of care when the contact with the health system may be episodic, or when the child is seen by different providers. They provide the caregiver with important time trends which may have clinical or epidemiological importance. Some child records are designed to continue through to adolescence, while others focus on the child up to school age, followed by a different record. Mother carried child health records ("road to health cards") provide the mother with all data needed for care during childhood and should continue through to adolescence, and be designed to include all basic information on one folded hard paper record, preferably with a plasticized envelope or cover pages. Many designs are possible, but timelines and visually easy formats enhance ease of use by providers. Maintenance of a child health record by the mother is important for their successful use.

Physical development in terms of weight and height for age and psychosocial developmental indicators should be carefully documented during infancy at the

TABLE 6.8   Components of a Child Health Record (Infancy, Toddler, Preschool, and School)

I. Personal data	II. Pregnancy/birth	III. Risk factors
Mother: name, age, family	Gestation	LBW
status, education, occupation	Place of birth	Previous care
Father: : name, age, family	Date of birth	Birth order
status, education, occupation	Birth weight	Family status: single parent
Address, telephone number	Delivery by	Social situation
Social security (ID) number	Condition at birth and at 1 and	Home conditions
Home situation: parent(s), sibs	5 minutes of age: Apgar	Family medical history
		Overall risk assessment
IV. Growth patterns: NCHS	**V. Immunizations dose,**	**VI. Nutrition, age when added**
growth curves in graphic	**date, reactions**	Breast-feeding: duration
chart form	BCG	Formula
Weight for age	Hepatitis B	Vitamins A and D
Height for age	DPT	Iron
Weight for Height	OPV	Supplements: solid foods,
Head circumference	IPV	cereals, fruit, vegetables,
	MMR	eggs, meat
	Haemophilus influenzae B	
	Varicella, Rotavirus	
VII. Intercurrent illnesses	**VIII. Development**	**IX. Medical examinations**
Diarrheal: age, duration	Hearing	Birth
Respiratory: age, duration	Vision	6 weeks
Hospitalizations	Responsiveness	3 months
Surgery	Relates with family	6 months
Others	Grasp, smile, grab	12 months
	Roll over	24 months
	Sit, crawl	36 months
	Stand, walk	48 months
	Talk: words, sentences	Preschool
	Other milestones	School ages
X. Summary of risk factors	**XI. Laboratory results**	**XII. Continuing notes**
Social	PKU	Document, developments, routine
Family	Hypothyroidism	examinations, illnesses, and
Genetic	Hemoglobin	care given
Medical	Others	
Other		

time of visits for immunization. Growth patterns are good health markers. The growth curves recommended by the World Health Organization are those developed by the U.S. National Center for Health Statistics (NCHS), based on a healthy U.S. white middle-class population. This provides an optimal or "gold standard" for use internationally, as opposed to local standards which may represent less than optimal child health and development. The NCHS is developing a revised set of standard growth curves based on data from a wider population sample. Most importantly it provides an external standard for comparison of change over time in the local setting.

While there is controversy on use of international or local standards (discussed in Chapter 8), there is full agreement on the need for growth monitoring and early intervention where an infant or a group of infants show signs of growth faltering or failure-to-thrive (FTT). Growth faltering occurs not only due to food insufficiency, but also as a result of intercurrent illnesses, which themselves can be more severe when nutrition status is compromised, such as the relationship of high case fatality rates from measles and other infections when vitamin A deficiency is present. FTT is a medically urgent situation requiring full evaluation and intervention as rapidly and consistently as possible.

THE PRESCHOOLER (AGES 1–5 YEARS)

The preschool child is undergoing rapid growth and development. The American Academy of Pediatrics recommends health assessments at 12, 15, and 18 months, and at 2, 3, 4, and 5 years (Committee on Psychosocial Aspects of Child and Family Health, 1985–198, American Academy of Pediatrics, 1988). The examination includes assessment of development status and physical examination as well as parental concerns and skills. Hearing should be checked along with signs of strabismus and general medical examination. The examination includes behavioral assessment based on both interview and observation. The care provider should use the visits to assess parental concerns and convey information on the child's health needs. Abnormal findings on physical or developmental examination should lead to referral to a clinical or child development assessment service. Hemoglobin (or hematocrit) and urinalysis should be checked at least once from 18 months to 5 years.

Counseling on common child care problems (e.g., nighttime crying), safety issues, and stimulation of the child should be discussed with the parent at each of these visits. Dangers of accidents are serious at these ages. Prevention of accidental falls, poisoning, aspiration of food, electrocution, burns, and scalding requires diligence and safety planning in the home. Parents should be educated on how to explain safety to their children as well as how to child-proof their homes (using childproof latches on cabinets and stairways, covering electrical outlets, safe storage of common household chemicals, or removal of poisonous houseplants). Signs of child abuse or neglect, especially injuries without adequate explanation, should be reported to police or welfare agencies. Assessment and counseling regarding appetite problems, toilet training, separation anxiety, demanding or obstinate behavior, relationship with parents and siblings, nutrition, self-feeding, and psychomotor development are all part of well-child care.

The physical and emotional well-being of the child up to school age is a key determinant of his or her health and capacities afterward. Completion of immunization, adequate nutrition with essential vitamin and mineral supplementation, and prompt, adequate treatment of intercurrent infections are essential. Warmth and responsiveness, including speaking and reading to the child to promote interest in books, pictures, and music are part of needed stimulation and play. Even warm and caring parents need guidance, as well as advice and support from the health provider, to promote optimal development for the child.

The community and national levels have a substantial role to play in promotion of safety and health for toddlers. Mandatory safety practices for children can save many lives and reduce child morbidity. These include such factors as correct use of safe child car seats, childproof medication or poisonous material containers, lead abatement programs, regulation of child clothing standards to prohibit inflammable materials, toy safety standards, supervision of nursery schools, and teacher qualifications.

Child abuse is an increasingly recognized hazard at all ages of childhood and in all socioeconomic groups. Preventive measures include public awareness through education and now mandatory reporting by teachers, social workers, and doctors of signs of abuse. Punishments of abusive parents or guardians, including imprisonment and mandatory treatment, are increasingly applied, but hidden physical, mental, or sexual abuse is far commoner than previously recognized. Home, school, and playground safety involves a wide range of community activities ranging from mandatory use of helmets for bicycle riding and many sports, proper use of car seats and seat belts for children in motor vehicles, to measures to reduce access of children to firearms.

In most industrialized societies, a large proportion of women work outside the home, so that child care includes preschool centers in which health factors can be both positive and negative. The potential for spread of infectious diseases among child care center attenders is high, and outbreaks of food-borne disease can also occur. In the United States, children in licensed day care centers have 3–4 times as much diarrheal disease as children not in organized child care centers. A Finnish study (1984–1989) showed that between 41 and 85% of colds, otitis media, and pneumonia in 1-year-old children were due to the children attending day care centers as opposed to family day care. On the other hand, children attending well-supervised preschool day care centers benefit from interacting with teachers and other children to develop social skills and to receive nutritious meals and professional attention. Preschool care centers should provide education in hygiene, nutrition, and health services including updating of immunization and developmental assessment. Services such as these may help parents to assess and meet the emotional and health needs for their children.

The value of preschool education, especially for children from socioeconomically deprived situations, was demonstrated by follow-up studies of children who had been in the U.S. Headstart program during the 1960s War on Poverty program. Disadvantaged children participating in Headstart were found to perform significantly better in school and to be less likely to have a criminal record or to be on social welfare than matched children not in the program. This resulted in benefits from reduced costs for special education and public assistance.

Preschool examination and screening for developmental, cognitive, and behavioral problems can be helpful to parents and teaching personnel. A variety of screening tests are available and used, but they are controversial as to their validity and predictive value. They may be culture specific and not necessarily valid in all societies without adaptation. Support, stimulus, and guidance can be culture specific, but are needed by the growing child.

SCHOOL AND ADOLESCENT HEALTH

The hazards that face schoolchildren include many public health issues. A central goal of public health regarding this young population is life skill training to promote healthful practices and avoidance of risk behavior. Public health efforts to reduce the toll of teenage premature loss of life (Table 6.9) must be in concert with educational, social, and police authorities. Efforts to convey a message to these risk groups should involve popular figures of the sports and popular music fields recruited to stress a positive image of teen adjustment and transition to adulthood. Research is needed to develop educational approaches for high risk teenagers and young adults.

A school health program should ensure a safe and healthful environment for the children, including sanitation, safety from violence, temperature control, and physical facilities for study and recreation. School-aged children need physical activity for their healthful development as an integral part of a comprehensive school health education program. School meal services contribute to good child and adolescent nutrition, learning, and performance.

Health monitoring includes ensuring full immunization at school entry and maintenance of its adequacy by appropriate booster doses at the elementary, secondary, and postsecondary stages. Monitoring of growth, vision, hearing, scoliosis, and skin testing for tuberculosis are frequent elements of school health programs.

The potential for learning about health in the school setting should be exploited by the public health system by encouraging the educational authorities to include health as part of the regular curriculum and preparing the teachers on how and what to teach in this area. During the early school years, the areas of study can include safety (fire and traffic), oral health (fluorides, brushing, and flossing),

TABLE 6.9 Total Mortality Rates for Children and Young Adults,[a] United States, and Percentage Change, 1950–1996

Age group	1950	1960	1970	1980	1990	1996	% change 1950–1996
1–4	139	109	85	64	47	38	−73
5–14	60	47	41	31	24	22	−63
15–24	128	106	128	115	99	90	−30

Source: *Health, United States,* 1998.
[a]Rates per 100,000 population in each age group (rounded).

healthy eating habits, hygiene, how the body functions, and healthy living (sleep and exercise). Topics for continuing education of teachers include assessment of child abuse, suicidal or violent tendencies, nutritional status, and substance abuse. Violence prevention has become an essential topic in educational systems at the primary as well as secondary levels.

Secondary school health programs include attention to personal behavior issues, including personal and family communication, sexual relationships, the right to say no (to peer pressure in situations ranging from alcohol and drug use to sexual activity), parental responsibilities, birth control, and STD and HIV prevention. Accident prevention is also vital in this population, including safety in sports, use of vehicles, work, and safety as a life habit.

Preadolescent and adolescent children are subject to social pressures that promote early sexual activity, smoking, substance abuse, eating disorders, accidents, suicide, and violence. The school is an important part of their life and has the potential to serve a socially educational role for attitudes and practices that may be lifelong. An activist approach to school health promotion should be integral to the school curriculum. School health committees of students, teachers, and parents can serve to promote awareness and program development, with community participation.

Special needs for those with physical or mental handicaps should be identified and recognized as part of the regular duties of the school. Teaching staff should be trained and facilities modified to accommodate children with special problems and to be able to recognize physical and social ill-health, in a wide range of areas from nutrition to child abuse.

BOX 6.7 CONTENT OF A SCHOOL HEALTH PROGRAM

1. Healthful, clean and safe school environment;
2. Health as topic in school curriculum;
3. Emergency or urgent health care in the school;
4. Physical education/fitness;
5. Education for special or handicapped children within regular school program;
6. Training and supervision of teaching staff in health content of teaching program;
7. Nutrition education and eating disorders prevention and case detection;
8. School lunch and breakfast programs;
9. Family life and sex education/parenthood/birth control;
10. STD and HIV education and prevention;
11. Violence and sexual assault prevention and reporting;

12. Safety orientation—safety and prevention of fires, and trauma as pedestrians, car, motorcycle, and bicycle operators or passengers, accidents at home, swimming and water safety;
13. Awareness of hazards of substance abuse, including smoking;
14. Personal hygiene and grooming;
15. Physical education and fitness habits;
16. Training in first aid/cardiopulmonary resuscitation;
17. Communicable disease control;
18. Mental and emotional health—suicide and violence prevention;
19. Understanding the health of a community;
20. Understanding health in the family (e.g., care of the elderly);
21. Understanding the environment and health;
22. Promotion of voluntarism in health.

Source: Derived from committee on School Health. American Academy of Pediatrics Website http://www.aap.org and www.schoolhealth.org/cshcap.htm and American School Health Association http://www.ashaweb.org/

The health of school-age children, adolescents, and young adults has improved in most countries as sanitation, vaccination, and nutrition have developed. Mortality rates overall have declined for all these age groups, but more so for preschoolers and primary school levels (Table 6.9). Teenage and young adult health has also improved, but the problem of trauma in its various forms asserts itself and causes much loss of young life.

Adolescence is a difficult and dangerous period of transition from childhood to adulthood. The teenager feels insecure, yet demands independence and responsibility for his or her own well-being. The mix provides for a stormy personal and family transition, with important physical and mental health dangers. Violence, accidents, and suicide are major causes of death in this age group, especially for males. Sexual activity causes dangers of teenage pregnancy, STDs, and HIV infection, especially when illicit drug use is involved. Developing individual responsibility in each of these categories is required of each individual, with the support of the family, the educational system, and the public health system.

Table 6.10 shows a decline in mortality of adolescents and young adults (ages 15–24) in the United States. Although mortality rates declined for both black and white persons in this age group, a sharp rise in mortality from homicide occurred, especially in black youths in the late 1980s. Failure to achieve the U.S. national health target for a mortality rate in the 15–24 age group (<85 per 100,000 by the year 1990) is primarily due to violence. Rates of mortality from motor vehicle accidents, homicide, and suicide have declined in the 1990s, but these factors still constitute major health problems associated with risk-taking behavior of teenage and young adult males (see Chapter 5).

TABLE 6.10   Age-Specific Mortality from Suicide, Homicide, and Motor Vehicle Accidents,[a] in Males Ages 15–24, United States, 1980–1996

Group	1950	1960	1970	1980	1985	1990	1993	1996
White males								
MVA	58	63	75	74	57	53	44	42
Homicide	4	4	8	16	11	15	17	14
Suicide	7	9	14	21	22	23	23	21
Black males								
MVA	42	46	58	35	32	36	34	35
Homicide	59	46	103	84	66	138	167	123
Suicide	5	4	11	12	13	15	20	17

Source: *Health, United States, 1998.*
[a]Rates per 100,000 population. Homicide includes "legal intervention."

Health targets for the United States for the year 2000 include reducing suicide mortality to 11 per 100,000 in the age group 15–24, primarily by health promotion and education, but also by community policing, and stricter weapons prevention programs in schools. Suicide is the seventh leading cause of death of males in Canada, nearly as high as motor vehicle deaths and more than deaths from diabetes. Male young adults are the highest risk group for suicide in North America and Europe, where rates in some countries, such as Hungary, Finland, and Sweden, are extremely high.

Smoking, alcohol, and drug abuse are among the most common contributing factors for disease and early death in modern societies. These personal behaviors or "lifestyle factors" which cause personal ill-health and early death constitute an enormous burden on society. They usually begin in late childhood, preadolescence, or adolescence, and their health effects continue into adulthood. Youth risk behavior is an important aspect of monitoring health status of a population, as reported by periodic survey data (Centers for Disease Control, 1996).

Smoking.   Smoking is a major contributor to cardiovascular diseases, cancer of the lung, and other diseases. The U.S. Surgeon General's Report on Smoking and Health issued in 1964 highlighted the accumulating evidence of the central role of smoking in the causation of chronic diseases. Subsequent reports addressed the continuing consequences of smoking and the prevalence of smoking among different population groups. Nearly all first use of tobacco occurs before age 18, and most adolescent smokers are addicted to nicotine. In 1995, one-sixth of students in grades 9–12 in the United States were regular cigarette smokers. There are identifiable psychosocial risk factors in smoking. Community-wide projects showed that public health efforts can successfully reduce adolescent use of tobacco. A 1997 Youth Risk Surveillance System Survey carried out by the CDC in all

50 states and the District of Columbia showed lifetime smoking in over 70%, current smoking in 36%, and frequent cigarette use in 17% of high school students. Current cigarette smoking increased from 28% in 1991 to 43% in 1997. Over 42% of high school students used some form of smoking in the 30 days prior to the survey (Centers for Disease Control, 1998).

Alcohol Abuse.   Alcohol use has been part of human culture since early civilization. Its use in excess causes serious health problems to the individual and society. The effects of alcohol abuse can be acute, chronic, or relate to the dependence itself. Binge drinking (five or more drinks per occasion) is a threat to the drinker, friends, and the community by associated violence, suicide, road and work accidents. Mortality from all causes is elevated in alcoholics. In the United States, alcohol-related deaths were 5% of all deaths in 1988, representing 1.5 million YPLL (years of potential life lost) to age 65 and 3 million total YPLL. Alcohol affects every organ in the body, and alcohol use is a compounding risk factor for a wide variety of diseases including hypertension, stroke, coronary heart disease, and cancers of the liver, esophagus, larynx, lung, stomach, large colon, and female breast. It increases reproductive disorders including amenorrhoea, anovulation, early menopause, and poor outcome of pregnancy with low birth weight or fetal alcohol syndrome in newborns. In 1997, 52.7% of senior high school students reported drinking alcohol, with 31.3% at a heavy level of consumption (5 or more drinks in a row, once or more during the previous 2 weeks).

Drug Abuse.   Drug abuse, while not a new problem, has become extremely widespread in most industrialized countries. Adolescence is a period of experimentation, intense social pressure, and risk-taking behavior. With widespread experience and availability of illicit drugs and other intoxicants in the United States, drug-related deaths among young people increased from 3.8 per 100,000 in 1987 to 4.1 in 1989. Cocaine-related emergency room episodes in the United States increased from 29,000 in 1985 to 80,000 in 1990 and to 142,000 in 1995, of which 2000 were for ages 6–17 and 22,000 for ages 18–25 (*Health, United States,* 1998). Adolescents use such drugs as glue (or other inhalants), marijuana, and "hard drugs" including heroin and crack. In 1997, 23.7% of U.S. senior high school students, and 10.2% of eighth graders, had used marijuana at least once in the previous month. Other drug use reported by high school seniors in the past month in this survey included cocaine (2.3%) and inhalants (2.5%).

Drug addiction is a compulsion to use a substance and obtain it by any means with a need to increase the dosage to obtain the desired effect. There can be a physiological and/or psychological dependence on the effects of the substance. The addiction has a detrimental effect on the family and the community through increased, stress, violence and crime. Drug use behavior can be classified as experimental, recreational, circumstantial, intensified, and compulsive. The experimental version is most common and motivated by curiosity and a desire to experience an altered mood state. Recreational use is voluntary but patterned in social settings where the user is not dependent on the drug. Circumstantial use may be related to certain situations, such as amphetamine use by students during exams or truck drivers on long drives. Intensified drug use occurs daily, motivated by stress and/or the wish to maintain level of performance, by use of sedatives, barbiturates, or tranquilizers. The individual remains integrated within the social and economic context. Compulsive use is of high frequency, with a dominant psychological and/or physical dependence. These are hard-core drug users whose lives are controlled by the dependence on and financing of the habit. This leads to even more serious health problems, caused by exchange of sex for drugs, lack of condom use, sharing of needles, poor hygiene, and poor nutrition. Transmission of hepatitis, STDs, and HIV are common. Arrest, imprisonment, injury, or death by violence in drug-related crime is also common.

Sexual Risk Behavior. Teenage sexuality risks include unplanned pregnancy and parenthood, STDs, and inadequate preparation for adulthood. The educational role in prevention of these is a vital element of a school health program. National efforts are required to control exploitation of children for prostitution, to provide care for runaway, homeless, and street children, and to regulate standards of care for orphans or handicapped children in institutions. These problems are particularly acute in South American and southeast Asian countries. The HIV/AIDS epidemic is taking an enormous toll among teenagers and young adult sex workers, truck drivers, military personnel, and increasingly in the general community.

Teenagers indulge in sexual intercourse, with 53% of U.S. high school students reporting having had sexual intercourse at least once in their lives, 18% having had four or more sexual partners, and 38% currently sexually active. Among these, just over half used condoms and 17% used birth control pills. Risks of STDs and HIV are commonly associated with high risk sexual behavior and drug abuse.

Dietary Risk Behavior. Only 27% of students had eaten the recommended dietary daily allowances (RDA) level of fruits and vegetables during the 24 hours previous to a 1995 U.S. survey. In addition, 28% thought of themselves as fat and 41% were currently trying to diet. Vomiting or laxative use to lose weight was nearly 8% among girls and over 2% among boys. Dietary deficiency conditions, especially iron deficiency anemia, are common in adolescent girls due to dieting behavior and extra need for iron to compensate for menstrual losses.

Eating disorders are a common form of behavioral disorder especially among teenaged girls influenced by the fashion model body type. They enter a vicious cycle of anorexia nervosa or bulimia which can be fatal, or in less severe forms extremely traumatic to the teen and family. These conditions may affect as many as 5% of college-aged females in the United States, and they are an occupationally associated condition among some female athletes and professionals such as ballet or modeling. Overeaters who take up obesity as a lifestyle also suffer the melancholies of this vulnerable age group. Prevention includes education and early case finding with suitable counseling and group therapy.

Physical Activity.   Nearly two-thirds of U.S. students had engaged in strenuous physical activity on 3 or more days of the previous week, 60% were enrolled
in physical education classes, and half were involved in team sports at their school.
Obesity in children and teenagers is increasing; for boys aged 6–8 years, the percentage of those overweight increased from 1963–1965 to 15.4%, in 1988–1994,
with similar, if less pronounced, patterns for those in the age groups 9–11, 12–14,
and 15–17. For girls, the increase in those overweight was similar to that of boys,
except for the age group 15–17 where the increase was less. These patterns seem
to be associated with increases in the passive activity of television watching and
with less activity in spontaneous or organized sports.

Violence and Gang Behavior.   Teenage violence in developing and industrial
countries is a growing public health hazard. Teenage alienation, availability of
drugs, weapons, and violent cultural icons provide breeding grounds for gangs and
violence, or individual violent outbursts which can lead to random or organized
shootings or terrorist outrages.

ADULT HEALTH

The health of adults between 25 and 64 years of age is important to the well-being of both the family health and society. The higher rate of death for males in this
age group creates a social imbalance with a predominance of females in the elderly
age group. Premature death or disability of males often results in loss of a major part
of family income, as well as productivity in the society. Much of the excess loss of
males can be prevented by currently available knowledge and techniques in clinical
medicine and public health. This has already been seen in reduced age-specific mortality from heart disease and stroke. Adult males should be targeted by coordinated
efforts of both preventive and curative services to reduce risk and promote health.

Risk factors that contribute to disease and mortality are both intrinsic and external. Intrinsic factors include age, sex, and genetics. Extrinsic factors include
personal lifestyle and the environment. Improved health for the adult population
requires a wide range of primary and secondary prevention programs focused on
extrinsic factors that can reduce personal risks and rates of mortality from chronic diseases and trauma (see Chapter 5). Programs to reduce morbidity and mortality from the chronic diseases include the following:

1. Smoking cessation;
2. Healthful diet (i.e., low in fat, moderate in carbohydrates, high in fruit,
   vegetables, and fiber);
3. Vitamin and mineral supplementation (e.g., calcium, iron, vitamins A, B,
   C, D, E, and others);
4. Daily physical activity (i.e., 30 minutes of moderately intense physical
   exercise 5 days per week;

   5. Road safety (e.g., seat belts, moderate speed enforcement, non use of al-
      cohol with driving;
   6. Workplace safety (e.g., safety helmets and shoes, preventive work site in-
      spections);
   7. Reducing violence (e.g., gun control, adolescent recreation activity pro-
      grams);
   8. Medical examinations for selective screening (e.g., diabetes, glaucoma,
      mammograms, Pap smears);
   9. General screening for and management of hypertension;
  10. General and high risk group screening for and management of cancer;
  11. Management of chronic diseases such as diabetes mellitus and glaucoma;
  12. Support services for mental health and social stress;
  13. Good management and prevention of infectious diseases;
  14. Immunization with dT for all adults, influenza and pneumococcal vac-
      cines for risk groups (i.e., persons over 65 and those with chronic dis-
      eases);
  15. Social and recreational activity and support groups;
  16. Employment, economic opportunity, and social equity.

The health care system has the responsibility of promoting the primary prevention
and screening programs needed to reduce the burden of disease as it affects the
adult population.

WOMEN'S HEALTH

   Women's health is a matrix of many factors relating to fertility, sexuality, and
societal conditions that affect health. Traditionally, in the years between menarche
and menopause, women's health needs have largely related to fertility. With higher levels of education and participation in the work force and other societal
changes, women's health increasingly consists of issues beyond those of fertility
above (Table 6.11).
   Since the 1980s, women's health needs have received increased attention from
the public health sector. Women are the largest consumers of health care, as well
as the largest group of health providers. Women live longer than men and visit
physicians 25% more than men and are hospitalized 15% more frequently. Lung
cancer has replaced breast cancer as the first cause of women's cancers in many
industrialized countries. Despite declining mortality rates, cardiac disease in
women is more common than previously thought. Indeed, stereotypes regarding
women as having symptomatic coronary heart disease less frequently than men
may lead to the underassessment of the severity of this problem.
   The life cycle of women between menarche and menopause involves psychosocial and physiological functions of fertility. Menarche, premenstrual syndrome
(PMS), and menopause are all associated with physical and psychological stress

TABLE 6.11   Women's Health Risks and Prevention by Age Group

Age	Major health risk factors	Screening/preventive activities
18–34	Nutrition, iron deficiency anemia, calcium, and micronutrient deficiencies Eating disorders, body self-image Marriage and parenthood Pregnancy and child care Single parenthood Physical fitness Self-reliance, self-determination Sexuality and birth control STDs and HIV Safety, unwanted sexual advances, assault, rape Safety from domestic abuse, physical and mental Physiological demands of menstrual cycle, PMS Smoking, alcohol, and drug use Poverty and economic status	Nutrition counseling, supplements Education, family, and social support Birth control education and access Safe sex education Prenatal care and high risk assessment Safe delivery Nutrition, exercise, weight control Self-defense Smoking cessation Premarital counseling Physical exercise Job training, academic work Support groups Shelters for abused, raped women Police, social service, and teacher sensitization
35–44	Late pregnancy Child rearing problems Marital stress, separation, divorce Diet Physical fitness Domestic abuse Obesity Cancers (breast, cervix, lung) Menopause	Counseling and support groups Screening for breast cancer, cervical cancer Physical exercise, weight control, fitness Screening for obesity, hypertension, cholesterol Diet counseling, weight control Preparation for menopause (counseling, hormonal therapy Immunization (dT)
45–64	Menopause Cancer (lung, breast) Diabetes, hypertension Physical fitness Depression Change in employment status Cardiovascular diseases Widowhood, divorce	Periodic physical examinations Screening for diabetes, hypertension, bone density Screening for breast, cervial, colonic cancer Screening for thyroid and vitamin status Counseling for life crises Immunization (dT), influenza Physical activity, regular exercise Retirement planning

Source: Adapted from U.S. Preventive Services Tak Force, 1996; Canadian Task Force on the Periodic Health Examination, 1994.

of clinical importance as well as public health significance. When education and awareness of women's physiological needs are accepted by the family, co-workers, and society in general, there is a greater likelihood of a more appropriate response by women suffering from these effects. Other women's health issues of public health significance include iron deficiency anemia, eating disorders, and nutrition (see Chapter 8).

Women's health is related both to biological and social factors. Postmenopausal women are subject to diseases for which they had previously been protected by hormones, such as cardiovascular diseases and osteoporosis with its accompanying fractures. The higher incidence of mental illness among women, in particular anxiety, depression, and eating disorders, relates to self-image, economics, and society's varying pressures on women. Postpartum and menopausal depression, the pressures of child rearing and work, and long periods of loneliness as a result of family breakdowns and widowhood all contribute to an excess of mental disorders. As women have become a substantial part of the workforce, employment and career are major sources of income, self-esteem as well as stress. Loss of employment can have negative effects on health for both financial and psychological reasons.

Screening for specific female cancers such as breast, ovary, uterus, and cervix are important public health programs, leading to early diagnosis and treatment. Sexual abuse, violence, single parenting, and widowhood are major factors that determine the content of health care services needed for women in the community. Females have been underrepresented in chronic disease research, which has tended to center on male subjects. Generalizations regarding treatment or prevention of a disease may not always be appropriate when they are based on male-only subjects. This has been especially problematic in heart disease, often presenting without clear symptoms, and stroke-related issues, with hypertension under-managed.

Women's health issues relate to their role in society in terms of education, employment, and equal pay for equal work. Attitudes, customs, laws and their enforcement regarding women's rights are extremely variable in different societies around the world, and the public health implications are widespread. In developed countries, a majority of people living in poverty are women and their children.

In developing countries, women's health involves a wide range of problems including high rates of pregnancy and maternal mortality, STDs and AIDS, and lack of access to family planning. In some parts of Africa, up to one-third of pregnant women are HIV positive. The World Health Organization estimates that about one-quarter of the world's women are subject to violence and abuse in their own homes, with rates over 50% in Thailand, Papua New Guinea, and Korea, and as high as 80% in Pakistan and Chile. WHO also estimates that between 85 and 115 million women in developing countries have undergone female genital mutilation (FGM) and some 2 million young girls suffer this procedure annually. FGM can cause hemorrhage, shock, and infection in the short term. Long-term effects include pelvic inflammatory disease, infertility, psychological damage, and sexual dysfunction. It may also contribute to obstructed labor, a common contributor to maternal mortality. Prevention of FGM will require basic changes in social attitudes toward women, their education, and place in society.

The Beijing Conference on Women in 1995 included health and reproduction as among those issues of most concern to women. Despite major differences in points of view between conservative religious representatives and liberal delegates on such issues as birth control and sexuality, a consensus emerged that women have rights to control their own bodies in these crucial areas. Violence and lack of education and economic development were also highlighted and recognized as impinging on the health status of women in both developed and developing societies.

Women's health is affected by socioeconomic status, education, information, societal equality, job opportunities, and women's input into health and social policies. Women are the primary caregivers in the family unit in most societies, but societal patterns may adversely affect a woman's health. Women's place in society affects many issues, ranging from fertility to public support for child care for working mothers. Some of these issues are politically contentious, with political, religious, and other social implications. Public health is closely involved in these issues, with responsibilities of advocacy, documentation and innovative leadership.

MEN'S HEALTH

Health care for the adult male has not received as much attention as women's health issues. Men are more vulnerable than women to early mortality, have an average life expectancy of some 8 years less than women, and are more subject to risk for early disease of a wide variety of conditions and disabilities from cardiovascular disease, cancers of specific sites, violence and trauma, suicide, and occupational hazards (Table 6.12).

Some contributing factors to the lower life expectancy for males than females may include less attention to self-care and professional care, less social and psychological adaptability, more risk taking behavior, and difficulty in transition from middle to old age. Emotional and psychosocial stresses on the middle-aged male (ages 45–64) may be very important in the morbidity and mortality patterns of a population.

Employment and career traditionally are central to the male's self-esteem. Feelings of competition with more aggressive younger colleagues, male and female, in a youth-oriented society may adversely affect the middle-aged male. Personal failure on the job or unemployment due to obsolescent industries, skills, or major economic change may result in stagnation at work, loss of self-esteem, and intrafamily strife that may lead to breakdown of physical and mental health. Social and economic changes produce loss of jobs and high rates of mortality among middle-aged men. Soaring rates of mortality from all types of chronic disease in Russia and other states of the former Soviet Union in the early 1990s reflects the social impact on physical health in a society in rapid transition. Further research in the psychological, sociological, and anthropological spheres is required.

Public health and clinical services should be aware of and address self-image, social role, and occupational issues related to men's health. The health care provider should utilize health education and promotion within the community to reduce risk factors in the male population.

TABLE 6.12   Men's Health Risks and Prevention by Age Group

Age	Major health risk factors	Screening/preventive activities
20–34	Trauma, accidents, homicide, violence	Nutrition counseling, supplements
	Suicide	Enforcement of drinking-driving, seat belt,
	Alcohol, drug abuse	motorcycle, and bicycle helmet laws
	Sexual problems, STDs	Sexual practices education
	Emotional problems	Anti smoking education
	Smoking	Peer group support
	Fatty diet—cholesterol	Exercise, fitness, and diet
	Lack of exercise	Premarital screening/counseling for
	Marriage and parenthood	genetic disorders
		Preparation for fatherhood
35–44	Diet	Screening for obesity, inactivity, BP,
	Hypertension	cholesterol
	Elevated cholesterol	Immunization (dT)
	Lack of exercise	Exercise, diet, and weight control
		Counseling for life crises (occupational,
		marital, family)
45–64	Cardiovascular disease risk factors	Periodic physical examination
	Hypertension, diabetes	Screening for hypertension, CVD, diabetes
	Cancer—colorectal, prostate	Screening for colorectal, prostate cancer
	Glaucoma	Counseling in life crisis, group therapy
	Benign prostatic hypertrophy	Immunization (dT, influenza)
	Changes in employment status	Early retirement planning
	Depression	Physical activity

Source: Adapted from U.S. Preventative Services Task Force, 1966; Canadian Task Force on the Periodic Health Examination, 1994.

HEALTH OF THE ELDERLY

Improving levels of health of the population result in an increasing elderly population in good physical, mental, social, and financial well-being in industrialized countries. With the changes in disease patterns in the population, more people are living and remaining relatively disease-free longer. Many life threatening conditions that occurred during middle age are now postponed to much later in life. As a result, it is not uncommon for the elderly to be well and free of major disease processes.

The elderly (65 years and older) comprises the fastest growing group in the population of the industrialized countries. As the mortality rates in adult and middle-aged years has declined, more people survive into the over 65, the over 75, and the over 85 years old groups. The elderly population constitutes 3.5% of the population in developing countries and up to 20% in developed countries. In the developed countries this proportion is growing and will reach 30% in some countries,

TABLE 6.13   Life Expectancy at Birth and at Age 65, Selected Countries, 1994

Country	Male life expectancy (years)		Country	Female life expectancy (years)	
	At birth	At age 65		Age 0	Age 65
Japan	76.6	16.8	Japan	82.5	21.3
Sweden	76.1	16.0	France	81.5	21.0
Greece	75.2	16.2	Switzerland	81.9	20.5
Switzerland	75.1	16.1	Sweden	81.4	19.8
Australia	75.1	15.8	Spain	81.1	19.6
Israel	75.1	15.8	Canada	81.0	19.9
Canada	74.8	15.8	Australia	80.9	19.7
Netherlands	74.6	14.8	Italy	80.7	19.3
Norway	74.2	14.8	Netherlands	80.4	18.8
Italy	74.0	15.5	Norway	80.3	18.8
England and Wales	73.8	14.4	Greece	80.2	18.5
France	73.8	16.2	Finland	80.2	18.5
Spain	73.7	15.8	Austria	79.8	18.6
Singapore	73.5	15.2	England and Wales	79.7	18.6
Austria	73.3	15.1	Northern Ireland	79.7	18.6
New Zealand	73.3	14.7	Germany	79.6	18.4
Germany	73.0	14.7	United States	79.0	19.0
Finland	72.8	14.7	Singapore	79.0	18.3
Denmark	72.7	14.1	Puerto Rico	78.9	19.4
Ireland	72.6	13.6	New Zealand	78.9	17.7
Northern Ireland	72.5	13.6	Israel	78.9	17.7
United States	72.4	15.5	Portugal	78.6	17.9
Portugal	71.5	12.7	Ireland	78.2	17.3
Scotland	71.4	13.1	Denmark	77.9	17.6
Chile	71.4	14.7	Scotland	77.7	17.3
Puerto Rico	69.6	16.3	Czech Republic	76.6	16.2
Czech Republic	69.5	12.9	Slovakia	76.5	16.5
Slovakia	68.3	12.9	Chile	76.3	17.6
Poland	67.4	12.7	Poland	76.0	16.4
Bulgaria	67.1	12.7	Bulgaria	74.7	15.4
Romania	66.0	12.7	Hungary	74.3	15.7
Hungary	64.9	12.0	Romania	73.3	15.0
Russian Federation	57.7	10.6	Russian Federation	73.1	15..6

Source: *Health, United States,* 1998.

with the very elderly (>75 years) being the most rapidly growing element of the population in many industrialized countries. The aging of the population has important implications for health services since the elderly are large consumers of health care, paid for by a shrinking work force.

Internationally, there are wide variations in life expectancy (Table 6.13) and in care programs for the elderly. The tradition of caring for the elderly in the family context is suffering in the industrialized societies where both men and women work, housing conditions may be crowded, and tolerance for the elderly reduced.

This, together with a growing elderly population, has increased the reliance on long-term care in institutional settings. Offsetting this trend in western countries has been the development of adequate pensions, social benefits, universal health care, home care, and a delay in disease to a later time in life.

Biological aging is measured by various functional abilities and performance of the individual, not necessarily reflected by chronological age. Aging carries with it social, occupational, psychological, financial, and physical change. These all directly impact health and the appropriate health and support needed by the elderly to sustain themselves. These factors can also interact with one another, causing or compounding health problems of the elderly.

Physical and mental deterioration are associated with aging, frequently causing

---

BOX 6.8 POTENTIAL ADVERSE CHANGES AFFECTING THE ELDERLY

1. Social/occupational/economic status
   a. Retirement;
   b. Widowhood and bereavement;
   c. Relocation;
   d. Loss of friends and family;
   e. Loss of financial security;
   f. Loss of professional status and self-esteem;
   g. Poor nutrition;
   h. Physical inactivity.
2. Physiological
   a. Hormonal changes;
   b. Onset of non-insulin-dependent diabetes;
   c. Hypertension;
   d. Thyroid dysfunction;
   e. Osteoporosis;
   f. Decreased absorption (e.g., vitamin B).
3. Pathophysiological
   a. Chronic disease of one or more organ systems;
   b. Medical/surgical conditions, medication, or support services;
   c. Handicaps limiting mobility, activities of daily living.
4. Mental
   a. Loneliness and depression;
   b. Loss of memory;
   c. Senility and agitation;
   d. Isolation from children, family, friends.

Source: Derived from National Institute of Aging: Improve Health and Quality of Life of Older People Website http://www.nih.gov/nia/plan/stratplan.htm

heavy burdens on the family unit and on social and health support systems. Chronic diseases, such as diabetes, cardiovascular diseases, and cancer, are increasingly common with advancing age. Physical limitations can affect social interaction and mental status, with depression affecting physical abilities. Alzheimer's disease causes serious mental deterioration in people over age 50 with rates increasing with age. Parkinson's disease produces progressive physical muscular rigidity and limitations of movement (see Chapter 5).

Health Maintenance for the Elderly

Public health attempts to promote well-being of the elderly by encouraging a healthful life style and access to good health care. This involves nutrition, physical fitness, recreation, work or daily activities, a positive family life, social and religious participation, and an active sexual life, all in keeping with the physical and emotional capability of the individual.

Good nutrition of the elderly, vital for a healthy old age, can be impaired by financial, social, and psychological factors. In many industrialized countries the elderly are protected by good pensions from their employment and social security, but there are also those with inadequate financial support who live in poverty. Isolation, loneliness, passivity, and malnutrition become a lifestyle which produces illness. Many elderly live on very severe budgets allowing little for food purchases. In addition, some have poor dental health. These are common contributing factors to the "tea and toast" syndrome of semistarvation that becomes a way of life for this population group. Even a well-off, physically capable elderly person, living alone, may lack the appetite and incentive to prepare a properly balanced diet for themselves. The situation is especially difficult when the person is restricted in activity and unable to move about. Vitamin deficiency conditions are common, especially of vitamins B and D, the latter mainly in winter months. Individual assessment by any caregiver should bear in mind the potential for low-level malnutrition among the elderly, especially those who are physically or mentally frail, or have severe dental problems. Community or voluntary programming to assist the elderly in nutrition and activity is a crucial contributor to maintaining the person in an independent life situation. The balanced diet for the elderly person is the same as for the adult, but must take into account the lower calorie output in activity and less exposure to the sun, especially in winter.

Fortification of basic foods is perhaps the most effective public health intervention for prevention of micronutrient deficiency in this, and other vulnerable population groups. Food fortification not only makes essential nutrients readily available to the entire population but also supplies a significant portion of the recommended daily allowances of the elderly, including iron, iodine, and vitamins B, C, D, and E. Daily vitamin supplementation is increasingly common and is gaining support in geriatric and nutritional professional circles. In 1998, the U.S. FDA recommended routine vitamin B complex supplements for the elderly. Annual immunization for influenza and periodic immunization against pneumococcal pneumonia (every 6 years) and dT (every 10 years) should be standard care for the elderly.

BOX 6.9 EUROPEAN AND UNITED STATES
HEALTH OBJECTIVES FOR THE ELDERLY
FOR THE YEAR 2000

WHO Europe

1. Ensure equity in health by reducing the gap in health status between countries and groups within countries;
2. Add life to years by ensuring full development and use of physical and mental capacity to derive full benefit and to cope with life in a healthy way;
3. Add health to years by reducing disease and disability;
4. Add years to life by reducing premature deaths and thereby increasing life expectancy for adults over age 65.

United States

1. Vigorous exercise—20% will engage in vigorous exercise three times a week for 30 minutes each occasion;
2. Muscle tone and endurance—50% will participate in physical activities that promote and develop muscle tone and endurance;
3. Flexibility—50% will regularly participate in physical activities that promote flexibility;
4. All will have more knowledge, positive attitude, and greater practice of physical exercise on a regular basis.

Physical fitness is an important preventive measure in preparation for advanced age and to maintain health when reaching that state. Regular physical activity, individually or in groups, tailored to the individual's capacity, is helpful in promoting good appetite, sound sleep, and good physical appearance. It can help prevent lethargy, apathy, and atrophy. Mutual support in group physical activity is part of regular socializing and mental well-being. The individual, even handicapped elderly persons, can also carry out physical exercise seated or lying down.

Social and family networks are important to the elderly and contribute to their sense of acceptance and well-being. Familial and social relations have direct health benefits, including reduced institutionalization. Mental health is strongly affected by the elderly person's self-perception of his or her role in the family and society. Recreation and social activity, including sexual relations, are part of the life of an older person. Recreational and social facilities designed to stimulate often socially isolated elderly persons to participate in recreational and support activity is dependent on development of a complex of adequate support systems such as rehabilitation, social security, transportation, and recreation facilities, all user-friendly and readily accessible to the elderly user.

Prevention and Screening Services Needed by the Elderly

Prevention services for the elderly begin well before age 65. A health-maintenance approach to preserve the well-being of the elderly involves preparation for a healthy old age, and requires self-care and a preventive-oriented approach begun in earlier years. This involves early case finding of chronic disease and care of existing diseases in order to prevent debilitating complications. Continuous contact and support of the elderly by health providers can prevent or alleviate complications from medication errors, misunderstanding of health needs, inadequate nutrition, and social isolation.

BOX 6.10 COMMUNITY HEALTH NEEDS OF THE ELDERLY

Community programs for the elderly should promote a wide range of knowledge and self-care and support services needed to prevent premature onset and progression of debilitation due to chronic disease and deterioration of general functioning of the elderly person, with the goal of helping elderly persons to function as independently as possible, consistent with safe and healthful conditions of a caring society.

1. Preventive self-care: healthy nutrition, regular exercise and exposure to the sun;

2. Social contact: regular contact with family, friends, and social support systems (e.g., church, ethnic, recreation, and social clubs);

3. Health education: to promote community awareness, and knowledge among the elderly;

4. Medical care services: preventive, diagnostic, treatment, hospitalization, and rehabilitation care;

5. Nutritional support programs: to provide assistance with counseling, home-delivered "meals-on-wheels," and group meals in senior citizen centers to promote socialization;

6. Injury prevention programs: inspect homes and provide safety devices such as nonslip carpets, railings, shower aids, bath mats;

7. Medical devices loan service: a service to provide and maintain medical aid devices, such as wheelchairs, kitchen and bathroom assisting devices;

8. Home care: organized nursing, physiotherapy, shopping, cleaning, and other services to assist handicapped or frail elderly persons to remain at home;

9. Hospital care: accessible but kept as short as possible to avoid infections and other complications;

10. Nursing homes: accredited facilities providing nursing and other care for elderly persons not able to live independently and requiring daily nursing care;

11. Supervised housing: group housing with supervisory nursing care, communal meals, recreational activities;

12. Recreation and occupational therapy: provided at community centers or in the home;

13. Volunteer work programs: volunteer traffic control, public garden maintenance, mutual help organizations, and community health workers;

14. Emergency call service: beeper service in case of emergency;

15. Security and safety measures installations: bathtub grips, smoke detectors, safety locks, and window bars;

16. Home help service: volunteer services to assist in maintaining independent homes by regular shopping, cleaning, and maintenance;

17. Home health aide service: cleaning, bathing, light housework, shopping, laundry, meals-on-wheels;

18. Mutual call service: telephone monitor services to maintain contact with elderly living alone.

Communication is vital to the life and health of the elderly. The availability of an emergency communication system may be lifesaving, as well as providing the frail elderly person with confidence. A telephone can be used for personal contact and social support as well as for emergencies and contact with medical personnel, providing a sense of security as well as contact with family and others outside the home.

Transportation for the elderly or the handicapped enables them to have access to medical care, social activities, shopping, and other activities of daily living. Bus companies operated by municipalities often make special arrangements for the elderly. They may provide special routes including home calls and lower costs of public transportation for the elderly. Ramps allow easy access for wheelchairs and walkers to public, residential, and commercial buildings, making continued participation in community life possible for the handicapped and elderly.

Relocation and transition are part of the life of the older person. Retirement from work brings with it the potential for rest and recreation or the possibility of isolation and depression. Training people for transition in life may be as important as their physical well-being. Death and bereavement are also part of this process of transition and require organized community, as well as family, support.

Organized community systems for assistance and interventions, based on community or social networks, require public health activities and support. Publication of needs, professional support for resource development, as well as direct provision of services are all part of public health. Public health plays an advocacy and promotion role, but many services will be provided by other agencies, such as hospital-based home care programs or nongovernmental voluntary organizations that provide support services for the elderly.

Finding methods of assisting the elderly to remain functional and adjusted in their own homes and in the community is essential to preserve the capacity of the health system to meet the needs of this population group. Such measures can be simple helping services, such as shopping or housecleaning. They may be devices to help in activities of daily living such as adaptations to stoves, wall grips for bathtub and toilet, or safety measures such as banisters, slip-free carpets, alarm systems, heating devices, safety locks, and police protection from vandals and burglars.

Finances in old age may be a serious burden for the younger generation. Children may need to provide financial assistance to the older member of their family who may have an inadequate pension or national social security system allocation. This can affect family relationships, living conditions, nutrition, medical care, social contact, and many other aspects of life. National social security systems have been developed in most industrialized countries to provide income security for the elderly, through payment from wages during the working years. Many social security systems are under pressure financially and politically. Increased life span and fewer births have created a situation in which the labor force may become smaller than the dependent members of society. Crises in social security systems may jeopardize the standards of living and security of the elderly in countries which today provide good income and other support systems for the elderly.

The elderly in many countries comprise a large percentage of the population and constitute a powerful political force. This has created a situation where the politician must take into account the special needs of the elderly. The elderly are strong in the political process, in professional organizations, and in municipal, provincial, and national parties. This "gray power" is a factor in western European and North American politics where the over-65 age group constitutes 15% or more of the total population, and an even larger percentage of the adult and political active population.

The economic aspects of aging of the population are their pension support and increased health care needs. On the other hand, the elderly are consumers whose accumulated savings or social benefits employ many persons as producers and care providers. The elderly also constitute a social asset to a country, not only as part of extended family networks, but also in their potential for volunteer work in the community.

Despite, or perhaps because of, their use of health resources, the elderly are healthier than ever before, with many continuing in the labor or volunteer work forces. Further, the elderly are an important consumer group and tend to have the time, money, and health to play important economic and social roles. A healthy elderly population should not be seen as a burden but as a vibrant contributing part of the family and society generally. The New Public Health seeks to improve the health status of the elderly and to ensure the provision of adequate support and health care services to assist the elderly to function independently, as long as possible, in their own homes. Society and the New Public Health should promote volunteer and self-help networking among the elderly to provide support during critical times such as illness or injury, following hospitalization or periods of emotional stress and depression.

SUMMARY

Individuals live much of their lives in some form of family unit, but they also pass through stages of vulnerability as individuals and as population groups with common health problems. Traditionally public health as paid great attention to some groups because of their particular vulnerability, as in maternal and child health. The benefits to society as a whole are great where such programs are well developed. Middle-aged men and women are important target groups for primary and secondary preventive programming in preparation for old age. The elderly also need special attention in public health, because of increasing numbers, changes in society, and the need to find effective ways of promoting their health as a supplement or, it is hoped, in place of costly services for unattended health problems. Preventive care for the elderly to sustain health can prevent unnecessary or premature dependency on high cost medical or nursing care institutions.

Each age group, from the newborn to the elderly, has specific problems and concerns that need to be addressed by the health care system. There are many medical, economic, and ethical issues involved in these aspects of public health. Public health needs to continuously monitor the health and social condition of the family as a key part of its overall individual care and population-orientated responsibility. Failure to do so leads to excess premature mortality and a costly burden on medical and hospital care to repair damage already done.

The New Public Health approaches the family unit both as a resource and a target group needing preventive and curative services at different stages of life. Family members may be cared for by different service providers, but there is a functional and economic relationship among those services. Inadequate prenatal care will increase the chance of poor results in infant health that can have long-term effects on the potential of the child. Inadequate support for the family struggling to cope with a chronically ill child or parent can result in unnecessary and damaging institutionalization. Lack of health promotion in nutrition, safety, and other community health issues may produce premature death of a parent, most often the male, with serious economic and social as well as emotional consequences for remaining members of the family.

The holistic approach of the New Public Health when applied to family health also addresses social and economic issues that affect or prejudice family function. Unemployment and poverty promote family distress and crises with long-term consequences for all members. Planning and resource allocation need to address this complex matrix of health in the family context.

This approach is based on unifying factors, such as integration of various service systems and new kinds of linkage between records or a new health provider, to assist the family to cope with normal family health events and the additional burdens of chronic disease. The family physician in an ideal sense should be supported in a team approach with a family nurse to help monitor and support families. Together, they can assist the family in coping with health problems of individuals within the family.

ELECTRONIC MEDIA

American Academy of Pediatrics http://www.aap.org/
American College of Obstetrics and Gynecology http://www.acog.org/
Beijing Conference on Women 1995 http://www.ijc.org/beijing
March of Dimes Foundation http://www.modimes.org/
National Institute of Child Health and Human Development http://www.nichd.nih.gov/
United Nations Population Division http://www.popin.org/pop1998/4.htm
United Nations Population Fund http://www.unfpa.org/
World Health Organization, Division of Child Health and Development http://www.who.org/int/chd/
World Health Organization, Division of Women's Health and Development whd@who.ch
World Health Organization, Family Planning and Population http://www.who.org/rht/fpp/

RECOMMENDED READINGS

American Academy of Pediatrics: Work Group on Breast-Feeding. 1997. Breast-feeding and the use of human milk. *Pediatrics,* 100:1035–1039.
Angastiniotis, M., Kyriakidou, S., Hadjiminas, M. 1986. How thalassemia was controlled in Cyprus. *World Health Forum,* 7:291–297.
Baker, J. P. 1994. Women and the invention of well child care. *Pediatrics,* 94:527–531.
Caravella, S., Clark, D., Dweck, H. S. 1987. Health codes for newborn care. *Pediatrics,* 80:1–5.
Heise, L. 1993. Violence against women: The hidden health burden. *World Health Statistics Quarterly,* 46:78–85.
Hilgartner, M. W. [editorial]. 1993. Vitamin K and the newborn. *The New England Journal of Medicine,* 329:957–958.
Kendig, J. W. 1992. Care of the normal newborn. *Pediatrics in Review,* 13:262–268.
Lozoff, B., Brittenham, G. M., Wolf, A. W. 1987. Iron deficiency anemia and iron therapy effects on infant developmental test performance. *Pediatrics,* 79:981–995.
Paneth, N. [editorial]. 1990. Technology at birth. *American Journal of Public Health,* 80:791–792.
Rodriguez-Trias, H. [editorial]. 1992. Women's health, women's lives, women's rights. *American Journal of Public Health,* 82:663–664.
Rowland, D. 1992. A five-nation perspective on the elderly. *Health Affairs,* 11:205–215.
Wegman, M. E. 1996. Infant mortality: Some international comparisons. *Pediatrics,* 98:1020–1027.

BIBLIOGRAPHY

American Academy of Pediatrics/American College of Obstetricians and Gynecologists. 1992. *Guidelines for Perinatal Care.* Third Edition. Washington, DC: ACOG, and Elk Grove Village, IL: American Academy of Pediatrics.
American Academy of Pediatrics: Committee on Environmental Health. 1998. Screening for blood lead levels. *Pediatrics,* 101:1072–1078.
American Academy of Pediatrics: Committee on Nutrition. 1989. Iron fortified formulas. *Pediatrics,* 84:1114–1115.
American Academy of Pediatrics: Committee on Nutrition. 1992. The use of whole cow's milk in infancy. *Pediatrics,* 89:1105–1109.
American Academy of Pediatrics. 1996. The role of the primary care pediatrician in the management of high-risk new-born infants (RE 9636). *Pediatrics,* 98:786–788.
American College of Obstetricians and Gynecologists. 1989. *Standards for Obstetric and Gynecologic Services.* Seventh Edition. Washington, DC: ACOG.

Belsey, M. A. 1993. Child abuse: Measuring a global problem. *World Health Statistics Quarterly,* 46:69–77.

Beresford, S. A. 1994. How do we get enough folic acid to prevent some neural tube defects? *American Journal of Public Health,* 84:348–350.

Canadian Task Force on the Periodic Health Examination. 1994. *The Canadian Guide to Clinical Preventive Health Care.* Ottawa: Health and Welfare Canada.

Centers for Disease Control. 1995. Surveillance for anencephaly and spina bifida and the impact of prenatal diagnosis—United States, 1985–1994. *Morbidity and Mortality Weekly Report,* 44 SS-4:1–13.

Centers for Disease Control. 1995. Economic costs of birth defects and cerebral palsy—United States. 1992. *Morbidity and Mortality Weekly Report,* 44:694–699.

Centers for Disease Control. 1996. Youth risk behavior surveillance—United States 1995. *Morbidity and Mortality Weekly Report,* 45:SS-4:1–84.

Centers for Disease Control. 1997. Children with elevated blood lead levels attributed to home renovation and remodeling activities—New York. 1993–1994. *Morbidity and Mortality Weekly Report,* 45:1121–1123.

Centers for Disease Control. 1998. Tobacco use among high school students—United States, 1997. *Morbidity and Mortality Weekly Report,* 47:229–233.

Centers for Disease Control. 1998. Recommendations to prevent and control iron deficiency anemia in the United States. *Morbidity and Mortality Weekly Report,* 47:RR-3:1–30.

Centers for Disease Control. 1999. Prevalence of selected maternal and infant characteristics. Pregnancy risk assessment monitoring system (PRAMS). *Morbidity and Mortality Weekly Report,* 48 (SS-5):1–43.

Committee on Psychosocial Aspects of Child and Family Health, 1985–1988, American Academy of Pediatrics. 1988. *Guidelines for Health Supervision I, II.* Elk Grove Village, IL: American Academy of Pediatrics.

Cooke, R. E. 1993. The origin of the National Institute of Child Health and Human Development. *Pediatrics,* 92:868–871.

Haddad, J. G (editorial). 1992. Vitamin D—Solar rays, the milky way or both? *The New England Journal Medicine,* 326:1213–1215.

Hughes, J. G. 1993. Conception and creation of the American Academy of Pediatrics. *Pediatrics,* 92:469–470.

Oski, F. A., Honig, A. S., Helu, B., Howanitz, P. 1983. Effect of iron therapy on behavior performance in nonanemic, iron-deficient infants. *Pediatrics,* 71:877–880.

Royston, E., Armstrong, S. (eds). 1989. *Preventing Maternal Deaths.* Geneva: World Health Organization.

Stopp, G. H. 1994. *International Perspectives on Healthcare for the Elderly.* New York: Peter Lang.

Thompson, R. S., Rivara, F. P., Thompson, D. C. 1989. A case control study of the effectiveness of bicycle safety helmets. *The New England Journal of Medicine,* 320:1361–1367.

UNICEF. 1996–1999. *The State of the World's Children 1996, 1997, 1998, 1999.* New York: United Nations Children's Fund, Oxford University Press.

United States Preventive Services Task Force. 1996. *Guide to Clinical Preventive Services,* Second Edition. Report of the U.S. Preventive Services Task Force. Baltimore: Williams & Wilkins.

Williams, C. D., Baumslag, N., Jelliffe, B. 1994. *Mother and Child Health: Delivering the Services,* Third Edition. New York: Oxford University Press.

World Health Organization. 1983. Community control of hereditary anaemias: Memorandum from a WHO meeting. *Bulletin of the World Health Organization,* 61:63–80.

World Health Organization. 1992. *The Prevalence of Anemia in Women: A Tabulation of Available Information,* Second Edition. Geneva: World Health Organization.

7

SPECIAL COMMUNITY
HEALTH NEEDS

INTRODUCTION

In any society there are groups of persons who require special attention from a health care system. They include those with mental health problems and mental retardation, native peoples, refugees, transient and homeless populations, the military, and imprisoned persons. There are also special needs, such as dental health and emergency health organization, that are required by the population as a whole. Traditionally, these special needs were dealt with through separate services which segregated them from the general health care system. Even after the initial reasons for separation are no longer relevant, tradition or vested interests of the systems sometimes continue this separation.

The New Public Health emphasizes the importance of seeing the individual with special needs and the special needs of the population in the context of the community and national health systems. This chapter addresses these special health needs, affecting particular subgroups and the population as a whole. Some of the interventions needed to protect the health of particular groups are directed specifically toward the group at risk. Other activities are directed at the general population since all are at risk, such as for development of mental health problems. The New Public Health advocates attention to the needs of these, often less privileged populations and subjects, and seeks to assure adequate attention to their special needs groups for preventive and long-term care needs, in the context care for the total population groups and all other health topics.

MENTAL HEALTH

"Mental health is a complex phenomenon which is determined by multiple social, environmental, biological and psychological factors and depends in part on public health efforts to control neuropsychiatric disorders including depression, anxiety disorders, schizophrenia, dementia and epilepsy" (World Health Organization, 1996). The WHO estimates that 1.5 billion people, three-quarters of whom live in developing countries, suffer from one or more neuropsychiatric illnesses. At the same time, great progress in understanding behavioral, neurological, and sociological factors, as well as in new methods of preventing and managing mental and neurological illness with effective medications and other methods of intervention, makes this area of health need an important element of the New Public Health.

Mental illness accounts for a large part of the total burden of disease and disability in any society. The contribution of mental illness to individual and family suffering and societal distress as well as their direct and indirect costs are enormous. Hundreds of millions of persons worldwide are affected by some form of mental disorder, varying from the relatively mild to the totally disabling.

Other definitions of mental health focus on functionality of the individual. The Canadian Department of Health and Welfare (1988) defined mental health as "the capacity of the individual, the group and the environment to interact with one another in ways that promote subjective well-being, the optimal development and use of mental abilities, the achievement of individual and collective goals consistent with justice and the attainment and preservation of conditions of fundamental equality." Donaldson and Donaldson define mental health as "a state in which a person is able to fulfil an active functioning role in society, interacting with others and overcoming difficulties without suffering major distress, abnormal or disturbed behavior" (Essential Public Health Medicine, 1993).

Short- or long-term mental and emotional problems may affect everyone to some degree during his or her life. These include a wide spectrum of conditions: anxiety, depression, isolation and loneliness, psychotic states, obsessive-compulsive disorders, behavioral and eating disorders, drug abuse, delinquency, suicide and violence, alcoholism, and intrafamily physical and mental abuse. These conditions affect the physical and social well-being of the patient, the family, and the community. Services based on waiting for the presentation with full clinical manifestation often lead to long-term institutionalization as opposed to an early crisis intervention or community-based approach.

Psychiatric conditions have traditionally been classified as psychotic or neurotic and by mode of presentation.[1] Psychoses are major mental illnesses characterized by severe symptoms such as delusions and hallucinations. These are divided into organic and functional psychoses, the organic being caused by a demonstrable physical abnormality, while the functional has no physical disease demonstrated. The functional psychoses include schizophrenia and affective or mood disorders.

Neurotic conditions can vary in severity, but they generally reflect an exaggerated response, such as anxiety or obsessional thoughts, to normal life events. These are sometimes divided into anxiety neuroses, obsessive-compulsive neuroses, hys-

---

[1] A British classification in the 1970s divided hospital admissions into depressive illness, schizophrenia, personality disorder, neurosis, and mania.

teria, and depression. Mental retardation and personality disorders have traditionally been considered separately from mental illness because they begin in early life
or adolescence, whereas mental illness has a recognized onset after a period of normal functioning in adult life, but they are increasingly seen as part of a whole spectrum of community health problems and program needs.

BOX 7.1 VULNERABLE GROUPS AND FACTORS OF MENTAL DYSFUNCTION

Children
: Genetic factors, birth injury, fetal alcohol syndrome, low birth weight, prenatal drug exposure
  Nutritional deficiency: iodine, iron
  Poverty and psychosocial deprivation
  Abuse, violence: actual or exposure to it in the home or environment
  Infections (e.g., viral encephalitis)
  Toxic exposure (e.g., lead)

Adolescents
: Sexual maturation and associated stress
  Family stress, abuse, violence, poverty
  Peer pressure, fear of failure
  School, career, occupational expectations
  Violence, trauma, stress
  Body image: fear of obesity
  Alcohol and drug abuse

Adults
: Alcohol and drug abuse
  Women in relation to fertility, pregnancy
  Parenthood, especially single status, poverty
  Abuse, physical, sexual, and psychological violence against women
  Occupation-related stress, unemployment
  Fear of aging, menopause
  Loss of reproductive function and virility
  Job and status loss, loss of self-esteem

Elderly
: Loss of spouse, friends, home
  Poverty and isolation
  Retirement and loss of occupational status
  Deterioration of mental and physical powers
  Malnutrition: "tea and toast" syndrome
  Abuse, violence
  Loss of independence
  Fear of dying and long process of deterioration

Mental health is influenced by many factors which are both intrinsic and extrinsic to the individual. Genetic heritage and family history of mental ill health may be present but brought to overt symptomatology by external factors. Stress contributes to behavioral and functional problems in families and communities. Life events such as illness, bereavement, unemployment or retirement, economic stress, dysfunctional families, family violence, infidelity, or divorce may trigger a social or mental breakdown and thus constitute risk factors which may help in defining preventive and intervention programs.

Emotions greatly influence physical and mental health, by stimulating physiological and psychological responses. Some responses to stress are beneficial in preparing the individual to resolve personal problems in life, but prolonged stress may become a factor in chronic emotional status and may become a strain on physiological systems. The result is social pathologies that impair the individual and family foundation for physical and mental health.

Historical Changes in Methods of Treatment

Traditionally mental illness has been stigmatized with superstition, brutal management, and treatment in isolation from the community. The methods of treating the mentally ill consisted of removal from the community into long-term institutionalization, using legal and physical restraints, and various severe forms of physical shock treatment. In the late eighteenth century pioneering reforms, by Vincenzo Chiarugi in Italy, William Tuke (and the Quakers) in England, and most influentially by Philippe Pinel in France, led to stopping the practice of chaining, starving, and beating patients in mental asylums. Psychiatric asylums, however, grew to be large isolated facilities for institutional care of the insane, usually under appalling conditions, and were the standard care of the mentally ill until well into the twentieth century.

In the United Kingdom, during the nineteenth century, local authorities were

---

Box 7.2 FAILINGS OF TRADITIONAL MENTAL HEALTH SYSTEMS

"Existing systems for the delivery of health care, including mental health care, have largely failed to meet the needs of most of the world's population. Many of the systems are centralized, hospital-based, and disease oriented, with care delivered by medical personnel in a one-to-one doctor/patient relationship. Such care is often inconsistent with the principle of social equity, particularly in developing countries."

Source: World Health Organization. 1990. *The Introduction of a Mental Health Component into Primary Health Care*. Geneva: WHO.

encouraged to build mental asylums and to supervise the notorious private asylums. In the Mental Treatment Act of 1930, community psychiatric clinics were established providing some alternative to the grimness of mental hospitalization. In 1948, the U.K. mental hospitals came under the National Health Service.

By the 1950s mental hospital beds in the United States and many other countries equaled the number of acute care beds. Hospitalization for a mental illness was long term and often lifelong, with repeated readmissions and custodial treatment with little optimism for improvement or release. Therapeutic measures relied heavily on custodial care with heavy sedation, insulin, and electroshock and even lobotomy as common forms of therapy. Reduction of tertiary syphilis and other organic causes of mental illness and development of the psychotropic drugs made possible major changes in custodial policies.

Mental health services based mainly on custodial care were felt to be costly and ineffectual service. Large custodial mental hospitals of the past consumed significant health resources. In many countries they were mainstays of mental health services with over 5 beds per 1000 population and large total expenditures. They had few cures and produced much long-term damage in the form of institutionalized patients unable to return to normal society.

The growing strength of advocates for mental health reform, called the mental hygiene movement, and passage of the National Mental Health Act in 1946 in the United States instigated new directions in mental health care. The U.K. Mental Health Act of 1959 encouraged a rapid reduction in mental hospital beds from 152,000 in 1952 to 98,000 in 1975, and to 59,000 by the beginning of the 1990s. During this period psychiatric units in general hospitals increased rapidly for short-term admissions. The concern that mentally ill persons are among the increase in the homeless population in the United Kingdom as well as in the United States is leading to a reappraisal of mental health policies.

Since the 1960s, release from institutional care and the return of large numbers of patients to the community required support by effective medication, follow-up services in the community, and other forms of therapy, as well as backup hospitalization for short- or even long-term care. Where community services are inadequate, this policy can contribute to increasing numbers of homeless mentally ill persons who are unable to cope in modern society. Hospital admission rates for mental health in the United States during the period 1969–1993 declined in state and county mental hospitals but increased in short-term admissions to acute care general hospitals and private mental hospitals (Table 7.1).

Mental Health Epidemiology

Mental disorders are common in all societies; an estimated 10 million adults and 4 million children in the United States are living with serious mental disorders, not including substance abuse (Mental Health, United States, 1996). Psychiatric epidemiology has developed as a scientific discipline over the past several decades with studies of prevalence, distribution, causes, and consequence of mental illness.

TABLE 7.1   Mental Health Inpatient Admission Rates,[a] United States, 1969–1992

Admissions by type or hospital	1969	1975	1983	1988	1992
Admissions to state or county hospitals	244	205	46	125	109
Acute general hospital (nonfederal)	240	257	337	359	377
Veterans hospitals psychiatric services	68	86	64	101	72
Private psychiatric hospitals	46	59	71	156	186
Total (including other categories)	644	737	701	819	830

Source: *Health, United States,* 1990 and 1996–1997.
[a]Rates per 100,000 civilian population.

In the mid-1950s a Midtown Manhattan study defined the population as follows: well 19%, mild symptoms 36%, moderate to marked symptoms 45%, and severe to incapacitated 10%. Studies carried out in other western countries show rates of clinical depression of between 4.5 and 7.2% in Finland and in cities such as Athens, Canberra, and Camberwell (United Kingdom). In 1985, mental illness accounted for 29% of all hospital bed occupancy in the United Kingdom and 4% of hospital admissions. Studies in the United Kingdom indicate that between 25 and 30% of patients visiting general practitioners have important or exclusively psychiatric causes for their presenting condition, even if the symptoms are primarily somatic.

The National Comorbidity Survey studied lifetime and 12-month prevalence of psychiatric disorders (DSM III) on a national probability sample of the adult population of the United States in 1990–1992. Nearly 50% of respondents reported at least one lifetime mental disorder, and at least 30% in the previous 12 months. The most common disorders were major depressive episodes, alcohol dependence, and social or simple phobias. More than half of all persons reporting lifetime disorders had three or more disorders, accounting for 14% of the total sample. Less than 40% of those with lifetime disorders were ever treated professionally. Women had higher rates of affective and anxiety disorders; men had higher rates of substance abuse and antisocial personality disorders. The prevalence of psychiatric disorders were higher than expected, with a high proportion not getting professional care. These findings suggest a need for widened outreach and integration of care for psychiatric needs within general primary care. Prevalence studies are fraught with difficulties in survey methods with recall and other biases in responses, diagnostic criteria, and representative sampling as well as in survey instrument design. Mental health epidemiology draws on data from national registries, as well as collaboration with social scientists, epidemiologists, biostatisticians, anthropologists, geneticists, and other disciplines to increase knowledge of contributory factors and to promote the search for causes.

The WHO estimates that at least 5% of the population of the European region suffers from serious diagnosable mental disorders (neuroses and functional psychoses), although prevalence estimates vary widely from study to study. At minimum, an additional 15% of the population suffer from less severe but partially incapacitating forms of mental distress. These affect their well-being and create the threat of more serious mental problems, such as severe depression, chronic psychiatric conditions, or psychoemotional problems, and life-threatening behavior, such as suicide, violence, and substance abuse. Western European countries have seen radical changes in psychiatric care over the 1980s and 1990s, with a process of "deinstitutionalization" that combines discharge of patients previously kept in long-term institutions with their reintegration into society. Development of community-based treatment and support services to promote independent living, vital social and employment skills, as well as backup support services has been part of this process.

The U.S. National Institute of Mental Health (NIMH internet http://www.nimh.gov) estimates that schizophrenia affects 2 million Americans, while 10 million are affected by bipolar (manic-depressive) disorders. Other conditions (phobias, post-traumatic stress disorders, and obsessive-compulsive disorders) together affect some 30 million, and a similar number are affected by Alzheimer's disease and other brain disorders. Eating disorders probably affect millions of teenagers, while substance abuse and associated comorbidity affect millions more. With a growing trend in managed care systems in the United States, and integration of mental health into general health systems under district health services in other countries, cost-effectiveness in care is increasingly a topic of concern for health economists and health systems managers.

The mission of the NIMH is to conduct research on the brain, behavior, and genetics, to develop new diagnostic and treatment methods, testing them in real world settings. It focusses on the neurosciences, especially at the molecular level (Nobel Prize, Julius Axelrod, 1970).

The WHO defines mental disorders as one of the major global public health problems, with tens of millions of cases (as seen in Table 7.2) accounting for approximately 10% of years of healthy life lost (DALYs) in 1990. Mental ill health is as relevant in developing as in industrialized societies, as a result of increasing life expectancy, and because of suffering from complex interactions between the biological, psychological, and social factors (e.g., aging, poverty, war and trauma, human rights violations, limited education, gender discrimination, malnutrition, and poverty).

The challenge of mental illness is partly one of changing priorities in health to include this set of conditions with equal priority to physical illness. An undefined burden of mental problems is the social and economic cost to families and society through illness, lost productivity, and personal financial outlays. A hidden burden is the social stigma and loss of human rights so commonly associated with mental illness. Rejection by society, families, employers, even caregivers compounds the isolation, humiliation, and pain suffered by the mentally ill, as well as their loss of earning power and independence.

TABLE 7.2   Global Prevalence of Mental and Behavioral Disorders, 1998

Disorder	Estimated prevalence (in 000s)
Anxiety disorders	400,000
Mood (affective) disorders	340,000
Alcohol dependence syndrome	288,000
Mental retardation (all types)	60,000
Schizophrenic disorders	45,000
Epilepsy	40,000
Dementia (including Alzheimer's disease)	29,000
Substance abuse syndrome	28,000
Suicide (attempted)	10,000–20,000
Suicide (completed)	835

Source: WHO, *World Health Reports,* 1998, and WHO Mental Health Program, 1999.

Mental Disorder Syndromes

Mental disorders present with a wide range of symptoms including personality change, confused thinking, abnormal anxiety, fear or suspiciousness, withdrawal from social contact, suicidal thoughts or actions, sleeplessness, change in eating patterns, outbursts of anger and hostility, alcohol or drug abuse, or simply incapability of dealing with daily activities, such as school, job, or personal needs. Mental disorders include a heterogeneous group of disorders ranging from exaggerated response to stressful events to altered mentation from specific neurological or genetic abnormalities (U.S. Department of Health and Human Resources, 1999). Chronicity is a problem which has, in the past, required long-term hospitalization. However, with improved medications and management in the community, hospitalization has been reduced as the primary method of treatment.

Organic Mental Syndromes. Organic mental disorders can produce a range of symptoms, such as decline of memory, comprehension, learning capacity, language, and judgment, including the ability to think and calculate, or severe dementia. Alzheimer's disease is the prominent condition in this category, but other causes of organic origin include traumatic brain damage, cerebrovascular accident, Parkinson's disease, alcoholism, Creutzfeldt-Jakob disease, HIV, postencephalitic disorders, syphilis and other dementias, and mental disorders due to physical brain disease. The WHO estimates 22 million persons suffer from such conditions internationally. Cognitive impairment in the elderly is less than 5% under age 75, but over 40% for those above age 80.

Prevention of brain injury, strokes, encephalitis due to vaccine-preventable diseases, management of alcoholism, and adequate nutrition reduces the prevalence of the dementias. Awareness and recognition of cognitive impairment in the elderly is an important function in primary care and specialized geriatric and psychiatric services. Because of increasing longevity, organic brain syndrome cases can place a great burden on families and the health care system. Support services for families caring for persons affected by organic brain syndromes are part of a comprehensive health program, and should include short-term respite care, home care, and long-term care services. Research into organic brain syndromes should be of high priority because of the personality destruction of an increasingly large group of the population, and the resultant effects on the individual and the family, as well as the costs of health services for this group.

Substance Abuse.   Substance abuse (mental and behavioral disorders due to psychoactive drug use) is intoxication with a substance that causes physical or psychological harm, impaired judgment, or dysfunctional behavior leading to disability and harming interpersonal relationships. Dependent syndromes are characterized by the presence of three of the following: a compulsion to use the substance, physiological withdrawal symptoms, tolerance of its effects, preoccupation with the substance, and persistence in using the substance despite negative effects. Substance abuse is associated with serious problems, including death from overdosage, crime to support the habit, STDs, AIDS and hepatitis transmission, imprisonment, social ostracism, and long-term brain damage.

Substance abuse is at pandemic levels. Drug dependence syndromes are estimated by the WHO to affect 28 million persons. Between 100,000 and 200,000 deaths occur from overdosage annually. Inhalation of volatile solvents (i.e., sniffing of glue, paint thinners, gasoline, and aerosols) among preadolescents is widespread and causes death and serious brain damage. Cannabis use is extremely widespread. New chemical formulations ("designer drugs") are appearing with death and brain damage effects. Use of opiates, cocaine, and psychotropic drugs affect all levels of society in developed and developing countries, influenced by urbanization and other social stresses and promoted by powerful economic and political interests in the international drug trade. Attempts to eliminate or control drug traffic have many similarities to international efforts to ban the slave trade in the late eighteenth and early nineteenth centuries, in that with both instances some governments covertly foster the trade, while others attempt to stop it.

Prevention should target vulnerable groups, especially young people, street children, and women drug users. Methadone substitution is widely used to reduce drug dependency. Needle exchange programs have been successful in reducing the spread of HIV and hepatitis B and C among intravenous drug users but are sometimes criticized as encouraging drug use. Detoxification and long-term treatment and follow-up programs are costly and frustrating, but they are better than the alternatives of disease, crime, social breakdown, and imprisonment now common in this group.

Alcohol abuse affects 120 million people internationally. Alcohol abuse, chronic alcoholism, and associated diseases of cirrhosis, cancers of various sites, and social breakdown are widespread in many countries, with resulting morbidity and mortality from trauma, violence, and child and spouse abuse. Alcohol abuse during pregnancy is associated with stillbirth, premature birth, and fetal alcohol syndrome.

Strategies for reducing alcohol abuse include raising the price and reducing the availability of alcohol, especially to adolescents, setting a minimum age for alcohol purchase, legislation and enforcement to curb driving while under the influence of alcohol, restrictions on promotion, marketing, and advertisement of alcohol, public education and awareness programs, individual counseling, group therapy, and self-help groups, as well as in-patient, out-patient, and rehabilitation programs.

Schizophrenia. Schizophrenia is a group of chronic conditions with episodes of psychotic illness with delusional hallucinatory thought or behavior disorders. It usually manifests itself around age 20, with distorted thinking, perception, and judgment. Symptoms may include excitement, withdrawal, or a catatonic state. Hallucinations can be visual or auditory.

The disorder appears in episodes lasting a few months with interval periods of normality but is a chronic illness. It occurs in both sexes and all social classes and affects about 1% of the adult population. It is characterized by chronicity with periodic need for hospitalization. Management with current medications improves the outlook for many patients.

The WHO estimates there are 45 million schizophrenics in the world with 33 million in the developing countries. In the United States, the direct cost of treatment is estimated to be close to 0.5% of the GNP. While thought to be biological in origin, this condition is affected in course and outcome by social and cultural conditions. Acute care and long-term follow-up require well-integrated community and hospital services. Neuroleptic drugs such as chlorpromazine, introduced in the early 1950s, greatly reduce symptomatology and enable patients to function in the community, especially with family support and suitable community-based services, with periodic short-term hospitalization if needed.

Mood Disorders. Elevated or depressed mood states affect some 340 million persons around the world at any given time, ranking as the fourth leading cause of the total burden of disease in developing counties. In the United States, the yearly cost of depression is estimated at $44 billion, equivalent to the cost of care of all cardiovascular diseases (WHO, *World Health Report 1997*).

Mood disorders range from manic to depressed conditions, often involving both mania and depression in sequence (manic-depressive or bipolar disorders). Depression is probably the commonest affective disorder, affecting approximately 5% of the population at any one time. Diagnosis of depression involves four or more of the following symptoms and signs: lost of interest or pleasure in normal activities, lack of emotional response, sleep disturbances (early waking, sleeplessness, or excessive sleepiness), depression that is worse in the morning, loss of appetite, weight loss of 5% of body weight in a month, loss of libido, and a psychomotor retardation (or agitation). Severity ranges from mild to severe. Depressive episodes may be recurrent and are a major risk factor for suicide and social breakdown. Severe depression is common in the elderly. Seasonal affective disorder (SAD) is a syndrome related to darkness during winter months and is common in northern countries with long winter nights and a closed-in lifestyle. It is associated with high rates of alcohol abuse, heart disease, and suicide.

Economic distress, unemployment, discrimination, and practices which limit women's rights are all contributors to mood disorders.

Women are more commonly affected by mood disorders, with depression especially among married women with children, related to social isolation and social devaluation of the role of the housewife. Postpartum depression can be a precursor to chronic depressive illness with a need for support and referral to prevent chronicity.

Antidepressant drugs provide an important advance in management of mood disorders, but should be accompanied by professional monitoring. In 1970, lithium was approved by the FDA for treatment of manic episodes, based on NIMH research, bringing major benefits to many persons suffering from bipolar disorders. Lithium and antidepressant treatment required close monitoring and supportive care by health providers. They help to reduce the disabling quality of mood disorders, reducing the economic burden of these conditions on society as well as the personal suffering of patients and families.

Primary care providers need to work jointly with mental health specialists in the same setting, to ease referral and consultation with patients unlikely to present to a separate mental health setting. Awareness of mood disorders among primary caregivers is of paramount importance in addressing this problem with understanding, patience, supportive therapy, and referral if chronic or leading to social breakdown at work or at home. Psychotherapeutic skills are important for primary care, but require support of specialized services. Community recognition is needed to address the widespread prevalence of domestic violence associated with mood disorders.

Neurotic (Anxiety and Dissociative) Disorders. This group includes a wide range of symptomatology and degrees of severity, including panic disorders, phobias, obsessive-compulsive disorders, anxiety conditions, and posttraumatic stress syndrome. Specific phobias include fear of crowds, public places, traveling, social situations, objects, animals, and closed spaces. Panic conditions are discrete episodes of intense fear, starting abruptly with physical symptoms that are inconsistent with the perceived threat of a specific situation or trigger event. Obsessive-compulsive disorders are repetitive, unpleasant obsessions, causing distress or interfering with normal functioning. They include compulsive behavior such as frequent hand washing, hair brushing, cleaning, and counting. Some obsessive-compulsive patients find relief with the new generation of medications. Stress reactions may be acute or occur long after the trigger event.

POST-TRAUMATIC STRESS DISORDERS. Post-traumatic stress disorders (PTSDs), with flashbacks and dreams of the event reawakening painful memories, usually occur within 6 months of the stressful event or time. These may include manifestations of depression or other affective or behavioral disturbances. PTSD, originally described as related to combat experience in the Vietnam War, is now also recognized as occurring in reaction to catastrophic events, such as violence, genocide, torture, disasters, and sexual abuse. Public and professional awareness and sensitivity are part of health system response, in preparation of psychological support as part of disaster planning or in response to catastrophic events such as hurricanes, earthquakes, terrorist bombings (e.g., Oklahoma City, 1995), or mass murder by deranged persons (e.g., Scotland and Australia, 1996).

DISSOCIATIVE (CONVERSION) DISORDERS.   Dissociative (conversion) disorders involve amnesia, stupor, or trancelike conditions, which have no physical cause but which are related to a specific trigger event. Preoccupation with a variety of physical symptoms, not explainable by detectable physical disorders, and refusal to accept medical reassurance are called somatoform disorders. Symptoms vary but usually involve gastrointestinal, cardiovascular, dermatological, genitourinary, or pain symptoms.

Behavioral Syndromes with Physiological Disturbance.   Behavioral syndromes include eating disorders, anorexia nervosa (self-starvation) and bulimia nervosa (self-induced vomiting or purging), abuse of nonaddictive substances (e.g., vitamins, antacids), sleep disorders, sleepwalking, and night terrors. Sexual dysfunction syndromes include loss of sexual desire or enjoyment, sexual aversion, failure of sexual response (male and female), orgasmic dysfunction, premature ejaculation, and painful intercourse.

Prevention requires public discussion, especially in the media, and awareness among family and caregivers concerning teenage psychological adjustment problems and life stress situations. The potential for conversion of normal anxieties into serious, even life threatening disorders, such as bulimia and anorexia nervosa, are especially important in adolescent care. Social norms, such as promotion of extremely thin models in advertisements, encourages these conditions. Middle-aged men, who may be under stress from insecurity of employment or loss of status, may present with sleep or sexual dysfunction and are at risk for physical disease such as premature coronary heart disease events. Unemployment, loss of marriage partner, or financial distress may trigger excessive psychological and physical responses that can be life threatening. Primary care services need to be oriented to detect and cope with potentially serious behavior syndrome situations in vulnerable groups and provide continuing support and referral services.

Personality Disorders.   Personality disorders include deviations of perception and interpretation of people and events, self-images, and affect (i.e., mood or responsiveness). They are associated with difficulty in control of impulses, gratification of needs and manner of relating to people and situations. The symptoms are persistent, inflexible, and cause distress not explainable by other mental disorders. Personality disorders range from the paranoid, or excessively suspicious, to the schizoid (i.e., emotional detachment, flattened affect, unresponsiveness, solitary life with fantasy and introspection). Emotionally unstable personality disorder involves impulsive behavior with anger and violence and difficulty maintaining a course of action not offering immediate rewards.

Borderline personality disorder includes impulsive qualities with disturbed self-image, emotional crises, threats or acts of self-harm, and chronic feelings of emptiness. The health system has a role in recognition and support of this group of disorders as it does for any significant physical illness; the consequences of such conditions can be serious for the individual, the family, and society. Early recognition may help to initiate professional care and support to help these persons through critical life periods. Mutual and family support groups with professional assistance may be of value.

Disorders of Psychological Development. Psychological developmental disorders include some level of impairment of speech, language, sound categorization, visual perception, attention, and activity control. They include childhood autism, that is, abnormal social communication, attachments, and play appearing before age 3 with lack of spontaneity, social and emotional reciprocity, and failure to develop according to mental age of peers. Early recognition and referral for specialist help requires awareness and cooperation between education and health systems.

Behavioral and Emotional Disorders of Childhood and Adolescence. Hyperkinetic disorders include inattention, overactivity, and impulsivity. They include a variety of attention disorders such as attention deficit disorder (ADD) and attention deficit hyperactive disorder (ADHD). Conduct disorders are characterized by aggressive behavior, repetitive behavior with temper tantrums, lying, stealing, use of dangerous weapons, and other unacceptable behavior.

Estimates suggest that 10–12% of children and adolescents suffer from mental disorders, including autism, hyperactivity, depression, developmental delay, behavior disorders, and emotional disturbances. A high percentage of children and adolescents who are dysfunctional with mental disorders do not receive appropriate therapy.

Suicide is the most serious outcome of mental disorders among adolescents, with 4660 suicides among people aged 5–24 years in the United States, with 30,903 completed suicides in 1996 for all ages (Health, United States, 1998). School systems and health care providers must be alert to the potential of teenage suicide. Provision of easily accessible care and specialist and family support services are needed at crucial times in the history of a disorder or disease that can lead to emotional breakdown or suicide.

Mental Health Goals for the Year 2000

The European region of the WHO has defined a series of mental health targets for the year 2000. These are designed to improve traditional mental health services, to reduce pathological and destructive behavior, to promote social well-being, and to reduce the prevalence of mental disorders. This is also meant to improve the quality of life of people with such disorders, and to reverse rising trends in suicide and attempted suicide. Other goals are as follows: improvement in societal factors (e.g., unemployment and social isolation) that put strain on the individual; improved access to measures that support people and equip them to cope with distressing or stressful events and conditions; improved access to supportive care, both formal and informal, for people with mental disorders, especially dementia; development of comprehensive community-based mental health services with a greater involvement of primary health care; special efforts to prevent health-damaging types of behavior such as substance abuse; and suicide prevention programs (World Health Organization, Mental Health Program, 1998).

United States targets for mental health relate to increasing awareness of mental health services and reducing risks within specific target groups within the population, such as young adults at risk for suicide, homicide, and other forms of violence. The U.S. mental health targets include

1. Increase access to support services;
2. Increase percentage of primary care providers taking careful history of medical and psychological coping skills to over 60%;
3. Increase proportion of those over age 15 who can identify support agencies for stress situations to >50%;
4. Reduce suicide rate for age group 15–24 to <11 per 100,000;
5. Reduce injury and death from child abuse by over 25%.

Controversies in Mental Health Policies

During the 1960s, depopulation of mental hospitals and reduction in traditional psychiatric hospitals took place in the United States, Canada, and the United Kingdom. Canada reduced its psychiatric bed capacity by 32,000 beds while increasing general hospital psychiatric beds. The general hospital units tended to treat mild psychiatric conditions, such as mild depression, while resources were not put into needed community-based programs to serve the more severely disabled.

Inadequate follow-up in the community increases the risk of the ex-psychiatric patient leading to exacerbation of symptoms, despondency, homelessness, alienation, and suicide. Community mental health programs should include case management, rehabilitation, housing programs, and other support services. These require well-organized systems to avoid poor coordination, losing patients from followup, lack of accountability, ineffective case management, high readmission rates, and inadequate linkages between hospital and community care systems.

Reviews of mental health programs in the past several decades have raised a number of issues, including concern for the civil rights of the mentally ill, the destructive effects of long-term institutional care, the availability of knowledge from the neurosciences and new medications that enable treatment in the community, and the high cost of institutional care. The search for evidence-based medical practice, managed care systems, and capitation payment for mental health care are part of continuing debate in mental health policy.

Community-Oriented Mental Health

Advances in drug therapy, together with concern for the negative effects of institutionalization and the low cure rates of large psychiatric hospitals, prompted the development of community-oriented mental health (COMH) programs. During the 1960s this model for delivery of community mental health services took the form of community mental health centers, but more recently there has been a trend to increase the involvement of primary care providers in mental health care.

Mental health services should be integral parts of other health and social services in the community. Patients may seek help at any one of the services, and appropriate care of the patient requires the intervention of many services. This requires development of intersectoral cooperation in health services and staff training in the elements of community mental health needs and services.

Community-oriented mental health is a programmatic approach providing links between primary health care with hospitals and long-term services, and it includes social support, rehabilitation, and preventive services. These services need to be developed and function as a network in order to provide the individual and the community with the appropriate level of care needed at a particular point or particular stage of an illness.

The development of COMH may require different approaches in a metropolitan area as compared with a small city or rural area. Services should be tailored to the setting and be comprehensive in scope to help the patient in acute need as well as one in need of long-term support services, preferably within the community. Crises in the form of patient or family breakdown may require short-term hospitalization in a community hospital setting, as an integral part of the service.

Community mental health involves primary care providers willing and able to recognize and manage mental health problems. Backup services of psychiatrists, psychologists, social workers, and perhaps paraprofessional community health

---

BOX 7.3 COMMUNITY MENTAL HEALTH POLICY

1. Target populations, including a commitment to care for the significantly disabled in the community to the maximum extent possible;

2. General hospital psychiatric units linked to community support, with greater diversification to include holding beds (e.g., 48 hour observation), day beds, crisis intervention, respite care services, and detoxification units;

3. Psychiatric hospitals provide for major psychiatric and behavior disorders (e.g., dementia, brain damage, paranoid and severely regressive schizophrenia); the supply of such beds is continuing to be reduced, but this requires well-funded and coordinated community-based services;

4. Continuity of care is essential to ensure adequate maintenance of the chronically mentally ill with highly individualized care, needs assessment, planning, and monitoring;

5. Attention to comorbidity, e.g., of drug abuse with underlying psychopathology;

6. Integration of mental health services with linkage to primary and other health care;

7. Consumerism and advocacy groups help define needs of the mentally ill in peer and family support, self-help groups, income maintenance, retraining and job placement, adequate housing, and social support;

8. Replacing dependency with independence promoting programs;

9. Psychiatric epidemiology provides greater evidence of prevalence of mental ill health and can compare the costs and benefits of different treatment approaches;

10. Community orientation includes outreach programs, supported housing, home services, patient and family support services with attention to the cultural and organizational needs of minority and other special needs groups;

11. Case management with multidisciplinary health care teams, with specialist and hospital backup;

12. Long stay accommodation in group residences with therapeutic services in the community in homelike units, day care.

Source: WHO, Mental Health Program, website, 1999.

workers trained to serve the mentally ill, particularly in minority groups or others with special needs, are needed. This requires active educational and organizational efforts to link traditional mental health services with primary care. Primary care physicians and other health personnel must be oriented and trained in the various mental dysfunctional conditions, such as mood and anxiety symptoms and disorders. Primary care providers and managed care programs need to assess the needs of risk groups, such as menopausal women and preretirement persons, as well as needed modalities of care that cannot be provided by mental health services alone. At the same time, traditional preventive and treatment services cannot function without backup mental health services. Each depends on the other. Building a community-oriented network approach to mental health may require finding new ways of working together, with multidisciplinary staffing either in the same setting or in close coordination.

Prevention and Health Promotion

Prevention requires working with risk groups and with groups with common problems where self-help or a support group may be the most effective form of therapy. This will encompass a wide range of resources which take into account that social, physical, and mental health are interactive. Identification of persons at risk for breakdown, such as suicide and violence, requires high levels of awareness by teachers, doctors, police, social workers, army personnel, employers, and the general public. Support groups take many forms including Alcoholics Anonymous and other groups based on its format, Al-Anon for family members of an alcoholic, drug rehabilitation programs, overeater and binge eater groups, Schizophrenics Anonymous, National Association of the Mentally Ill (NAMI) representing families of mentally ill, and bereavement groups.

Prevention programs attempt to reduce the possibility that mental dysfunction will proceed to more severe forms in which serious functional breakdown occurs. Social factors are vital in community in mental health practices. Prevention in mental health involves many aspects of patient/client care on the primary, secondary, and tertiary levels. Health promotion in the mental health field involves making the public aware of a healthy lifestyle, including activity, rest, recreation, and socialization, and the dangers of substance abuse. Specific programs for activity and socialization should be targeted at the young and elderly. Secondary prevention involves early detection and supportive treatment, and tertiary prevention requires the management of long-term mental illness with adequate support systems in the community.

Primary prevention includes prevention of nutritional deficiencies (i.e., iodine, iron, and vitamins), reduction of environmental contamination (e.g., lead), reducing social and educational deprivation (e.g., education and recreation for youth in high crime/drug abuse neighborhoods), promoting toddler and preschool enrichment programs, promoting family support systems (e.g., for single parent families), providing vocational and employment assistance (e.g., for teenage mothers), providing social support for the aged and handicapped (e.g., social security), targeting family and social violence (e.g., domestic, antifemale, and school violence), and raising public and professional awareness and cooperating (e.g., education and health systems).

Secondary prevention requires crisis intervention by trained professionals/paraprofessionals at the primary care level, social relief and support for abuse victims (e.g., shelters), early diagnosis of mental health disorders including screening, early referral to adequate care by police, courts, schools, hospitals, armed forces, and employers, effective treatment and follow-up including outreach and use of paraprofessionals in COMH teams, increased supply and accessibility of crisis intervention services at the community level, use of existing facilities such as hospitals for detoxification and crisis intervention, defining treatment goals and quality of care, and ensuring continuity of follow-up.

Tertiary prevention requires maintaining contact with clients to monitor medications, mood, function, family, and social relations; providing support, referral, assistance; assuring continuity in care and follow-up; assisting client contact with social support networks; protecting/rehabilitating the patient, family, and community regarding the effects of the disorder (e.g., drug addiction); providing support groups and structured rehabilitation; funding community-based group transitional residences, halfway houses; promoting independence and self support; and training and providing continuing education to caregivers, clients, families, and the community.

Mental health research is opening new approaches to management of clinical psychiatric conditions, along with a wider recognition in the community that mental illnesses are real and treatable conditions. Psychotherapeutic medications and psychotherapies are effective, helping people with mental illness by correcting abnormal brain function. More research is needed to further define the relationship between the brain and behavior. New knowledge also helps establish the role of genetic and environmental factors in shaping brain and behavioral function. Great progress being made in the neurosciences and genetics helps biomedical scientists to study the normal and pathological function of the brain and devise new methods of treating mental disorders. This will bring new hope to millions of persons and their families suffering the debilitating effects of mental illness.

MENTAL HANDICAP

The World Health Organization defines mental disability as a person handicapped because of learning disabilities. Internationally, both scientific study and health service practice have yielded a diverse terminology. It is therefore important to properly define the term mental handicap, including the basic concepts of intellectual impairment, learning disability, and mental handicap or retardation:

1. Intellectual impairment: Low levels of intelligence as measured by developmental tests (IQ). Ranges of impairment according to the ICD-10 are shown in Table 7.3.
2. Learning disability or dysfunction: Specific learning disorders, such as dyslexia, which are unrelated to intelligence, require specialist assessment and educational support.
3. Mental handicap: This is graded by social maladaptation, which relates to acceptance and accommodation of the handicapped in society.

The term mental handicap includes a wide range of causes and associated conditions. Among them are organic neurological impairment, genetic endowment, and developmental or educational deprivation. Other factors that greatly influence the social adjustment of the mentally handicapped include structure and orientation of services, professional attitudes and training, employment and training practices, social and cultural expectations, family and kinship structures, as well as legislation in relationship to health, welfare, education, and employment.

Many conditions that result in mental handicapping are preventable. Prevention can be categorized into three areas:

1. Preventive processes before conception include health education programs to minimize the birth prevalence of inherited diseases in identified families. This includes immunization against rubella, adequate nutrition, and smoking cessation prior to pregnancy, and genetic screening and prenuptial counseling and population screening. Other preventive measures directed at processes before conception

TABLE 7.3   International Classification of Diseases (ICD-10) Diagnostic Categories
for Mental and Behavioral Disorders, F00-F99[a]

Diagnostic category	ICD-10 code	Clinical features
Organic, symptomatic, mental disorders	F00–09	Mental disturbance due to brain damage, toxicity, or trauma (e.g., Alzheimer's, vascular dementia)
Mental and behavioral disorders due to psychoactive substance use	F1–19	Alcohol, opiates, sedatives, cannabis, cocaine, hallucinogens, volatile substances, multiple drug use, and other psychoactive substances
Schizophrenia, schizotypal, and delusional disorders	F20–29	Delusional psychotic disorders and schizoaffective disorders
Mood (affective) disorders	F30–39	Manic, bipolar, and depressive disorders
Neurotic, stress related, and somatoform disorders	F40–49	Phobias, anxiety disorders, obsessive-compulsive disorders, stress reactions, dissassociation, somatoform, and other disorders
Behavioral syndromes associated with psychological disturbances and physical factors	F50–59	Eating, sleeping, sexual, behavioral, and other disorders
Disorders of adult personality and behaviors	F60–69	A variety of specific personality disorder syndromes, including habit and impulse disorders, gender disorders, sexual preference disorders (pedophila, voyeurism, etc.)
Mental retardations	F70–79	Mild, moderate, severe, and profound
Disorders of psychological development	F80–89	Speech and language, scholastic, motor, and development disorders (e.g., autism)
Behavioral and emotional disorders with onset usually occuring in childhood or adolescence	F90–98	Hyperkinetic conduct, emotional and social dysfunction disorders
Unspecified mental disorders	F99	

[a]The previous American Psychiatric Association (APA) classification of mental disorders included
the major categories alcohol or drug abuse or dependence, phobias, major depression, obsessive-compulsive disorders, antisocial personality, panic disorders, cognitive impairment, schizophrenia, mania,
and somatization. The present classification (DSM-IV) was developed in cooperation with the WHO
and is close to the ICD-10 classification.

Source: American Psychiatric Association, in 1994, *DSM-IV: Diagnostic and Statistical Manual of
Mental Disorders,* Fourth Edition, in cooperation with the World Health Organization.

include the reduction of hazardous factors in the environment, such as drugs,
chemicals, and radiation exposures, and rubella prevention.

2. Preventive processes during fetal life and birth include prevention targeted
at the nutritional status of pregnant women, including folate supplements, prevention of iron deficiency anemia, and iodine deficiency before and during pregnancy; minimizing harm to the fetus by eliminating alcohol intake, drugs, and

smoking during pregnancy; screening to detect fetal anomalies associated with mental retardation; good prenatal and obstetrical care to avoid fetal damage, anoxia, and birth trauma; good neonatal care to avoid anoxia and vitamin K deficiency induced intracranial hemorrhages; fetal monitoring in labor.

3. Preventive processes after birth include vaccination against communicable diseases that may cause brain damage; screening for PKU, congenital and other inborn errors of metabolism, and hypothyroidism followed by case management; good infancy care and nutrition to avoid nutritional and emotional deprivation that can reduce psychomotor development; and strategies to reduce accidents and their impact.

Improved knowledge of nutrition and risk factors in pregnancy and technological developments are providing improvements in such areas as antenatal detection, treatment of fetal abnormalities, and genetic risk identification, so that mental retardation from preventable causes can be reduced in frequency. Down's syndrome can be reduced in terms of new cases by family planning and reduced late maternal age pregnancies and use of currently available screening methods of amniocentesis. Down's syndrome cases can also achieve higher levels of learning and social adaptation than was thought possible even a decade ago.

Adaptation of mentally handicapped persons to living and functioning in the community is part of rehabilitation. Political, community, and parental acceptance and support is essential to establish and operate successful community group living and working arrangements for this population group. Public health can address issues relating to adaptation in noninstitutional surroundings within the community. This requires advocacy and education among employers and the general public, and active promotion by the health community.

ORAL HEALTH

The term stomatology is used for dental health in some countries, derived from stoma, the Greek word for mouth. Oral health is an important element of general health status, affecting everyone. Poor dental health can cause pain and interfere with nutrition. Oral disease can progress to loss of teeth and costly treatment, yet much is preventable. In extreme cases, dental illness can result in osteomyelitis, brain abscesses, systemic infection, and death. Public health has a key role to play in oral health because it has important preventive methods available to it. These include health education for oral hygiene, fluoridation of community water supplies, and availability and assurance of quality standards in the dental and paradental professions.

Ancient societies faced dental problems and sought explanations and treatments based on practical experience, with explanations that "nematodes" grew in teeth. Egyptian mummies show evidence of dental treatment. Hippocrates and later Ibn Sinna (Avicenna) discussed hygiene for prevention of dental disease. In

1672 Russian *Lekars* or military doctors were trained in dental treatment for the army. Dental care was part of general medical care, but at the end of the seventeenth century it separated into its own field. Pierre Fauchard, a French surgeon (1678–1761), described 130 diseases of the mouth and teeth and is considered the father of modern dentistry.

In 1736, artificial golden dental caps were first used. Silver amalgam was first used for filling cavities in 1819. Horace Wells introduced nitrous oxide for dental anesthesia in 1844, which, along with ether, was quickly adopted by the medical community for general surgery. Special cements for filling cavities were developed in 1855. The dental drill was invented in 1870. Four years later fluoride was discovered to prevent tooth decay. In 1942 the U.S. Dental Health Officer reported on studies in 13 cities which determined a relationship between fluoride use and reduced caries, with excessive levels of fluoride causing fluorosis. In 1946 a study was taken in Grand Rapids, Michigan, Kingston, New York, and Brantford, Ontario, whereby fluoride was administered through the community water systems. The results were then compared to cities wherein fluoride was not dispensed. The study showed that fluoridated drinking water reduced caries by 48 to 78%. By 1999, 144 million Americans were using fluoridated community water, with fluoride levels adjusted to the optimal 0.7–1.2 parts per million.

In developing countries, as standards of living rise, increased use of sugars and candies cause a deterioration in children's dental health. Use of sugar in water or tea as a baby pacifier is common in some parts of the world and should be discouraged. Education of parents to prevent this should be part of health promotion.

The major issues in dental public health are dental caries, periodontal disease, malocclusion, and oral cancer. All of these contribute to loss of dentition with effects on general health status. The important interventions to reduce caries and periodontal disease are fluoridation of community water supplies, education in oral hygiene, reduction of sugar intake in children, regular dental care, and dental sealants. Feeding babies sugared water or tea and allowing babies to go to sleep with a bottle in the mouth can cause serious tooth decay. Self-care by brushing teeth after meals, regular dental flossing, and regular dental checkups are major aspects of dental health and should be part of school and home health education.

Fluoridation

Fluoridation of community water supplies reduces the number of caries and extractions in both children and adults by some 60%. Fluoride is present naturally in most water supplies. When adjusted to a level of 1 part per million, it prevents dental decay. This is one of the most effective public health interventions available. Fluoridation must be a controlled procedure because excessive fluoride in drinking water can cause fluorosis, which results in staining and fragility of the teeth. Good public health practice should therefore ensure adequate fluoride levels to reduce dental caries but avoid excessive levels of fluoridation.

Other methods of giving fluoride include tablets, fluoride rinses, fluoride in toothpaste, or fluoride enrichment of salt or milk. Fluoride tablets are a useful preventive adjunct to child health in areas without adequate fluoride in the water, but the most cost-effective method is by fluoridation of water supplies. The cost of fluoridation of water supplies depends on the size of the community served, but in the United States this varies from 12 to 21 cents per person per year in communities over 200,000 population, to $0.60–5.41 for smaller communities.

Fluoridation has been the subject of political and emotional controversy in many areas. This has prevented the institution of this preventive health measure universally. As time goes by, it is being more widely implemented. Fluoridation has been adopted by most of the major cities in the United States. Sixty percent of the U.S. population is served by fluoridated water supplies. The United States health targets for the year 2000 include one in which at least 75% of the population's community water systems should be receiving optimally fluoridated water (from 62% in 1989). By the early 1990s, more than 20 states had already reached that objective. A second goal is that 50% of all schoolchildren living in fluoride-deficient areas should be served by optimum fluoridated school water supplies. A final target is that 65% of all schoolchildren should be proficient in personal oral hygiene practices and should receive other needed preventive dental services.

Progress in the reduction of the prevalence of dental caries has been occurring over the past several decades in countries where fluoridation has been adopted on a wide scale. Worldwide, approximately 210 million persons consume drinking water with optimal levels of fluoride. Health education, fluoride tablets or rinses when fluoridation is not available, and access to dentists for assessment, cleaning, and treatment are also an essential part of public health. More recently, topical application of fluoride and use of plastic dental sealants, developed in the 1970s, to reduce exposure of pits and fissures in teeth to cariogenic organisms has been shown to be highly effective.

Periodontal Disease

Disease of the oral tissues that support the teeth is described as periodontal disease. The accompanying inflammation of the gums or gingival tissue extends to the periodontal ligament and causes the loss of the supporting bone. If control is not imminent, teeth become loose and must be extracted. The major factor in the development of periodontal disease is poor oral hygiene, especially plaque formation. Prevention of this disease should include such measures as tooth brushing and flossing, along with use of anti-plaque rinses.

Dental Care

Expenditures on dentist services in the United States increased from $2 billion in 1960 to $13.3 billion in 1980 and to $47.6 billion in 1996. However, as a percentage of total national expenditures on health, dental services declined from 7.3% in 1960 to 5.4% in 1980 and 4.6% in 1996. With the dramatic decline in dental caries in the child population of industrialized countries since the 1960s, there has been a tendency to increasing use of costly dental procedures. At the same time there has been a reassessment of the numbers of dentists required, whereas in the 1960s there was a continuous chorus of demand to increase the size and number of dental schools in western countries. Dental care is costly, and oversupply of manpower does not necessarily lower prices by competition.

The supply of dentists is affected by demand for services, and between 1985 and 1996 the number of dental schools in the United States declined from 60 to 53, and the number of graduates from 5400 in 1985 to 3700–3800 per year in the period 1993–1996.

Visits to dentists for preventive, cleaning, or restorative work is limited by economic factors and shortage of personnel in many countries. In the United States, annual dental visits for persons over age 25 increased from 54% in 1963 to 61% in 1993, varying by age group, years of education, poverty status, and gender (females visit dentists more than males). Dental-related illness accounted for 6.4 million days of disability, 14.3 million days of restricted activity, and a total of 20.9 million days of lost work days in 1988. Many countries with universal health plans do not include dental care as a covered benefit, largely because of its high cost.

Dental care programs in schools and through community organizations should include education, fluoride rinse programs where the water supply is inadequately fluoridated, and regular dental checkups and care. Dental sealants applied to children's teeth after eruption of the first (age 6–8) and second molars (age 12–14) is a safe and highly effective means of preventing caries and should be used especially in areas without fluoridation.

In countries with dental health as a part of the national health program, such as the United Kingdom, priority on prevention by fluoridation of community water supplies and school fluoride rinse programs supplemented by dental sealants could reduce the need for restorative dentistry and allow for reducing dental manpower and costs. Dental nurses providing treatment, oral hygiene, and education in school-based programs have been successfully used in New Zealand and parts of Canada, bringing dental care to children otherwise unchecked and unserved by private dentistry.

Oral Cancer

The control of oral cancer is primarily through early detection, as well as educating the public and training quality professional dentists and other health providers regarding its epidemiology and symptomatology. All suspicious lesions should be biopsied, especially in high risk patients such as males above the age of 40 who smoke and/or drink heavily. Health promotion emphasizing the elimination of risk factors is most important. Risk factors include smoking (especially pipe and cigar smokers), heavy drinking of alcohol, use of chewing tobacco, and poor oral hygiene.

PHYSICAL DISABILITY AND REHABILITATION

Many people with disabilities need long-term care. These needs will only partially be met in a hospital environment; most care will be delivered within the community by families. Professional health care providers such as physiotherapists, occupational and speech therapists, as well as medical and nursing personnel should be part of the support services. The provision of high quality rehabilitation services in a community should include the following:

1. Conducting a full assessment of people with disabilities and suitable support systems;
2. Establishing a clear care plan;
3. Providing measures and services to deliver the care plan.

With so many different handicapping situations, this field is by its very nature both extensive and complex. In most countries services and legislation for the disabled have developed in a piecemeal fashion, often as the result of lobbying by social reformers. In more recent years this lobbying has been by organized groups of the disabled themselves.

The 1975 United Nations' Declaration of the Rights of Disabled Persons, the WHO's 1981 International Year of Disabled Persons, and the United Nations' Decade of Disabled Persons 1983–1992 reviewed legislation and programs for the disabled in member countries of the WHO European Region setting a benchmark in progress in the field.

The WHO estimates that there are 450 million disabled persons in the world, or some 10% of any given country's population. This is generally equivalent to the size of the population 65 years of age and over or to the 0- 4-year-old population in developed countries. The elderly and the under 5 population receive a great deal of attention in the delivery of health services and social programming. Despite progress made, the disabled are still not as high on the social agenda as these two groups in most countries. The WHO has undertaken promotion of political and professional awareness of the scope of the problem and unmet needs in the area, with a stress on prevention and rehabilitation.

Principles that are common to many countries in Europe include a widespread recognition that prevention of disability is one of the most essential elements of a comprehensive approach, incorporating the classic public health definitions: primary prevention (i.e., stopping the disease or injury from occurring at all), secondary prevention (i.e., if the event occurs, stopping or reducing the severity of complications which lead to disability), and tertiary prevention (i.e., restoring the injured person to maximum feasible physical, psychological, and social functioning). Prevention of handicaps from birth injury, prematurity, failure-to-thrive, as well as injury in accidents are among the most important issues to address.

Many countries have developed a comprehensive range of laws, activities, and programs for the disabled that include prevention of disability, reduction of the severity of complications, and physical, social, and employment rehabilitation, as well as support or compensation systems. In most countries there are difficulties coordinating policies, services, and support systems.

The handicaps commonly dealt with in most countries include those due to medical, accidental, or occupational ill-health or injury: veterans and civilian war victims, the hard of hearing, the blind, the mentally handicapped, mentally ill, and

---

BOX 7.4 PRINCIPLES OF A
NATIONAL PUBLIC HEALTH PROGRAM
FOR DISABILITIES

1. Political commitment of national, state, local governments;
2. Multisectoral involvement of relevant government agencies, NGOs, media, professional, judiciary, church, academic, private sector;
3. Use of legislative, regulatory, and moral suasion powers;
4. Comprehensiveness of approach;
5. Prevention-oriented approach;
6. Restoration-oriented approach for physical, psychological, and social function;
7. Promotion of full participation and equality of handicapped in society;
8. Promotion of acceptance of the disabled in society;
9. Promotion of community versus institutional care;
10. Reallocation of resources and expenditures to community-based services;
11. Local service responsibility;
12. National coordination, support, and financing;
13. Participation of the disabled in decision-making regarding their needs;
14. Promotion of education, employment, housing, and support services;
15. Defined legal, financial, and social rights;
16. Social benefits and compensation without litigation.

---

others. National programs in addition to medical, rehabilitative, and preventive health services include income maintenance and compensation, legal rights, normal and special education, vocational training, work in protected or normal settings, housing and the physical environment, communication and transport, leisure and sport, and self-help groups.

Prevention of road accidents, work place accidents, and toxic exposures are all part of national agendas. Some countries have clear objectives with measurable targets specified for deaths, but also reflecting injury, in multiyear programs. The WHO European Region in 1985 stated, "By the Year 2000 deaths from accidents in the European region should be reduced by at least 25% through an intensified effort to reduce traffic, home and occupational accidents" (WHO, European Region, 1985)

The United Kingdom stresses not only extensive social security legislation and support systems, but also the vital role of prevention. This ranges from nutrition and vitamin/mineral enrichment of basic foods to smoking reduction, prevention of low birth weight and perinatal mortality/morbidity, infectious disease control, workers' health and safety, accident prevention, and mental health and disability legislation and compensation. The National Health Service has moved toward decentralized management through district health systems with national support. Organization of services for the disabled has evolved in this direction as well, reflecting a return to the principles of the local administration of the Elizabethan Poor Laws.

The Danish program for the handicapped has evolved through social legislation over the past 100 years, especially the Public Assistance Act of 1933. A wide range of national legislation and programs have been implemented for the handicapped. As in the United Kingdom, the Danish trend has also been to return responsibility for management of social and health programs from the central government to the county and local governmental levels, with support and service backup from national organizations and the national government.

The Finnish low birth weight rate of 4%, down from 5% over the past 10 years, is due in part to the 99% attendance at their maternal and child centers. Finnish pregnant women average 12 prenatal and postnatal visits. A maternity benefit of $125 is paid only to those who attend prenatal care before the end of the fourth month of pregnancy.

France has an impressively comprehensive program which also places great importance on prevention. As an example, there is much stress on prevention of low birth weight and associated morbidity by requiring early prenatal care for maternity benefits and 20 obligatory examinations for children from birth up to the age of 6 years, including 9 during the first year of life. Screening for metabolic disorders at birth is now augmented by screening for thalassemia, hemophilia, and sickle-cell disease. Great importance is placed on integration of disabled children in regular schools and the social integration of handicapped adults. Income support systems are tailored to promote these objectives. For example, in order to qualify for birth and child financial benefits, prenatal and infant visits to preventive care services are obligatory in France.

The situation of the handicapped in Russia deserves review jointly by the Ministry of Labor and Social Affairs and the Ministry of Health. The major issues in disability prevention relate to the high rates of violence, suicide, homicide, poisonings, and industrial and road accidents. Besides causing high rates of mortality they cause high rates of injury and disability. Prevention of low birth weight (now reportedly 7%) and perinatal mortality and morbidity would help to reduce Russia's infant mortality rate, now double those of other industrialized nations, as well as prevent disability.

The U.S. health target for 1990 for vehicle fatality rates was no greater than 18 per 100,000 population. This target was met, as the rate has fallen steadily from its peak of 27.4 per 100,000 in 1970 to 18.8 in 1985 and 16.2 per 100,000 in 1996. Death from homicide, in contrast, increased from 5.4 in 1960 to 10.7 in 1993, then declined to 8.5 per 100,000 population in 1996. Firearms related death rates increased from 1.8 per 100,000 in 1980 to 15.6 in 1993, but then fell to 12.9 per

TABLE 7.4   Burden of Injury, United States, 1995

Episode	Number
Deaths	147,891
Hospital discharges	2,591,000
Emergency room visits	36,961,000
Reported injury episodes	59,127,000

Source: *Health, United States,* 1996–1997.

100,000 in 1996. Monitoring of injuries in the United States uses multiple systems including the National Center for Health Statistics, the National Vital Statistics System, the National Hospital Ambulatory Medical Care Survey, the National Health Interview Survey, the National Hospital Discharge Survey, and the Centers for Disease Control and Prevention. The burden of injury in the United States (Table 7.4) is substantial.

Progress is being made in developing national programs in prevention, care, and services for the disabled. Sharing experiences from one country to another can help in defining how health and social services can be more effectively arranged both to prevent disability and to integrate the disabled into society. Review of legislation and service organizations as well as the incorporation of new medical, therapeutic, and technologic innovations are part of that process.

SPECIAL GROUP HEALTH NEEDS

In every society there are groups of people with special health needs. Indeed, a society is often judged on how it cares for its native peoples, its prisoners, its homeless, and others. They may be set aside from the mainstream of the society as a whole by historic, ethnic, legal, or economic circumstance. They may require special attention to deal with their problems of health because they are either more dependent, vulnerable, or less able to access services based on traditional medical practice, or simply because their needs are greater or more specialized than those of the general population. The New Public Health has an advocacy and pioneering role here, as it does for special needs of the whole population.

GAY AND LESBIAN HEALTH

As many as 10% of the population is estimated to be gay, lesbian, or bisexual. Accurate rates are difficult to determine; societal stigmatization of homosexuality may lead to underreporting, whereas nonrandom convenience studies (e.g., patients at STD clinics) may overestimate true prevalence. Furthermore, sexual orientation may not always correlate with sexual behavior (e.g., 70% of gay men report having sex with married men and 45% of lesbians report having sex with men). Regardless of the exact figures, gay men, lesbians, and bisexuals represent a significant proportion of the population. In heterogeneous societies, such as the United States, the gay/lesbian population reflects the diversity of the population-at-large, in terms of ethnicity, religion, socioeconomic status, and geography. Thus, health care needs deriving from homosexual behavior must be addressed in their appropriate social context.

Specific health problems related to same-sex behavior include high incidence of STDs, gastrointestinal infections, and hepatocellular and anal cancer among gay men, related to oral–anal, oral–genital, and receptive sexual intercourse. Health problems of particular concern to lesbians include high incidence of breast, ovarian, and endometrial cancer related to low rates of parity and low rates of breast-feeding and oral contraceptive use. Lesbians tend to visit gynecologists less frequently, and may therefore not receive important tests such as Pap smears or mammograms. Antigay violence is another serious health issue for both gays and lesbians, as are high rates of smoking, alcoholism, substance abuse, depression, suicide, and cardiovascular disease, all possibly related to the stress of living in a homophobic society.

Among the most significant and relatively easy to improve health issues for gay men and lesbians is the lack of preventive services and early treatment due to negative experiences with the health care establishment and fear of stigmatization by the medical community. Studies have shown that only 10–40% of primary care physicians routinely take a sexual history from a new adult patient and only 30% feel comfortable with gay patients. Over 60% of Gay and Lesbian Medical Association members surveyed felt that homosexuals who revealed their sexual orientation risked substandard care as a result. The overall outcome is a failure to screen, diagnose, and treat important medical problems of gay and lesbian patients.

The AIDS epidemic has brought sexual behavior issues to the fore, especially for the homosexual population. However, with only an estimated 10% of U.S. students receiving comprehensive sex education and fewer than 5% of primary care physicians routinely obtaining information regarding high risk activities, gay and lesbian organizations themselves have often had to fill the vacuum. Significant successes have been recorded (e.g., widespread adoption of condom use among gay men in the 1980s), but educational programs face enormous ongoing challenges.

One target group of particular concern is the gay and lesbian adolescent population. According to the American Academy of Pediatrics (1993), these teens:

> are severely hindered by societal stigmatization and prejudice, limited knowledge of human sexuality, a need for secrecy, a lack of opportunity for open socialization, and limited communication with healthy role models. Subjected to overt rejection and harassment at the hands of family members, peers, school officials, and others in the community, they may seek but not find, understanding and acceptance by parents and others. . . . Such rejection may lead to isolation, runaway behavior, homelessness, domestic violence, depression, suicide, substance abuse, and school or job failure. Heterosexual or homosexual promiscuity may occur, including involvement in prostitution (often in runaway youths) as a means to survive.

Thus, with the new generation as with the previous one, ostracism by society leads to those very behaviors which put both homosexuals and heterosexuals at increased risk for social isolation and the spread of diseases, including hepatitis B and AIDS. Breaking this vicious cycle will require a concentrated multidisciplinary effort, in which public health workers, primary care providers, and gay and lesbian organizations all have pivotal advocacy roles to play.

NATIVE PEOPLES' HEALTH

Both the United States and Canada provide health care for native populations as special categories under direct provision of care by federal government agencies. These populations are often segregated from the general population, with different schools, legal systems, and health care systems.

The Canadian Indian population in the sixteenth century was approximately 222,000 persons, but disease and famine accompanying European immigration reduced this to 102,000 by 1867 (the year of Canada's confederation). By 1941, the number of aboriginal people was between 100,000 and 122,000, but by 1988 had reached 443,884. There has been a renewal of tribal energy following disillusionment with the experience of absorption into the general society, which was marked by tragedies of epidemic alcoholism, violence, drug abuse, prostitution, and social breakdown.

Health status indicators in Canada's native peoples indicated widespread alcoholism in the form of binge drinking among all age groups from children and adolescents upward. Glue sniffing, diabetes, hypertension, and hospitalization rates are about twice the Canadian averages. Over the past several decades there has been a decline in mortality rates among Canadian Indian and Inuit people, but the gap with the overall population remains high. A nutrition survey carried out in 1973 and subsequent follow-up studies showed marked levels of micronutrient deficiency conditions among Canadian Indians and Inuit, with inadequate intakes of vitamins A and D, iron, and calcium and high levels of iron deficiency anemia and rickets. In 1990, the relative mortality rates as compared to the total population varied from 5:1 in infancy to 1.1:1 for those over age 65. The major causes of death were, in order of decreasing frequency, injury and poisoning, diseases of the circulatory system, neoplasms, and diseases of the respiratory system. Suicide rates for those aged 15–24 were five to six times the Canadian rates for this age group.

In 1997, infant mortality in the Canadian aboriginal population was twice the national average, and life expectancy at birth was 8 years less than the national level. Chronic illness, including diabetes, cardiovascular disease, cancer, and end-stage renal disease, as well as infectious diseases including tuberculosis, STDs, and hepatitis are all much commoner among native peoples than the general population. Poverty, physical violence, and alcohol and drug abuse are widely prevalent.

Indian and Inuit peoples come under direct federal responsibility for health and are not part of Canada's universal health insurance program operated by each province. Services are provided by the Medical Branch of a federal department. Recent trends to decentralize management of services to the tribal council have been matched by proposals for constitutional change making aboriginal communities recognized levels of government. This has, however, not been adopted. The social problems of unemployment, poor education, alcohol abuse, family breakdown, violence, and welfare syndromes plague Canadian Indians who move to the cities perhaps even more than those who remain on the reservation.

The violent record of the United States in its wars against the Indians was followed by forcible concentration on poor land reservations and lethal administrative practices far more harsh than the Canadian record. However, provision of care and prevention of social breakdown may be even less effective in Canada than in the U.S. in the latter part of the twentieth century.

The history of U.S. government provision of health care to American Indian and Alaska natives is equally troubling. Since the early nineteenth century, care for Indians was provided by the federal government, namely, by military doctors up to 1849, then by the Bureau of Indian Affairs as part of the Department of the Interior until moved to the U.S. Public Health Service in 1954. The Indian Health Service (IHS), formed as a national health service for Indians, is administered through regional offices, providing primary and specialty care with increasing emphasis on community participation and control.

The U.S. population of Native Americans and Eskimos in 1996 totaled 2.3 million persons. The IHS provides care to some 60% of this population. Gains in health status in reduced infant and general mortality still leave the native peoples in poorer health than the general U.S. population. Poor nutrition, unsafe water supplies, isolation and poor transportation, inadequate waste disposal, and high rates of obesity, alcohol abuse, and violence shorten life expectancy. The IHS operates a network of 43 hospitals and over 110 health centers and health stations. Expenditures for Indian health care, however, are 60–65% of per capita national average expenditures. Life expectancy for Native Americans improved from about 60 years in 1950 to 73.2 years in 1989–1991. Improved infant mortality and a reduction in deaths due to tuberculosis, gastroenteritis, and acute respiratory disease have been accompanied by a reduction in alcohol and violence related mortality. Rates of diabetes (type 2) are high and increasing, as compared to the non-Indian population. Fetal alcohol syndrome (FAS) is a large problem for Native Americans, with an estimated incidence of 2.7 per 1000 live births. The rate of FAS for American Indians is higher than that for any other racial or ethnic group in the United States. Increasing tribal management of health services has been accompanied by increasing tribal activity to reduce alcohol abuse and violence. Health goals for the year 2000 include reducing rates of cancer, infant mortality, and alcoholism.

The aboriginal populations, initially decimated by famine and acute infectious disease and currently by alcohol abuse, violence, and diabetes-related conditions, have a health status that remains well below the health of both Canadian and American populations. Federal administration and separateness from general health services have been subject to much criticism in both countries. Integration with other health services and decentralization of management with community involvement are current trends. Federal financing of this service has stabilized, while increasing local management leads to increased expenditures by this population group in the private medical market.

The prospects of the native population in the United States, Canada, and elsewhere will depend on social and economic development, with a large measure of self-government. In 1999, the Canadian government established a self-governing Inuit (Eskimo) territory which has authority for health, social services, education, taxation and economic rights (e.g., minerals), and many other areas accorded to Canadian provinces. This will provide an important test for this concept.

Similar problems appear among aboriginal populations and minority groups in other countries, including Australia, New Zealand, Peru and other South American countries. Papua-New Guinea, Taiwan, central Africa, and the former Soviet Union. Australian aboriginals suffer from conditions of high rates of cardiovascular disease, diabetes, end-stage renal disease, alcoholism, glue sniffing and drug abuse, rheumatic fever, infant and child mortality, and general social breakdown. The poor health of native peoples is a blight on the records of many countries without clearly defined health responsibility, authority, and programming. The issue of health of aboriginal peoples is a serious challenge to the New Public Health in finding ways to provide adequate preventive and curative care, and more importantly to reduce social gaps engendering apathy and social decay.

PRISONERS' HEALTH

Since the mid-1980s, there have been extraordinary changes in the condition of jails and the health of prison inmates in the United States. The increase of urban decay, illicit drug use, and poverty and associated epidemics has had a great impact on incarcerated Americans. Prison medical services have been transformed into outposts faced with the challenge of meeting near impossible demands.

Prisons are a growth industry in the United States, with over 1.6 million prisoners in 1997. Between 1985 and 1996, the population in federal and state prisons increased from 480,568 to 1.1 million persons, while inmates of local jails rose from 256,615 to 507,044. The total incarceration rate increased from 313 to 615 per 100,000 population over this time period. The demographic characteristics of the prison population are skewed from that of the general population. Approximately 6% of white males are incarcerated and 23% of black males are under the supervision of the correction system. About 47% of prisoners are African-American, and a large number are young and poor.

In 1970, concern for the poor state of prisoner health led the American Medical Association and the American Public Health Association to conduct reviews of prison health services, recommending guidelines which provided state prison services with standards to meet and resulted in improved services in many states. Prison health services took on a more professional role, with less reliance on part-time practitioners. Voluntary accreditation of services by the American Council on Health Services Accreditation serve as an external review procedure.

Persons in correctional institutions are at increased risk for tuberculosis because of the high prevalence of HIV and hepatitis B infections and latent TB, over-crowding, poor ventilation, and the frequent transfer of inmates within and be-tween institutions. The recent emergence of multidrug-resistant TB as an impor-tant opportunistic infection of HIV-infected persons underscores the need for improving infection control practices. The increase in tuberculosis in Russia since 1990 is in part related to large-scale release of prisoners following perestroika in the 1980s. Specific control measures in correctional facilities should include the following:

1. Regular and systematic screening of inmates and staff for HIV and TB. Those that test positive should be eligible for preventive therapy.
2. Rapid identification, isolation, and treatment of suspected cases of TB.
3. Directly observed therapy as well as rigorous follow-up and record keep-ing to ensure completion of treatment.
4. Follow-up to assure continuity of care both inside and outside the correc-tional facility.

Prisoners are also at risk for sexually transmitted diseases. Both male and fe-male prisoners may be subject to physical or sexual assault or intimidation by fel-low prisoners and staff. Prison medical services should have screening and treat-ment capacity as well as powers to assure protection of vulnerable prisoners, such as segregation of young prisoners from potentially violent or long-term inmates.

In the 1990s, U.S. federal legislation mandating life sentences for repeat of-fenders ("three strikes and out") increased the number of elderly prisoners ineli-gible for parole. As larger numbers of inmates spend longer periods in prison, there is a need to regard prisoner's health as part of that of society as a whole.

Short- and long-term health risks for prison staff requires the attention of prison health services. Alcoholism, smoking obesity, and job stress are linked to family breakdown and early mortality from cardiovascular diseases especially. Helping staff deal with the stress and latent violence on the job should work to reduce per-sonal risk and propensity to staff-initiated violence.

The need for mental health services in prison is great and growing. The defi-ciency of community mental health services has accelerated the movement of in-dividuals from the street to the penitentiary. A study by the National Association for Mental Health found that 25% of prisoners admitted were psychotic, with an-other 14% manifesting some psychotic symptoms. Mental illness may be respon-sible for a large part of violent behavior and violent crime. There is a need to com-bine jail treatment facilities with rehabilitation programs.

Ethical issues in prisoner health include torture, rape, murder, starvation, un-ethical medical experimentation, and inadequate medical and psychiatric care. The Nuremberg and Helsinki codes of medical conduct in medical experimentation are internationally accepted standards. The Geneva Convention related to war and military occupation conduct are also accepted standards often ignored, but still a basis for judgment. Abuse of prisoners in civil war situations, such as in Bosnia in the mid-1990s, reached genocidal levels with massacres, starvation, and rape applied widely. The health community in any country or conflict situation has a professional obligation to monitor care of prisoners and play an advocacy role in times of peace and of civil war.

MIGRANT POPULATION HEALTH

In many countries large numbers of persons and families move with seasons for farm or other temporary work. They lack stable social environments and support systems and are often dependent on exploitative employers willing to provide only subsistence wages and a possibly hostile surrounding community. This group includes migrant and seasonal farm workers, miners, foresters, and building workers who often live away from their families for extensive periods.

In the United States, this phenomenon is widespread. The Migrant Health Program of the U.S. Department of Health and Human Services estimates the size of the population of migratory workers at 1.5 million, and another 2.5 million persons work as seasonal farm workers. The majority are married with children, so the total size of the population including dependents is much larger. Most of these families are living below the poverty line. Many are illegal immigrants, subject to exploitation and abuse by employers or contractors. They are exposed to poor sanitation and housing conditions, pesticides, and infectious diseases.

Migratory workers in the United States are of mixed ethnic origin, including white Americans, Hispanics, Haitians, and Jamaicans. Immigrant groups, such as Haitians, Mexicans, and Filipinos, have high rates of tuberculosis, and their work in agriculture may lead to disease transmission to other farm workers. Migrant farm workers are approximately six times more likely to develop TB than the general population of employed adults.

Migrant farm workers may bring their families with them to poor housing encampments operated by their employers with unsanitary conditions and limited access to food and other necessary supplies. Arrangements for health care, schooling, and recreation are likely to be inadequate. Migrant workers in urban settings are frequently single men, who may have left families at home and who migrate from declining rural regions in the same country or who are brought into a foreign country for specific work in agriculture, construction, or mining. Women workers are often brought in for housekeeping, personal caregivers or as sex workers. Professional migrant workers in nursing medicine, computers, and other skills are also common in many parts of the world. People living in a foreign culture as contract workers are subject to abuse and exploitation as well as social isolation with associated mental stress, violence, and STDs.

Migrant workers are more likely to live in poverty, and to suffer unsanitary and crowded living conditions, with lack of medical care. In addition, they generally lack the necessary documentation for establishment of legal residence status. Access to health insurance and regular medical care, especially for children and women is often one of the most pressing issues, especially for illegal immigrants. Nutritional problems of iron deficiency and stunted growth are common as well as micronutrient conditions,. Dental disease is common in migrant families, and access dental care is usually lacking. Injury and deaths from work accidents are high among farm workers. Agricultural workers are the occupational group having one of the highest mortality rates (37 per 100,000 agricultural workers) from work related injuries (see Chapter 9).

High prevalence of syphilis, HIV infection, and TB among migrant workers, underscores the need for public health professionals who are trained to respond to health care needs within the migrant worker population. These studies led to development of cross-training for public health workers on STDs (including HIV infection), TB, and other communicable diseases among migrant farm workers.

Migrant workers are an international phenomenon. In Europe, large numbers of "guest workers" provide unskilled labor in agriculture and the building trades primarily to the wealthy economies which over several decades imported workers from poor countries in North and sub-Saharan Africa or eastern Europe. This has created long-term social and political problems for workers and their families, who are often subject to ethnic resentment, violence, and cultural adaptation difficulties. Public health agencies should supervise living conditions and arrange preventive and treatment services with screening for common diseases in this population group.

HOMELESS POPULATION HEALTH

Homelessness due to poverty and lack of low rental permanent housing has become a common problem in urban centers, in both industrialized and a long standing issue in developing countries. It is increasingly capturing attention in the media because of the contrasting lifestyles of the homeless and the wealthy population. Depending on how "homeless" is defined, the WHO estimated in *The World Health Report 1997* that there are between 100 million and 1 billion homeless people worldwide (World Health Organization, 1997). An estimate suggests that as much as 7.4% of the U.S. population has experienced homelessness at some time in their lives. Long-term street dwellers are at great risk for health problems.

Health care, especially preventive, is not a priority for homeless people. Instead, many of their resources, such as time and money, go toward securing a safe, warm, and dependable place to sleep, toward purchasing food, and often toward drugs and alcohol.

The global pandemic of homelessness is the result of the lack of affordable housing, unemployment, and the deterioration of social services and support. The homeless do not "choose" to live on the streets, rather thay are often victims of circumstance. It is often argued that homelessness is the result of deinstitutionalization from mental institutions, and some estimates of the percentage of homeless individuals suffering from mental illness range from 25 to 50%. It is difficult to gauge whether homelessness causes mental illness or vice versa, but it is clear that the two combine to produce a vicious cycle in which individuals face social isolation, inadequate nutrition, and hygiene barriers to housing and employment, which together serve to perpetuate illness and insecurity.

Surveys of health problems faced by the homeless show that rates of both chronic and acute diseases are extremely high. With the exception of stroke, cancer, and obesity, homeless people are far more likely to suffer from every category of chronic health complication. Because of the transient nature of the population, conditions which require vigilant care and attention, such as TB, are rampant. This, of course, has grave dangers for the public-at-large. Along with prisons, homeless shelters account for a large percentage of TB cases in the industrialized countries. Other health problems include death by freezing or violence, frostbite, leg ulcers, burns, respiratory infections, STDs, including HIV, and trauma from muggings, beatings, and rapes. Homelessness precludes good nutrition, shelter, warmth, security, personal hygiene, and basic first aid. The homeless have poor access to care until the medical condition is dire.

As the members of this group are often uninsured, access to health care services is compromised by cost and distance. One of the most effective means of serving this population is through mobile health units. In Washington, D.C., among other U.S. cities, "health vans," coupled with mobile "soup kitchens," operated by charitable organizations, provide nutritional and medical services to the homeless. If conditions are severe enough to warrant more substantial care, the individual is encouraged to enter a city or private homeless facility, or may be referred for medical care. These provide medical care, a hot shower, a place to stay the night, and hot meals. The role of NGOs and charitable organizations is important in helping the homeless to survive. Clearly, though, these are only temporary solutions to the millions of homeless people worldwide living in dangerous conditions.

Homeless female-headed families with young children may occur from family breakdown due to an abusive spouse, mother's use of drugs and alcohol, pregnancy, migration, or the collapse of an economic base of the family with eviction from a stable home. Homeless runaway children or teenagers are vulnerable to serious social and health consequences, including violence, substance abuse, suicide, and high risk behavior, resulting in unwanted pregnancies, STDs, or HIV infection. In many developing countries, homelessness and poverty provide a source for sale of babies, organs, and child prostitutes.

Homelessness is a public health problem, mainly to the homeless themselves, but also affects the community around them. Solutions, like those for many other social problems, are complex and costly. The root issues lie at conquering the common characteristics of homelessness: poverty and the inadequate supply of permanent housing. Reintegrating the homeless into mainstream society requires community-based treatment and social services. This means client engagement, case management, housing options, long-term follow-up, and support. Vocational training and employment opportunities may be the key. Until such help is available, homelessness will continue to cause much suffering as well as strain the charitable and public health systems which try to care for this population.

REFUGEE HEALTH

An estimated 35 million persons have fled their countries as refugees or have been displaced internally in the mid-1990s. Of some 20 million cross-border refugees, about 14 million are in Africa, southwest Asia, or the Middle East. The mass exodus of people from their homelands can be the result of ethnic or religious persecution, political conflict, war, civil strife or political conflict, environmental degradation, or conflict over economic resources. Annually, millions of persons abandon their homes, farms, regions, or countries in search of food and water, employment, or security.

The ability of a country to cope with a sudden mass refugee situation depends on the function of the state infrastructure and basic services, the security situation, and prior planning for disasters. Appropriate and timely response depends on local strengths in these areas, with international assistance as a backup for needed supplies and services. Effective relief services must be primarily based on the establishment of security, shelter, clothing, sanitation, safe water and food supplies, family reintegration, personal preventive services, and treatment services. Education, family planning and longer term health services are part of any relief program when the refugee situation continues.

The United Nations High Commissioner for Refugees (UNHCR) attempts to monitor and coordinate refugee assistance needs and relief services. The UNHCR estimates the global refugee population at the end of 1997 at some 12 million persons, a decline of 1.2 million from 1996, mainly in Africa. Almost 900,000 refugees were repatriated during 1997, spontaneously and with UNHCR assistance. The 1999 Kosovo refugee crisis added almost 1 million persons to the roster of refugees as a result of ethnic cleansing policies of the government of Serbia.

The acute phase of a refugee situation, discussed later in the section on manmade or natural disaster relief, may become a long-term situation which imposes a different set of problems. The drama of the acute phase may become an international media event with many countries and NGOs offering assistance, relief supplies, and financial contributions. Longer term refugee situations require help which places more responsibility on the UNHCR and NGOs such as the International Red Cross, Médicins sans Frontières, Catholic Relief Services, and others (see Chapter 2). Long-term refugee centers require all the basic services of any population group, with the additional factor of the risk of collapse of fragile sanitation, nutrition, or health services following political change or further disaster. The Kosovo crisis of 1999 has brought together international efforts of the UNHCR, the North Atlantic Treaty Organization, and many international NGOs, with bilateral and public assistance with finances and essential survival matériel, but the immediate solution to the massive displacement is was resolved through military and political action to force the home country to allow repatriation of the refugees, although long-term political solutions are hard to foresee.

MILITARY MEDICINE

Military medicine has played an important role in the development of surgery and related skills, and public health. Roman military success was aided by skill in preventive health measures through hygiene and camp discipline no less than in care of wounds, basic military organization and discipline. Armies and navies depend on medical and nutritional support to maintain health and ability to perform missions assigned. The mandatory use of limes for British sailors following the epidemiological breakthrough of James Lind established the necessity of nutritional support and discipline to maintain function and competence of the individual and the unit. The American army adopted mandatory vaccination soon after Jenner's method became known. Innovations in public health by military personnel during the nineteenth century were numerous, from Ronald Ross' work on the malaria parasite to the conquest of yellow fever by the U.S. Army in Cuba in 1901.

Armed forces are part of national responsibility, and health issues related to the military have an important place in public health. Armed forces seek to recruit people in apparent good health, but conditions of service may cause disease both in peace and in wartime. Military medicine emphasizes prevention to maintain the health of personnel who will be placed in hazardous conditions for disease and injury.

Death from disease outweighed deaths in battle in most armies until the early part of the twentieth century, as seen in data from the U.S. Army from the Civil War to the Vietnam War (Table 7.5). Awareness of the predominance of disease in army casualties was a major contribution of Florence Nightingale from the Crimean War. Military medicine has contributed to the development of emergency

TABLE 7.5   Number and Percent of Deaths from Disease and Battle Injury, United States Army, 1860–1975

War	Disease (D)	Battle injury (BI)	D/BI%
Civil War (North) 1861–1865	199,720	138,154	145
Spanish American War, 1898	1,939	369	525
Phillippines, 1899–1902	4,356	1,061	410
World War I, 1917–1918	51,447	50,510	102
World War II, 1941–1945	15,779	234,874	7
Korean War, 1950–1953	509	27,704	2
Vietnam War, 1961–1975	1,433	30,900	5

Source: Adapted from Legters and Llewellyn, 1992. Military medicine, In (Last, J. M., Wallace, J. B. Eds.). *Public Health and Preventive Medicine.*

medical care, bringing enormous benefit for the civilian sector, including innovations in medical adaptations from military technology such as ultrasound.

Protection of troops from disease requires assurance of immunization to prevent diseases that could easily be transmitted in barracks or shipboard living conditions. This may include updating childhood vaccinations with boosters of diphtheria and tetanus and vaccinating against hepatitis B, influenza, polio, measles, mumps, rubella, meningococcal meningitis, hepatitis A, anthrax, and others. Antimalarials and other preventive measures are used depending on location of service.

Nutrition, protection from food-borne, waterborne, and vector-borne diseases, and prevention of training and motor vehicle accidents, suicides, exposure to toxic materials, contact with STDs, and HIV exposure are among the many issues of peacetime armed forces. Violence and brutality are frequent issues in the training period, as are suicides. Officer and noncommissioned officer vigilance, accountability, and medical surveillance are required.

Prevention of war is the surest method of preventing related military and civilian death and injury. In war, battle casualty prevention depends on armaments, skill in their application, discipline, and leadership. Medical support in the field, evacuation and triage, and rapid transfer with life support systems to medical centers is crucial to keep fatality:injury rates low and save many young lives. Organization of medical services from the medic on the battle field to the evacuation post and the base hospital requires skilled organization and management with assurance of adequate supplies of everything from water and food to diagnostic and treatment resources.

Casualty management to conserve fighting strength is based on prevention and collection, treatment, and evacuation of casualties in a manner that supports morale of the troops and prevents complications or death following injury. Triage or sorting is based on providing the best care one can for the most patients under the circumstances. Triage categories include the following: urgent cases require airways, chest tubes, hemorrhage control, and replacement therapy to prevent immediate death; immediate cases are life threatening wounds temporarily stable but requiring surgery within a short time and a good chance of recovery; delayed cases are injuries that can be cared for successfully 8–16 hours after the injury; minimal, or superficial, wounds require minor surgery, fractures setting, or observation; expectant cases are mortal wounds with little chance of survival.

Conditions of potential atomic, biological, and chemical warfare require special preparation, and above all prevention. The potential exists, and preparation for its effects on the health of combat units and civilian populations are part of this special field of military medicine and, indirectly, public health.

HEALTH PROTECTION IN DISASTERS

Natural and man-made disasters are frequent occurrences (Table 7.6) and have important implications for public health. Natural disasters are naturally occurring

TABLE 7.6   Recent Disasters—Manmade and Natural

Location	Type	Year	Effects
Guatemala	Earthquake	1976	23,000 deaths, 77,000 injured, 1.2 million affected
Serveso, Italy	Chemical factory explosion	1976	17,000 evacuated
Cambodia	Genocide, political	1979	1–2 million deaths
Bhopal, India	Chemical leak	1984	2000 deaths, 70,000 evacuated
Mexico	Earthquakes	1985	10,000 deaths, 60,000 homeless
Colombia	Volcano	1985	23,000 deaths, 200,000 homeless
Chernobyl, Ukraine	Nuclear reactor meltdown	1986	30 dead, 100,000 evacuated, long-term effects (cancer, birth defects) unknown
Bosnia, Herzogovina	Civil war, genocide	1993–1995	Tens of thousands of casulties, mass rape and murder, breakdown of civilian services
Rwanda	Genocide, tribal	1994	Up to a half million deaths
Kobe, Japan	Earthquake	1995	6000 deaths
Caribbean	Hurricane Gordon	1995	11,000 deaths in Haiti, Cuba, Jamaica, and Dominican Republic
China	Flood	1998	4150 dead, 18.4 million displaced, and 180 million affected
Nicaragua	Hurricane Mitch	1998	10,000 dead, thousands missing, 120,000 homeless
Kosovo	Civil war, ethnic cleansing	1999	Some 1 million persons forcibly displaced, with mass murder, community destruction

Source: Adapted from Menu, J.-P. (ed). Emergency and humanitarian action. *World Health Statistics Quarterly,* 1996, and 1999 Internet website http://www.reliefweb.int

extreme events that can cause excess morbidity, mortality among the population, and damage to the physical environment. These include earthquakes, floods, hurricanes, droughts, blizzards, and volcanic eruptions. The term disaster also includes a wide range of events including war, industrial accidents, terrorist incidents, and others. Disasters are often classified into "man-made" and "natural" and "sudden" or "slow" onset. The distinctions are often blurred, since a natural disaster may be the result of inappropriate policy decisions such as a drought occurring in an area made vulnerable by political or policy actions. In semiarid areas with a delicate food balance and chronic undernutrition, a modest natural disaster may tip the balance and cause wide-scale suffering and malnutrition continuing over an extended period of time.

The accumulated experience of disaster relief management has been enhanced in the past several decades by improved technology in communications and transportability of air-mobile supplies of tents, blankets, food, chlorinators, generators, heavy equipment for lifting debris, as well as field hospitals, and medical supplies such as sterilizing and trauma equipment, antibiotics, vaccines, and oral rehydration therapy.

Disasters require a highly professional and aggressive response based on intersectoral cooperation. This can be a matter of life and death to large groups of people. Disasters can occur in the midst of an urban metropolis or in a remote jungle. The details will differ, but meeting basic human needs is common to both settings. The issues of limiting further death and injury, protection, safety, water, food, and shelter come first in priority. Organized prevention in the form of sanitation and management of the most common health problems such as diarrheal diseases, acute respiratory diseases, measles, and other diseases associated with poor sanitation (e.g., hepatitis, typhoid, cholera, and gastroenteritis) is needed. Dysentery, malnutrition, and respiratory infection outbreaks can kill large numbers of debilitated children and the elderly. Posttraumatic syndrome can be dealt with as part of a community-oriented support program with involvement of the population affected as health aides.

Epidemiologic monitoring of death, injury, and disease is usually difficult because of the chaotic events associated with disasters. Monitoring of the process and epidemiologic patterns provides information for intervention and lessons which can be important in planning and management of future crises. Monitoring admissions to hospitals indicates morbidity patterns for specific diseases such as typhoid fever and viral hepatitis. Such was the case during the 1980 earthquake in southern Italy causing water contamination. Cholera is also a serious danger in disaster situations as occurred in the Rwanda refugee disaster. Contamination of limited water supplies makes this one of the most urgent aid issues, with pumps and portable chlorinators as basic equipment needs. Tents and food distribution facilities may need to be protected by armed guards.

Intervention by aid agencies should include the potential for vaccination against measles and poliomyelitis and sustaining regular DPT vaccination, among with vitamin and iron supplements if the return to normal conditions is delayed over months. Typhoid and salmonellosis are combatted by sanitary and case management approaches. The absolute necessity of potable water and oral rehydration therapy on a large scale was demonstrated in the Rwanda disasters of 1993–1994.

Over longer periods, nutrition monitoring of child weight and height for age may be done on sample populations to assess the effects and changing situation. Disaster relief should include the possibility for vitamin and iron supplements for children to offset the damage of food deprivation. Provision of food staples is a high priority along with shelter and water, but the bulk shipment required may overwhelm even the ability of developed countries to cope with local disasters.

Secondary damage from unsanitary conditions can spread infectious disease. This is aggravated by food deprivation and lack of water and sanitary facilities. The large numbers of refugees from civil war and ethnic massacres seen in the 1994 Rwanda refugee situation resulted in massive loss of life. International or local relief efforts may be overwhelmed and hampered by lack of coordination. Planning refugee camps in disaster situations must take into account natural drainage, water sources, access roads, separation of sanitary facilities including latrines and garbage disposal with safe areas for food and water distribution.

Hurricane Opal struck the Florida coast in October 1995 and caused 27 deaths of which 13 were due to falling trees, 4 due to carbon monoxide poisoning, 3 from house fires, 7 from motor vehicle crashes, and 5 during the postimpact phase from drowning, electrocution, and other causes. Better coordination of storm warnings could have reduced deaths at sea; warnings to remain off the roads and curb use of vehicles in danger areas may have reduced road crash deaths. Emergency information to promote safe evacuation, safe havens, risk avoidance, and assurance of support services are part of public health action in emergency situations, in conjunction with other local, state, and federal agencies.

In late 1996, a massive refugee crisis emerged in Rwanda and Zaire, with complex ethnic and political origins, which resulted in death on a massive scale due to genocidal military actions as well as dehydration, disease, and starvation. International intervention involving military field hospitals, aid, and health organizations were sporadic and insufficient. NGOs such as Medicins sans Frontieres and U.N. aid organizations are as usual on the ground first, but without strong international political and military backing, all efforts will be limited. Medical interventions with field hospitals capture media attention, but efforts to promote basic organization for shelter, safe water, and sanitation receive fewer resources and less support.

In 1998, 37 major disasters including floods (15), earthquakes (6), volcano eruptions (1), dam bursts (1), hurricanes/cyclones (6), forest fires (1), and droughts (3) occurred in all parts of the world entailing loss of life and population displacement with heavy costs in damage to property and other economic costs. In 1999, 13 such events were reported up until March. The flood in China during the summer of 1998 resulted in 4,150 deaths, 18.4 million persons evacuated, and a total of 180 million persons affected. The economic loss was estimated at over $26 billion (website http://www.reliefweb.org) Two 1999 earthquakes in Turkey caused massive destruction and loss of life, in part due to poor construction standards in a known earthquake zone.

Prevention of human injury and death in disasters must take into account experience in a given area. Earthquakes in California are a serious threat and have led to strict building codes that reduce the damage and secondary loss of life. Monitoring can give warning of potential disasters such as hurricanes, and preparation of evacuation plans and facilities make a major difference in disaster management. Zoning laws and building code enforcement can prevent many deaths when earthquakes strike in urban areas. Flood control measures in areas traditionally threatened are worthwhile investments to prevent the massive damage to property and homes. Storage of appropriate equipment and supplies and training of manpower for such emergencies is part of prudent public administration, and is important for public health as well.

Documentation of disaster experience is essential to improve efforts in future

BOX 7.5 DISASTER PLANNING ISSUES

1. Treat and evacuate the injured;
2. Limit further death, injury, and disease;
3. Assure safety/protection/security/public order to coordinate interventions of police, army, official health agencies, local and international NGOs;
4. Provide safe water, food, warmth, and shelter;
5. Provide sanitary facilities and prevent environmental hazards;
6. Promote epidemic control/prevention and treatment of communicable diseases (e.g., diarrheal disease, acute respiratory infections, measles, hepatitis, malaria);
7. Provide ongoing medical care for the injured and sick;
8. Mobilize and coordinate all official and voluntary local, state, national, and international aid;
9. Prevent malnutrition, provide micronutrient supplements;
10. Provide maternal and child care for pregnancy, delivery, infancy, and childhood;
11. Monitor disease and epidemiological surveys;
12. Mobilize health aides among the affected population;
13. Promote early restoration of normal functions (e.g., family, health care, work);
14. Prevent posttraumatic stress syndromes;
15. Assess, evaluate, monitor, and report the process and lessons learned;
16. Survey, document, publish, and follow up;
17. Educate (e.g., provide temporary preschool and school activities);
18. Employ able-bodied persons (e.g., promote participation in refugee care activities);
19. Promote rapid return to homes and rehabilitation;
20. Review experience and plan for potential future disasters.

situations, and although lessons may be absorbed slowly, they are part of the development of public health. Disaster planning is an essential component of public health agencies at the national state/provincial/regional levels and at local health authority levels. Coordination with police, army, civil defense, fire service, local and state disaster planning, hospitals, and many other local agencies as well as international relief agencies is fundamental to coping with disasters. Preparation of protocols for operating procedures with drills may make a large difference in outcomes. Although disaster situations are unique they have common features relevant to planning, adapted from documentation of trial, error, and experience.

SUMMARY

A society is often judged on how it treats its minorities, its poor, its prisoners, and its refugees as much as how it cares for the main population groups. All such groups need special attention because they are people in need, but also because they can affect the health of others, as in the case of tuberculosis spread to prison guards in the United States, and the spread of tuberculosis in the general population by ex-prisoners in the former Soviet Union.

Public health agencies are often the advocates and pioneers in implementing programs for such groups. These may be special needs of the whole population or needs of special groups in the population. The public health approach emphasizes prevention in all its phases, and preparation for foreseen and unexpected emergency health situations, in coordination not only with health service systems, but also many other agencies in society.

Preparation for handling emergencies is an important challenge to all elements of a health system, from sanitation and pest control to tertiary care neurosurgery. In such a situation, there will be little question of the need for all elements to work together to treat the injured and prevent further damage. This requires preparation as well as improvisation at the time of the event. Mental and dental health are two areas where community care and prevention are established as effective measures but generally not well linked to the mainstream of organization and funding of health activities.

Special groups such as refugees, prisoners, or those suffering from natural and man-made disasters provide living proof of John Donne's famous phrase, "No man is an island unto himself alone." Today's prisoner or refugee can become tomorrow's free citizen infected with HIV or tuberculosis. Disasters can affect anyone, and the role of health care providers and public health agencies becomes all too obvious and the need for preparation reinforced. The New Public Health seeks to use the resources of prevention and care more effectively for what have been treated as marginal populations or issues.

ELECTRONIC MEDIA

American Psychiatric Association, http://www.apa.org/
American Dental Association, http://www.ada.org/
Disasters, http://www.reliefweb/
Disaster medicine, http://wwwnotes.reliefweb.int
Indian Health Service, U.S. Department of Health and Human Services, http://www.ihs.gov/
Mental health, http://www.nimh.nih.gov/, http://www.who.int/msa/, and http://www.medscape.co/
   Medscape/MentalHealth/public
Migrant health, http://www.bphc.hrsa.dhhs.gov/uhsc/Pages/about_nhsc/3D1b_migrant.htw
Migrant health, http://www.aap.org/policy? and http://www.bphc.hrsa.gov
National Institute of Mental Health, http://www.nimh.nih.gov/
National Institute of Dental and Craniofacial Research, http://www.nider.nih.gov/
Native peoples' health, http://www.cdc.gov/ and http://www.aaip.com/
Prisoner's health (U.S.), http://www.ojp.usdoj.gov:80/bjs/pub/pdf/cpius95.pdf

United Nations High Commissioner for Refugees, http://www.unhcr.ch
Women's health, http://www.ahcpr.gov/research/womenres.htm
World Health Organization, Mental Health Program, http://www.who.int/msa/

Note. For full listing of electronic resources related to mental health, see *American Journal of Public Health* 1999;89:1440.

RECOMMENDED READINGS

Breakey, W. R. (editorial). 1997. It's time for the public health community to declare war on homelessness. *American Journal of Public Health,* 87:153–154.

Centers for Disease Control. 1992. Famine-affected, refugee and displaced populations: Recommendations for public health issues. *Morbidity and Mortality Weekly Report,* 41(RR-13):1–76.

Centers for Disease Control. 1992. Public health focus: Fluoridation of community water supplies. *Morbidity and Mortality Weekly Report,* 41:372–375.

Centers for Disease Control. 1993. Public health consequences of a flood disaster—Iowa, 1993. *Morbidity and Mortality Weekly Report,* 42:653–656.

Centers for Diseases Control. 1993. Prevalence of work disability—United States, 1990. *Morbidity and Mortality Weekly Report,* 42:757–759.

Centers for Diseases Control. 1993. Reduction in alcohol-related traffic fatalities—United States, 1990–1992. *Morbidity and Mortality Weekly Report,* 42:905–909.

Centers for Disease Control. 1994. Core public health functions and state efforts to improve oral health—United States, 1993. *Morbidity and Mortality Weekly Report,* 43:201, 207–209.

Centers for Disease Control. 1996. Deaths associated with Hurricanes Marilyn and Opal—United States, September–October, 1995. *Morbidity and Mortality Weekly Report,* 45:32–38.

Durkin, M. S. (editorial). 1996. Beyond mortality—Residential placement and quality of life among children with mental retardation. *American Journal of Public Health,* 86:1359–1360.

Easley, M. 1990. The status of community water fluoridation in the United States. *Public Health Reports,* 105:348–353.

Elias, C. J., Alexander, B. H., Sokly, T. 1990. Infectious disease control in a long-term refugee camp: The role of epidemiologic surveillance and investigation. *American Journal of Public Health,* 80:824–828.

Gaiter, J., Doll, L. S. (editorial). 1996. Improving HIV/AIDS prevention in prison is good public health policy. *American Journal of Public Health,* 86:1201–1203.

Johnston, I. 1992. Action to reduce road casualties. *World Health Forum,* 13:154–162.

Jorgensen, J. G. 1996. Recent twists and turns in American Indian health care. *American Journal of Public Health,* 86:1362–1364.

Kessler, R. C., McGonagle, A., Zhao, S., Nelson, B., Hughes, M., Eshelman, S., Wittche, H. U., Kemder, K. S. 1994. Lifetime and 12 month prevalence of DSM-III-R psychiatric disorders in the United States: Results from the National Comorbidity Study. *Archives of General Psychiatry,* 51:8–19.

Kunitz, S. J. 1996. The history and politics of U.S. health care policy for American Indians and Alaska natives. *American Journal of Public Health,* 86:1464–1473.

Neugebauer, R. 1999. Mind matters: the importance of mental disorders in public health's 21st century mission. *American Journal of Public Health,* 89:1309–1311.

Phelan, J. C., Link, B. G. 1999. Who are the "homeless": reconsidering stability and composition of the homeless population. *American Journal of Public Health,* 89:1334–1338.

Shenson, D., Dubler, N., Michaels, D. 1990. Jails and prisons: The new asylums? *American Journal of Public Health,* 80:655–656.

Susser, E., Valencia, E., Conover, S., Felix, A., Tsai, W-Y., Wyatt, R. J. 1997. Preventing recurrent homelessness among mentally ill men: A "critical time" intervention after discharge from a shelter. *American Journal of Public Health,* 87:256–262.

Teplin, L. A. 1990. The prevalence of severe mental disorder among male urban jail detainees: Comparison with the Epidemiologic Catchment Area Program. *American Journal of Public Health,* 80:663–669.
Ustün, T. B. 1999. The global burden of mental disorders. *American Journal of Public Health,* 89:1315–1318.

BIBLIOGRAPHY

American Academy of Pediatrics. 1993. Homosexuality and adolescence. *Pediatrics,* 92:631–634.
American Medical Association Council on Scientific Affairs. 1996. Health care needs of gay men and lesbians. *Journal of the American Medical Association,* 275:1354–1359.
American Psychiatric Association. 1994. *DSM-IV: Diagnostic and Statistical Manual of Mental Disorders,* Fourth Edition. Washington, DC: The American Psychiatric Association.
Braddock, D. 1992. Community mental health and mental retardation services in the United States: A comparative study of resource allocation. *American Journal of Psychiatry,* 149:175–183.
Buckner, J. C., Trickett, E. J., Corse, S. J. 1985. *Primary Prevention in Mental Health: An Annotated Bibliography.* Bethesda, MD: National Institutes of Mental Health.
Eaton, W. W. 1985. Epidemiology of schizophrenia. *Epidemiologic Reviews,* 7:105–126.
Fisher, W. H., Geller, J. L., Altaffer, F., Bennett, M. B. 1992. The relationship between community resources and state hospital recidivism. *American Journal of Psychiatry,* 149:385–390.
Garrison, C. Z., McKeown, R. E., Valois, R. F., Vincent, M. L. 1993. Aggression, substance use and suicidal behaviors in high school students. *American Journal of Public Health,* 83:179–184.
Hveman, M. J. 1996. Epidemiologic issues in mental retardation. *Current Opinion in Psychiatry,* 9:305–311.
International Helsinki Federation for Human Rights. 1999. *Annual Report on United States, 1997.* Vienna, Austria: IHF.
Kemp, D. A. (ed). 1993. *International Handbook on Mental Health Policy.* Westport, Connecticut: Greenwood Press.
Leaning, J., Briggs, S. M., Chen, L. C. 1999. *Humanitarian Crises: The Medical and Public Health Responses.* Cambridge, MA: Harvard University Press.
Leff, H. S., Lieberman, M., Mulkern, V., Raab, B. 1996. Outcome trends for severely mentally ill persons in capitated and case managed mental health programs. *Administration and Policy in Mental Health,* 24:3–23.
Legters, L. J., Llewellyn, C. H. 1992. Military medicine (Chapter 71). In Last, J. M., Wallace, R. B., *Macxy-Rosenau-Last, Public Health and Preventive Medicine,* Thirteenth edition. Norwalk, CT: Appleton & Lange.
Lieberman, E. J. 1975. *Mental Health: The Public Health Challenge.* American Public Health Association. Washington, DC.
Logue, J. N. 1996. Disasters, the environment, and public health: Improving our response. *American Journal of Public Health,* 86:1207–1210.
Menu, J-P. (ed). 1996. Emergency and humanitarian action. *World Health Statistics Quarterly,* 49:165–242.
Nease, D. E., Volk, R. J., Cass, A. R. 1999. Investigation of the severity-based classification of mood and anxiety symptoms in primary care patients. *Journal of the American Board of Family Practice,* 12:21–31.
Olfson, M. 1990. Assertive community treatment: An evaluation of the experimental evidence. *Hospital and Community Psychiatry,* 41:634–641.
O'Neill, D. M. 1996. Measuring changes in resource use in state hospitals, 1969–1990: The effect of alternative deflators. *Administration and Policy in Mental Health,* 24:3–23.
Post, B. 1997. The underserved and desperate: Native health, its time for action. *Canadian Medical Association Journal,* 157:1655–1656.

Sandhaus, S. 1998. Migrant health: A harvest of poverty. *American Journal of Nursing,* 98:52–54.

United States Census Bureau. 1998. *Statistical Abstract of the United States, 1998.* Washington, DC: U.S. Census Bureau.

United States Department of Justice. 1997. Prisoners in 1996. *Bureau of Justice Statistics Bulletin,* June: 1–8.

World Health Organization. 1985. Targets for Health for All: Targets in Support of the European Regional Strategy for Health for All. WHO, Regional Office for Europe, Copenhagen.

World Health Organization. 1991. *Health for All Targets: The Health Policy for Europe. European Health for All Series No. 4. Copenhagen: World Health Organization, Regional Office for Europe.*

World Health Organization. 1997. The World Health Report 1997. Conquering Suffering, Enriching Humanity, Geneva: WHO.

World Health Organization. 1998. *The World Health Report, 1998: Life in the 21st century: a Vision for All.* Geneva: WHO.

8

NUTRITION AND
FOOD SAFETY

INTRODUCTION

Nutrition has a direct affect on growth, development, reproduction, and the physical and mental condition of the individual. Nutrition is one of the most important single factors for the health of an individual or a community, and is, consequently, a fundamental issue in modern public health. People require nutrients to provide heat and energy (carbohydrates, fats, and proteins), regulate body processes (water, minerals, fiber, vitamins, proteins, and essential amino acids), and build and renew body tissue (water, proteins, and mineral salts). The nutritional status of an individual and society is influenced by the supply, quality, distribution, access to, and cost of foods. It is also influenced by knowledge, attitudes, beliefs, and practices regarding essential and balanced nutrition. Food policy at the individual and government levels, eating habits, as well as economic and technical factors all contribute to the nutritional state of the public, and the individual person.

Improved nutrition has made a major contribution to better health in past centuries. In the twentieth century nutrition emerged as a basic and applied science. Knowledge of the elements of proper nutrition and its role in prevention of deficiency diseases or chronic diseases has played a vital part in the development of modern public health. And, despite rapid population growth, food production and average food consumption has improved steadily worldwide.

Nevertheless, malnutrition is widely prevalent throughout the world. Developed countries struggle with problems of inappropriate and excessive nutrition which can lead to chronic diseases, which developing nations also face along with diseases of nutritional deficiencies. Rising standards of living in many developing countries have brought diseases of modern living to prominence as communicable disease are increasingly brought under control. Subpopulations within rich and poor nations alike suffer from a broad spectrum of nutritional diseases.

Public health attempts to ensure that all groups in the population have an adequate, but not excessive, intake of the basic food groups, essential vitamins and minerals for growth, health maintenance, and physical activity. This includes recommendations of daily human needs for nutrition and energy, which vary according to age, sex, body size, level of activity, individual health status, and environmental conditions. It also requires monitoring of the nutritional status of the population and its subgroups.

DEVELOPMENT OF NUTRITION IN PUBLIC HEALTH

The steady improvement in life expectancy seen in Europe and North America in the seventeenth to the nineteenth century probably had as much to do with improved nutrition as with improved sanitation. Pioneering epidemiologic studies of James Lind in the mid eighteenth century and that of Joseph Goldberger in the early twentieth century developed nutritional epidemiology. They each established proof of deficiency conditions that met the Koch–Henle criteria of causation in epidemiology.

Just as Snow's work on cholera preceded Koch's discovery of the *Vibrio cholerae* organism by 30 years, so Lind's work on scurvy preceded the isolation of ascorbic acid by more than a century. Antoine Lavoisier (1743–1794) in Paris developed basic concepts of metabolism, measuring oxygen consumption and carbon dioxide production, and is called the "father of the science of nutrition." Justus von Liebig (1803–1873) demonstrated that fat, protein, and carbohydrates are burned in the body and developed methods of analysis of composition of foods, body tissues, urine, and feces. In the mid nineteenth century, scientists in France, Germany, and later in Britain and the United States made rapid advances in the chemistry of oxygen, carbon dioxide, calcium, iodine, and iron. In 1897, Christian Eijkman demonstrated the origin of beriberi by showing the disease in populations eating polished rice and its absence when rice was eaten with its husk (Nobel prize, 1982).

In the early years of the twentieth century animal studies showed diets consisting of only pure protein, carbohydrates, and fats led to failure-to-thrive, sickness, and death. This led Casimir Funte to the idea of "vital amines" missing in the diet, later termed vitamins, and the isolation of chemical substances with antineuritic and antiberiberi properties. The delineation of the vitamins necessary to prevent disease and promote health produced immeasurable improvement in human health where the knowledge was applied.

The development of applied nutritional public health interventions in the United States was led by the federal Departments of Agriculture and Education, but the Department of Health and Human Services also contributed greatly to this evolution. The investigation of pellagra, the establishment of national recommended dietary allowances and national food supplementation, and school lunch programs established nutrition as a central issue in public health in the United States. In many parts of the world application of the knowledge of the fundamental importance of essential vitamins and trace minerals has still to be implemented as applied nutritional public health measures.

NUTRITION IN A GLOBAL CONTEXT

In 1997, "about a quarter of the world population lives in dire poverty and in many regions it is increasing. There is also a widening gap between the living standards of this quarter and those of the more privileged who enjoy rising standards of living" (World Health Organization, 1998). The people most affected by poverty are women, children, and the elderly, and poor nutrition is one of the key mediators of poverty and poor health. The effects of maternal malnutrition on newborns can be lifelong. Micronutrient deficiencies, despite their preventability by low cost public health measures, are still widespread. Yet, while millions are starving or in a chronic state of undernutrition, there is enough food in the world to feed everyone, so that national strategies for combating the effects of poverty are needed based on good governance, suitable legislation, participation by all in society, and provision of effective basic services. Poverty, unequal distribution of income and food, ignorance, low education levels, large family size, acute and chronic illness, and government inaction are major contributors to world hunger (Table 8.1).

The cumulative effects of poverty, population, environmental degradation, overfarming of land, erosion of topsoil, harmful agricultural practices, and inadequate storage and transportation are vital in global health. Malnourished people, especially children are susceptible to diseases including diarrhea, acute respiratory infections, and many which are vaccine-preventable.

Frequent pregnancies among uneducated women, poorly nourished and anemic, contribute to high maternal mortality rates and low birth weight babies, who are then vulnerable to infections. Poor quality breast milk affects the infant already vulnerable to increased chance of infection. Inadequate intake of essential vitamins and trace minerals, or poor absorption of nutrients, adversely affects tissue metabolism, the oxygen carrying capacity of blood, and the formation and function of muscle and brain tissue, as well as growth of long bones. These all contribute to poor productivity and economic development. This complex is worsened by movement of farm labor to the crowded city slums where poor sanitation perpetuates the cycle of ill health.

Even wealthy societies contain population groups at risk for inadequate or "subnutrition," often masked by apparent "middle-class" status with the elderly, the working poor, teenagers, women, and toddlers affected by dietary inadequacies and especially micronutrient deficiencies with long-term health effects. These issues, too, require public health and governmental interventions. Especially important are education about good dietary practices as well as assurance of food quality, including fortification with essential minerals and vitamins.

TABLE 8.1   Nutritional Condition in Developing and Developed Countries

Issues	Factors in developing countries	Factors in developed countries
Poverty	Lack of resources or access to healthy foods; traditional crops and foods inadequate for healthy diet	General prosperity but poverty in significant minorities; supplemental programs (e.g., WIC, food stamps, school lunches) offset deprivation
Educational Deprivation	Lack of knowledge of good nutrition	High awareness of good nutrition, but obesity common, especially among the poor
Micronutrient deficiencies	Widespread deficiencies of iron, iodine, and vitamins A, B, C, and D; inadequate food processing technology	Food fortification with iodine, iron, vitamins, A, B, C, and D in the U.S., Canada, and some others
Malnutrition–infection cycle	Poor sanitation; lack of control of vaccine- preventable diseases, parasitic, diarrheal, respiratory diseases; HIV, tuberculosis, malaria all harm nutrition status	High levels of sanitation, vaccination coverage and hygeine with benefit to nutritional state; high risk subgroups (HIV tuberculosis, drug abusers, the homeless)
Food security	Harmful agriculture practices; overgrazing and overfarming; soil erosion, small plots; inadequate and wasteful storage and distribution facilities; food costly relative to income	Science-based technological agriculture; high productivity, abundant produce, variety, food cheap relative to family income
Unhealthy diets and noncommunicable diseases	Undernutrition with poverty, inadequate calories per person; middle class overnutrition with high rates of cardiovascular diseases, diabetes	Overnutrition with excess sugar and animal fats; high rates of cardiovascular disease; pockets of poverty and nutritional inadequacy; cardiovascular disease mortality falling
Breast-feeding (capacity to care)	Common and prolonged but with poor supplementation practices	Increasing and well supplemented
Food quantity quality, variety, cost	Contamination, waste, and destruction common; and lack of supply and variety of vegetables, fruit, protein; costly, wasteful	Good supply and quality; good distribution, and marketing systems, labeling, regulation, and supervision; comparatively inexpensive
Monitoring	Monitoring supply, distribution, intake of food; growth, anemia, and intake studies needed	Nutrition surveys
Need for national policies and objectives	Prevent deforestation and land loss; rural poverty, lack of education; lack of credit and agricultural support; poor and harmful farming practices; pesticides used dangerously	Government promotes agriculture, research, support systems, transport, and marketing; limit use of pesticides; industrialized farming

Source: Adapted from *International Conference on Nutrition: Nutrition the Global Challenge.* World Health Organization and Food and Agricultural Organization, 1992.

NUTRITION AND INFECTION

The infectious disease–malnutrition cycle causes millions of deaths of children yearly from preventable causes. Control of measles and wide use of oral rehydration therapy alone would save hundreds of thousands of lives and improve the nutritional status of millions of children. Malnutrition and infection interact, each exacerbating the other. Infections such as measles may have a case fatality rate of 2 per 1000 in an industrial country but 50 or more per 1000 in a developing country where there are widespread deficiencies of essential nutrient elements such as vitamin A. Conversely, even a relatively common infection can adversely affect the nutritional state, growth pattern, and resistance to further infection of a child. The time required to make up nutrient losses from infection following recovery may be two to three times as long as the duration of the infection itself.

Normal growth in weight and height indicate that a child is more likely to resist infection or prolongation of infections that occur. Child health survival strategies rest on the twin pillars of infectious disease control and nutritional adequacy. Nutritional deficiency may also be a factor in reduced resistance to infection in the elderly and the immunocompromised, as both social groups may be socially and economically marginalized in industrialized societies.

FUNCTIONS OF FOOD

Consumed food substances provide varying levels of energy and essential requirements for growth and maintenance of body functions. Exercise and moderate eating habits maintain body weight and reduce the risk for chronic diseases associated with excess body fat, such as diabetes, hypertension, cardiovascular disease, and some cancers.

Nutrients have specific roles within the body, but their functions are interdependent. The diet of the individual determines their availability. Consequently, it is important to identify the sources of proteins, carbohydrates, fats, vitamins, and minerals available in common foods in a target community. The six important nutrient groups are the water and fat soluble macronutrients (carbohydrates, fats and proteins), the micronutrients (trace minerals and vitamins), and water. The macronutrients provide energy, essential and other amino acids, and essential fatty acids. Micronutrients are required for utilization of that energy.

The body processes foods into simpler forms in order to absorb them by digestion through a continuous mechanical and chemical process in the digestive tract. Foods are first ground up by chewing of food, requiring good dentition. Mixing of the food with saliva and swallowing brings the partially digested food to the stomach and small intestine, where it is acted on by enzymes. These enzymes break the food down into smaller and smaller fragments which can then be absorbed through the walls of the small intestine to enter the bloodstream. Disease of the gastrointestinal tract can interfere with this process.

Pancreatic enzymes are released into the small intestine as proteases (which split proteins), amylases (which split polysaccharides), and lipases (which break down fats). Carbohydrates are absorbed as sugar and stored to provide energy, and stored in the liver as glycogen, which is released into the bloodstream to sustain sugar levels. The liver also stores fat soluble vitamins and manufactures enzymes, cholesterol, proteins, vitamin A, blood coagulation factors, and bile salts which are released into the intestine to help in absorption. The pancreas also produces insulin, vital for control of blood sugars.

Composition of the Human Body

The human body is composed, by weight of approximately 62% water, 17% protein, 13% fat, 6% minerals, and 2% carbohydrate. Body composition can vary with stress and nutritional status. Food deprivation uses body stores. During starvation, depletion of carbohydrates is made up by synthesis from reserves of fat and protein. Depletion of up to 10% of total body water can occur without serious risk, but in small children the margin of safety is smaller.

HUMAN NUTRITIONAL REQUIREMENTS

Calculation of the appropriate amounts of nutrients for age and sex groups is a complex task. Energy from food is converted into mechanical work (up to 25%) and growth, dissipated as heat, and used in maintaining the basic functions and temperature of the body. One calorie is the energy required to raise 1 gram of water by 1 degree Celsius. Food energy is expressed in kilocalories (1000 calories). The international unit of energy is a joule, but the more commonly used measure in nutrition is the kilocalorie, or Calorie (1 kcal = 1 Cal = 4.1868 kilojoules, kJ). Need for energy intake varies with body size and is increased by activities of work and recreation. Persons with a sedentary lifestyle and average body size will need less food intake than those with moderate or high levels of activity, or greater body size to maintain a status quo.

Both deficiency and excessive dietary intake of any nutrient can cause disease or death. The range of intake needed for optimal physiological function depends on age, body size, gender, and activity level, as well as on pregnancy, disease, or injury. The range of intake needed for optimal health (shown in Fig. 8.1) emphasizes that there are subclinical phases of basic undernutrition and overnutrition. Table 8.2 lists the essential nutrients and their functions in the human body.

Carbohydrates. Carbohydrates are a major source of energy (4 kcalories per gram) used for metabolic processes and for producing cellular substances including enzymes and cell membranes. Carbohydrates are classified as monosaccharides or disaccharides (simple carbohydrates) or as polysaccharides (complex carbohydrates).

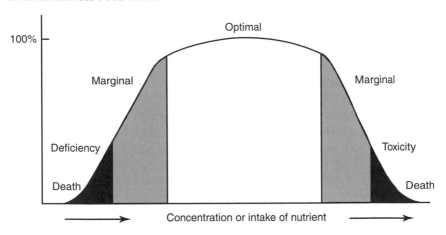

FIGURE 8.1   Range of nutrient intake for optimal health. For nutrients and energy sources, there is a range of intake that confers optimal physiological function. Below this range, deficiencies can cause disease or death. Excessive intake also can lead to increasing symptoms of toxicity. The optimal range varies for each nutrient and is affected by many individual and environmental factors. (Reprinted from Mertz W. 1981. The essential trace elements. *Science,* 213:1332–1338 with permission; from *Surgeon General's Report on Nutrition and Health, 1988*)

Monosaccharides are found as glucose and fructose in fruits, vegetables, and honey. They are simple sugars that can be absorbed in the gut without any digestive process. Disaccharides are made up of two monosaccharides and are commonly found in fruits and vegetables, including sugar beets and sugarcane.

Polysaccharides are larger molecular structures of monosaccharides linked together. Disaccharides and polysaccharides must be broken down into monosaccharides before they can be absorbed. Excess glucose is stored in the liver as glycogen or in fat tissue of the body. Diets rich in complex carbohydrates (polysaccharides) and fiber are associated with reduced risk of cancer and cardiovascular diseases.

Proteins.   Proteins are large molecules made up of chains of amino acids that are broken down by the digestive process into their component units (1 gram protein yields 4 kcalories). There are 20 common amino acids in biological materials needed by the body. Humans lack the ability to synthesize at least nine amino acids, and it is therefore essential to obtain them in protein from animal sources or combinations of foods, such as legumes and cereals, in the diet. Young children and adolescents need protein for their growth spurts. Proteins function in the body as structural components of cells and tissues, enzymes which act as catalysts for chemical reactions, and hormones which act as chemical messengers. Lack of protein and calories in the diet is called Protein Energy Malnutrition (PEM).

TABLE 8.2   Essential Nutrients and Their Functions

Dietary components	Types	Functions
Carbohydrates (4 kcal/g)	Sugar and starches	Provides efficient source of energy; water soluble, easily transported and available in tissue fluids; should comprise 40–85% of energy intake
Proteins (4 kcal/g)	Essential amino acids	Provides amino acids and building material for all body cells, especially muscle and bone; should comprise 10–15% of energy uptake
Fats and oils (9 kcal/g)	Essential fatty acids (linoleic and linolenic acids)	Concentrated energy source; transports fat-soluble vitamins (A, D, E, and K); enhances flavor; should comprise 25% of diet, mostly nonanimal sources
Vitamins	Fat soluble, water soluble	Healthy body tissue
Minerals	Calcium, chloride, phosphorus, potassium, sodium,	Body fluid and electrolyte balance; bones and teeth
Trace elements	Chromium, cobalt, copper, fluoride, iodine, iron, manganese, selenium, zinc, and others	Essential for building healthy body tissue, blood, and hormones
Fiber	Vegetable matter	Food bulk and prevention of cancer
Water	—	Fluid and body tissue balance; conveyer of food and water-soluble vitamins

Source: Adapted from *Surgeon General's Report on Nutrition and Health, 1988.*

Fats and Oils.   Foods of animal and plant origin include a variety of substances known as fats and oils (lipids) that are soluble in organic solvents but not in water. Dietary fats are broken down in the gut for absorption into the body to provide energy and fatty acids needed for many physiological functions. They provide a concentrated form of energy (9 kcal per gram as compared to 4 for proteins and carbohydrates, and 7 for alcohol). They also provide essential fatty acids needed for production of hormones, cell membranes, and other substances. Linoleic and linolenic acids are essential fatty acids which cannot be synthesized in the body but can be retrieved from animal fats.

Fats or lipids stored in body fat tissue insulate and protect vital organs, insulate the body against heat loss, and provide energy during periods of reduced consumption or greater body need of energy in periods of growth, illness, or injury. Fats provide for production of bile acids needed for absorption of fat-soluble vitamins (A, D, E, and K).

Dietary fats consist of mixtures of saturated, monosaturated, and polyunsaturated fats, depending on their chemical structure. The degree of saturation of fatty acids is based on the number of double bonds in the side chains of molecules made up of carbon, hydrogen, and oxygen atoms. Fats from animal sources (meat, poultry, fish, and dairy products) are mostly saturated fats (i.e., contain no double bonds). Fats from plant sources such as sunflower, olive, or peanut oils are monounsaturated (i.e., contain side chains with one double bond) and are preferable to saturated or polyunsaturated fats. Whereas coconut and palm oils are high in saturated fats, fish are good sources of unsaturated fats.

Foods of animal origin contain high amounts of saturated fats such as triglycerides and cholesterol. Cholesterol, which can also be synthesized in the body, is needed for synthesis of sex hormones, vitamin D, and cell membranes. Excess dietary intake of triglycerides and cholesterol increases the risk for atherosclerosis, cardiovascular diseases, diabetes, and some forms of cancer.

Vitamins. Vitamins are organic compounds are essential in small amounts for specific functions of the body for health, growth, reproduction, and resistance to infection (Table 8.3). They differ in physical and chemical properties and in biological functions. Vitamins function in highly specialized metabolic processes. They cannot be synthesized in sufficient quantity by the body alone and must be obtained from the diet or from supplements.

Vitamins are classified based on solubility, either in fat or water. Fat-soluble vitamins (A, D, E, and K) are found in high concentrations in the fatty portions of food. Excretion of excess intake of this type of fat is minimal. Vitamin C and those of the B-complex group are water soluble and should be supplied in adequate amounts in the daily diet, as they are easily excreted.

Storage of vitamins in the body for the water-soluble vitamins is limited, so that regular sources are even more essential than for the fat-soluble vitamins, which are stored in body fat and the liver. Deficiencies of either in the diet leads to depletion of body stores, followed by nonspecific symptoms (fatigue, confusion, weakness, neuritis, and reduced resistance to infection) before classic deficiency can be recognized clinically. A deficiency condition of even one vitamin can jeopardize health.

Vitamins are present in natural foods, and an appropriate diet should supply most vitamin needs; however, since eating such a balanced diet is often problematic, food enrichment or supplementation is necessary, especially for vulnerable groups, e.g., children, the aged, adolescents, and institutionalized patients. Enrichment means replacing nutrients in foods to levels found in the natural product before processing. For example, enrichment of white flour should replace the 22 natural elements that are normally present in whole grains but are removed in processing, including B vitamins, vitamin D, calcium, and iron salts.

Minerals and Trace Elements. Minerals are distributed in a variety of foods, but they usually are present in limited amounts. Diets must contain a sufficiency and variety of foods to meet daily requirements. Eighteen known minerals are required for body maintenance and regulatory functions. Of these, recommended

TABLE 8.3   Essential Vitamins

Vitamins	Source and activity in body	Deficiency condition
Water-Soluble Vitamins	Absorbed in intestines, excreted in urine, so very large amounts are required to produce overdosage; body needs adequate daily intake or tissue depletion occurs within weeks or months; they are essential for enzymes to catalyze biochemical reaction in energy production, biosynthesis, and nervous system development and maintenance	
Vitamin B complex	Thiamine (B1), riboflavin, niacin, pyridoxine (B6), cobalamin (B12), and folic acid; sources are whole grain cereals, legumes, leafy vegetables, meat, and dairy foods	Loss of memory, mental confusion; occurs with chronic illness, alcoholism, dietary restriction
Thiamin, vitamin B1	Part of enzyme systems for release of energy from carbohydrates	Beriberi, anorexia, emotional lability, depression, fatigue, constipation; cardiomyopathy, cardiac failure, polyneuritis; Wernicke's encephalopathy; Korsakoff's psychosis
Riboflavin, vitamin B2	Enzymes for metabolism of protein and carbohydrates	Skin and mucous membranes dry, stomatitis, photophobia, blurred vision, polyneuritis
Pyridoxine, vitamin B6	Part of enzyme process in protein and carbohydrates	Irritablity, depression, muscle weakness, cardiomyopathy, liver damage, prevents neuropathy in isoniazid therapy for tuberculosis
Niacin, nicotinic acid	Maintains normal gastrointestinal and nervous system	Pellagra, gastronintestinal, skin, and neurological changes, depression, psychosis, neuropathy
Folic acid, folate	Red blood cell formation	Megaloblastic anemia of pregnancy; neural tube defects (NTDs, spina bifida, and anencephaly)
Vitamin B12, cobalamin	Source only in foods of animal origin; essential for red blood cell formation	Macrocytic anemia; mental retardation
Vitamin C, ascorbic acid	Source fruits and vegetables; needed to form, maintain intercellular substances	Scurvy, poor bone and cartilage anemia, stunting, infections, bleeding, weakness, death
Fat-Soluble Vitamins	Found mainly in fat component of food; absorbed, transported along with fat; requires bile and dietary fats; stored in body fats and takes longer to deplete than water-soluble vitamins	
Vitamin A, retinol	Found in yellow vegetables, essential for epithelial cells of mucous membranes; regulation of vision in dim light	Night blindness, loss of color vision, dryness, corneal ulceration and scarring, blindness; poor bone and tooth fomation; susceptability to infection and survival

(continues)

TABLE 8.3   Essential Vitamins (*continued*)

Vitamins	Source and activity in body	Deficiency condition
Vitamin D calciferol	Produced in skin by exposure to sun, found in enriched foods, milk pro-ucts; enhances calcium and phosphorus utilization and bone growth	Rickets, stunting, soft bones, bowed legs, carious teeth; osteomalacia, soft bones, fractures, muscle pain in adults; contributes to osteoporosis
Vitamin E, tocopherpl	Found in leafy green vegetables, legumes, nuts; protects fat from oxidation (antioxidant) and red cell breakdown	Low birth weight; hemolysis of the newborn; degenerative disorders; probable role in heart disease
Vitamin K, phylloqui-none (K1) menanquinone (K2)	Spinach, cabbage, cauliflower; formation of prothrombin	Hemorrhagic disease of newborn,

Source: Adapted from Canada's Food Guide Handbook; Passmore and Eastwood, 1986, and *The Surgeon General's Report on Nutrition and Health, 1988.*

dietary allowances (RDA) have been established for seven: calcium, iodine, iron, magnesium, phosphorus, selenium, and zinc. Other active minerals in the body include: sodium, chloride, potassium, chromium, cobalt, copper, fluoride, manganese, molybdenum, sulfur, and vanadium. Sodium, potassium, chloride, and calcium are particularly crucial for electrolyte balance in the blood and body tissues. Essential trace minerals also include boron, silicon, nickel, and arsenic for optimal growth and membrane function. The body needs a small but continuing intake of these elements for its structure and function. If metabolic needs are not met, a deficiency ensues. Deficiency disorders vary with the mineral element involved, the duration and extent of dietary intake deficiency, and depletion of body stores (Table 8.4).

GROWTH

Growth is not a steady progression, but a process during which nutrition requirements are determined by a genetic timetable, affected by nutritional intake and health status. Optimal growth occurs only if the organs and tissues receive the nutrients needed for synthesis of proteins and other molecules. Insufficient energy and protein (Protein Energy Malnutrition or PEM) are the commonest causes of failure to grow, but deficiencies of essential minerals and vitamins also adversely affect growth. Deficiencies of iodine slow thyroid hormone production and cause developmental effects. Lack of vitamins A or D or minerals such as iron, iodine, calcium, and phosphorus adversely affect growth and development of epithelial cells, bone, and red blood cells.

TABLE 8.4   Essential Minerals

Mineral	Activity in body	Effect of deficiency	Food source
Calcium	Builds and maintains bone stucture and teeth; muscle and cardiac function, blood coagulation, neuromuscular irritability	Poor bone and teeth formation; rickets in children, osteoporosis in elderly	Fortified milk, hard cheese, egg yolk, cabbage, clams, cauliflower, soybeans, spinach
Iron	Constituent of hemoglobin, muscle, and bone; carries oxygen in red blood cells	Iron deficiency anemia, fatigue, poor linear growth in infants, psychomotor deficiency affecting school and work performance	Liver, red meat, turkey, legumes, egg yolk, peaches, apples, raisins, prunes, molasses
Iodine	Constituent of thyroxine, normal thyroid function needed for growth and mental development	Stunting, retardation, cretinism	Iodized salt, seafood
Phosphorus	Builds and maintains bones, teeth, cells, body fluid	Poor teeth, stunting, rickets	Milk, cheese, egg yolk, meat, legumes, cereals, nuts, vegetables
Chloride	Needed to maintain acid-base and osmotic balance in body fluids	Lost in diarrhea, with dehydration, shock, and death in children or elderly from diarrheal diseases	Milk, salt, fish, cheese
Copper	Needed for the central nervous system, hemoglobin formation	Anemia, liver function, metabolism of ascorbic acid	Seafood, nuts, whole grain cereals, liver, meat, legumes, vegetables
Fluoride	Strengthens tooth enamel, bone formation	Dental caries and osteoporosis in the elderly	Fluoridated drinking water, fluoride rinses
Magnesium	Constituent of bones and and teeth, enzymes; needed for cardiac and neurological functions	Cardiac arrhythmia, nervous irritability	Same as diet for calcium and phosphorus
Sodium	Intracellular and extracellular fluid balance, muscle and nerve irritability	Fluid loss, circulatory collapse	Table salt, meat, milk
Potassium	Electrolyte imbalance, cardiac arrhythmia	Fluid loss, circulatory collapes, and muscle irritability	Vegetables, cereals, fruits, bananas, melons
Selenium	Antioxidant	Changes in biochemical systems cardiomyopathy, cardiogenesis, liver damage	Meat, seafood, cereals, grains

Source: Adapted from *The Surgeon General's Report on Nutrition, 1988.*

MEASURING BODY MASS

The body mass index (BMI) is a standard method of measuring body size. It encompasses in one number, summarizing nutrition status height and weight. BMI is useful clinically for the individual patient and as a description of nutritional status of a community based on survey data. It is calculated as follows:

$$BMI = \text{body weight in kilograms}/(\text{height in meters})^2$$

or

$$BMI = \text{weight in pounds} \times 703/(\text{height in inches})^2.$$

Obesity is defined for males and females as a BMI above 30; overweight, BMI = 25–30; *undernutrition* is defined as a BMI below 18.5.

Body weight is taken as usual weight and expressed as a percentage of desirable. The BMI is a convenient measure. Ranges of BMI shown in Table 8.5 are arbitrary, but these provide a useful categorization which may need to be augmented by clinical and other anthropometric measures. The National Heart, Lung, and Blood Institute revised its standards in 1998, defining a BMI of over 25 as overweight and over 30 as obesity. This defines 55% of the U.S. population as overweight or obese, and at increased risk for hypertension, diabetes, and heart disease. These new standards bring the U.S. standards into conformity with those issued by the WHO. Undernutrition is graded with a BMI of 18.4 or under.

RECOMMENDED DIETARY ALLOWANCES

In 1941, the Committee of Food Nutrition of the National Research Council in the United States developed recommended daily allowances (RDAs, later renamed daily dietary allowances) in response to a request from the U.S. Council of National Defense, which was concerned over possible food shortages and nutritional ill effects on the health of the population during World War II.

The RDAs represent the levels of essential nutrients needed to adequately meet the nutritional needs of virtually all healthy persons. They are used for planning for national emergencies and for the needs of institutionalized or socially deprived persons. The RDAs are modified in light of expanding knowledge of nutrition and have become a guideline for proper nutrition for the health-conscious individual.

TABLE 8.5 Body Mass Index and Grades of Chronic Protein-Energy Malnutrition and Obesity

Grade	Body mass index
Obesity	>30
Overweight	25–30
Normal	18.5–25
Protein-energy malnutrition*	
I	17–18.4
II	16–16.9
III	<16

Source: National Heart Lung and Blood Institue, 1998
*Shils *et al.,* 1994.

TABLE 8.6   Recommended Dietary Allowances (RDAs) by Age and Sex,[a] United States, 1993

Group, age (years)	Weight (kg)	Height (cm)	Average energy needs (kcal)	Protein (g)	Fat-soluable vitamins (A, D, and E)	Water-soluble vitamins (C, thiamin, riboflavin, niacin, B6, folate, B12)	Minerals (calcium, phosphorus, iron, magnesium, zinc, iodine)
Infants							
0–0.5	6	60	kg × 115 (95–145)	Wt × 2.2	A 410 μg D 10 μg E 4 μg	Vit C 35 mg Thiam 0.4 mg Ribofl 0.5 mg Niac 7 mg B6 0.5 mg Folic 38 μg B12 0.1 μg	Ca 450 mg P 300 mg Mg 60 mg Fe 13 mg Zn 4 mg I 45 μg
0.5–1.0	9	71	kg × 105 (80–135)	Wt × 2.0			
Children							
1–3	13	90	1300 (900–1800)	23	A 500 μg D 10 μg E 7 μg	Vit C 45 mg Thiam 0.4 mg Ribofl 1.2 Niac 14 mg B6 1.3 mg Folic 120 μg B12 3 μg	Ca 800 mg P 800 mg Mg 200 mg Fe 10 mg Zn 10 mg I 80 μg
4–6	20	112	1700 (100–2300)	30			
7–10	28	132	2400 (1650–3300)	34			

Males							
11–14	45	157	2700 (2000–3700)	45	A 1000 µg	Vit C 55 mg	Ca 1000 mg
15–18	66	176	2800 (2100–3900)	56	D 8 µg	Thiam 1.1 mg	P 1600 mg
19–22	70	177	2900 (2500–3300)	56	E 10 µg	Ribofl 1.0 mg	Mg 300 mg
23–50	70	178	2700 (2300–3100)	56		Niac 14 mg	Fe 15 m
51–75	70	178	2400 (2000–2800)	56		B6 2 mg	Zn 10 mg
76+	70	178	2050 (1650–2450)	—		Folic 400 µg	I 135 µg
						B12 3 µg	
Females							
11–14	46	157	2200 (1500–3000)	46	A 1000 µg	Vit C 55 mg	Ca 1200–800
15–18	55	163	2100 (2100–3000)	46	D 5–10 µg	Thiam 0.5 mg	P 1200–800
19–22	55	163	2100 (1700–2500)	44	E 8 µg	Ribofl 1.2 mg	Mg 300 mg
23–50	55	163	2000 (1600–2400)	44		Niac 14 mg	Fe 18 m
51–75	55	163	1800 (1400–2200)	44		B6 2 mg	Zn 15 mg
76+	55	163	1600 (1200–2000)	—		Folic 400 µg	I 150 µg
						B12 3 µg	
Pregnant			+300	+20	A + 200– 400 µg	Vit C +30 mg	Ca +400 mg
Lactating			+500	+30	D + 1+5 µg	Thiam +0.4 mg	P +400 mg
					E + 2-3 µg	Ribofl +0.4 mg	Mg +150 mg
						Niac +4 mg	Fe +30–60 mg
						B6 +0.5 mg	Zn 15– 10 mg
						Folic 400 µg	I 125°50 µg
						B12 +1µg	

[a]Average requirements given for vitamins and minerals; ranges given for energy requirements and for pregnant and lactating women only.

Source: From the *Surgeon-General's Report on Nutrition and Health, 1988*; and National Academy of Science, 1980, Recommended Dietary Allowances.

They serve to remind us of what should be included in our regular diet, and assist in preventing excessive intake of essential nutritional elements that could be potentially harmful if taken in exaggerated amounts. The U.S. National Academy of Sciences (NAS) revised the RDAs in 1989 and again in 1997; they are now classified as Dietary Reference Intakes (DRIs). DRIs have been established for 13 vitamins along with 15 minerals and other elements, as well as protein (Nutrition Health Reports website, 1999).

The DRIs serve as nutritional guidelines and are useful in standardizing food manufacturing package labelling and education practices. They have become an essential part of both clinical nutrition and public health nutrition standards internationally. The RDAs are average figures, and, although age and sex specific, they do not take into account periods of illness, injury, or physical stress, nor the elderly. There are, as in many other areas of public health, different points of view on the importance of DRIs. Intake of enriched foods and additioinal vitamins or minerals as supplements are thought by some to have the potential for excessive intake if multiple sources of intake are used.

The NAS sets DRIs, which are used by government, industry, researchers, and clinicians for individual and public health purposes. As an example, NAS recently recommended folic acid and vitamin B supplements for routine use to prevent birth defects and micronutrient deficiency conditions among vulnerable groups in the U.S. population.

Malnutrition is a pathological state caused by a relative or absolute deficiency or excess of one or more essential nutrients, the clinical results being detectable by physical examination or biochemical, anthropometric, or physiological tests. The types of malnutrition identified by Derick Jelliffe are underweight/starvation, overweight/obesity, specific deficiency, and imbalance.

DISORDERS OF UNDERNUTRITION

Undernutrition includes at least 25 different deficiency diseases resulting from lack of one of the essential nutrients, proteins, vitamins, fats, or minerals (Table 8.7). A person is more likely to suffer from multiple deficiencies as opposed to deficiency in only one nutrient. Deficiency disorders may be subclinical and not show characteristic clinical features.

The WHO and UNICEF estimate that over 200 million children in developing countries are malnourished and that malnutrition contributes to more than half of the nearly 12 million deaths of children under age 5 in developing countries each year. Malnourished children suffer loss of physical and mental capacities. They are subject to illness and complications or death from diseases that would otherwise be less dramatic. Much of this is due to early cessation of breast-feeding, lack of adequate food supplies, ignorance, and failure of public health systems to fortify foods or provide supplements to vulnerable groups. But the main cause is political chaos or repression, with indifference to poverty and its consequences (UNICEF, 1998).

TABLE 8.7   Terms and Syndromes of Undernutrition

Term syndrome	Condition
Low birth weight (LBW)	Birth weight of 2500 g or less; %LBW is a good indicator of nutrition status of a population
Marasmus	A wasting disease, due to lack of energy and protein; usually occurs between 6 and 18 months with a combination of maternal neglect and non-breast-feeding; occurs in urban slums; infant looks like a wizened old person, skin and bones, with recurrent infections; causes permannent impairment of brain development; needs warmth, attention, adequate nutrition, and maternal education, prevention of infectious diseases
Kwashiorkor ("deposed child")	Calorie, vitamin, and protein deficiency commonly affects children aged 2– 4 deposed from the breast; when the next child is born and insufficient food, triggered by infection (e.g., measles); depigmented flaky skin and hair; poor appetite, low serum albumin, edema or swelling of abdomen, enlarged liver in a child who is thin in trunk, arms, and legs
Protein-energy malnutrition (PEM)	Person does not get enough food, calories, or proteins needed for normal growth and energy or normal human activities; low birth weight and height for age; fatigue, poor work or school capacity
Specific deficiency conditions	Vitamin A deficiency—xerophthalmia
	Vitamin B deficiency—pellagra
	Folic acid deficiency—anemia and congenital birth defects
	Vitamin C deficiency—scurvy
	Vitamin D deficiency—rickets
	Vitamin K deficiency—hemorrhagic disease of newborn
	Iron deficiency anemia (IDA)
	Iodine deficiency disorders (IDD)
Low weight for age (wasting)	Indicator of current calorie and protein undernutrition (chronic or acute or both)
Low height for age (stunting)	Low height for age compared to reference population indicates chronic undernutrition at some time during the growth years, but may not be currently underdeveloped
Low weight for height	Thin person whose weight is low for his height indicating current undernutrition
Infection–malnutrition cycle	Interaction of nutrition status, susceptibility to infection with effects of infectious disease on nutrition status

Source: Williams, C. D., Baumslag, N., and Jelliff, D. 1994. *Mother and Child Health: Delivering the Services*. Third Edition. New York: Oxford University Press.

Failure of the political systems to sustain or promote conditions to ensure adequacy of food supply or distribution systems may cause widespread undernutrition or famine. Chaos in a political system produces the conditions in which food production, storage, and marketing break down and prove inadequate for growing populations. Hunger is particularly common in populations living in sub-Saharan Africa, the Indian subcontinent, and southeast Asia, with significant pockets of undernourished populations in all communities throughout theworld.

Underweight: Protein-Energy Malnutrition. The state of being underweight, or protein-energy malnutrition (PEM), is the result of inadequate consumption of nutrients (measured in calories or joules) to meet the basic needs of the human body. The main characteristics are weight loss and wasting of body fat and muscle mass, low weight for height, and low height for age. Severe forms include a spectrum of failure-to-thrive, marasmus, or kwashiorkor in children and starvation in adults. Weight loss in adults may be the result of loss of appetite, fasting, anorexia nervosa, persistent vomiting, inability to swallow, incomplete absorption, and increased base metabolic rate such as in prolonged fever, hyperthyroidism, cancer, diabetes mellitus, or other medical conditions. Chronic underweight in developed countries can occur in high risk population groups because of poverty, illness, or ignorance of appropriate diets.

In young children, PEM may be due to an infection and/or a lack of food. In developing countries, undernutrition due to poverty or failure of the food supply is the single most prevalent public health problem, particularly in infants and young children. A malnourished child is more vulnerable to infection, with lowered resistance and reduced immunity. A child with an infection then becomes even more undernourished and may suffer long-term growth retardation as a result.

In starvation there is a compensatory reduction in metabolic rate, slowed and weak pulse, lowered blood pressure, loss of body fat, muscle wasting, decreased muscle tone, loss of skin elasticity, mental dullness, and easy fatigue. The symptoms of specific deficiencies in vitamins and minerals are likely to be minimal. Recovery of weight loss after the start of feeding is slower in adults than in children. Starvation is more likely to affect children, women, and the elderly, with infants and toddlers especially vulnerable.

Failure-to-Thrive. Failure-to-thrive, or growth retardation, describes the failure of growth in keeping with age compared to a standard growth pattern. This is particularly common in inadequately fed babies in developing countries. Failure to recover growth following an illness such as diarrhea, acute respiratory infection, or measles is also common.

Marasmus. Marasmus is a severe failure-to-thrive condition due to marked deficiency of caloric and protein intake and general deprivation. It occurs in the 3–9 month age group, commonly due to early weaning with resultant starvation of the infant. The child appears wasted and irritable with depletion of subcutaneous fat and muscle tissue which are burned up to maintain blood glucose.

Kwashiorkor. Kwashiorkor is a severe form of undernutrition, primarily of protein intake. It occurs in infants and children up to about age 6. This widespread protein deficiency syndrome occurs in young children who have been weaned, often after the birth of a new child, to a diet high in carbohydrates and low in protein, and the condition is often aggravated by an infectious disease. Kwashiorkor is characterized by retarded growth and development, apathy, gastrointestinal irritability, depigmentation of hair, edema resulting in a swollen abdomen, fatty infiltration of the liver, and dry skin. Treatment consists of establishing adequate dietary intake and balance. Untreated, the mortality of this condition is high.

VITAMIN AND MINERAL DEFICIENCY CONDITIONS

Deficiency conditions for one or more of the essential nutrients are an important in public health. In developing countries incidence varies according to urban/rural residence and social class, but among the majority of rural and urban poor they are major factors in excess morbidity and mortality. In developed countries, the predominant nutritional problem is obesity, but significant pockets of deficiencies exist among special population groups.

Even in developed countries, specific vitamin and mineral deficiency states can be widespread. The Canadian National Nutrition Survey in the early 1970s showed significant vitamin and mineral deficiencies in different parts of the country, especially in native peoples, teenagers, women, and the elderly. Many developed countries have taken steps to reduce these problems through improved standards of living and the implementation of policies of vitamin and mineral enrichment of basic foods.

The Report of the President on Health in Russia in 1992 indicates there is widespread iodine deficiency disease, iron deficiency anemia, and even protein-energy malnutrition. The situation in Russia is made worse by the current economic crisis, with large segments of the population lacking funds for basic nutrition. Consequently, there exists a danger of frank malnutrition among vulnerable groups.

Vulnerable population groups include pregnant and lactating women, infants and toddlers, and the elderly, especially those in poverty groups. Other groups at risk include alcoholics, persons with chronic or frequent infection, such as AIDS patients and persons with chronic diseases, or those who restrict themselves to certain foods, such as vegetarians. Vegetarianism is compatible with good health, provided it is done with good nutritional advice.

Vitamin A Deficiency

Vitamin A is essential for normal vision and ocular function, because of its primary role in forming visual pigment. The dietary sources of vitamin A are animal products, such as egg yolk, liver, dairy products, and breast milk, and plants containing carotenoids, such as dark green leafy vegetables, yellow and reddish fruits, and red palm oil. Symptoms of vitamin A deficiency include growth retardation, alterations in the differentiation and morphology of epithelial and mesenchymal tissues, and impaired vision.

Vitamin A deficiency decreases resistance to infections and increases the severity, complications, and risk of death from various diseases. It also results in night blindness and xerophthalmia (drying out of the eyes, leading to scarring). This nutritional deficiency is widespread among children in developing countries, and the problem is exacerbated by a tendency by some to withhold vegetables from children for cultural or other reasons. Vitamin A deficiency is associated with increased mortality from measles, and high doses of vitamin A should be given in treatment and prophylaxis during measles outbreaks.

The WHO and UNICEF estimate that vitamin A deficiency affects very large numbers of children:

Deficient intake (subclinical)	562 million
Deficient intake (clinical susceptibility to infection)	231 million
Night blindness	13.5 million
Xerophthalmia	3.1 million
Severe eye damage/blindness	0.5 million

Preventive measures for populations identified as experiencing early or marginal vitamin A deficiency can markedly reduce the risk or mortality from intestinal and respiratory infections and measles. Methods for improving vitamin A status include periodic distribution of large-dose capsules appropriate for age, fortification of readily consumed dietary staples, and increased intake of vitamin A-rich foods. Adding vitamin A, along with vitamin D, to margarine, milk, and dairy products was practiced in the United States, Canada, and the United Kingdom during World War II. The practice is mandatory in Canada and nearly universal in the United States.

Cumulative evidence since the mid-1980s has reinforced the importance of vitamin A in combatting infections, so that food fortification is recommended especially in developing countries or for populations subject to high infection rates, such as HIV positive persons. Routine vitamin A supplementation for children is now recommended by WHO in conjunction with immunization programs (EPI plus, see chapter 4). Massive doses of vitamin A are used to treat children with measles with great benefit in reducing mortality and complication rates.

Fortification of sugar with vitamin A has been implemented in a number of countries in South and Central America. The Philippines fortified margarine with vitamin A during 1998. Indonesia is using a variety of techniques to increase supplementation and fortification to reduce deficiency conditions. Supplementation policies of giving vitamin A to children and postpartum mothers was carried out in 78 countries in 1996, and this movement is spreading with WHO and UNICEF promotion.

Vitamin A toxicity may occur when more than three times the RDA is consumed, usually from medication overdose, especially among pregnant women, alcoholics, or persons with chronic liver conditions. This can result in hyperkeratinosis (i.e., orange skin), nausea, vomiting, birth defects, and neurological, gastrointestinal, and dermatologic symptoms.

Vitamin D Deficiency (Rickets and Osteomalacia)

Rickets from vitamin D deficiency was a common disorder in many parts of the developed world well into the twentieth century. It constitutes one of the important diseases of infancy because of its serious complications, including disorders of long bone growth, bowing of legs, pelvic deformity, and, in extreme forms in infants, tetany and convulsions. The vitamin D content of human milk is extremely low, and the exclusively breast-fed infant not adequately exposed to sunlight may develop clinical rickets. In adults, malabsorption or poor dietary intake of vitamin D can result in osteomalacia and osteoperosis with fragile bones and frequent fractures. Among the elderly, who are closed in especially during winter months, and among homebound or institutionalized patients, vitamin D deficiency is common.

There are few natural sources of vitamin D, but deficiency is preventable by exposing the skin to the ultraviolet rays of the sun. Prevention became common through the practice of giving children cod-liver oil, a successful antirachitic measure, until replaced by use of vitamin D supplements for infants. In colder climates or in foggy locations where exposure to the sun may be limited, rickets was common. However, seasonal deficiency of vitamin D can occur even in locations with ample sunlight.

Rickets remained widespread until fortification of milk was introduced in the 1940s during World War II in Britain and North America. Rickets prevalence in the United Kingdom, especially in the industrial cities in northern England and Scotland, declined dramatically. The 1971 Canadian national nutrition survey found significant vitamin D deficiencies in certain age, sex, ethnic, and geographic population groups. Although antirachitic procedures (cod-liver oil and adding vitamin D to milk) were routine during the 1940s, the practice waned in the 1950s and 1960s. Hospitalizations for rickets in Montreal followed the abandonment of vitamin D milk enrichment.

Addition of vitamin D to milk, nearly universal in the United States, was made mandatory in Canada in 1979 and rickets disappeared. No incidents of vitamin D toxicity were reported. Vitamin D toxicity, due to human error in formula preparation or from multiple source supplementation, can cause failure-to-thrive, nausea, vomiting, and weakness.

Even in a sunny climate, there are seasonal variations in vitamin D availability, and social customs of keeping infants overly wrapped and out of the sun may cause rickets in some population groups. Studies of vitamin D levels among institutionalized elderly found low levels, and among elderly in the community vitamin D levels were reported to be low in the winter months.

Prevention of vitamin D deficiency conditions should include routine supplements of vitamin A and D for infants from 1 to 12 months (and up to 24 months). Fortification of baby formulas and cereals is common. Standard textbooks of pediatrics and of public health take the same position with respect to rickets (vitamin D deficiency): vitamin D deficiency is a problem in childhood and in the elderly that cannot be addressed by providing vitamins to high risk groups, so that milk fortification is justified. Fortification of milk products is one of the most important public health nutrition measures used. Failure to promote its application widely internationally has been an important lapse of the international health community.

Vitamin C Deficiency (Scurvy)

Scurvy is a deficiency disease due to lack of vitamin C (ascorbic acid) in the diet. It was common among seamen (see Chapter 1) and others deprived of fresh fruits and vegetables. Vitamin C deficiency causes skin lesions, weakness, fatigue, weight loss, muscle pains, susceptibility to infection, hemorrhages, debility, and even death. It can occur at any age due to inadequate diet. Infantile scurvy formerly appeared in bottle-fed infants, but with advent of vitamin-fortified infant formulas this has become rare in industrialized countries. Infants should have a source of vitamin C from the first month of life, such as in orange juice. During infections, increased vitamin C intake is recommended. Vitamin C is widely recommended as a preventive measure to reduce cholesterol levels and prevent cancer, but this has not been conclusively established.

Vitamin K Deficiency (Hemorrhagic Disease of the Newborn)

A deficiency of vitamin K occurs in the newborn, causing impaired production of prothrombin, a key blood clotting factor. Lack of prothrombin shortly after birth can cause hemorrhagic disease of the newborn (HDN; see Chapter 6). Secondary HDN can occur weeks later. To prevent the deficiency, a single injection of vitamin K is administered at birth and vitamin K is added to fortified baby formulas and cereals.

Vitamin B Deficiencies

Niacin Deficiency (Pellagra).   Niacin (nicotinic acid) is essential for specific oxidation–reduction reactions in the body. A deficiency of niacin causes diarrhea, dermatitis, and dementia. Pellagra was established as a nutritional deficiency condition (and not infectious) in investigation of the condition in orphanages and hospitals in the southern United States in 1917–1922 (see Chapter 1). Pellagra is prevented with adequate dietary intakes of niacin or niacin substitutes contained in vitamin-enriched bread, liver, meat, fish, poultry, potatoes, green vegetables, peanuts, and cereals. Niacin also exerts beneficial effects by reducing blood lipids and raising high density lipoproteins (HDL), slowing atherosclerosis and coronary artery lesions.

In 1998, the National Science Foundation issued a recommendation for multiple vitamin supplements for all adults.

Thiamine (B1) Deficiency (Beriberi).   Thiamine deficiency causes derangement of carbohydrate metabolism. "Dry beriberi" results in neurological symptoms and death. "Wet beriberi" is characterized by cardiac symptoms and disorders, including heart failure and death. This disease was common among prisoners of war in Japanese camps during World War II due to diets mainly of polished rice. Prevention requires a diet containing liver, glandular organs, yeast, wheat germ, whole wheat or cereals, unpolished rice, milk, legumes, soybeans, and peanuts.

Alcoholic (Korsakoff's Syndrome) or nutritional dementia (Wernicke's encephalopathy) may be a combination of B complex deficiency and constitute important public health problems. Wernicke's encephalopathy together with Korsakoff's syndrome are termed "cerebral beriberi." Enrichment of bread, breakfast cereals, and other flour-based products is practiced in Canada and the United Kingdom on a mandatory basis, and widely done in the United States, where it is as mandatory when the term "enriched" is used.

Folate Deficiency. Folic acid is required for normal blood formation and neurological health. Its deficiency is common among low socioeconomic groups, especially during pregnancy, infancy, and childhood. Deficiency is common in alcoholics, due to malnutrition or impaired absorption. Alcoholics with good diets are less likely to develop folate deficiency than those with poor eating habits. Folate deficiency constitutes a major health risk for alcoholics to develop severe neurological damage in the spinal cord, or in the optic or peripheral nerves.

Folic acid deficiency has been demonstrated to be a cause of neural tube defects (NTDs), an important congenital disorder (see Chapter 6). This condition ranges in severity from anencephaly to disabling defects of the spinal cord and is largely preventable by prenatal supplements of folic acid or food fortification as introduced as mandatory in the United States in 1998.

Folic acid supplementation may be important in prevention of coronary heart disease by lowering homocysteine levels, but this requires further substantiation (see Chapter 5).

Vitamin B12 Deficiency. Vitamin B12 deficiency causes enlargement of red blood cells with poor hemoglobin content (macrocytic anemia). Vitamin B12 deficiency may be due to malabsorption in the stomach (pernicious anemia) or from long-term deficiency of vitamin B12 intake from vegetarian diets which exclude eggs and dairy products. Pregnant women and lactating mothers, especially if vegetarian, need vitamin B12 supplements. Vitamin B12 deficiency can result in degeneration of the spinal cord, optic nerves, cerebral tissue, and peripheral nerves. Prevention is by promotion of healthy nutrition with foods rich in B complex vitamins and supplementation for infants in fortified cereals.

Iron Deficiency Anemia

Iron deficiency anemia (IDA) is the commonest nutritional deficiency condition in the world, widespread in both developing and developed countries. Iron plays a critical role in the transport of oxygen in human blood. Iron exists in two major forms in food. The first is heme iron, which is found only in animal sources. It is readily available, as absorption is not influenced by other constituents of the diet. The second type is inorganic iron, whose absorption is strongly influenced by factors present in foods ingested at the same time. The composition of a meal can influence the amount of iron that is absorbed. Consumption of foods containing heme iron will improve the absorption of nonheme iron. Vitamin C enhances the absorption of nonheme iron, but substances like tannin from tea by combining with the iron in the intestine to inhibit its absorption.

Iron deficiency anemia as a public health problem has received considerable attention since the 1980s by international organizations such as UNICEF and the World Health Organization. The issue is especially vital for developing countries, but even in developed countries iron enrichment of staple foods is an important public health measure. Iron deficiency anemia has been extensively documented, particularly in infancy, among pregnant women, in the adolescent and adult populations, as well as among the elderly. A 1985 study of presumably healthy adult blood donors in Jerusalem found 21% with low serum ferritin levels and some 30% with low ascorbic acid levels.

In developing countries, two-thirds of children and women of childbearing age are estimated to suffer from iron deficiency; one-third or more of them have the more severe form of the disorder, anemia. Symptoms include listlessness and fatigue. Low levels of iron are associated with often irreversible damage to brain development. Studies of the effects of iron deficiency anemia have shown diminished psychomotor performance suggestive of low levels of brain dysfunction or damage resulting from iron deficiency, even in the absence of anemia. The intellectual development of children and the physical activity of both adults and children are impaired in those suffering from anemia. There is a direct correlation between low hemoglobin levels and the prevalence of diarrheal and respiratory diseases which result from an impaired immune system. Lack of recovery in many children even after iron supplementation underscores the importance of preventing iron deficiency through dietary manipulations and health education.

The International Society for Prevention of Iron Deficiency Anemia promotes preventive approaches, such as iron fortification of basic foods (bread, sugar, salt) and routine supplementation for infants and pregnant women, to prevent what it defines as the most widespread nutritional deficiency in both developed and developing countries. Globally, nearly 2 billion people are anemic and 3.6 billion are iron deficient.

Iodine Deficiency Diseases

Iodine is an essential element in nutrition. Insufficient iodine in natural sources causes clinical or subclinical thyroid disorders. Deficient or irregular supply of iodine damages fetal development and produces fetal hypothyroidism, with low levels circulating thyroid hormones and urinary iodine, causing varying degrees of brain damage in infants, including cretinism. Excess prevalence of goiter, or enlargement of the thyroid gland, indicating suboptimal thyroid function, has been documented in large areas of the world where there are low levels of ground and surface water iodine.

Iodine deficiency affects an estimated 1.5 billion persons, with 844 million having goiter, 49.5 million cretinoids, and 16.5 million cretins (WHO, 1998). Prevention of iodine deficiency, pioneered by David Marine and David Cowie following a series of studies from 1910 onwards, leading to iodization of commercial table salt in 1924 (Morton's Iodized Salt, see Chapter 1). By 1930, most of the salt consumed in the United States was iodized, and goiter had largely disappeared even in previously endemic areas.

Iodination of salt as a preventive measure has become standard public health practice in many countries, since World War I. It has been compulsory in Canada since 1979, and widespread in western Europe, although not universal and not always at effective levels.

In the 1980s the World Health Organization expressed growing concern with the widespread nature of iodine deficiency disorders in large areas of the world, affecting an estimated 2.3 billion persons especially in China, the former Soviet Union, southeast Asia, and many developing countries. In 1986, the World Health Assembly called on all nations to introduce iodination of salt or other appropriate technology to reduce this silent pandemic.

Australian scientist and public health advocate Basil Hetzel demonstrated that insufficient iodine during fetal development adversely affected brain development and promoted WHO and other international organizational response to the issue of IDD.

Iodine deficiency was described as follows in a 1993 editorial in *The New England Journal of Medicine:* "The most important effects of iodine deficiency are on the developing central nervous system, and they form a continuum from mild intellectual impairment to full blown cretinism." Prevention of iodine deficiency is "best achieved through the iodization of salt on a national scale at a level of 1 part iodine to 10,000–20,000 parts of salt" (Last, J. M., 1986). A global effort for prevention of IDD is underway.

UNICEF, the International Council for Control of Iodine Deficiency Disorders, the European Thyroid Association, Kiwanis International, some countries (e.g., Canada), and the World Bank have called for national and international action to control this widespread public health problem. The World Summit for Children called for universal iodizing of salt with a target of 95% iodination in each country by 1995. By 1994, 94 countries had national plans for iodizing salt, with 58 countries, including almost 60% of the world's children, on schedule. Progress is being made, but inertia and complacency are still barriers to achieving this goal.

Osteoporosis

Osteoporosis is an important chronic condition, discussed in Chapter 5, with serious health consequences, particularly among older women in terms of fractures of the hip, the spine, and forearm. Hip fracture mortality is between 12 and 20%, and many survivors are institutionalized as a result of complications. Osteoporosis is preventable to a large degree by adequate calcium and vitamin D supplementation and exposure to exercise and sunlight from the early years of life. Fluoridation of water may have beneficial effects. Exercise helps to increase bone mass. Because of the increasing population of elderly persons, especially women, osteoporosis and its complications are important problems for preventive, curative, and rehabilitative services as well as an issue for health promotion in terms of dietary and other self care and preventive measures in individual medical practice as well as in public health. Food enrichment with calcium and vitamin D play an important role in reducing the severity of this condition currently affecting some 28 million Americans (National Osteoporosis Foundation, 1999).

EATING DISORDERS

Eating disorders are an important health risk, primarily for teenaged girls, and an occupation-related health risk associated with sports, ballet, and modeling. These disorders can be communicated between people by group pressure of fashion, precedent, and close contact among teenaged girls and young women. Fashions in a society have great influence on vulnerable adolescents who can enter a cycle of self-denial or purgation that may be extremely destructive and even fatal.

Anorexia Nervosa.　Anorexia nervosa is a self-imposed severe dietary restriction that may be chronic or acute. This condition is most common among teenage girls and occupational groups including models and ballet dancers. Eating disorders reportedly affect some 5% of female college students in the United States. Anorexia is often associated with laxative and diuretic abuse or excessive exercising. Specific features common to anorexia nervosa are abnormal sensitivity to being fat and fear of losing control over the amount of food eaten; severe restriction of food intake and refusal to accept food; marked weight loss; cessation of menstruation; damage to dentition and chronic tooth pain; and more serious consequences including liver and heart muscle damage. The psychological features of rigidity, perfectionism, and fear of obesity preceding the condition are usually resistant to treatment. Hospitalization is commonly required with strict regimens of antidepressant and behavior therapy. Anorexia nervosa may lead to death by self-starvation, or suicide.

Bulimia or Binge Eating.　A binge-eater has a compulsion to eat large quantities of food within a short period of time, usually 2 hours or less, with deliberate vomiting to expel the food to lose or maintain their weight. Many features are common to both bulimia and anorexia nervosa. The physical effects of this form of eating disorder can be severe (such as esophageal damage from repeated vomiting) but usually less so than anorexia nervosa. The presence of depression may require antidepressant therapy supported by psychotherapy techniques or behavior modification therapy and hospitalization. Like anorexia, this disorder is most common in societies that promote images of beauty as thinness through advertisement and related social pressures affecting the psychologically vulnerable with poor self-images.

DISEASES OF OVERNUTRITION

Though media attention tends to focus on the extreme forms of malnutrition, the more common nutritional health concerns in industrialized countries are those relating to specific deficiencies and dietary imbalances. In the United States, *The Surgeon General's Report on Nutrition and Health, 1988* addressed the problems of nutrition in a developed country in the 1980s, with an emphasis on a type of malnutrition which could be described as overnutrition. The report also addressed the issues of specific vitamin and mineral deficiency states and food fortification. Diseases of overnutrition, discussed in Chapter 5, are of increasing importance in developing countries as well, as the diseases of improving standards of living affect growing middle classes and as social patterns, food, and eating habits switch from traditional styles toward western diets including excess calories, especially from fatty foods and red meat.

Overweight/Obesity

Economic development leads to changes in a population's diet. Dietary habits identified with an "affluent" or western lifestyle are characterized by an excess of energy-dense foods rich in fat and free sugars with a relative deficiency of complex carbohydrate food. Modest improvements in the economic state of a country are associated with an epidemiological shift characterized by a rise in the incidence of chronic diseases of middle and later adult life. In developing countries, these typically coexist with the traditional and persistent problems associated with nutritional deficiencies. Obesity in the industrialized countries is often associated with poverty, and its public health significance increasingly recognized.

Obesity is an excess of fat in the body caused when energy consumed has been greater than the energy expended. Nutrient excess may occur in the short term or over a long period (chronic nutrient toxicity). Excessive weight for body size (BMI, see above) is a public health problem because it is associated with premature death and is a risk factor for coronary heart disease, diabetes mellitus, hypertension, asthma, and gastrointestinal disorders. While there is evidence of a genetic predisposition for obesity, diet and other environmental factors, such as a sedentary lifestyle, can play a major role in excess body fat.

Restriction of caloric intake and increased exercise are advisable for those individuals seeking to lose excess weight. Obese individuals tend to be inactive; attempts to increase physical exercise have well-documented health benefits. Drugs and fad diets are treatment techniques that have the potential for damaging one's health, and any weight loss tends to be temporary. Surgical treatment may produce permanent weight loss but often with serious side effects, such as chronic diarrhea. Nutritional counseling and assistance in development of nutritious and low-fat diets will be more useful, especially when accompanied by exercise.

Obesity is extremely resistant to treatment, and thus primary prevention of obesity is a major public health target. This requires health education of mothers and very young children in proper nutritional practices. Eating and physical activity habits learned in childhood are very difficult, although not impossible, to change. Information about food choices should highlight the need to reduce fat and salt intake and increase dietary fiber and complex carbohydrates. Food labeling can serve to inform the individual of the caloric and fat content of items intended for consumption. The individual as well as governments, schools, parents, the media, and communities all have responsibilities in health promotion for prevention of obesity.

Diabetes

Diabetes mellitus, discussed in Chapter 5, is a chronic metabolic disorder found throughout the world. It develops in individuals who lack sufficient insulin production or whose insulin is impaired in function. As a result, they have reduced capacity to utilize glucose derived from carbohydrate foods or from body stores of glycogen. Type I, insulin-dependent diabetes (IDDM) or juvenile diabetes appears in childhood or adolescence due to failure of insulin production is not diet generated, but its management requires dietary controls as well as insulin. Type II or mature onset, nutritional, or non-insulin-dependent diabetes (NIDDM) occurs in middle adulthood or in the elderly, mostly due to excess dietary caloric intake, with saturated fats and low intake of dietary fiber, resulting in decreased insulin sensitivity and abnormal glucose tolerance. Genetic predisposition and possibly fetal influences result in a high prevalence of type II diabetes in some ethnic groups such as native Americans, Inuit, and Australian aborigines, but prevalence is magnified by societal factors including poor diet, lack of physical exercise, and alcohol abuse. This interacts with other risk factors, such as body weight, to instigate the onset of disease.

Type I and Type II diabetes both require careful nutrition. Since the discovery of insulin by Frederick Banting and Charles Best in Toronto in 1921, type I diabetes has been managed by regulation of blood sugar with daily insulin injections, with monitoring of blood sugar, diet, and exercise. Type II diabetes is controllable by diet and exercise but may also require medication to lower blood sugar. Approximately 80% of patients with type II diabetes are obese. Modern public health emphasizes weight reduction for obese diabetic patients to reduce their high risk of coronary heart disease and stroke. Nutrition plays a key role in the causation of type II diabetes, and in its management.

Cardiovascular Diseases

As discussed in Chapter 5, a strong relationship exists between diet, lifestyle, and the risk of cardiovascular disease (CVD), and especially between saturated fat intake and incidence of this disease. Cholesterol levels can be reduced by lowering dietary intake of foods high in cholesterol and by increasing foods with fiber and high density lipids, such as olive oil and avocado.

Mortality from coronary heart disease and strokes has declined dramatically in western countries, but strokes still occur in some half million persons in the United States each year, with some 150,000 deaths, large numbers of disabilities and associated health care costs (estimated at $11 billion in the late 1980s). Estimates of the incidence of strokes based on the Framingham study (Chapter 5) underestimate the true incidence since the study is primarily of a white middle class population; more representative population based studies report higher rates since strokes are commoner in black and hispanic populations.

Hypertension and diabetes are major risk factors for heart disease and stroke. Hypertension is associated with imbalanced nutrition, with excess fats and low intake of fruit and vegetables. Primary prevention is largely dietary, with weight loss, salt restriction, smoking cessation, and increased physical activity. Nutrition and diet play a major role in clinical care of the patient with cardiovascular diseases as well as in public health policy related to food, nutrition, and population awareness. Alcohol, especially red wines, in moderation, has a protective effect against coronary heart disease, but in excess contributes to increased rates. Clinical and health promotion intervention must also promote regular screening for elevated resting blood pressure, associated risk factors, and careful clinical management of hypertension to reduce risk of CVD. Antihypertensive medications play an important role in prevention of known complications of this condition.

Cancer

The relationship between specific nutrients and cancer is less well established than between diet and cardiovascular disease (Chapter 5). The relationship of food consumption data and cancer rate is supported by studies including the following: changing cancer rates in ethnic groups after migration with changes in dietary patterns; case–control studies of cancer patients and controls; prospective studies of populations with known dietary habits; and animal experimental data.

Cancers in which diet is implicated as an etiological factor include cancers of the oral cavity and pharynx, larynx, gastrointestinal tract, breast, liver, pancreas, lung, endometrium, cervix, and prostate (Table 8.8). It is currently accepted that there is a relationship between diet and specific cancer sites. The strongest dietary associations are high fat intake with cancers of the prostate and colon; high body weight with cancer of endometrium; alcohol with cancer of the esophagus; and smoked, pickled, or salted foods with cancer of the stomach. The protective effect of a diet high in fruit, whole grains, and vegetables on colon and other cancers are possibly due to vitamin A and C and their antioxidant effects.

TABLE 8.8   Association between Selected Dietary Factors and Specific Cancer Sites[a]

Cancer site	Body weight	Fat	Fiber	Fruit/ vegetables	Alcohol	Smoked, salted, pickled foods
Lung	0	0	0	P	0	0
Breast	C	C	0	0	C	0
Colon	0	C*	P	P	0	0
Prostate	0	C*	0	0	0	0
Bladder	0	0	0	P	0	0
Rectum	0	C	0	P	0	0
Endometrium	C*	0	0	0	0	0
Oral	0	0	0	P	C	0
Stomach	0	0	0	P	0	C*
Cervix	0	0	0	P	0	0
Esophagus	0	0	0	0	C*	C

[a]C = contributes to development of cancer; C* = greater contribution to the development of cancer; P = protective effect against cancer; 0 = no known effect.

Source: World Health Organization, 1990; and *The Surgeon General's Report on Nutrition and Health 1988.*

Some epidemiologists attribute 30–40% of cancer in men and 60% in women as diet related. Since the 1960s, there has been a growing consensus that diet plays an important, although still undefined role in carcinogenesis. The "Mediterranean diet," which is low in total and saturated fat, high in green and yellow vegetables and citrus fruits, and low in alcohol, salt-pickled, smoked, and salt-preserved foods, is consistent with a lower risk of many of the currently major cancers.

Dietary guidelines for prevention of cancer currently focus on the following:

1. High intake of fruits, vegetables, and whole grains;
2. Low intake of salt-cured, pickled, and smoked foods;
3. Low animal fat consumption (with <30% of total caloric intake of fat from all sources);
4. Moderate alcohol consumption;
5. Moderate caloric intake and physical activity to reduce obesity.

Public health nutrition includes a wide ranging program of health education, and health promotion programs are effective primary prevention. Governments, especially departments of agriculture and finance, should take steps to ensure fruit and vegetable supply at low cost to the consumer. Decisions affecting the supply and pricing of foods interact with the interests of farmers, producers, agri-business, transport and storage, as well as marketers of food to the consumer and the food production industry. However, they exist within a social context in which public opinion and purchasing power affect the type of produce offered and available.

NUTRITION IN PREGNANCY AND LACTATION

Pregnant women and lactating women are in effect feeding two people. They must eat enough to meet their own requirements at a time of increased need, as well as provide for the growth of the fetus and infant. During pregnancy, a woman needs 300 extra Calories and an additional 20 g of protein per day. During breast-feeding, the mother needs an extra 500 Calories and 20 g of protein above her dietary needs based on height, weight, and activity levels. The needs for (many) vitamins and minerals are similarly higher during pregnancy and lactation (see Table 8.6).

Adequate nutrition and weight gain during pregnancy are important for the development of the fetus and for the woman's health. Weight gain of a mother is a good predictor of birth weight of her infant. Since infant birth weight is a determinant of potential for survival and future development, the recommended weight gain for the mother is important to achieve. Women who are of normal weight for their height or slightly overweight women have better pregnancy outcomes than those who are underweight.

Both caloric intake and nutritional quality need to be considered. A pregnant woman needs to increase her intake of folic acid, iron, and certain trace elements. This can be done through supplementation as well as food fortification. Other factors influencing weight gain include smoking, strenuous physical work, and chronic illness. The social pressure on women to be thin may make it difficult for some to allow themselves adequate weight gain. For further discussion on nutritional needs during pregnancy, see Chapter 6.

Mineral and vitamin supplementation is also important in pregnancy and lactation; iron, iodine, selenium, folic acid, and vitamins A, B, and C are especially important when these additional needs cannot be met from diet alone. As noted in Chapter 6, the need for folic acid supplements precedes pregnancy in order to prevent neural tube defects in the fetus, since the neural tube develops in the first weeks after conception when a woman may not realize she is pregnant. Since not all pregnancies are planned and prepregnancy preparation is not common, fortification of flour with folic acid has been made mandatory in Canada and the United States in order to prevent neural tube defects. Food enrichment is discussed later in this chapter.

Lactating women should continue iron, folate, and multiple vitamin supplementation. Return to prepregnancy weight is often a major preoccupation but should be secondary to meeting the calorie, micronutrient, and fluid needs of lactation and energy requirements of caring for a newborn.

PROMOTING HEALTHY DIETS
AND LIFESTYLES

Nutrition plays a central role in health of an individual and a population, making this a major function of public health. Nutrition is also a very personal matter and requires understanding on the part of each individual as well as the society which deals with food and agricultural policies, costs and infrastructure, cultural standards, food enrichment, and many other aspects of nutrition. Education of the public in nutrition is a part of creating a social awareness of nutrition and its role in health (U.S. Department of Agriculture, 1995).

Nationwide and community-based education to promote healthy eating patterns is an essential part of health promotion. Education for health can be promoted by departments or ministries of health, education, and agriculture, as well as by nongovernmental organizations. Availability, quality, variety, and cost of foods depend on national policy, economics and personal preferences, knowledge, and community patterns. A nutrition program should provide consumers with information regarding food selection and should consider why people make their choices and the individual's responsibility for health.

Healthy eating behavior is part of all stages of life. Beginning in kindergarden and grade school, children can be taught about the nutritional values of foods. School curricula and teacher training should provide sufficient background in nutrition to provide children with guidance in food selection and promote desirable food habits. Child and adult education programs offer opportunities to emphasize the value of nutritious, balanced, and adequate diets for child and maternal health.

Dietary diversification programs need to take into account cultural beliefs about appropriate foods for different age groups, economic barriers to a healthy diet, and the availability of food types. The use of mass media to inform populations of food choices should not be overlooked.

Food support programs should be use to promote healthful nutrition. Community promotion of healthful nutrition can be implemented through schools and through health and social services especially for vulnerable groups, including women at the age of menopause, single parent families, the elderly, homeless, and the immunocompromised.

DIETARY GUIDELINES

Dietary guidelines provide a basis for individual and community education regarding healthful nutrition. The pyramid of recommended daily nutrition (Fig. 8.2) provides a guideline for healthy eating. Adolescents, pregnant, and nursing women require larger amounts than adults in all categories. The elderly need to include in their diet eight 8 ounce servings of water because they have a reduced thirst mechanism and need to drink regularly. Preference should be given to the dark green and yellow vegetables; dairy foods selected should be low fat, and the meat. Poultry, fish, and lean meat should be emphasized. Breads should be whole grain and enriched, if possible, with minerals and vitamins. The elderly should also use daily supplements of calcium and vitamins D and B12, since it is difficult to get a sufficient supply of these important nutrients from foods or sunlight alone.

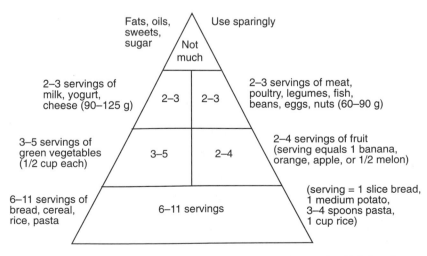

FIGURE 8.2   Recommended daily servings of food. Serving sizes (~75–100 Cal) are in parentheses. (Source: Willett, 1994.)

VITAMIN AND MINERAL ENRICHMENT
OF BASIC FOODS

Health education regarding good nutrition practices alone is not sufficient to prevent significant manifestations (borderline or clinical) of vitamin and mineral deficiency states in vulnerable sections of the population even in wealthy countries. It is a public health responsibility to ensure that all people get an adequate basic vitamin/mineral daily intake even if their food budget, or access knowledge, is limited. This can best be assured through the appropriate vitamin and mineral enrichment of basic foods, such as bread, milk, and salt. Food enrichment to provide basic essential trace elements is standard practice widespread in North America and the U.K. It must become a basic component of modern public health in nations where it has not yet been implemented.

In the United Kingdom, enrichment of white flour with two vitamins, thiamine and niacin, as well as two minerals, iron and calcium, has been compulsory since World War II. As a result of a 1971 national Canadian nutrition survey, the Canadian federal government's Food and Drug Directorate passed regulations which made it illegal to sell milk products without vitamin A and D, bread without iron, vitamins B, and niacin, or salt without iodine, each to specified levels. In the United States, enriched bread is permitted and defined in FDA regulations, and enrichment is very widely practiced, with most breads enriched in keeping with these regulations.

In the early 1990s studies in the United Kingdom showed the protective effects of folic acid, if taken prior to and during pregnancy, against neural tube defects (NTDs), including anencephaly and spina bifida. This important finding raised new possibilities for prevention of birth defects by nutritional means. For planned pregnancies among educated women, this can be done through education and physician advice and taking of folic acid supplements prior to and during pregnancy. However, this is not practical for the majority of pregnancies. The alternative is addition of folic acid to a common food such as bread to reach the population at risk to prevent these fatal or difficult to manage birth defects. As a result, folic acid fortification of enriched flour is mandatory in the United States, Canada, and the U.K.

Controversy in Food Enrichment

Food enrichment in the United States, Canada, and U.K. model is not widely practiced internationally. Inaction where food enrichment is concerned is based on inertia, tradition, lack of a strong professional lobby group on health/nutrition issues, interpretation of food fortification as an invasion of personal rights, and the lack of perfect evidence of deficiency states. Costs are not the issue since the cost of enrichment of a loaf of bread with iron and B vitamins such as niacin is negligible, and the vitamin A and D enrichment of milk and iodination of salt are of similar magnitude. The population-based strategy is justified because fortification is inexpensive, and it is harmless to others.

Is food enrichment an infringement of personal rights, or a manifestation of "paternalism versus liberalism"? It is the most liberal countries, including Canada, the United States, and the United Kingdom, that have adopted food enrichment for the common good, whereas the tradition-bound, more centralistic, less liberal countries have persisted in ignoring this issue. Chlorination and fluoridation of community drinking water are relevant precedents with positive experience in the western world. The philosophy of health promotion is based not only on prevention of disease, but also on ensuring an optimal health environment particularly for the vulnerable groups in society (see Chapters 7 and 15).

Mandatory enrichment should be supported by mandatory labeling requirements, monitoring of levels of vitamin and mineral enrichment, and continuing health education on appropriate eating habits. National monitoring of the nutrition status of the population should be carried out in sentinel population groups through appropriate measurements (i.e., hematologic, biochemical, and anthropometric).

Although malnutrition is widespread throughout the world, it is most common in the form of micronutrient deficiencies. These are not only health problems, as defined by the World Health Organization and UNICEF, affecting well-being and but also the economic growth potential of a population. The World Bank has increasingly accepted the importance of public health nutrition as a basis for infrastructure and economic growth, with nutrition interventions being among the most cost-effective public health measures. These include nutrition education (including promoting breast-feeding), micronutrient fortification, micronutrient supplementation, food supplementation, food price subsidies, and control of parasitic diseases (World Development Report, 1993).

FOOD AND NUTRITION POLICY

A national food policy aimed at prevention of conditions related to nutritional deficiency or excess can only succeed with continuous intersectoral cooperation. Regulation of food products by public health authorities involves national and provincial governments, as well as the local health authority. The food processing industry must be educated to compliance and self-regulation as part of a corporate culture working within adequate government regulation and educated consumer demand, with the competition of the marketplace serving a positive role. Active support is needed from government departments or ministries, food manufacturers and marketers, community leaders, health professionals, educators, women's groups, and the media in informing the public of the key issues.

The Evolution of a Federal Role

The U.S. federal government has been a force for improved nutrition for the past century. This included extension or outreach education programs as well as in food and agricultural research and development and in food supplementation programs. The federal role in nutrition research began in 1887 with the establishment of a nutrition laboratory which was the forerunner of the National Institutes of Health. The U.S. Department of Agriculture (USDA) instituted food monitoring programs in 1893 and extension services to promote nutrition education in 1914.

BOX 8.1 FOOD AND NUTRITIONAL POLICY: RESPONSIBILITIES OF NATIONAL AND INTERNATIONAL AUTHORITIES

International agencies: Establish and promote the science and practice of food production, processing, distribution, and marketing for adequacy of nutrition internationally. Agencies concerned include the United Nations Food and Agriculture Organization (FAO), the World Health Organization, UNDP, the World Bank, and many bilateral governmental aid agencies (e.g., the Agency for International Development, or AID).

National authority: Establish national standards and regulatory requirements in such areas as food supply and safety, labeling, fortification, and food content; responsible for regulating imported foods; establishing standards and enforcement of safety, packaging, and labeling; nutrition status monitoring; promotion of nutrition education and policies by government; coordination with the other departments of government (e.g., agriculture, trade, and industry) and international agencies.

State/provincial authority: Legislate ordinances that reinforce national standards; licensing of food establishments; supervisory and regulatory functions; coordination with agricultural and industry departments of government; nutrition education; liaison with private food production, processing, and marketing industry.

Local authority: Supervise local food producers; inspection of meat and dairy products and certain processed foods; monitoring of school lunch programs; provision of nutrition services and education; periodic inspection of public eating places; promotion of nutrition education, breast-feeding, and healthy weaning practices; provision of well child care including growth monitoring and nutrition education.

In 1917, national concern for nutrition increased as a result of findings of large numbers of draftees being rejected from military service because of poor nutritional status (e.g., goiter). The U.S. Food Administration was established to supervise food supply during World War I. National initiatives in nutrition increased during the Great Depression (1929–1936) with use of surplus agricultural commodities for food relief. This initiative later developed into routine school lunch programs on a national level continuing to the present time.

In 1927, the U.S. Food and Drug Administration (FDA) was established. During the 1930s, the USDA conducted the first nationwide food consumption surveys, and in 1939 a federal food stamp program was established. In the 1950s and 1960s, food stamp and lunch programs were expanded. In 1972, the USDA established the Women, Infants, and Children Program (WIC) providing food supplements to needy pregnant women and children at risk. Programs to provide meals for the elderly were also developed and the school lunch program expanded to include breakfasts in needy communities.

A 1940 report on malnutrition in the United States led to a national nutritional conference and to recommended dietary allowances (RDAs). In 1941 the FDA issued enrichment standards for flour and bread with vitamin B complex and iron, and in 1942 addition of vitamins A and D to milk and margarine was mandated. Following World War II, in 1946 a national School Lunch Program was established. In 1965, as a part of the War on Poverty, the Headstart program was initiated, the Food Stamp Act was passed by Congress, followed by the Child Nutrition Act and School Breakfast Program in 1966. The National Health and Nutrition Examination Survey (NHANES) I (1971–1974), NHANES II (1976–1980), and NHANES III (1988–1991) helped document the nutritional status of the U.S. population.

The WIC supplementary food programs for women, infants, and children and the elderly were initiated in 1972, and labeling of foods with added nutrients or nutrient claims was mandated. Enrichment of infant cereals and formulas with vitamins B1, B2, B6, and iron, with vitamin C in drinks and dessert powders, was required. In 1979, a Nutrition Policy Board and Surgeon General's publication *Healthy People* established Nutritional Objectives for the nation. In 1988, Dr. Everett Koop, The Surgeon General of the United States, issued *The Surgeon General's Report on Nutrition and Health,* a landmark document on nutrition in national health policy. Federal regulations were issued in 1993 under the Nutrition Labeling and Education Act, and in 1994 regulations were issued for labeling of meat and poultry. This was followed in 1996 by federal requirements for addition of folate to enriched flour, and in 1998 federal agencies recommended vitamin B supplements for older persons.

Nutrition Issues in Development Policies

Countries or regions that are poor and suffer from high rates of borderline or overt malnutrition must address the nutritional needs of the population as an essential part of planning for development. Economic development and nutritional status are directly related, as a malnourished population is not productive and cannot learn well. Efficient planning and policy making requires all governments to establish a cycle of assessment, intervention, and monitoring of the existing nutritional situation.

Piped or clean water and public sanitation and improved food safety help to reduce diarrheal disease which, especially in children, aggravates malnutrition. Loans and technical support for development of small agricultural, food production, and marketing businesses in rural and poor urban areas can improve food supply and food quality reaching a poor population.

Successful nutrition policy ultimately depends on consumer knowledge and access to quality foods at costs that are affordable. Fortification is an important supplement, but education of the producer, processor, marketer, and consumer are no less vital. Consideration must be given not only to the supply of food and nutrients but also to the short- and long-term effects of agricultural policies. Food security policies must aim to guarantee all families access to their minimum food requirements.

The cost to the health system of sickness associated with over or undernutrition alone justifies expenditures on educational effort to reduce to burden of disease or disability associated with inappropriate nutritional status.

The Role of the Private Sector and NGOs

The private sector grows, processes and markets food and thus plays a vital role in nutrition and food safety. This includes the farmer producer, but equally the food processor/marketing industry. The private sector responds to the consumer but can also create demand for new products with a health component by labelling, advertising and marketing practices.

Manufacturers of salt initiated ionization of salt in the United States, with widespread manufacture and marketing of iodized salt to prevent goiter was initiated in 1924. Baby food manufacturers enriched their products with vitamins and minerals without being required to by law. Where governments are reluctant to mandate food enrichment, private industry may already have taken the decision to do so. This can be a valuable marketing tool, and where government regulations stipulate permissable enrichment, private industry fulfills an important public health function by pursuing enrichment policies.

Nongovernmental organizations have played important roles in nutrition especially in the form of women's organizations. This was seen in the early part of the twentieth century in rural women's organizations in North America which promoted better nutrition through education and consciousness raising. Lobbying of government and private industry for constructive nutrition policies is also important for promoting better nutrition for vulnerable groups. Concern for nutrition of the elderly is an area where NGO support for nutritional assistance programs can be very effective. Women's magazines have also played an important role in disseminating information on healthful diets and raising interest in the relationship of nutrition and health.

The Role of Health Providers

Providers of care play a pivotal role in education of patients regarding appropriate nutrition during normal healthy periods and certainly during illness. The provider of health care has the trust of the patient and the professional knowledge to convey information to the patient in a time and manner that may be more effective than any other form of education. This includes the person at risk for development of ischemic heart disease, hypertension, and other diseases associated with nutritional factors, before and even moreso after the disease process has become clinically apparent.

Malnutrition is associated with an increase in morbidity from infectious and chronic diseases. Conversely, disease, whether infectious or chronic, can cause a

TABLE 8.9   Standards and Samples for Monitoring Nutrition Status of Population

Standards/samples	Factors
Reference standards	International reference population preferable to local standards to compare and monitor changes over time with optimal "ideal growth" population ("gold standard")
Representative samples	Epidemiologically representative sample of total population at risk
Sentinel populations	Monitor population in particular program such as food supplement program (e.g., WIC) or attendees of a mother and child preventive care program
Sentinel centers	Selected places of contact with population in which monitoring can be added to basic program such as in MCH centers, emergency rooms, doctors' offices, school

worsening of underlying malnutrition. Health providers need to be sensitive to these interactions and aware of their consequences. Prevention and management of infectious diseases can have an important effect on the nutritional status of children at risk for protein energy malnutrition, growth faltering, or failure-to-thrive. Identification of children at risk for malnutrition can result in additional attention and instruction of the mother regarding appropriate nutrition. Counseling and education for healthful nutrition according to age and needs is a key element of all patient care, as it is a public health issue.

NUTRITION MONITORING AND EVALUATION

Ongoing monitoring of the nutritional status of representative samples of the target population is an essential part of monitoring the health of a population (Tables 8.9 and 8.10). This may be done through population samples or at selected sentinel centers. Nutrition monitoring systems, such as the U.S. National Health and Nutrition Examination Survey (NHANES), enable governments to monitor trends in food intake of individuals sampled and trends in nutritional status and general health of the population. Study results can direct policy makers in dietary standards and recommendations for manufacturers, state and local governments, and the general public.

Surveys of food purchases and consumption through direct observation, and interviews of family members can provide valuable nutrition monitoring data, as can interviews with food marketing networks. Food supply and consumption tables can be calculated from total national data to estimate average consumption. Anthropometric measurements of height and weight, skin fold, and body mass indexes provide valuable data on nutrition status, as do hematologic (iron, ferritin) and biochemical indicators (cholesterol, lipids, fasting blood sugar, vitamin levels). Large-scale surveys are costly, so sentinel center studies in selected representative locations may be more practical and can provide time trends in successive cross-sectional studies.

TABLE 8.10   Measures for Monitoring the Nutrition Status of a Population

Assessment	Indicators
Indirect measures	
Education levels and cultural practices	Literacy, years of schooling, and especially female education levels are directly and indirectly mediated through improved family incomes and competencey in dealing with the challenges of child care; traditional dietary patterns and individual choices
Household income and food purchase surveys	Per capita GNP, average household incomes, and other indicators of purchasing power are derived from census data and special surveys, including purchasing practices
Food balance sheets	An indirect measure of nutrition state by aggregating national data on amounts of food products adding imports, subtracting exports; the balance is converted to calories, deriving per capita consumption which reasonably estimates actual average intake
Infant mortality, low birth weight	Where these rates are high, it is probable that malnutrition is widespread
Parasitosis	High rates of parasitosis contribute to malnutrition
Direct measures	
Clinical assessment	Hospitalizations with primary or secondary diagnoses serve as index cases of extent of nutritional problems such as rickets, diabetes
Anthropometric measures and surveys	Weight and height for age and weight for height using WHO/NCHS standard are used as the international "gold standard"; skin fold thickness of upper arm is also used; these indicate protein/calorie status of individual child or the population of children; body mass index is an important individual and population health measure
Biochemical measures and surveys	Examinations of body fluids including blood, urine are measures of nutrients or indicators of complex metabolic processes; anemia is measured by hemoglobin, hematocrit, serum iron, or feritin; thyroid function tests and iodine levels in urine measure iodine status; vitamin C in cells and serum, serum levels of vitamins B and D also are used; these are expensive tests not readily available for surveys—surrogate inexpensive field tests such as zinc protoporphyrin can serve as indicators of blood lead and or anemia; fasting bleed sugar levels can give an idea of prevalence of glucose intolerance
Dietary assessment	Dietary surveys in institutions or households are based on recall of intake over past 24 hours or 7 days, or on dietary records weighing food served and uneaten

Standard Reference Populations

Reference populations are used for comparison purposes and have been found to reflect the "optimal" growth of children who are not excessively ill and who receive what is currently thought to be "good nutrition." The data may be presented in chart form and as percentiles. Recorded weights or heights at specific ages for the individual or group of children are seen graphically at different ages.

The U.S. National Center for Health Statistics growth patterns data derived from a white middle class population sample is increasingly used internationally as the standard reference population. The NCHS standard was adopted by the World Health Organization so as to have an international standard reference population. The WHO concluded that child growth potential is similar across national and ethnic lines if nutrition is appropriate and intercurrent illness is not excessive.

There are criticisms of this approach, based on the feeling that local studies should be compared to local standards and conditions. However, this fails to provide for change over time, the possibility of endemic malnutrition, and the importance of an international standard of expected heights, weights, and head circumferences for comparison. The controversy will continue, but the use of an international standard reference population is widely accepted.

Anthropometric measures were first developed in the nineteenth century by Richer using skin fold thickness as an index of fatness. During World War I, Matiega developed ways of measuring the composition of the human body to assess the physical efficiency of soldiers. Anthropometry includes composition of the body at the atomic, molecular, cellular, tissue, and whole body levels. Whole body studies involve measurements of body weight and stature, including skin folds, circumferences, and bone measures. These studies have clinical as well as public health importance. The association between weight loss and disease outcome has become a useful predictor of prognosis in individual patient care.

TABLE 8.11   Classification Systems Used Internationally for Monitoring Child Growth and Protein–Energy Malnutrition

Classifications	Features
Gomez (weight for age using Harvard standard)	Those 90% or more of median considered normal; those 76–90% called mildly malnourished (first degree); those 61–75% of median as moderately malnourished (second degree); those 60% or less as severely malnourished (third degree), using Harvard standard population
Waterlow (weight for height and height for age)	Height for age identifies past malnutrition, and weight for height indicates current malnutrition; uses NCHS/WHO reference population
NCHS/WHO	Weight and height for age, weight for height, and head circumference standards based on NCHS survey finding of primarily white middle class children from Iowa; this is accepted by the WHO as an international standard
Shakir arm circumference (as % of standard correlates to weight as % of standard for age)	A well-nourished child has same arm circumference between 1 and 6 years; Shakir used tape measurement of arm circumference as a simple screening device for severely undernourished preschoolers; effective for case finding of moderate to severe malnutrition

Measures used widely in anthropometry include the following:

1. Height and weight to calculate body mass index (BMI);
2. Knee height to predict adult stature;
3. Circumferential measures, including head, mid upper arm, mid thigh, and mid calf;
4. Skin fold measures, including biceps, triceps, subscapular area (1 cm below lower angle of scapula), suprailiac (mid axillary 2 cm above iliac crest), thigh, and calf.

Such combinations of these are used to calculate the fat and fat free content of body mass, which is useful for clinical assessment and special studies of groups. Fat free body mass (FFM) indicates the muscle mass for nutrition studies or in patients in long-term nutrition follow-up and for teaching purposes.

Measuring Deviation from the Reference Population

Comparing the growth pattern of a study population to a standard provides a comparison similar to measuring blood cholesterol and comparing it to a normal range. A simple method of surveying a population for comparison with a standard, such as the NCHS/WHO reference population, is to record heights and weights of children presenting for immunization on a standard growth chart and compare the clustering of those observed to the international standard. This provides personnel in primary care a method of monitoring the population they serve. Table 8.11 lists classification systems used for monitoring child growth.

Z-scores are a widely used method of showing anthropometric data compared to a standard. A Z-score shows deviation from the mean in terms of standard deviations. Thus the birth weight or growth pattern of children may be compared to a norm in standard deviations and not absolute numbers. Observations representing 68% of a normal distribution curve are included within one standard deviation, while 95% are included within 1.96 standard deviations (see Chapter 3). Z-scores simplify graphic summation of the data in anthropometric studies.

FOOD QUALITY AND SAFETY

Food-borne disease is a major public health issue in all societies. In developing countries one of the major causes of morbidity and mortality, and even in the most advanced countries great care must be taken by governments, the food industry, and the consumer to prevent food-borne disease. Both government and the food industry have a vested interest in providing nutritious and safe food to the public at reasonable cost. Other concerns to private industry are attractiveness of food and its packaging, long shelf life, low rates of loss of food or spoilage, contamination, and deterioration. Because the food producer and the processing/ marketing industry wish to maximize their profit, self-policing is not sufficient,

TABLE 8.12   Food-Borne Disease Agents, Organisms, Chemicals, and Vectors

Agent	Organism/chemical	Food vector
Bacteria	*Campylobacter, Clostridium, E. coli, salmonella, Vibrio cholerae, Listeria*	Raw and processed foods contaminated with human, bird, or animal excreta; undercooking and contamination during preparation, storage, or serving
Viruses	Poliomyelitis, hepatitis A and E	Vegetables, vegetables irrigated with sewage, shellfish
Prions	Bovine spongiform encephalopathy (BSE)	Food of animal origin for cattle transmitting Creutzfeld-Jakob disease (CJD) to humans
Molds	Mycotoxins, aflatoxins, ochratoxins	Stored nuts and cereals
Protozoa	Amebas, giardia, cryptosporidia	Vegetables, fruit, water, unpasteurized milk
Helminths, parasites	Ascaris, fasciola, taenia, trichinella, trichura, opisthorchiasis	Vegetables, underciooked meat , raw fish from areas with contaminated water, untreated sewage, or irrigation water
Chemicals, agricultural and veterinary	Pesticides, fungicides, fertilizers, hormones, antibiotics	Contaminated food from abuse of chemical agents; widespread antibiotic use may engender allergies and organism resistance
Heavy metals	Mercury, cadmium, tin	Fish, seafood from water contaminated by industrial waste
Radionuclides	Radioactive materials	Foods contaminated from atmospheric fallout

Source: Adapted from World Health Organization and Food and Agriculture Organization, 1992.

and a strong governmental regulatory agency is necessary to protect the public interest.

Fruit and vegetables contaminated with enteric pathogens are common causes of disease. Ingestion of contaminated products results in frequent outbreaks of gastroenteric disease, and contributes to the background level of diarrheal diseases not necessarily identified as outbreaks.

Animals used for human food can be a source of transmission of communicable disease as well as chronic disease. This may be in the form of infectious diseases when food is contaminated with pathogenic organisms during growth, transportation, preparation, storage, or handling before consumption. Food poisoning or diarrhea can result from salmonella or campylobacter on chicken meat or eggs. Animals raised for food can harbor diseases such as brucellosis that can be transmitted to humans through consumption of unpasteurized milk products from infected animals. People can become seriously affected by toxins in food from animal origin, including meat, milk, and seafood. Diseases such as brucellosis, anthrax, and Rift Valley fever can be spread from infected domestic animals by direct contact of persons handling meat products. Table 8.12 lists various food-borne agents capable of affecting human health.

Animal husbandry practices are now part of public health concerns as are food processing, transport, and cooking. A large number of cases of dangerous *E. coli* infections resulting from contaminated meat in the late 1990s in the United States, Japan, and many other countries show the potential danger even in technologically advanced industrial countries. Use of animal materials rendered for animal feed is common practice instead of more expensive alfalfa. This raises serious health risks because of the salmonella and campylobacter it contains. In August 1997, tainted beef found in products of one company resulted in tens of thousands of tons of hamburger meat being condemned in the United States. A ban was imposed on British beef as a result of by the European Union in the late 1990s as a result of the bovine spongiform encephalopathy (BSE or mad cow disease) linked to Creutzfeld-Jakob disease cases in humans (see Chapters 4 and 5) due to animal feed using contaminated animal parts.

Hazards associated with foods require constant vigilance on the part of government, the food industry, and consumers. Cooperation between government and food producers, processors, and manufacturers is one of mutual interest as well as a regulatory relationship. Private industry has a major role to play in improving food supply, and nutritional quality both for commercial and legal liability reasons. The consumer is also responsible for safe food handling practices, in terms of hygiene, storage, and usage. Sources of health-related problems exist at the points of production, processing, transporting, storing, sale, preparation, and serving of food products. Outbreaks of food-borne diseases serve as a reminder of the need for careful inspection of fresh produce, meat and dairy products, and processed foods on both a national and local basis.

Hazards that occur in many agricultural regions include contamination by pathogens. It is not uncommon in developing countries for foods to be exposed to pathogens by the use of untreated or inadequately treated sewage water for irrigation, malfunctioning sewage systems, or deliberate use of "nightsoil" for fertilization. Such fruits and vegetables eaten raw can expose the consumer to a host of infectious diseases transmitted via the fecal–oral cycle. Destruction and contamination of food by ruminants, rodents, or other pests during growth, harvesting, storage, transportation, or processing of food can be major factors in safety and adequacy of food supply. Even in highly developed agriculture, contamination of food can occur at many stages in its processing. Imported parsley, basil, lettuce, and strawberries resulted in a number of multistate outbreaks of shigellosis and cyclosporiasis in the United States in the period 1997–1999 (CDC, 1999).

A second major concern is the hazard posed by ingestion of chemicals used in food production. Modern agriculture has come to rely on extensive use of chemical fertilizers and pesticides. These accumulate in the soil over time and form a residue on the produce grown. Pesticide residues on orchard and field crops and in processed foods include chlorinated hydrocarbons, dieldrin, and DDT. Pesticides can cause serious problems of toxic exposure to farm workers, contamination of

---

BOX 8.2 FOOD ADDITIVES

Food additives are substances used in production, processing, treatment, packaging, transportation, or storage of food. These include simple additives such as salt, baking soda, vanilla, or yeast, but also complex ones.

Additives are used in food: to maintain product consistency, to improve nutritional value (vitamins and mineral enrichment); to maintain palatability and retard spoilage; to provide leavening and prevent acidity/alkalinity); and to enhance flavor or desired color.

All additives are subject to FDA regulation, and some have been banned for harmful effects, including cyclamates (sweetener), cobalt salts (beer), tar derivative used for food coloring, some pesticides, fungicides, and herbicides, and polyvinylchloride in plastic containers.

---

rain runoff to surface waters or groundwater, and contamination of inadequately washed farm produce. Safety requires consumer self-protection by selection of foods and careful washing before use, but education in seeking effective alternatives and reducing use of pesticides is important primary prevention (see Chapter 9).

Growth enhancers, including antibiotics and hormones to give rapid and vigorous animal growth and milk production, can adversely affect milk and meat products. Prophylactic use of antibiotics may contribute to increasing the numbers of resistant organisms which can potentially endanger public health.

Food additives are substances intentionally added to foods to modify their taste, color, texture, nutritive value, appearance, and resistance to deterioration. Some have been in use for a long time, such as sugar, salt, and spices. Natural additives are important in processed foods; others have positive public health benefit, such as in added vitamins and minerals.

Concern over some additives centers on their toxic or carcinogenic effects, and more so when these additives are for cosmetic value only. In 1958, the Food Additives Amendment was added to the U.S. Food, Drug, and Cosmetics Act. This amendment requires food manufacturers to comply with FDA rulings on food additives. In the same year, the famous Delaney Clause banned any additive that has been shown to be carcinogenic in any amount from use as a food additive. As a result, there was large-scale testing of additives. Those additives already in use were categorized as "generally regarded as safe" or GRAS. Some of these were subsequently tested and found to be carcinogenic and removed from general use by FDA ruling (Box 8.2).

Methods of preserving foods, some as old as civilization itself, were developed to protect the quality of foods and ensure supply beyond the immediate period of harvesting of crops or foods of animal origin. They are meant to prevent microbial

TABLE 8.13   Major Food Processing and Preservation Technologies

Technology	Examples
Drying	Radiant (solar), spray, air, freezer, or vacuum drying
Fermentation	Yogurt, cheese, soya sauce, wine and beer, vinegar, sauerkraut
Chemical treatment	Salting, pickling, sugaring, smoking, fumigating, or treating with food additives
Cold treatment	Refrigerating, freezing
Irradiation	Microwave cooking, ionizing irradiation of meat, poultry, fish, spices, fresh fruit, vegetables, and prepared foods to destroy pathogenic organisms and those which cause spoilage

or chemical deterioration of food, without causing harmful effects to the safety, nutritional, or aesthetic value or shelf life of the foods. Techniques developed to protect and preserve the quality of food beyond the immediate period following production are vital to any population group. They are meant to prevent microbiological and chemical deterioration, without harmful chemical toxic, or carcinogenic effects. Some of the major categories of food preservation technology are shown in Table 8.13.

While government and industry have major roles in ensuring safe food, final responsibility rests with the consumer. Public education, knowledge, attitudes, and practices determine the safety of food served to the family. The consumer in a market economy can influence and even determine which goods to purchase, creating demands felt in the marketplace that guide producers and processors to change the content, packaging, labeling, advertising, and prices. Consumers can influence the market by refusing to purchase goods known to be problematic or lacking nutritional value, an effective part of quality control. Consumer advocacy can also play a role in food quality, as manufacturers and governments are sensitive to public expression of concern through the media on food quality. The Food and Drug Act setting up regulatory mechanisms in the United States was enacted in response to newspaper articles and novels (such as *The Jungle*, by Upton Sinclair) exposing the food quality in the United States in 1906. Consumer boycotts have also made producers very sensitive to public concerns.

Concern for public health aspects of food have increased with serious incidents of food poisoning from new variants of *E. coli* and salmonella resistant to available antibiotics, and as large sectors of the population are vulnerable to infection, especially immunocompromised persons such as HIV carriers and splenectomized patients. In the summer of 1997, 25 million pounds of ground beef were recalled in the United States because of suspected contamination with *E. coli* O157:H7 (FDA, 1998).

In the 1990s, despite safe water supplies, good hygiene, and food technology, food poisoning increased in many industrialized countries. Current estimates suggest that between 5 and 10% of the population of industrial countries are affected annually. In the United States, there are between 3.3 and 12.3 million cases and up to 3900 deaths per year from food poisoning. Food-borne and waterborne diseases are perhaps the most important public health problems of developing countries, with 1.5 billion cases and 3 million deaths annually from diarrheal disease. Consumer resistance to use of irradiation of food remains high, despite it being widely regarded as probably the safest and most reliable method of protecting food supplies from contamination. This resistance may decline as more instances of food contamination are reported in the public media.

Irradiation to improve food safety, and genetic engineering to produce food crops of higher quality and freedom from need of pesticides and chemical fertilizers are producing a new revolution in agriculture and food safety. They are also controversial and under critical review for safeguards and labelling requirements to protect consumer freedom to choose. Nevertheless, these are important new technologies which will be vital to meeting the needs of developing countries, as well as those of the industrial world in promoting good nutrition and food safety.

NUTRITION AND THE NEW PUBLIC HEALTH

Population and individual health are closely related to nutrition status. The individual chronically ill patient may experience a serious decline in his or her condition as a result of inadequate daily nutrition. The cost of hospitalization alone for such persons may outweigh the support services that should assist that person in maintaining adequate nutrition. Overt or subclinical deficiency conditions can cause widespread health damage if national health policies fail to take into account current knowledge and practices for their prevention. Failure to implement such policies engenders undernutrition with important health consequences. The success of developed countries in controlling micronutrient deficiency conditions has been achieved partly by social and economic development and improved pediatric care, but also by enlightened food policy with food fortification. Adoption of population oriented food policies is crucial for developing countries to cope with basic undernutrition for large vulnerable parts of the population.

Similarly, overnutrition and its associated diseases, such as diabetes, cardiovascular diseases, and cancer, are the cause of widespread pathology and use of costly health services, whereas education and other programs of intervention can have important preventive effects. These cannot be addressed solely as individual patient problems, but require a parallel population approach which involves a broad set of food and nutrition policies. The New Public Health takes a holistic approach to prevention and management of disease, as well as promotion of optimal health. Nutrition is central to that task, requiring well-planned interventions at the national, community, and individual level.

SUMMARY

While much of the glamour and stress in public health achievements over the past century has involved sanitation and control of communicable diseases, the contribution of improved nutrition and food safety to better health has been enormous. Mortality from all causes, including infectious diseases, began to decline in the seventeenth century, well before organized infectious disease control emerged, and that was, in part, a result of the agricultural revolution and improved food production.

Today, nutrition is one of the major public health issues of developed countries. The problems are largely of overnutrition or inappropriate balance and excessive caloric intake. In developing countries, mass nutritional deprivation is perhaps the most important public health problem, resulting in wide-scale deficiencies of calories, protein, and essential vitamins and minerals. Mid-level developing countries are experiencing rapidly increasing cardiovascular mortality due to excessive animal fat intake as living standards increase for part of the population, with widespread undernutrition for the majority.

Since the 1960s, the people in most industrialized countries have paid a great deal of attention to their personal nutrition, prompted by concern in the media and the health professions. The dietary approach to preventive health care has contributed to the dramatic drop in death rates from coronary heart disease, stroke, and cancer of the stomach in most western countries.

Monitoring nutrition status and developing national policies to assure adequate and high quality food are major governmental functions. The economic, agricultural and marketing systems all have roles to play in ensuring population nutritional health. Food fortification and promotion of healthful dietary habits are vital to prevent disorders due to deficiency and to excess nutrition. The supply and quality of food for growing populations with rising expectations will depend on new science and technology such as genetic engineering and food irradiation, despite current controversy and manipulation of legitimate public concerns. The clinician deals with the individual patient and public health with the community and nation. This duality is the New Public Health in which nutrition plays a central role.

ELECTRONIC MEDIA

Food and Agriculture Organization (FAO) home page, http://www.fao.org/

Health Canada, Review of Canada's Policies on the Addition of Vitamins and Minerals. April 1999, Canadian Food Inspection Agency via http://www.hc-sc.ca/food-aliment/

U.S. Food and Drug Administration (FDA) home page, http://www.fda.gov/

U.S. Department of Agriculture (USDA) home page, http://www.usda.gov/

USDA National Agricultural Library home page, http://www.nal.usda.gov/

USDA nutrient data and RDAs for specific foods can be found at
http://www.nal.usda.gov/fnic/foodcomp/ and http://www.nal.usda.gov/fnic/dga/rda.pdf/

USDA Nutrient Data Laboratory, Riverdale, MD, ndlinfo@rbhnrc.usda.gov/

U.S. Environmental Protection Agency (EPA) home page, http://www.epa.gov/

RECOMMENDED READINGS

Anderson, S. H., Vickery, C. A., Nicol, A. D. 1986. Adult thiamine requirements and the continued need to fortify processed cereals. *Lancet,* 2:85–89.

Centers for Disease Control. 1999. Achievements in public health, 1900–1999. Safer and healthier foods. *Morbidity and Mortality Weekly Report,* 48:905–913.

DeMaeyer, E. M., Dallman, P., Gurney, J. M., Hallberg, L., Sood, S. K., Srikantia, S. G. 1989. *Preventing and Controlling Iron Deficiency Anemia through Primary Health Care.* Geneva: World Health Organization.

Dunn, J. T. (editorial). 1992. Iodine deficiency—The next target for elimination? *The New England Journal of Medicine,* 326:267–268.

Foster, P. 1992. *The World Food Problem: Tackling the Causes of Undernutrition in the Third World.* London: Boulder Adamantine Press.

Gorstein, J., Akre, J. 1988. The use of anthropometry to assess nutritional status. *World Health Statistics Quarterly,* 41:48–58.

Rose, D., Oliveira, V. 1997. Nutrient intakes of individuals from food-insufficient households in the United States. *American Journal of Public Health,* 87:1956–1961.

Scrimshaw, N. S. 1992. Iron deficiency. *Scientific American,* 265:46–52.

Scrimshaw, N. 1995. The new paradigm of public health nutrition. *American Journal of Public Health,* 85:622–624.

Sidel, V. 1997. The public health impact of hunger. *American Journal of Public Health,* 87:1921–1922.

BIBLIOGRAPHY

Barkan, I. D. 1985. Industry invites regulation: the passage of the Pure Food and Drug Act of 1906. *American Journal of Public Health,* 75:18–26.

Beghin, I., Cap, M., Dujardin, B. 1988. *A Guide to Nutritional Assessment.* Geneva: World Health Organization.

Bothwell, T. H. E., Charlton, R. W. 1981. *Iron Deficiency in Women.* Washington DC: International Nutritional Anemia Consultative Group (INACG).

Centers for Disease Control. 1989. CDC criteria for anemia in children and child bearing-aged women. *Morbidity and Mortality Weekly Report,* 38:400–404.

Centers for Disease Control. 1994. Daily dietary fat and total food energy intakes—Third National Health and Nutrition Examination Survey, Phase 1, 1988–1991. *Morbidity and Mortality Weekly Report,* 43:116–117, 123–125.

Centers for Disease Control. 1994. Epidemic neuropathy—Cuba 1991–1994. *Morbidity and Mortality Weekly Report,* 43:183, 189–192.

Compston, J. E. 1998. Vitamin D deficiency: Time for action. *British Medical Journal,* 317:1466–1467.

Delange, F., Burgi, H. 1989. Iodine deficiency disorders in Europe. *Bulletin of the World Health Organization,* 67:317–325.

Dunn, J. T., Van der Haar, F. 1990. *A Practical Guide to the Correction of Iodine Deficiency.* Amsterdam: International Council for the Control of Iodine Deficiency Disorders.

Editorial. 1986. Prevention and control of iodine deficiency disorders. *Lancet,* 2:433–434.

Food and Agriculture Organization. 1987. *The Fifth World Food Survey.* Rome: Food and Agriculture Organization of the United Nations.

Gerstner, H. 1997. Vitamin A—Functions, dietary requirements and safety in humans. *International Journal of Vitamin and Nutrition Research,* 67:71–90.

Gibson, R. S. 1990. *Principles of Nutritional Assessment.* New York: Oxford University Press.

Health Protection Branch, Health and Welfare Canada. 1974. *Addition of Vitamin and Mineral Nutrients to Foods* 426:1–4. Ottawa: Health and Welfare Canada.

Henkel, J. 1998. Irradiation: A safe measure for safer food. *FDA Consumer,* 32:12–17.

Krebs-Smith, S. M., Cook, D. A., Subar, A. F., Cleveland, L., Friday, J. 1995. US adults' fruit and veg-etable intakes, 1980 to 1991: A revised baseline for Health People 2000 objectives. *American Journal of Public Health,* 85:1623–1629.

Marshall, J. R. (editorial). 1995. Improving American diet—Setting public policy with limited knowl-edge. *American Journal of Public Health,* 85:1609–1611.

Mertz, W. 1981. The essential trace elements. *Science,* 213:1332–1338.

Morley, J. E., Mooradian, A. D., Silver, A. J., Heber, D., Alfin-Slater, R. B. 1988. Nutrition in the elderly. *Annals of Internal Medicine,* 109:890–904.

National Academy of Sciences–National Research Council. 1968. *Recommended Dietary Allowances,* Seventh Edition, Washington, DC: Food and Nutrition Board.

National Research Council. 1989. *Diet and Health: Implications for Reducing Chronic Disease Risk.* Washington, DC: National Academy Press.

Palti, H., Tulchinsky, T. H. 1988. Anemia as a public health problem in Israel. *Public Health Reviews,* 16:215–250.

Passmore, R., Eastwood, M. A. 1986. *Davidson and Passmore Human Nutrition and Dietetics,* Eighth Edition. Edinburgh: Churchill Livingstone.

Pinstrup-Andersen, P., Pelletier, D., Alderman, H. (eds.). 1995. *Child Growth and Nutrition in Devel-oping Countries: Priorities for Action.* Ithaca: Cornell University Press.

Shils, M. E., Olson, J. A., Shike, M. (eds.). 1994. *Modern Nutrition in Health and Disease,* Eighth Edi-tion, Philadelphia: Lea & Febiger.

UNICEF. 1998. *The State of the World's Children, 1998.* New York: Oxford University Press.

United States Department of Agriculture. 1995. *Dietary Guidelines for Americans, Fourth Edition.* Washington, DC: USDA.

United States Department of Health and Human Services. 1988. *The Surgeon General's Report on Nutrition and Health 1988.* DHHS (PHS) Publ. No. 88-50210, Washington, DC: United States De-partment of Health and Human Services.

Weber, P., Bendich, W. A., Schach, W. 1995. Vitamin C and human health—A review of recent data relevant to human requirements. *International Journal of Vitamin and Nutrition Research,* 66:19–30.

West, K. P. (ed). 1993. *Bellagio Meeting on Vitamin A Deficiency and Childhood Mortality.* New York: Helen Keller Foundation.

Willett, W. 1990. *Nutritional Epidemiology.* New York: Oxford University Press.

Willett, W. C. 1994. Diet and health: What should we eat? *Science,* 264:532–537.

Williams, C. D., Baumslag, N., and Jelliffe, D. B. 1994. *Mother and Child Health: Delivering the Services, Third Edition.* New York: Oxford University Press.

World Bank. 1993. *World Development Report, 1993: Investing in Health.* New York: Oxford Uni-versity Press.

World Health Organization. 1990. *Diet, Nutrition, and the Prevention of Chronic Diseases.* Technical Series Report 797. Geneva: World Health Organization.

World Health Organization. 1998. *World Health Report 1998.* Geneva: World Health Organization.

World Health Organization and Food and Agricultural Organization. 1992. *Major Issues for Nutrition Strategies, 1992: International Conference on Nutrition.* Italy: FAO and WHO, 1992.

NUTRITION AND FOOD TECHNOLOGY JOURNALS

American Journal of Clinical Nutrition (American Society of Clinical Nutrition)

Food and Nutrition Digest (Kansas State University)

Food Technology

International Journal of Vitamin and Nutrition Research

Journal of Food Sciences

Journal of Nutrition (American Society for Nutritional Sciences)

9

ENVIRONMENTAL AND
OCCUPATIONAL HEALTH

ENVIRONMENTAL HEALTH: INTRODUCTION

A safe environment is fundamental to health. Air and water precede even food and shelter in a hierarchy of health and survival needs. Safe water supplies and waste management are still problematic aspects of public health and community hygiene. Contamination of the environment by biological, chemical, physical, or other disease-causing agents in the external environment and the workplace are major public health and political concerns as we approach the end of the twentieth century. Since the 1960s, a high degree of consciousness has developed regarding these problems. The growth of the concepts of right-to-know, consumerism, and advocacy in public health has led to greater sensitivity to these issues in many countries. Air, water, ground, and workplace pollution are issues of concern to the public, business, the media, nongovernmental organizations, and government and are part of the general culture of our times.

Occupational health developed as a separate area of concern from environmental health, but in recent years it has come to address issues in common. It is included as the second part of this chapter because of common advocacy, professionalism, technology, and regulatory approaches. The level of public response to environmental threats is illustrated by the groundswell of public opinion against nuclear testing and environmental decay. The twentieth century advocacy groups and reformers have made major contributions to public policy. These are akin to the achievements of eighteenth and nineteenth century reformers in the areas of abolition of slavery, humane treatment of prisoners and the mentally ill, working conditions in factories and mines, and general sanitary improvements (see Chapter 1).

Issues have become more complex and go beyond the prevention of disease and traditional public health. They enter the wider realm of human survival in the earth's natural environment. While the resources needed to reduce the environmental neglect from inadequate sanitation and high levels of pollutants in the air and soil are costly, the burden to society of environmental decay can be even greater in the long term.

Human society must order its affairs so that its use of natural resources does not deplete or overwhelm the self-sustainable capacity or natural regenerative powers of the environment. Environmental health is a central issue in the New Public Health in that degradation of the environment creates preventable disease and irreversible loss to society.

ENVIRONMENTAL ISSUES

The environment and society interact and are mutually dependent. The ecological issues that face the world include those which can be addressed locally and nationally and others which require concerted international cooperation. Local action is part of global responsibility. Local issues require close cooperation among different agencies of government at all levels and with nongovernmental organizations, the media, and voluntary groups interested in promotion of a healthful environment. Unrestrained population growth and rising standards of living undermine local and international efforts to maintain a balance between nature and human society.

Issues confronting global society include those that relate to poverty and excess population growth in the poorest countries. Among the long-range issues confronting many countries are water supplies and their quality, which are endangered

---

BOX 9.1 GLOBAL ENVIRONMENTAL CHALLENGES FOR THE TWENTY-FIRST CENTURY

1. Population growth;
2. Economic growth;
3. Food production and distribution;
4. Energy and resource depletion;
5. Soil erosion/desertification;
6. Deforestation;
7. Water supplies/shortages;
8. Air pollution;
9. Chemical/toxic wastes;
10. War/nuclear threats/terrorism/armament costs;
11. Ozone depletion;
12. Global warming;
13. Social, economic, and political inequalities nationally;
14. Inequalities between industrial and nonindustrial countries.

Source: WHO. 1992. *Our Planet Our Health.* Report of the WHO Commission on Health and the Environment.

by overuse and the pollution of groundwater sources. Air and soil pollution, deforestation, and desertification require local, national, and international multisectoral cooperative planning and intervention.

Public consciousness regarding these issues has increased during the past several decades. Environmental concern has become an essential part of accepted public philosophy in many developed countries. Its place in developing countries is often of low priority, coming after the struggle to expand economically and the severe problems of population growth and basic services. Economic growth and health are closed related to agriculture, food supplies, and distribution systems, as well as preservation of agricultural land and rational use of energy. In the countries of eastern Europe, industrialization was given priority over all other issues and environmental concerns subordinated, so that accumulated environmental degradation is part of the long-range burden of post-Soviet societies. This also occurred in other countries during their industrial development and urbanization.

GEOGRAPHIC AND ENVIRONMENTAL EPIDEMIOLOGY

Geographic epidemiology is defined as the description of spatial patterns of disease incidence and mortality. It is part of descriptive epidemiology which generally describes the occurrence of disease according to demographic characteristics of the population at risk and in terms of place and time. Snow's description of cholera in London in 1854 and many other observational studies supported hypotheses that turned out in practice to be the case, even though the direct causal relationships were not demonstrable at the time.

Geographic epidemiology helps to generate hypotheses that can then be tested by more rigorous methods. Environmental and occupational epidemiology study disease in relation to environmental or work-related factors. In practical, everyday public health, the findings of a common point source of disease or death may lead directly to contaminated water, toxic exposure at a work site, or polluted air of a city, but may need case-control or other more formal studies for confirmation.

Epidemiological studies may describe in quantitative terms the relationship between the frequency of disease and the degree of exposure to a particular agent. Such studies are subject to errors in the measurement of exposure. Measurement of exposure by place of residence or work is only an approximation. Moreover, within the same community there will be wide variation in actual exposure levels to the toxic agent. The agent may affect different populations or subgroups differently.

ENVIRONMENTAL TARGETS

In 1985, the European Region of the WHO issued consensus statements on health targets for the year 2000; some of these are listed in Table 9.1. These targets represent a broad societal commitment to stop environmental degradation.

TABLE 9.1   Selected European Region WHO Environmental Health Targets and Issues,
1985–1997

18. Multisectoral policies to protect the environment	Coordination between agencies at international, national, regional, and local levels
19. Monitoring and control mechanisms for environmental hazards	Chemicals, ionizing radiation, noise, biological agents, consumer goods, risk assessment
20. Adequate supplies of safe drinking water	Quantity, quality of water; international, national programs, groundwater and surface water surveillance, quality control; water management standards
21. Protection against air pollution	Legislative administrative and technical measures to control indoor and outdoor pollution
22. Reduced risk of food contamination including harmful additives	Legislative, administrative, and technical measures to control food contamination and additives, and production, storage, transport, sale, and use
23. Eliminate risks of hazardous waste	Effective legislative, administrative, and technical measures for surveillance and control of dumped wastes
24. Healthy and safe urban environment	Housing and urban planning standards, waste disposal, water supply, recreation, open spaces, traffic control, waste disposal, and sanitation
25. Protection against work-related risks	Protection against biological, chemical, physical hazards; worker education, industry self-monitoring, and government regulation

Note: Review of progress up to 1997 indicate moderate cumulative progress has been achieved.
Source: WHO European Region. 1997. *Health in Europe.*

GLOBAL ENVIRONMENTAL CHANGE

Public health has traditionally placed high priority on sanitation, housing, and urban planning in the battle to reduce the burden of infectious disease. The Sanitary Movement of the nineteenth century had an enormous impact on the control of communicable diseases. The lessons learned in disinfection of water supplies and treatment of solid and liquid waste are still not applied universally. This includes not only the less developed countries, but middle-level developing and industrialized countries as well. The threat of local, national, and international disasters, including the reemergence of cholera in South America and Russia and giardia in a major outbreak in the United States, returned the classic issues of water quality to center stage in modern public health. Plague in India and Rift Valley Fever in Egypt in 1994 are dramatic events while malaria and dengue fever resurgence highlight even more the problems of vector control in modern public health.

The concept of environmental health has been widened in recent decades by the spectrum of global changes to the environment as a result of man-made environmental pollution and natural events such as volcanic eruptions. The greenhouse effect is the warming of the global environment and occurs partly as a result of the thinning of the ozone layer of the stratosphere. Disposal of toxic and radiologic waste constitutes very difficult public health challenges in many countries. Land degradation, loss of topsoil, deforestation, groundwater depletion, and acidification of water and soil are all challenges in environmental health for the new century.

The effects of global environmental changes cannot be predicted with certainty, but they require an international response rather than local efforts alone. Poverty, low levels of education, and rapid population growth in the poorest countries with limited food production potential stand in contrast to high levels of consumption and energy use and low rates of population growth in the industrialized countries. Many environmental issues involve more than one country, partly because of transportation of waste products or hazardous materials from one country to another, by wind, water, or deliberately by man. Economic concerns include the destruction of fishing stocks, damage to forests, and more global concerns of ozone depletion, global warming, and ocean pollution. Intersectoral cooperation within a country, and international cooperation and regulation to reduce pollution of common waters in seas, lakes, and rivers shared by more than one country, are part of a broad New Public Health agenda.

Sulfur and nitrogenous oxides from fossil fuel electric power plants can travel long distances after being released from tall chimneys. Such pollutants falling as acid precipitation have led to the destruction of forests in central and eastern Europe. Acid rain generated in some European countries falls in others, affecting waters and animal life in addition to forests. Reduction of acid rain has been achieved during the 1980s in North America by greater selectivity in fossil fuels, and the result is reduced damage to forest and water sources.

The release of various organic solvents, called chlorofluorocarbons (also known as freons or CFCs), used in cooling systems, refrigerators, and consumer aerosol products, causes damage to the earth's ozone layer. This permits entry of ultraviolet (UV) light that was formerly excluded into the earth's atmosphere. UV light causes a rise in skin cancer and cataracts in humans. Substitution for freons is vital to reduce damage to the ozone layer. This can be achieved on an individual level by use of water-based paints and chemical products in daily life. The search for substitutes for refrigerants and toxic chemicals to replace those that damage the environment, and exposed workers, has, along with regulation, become the hallmark of environmental and occupational health.

Greenhouse gases are built up in the atmosphere by carbon dioxide emissions from fossil fuel combustion. These gases block infrared radiation from Earth's surface, leading to trapping of heat. This effect resembles the use of glass or plastic covers to retain heat in a greenhouse. This global warming effect may have long-term serious consequences for Earth's thermal balance. The effects on the polar ice caps can lead to global changes in the level of oceans. Reduction of the greenhouse effect requires international, national, and individual effort, and especially an environmentally conscious public.

Hazardous wastes are being exported from developed to developing countries, and this is potentially solvable by heightened national awareness and stronger international conventions with publicity and fines imposed by international courts against offending firms or nations. In a global economy all of these factors link up with effects on the physical environment as well as on working conditions and many social/political factors, such as the widening gap between rich and poor.

COMMUNITY WATER SUPPLIES

Freshwater is an increasingly scarce resource and can be a serious danger to health. Waterborne disease is still among the major causes of death in developing countries, which often lack adequate supplies of water. In both developed and developing countries, pollution control, reuse of wastewater, and water planning are vital to the national economy and public health.

The International Decade for Drinking Water and Sanitation in the 1970s and early 1980s promoted national, binational, and international efforts to improve community water supplies, sanitation, drainage, education, and hygiene. Implementation of appropriate technology for maintaining water and sanitation infrastructure was emphasized. Safety of community drinking water, as defined by the World Health Organization, requires a combination of standards and protection of raw water sources from contamination. Treatment of community water supplies requires sedimentation, coagulation, filtration, chlorination, and continuous monitoring. High standards of construction and maintenance of water distribution systems are needed whether at the village well or the municipal water supply system. Filtration removes solid and suspended particles, improving the quality of surface source water, and disinfection by chorination effectively kills most microorganisms.

Coverage and protection of reservoirs and canals is also beneficial in improving the security of water sources and in preventing contamination from natural sources, including birds, animals, and vegetation. Community water regulation and enforcement require both physical treatment and disinfection to protect the public against microbiological, chemical, and other health hazards. Agricultural runoff of pesticides and animal wastes are also important contaminants of water sources.

The Clean Water Act (CWA) of 1977, amended the 1972 Federal Water Pollution Control Act. At that time some of the Great Lakes were seriously polluted as were many of the major rivers of the country. The CWA set new U.S. national standards and regulatory mechanisms at federal, state, and local levels of government. It increased regulatory powers to "restore and maintain the chemical, physical, and biological integrity of the Nation's waters." Regulations under the Act permitted effective action against industrial and other polluters and allowed for controls to be established where multiple municipalities were involved in a river or regional water system. This led to steady improvement in water quality of lakes, rivers, and groundwater sources throughout the country, but in 1999, 40% of U.S. waterways were below EPA standards (EPA, 1999).

Concern about potential carcinogenic effects of trihalomethanes may cause withdrawal of mandatory chlorination of surface waters. The absence of adequate disinfection with chlorine increases the risk of serious waterborne disease outbreaks such as the wave of cholera epidemics in South America from 1991 to 1999. New standards may require time to be implemented because of prevailing conservative professional and public attitudes and the cost of treatment plants. In Israel, for example, opinion gradually shifted to mandatory chlorination policy. This was due to a number of factors: increased public and news media awareness of drinking water quality, a greater recognition at the leadership level of the Ministry of Health of the importance of preventive and environmental factors in enteric disease, an increasing presence of younger, better trained sanitary engineers willing to challenge previously accepted dogmas, and persuasive documentation of the impact of contaminated community water supplies on the infectious disease burden of the country.

Waterborne Diseases

Waterborne disease may be so common as to escape detection in point outbreak form. This seems to be the case in many countries, where hepatitis (especially hepatitis A and E) is endemic and where incidence of gastroenteritis from shigella and *E. coli* remains high. In industrialized countries, waterborne disease outbreaks have become uncommon events because of high levels of water management. Water contamination and enteric disease can occur from organisms for which routine testing is not currently practiced. For example, testing for rotaviruses (which cause enteric disease) and organisms such as campylobacter and giardia is not done routinely, but water is tested if there is a suspicion of contamination. Safe water requires physical treatment as well as disinfection of all surface water sources. Effective water management can reduce the burden of gastroenteric disease even in a relatively developed country.

Israel, developing rapidly in the 1960s, built a national water carrier to distribute unfiltered chlorinated water to communities and for agriculture. Local groundwater was not necessarily chlorinated, and sewage system development was inadequate. In the 1970s and 1980s, Israel experienced large numbers of waterborne disease outbreaks. A 1985 outbreak resulted from the contamination of groundwater sources by sewage from a nearby sewage pipe accidentally broken during roadwork, resulting in 9000 cases of shigellosis, 49 cases of typhoid fever, and 1 death. Introduction of mandatory chlorination in 1988 in Israel resulted in a substantial improvement in the quality of community water supplies and greatly reduced the incidence of waterborne disease outbreaks and the total burden of diarrheal disease. At the same time, there was a marked reduction in the overall burden of enteric diseases in the country, including hepatitis A; however, primarily foodborne salmonellosis continues to increase.

In the United States during 1995–1996, 22 waterborne disease outbreaks due to contaminated drinking water with 2567 cases, largely due to giardia. In 1993, cryptosporidium contamination of water sources caused waterborne disease outbreaks in Milwaukee and elsewhere. Some waterborne organisms such as giardia and cryptosporidium constitute a special risk for persons with compromised immune systems, including cancer patients being treated with chemotherapy, HIV positive persons, and patients following transplantation being treated with immunosuppressants. Outbreaks of giardia and cryptosporidium in the United

BOX 9.2 INTERNATIONAL WATER MANAGEMENT STANDARDS

1. International: the United Nations and the World Health Organization promoted the International Decade for Drinking Water and Sanitation and promulgated clear standards of water quality for community water supplies (1958, 1963, 1971, 1984, and 1997).
2. National, state, and local authorities: policy commitment, funding, and professional departments for supervision of community water systems.
3. Municipal water systems: water management and testing varies according to the quality of the source water, and methods of treatment include
   a. High standards of acceptability of source surface water;
   b. Physical treatment (coagulation and filtration);
   c. Disinfection (chlorination)—routine and mandatory;
   d. Maintaining and monitoring of residual chlorine;
   e. Construction and maintenance of water storage and distribution systems;
   f. Monitoring of enteric disease;
   g. Investigation of suspect waterborne disease outbreaks;
   h. Continuous monitoring by bacteriologic and chemical testing;
   i. Assurance of safe distance between sewage and water pipes;
   j. Integrity of water distribution system against inflow.
4. Village wells:
   a. Protection of wells from human and animal wastes;
   b. Regular or periodic chlorination.
5. Sanitary education: at all levels of society including governments, NGOs, intersectoral cooperation, public, professional communities, and in schools.

States have raised concerns because these organisms are not efficiently eliminated by standard water treatment, and are not routinely tested in regular water-sampling monitoring.

The 1993 Milwaukee outbreak was the largest reported waterborne disease outbreak in U.S. history, with approximately 403,000 ill persons of which 4400 required hospitalization. Attack rates were as high as 50% in some parts of the city. Cryptosporidium, in addition to being transmitted from person to person and from animals to humans, can also be transmitted in swimming pools. *Cryptosporidium* is reportedly present in 65–87% of surface water samples tested in the United States.

Early detection by laboratory diagnosis requires preparation of laboratories for identification of these organisms. Regular testing of the community water supply at its origin and within the supply system is essential to monitor water safely. The presence of coliform bacteria indicates fecal contamination and potential hazards, warning sanitation officials that other more dangerous organisms, such as dysentery bacilli or enteric viruses such as hepatitis, may be present. Testing for crypto-sporidium, giardia, and viruses is difficult, costly, and insensitive; therefore, routine testing is not done. Chlorination and filtration may not be sufficient to prevent waterborne disease transmission of these organisms. This is a problem for sanitary control. New methods of testing and disinfection of water supplies need to be devised. At present, filtration and chlorination remain the basic methods of assuring safe community water supplies, supplemented by boiling of suspect water during outbreaks of disease.

Standard water treatment processes remove solid and suspended material, bacteria, and odors from water and have been outstandingly successful in reducing waterborne disease. New concerns over chemical contamination of community water supplies have become prominent in recent decades. Heavily polluted waters have been linked to neurological damage and cancers of the bladder, intestinal tract, liver, and kidney. The U.S. Safe Drinking Water Act of 1974, as amended in 1996, establishes criteria for monitoring of public watersystems for microbiologic, chemical, and other contaminants. It defines maximum contaminant levels (MCLs) for specified chemical pollutants. By 1989 the U.S. Environmental Protection Agency (EPA) had defined MCLs for 83 pollutants, out of some 700 organic, inorganic, biological, and radiological contaminants detected in water supplies around the country. This is an area of public health concern that still requires much epidemiologic and sanitary engineering research.

Right-to-know laws, a critical investigative press, and an environmentally con-

---

BOX 9.3 WATER CONTAMINANTS UNDER UNITED STATE ENVIRONMENTAL PROTECTION AGENCY REGULATION

Microbiological contaminants: turbidity, total coliforms, viruses, *Giardia lamblia, Cryptosporidium, Legionella;*
Volatile organic chemicals: trichlorethylene, tetrachloroethylene, carbon tetrachloride, vinyl chloride, benzenes, ethylene, and ethane compounds;
Synthetic organic compounds: pesticides (lindane, endrin, 2,4-D), trihalo-methanes, dioxin, polychlorinated biphenyls (PCBs);
Inorganic chemicals: arsenic, barium, cadmium, chromium, lead, mercury, nitrates, nitrites, selenium, nickel, silver, fluoride, copper, zinc, thallium, beryllium, cyanide, aluminum;
Radiological contaminants: radium, alpha and beta particle activity, natural uranium, radon.
Note: Not all contaminants have currently defined maximum containment levels (MCLs).

Source: Nadakavukaren A. 1990. *Man and Environment,* Third Edition, Prospect Heights, IL: Waveland Press, and website www.epa.gov/history/topics/sdwoi.htm

scious public are fundamental to prevent serious ecological degradation. Environmental activism has made important contributions to public health, but such activism can be a two-edged sword. One example is the excessive zeal focused on the environmental impact of chlorination and its by-products. Trihalomethane, produced by the combination of chlorine and nitrogenous material (chloroform, bromoform, bromochloromethane, and chlorodibromomethane) in unfiltered surface water, can reach levels that are carcinogenic. However, levels within accepted norms are considered safe, that is, total trihalomethane levels under 0.10 mg/L (EPA, Dec. 1998).

Some opposition to disinfection by chlorination has been based on this concern and led to the spread of cholera in South America in the period 1991–1995 (see Chapter 4). The offset of benefits against risks has resulted in current professional consensus that this is not a reason to cease chlorination, but rather provides additional justification for physical treatment of raw water before chlorination to reduce the nitrogenous material content and thereby reduce the combination with chlorine which produces trihalomethanes, improving water potability and clarity.

Developmental programs including local and large-scale dam projects can have negative health effects by providing a hospitable environment for vectors for diseases such as malaria, schistosomiasis, and onchocerciasis, resulting in resurgence of diseases once controlled. Planning of development projects must take into account the potential ecological effects and the needed control measures to prevent greater health damage than benefit.

SEWAGE COLLECTION AND TREATMENT

Sewage collection and treatment, along with filtration and disinfection of drinking water, have made enormous contributions to improved public health, perhaps even more than the use of modern medicines and vaccines. Sewage contains bacteria, viruses, protozoa, and other pathogens that can cause serious disease. Collection of sewage prevents surface environmental contamination and seepage into groundwater and contamination of local water sources. The purpose of sewage treatment is to improve the quality of wastewater to a level where it can be discharged into a waterway or prepared for reuse for agriculture without damaging the aquatic environment or causing human health problems in the form of waterborne disease. This requires killing the pathogenic organisms present in the sewage.

Primary treatment of community wastewater begins with the removal of solids from the wastewater. This is done by several mechanical processes of screening and sedimentation. The wastewater is passed through screens to remove large solid objects and then through grinders to further break up the solid wastes. The wastewater then flows at reduced velocity through a grit chamber where sand, gravel, and other inorganic materials settle out. Air is injected into the tank to remove trapped gases and to maintain an aerobic environment. The wastewater then flows into secondary settling tanks for further sedimentation of solid particles. Primary treatment removes just over half of the suspended material and particles. The wastewater is then ready for secondary treatment.

Secondary treatment of wastewater is based on biological treatment assisted by mechanical methods, accelerating the natural decomposition of organic wastes. Aerobic microorganisms are used in the presence of an abundant oxygen supply to decompose the organic material into carbon dioxide, water, and minerals. The wastewater is sprayed over trickling filters or beds of crushed stone covered with a slime containing various types of microbes. These microbes absorb the organic material and act to break it down into its various components. The sewage is then processed by the activated sludge method, carried out by introducing bacteria-containing sludge into a tank of wastewater along with compressed air. The waste is then agitated and mixed for 4–10 hours. The microbes are adsorbed to suspended particles and oxidize the organic material. Afterward the sludge, consisting of masses of bacteria, settles out into the tank. The sludge is then removed and recycled into the next tank load of wastewater.

Following primary and secondary treatment, the suspended material and the BOD (biochemical oxygen demand) are reduced by some 90%. This process depends on temperature, which affects the metabolic rate and activity of the organisms needed to break down the suspended organic material. Secondary treatment is most effective against protozoa, worms, and bacteria, but less effective against viruses, heavy metals, and other chemicals. Since 1988, all sewage plants in the United States are required by federal regulations to provide at least secondary treatment.

Tertiary treatment is required if the wastewater is to be recycled for the purposes of drinking, irrigation, or recreation. Tertiary treatment includes a combination of physical, chemical, and biological processes to reduce the particles and BOD to less than 1% of those of the original wastewater. The process includes chemical coagulation, filtration, sedimentation, activated carbon adsorption, oxygenation ponds and aerated lagoons, osmosis, ion exchange, foam separation, and land application. All of these processes remove different pollutants present in the wastewater, in particular tiny particles of suspended organic matter. They also remove synthetic chemicals, ammonia, nitrates, phosphates, and dissolved organic materials.

Disposal of the sludge remaining after sewage treatment by incineration or ocean dumping is environmentally problematic. Use of the sludge for compost in agriculture or gardening is increasing, but contamination may enter the food chain and create another hazard. Sludge disposal should be carefully regulated.

Disinfection is the final stage, accomplished by introducing chlorine into the water so that there is a residual level of chlorine to protect the water from contamination in the water storage and distribution system. In many countries or regions, lack of sufficient local water supplies for community agriculture and industrial use necessitate recycling of wastewater as part of the process of water conservation. Supplementation of water sources by desalination and recycling will be increasingly important as population growth, increasing standards of living, and pressures of agricultural and industrial contamination of water sources increase. New technology in membrane filtration offers hope to improve the efficiency and economics of this sector of the ecological sciences.

SOLID WASTE

The disposal of solid waste has been a problem faced by human society from prehistoric to modern times and will only increase so in the future. With the growth of cities, disposal of refuse took on more significant health importance. In biblical times Jerusalem burned its garbage in a valley outside the city walls (Gehennam, a term later adopted for "Hell"). The Greek city-states had ordinances against dumping refuse in or near cities, providing waste disposal sites for this purpose. In medieval European cities, garbage, human, and animal wastes were discarded into the streets and areas surrounding the home. In the thirteenth century Parisians were forbidden to throw waste on the streets and had to dump it outside the walls of the city. In 1388, the English parliament prohibited waste disposal in public waterways. During the industrial revolution, medieval cities evolved into working class slums. Crowding, poor housing, and poor sanitation forced municipal governments to organize measures to reduce the nuisance and health hazards of solid waste.

Waste management continues to be a problem in the present as greater amounts are generated by the affluent lifestyles of the population of industrialized countries. In developing countries where rural to urban population shift is underway on a massive scale, rapid population growth, crowding, and slums increase the burden of solid waste disposal. During the 1980s and 1990s, return, recycling, and reuse of waste products entered the popular culture in some countries and is beginning to have an impact on reducing landfill needs.

Sources of solid waste include agriculture, mining, industry, and urban waste. In the United States, 95% of solid wastes are from agriculture, mining, and industry. The remainder is from household waste, which generates 150–180 million tons of solid waste annually. This is the equivalent of 4 pounds of refuse per person per day. Municipal waste collection and disposal are serious problems involving high costs and a serious public health burden if not done well.

Waste management involves a variety of techniques, including recycling, composting, incineration, and land refill. Each has its advantages and disadvantages. The techniques are part of the engineering of community infrastructure. Seawater dumping is still practiced in some countries, but increasing global concern about the effects of such practices on the ecology of the lakes and oceans makes this solution unacceptable. Waste disposal alternatives include sanitary landfill, composting, incineration, and reclamation through recycling and reuse.

Landfill is the major method of solid waste disposal. This involves spreading garbage in layers 8–10 feet deep and covering them with a thin layer of soil. This method is adequate if well planned and supervised. The problem of seepage of toxic materials from landfill sites and potentially explosive gas accumulation requires careful assessment of landfill sites and limits their potential. Limited possibilities for suitable landfill locations in large urban concentrations makes this method of disposal a serious urban planning problem. Sanitary landfill is expensive because of the cost of collection and transportation, the land value, and the manpower required. Landfill under sanitary conditions requires compaction of waste and covering by well-spread yet compacted earth far from ground water and surface water. Sites must be fenced to prevent scavenging by people, animals, and off-hours dumpers. It should have well-paved and well-drained access roads. The landfill should be away from residential areas and should be well maintained and tidy. It should be seeded in completed areas with grass and trees to control erosion and should be maintained by well-trained sanitarians. Methane gas produced by anaerobic decomposition in a landfill can be recovered for use.

Composting or conversion of waste products into topsoil can be applied at the household and municipal levels. By-products of wood and food processing can be composted and used to reduce soil pollution from petroleum-based products. This involves separation of nonbiodegradable materials from biodegradable materials and treatment of the materials to break down organic waste. Decomposition at high temperatures (140°F) kills flies, weed seeds, and potentially pathogenic organisms. In closed systems with forced draft aeration this process can be accomplished in a few days. However, with passive methods it takes many months. After further treatment of "curing" and screening or grinding, an excellent soil conditioner can be produced which can be used to enhance agricultural or horticultural work such as in nurseries, public gardens, and parks. Incineration is attracting wide interest, but this is limited by high capital cost and the possible release into the atmosphere of potentially toxic materials such as dioxin and heavy metals. Meticulous maintenance requirements are needed to properly mix the materials for clear burning at high temperatures. In addition, there is the residual problem of disposal of ash that in itself is toxic. Waste-to-energy incineration reduces the volume of waste products by 80–90% and produces energy which can generate electricity and replace fossil fuels. In Japan and western Europe, 30–40% of solid waste is incinerated in waste-to-energy plants.

Using garbage as feed for pigs is no longer acceptable because of the problem of meat contaminated with trichinosis (pork tapeworm). However, the practice is returning on an experimental basis with scraps ground and steamed prior to being used as animal feed. Control of the use of animal parts for animal feed is now being re-evaluated and more intensively regulated following the Bovine Spongiform Encephalopathy (BSE) experience in the United Kingdom and Europe in the 1990s (Chaper 4).

Recycling and waste reduction are methods gaining wide support. Reducing use of disposables (e.g., packaging materials, disposable diapers) requires an ecologically conscious public and municipal, nongovernmental, or volunteer collection systems. Scrap metal, paper, glass, and plastic recycling can be commercially successful. Industry and commercial enterprises can be convinced to reduce use of bulky packaging materials and to adopt "ecologically friendly" practices. Plastics and rubber tires are also recyclable in economically valuable ways. Ecological consciousness is fundamental to the success of such practices.

Recycling of iron, steel, and aluminum in the United States in 1991 was a substantial part of total new production of these metals. Over 11 million vehicles are recycled, supplying 37% of all ferrous scrap. Use of recycled iron and steel reduces air pollution by 86%, water pollution by 76%, and solid wastes by 105%, as compared to production from new ores. Similar benefits accrue from recycling of aluminum scrap. Production of both steel and aluminum from virgin ore are both very polluting and energy intensive.

TOXINS

A toxin is a substance in the environment with the potential for causing human disease or injury. Toxicology is the study of such substances and their effects on humans. All chemicals are toxic under some conditions, depending on the dose, concentration, and the threshold or sensitivity of a given species for that substance.

The range of chemical toxins and methods of classifying them are shown in Table 9.2. The factors that affect the toxicity of an agent, in addition to the extent and duration of exposure, include host factors (e.g., age, gender, fitness, previous

BOX 9.4 BASIC CONCEPTS OF TOXICOLOGY

Bioavailability: the ability of a substance that enters the body to be liberated from its environmental matrix (water, tissue, soil) and to enter the circulation of the host.

Dose–response relationship: the relationship between the quantity of a toxicant received by the host and the probability of an effective concentration at the vulnerable site.

Intermediary metabolism: the metabolic changes that a chemical undergoes once it reaches the cell of the body, usually in the liver. The substance may be detoxified to benign compounds, or may be converted to biologically harmful metabolites.

Mechanism of action: the way the toxic substance acts on a cellular or subcellular level to disrupt the living organism. Some toxic agents are metabolic poisons; others act on cell membranes, interfere with chemical reactions, or bind to nucleic acids.

Susceptibility: the ability of a living thing to be harmed by an agent, which may be influenced by age, sex, genetic disposition, nutrition, prior exposure, immune state or general health, stress, location at work, airflow, temperature, and humidity.

Threshold: the lowest does of a chemical that has a detectable effect.

Toxic effect: damage to an organism as measured in terms of loss, reduction, or change of function, clinical symptoms, or signs. Effects may be adverse in one person and not in others.

TABLE 9.2   Classification of Toxic Agents in Environmental and Occupational Health

Classification	Subgroups
By structure: organic, inorganic	Organic: aromatics (e.g., benzenes), polyaromatics, amines, ethers, ketones, alcohols
	Inorganic: Anions, cations, heavy metals, metalloids (e.g., selenium)
By chemical type	Organochlorines, organophosphates, halogenated aliphatic hydrocarbons, halogenated ethers, polychlorinated biphenyls, monocyclic aromatic hydrocarbons, phthalate esters, polycyclic aromatic hydrocarbons, nitrosamines, metals and inorganics, others
By source	Natural: plants, bacteria, fungi
	Synthetic: industrial reagents, products, or by-products; pharmaceuticals
By use	Pesticides, solvents, paints, dyes, coatings, detergents, cleansers, pharmaceuticals
By action	Enzyme damage, metabolic poisoning; binding of macromolecules (e.g., DNA); cell membrane damage; sensitization, irritation
By target organ	Affecting nervous system, blood, kidneys, liver, lungs, skin, metabolic processes; reproductive system and genetic effects, teratogens; carcinogens

Source: Adapted from Last, 1992.

exposure), environmental factors (e.g., temperature, air flow), and the nature of the toxic agent (e.g., physical and chemical properties). Toxicology is an important part of environmental and occupational health; further reference will require a specialized text, and appropriate internet websites (see bibliography).

Toxic Effects on Fertility

Toxins can adversely affect fertility, pregnancy, and early or later child development. Reproductive potential can be adversely affected by reduced male reproductivity, such as by exposure to the pesticide dibromochloropropane (DBCP). Other chemicals have been implicated in increased abortion rates among exposed pregnant women. Birth defects or teratogenesis occurred with exposure to thalidomide. Other chemicals relate to low birth weight and toxicity in newborns. Exposures to other chemicals such as lead, produce brain damage in children.

Teratogens are substances that cause birth defects, diseases, or abnormalities in the embryo or fetus either by disturbing maternal homeostatis or by acting directly on the fetus. Birth defects historically were attributed to retribution for sin, witchcraft, or moral or physical defects in the mother. Scientific knowledge of genetic disorders has grown since the 1940s, and many agents have been shown to cause birth defects. Such agents act on fetal development and not on genetic DNA, so that a threshold effect is assumed, that is, the effect occurs only if the causative exposure is above a certain threshold. Some currently known teratogenic agents and their effects are shown in Table 9.3.

TABLE 9.3   Some Teratogens and Their Effects on the Fetus and Newborn

Teratogen	Effects on fetus and newborn
Maternal infections	
Rubella	Congenital rubells syndrome, deafness, cataracts, heart defects
Syphilis, herpes simplex	Mental retardation, microcephaly
Cytomegalovirus	Infected kidney, liver, lungs
Toxoplasmosis	Central nervous system lesions
HIV	HIV neonatal transmission
Nutritional deficiency	
Protein deficiency	Abortion, prematurity, low birth weight
Folic acid deficiency	Anencephaly, spina bifada
Ionizing radiation	
X-rays, or nuclear radiation or fallout	Central nervous system disorders, microcephaly, mental retardation
Drugs	
Alcohol	Mental retardation, microcephaly, facial defects
Cocaine	Prematurity, retardation, addiction
Thalidomine	Phocomelia (i.e., small deformed limbs)
Dilantin, valproic acid	Heart malformations, cleft palate, retardation, microcephaly
DES (diethylstilbestrol)	Vaginal cancer in girls, genital deformities in boys
Anesthesia	Miscarriages, structural deformities
Barbiturates	Heart defects, microcephaly, retardation
Chemicals and heavy metals	
Methyl mercury, lead, cadmium	Miscarriages, mental retardation, neurological disorders
Dioxin	Physical deformities, miscarriage
Cigarette smoke, direct and "second hand smoke"	Miscarriage, prematurity, low birth weight

Source: Adapted from Nadakavukaren, 1990.

Toxic Effects of Lead in the Environment

In the United States in the 1920s, tetraethyl lead use in fuel was promoted to improve automobile performance. This caused a long struggle between public health and regulatory agencies and industry. Industry won, and leaded gasoline was used well into the 1960s and is still available in many parts of the world. Alice Hamilton during the 1920s investigated the widespread use of lead in industry and successfully lobbied for legislative changes to increase surveillance and improve safety by reduced exposure. Community exposure to lead was identified as a public health problem in the 1960s when trace quantities were found in food, beverages, soil, and air. Lead used in fuels for cars and in lead-based paints manufactured from the 1920s to the 1960s was an important source of community exposure.

Children were exposed to lead from the environment, mainly from automobile and industrial emissions, as well as sources in the home from lead-contaminated paint. Clinical effects appeared particularly in children and at lower blood concentration levels than previously thought to be significant. "Acceptable" levels were lowered and lead abatement programs introduced. These programs were es-

---

BOX 9.5 LEAD ABATEMENT IN THE
UNITED STATES, 1977–1994

Studies in the United States based on the NHANES II (Second National Health and Nutrition Examination Survey) showed that blood lead levels declined by 78% from 1976–1980 to 1988–1991. Among children 1–2 years of age, blood lead levels (BLLs) declined by just over 88%. This reduction was due to a number of factors, including the following:

1. Reduction by 99.8% of lead use in gasoline from 1976 to 1990;
2. Reduced use of food and soft drink cans containing lead solder, from 47% in 1980 to 0.9% in 1990;
3. Reduced use of lead-based house paint;
4. Promulgation of national standards for lead exposure in industry;
5. A ban on the use of lead soldering on household plumbing;
6. Screening of children, and intervention where elevated BLLs found;
7. Lead abatement by county health departments through removal of lead-based paint in older housing;
8. Increased provider and parental awareness of lead-induced permanent brain damage hazard;
9. Strong positions of the Centers for Disease Control, American Academy of Pediatrics, state and county health departments, and child advocacy organizations;
10. Public awareness.

Source: Centers for Disease Control. 1999. *Morbidity and Mortality Weekly Report,* 48:461–469.

---

pecially needed in urban slum areas where children were exposed to lead-based paints in older homes and heavy urban traffic and were found to have high blood lead levels (BLLs) with risk of brain damage as a result. In 1992, the American Academy of Pediatrics adopted even lower BLLs levels as danger signs of lead toxicity sufficient to cause brain damage in children.

Safe levels of lead are a subject of controversy. There may be no harm-free body level of lead, and current recommendations include routine testing of infants and young children as well as exposed workers, along with reduced emission levels and industrial or home use of lead or lead-containing products.

AGRICULTURAL AND
ENVIRONMENTAL HAZARDS

Pesticide and herbicide use to increase agricultural production is a worldwide phenomenon. Resistance to widely used chemicals has developed so that there is a continuing search for new chemicals. Excess use affects the ecosystem by the buildup of pesticides in the food chain and in groundwater, and the long-term effects may be serious.

Short-term exposure to agricultural chemicals may result in acute poisoning, especially in developing countries where this is estimated to affect some 3 million persons with 220,000 deaths annually. Suspected concentration of pesticides in breast fat tissues may be linked with excess breast cancer. Pesticide use in North America and the former Soviet countries is high but has declined during the 1980s, while their use in western Europe exceeds both and is increasing. Widespread use of pesticides in developing countries is often poorly supervised, and pesticide poisoning episodes are common.

The use of herbicides and pesticides within the recommended limits of the *Codex Alimentarius* (joint foods standards manual of the Food and Agriculture Organization, or FAO, and the WHO) and methods recommended by the International Code of Conduct on the Distribution and Use of Pesticides are considered safe. Current recommended practice is to reduce the amounts of pesticide and herbicide use, accompanied by care and safe use and storage practices to reduce the chance of acute poisonings. Alternative agricultural methods, using little or no chemicals, are the subject of wide research and experimentation.

AIR POLLUTION

The External Environment

Air pollution is contamination of the air by smoke, solid material, or chemicals that cause health and ecological damage to the community and the environment. It includes the oxides of sulfur and nitrogen spread locally and over long distances, including between nations. The effects are increasingly important as the use of fossil fuels has grown for the internal combustion engine, heating, and power generation. Coal fuel used in homes created the terrible air pollution of nineteenth and early twentieth century London. This has subsided since the 1950s with reduction of brown soft coal usage in individual homes. Use of coal fuel-energy plants in central and eastern Europe has created a gray zone of air pollution carried long distances, destroying forests and creating serious damage to the human environment and health hazards to large population groups. Similarly, extensive damage has occurred in Canadian forests from acid rain originating in the United States.

Large modern fossil fuel plants built near population centers may use high chimneys to disperse the effluent. This reduces exposure of the adjacent population but contributes to long-distance pollutant effects, carrying sulfur and nitrogen oxides to forests and bodies of water creating sulfuric and nitric acid or acid rain. Acid precipitation affects rivers, streams, and lakes, many already burdened with sewage effluent and pesticide runoff, damaging the ecosystem and animal and plant life. The effects on human health are not easily measurable in a directly attributable way, but environmental damage affects the quality of life. International transmission of environmental damage is seen in nearly a quarter of Europe's forests from acid rain originating in eastern European countries with poor emis-

---

BOX 9.6 UPTON SINCLAIR—THE JUNGLE

"A full hour before the party reached the city they had begun to note the perplexing change in the atmosphere. It grew darker all the time, and upon the earth the grass seemed to grow less green. Every minute as the train sped on, the colors of things became dingier; the fields were grown parched and yellow, the landscape hideous and bare. And along with the thickening smoke, they began to notice another circumstance, a strange, pungent odor. . . . It was now no longer something far off and faint, that you caught in whiffs; you could literally taste it as well as smell it."

Source: Sinclair U. 1906. *The Jungle.* Classic Series, 1965. New York: Airmont Co., p. 31.

---

sion control standards. The range of damage measured by the percentage of dead and dying trees varies from over 50% in some eastern European countries (the Czech Republic and Poland) to over 24% in central and western European countries (Denmark, Norway, the Netherlands, and Germany). In the 1979 Convention on Long-Range Trans-Boundary Pollution, European states agreed by 1993 to reduce by 30% emissions that could cross international boundaries.

Air pollutants can enter the food chain by contaminating fish, fowl, and livestock. Changes in the acidity of water can have further effects by corrosion of water pipes, affecting the lead, mercury, aluminum, cadmium, or copper content of drinking water. Acidified metals may cause chronic conditions such as chronic obstructive lung disease and asthma as well as specific chemical toxicity. These are difficult to measure epidemiologically, so that regulation of source emissions are set as proxy measures for preventable exposure to unhealthful contaminants.

Particulate matter in air pollution has both physical and chemical effects on the nasopharynx and respiratory tract. Excess cancer of the respiratory tract and chronic obstructive lung disease can be demonstrated in exposed populations. A variety of syndromes are associated with specific respiratory irritants, such as coal dust (miners lung), cotton dust (byssinosis), and others among exposed occupational groups. A study of regional cancer rates in Israel in the 1980s showed an excess of cancers of the nasopharynx and respiratory tract in persons living in an area exposed to high levels of silicate materials in emissions from a local cement plant. Geographic cancer epidemiology in the United Kingdom shows higher levels of many diseases in terms of SMRs in urban or other polluted areas in Britain, correlating excess morbidity with excess air pollution (Chapter 3).

The London "killer fog" incident in 1952 implicated in up to 4000 deaths, raised international concern over the deadly effects of critical levels of pollution as well as the long-term effects. In Britain this led to controls on use of soft coals for home fires and gradual reduction of the Victorian levels of smog that had fouled ambient air quality in British industrial and commercial centers. A similar inversion in 1948 in Donora, Pennsylvania, affected over 40% of the population of 14,000 with 20 dealths. A smog crisis in New York City in 1966 occurred a month before the third National Conference on Air Pollution, followed by a series of smog crises in California.

Localized air pollution is largely generated by the automobile as well as by industry. Pollution of urban areas with lead, SO_2, and nitric oxide (NO) has been reduced where catalytic converters and unleaded gasoline are compulsory. However, the beneficial effect is reduced simply by the increase in the number of automobiles, as shown in the southern California experience. During the 1960s and 1970s, there was a growing sense of crisis in pollution of the environment in the United States. Until the 1970s, solid and liquid industrial wastes were dumped or discharged indiscriminately with volatile chemicals contaminating water sources and the air. Pollution of lakes and rivers, and poor air quality in the cities led to a series of federal legislative acts, including establishing of the Environmental Protection Agency (EPA) in 1970, Motor Vehicle Air Control Act of 1967, the Air Quality Act of 1967 and the more effective Clean Air Act of 1970, the Clean Water Act of 1977, and the Safe Drinking Water Act of 1974 (amended in 1996).

Traffic congestion in modern cities exposes car occupants and pedestrians to exhaust fumes containing particulate matter and air pollutants. Pollutants may act to compound the ill effects of other risk factors such as smoking. Los Angeles is subject to heavy pollution and temperature inversions, producing harsh conditions for those prone to chronic bronchitis, asthma, and chronic obstructive lung disease.

A study in Los Angeles showed that an increase of 10 parts per million of CO levels in the air was associated with a 37% increase in hospital admissions. Other large cities that have not implemented such pollution controls, such as Mexico City, continue to have serious air pollution levels.

Where the numbers of cars and trucks increase but these measures are not required, the pollution can have alarming effects on child health in the form of increased blood lead levels and respiratory tract damage from other chemical and particulate pollutants. Carbon monoxide (CO) blocks red blood cell uptake of oxygen and can reduce the oxygen carrying capacity of blood. In vulnerable groups, such as children, the elderly, pregnant women, and the immunosuppressed, this can have serious deleterious effects on psychomotor function. Polycyclic hydrocarbons released from car emissions and other sources are carcinogens. Nitrogen oxides (NOx) affect the terminal respiratory tractalveoli, increasing susceptibility to lower respiratory tract infection in children. Ozone (O_3) and secondary pollutants affect ultraviolet (UV) light absorption, increasing skin cancer incidence. Ozone can travel hundreds of kilometers, causing clinical effects close to and well away from the site of the traffic. Carbon dioxide (CO_2) affects global warming with potentially important effects on world climate and water supplies. The health and environmental effects of air pollutants are summarized in Table 9.4.

Emission control through regulation and new technology should be seen in the context of overall transportation policy. Policy in transportation has long-term effects in determining degrees of air pollution, land use, and trauma from motor vehicle crashes. A full accounting of the costs of morbidity and mortality associated with air pollution and traffic accidents should be included in cost-effectiveness

TABLE 9.4   Air Pollutants: Sources, Health and Environmental Effects, and Changes in Emissions and Air Concentrations, United States, 1988–1997[a]

Pollutant	Environmental and health effects	Decrease (%), 1988–1997 Concentrations	Emissions
Carbon monoxide (CO), from incomplete combustion of motor vehicle fuel	Interferes with hemoglobin oxygen carrying capacity causing slowing of thought, responses, headaches; may cause unconsciousness and death; long-term exposure adversely affects ischemic heart disease, fetal development	38	25
Nitrogen oxides (NO_x), normal by-product of fuel combustion	Irritates lung tissue, hay fever, asthma, allergies, bronchitis, pneumonia; constitutes one-third of acid rain and greenhouse gases	14	1
Hydrocarbons (organic compounds, e.g., benzene), from vehicle and industrial emissions	Associated with cancer, leukemia, and impotence		
Airborne particulates, from vehicle (lead 80%, chromium) and industrial emissions (PM-10)	Lead impairs mental development of fetus and infant; heavy metals are respiratory irritants, carcinogens, and toxins affecting nervous, circulatory, and reproductive systems	Pb 67	44
Ozone (O_3) produced by sunlight acting on ambient NO_x and volatile organic compounds	Irritates eyes, ears, nose, throat, and lungs causing coughing, headaches, reduced resistance to infectious diseases; Aggravates asthma, bronchitis, heart disease; acid rain can destroy forests	16–19	20
Carbon dioxide (CO_2), from motor vehicle exhaust fumes	No direct health effects, but important in "greenhouse effect" accelerating global warming	na[b]	na
Sulfur dioxide (SO_2) from sulfur in fuels (oil, coal)	Irritant exacerbates bronchial and lung problems; major source of acid rain	39	12

Source: Office of Air Quality Planning and Standards, 1998. Latest Findings on National Air Quality: 1997 Status and Trends. Research Triangle Park, NC: Environmental Protection Agency, EPA-454/F-98-009, December 1998.

[a]In 1997, the EPA revised ozone (O) and particulate matter (PM) national air quality standards. PM size was reduced from 10 μm (PM-10) to 2.5 μm (PM-2.5). Air quality data were derived from averaging thousands of yearly measurements from thousands of monitoring sites around the United States. Emission data were derived from engineering estimates of tonnage of pollutants released into the air annually, and emission monitoring of SO_2 and NO_x from electric utility plants.

[b]na, Not available.

studies of rail versus road transport, especially in crowded urban communities and in countries with limited land space.

The U.S. federal Clean Air Act of 1970 established air quality standards for major pollutants such as NOx, CO, SO_2, ozone, asbestos, dioxin, and other toxic air contaminants (TACs). Improving enforcement, especially of automobile emissions, has led to improved air quality in many parts of the country. This federal legislation is implemented at the state level. It sets standards for ambient air quality, automobile emissions, and emission by stationary facilities, such as power plants and factories. Such standards are also being implemented in many other countries.

The Clean Air Act Amendments of 1990 listed 189 hazardous air pollutants (HAPs) for which Congress mandated EPA to issue standards. These include asbestos, dioxin, diesel, and many other potentially toxic agents, including latex, which has been identified as a factor in asthma. The FDA is continuing to develop standards for other HAPs. The Clean Air Act as amended provides for state agencies to regulate local air districts.

The California Air Quality Management Board regulates local air quality management boards (e.g., of southern California) that carry out a certification process of local industry through regional Air Quality Management Boards. This board has powers to sanction changes in industrial practices in any given industry by attributing its component of ambient air pollution, with the potential for closing an offending industry. As a result, California was able to reduce air pollution dramatically since the mid 1980s, with only one major smog alert in 1997 compared to 66 in 1987 in Los Angeles. Technological innovations are becoming standard in some new automobiles at the end of the 1990s that will further reduce emissions and increase air mileage per gallon of gasoline. Other innovations incorporate hydrogen fuel cells and hybrid and electric vehicles that will release nearly zero pollutants.

Diesel air pollutants became the subject of scrutiny and decision by the state Air Resources board which carried out a meta-analysis and defined diesel pollution as a health hazard. This requires use of best available control technology (BACT) to reduce emissions, from the defined "acceptable risk" of 10 excess cases of cancer per million population. In the case of diesel emissions the excess rate was determined to be 100 times in excess of the acceptable rate. Industry opponents to this step raise the specter of horrendous economic effects of such decisions, but the BACT approach minimizes this potential harm to the economy. At the same time the process of identifying the issue spurs industry to seek out technological solutions that are compatible with greater efficiency in the long run.

Methyl *tert*-Butyl Ether

Methyl *tert*-butyl ether (MTBE) is a synthesis of methanol and isobutylene, developed as an additive to gasoline to improve octane performance, and was widely adopted in the United States, especially in California in place of ethanol used in other states. Ethanol is a farm product that was encouraged by the U.S. Department of Agriculture as a new economic opportunity for farmers and relatively nontoxic agent environmentally. Some gasoline producers opted to use MTBE instead as it is produced and promoted by the chemical industry. MTBE is a volatile, ether-based chemical agent that when found in drinking water gives a bad taste. MTBE was adopted widely without adequate testing for potential toxic effects and has come under scrutiny. Evidence of carcinogenesis in rats raised concerns that MTBE may have this effect in humans, and is a public health risk, especially to drivers, gasoline station attendants, and refinery workers exposed to high levels of

MTBE. The effectiveness of MTBE in promoting clean burning of gasoline and reducing exhaust pollutants has also been questioned. MTBE has been found in 3.4% of water districts in California. Some 50% of drinking water wells in Santa Monica, California, were closed due to MTBE contamination in 1995.

The MTBE case showed the EPA allowing use of a chemical substance widely used in industrial settings as a gasoline additve and thus present in automobile emissions. A more environmentally safe substance from the farm industry was allowed to be replaced even before environmental concerns brought the issue under public scrutiny. The American Public Health Association has since 1996 called on the FDA to ban MTBE as a hazardous chemical, to place restrictions on the use of gas powered boats on lakes and rivers, and to return to ethanol based fuel additives. An EPA-appointed committee in 1999 recommended reduced use of MTBE and substitution of ethanol additives, as well as major efforts to reduce MTBE contamination of ground and surface water supplies (EPA, 1999).

INDOOR POLLUTION

Contaminants within private dwellings may be a greater health hazard than external pollution. Indoor pollution particularly affects women, the very young, the ill, and the elderly because they usually spend more time within the home. Increased insulation, window layers, sealed doors, and smoking all contribute to increased concentration of indoor pollutants, including benzene, formaldehyde, carbon monoxide, and radon gas, as well as bacteria, fungi, and viruses. Smoking is a widespread habit, and passive smoking, or inhalation of smoke generated by other persons, is a long-term health hazard.

Wood and its waste products, vegetable matter, and animal dung are sometimes referred to as bamboo fuels. These are not efficient fuels in terms of heat produced per unit mass as compared to fossil fuels. These fuels are used extensively in rural areas of developing countries because they are cheap and widely available, but they require much time to gather and lead to deforestation with other damage to the environment. Often primitive stoves are used, which result in high levels of continuous daily indoor pollution due to poor ventilation, as well as creating fire hazards. Carbon monoxide poisoning during sleep is also a danger. Approximately half of the world's population depends on such fuels for their daily needs.

The dangers associated with use of bamboo fuels include fires, smoke inhalation, and chronic indoor pollution. These fuels release many chemical compounds including suspended particulate matter, carbon monoxide, nitrogen and sulfur oxides, aldehydes, hydrocarbons, benzene, phenols, and complex hydrocarbons. Women in India show high rates of right heart failure (cor pulmonale) from cooking stove fumes. Technological development of more efficient wood stoves would reduce the problem; however, other forms of energy are more efficient and less damaging to health in the home and to the environment.

Indoor pollution from materials used in construction are important health problems. Asbestos in the home may contribute to mesothelioma and lung cancer. Lead paint in the home increases the hazard of lead toxicity among young children, which is associated with brain damage. Unsafely packaged household chemical solvents and mold in the home contribute to poisonings as well as asthma morbidity and mortality.

Radon Gas

Radon is a very heavy gas producing alpha particles. Radon originates in the natural radioactive decay of uranium from soil and rocks such as granite, shale, and phosphate and is present as a gas in ground crevices, dissolved in water, or dispersed in open air. It seeps into homes via basement cracks, and into well water and point sources which enter homes. Radon by-products enter the lungs. Radon was first detected in homes in the United States in 1984 near Philadelphia. Early investigations showed in-home radiation exposure as high as the equivalent of 455,000 chest X rays. Further investigation revealed that sections of eastern Pennsylvania, New Jersey, and New York lie over uranium-rich geological formations that result in high levels of radon contamination.

The U.S. Environmental Protection Agency in 1988 advised that all homes be checked for radon levels. Inexpensive home radon detectors are available that meet EPA standards. In 1988, the EPA estimated that radon contributes to between 7000 and 30,000 cases per year, or up to 10% of all lung cancer deaths in the United States. Hundreds of thousands of U.S. citizens receive as much radiation as did persons living near the Chernobyl plant at the time of the nuclear accident in 1986. Cigarette smoking has a synergistic effect, enhancing the radon-related risk of lung cancer by a factor of 10. Radon reduction can be carried out in high risk homes by carefully planned sealing of identified sources, ventilation, and fans for high radon basements.

Outdoor–Indoor Pollutants

Carbon monoxide, nitrogen oxides, chemicals, and particulate matter are common outdoor pollutants that can accumulate in homes with kerosene and wood stoves, attached automobile garages, or cigarette use. Passive smoking can expose the nonsmoker to benzene and other carcinogens. Formaldehyde is produced from insulation material, plywood, and floor coverings, especially in mobile homes. Chemical fumes from household products, such as disinfectants, solvents, hair sprays, furniture polish, and dry cleaning solvent, also pollute the home atmosphere as well as provide a potential for childhood poisonings.

Carbon monoxide poisoning from home heaters where there is inadequate ventilation causes 100 deaths per year in the United Kingdom. Some deaths from carbon monoxide poisoning may be attributed to heart disease and can only be diagnosed affirmatively by measurement of carbon monoxide in the air or blood carboxyhemoglobin levels.

Biological Pollutants

Bacteria and fungal spores can enter a building and infect inhabitants, usually through the ventilation system, as is the case with Legionnaires' disease (see Chapter 4). Occupants of a building may suffer from allergies due to fungal spores, mites, animal dander, and feces of roaches or mites. These allergies are more likely to occur in buildings using humidifiers or vaporizers with stagnant water, which favor bacterial and fungal growth.

Sick Building Syndrome

The term sick building syndrome used to describe a common symptomatology (headache, eye and nose irritation, dizziness, fatigue, wheezing, or recurrent respiratory infections) among persons working in a specific building. This may occur as a result of a poor ventilation system that fails to provide sufficient fresh air, specific microbiological pollution of a ventilation or humidifier system, vehicle exhaust entering a ventilation intake vent, ozone emissions from photocopying machines, formaldehyde in wood paneling or furniture, or cigarette smoke.

Nonresidential buildings are often sealed with ventilation provided by mechanical means, so that such conditions can occur when the ventilation system is inadequate. Building codes should specify minimum levels of outside air admission, acceptable levels of oxygen, carbon monoxide, and carbon dioxide, odor dilution, and adequacy of ventilation equipment.

HAZARDOUS OR TOXIC WASTES

Toxic materials used in industrial processes can cause ill-effects in workers exposed to the material at the site of production and in storage, transport, and use of the materials. They can also cause harmful effects on persons living near to the material, as well as long-term effects on the environment. Case studies of serious environmental pollutants and their effects on health demonstrate the problems involved.

Hazardous wastes are defined as any discarded material that may pose a substantial threat to human health or the environment when improperly handled. They include toxic wastes such as arsenic, heavy metals, and pesticides which can cause acute or long term health problems; ignitable wastes include organic solvents, oils, plasticizers, and paint waste; and corrosive wastes (with a pH of under 2 or over 12.5) which can eat away metal containers or living tissue.

Reactive wastes include obsolete munitions and acids which react with water or air to produce explosions or toxic fumes. Radioactive and infectious wastes from hospitals are also hazards to public health. Hospital wastes took on new importance with the dangers of transmission of hepatitis B, HIV, and drug-resistant microorganisms in contaminated materials. The problem caught worldwide attention in the late 1980s when waste material from hospitals washed onto beaches in the United States. Prevention, and waste site remedies have gained wide attention by federal, state, and local government, and by industry, because of media and public concern generated by episodes such as Love Canal.

Love Canal.   The Love Canal episode in the late 1970s served to mobilize public opinion and awareness of environmental health in the United States. In the

1890s, Mr. William T. Love built a canal bypassing Niagara Falls with the intent of building an industrial city using inexpensive hydroelectric power. The project failed and the canal was abandoned and the land sold at public auction. In 1942, the Hooker Chemical Company (subsidiary of the Occidental Petroleum Co.) received permission to use the canal to dump chemicals from its several plants in the area. Up to 1953, 21,000 tons of chemical wastes (acids, alkalis, solvents, chlorinated hydrocarbons, etc.) were disposed of at the site, when it was covered by land fill. Despite warnings, the land was sold and over 1,000 homes, apartments, and schools were constructed along the covered canal.

From the 1950s, local residents complained of foul odors and chemicals oozing from the covered canal were found. In 1978, pressure from local congressmen and news media prompted investigation by the U.S. Environmental Protection Agency and the New York State Department of Health. Over 200 different chemicals were identified, including dioxin and 12 known or suspected carcinogens, mutagens, and teratogens. The New York State Commissioner of Health proclaimed an imminent health peril and called for evacuation of pregnant women and children under age 2. Over 1000 families were evacuated and 300 homes demolished at public expense. Work at the site to contain the chemicals and prevent seepage and groundwater contamination cost over $180 million. This could have been prevented by the Hooker Chemical Company by a $2 million investment at the time of the disposal.

Epidemiologic studies of the exposed residents showed that they experienced statistically significant elevated rates of miscarriage, birth defects, and chromosomal abnormalities, but the studies' methods and conclusions remain controversial. This episode focused national concern on the approximately 16,000 hazardous waste sites throughout the United States. In 1980, Congress established a "Superfund program", funded by a federal tax on the chemical and petroleum industries, to locate, investigate, and cleanup the worst sites nationally

Minimata Disease. Minimata disease is a chronic neurological disorder caused by methyl mercury. The disease was first reported near Minimata Bay in Japan in 1968. Mercury oxide was being discharged from a chemical plant into the waters of the bay. It was converted to an organic form, methyl mercury, by organisms in the mud and slime of the bay floor. This organic form entered fish where it was concentrated and poisoned consumers of the fish. As of 1990, 2248 cases were reported with 1004 deaths. Compensation, cleanup, and damages cost hundreds of millions of dollars. This episode also served to mobilize international public opinion to the dangers of toxic waste disposal as a health hazard. Mercury poisoning of fish is a recurrent phenomenon where industrial wastes discharged into rivers, lakes, and the sea enter the food chain, and humans are affected through fish consumption. In 1999, people living in remote areas of Brazil were found to have methyl-mercury poisoning, probably from fish contaminated by methyl mercury used to purify gold and thus contaminating rivers.

Toxic Waste Management. Pollution prevention at the workplace has become part of management processes as industry responded to increasing federal and state regulation and as the public demand for greater corporate responsibility led to increasing punitive litigation. In 1986, the federal Office of Technology Assessment published *Serious Reduction of Hazardous Waste*. An OECD (Organization of Economic Cooperation and Development) 1992 publication called on workers to play a greater role in pollution prevention. The chemical industry responded with the idea of total quality environmental management (TQEM), adopting pollution prevention as integral to industrial management. Companies like 3M, Monsanto, and Rhone-Poulenc and industrial associations (the Chemical Manufacturers Association) undertook environmental prevention policies. Community activism helped industry to respond positively and openly to environment hazards in their communities. The search for safe alternatives to toxic chemical waste management can be costly but may save a company large outlays in fines, litigation, and damage to corporate image. The issue is now very broadly shared between government, private industry, workers, and community involving planners, scientists, engineers, regulators, residents, as well as environmental organizations and consumers. The EPA is currently promoting waste minimization of persistent bio-accumulative and toxic (PBT) chemicals from industrial sources. This includes source reduction and recycling aimed at reducing hazardous waste products (EPA, 1998. Website http://www.epa.gov/wastemin).

RADIATION

In 1895, Wilhelm Roentgen's discovery of X rays opened an important contribution to medical science, as well as opening the fields of physics and chemistry to radioactivity. Ionizing radiation includes particulate radiation of alpha and beta particles, as well as electromagnetic X rays and gamma rays. Alpha particles are easily stopped by a thin sheet of paper, while beta and gamma radiation can penetrate barriers both inside and outside of the body. Ionizing radiation can dislodges atoms or parts of atoms and destroy chemical bonds. This can adversely affect living organisms, especially vulnerable fetal cells, resulting in mutations or carcinogenesis.

Ionizing Radiation

Ionizing radiation includes high-energy electromagnetic radiation, such as X rays and gamma rays, that are of shorter wavelength and higher energy than ultraviolet or visible radiation. It also includes high energy particles such as electrons, neutrons, protons, and alpha particles. Excessive exposure to these forms of radiation have early and late effects depending on dose and the tissue exposed. Early effects of exposure to high doses of radiation may be fatal due to acute damage to the gastrointestinal, erythropoietic (blood-forming), and central nervous systems. Late effects include malignant disease such as leukemia and birth defects. The principal sources of radiation exposure for the general public include natural background radiation from radon (55%) outer space (8%), the Earth (8%), inhaled or ingested materials (11%), medical exposures (15%), discharges from the nuclear sources (1%), and consumer products (3%). Variation in background exposure from natural sources, home construction materials, and geographic location can be quite high.

Shorter exposures with high dosage is far more serious than long-term low dose exposure. Radiation sickness of people exposed to radiation from the atomic bomb explosions at Hiroshima and Nagasaki and nuclear accidents ranged in severity with a variety of short- and long-term responses. The long-term responses have been less severe than originally feared.

Ionizing radiation of humans can act as a mutagen, a carcinogen, and a teratogen. It can cause cataracts, impaired fertility, premature aging, and skin damage. Radiation-induced cancer can occur as little as 2–5 years after exposure or following a latency period of up to 25 years after exposure. Greater risk occurs for those exposed *in utero*. X-ray-induced disease from excess exposure, faulty equipment, or human error is a hazard of medical care. There is perhaps no safe exposure to ionizing radiation beyond atmospheric background, and any extra exposure should be limited, with prudent exposure to X rays and limited exposure to atomic radiation from domestic or military uses.

Nonionizing Radiation

There are two types of nonionizing radiation, optic and some electromagnetic fields. Optic radiation inclues ultraviolet and infrared. Electromagnetic fields, such as those induced by microwave or radio frequencies, are described in terms of wavelengths or frequency. The harmful effects of nonionizing radiation are of three main types: photochemical (sunburn or snow blindness), thermal, and electrical.

The health effects of ultraviolet (UV) radiation include increasing incidence of squamous and basal cell carcinoma and melanoma of the skin, a highly malignant cancer. This kind of radiation is associated with excess exposure to the sun, which, in addition to these skin cancers, causes skin and eye burns, cataracts, reduced immunity, and damage to blood vessels. Infrared radiation exposure over long periods is associated with increased risk for cataracts, impaired fertility, and tissue damage.

Long-term exposure to high voltage power lines and radio and radar transmitters is suspected to be associated with increased risk of cancer, but this has not yet been proved. Microwave exposures at high levels can damage vulnerable tissues, but the level of dangerous exposure has not yet been conclusively determined. Some people have recently become concerned about the effects of long-term cellular phone use. Lasers are pulsed electromagnetic waves used increasingly in medicine and industry. Lasers not intended for medical use (and misused medical lasers) can cause irreparable retinal damage and severe burns.

Use of low dose irradiation in production, processing, and handling of foods to prevent food safety hazards is being widely supported by professional organizations. It provides an important adjunct to sanitation and good manufacturing practices to reduce morbidity and mortality associated with food-borne diseases even in industrialized countries. More than 40 years of research and use in the United States and many other countries have demonstrated the effectiveness and safety of low dose irradiation. This is rapidly becoming an essential part of public health protection from food-borne disease in the United States and internationally, although public acceptance is still problematic.

ENVIRONMENTAL IMPACT

The U.S. National Environmental Policy Act (NEPA), passed in 1970, made protection and restoration of the environment a matter of national policy. NEPA required all federal agencies to take environmental considerations into acount in decision-making processes and program implementation. Environmental impact statements are required for major construction and public works programs, delineating positive impact, possible adverse effects, alternatives, and any irreversible effects. This legislation resulted in changes in many national projects and promoted a governmental regulatory approach to supervision, control, and prevention of pollution with materials and processes that could harm human health and the environment.

Emergency Events Involving Hazardous Substances

Since World War II, there has been a rapid increase in the number of chemicals developed and used worldwide. More than 60,000 chemicals are available, with some 600 new substances produced every year, an unknown number of which are hazardous. The health effects resulting from the release of a hazardous substance are often unknown. A hazardous substance release is defined as the uncontrolled or illegal release or threatened release of chemicals or their hazardous by-products.

Reportable events are defined as those events in which the substances need to be removed or cleaned up. Plant management is liable for damages due to negligence in both civil and criminal law. Where community exposure occurs from negligence, accident, or natural disaster, a public health emergency response is required, based on prior preparation.

In the United States between 1988 and 1992, 34,575 toxic chemical accidents were reported, of which 2186 involved deaths, injuries, or evacuations. These accidents released 680 million pounds of toxic chemicals into the environment. Almost two-thirds of these incidents involved one of 15 chemicals with polychlorinated biphenyls (PCBs) heading the list followed by anhydrous ammonia, sulfuric acid, chlorine, hydrochloric acid, ethylene glycol, sulfur dioxide, radioactive materials, and hydrogen sulfide. The potential for chemical disasters requires a fundamental preventive approach by industry with supervision by federal and state regulatory agencies. In the United States the major federal agencies involved are the Environmental Protection Agency (EPA) and the Occupational Safety and Health Administration (OSHA).

Internationally, a number of major disasters have occurred in recent decades. In Seveso, Italy, in 1976, an explosion in a chemical factory resulted in 17,000 persons being evacuated and many terminations of pregnancy among exposed women. In 1984, a sudden release of highly toxic methyl isocyanide from a chemical plant in Bhopal, India, caused thousands of deaths, blinding or permanently injuring several thousands more, requiring evacuation of an estimated 300,000 persons living in adjacent neighborhoods. The tragedy at Bhopal led to greater recognition by policy makers and the public that the potential for toxic accidents is anywhere, any time, and any place and not just in developed countries. Transfer of hazardous occupations and industries to less developed areas is a growing issue.

Nuclear and chemical disasters have become a major element in disaster planning for corporate, investor, and occupational and environmental health agencies, as well as for communities adjacent to chemical production, storage, or transportation. Emergency responses to chemical, radiation, or biological catastrophes involve specialized expertise, based on common principles of prevention, monitoring, and crisis management. These include prior emergency planning, speed, coordination of civil and military resources, skilled professional teams, providing information to meet the public concern to know, logistical, medical, and laboratory support, on-site case management and evacuation, investigation of causes, and continuous teamwork among all involved agencies.

The *Exxon Valdez* was a large oil tanker that ran aground in Alaska in 1989, spilling large amounts of crude oil in Prince William Sound. Cleanup efforts required enormous amounts of money, and the spill became a cause celebre for the environmental movement. The incident highlighted the importance of monitoring of seagoing chemical and fuel vessels. New cleanup techniques have been researched and applied in recent years. This event was precedent-setting due to the public and legal recognition that personal responsibility lies with the ship's captain and fiscal responsibility of the company which owns the ship for cleanup and other costs to reduce the environmental damage. In response, the Coalition for Environmentally Responsible Economics (CERES) of investment fund advisors and social advocates established the "CERES principles" demanding environmental monitoring of corporations regarding energy use, public disclosure, damage compensation, sustainable use of natural resources, and environmental representatives on boards and in management of corporations.

Man-made Disasters, War, Terrorism

The man-made disaster of war has used chemical, biological, and nuclear methods of destruction as well as traditonal methods of warfare including economic blockade. War's offspring, terrorism, has used chemical armamentaria and may use biological or even nuclear destruction sooner or later. All can cause great damage to human life and constitute a part of public health responsibilities at the end of the twentieth century. When disasters occur, lessons can be learned to improve services for the next disasters, be it natural or man-made (see Chapter 7).

Since World War I, poison gas has been used as a weapon against both front-line troops and against civilian populations. This practice has continued almost to the end of the twentieth century. Poision gas was used with deadly efficiency by the Nazis in the Holocaust of the Jews in World War II. Egypt used poison gas in its war in the Yemen in the 1960s, and in the 1980s Iraq targeted Kurdish civilian

BOX 9.7 RECENT NUCLEAR ACCIDENTS:
THREE MILE ISLAND AND CHERNOBYL,
1979 AND 1986

Three Mile Island: In 1979, the nuclear plant at Three Mile Island in Pennsylvania suffered a near disaster which devastated the plant but did not release nuclear material. It led to review of safety procedures and heightened public concern as to the overall safety of nuclear energy facilities.

Chernobyl: In 1986, a nuclear energy plant located at Chernobyl in the Ukraine suffered a meltdown which breached the integrity of the containment vessel and caused a massive explosion of the reactor. A series of staff errors led to loss of control of the reactor with power levels soaring to 120 times the normal, rupturing of the fuel rods, and vaporization of the cooling system. A steam explosion then blasted open the 100-ton concrete slab covering the reactor, setting uncontrollable fires. Despite valiant attempts by emergency personnel and staff, the fires could not be controlled immediately. Air dropping of sand, clay, limestone, and lead led to fire control, but the heat of the reactor and radiation could not be reduced for many days. Immediate deaths included 33, mostly among the firefighters, with 237 suffering acute radiation poisoning. Evacuation of an area of 19 square miles included 135,000 people.

The nuclear fallout material carried in a 1900-foot plume, including iodine-131, cesium-137, and xenon isotopes, spread across much of Europe. Fallout reached some 20 countries, and an international public health threat of major proportions occurred. At 10 years after the incident, there was a highly significant increase in thyroid cancer in children in the three affected countries, Ukraine, Belarus, and Russia. The long-term effects in terms of increased cancer and birth defects are hard to assess, but current estimates are of 500 (1–2%) additional cancer cases among 100,000 persons exposed to 10–20 rads. The actual increase in incidence of thyroid cancer, other cancers, and birth defects and the general impact on health will only be determined by careful epidemiologic follow-up of the exposed population over many years. The economic impact of the disaster is estimated at over $19 billion and replacement of the plant at a similar sum. Close to the tenth anniversary of the Chernobyl disaster, a second nuclear leak nearly occurred due to human error. The Ukrainian government reopened the second reactor in 1999.

Source: International Conference One Decade after Chernobyl. Sponsored by WHO, International Atomic Energy Agency, and other international agencies. Austria, 1996. Available at website http://www.iaea.or.at/worlatom/this week/preview/chernobyl/conclsn9.html

villages killing thousands of men, women, and children. Defoliants (Agent Orange) used widely during the Vietnam War are believed to have long-term effects on Vietnamese civilians and on exposed military personnel.

During the Gulf War of 1991, the potential use of poison gas in long-range rockets on civilian populations targets was narrowly averted. Several years later, thousands of U.S. service personnel reported a variety of neurological symptoms and general fatigue. By 1996, these cases were acknowledged by the Department of Defense as possible long-term sequelae of accidental exposure by troops to toxic agents following destruction of Iraqi chemical weapons, or due to antidotes taken for potential gas warfare exposure (Soman). In 1995, a chemical attack with an extremely dangerous chemical warfare agent (sarin) was carried out by an extremist cult in Japan on subway passengers in Tokyo, resulting in 12 deaths and 3000 injuries with hundreds of hospitalized persons.

Terrorist bombing incidents occurred in many parts of the world during the 1990s. The 1995 bomb detonated by domestic terrorists in a federal building in Oklahoma City in the United States killed over 160 persons. Terrorist bombings of a U.S. military living complex in Saudi Arabia, Israeli buses in 1996, U.S. Embassies in Africa in 1998, and Moscow apartment buildings in 1999 caused large numbers of deaths and injuries. Each incident resulted in national concern over the threat of terrorist action causing mass casualties. The fear increased when people realized that many bombs can be made with easily obtainable chemicals or explosives. Destruction of pipelines and oilfields caused extensive environmental damage in the aftermath of the Gulf War.

Residual land mines left in place without markings after warfare or in preparation for it cause continuous loss of life and limbs often among farmers and children. Land mines are present in many areas of conflict past and present in the millions, and cleanup is dangerous and costly. In 1997, approximately 800 persons died and 1200 were injured, one-third of whom require amputation, each month from land mines. Land mines limit land and water use by limiting access and have serious economic consequences for farmers. An international movement to ban the use of land mines has gained international prominence with support given to it by Princess Diana and by awarding the 1997 Nobel Peace Prize to Jody Williams, founder of this movement. Prevention is by raising awareness and political action to prevent land mine use and support efforts for land mine clearance. The United States is one of the few countries to refuse to sign this international accord.

The potential for intentional, negligent, or accidental disasters, whether manmade or natural, is a real and present danger requiring health officials to coordinate with civil defense and military authorities to prepare disaster plans and continuously exercise to prepare for such events. Planning can greatly reduce the number and severity of casualties of toxic chemical disasters.

Preventing and Managing Environmental Emergencies

Public health has an important role to play in prevention, management, and mitigation of the effects of man-made and natural disasters. The U.S. Congress passed the Emergency Planning and Community-Right-to-Know Act (EPCRA) of 1986 following the 1984 Bhopal disaster. This legislation established state and local agencies for chemical emergencies. It requires facilities that handle hazardous chemicals to make information available to the public.[1] It also requires steps to be taken in preparedness for possible chemical accidents. This involves a holistic approach integrating technology, procedures, and management practices. The first responsibility lies with management, which must have a high level of awareness and commitment to accident prevention and safe practices. The range of industries-at-risk is very broad in modern societies. It includes local dry cleaners and furniture manufacturers as well as the chemical industry. The right-to-know extends from governments, professional societies, trade associations, labor unions, the research community, the news media, and environmentalists as well as the general public. The right-to-know has become the need-to-know.

Environmental emergencies occur from release of chemicals or radiation into the air. Inhalation and fallout effects downwind of the site depend on weather conditions and dispersion of the smoke plume. Clinical management of exposed civilians and emergency personnel is an activity of health management that involves organizing triage and transportation services at the site of the disaster. The decision to evacuate civilians is often made with limited information but must take into account the potential for exposure during evacuation weighed against the protective effect of sealing homes and staying indoors.

The team needed to manage such a situation involves public health, occupational health, and epidemiology investigators as well as police, fire services, civil defense, armed forces, chemical warfare units, and psychological staff. Postdisaster recovery planning is part of the planning process. Disaster planning is discussed in Chapter 7. Long-term effects include post-traumatic stress disorder (PTSD) which can result in serious psychological dysfunction in affected individuals. PTSD can be alleviated by early psychological support for victims of mass disasters at the site and at evacuation or follow-up centers and should be part of emergency care planning.

Rapid risk assessment involves weighing the hazard, exposure potential, dose–response, as well as both short- and long-term risks. Command centers and designated leaders are needed to maintain control of the multitude of needs for information, coordination between agencies, and the distribution of resources to areas of greatest need. Long-term epidemiologic assessment may be necessary for legal and compensation purposes, as well as for training and preparation for future events.

Advocacy is a public health function, and environmental and safety issues are areas where advocacy can bring important public benefit. Leadership in defining public health problems and in defining necessary action to reduce risk factors, or short- or long-term ill-effects, requires skill in interpretation of epidemiologic events and studies, providing perspective for policy makers for addressing those issues.

---

[1]Information and technical assistance bulletins are available by contacting the following: Emergency Planning and Community-Right-to-Know Information Service, U.S. Environmental Protection Agency, OS 120, 401 M Street SW, Washington, DC. See website Nov. 18, 1999: http://www.epa.gov/swercepp/crtk.html

BOX 9.8 EMERGENCY PROCEDURES FOR HAZARDOUS SUBSTANCES, CHEMICAL OR RADIATION DISASTERS

1. Contain and reduce spread of toxin;
2. Inform the community
3. Notify municipal, state, and federal emergency organizations;
4. Minimize exposure by evacuation or ensuring the potentially exposed population remain inside, with the affected areas sealed off and quarantined;
5. Identify, decontaminate, and triage exposed persons;
6. Measure exposure and reaction;
7. Determine causative agents and antidotes;
8. Initiate antichemical procedures for exposed including removal of clothing, showers, antidote;
9. Coordinate on site triage and evacuation for medical care;
10. Ensure medical or hospital care for exposed persons;
11. Provide accurate public information;
12. Promote health and supportive care at evacuation sites;
13. Investigate—professional and criminal;
14. Compensate the injured or displaced;
15. Pursue civil and criminal charges against negligent management persons and corporations;
16. Provide documentation and recommendations from lessons learned;
17. Review procedures and revise disaster plan operation;
18. Promote public and professional discussion.

Note: See websites http://www.epa.gov/superfund/programs/er/hazsubs/ and http://www.epa.gov/swercepp

ENVIRONMENTAL HEALTH ORGANIZATION

The World Health Organization Commission on Health and Environment report (1992) represents a consensus documentation of international environmental health issues. This commission, chaired by Simone Weil of the European Parliament included many distinguished scientists, professional leaders, and international organizations. This represented a strong international consensus on joint action to prevent and clean up environmental degradation that had occurred in Europe over several decades.

National organization for environmental health can take various forms. In the past it was common for ministries of health to have environmental health departments, but in recent years this has increasingly moved to ministries of the environment. In 1970, the United States established the Environmental Protection Agency as the head federal agency reporting to the President to coordinate the administration of a wide range of environmental health problems, because of concern with environmental decay and fragmentation of government efforts to regulate and clean up the chaotic situation. The EPA sets standards and regulations for a variety for legislation pertaining to the environment, such as air and water pollution, solid and hazardous waste management, noise, public water supplies, pesticides, and radiation. Despite the growth of the EPA and its control of a Superfund to reduce toxic and other waste sites, inter-agency coordination is complex. In the U.S. federal government, a wide variety of agencies located in different government departments have responsibilities related to the environment. The substantial environmental progress made in the U.S. in the past 25 years is outlined in Table 9.5.

OCCUPATIONAL HEALTH: INTRODUCTION

Occupational health is the promotion and maintenance of the highest physical, mental, and social well-being of workers in all occupations by preventive departures from health, controlling risks, and adapting of work to people and people to their jobs (International Labour Organization and WHO, 1950). Diseases related to occupations, always an essential part of public health, increasingly relate to environmental health, but to other fields as well. The worker is also a member of a family and a breadwinner, so that the health of the worker is related to family health. The worker is concerned not only with what happens at the place of employment but also with hazardous agents he or she might accidently bring home. The retired or laid-off worker is worried about well-pensioned and honorable retirement. Occupational health in this wider context has an important place in the New Public Health.

DEVELOPMENT OF OCCUPATIONAL HEALTH

Occupational health is one of the oldest sectors of public health, dating back to Roman times. Documentation of occupational diseases was begun in 1700 by Ramazzini. Historic examples of work-related health hazards and diseases include scurvy among sailors, cancer of the scrotum specific to chimney sweepers in nineteenth century England, black lung in coal miners, mercury poisoning in hat makers, byssinosis in cotton mill workers, and mesothelioma in asbestos workers. The list is long and extends to musculoskeletal injuries and hepatitis B in hospital workers, spinal disorders in typists, and medial neuritis in computer workers (carpal tunnel syndrome). Interventions vary widely from the banning of asbestos usage to modifying the office work environment through better chairs, exercise breaks, and ergonomic training of workers.

During the early part of the nineteenth century, the harsh working conditions of children, women, and workers led to parliamentary action to regulate mines and

TABLE 9.5 Environmental Health Progress, United States, 1970–1995

Medium	Example Achievements	Remaining Challenges
Air quality	90 large cities had 72% reduction in number of days air unhealthy due to ozone (Los Angeles 33% reduction) (1985–1994) Total emissions of six common air pollutants declined by 24% Particulate matter emissions declined by 78% Lead emissions declined by 90% Power plant SO_2 emission reduced by half Ozone-depleting hydrocarbons fell >60% (1987–1993)	60–70 metropolitan areas (62 million population) fail to meet air quality standards for 1 or more pollutants Total emissions of NOx due to motor vehicles and coal fired plants increased by 14% Indoor air pollutants still hazardous Earth's protective ozone layer and greenhouse gases still global hazard
Water quality	60% of lakes and streams now clean for swimming and fishing Ocean dumping of sewage sludge, industrial, medical, and plastic waste banned Wastewater standards developed for >50 different industries 57,000 industrial facilities now under pollution control Improved drinking water standards Waterborne disease reduced by 200,000–470,000 cases annually Thousands of communities upgraded sewage treatment	Over 40% of lakes, streams, and estuaries still unfit for fishing and swimming Wetland destruction continues at 70,000–90,000 acres per year Pesticide runoff now the leading cause of water pollution Local water contamination occurs with new organisms not readily detected on routine monitoring nor destroyed by routine treatment, e.g., cryptosporidium outbreak in Milwaukee in 1993 Need better coordinating with other federal and state agencies, tribes
Waste, toxics, and pesticide management	Toxic air emissions fell by 39% Reduced toxic discharge into water by 13% Toxic substance disposal in deep wells fell by 57%, in landfill sites by 44% Over 230 pesticides were banned or eliminated, including 20,000 pesticide products; reduced pesticide use, safer use of, and new safer pesticides Safer working conditions for 4 million agricultural workers Clean up of 141,000 underground storage tanks since 1990 Of 1,300 Superfund danger sites, 95% were partially or fully cleaned up (349 or 27%)	Toxic Release Inventory (TRI) program just beginning Need to empower public action against offenders Need improved public access to information Legal barriers to an integrated approach—facility, industry, and community-wide

Source: Environmental Protection Agency. *Twenty-five years of Environmental Progress at a Glance.* Website: http://www.epa.gov/25year/intro.html

factories, improving conditions generally. The first factory inspectors in the United Kingdom were appointed in 1833 to administer the provisions of the Factory and Workshops Acts. In 1898, Thomas Legge became the first medical doctor appointed to the post of Chief Factory Inspector in the United Kingdom. He articulated the basic public health approach to worker's health and established the principle that management is responsible for the health of its employees. These issues are termed Legge's Axioms and are still relevant to the field of occupational health today (Table 9.6).

Government responsibility for setting standards, monitoring, intervention, and regulating compensation grew slowly over the past century. Case reports, epidemiologic studies, and advocacy regarding the effects of lead, asbestos, vinyl chloride, silica, and dust fibers led to steps to reduce the hazards to workers and provided the professional support for legislative initiatives. International standards developed by the League of Nations, the International Labour Organization (ILO), and other international organizations promoted development of this field.

THE HEALTH OF WORKERS

The health of workers is subject to normal health threats for the adult population, but there are specific threats to health associated with the work situation.

TABLE 9.6   Thomas Legge's Axioms on Worker's Health and Modern Equivalent

Legge's axioms	Modern version
1. Unless and until the employer has done everything—everything means a good deal—the workman can do next to nothing to protect himself . . .	Don't blame the victim; the health of workers is the responsibility of management
2. If you can bring an influence to bear, external to the worker, that is one over which he can exercise no control, you will be successful; if you cannot or do not, you will never be wholly successful.	Structural change is best
3. Practically all industrial lead poisoning is due to inhalation of dust and fumes; and if you stop their inhalation you will stop the poisoning.	If you stop the exposure you stop the poison
4. All workmen should be told something of the danger of the material with which they come into contact and not be left to find out for themselves—sometimes at the cost of their lives.	Workers have the right to know of potential hazards to their health at their place of employment
5. External influences depend on the will or whim of the workers to use them—respirators, goggles, and washing conveniences.	Educate the workers about risk reduction.

Source: Hunter, D. 1969. *The Diseases of Occupations.* Fourth Edition. London: The English Universities Press Ltd. and Harrington, J. M. 1999. 1998 and beyond—Legge's legacy to modern occupational health. *Annals of Occupational Health,* 43:1–6.

Workers exhibit lower death rates than the general population. A worker population is demographically different than the general population and even epidemiologically different than a population matched for age and sex. This is due to the fact that there is a process of selection of workers which excludes the severely ill and disabled from employment. The selection process continues with attrition of unhealthy persons from the work place. This is termed the Healthy Worker Effect and is a factor to be considered in occupational health studies and practice. Death rates or other population-based norms from the general population may be inappropriate for comparison if this effect is not taken into account. Case matching or control studies may be needed to accommodate for this phenomenon. Other population groups such as immigrants or refugees go through similar selection, where only the healthy may be included or survive.

THE BURDEN OF OCCUPATIONAL MORBIDITY AND MORTALITY

In the United States, the work force is made up of 110 million persons. Premature disease, injury, and death related to occupational exposures are a major burden on the economy and the health system. Over the period 1980–1994, a total of 88,622 workers died in the United States from work-related injuries and an additional 60,000 died from occupational diseases. In 1992, the costs of such injuries were an estimated $145 billion.[2]

Deaths due to work injuries declined from 7400 in 1980 (8.9 per 100,000 workers) to 6250 in 1985 (7.0 per 100,000) and 5714 in 1989 (5.6 per 100,000) and 5406 (or 4.4 per 100,000) in 1994. Deaths from work-related disease are estimated at 100,000 annually. The largest numbers of deaths occur in the following industries: construction (18.2%), transportation/communications/public utilities (17.7%), and manufacturing (14.0%). The decrease in occupation related deaths over the period 1980–1994 is related to the cumulative effect of increased awareness and regulation of worksite dangers and toxins, as well as new technology and mechanization, changes in the economy, and workforce distributions.

Even though occupational deaths have been declining, permanent impairment suffered on the job grew during the 1980s. During 1987, permanent impairments were suffered by 70,000 workers, and the number of totally disabled increased to 1.8 million. Further, there is an increasing rate of lost workdays. The decline in mortality and increase in work injury may be due to improving care of the injured or to a real increase in the number of injuries.

In 1996 in the United States there were 3.1 nonfatal injuries with lost work days per 100 employees in the private sector, a reduction from 3.9 in 1990. Substantial

---

[2]Figures from the National Institute of Occupational Safety and Health (NIOSH) Web site http://www.cdc.gov/niosh/homepage.html, and Centers for Disease Control. 1998. Worker's Memorial Day—April 28, 1998. *Morbidity and Mortality Weekly Report,* 47:297.

TABLE 9.7 Occupational Injury Death Rates (per100,000 workers) by Industry, United States, Selected Years, 1980–1993

Industry	1980	1985	1990	1993	Δ%
Total civilian workforce	7.6	5.8	4.6	4.2	−44.7
Mining	43.8	30.0	30.0	25.4	−42.0
Agriculture, fishing, forestry	24.4	23.7	18.0	18.5	−24.2
Construction	21.3	16.6	14.0	11.8	−44.6
Transport, communication, public utilities	21.2	15.7	10.4	10.1	−52.4
Public administration	7.7	6.4	3.8	4.2	−45.5
Manufacturing	4.7	4.0	4.0	3.6	−23.4
Wholesale trade	4.4	2.8	3.6	3.6	−18.2
Retail trade	3.7	2.7	2.8	2.9	−21.6
Services	2.4	1.8	1.5	1.4	−41.7

Source: *Health,* United States, 1998

improvements are seen in the more dangerous occupations such as agriculture, fishing and forestry, mining, construction, and manufacturing during the 1990s, as seen in Table 9.7 in the United States from 1980 to 1996.

The 10 most frequent work-related diseases and injuries are

1. Lung disease;
2. Musculoskeletal injuries;
3. Cancers;
4. Severe trauma;
5. Cardiovascular disorders;
6. Disorders or reproduction;
7. Neurotoxic disorders;
8. Noise-related hearing loss;
9. Dermatologic conditions;
10. Psychological strain and boredom.

Injury surveillance in the United States is maintained by the CDCs National Institute of Occupational Safety and Health.[3]

INTERNATIONAL ISSUES IN OCCUPATIONAL HEALTH

Occupational health has become an international issue as the global economy transfers manufacturing from one country to another with great speed and ease.

[3]The National Electronic Injury Surveillance System (NEISS) is maintained by the federal Consumer Product Safety Commission (CPSC). It collects data since 1981 for surveillance of work-related injuries treated at 65 of 91 hospital emergency departments selected from a stratified sample of all hospitals in the United States (*Morbidity and Mortality Weekly Report,* 47:302–306, 1988).

BOX 9.9 OCCUPATIONAL HEALTH ISSUES
IN THE GLOBAL ECONOMY

1. Technology transfer from industrial to developing countries or areas within a country;
2. Child labor in developing and developed countries;
3. Pesticide overuse, toxicity, and food contamination;
4. Ecological damage from toxic wastes spills and waste disposal;
5. Toxic waste transfer from industrial to developing countries;
6. High technology industrial toxic wastes;
7. Nuclear energy, accidents, and wastes;
8. Technological and professional common interest between occupational and environmental health;
9. Poor safety and control standards in former Soviet and developing countries;
10. Poor wages, psychological stress, boredom, and shift work;
11. Management negligence and lack of accountability for workplace safety;
12. Governmental negligence and corruption in developing regulatory role;
13. Inadequate health and safety measures in developing countries;
14. Widening income gap between upper and lower income groups.

This is often motivated by lower wages, but also by lower occupational and environmental regulatory controls, and less stringent or nonexistent legal protection against toxic exposures and child labor in developing countries. Transfer of occupational hazards from industrialized to nonindustrial countries has become an issue in international cooperation and trade agreements. Developed countries have stricter environmental regulations and worker organization than developing countries that are anxious for job-producing industry at any price.

NATIONAL AND MANAGEMENT RESPONSIBILITIES

In the United States, workers' health benefits cost more than the steel to make a car. As a result, there is a growing interest on the part of management and of workers in promoting workers' health through improved nutritional monitoring of canteens and cafeterias, antismoking activities, and physical fitness programs. The management interest in a healthier work force to restrict rising health care costs is part of the modern corporate culture. The primary responsibility, however, legally and morally, lies with management, in addition to protecting the worker by monitoring risks, providing a safe environment and provision of care at the time of injury.

Occupational injuries and illnesses are social as well as engineering and man-

---

BOX 9.10 PRINCIPAL TASKS OF
OCCUPATIONAL HEALTH

1. Anticipation: dealing with potential disease and injury to include preparation for prevention as facilities are planned or renovated;
2. Surveillance and monitoring assuring timely and accurate identification, reporting, and recording of occupational disease and injury; medical surveillance: passive or active and industrial hygiene and safety;
3. Right-to-know: for workers, health professionals, community at large;
4. Epidemiologic analysis: analyzing collected data—linking exposure to outcome data helps to locate trends, clusters, associations, and causes of disease and injury for more in-depth investigation and prevention;
5. Exposure reduction: minimizing toxic exposure, to prevent approaching or exceeding established limits;
6. Substitution: substituting less toxic substances;
7. Awareness: promoting awareness at government, management, community, worker, and consumer levels;
8. Government regulation: on-site supervision by regulatory agencies; publication of standards of exposure and "good practice;"
9. Compensation: compensating for illness and loss of life related to work accidents, toxicity, and stress;
10. Management–worker cooperation: recognizing that worker participation in health and safety is of mutual benefit.

Source: Weeks, J. L., Levy, B. S., Wagner, G. R. (eds.) 1991. *Preventing Occupational Disease and Injury.* Washington, DC.: American Public Health Association.

---

agement concerns. Compensation, litigation, class action suits, and union action are all associated with increasing awareness of toxic and trauma effects on workers, and court decisions regarding management liability. The field is made more complex because some occupational illness may occur long after the exposure: silicosis, asbestos-related mesothelioma, and asbestosis may develop after a long latency of up to 20–30 years following exposure. Follow-up of exposed workers may be difficult, and issues such as compensation may also be complicated. Occupational health involves a governmental regulatory function and legislated responsibility to protect workers from toxic or physical risks at the work site.

Standards and Monitoring

Monitoring of occupational health involves a set of activities designed to increase the safety and protection of the workers. It involves a number of parallel services to promote the health of the individual worker and the safety of the work environment and should be coordinated in an overall strategy.

In the United States prior to 1970, prevention of occupational injuries, death, and disease was governed by state and local government or market forces. Federal initiatives to raise standards of occupational health and safety were mandated in the Occupational Safety and Health Act of 1970, which established two government agencies to implement the Act, the Occupational Safety and Health Administration (OSHA) and the CDC's National Institute of Occupational Safety and Health (NIOSH). OSHA is responsible for promulgation and enforcement purposes, within the U.S. federal Department of Labor. OSHA sets standards based on consensus derived from professional organizations in consultation with labor, industry, and health authorities, meant to promote safety and reduce risk for employees and set performance standards for employers. NIOSH was established to conduct research related to the objects of the Act for occupational disease, particularly those derived from exposure to toxic physical and chemical agents.

The Act provides an environment for regulation and study of occupational health issues including public petitions, court decisions, and new research findings used to formulate priorities for standards development. Monitoring is done by a combination of federal, state, and local health authorities with participation of professional and industrial organizations. The legal responsibility for worker safety and health is placed with the employer (Table 9.8), but worker awareness and participation in safety programs is vital to a successful approach.

OCCUPATIONAL HEALTH TARGETS

The U.S. Surgeon General's report *Healthy People 2000* formulated a number of targets for occupational health and safety issues (Table 9.9). These are national targets that are also being adopted by state departments of health and have organizational as well as legal implications.

TOXICITY AT THE WORKPLACE AND IN THE ENVIRONMENT

Toxic substances are widely used in industry, not only in manufacturing but also in services such as laboratories, and they constitute a major concern of both occupational and environmental health. Extensive information is published on toxic substances by the World Health Organization and the Centers for Disease Control.[4]

Much of the concern of occupational health has been on detection, prevention, and reduction of exposure to toxic materials at a workplace, but more recently also with contamination of the surrounding environment. The scientific knowledge of

---

[4]Toxic Substances and Disease Registry, Division of Toxicology, Centers for Disease Control, 1600 Clifton Road N.E., Atlanta, GA 30333.

TABLE 9.8   Management and Governmental Responsibilities in Worker's Health

Management responsibility	Governmental responsibility
Substitute less dangerous materials	Legislation: substitute, ban, define legal responsibility (civil and criminal) and compensation
Enclose/separate	Regulate to set and enforce standards for toxic emissions and controls
Process exhaust	Litigation: civil suits versus compensation
General ventilation	Test environment and workers, with notification of test results
Good housekeeping	Label hazardous materials, regulate labeling and disposal
Monitor health of workers	Monitor health of workers
Personal protection	Educate managers and workers
Investigate	Research: scientific and operational
GMP (good manufacturing practices)	Compensate for income loss and health damage

toxins used in occupational setting and their sources, uses, effects, actions, and target organs is extensive. Factors that affect the toxicity of an agent, in addition to the extent and duration of exposure, include host factors such as age, gender, fitness, previous exposure, and compounding risk factors such as smoking. Environmental factors include temperature and air flow as well as the physical and chemical properties of the toxic agent. A number of examples of toxic substances and the history of control measures for them illustrate the complexity of this problem.

Lead.   Lead is a mineral with thousands of applications because of its plasticity and its softness. Lead poisoning has been a worker hazard since ancient times. Lead enters the body through inhalation and ingestion, affecting the gastrointestinal, nervous, hematologic, and circulatory systems. It is associated with intestinal colic, encephalopathy, delirium, and even coma in its acute forms. Chronic forms of plumbism or lead poisoning can be seen as mental dullness, headache, memory loss, neurological defects (wrist drop), anemia, and a blue line on the gums.

Lead toxicity has been a traditional health problem of glaziers and potters because of lead use in the manufacturing process. Wines or rum produced and stored in lead containers or in pewter (lead–tin alloy) utensils were known to be associated with the "dry gripes" in the seventeenth and eighteenth centuries. The Devonshire colic, described in 1776 by George Baker, was widespread for more than 100 years in parts of England where cider was made and stored in lead containers.

Lead toxicity and excess exposure in the workplace remains a problem in the United States. Lead-induced hypertension, neuropathy, carcinogenesis, reproductive damage for men, and abortion for women are the major toxic effects. The 1995 blood lead surveillance by the CDC's NIOSH Adult Blood Lead Epidemiology and Surveillance Program, which monitors elevated blood levels among adults, reported a continuing hazard of work-related exposures as an occupational hazard in the United States. Studies of lead exposure in industrial settings in the United

TABLE 9.9   United States Health Targets in Occupational Health for the Year 2000

Subject	Previous	Target
Reduce death from *work-related* injuries	6 per 100.000 full-time workers (1983–1987)	<4 per 100,000
Reduce *work-related injuries* resulting in medical treatment, lost time from work, or restricted work activity	<7.7 per 100 full-time workers in 1987	<6 per 100
Reduce *hepatitis B infection* by increasing immunization levels to > 90% among occupationally exposed workers	6200 cases in 1987	<1250 cases
Increase proportion of work sites with 50 or more employees that mandate use of *occupant protection systems* (such as seat belts) during work related motor vehicle travel		>75%
Reduce proportion of workers exposed to average daily *noise levels* that exceed 85 decibels		<15%
Eliminate exposures that result in workes having *blood lead concentrations* greater than 25 μg/dl of whole blood		
Implement statewide *occupational safety and health plans* for the identification, management, and prevention of work-related diseases and injuries	10 states in 1989	50 states
Establish exposure standards adequate to prevent the major occupational lung diseases to which worker populations are exposed, including byssinosis, asbestosis, coal workers' pneumoconiosis, and silicosis		50 states
Increase the proportion of work sites with 50 or more employees that have implemented *programs on worker health and safety*		>70%
Increase the proportion of work sites with 50 or more employees that offer *back injury* prevention and rehabilitation programs	28.6% in 1985	>70%
Establish either public health or labor department programs that provide *consultation and assistance to small businesses* to implement safety and health programs for their employees		50 states
Increase the proportion of primary care providers who routinely elicit occupational health exposures as a part of patient history and provide relevant counseling		>75%

Source: U.S. Surgeon General, 1992, *Healthy People 2000: National Health Promotion and Disease Objectives.*

States have shown widespread exposure above permissible exposure limits. This includes the traditional high exposure industries such as primary and secondary lead smelting, battery and pigment manufacturers, brass/bronze foundries, and 47 other industries. High exposure job titles throughout industry were painters. Oc-

---

BOX 9.11 ALICE HAMILTON AND
TETRAETHYL LEAD

Alice Hamilton, a pioneering researcher and public health advocate in the 1910s and 1920s, with others, demonstrated workplace hazards to toxic substances such as white phosphorus used in match production, lead additives to gasoline, and radium in watch dials. Tetraethyl lead (TEL) was produced and promoted by DuPont, despite being identified as hazardous. Hamilton and others strenuously opposed its use, but TEL use expanded, and with it her research on behalf of state and federal government commissions. Environmental lead toxicity increased until the 1970s when further research revealed the extent of the problem and its public health effects, especially on children. Hamilton's work set standards for toxicology research in occupational and environmental health that led to the regulatory successes of the 1970s in the United States.

Source: Rossner, P., Markowitz, G. 1985. A "Gift of God"? The public health controversy over leaded gasoline in the 1920s. *American Journal of Public Health,* 331:161–167; and CDC. 1999. Improvements in workplace safety—United States, 1900–1999. MMWR, 48: 461–469.

---

cupational exposure continues to be an important source of lead toxicity. OSHA standards promulgated in 1978 came at a time when lead prices dropped, reducing the number of producers and the degree of compliance overall.

Concern for lead toxicity evolved from a strictly occupation-related toxicity to an environmental one in which both the exposed worker and the general population are adversely affected by this widely used metal. In 1997, the CDC adopted a BLL standard of <10 micrograms per deciliter (<10 μg/dl), a level at which a negative effect on cognitive development is recorded. Between 1976–1980 and 1980–1991, geometric mean BLLs of persons aged 1–74 in the U.S. declined from 12.8 μg/dL to 2.9 μg/dL, and even further in 1991–1994 to 2.3 μg/dL (NHANES surveys).

Despite major improvements (see Box 9.5), some 1.7 million children aged 1–5 in the United States still have BLLs above 10 μg/dl. Further progress in BLLs will require reduction in lead hazards in housing and reduced contact with lead-contaminated dust, house paint lead, and work-site exposure. Work-related and environmental lead exposures continue to be public health problems in the United States, requiring continued diligence on the part of pediatricians and internists as well as occupational and public health workers.

The Upper Silesia region of Poland with its capital Katowice and a population of 4 million is the site of many nonferrous metal plants, especially using lead and zinc. In the Katowice district there are four such plants, two of which are more than a century old and have a high output of atmospheric lead, and two built in the 1960s with inadequate pollution control equipment. Although emissions of lead and cadmium from one plant reportedly fell during the late 1980s, high levels of blood lead and cadmium are found in children, and soil contamination is extensive, including high levels of contamination of vegetables. This problem is widespread in eastern Europe.

Reduction in lead exposures has been achieved in the United States by a combination of legislation and professional and social pressures, resulting in the adoption of lead-free gasoline, removal of lead from paints, and its substitution in many industrial practices. Awareness and active lobbying by public health minded groups has had a beneficial effect in reducing lead toxicity in the community and workplace. The American Public Health Association continues its concern that 4.4% of U.S. children age 1–5 have blood lead levels above 10 μg/dl, and it promotes a wide ranging program of further abatement of lead paint hazards including litigation against manufacturers as well as community-based prevention and health education programs.

Asbestos. Asbestos-related disease is an occupational and public health problem that grew from the rapid increase in the use of asbestos during World War II. It left a legacy of death and disease that only became apparent many years later. Fibrotic lung disease resulting from asbestos exposure was called asbestosis by W. E. Cooke in 1927. A subsequent British government investigation of the subject reported to Parliament that inhalation of asbestos dust over a period of years results in the development of a serious type of fibrosis of the lung and recommended dust suppression measures. This was followed by many case reports and the wide recognition of the health hazards associated with asbestos exposure. During World War II, the U.S. Navy issued minimum requirements for safety in shipyards contracting for naval work, involving some 1 million workers.

The first reports of an association between asbestos and lung cancer began to appear in the 1930s. Studies by Irving Selikoff in 1965 in New York reported high rates of lung cancer in several large population groups of ex-shipyard workers. Selikoff and colleagues also showed a synergistic relationship between asbestos exposure and cigarette smoking (Table 9.10), namely, a greater risk of

TABLE 9.10 Lung Cancer Death Rates (Age Standardized)[a] for Workers Exposed to Asbestos Dust and Cigarette Smoking with Controls

Group	Asbestos exposure	Cigarette smoking	Death rate	Mortality difference	Mortality ratio
Control	No	No	11.3	—	1.0
Asbestos workers	Yes	No	58.4	+47	5.2
Control	No	Yes	122.6	+111	10.9
Asbestos workers	Yes	Yes	601.6	+590	53.2

Source: Selikoff, I. J. 1986. Asbestos-associated disease. In Last, J. M. (ed.). *Maxcy-Rosenau: Public Health and Preventive Medicine.* Twelfth Edition. Norwalk, CT: Appleton–Century–Crofts.

[a]Rates per 100,000 man-years, age standardized, on 12,051 asbestos-exposed workers followed prospectively between 1967–1976. Controls included 73,763 similar men in a prospective study of the American Cancer Society for the same decade. Number of lung cancer deaths are based on death certificate information.

BOX 9.12 RAMAZZINI ON SILICOSIS, 1700

"We must not underestimate the maladies that attack stonecutters, sculptors, quarrymen and other such workers. When they hew and cut marble underground or chisel it to make statues and other objects, they often breathe in the rough, sharp, jagged splinters that glance off, hence they are usually troubled with cough, and some contract asthmatic affections and become consumptive."

Source: Ramazzini, B., De Morbis Artificum Diatriba, 1700, as quoted in Hunter, D. 1969. The Diseases of Occupations, Fourth Edition. London: English University Press.

---

lung cancer with heaver smoking and a reduction in risk following cessation of smoking.

The United States Toxic Substances Control Act of 1976 placed the responsibility for harmful chemicals, including asbestos, on those who would profit from their sale. The long time lag between the first reports of asbestos-related disease followed by definitive studies and implementation of control measures raised questions as to the way in which occupational health functions. As a result of these studies and the regulatory responses by federal legislators, there was a fourfold reduction in asbestos usage in the United States from 1972 to 1982. In 1986, the U.S. Asbestos Hazard Emergency Response Act reinforced federal regulation of asbestos usage.

Asbestos exposure is accepted as the cause of mesothelioma, a highly malignant cancer of the chest or abdominal lining. The latency period may be 20–30 years or more, and the risk of the disease is related to the extent of exposure. The exposure may occur in asbestos-cement production, shipyard workers, garage workers exposed to brake linings, plumbers, and construction workers using asbestos-based products. During the 1980s concern was expressed that asbestos was being exported to developing countries lacking the regulatory mechanisms of the developing world. In the 1990s there is still some concern that asbestos products manufactured in developing countries are being imported to developed countries. In 1999, the European Union effectively banned the use of asbestos products.

Silica. Silicosis is one of the oldest known occupational diseases, affecting miners in particular. It was described in ancient Greece and Rome as the "fatal dust." Silica occurs in minerals and rocks throughout the world either as free silica or combined in quartz, flint, or sandstone. Mining, tunneling, stone cutting, quarrying, iron and steel works, sandblasting, brick making, polishing of stone, glass, metals, and many other industries cause the exposure of workers to inhalation of silica dust.

Silicosis is a condition of massive fibrosis of the lungs resulting from prolonged inhalation of silica dust. It is classified as a pneumoconiosis, a general inflammatory fibrotic lung condition caused by inhalation of dust particles. This condition can progress through mild symptoms to shortness of breath, with radiologic evidence of pulmonary consolidation and concomitant tuberculosis. Silica has not been proved to be a carcinogen.

Studies of hard coal miners in the nineteenth century documented the effects of silicosis. By 1918, English workers could receive disability compensation for silicosis and tuberculosis. In the 1920s and 1930s in the United States, studies showed silicosis in cement production workers, anthracite miners, tunnel workers, lead–zinc miners, and other hard rock miners. In the mid-1930s, an estimated 700 U.S. workers died as a result of construction of the Hawk's Nest Tunnel in Gauley Bridge, Fayette County, Virginia, leading to compensation laws covering workers with silicosis. At present there is still a controversy as to legally enforceable standards, and the problem remains difficult to prevent.

Cotton Dust (Byssinosis).   Cotton dust has been a common cause of chronic obstructive lung disease among long-term workers in textile industries, widespread in the United States until the 1960s. OSHA promulgated new standards in 1978 based on assessment of the potential of improved ventilation and filtration, and improved machinery use. The industry at that time was in the process of replacing old equipment with modern and more automated machines, which gave improved production speed, more effective use of floor space, reduced labor input, and a higher quality product, along with lower dust levels. The technical and economic feasibility of the higher standard was correct, and compliance by industry exceeded early expectations at about one-third of anticipated costs.

Vinyl Chloride.   Vinyl chloride is a colourless, flammable gas with a faintly sweet odour. It is an important component of the chemical industry because of its flame-retardant properties, low cost, and many end product uses. It is also a carcinogen causing liver, brain, and lung cancer, as well as spontaneous abortion. Vinyl chlorides are dangerous primarily when inhaled or ingested. Vinyl chloride usage increased since the 1930s and more dramatically after the end of World War II until the 1970s. In the 1960s, polyvinyl chloride (PVC) was shown to be associated with Raynaud's phenomenon and later with malignancies, including hemangiosarcoma of the liver.

The carcinogenicity of PVC was established as a result of the review of all evidence in 1974 by the U.S. Office of Technology Assessment and OSHA. Scientists concluded that there was no safe level of exposure to vinyl chlorides. OSHA adopted 1 part per million as the maximum possible dose. While the risk assessment issues are still controversial, reduction of exposure to workplace carcinogens such as vinyl chloride is the accepted standard of modern occupational health.

Despite vigorous opposition by the industry to this reduction in PEL, full compliance was achieved within 18 months by improving ventilation, reducing leaks, modifying reactor designs and chemical pathways, and using greater automation of the process. Even more effective was a major improvement in the production of PVC using less vinyl chloride. The costs to industry of reducing exposure levels were less than 25% of anticipated costs because of unanticipated innovations in the production process.

Agent Orange.   Agent Orange is a herbicide used widely by the U.S. armed forces in the Vietnam War to defoliate large areas of that country. This agent includes dioxin and is carcinogenic. High levels of dioxin have been found in breast milk, adipose tissue, and blood of the Vietnamese population. Even though sampling has not been systematic, studies carried out between 1984 and 1992 show that high levels of dioxinlike contaminants (TCDD) or 2,3,4,8-tetrachloro-dibenzo-*p*-dioxin are seen in blood samples of the Vietnamese population exposed to Agent Orange during the war.

Studies of effects among American veterans of the Vietnam War have not produced convincing evidence of long-term effects. Additional studies will be needed to verify effects such as increased cases of cancer or birth defects. However, court and compensation decisions have been made in favor of veterans exposed to Agent Orange despite inconclusive epidemiological evidence of its ill effects on health.

WORKPLACE VIOLENCE

Violence is endemic in many societies and affects many organizations and institutions. Violence has become a leading cause of fatal injuries in the workplace. Violence in the health setting has an extensive history, with the first documented case in 1849 when a patient fatally assaulted a psychiatrist in a mental health care facility. Since then there have been many other studies reporting assaults, hostage taking, rapes, robbery, and other violent acts in the health care and community settings. During the 1990s, homicide became the leading occupational cause of death for females and the second leading cause, after motor vehicle crashes, for men in the United States.

Murder of convenience store employees in the U.S. has become a major occupational problem. NIOSH issued new guidelines for addressing this problem in April 1998. Response from the association of operators has opposed standards such as installation of bulletproof glass and television monitors and nighttime double staffing that have been demonstrated to reduce violence and death in armed robberies. Shocking incidents of violence and homicide have occurred in which bombs and handguns were used in assassinations of health workers in clinics carrying out abortions, while assault and murder of health workers occured in hospitals and other settings.

Homicide at work has only recently been addressed as an occupational hazard, and research in this area is in its infancy. No universal standards exist to protect workers from work-related violence, and no policy has been created to protect workers. Preventing violence in the workplace is essential and must be addressed at the national level. Currently the California Occupational Safety and Health Authority has promulgated guidelines with emphasis on preventing violence before it occurs, by developing an effective policy to ensure workplace safety. Management and workers organizations as well as the health system share responsibility. Prevention of drug, alcohol, and sexual abuse or exploitation at work are vital to prevention of workplace violence.

OCCUPATIONAL HEALTH
IN CLINICAL PRACTICE

The clinical physician should be aware of the patient's occupation and previous work history. The inclusion of questions related to past employment (Table 9.11) may be crucial in investigation of patients and without which it might be impossible to find the cause of the disease. The health care provider should be aware of industries in the community and their potential hazards. The clinician is particularly important because he or she may be the first to see index cases of toxicity. This requires simple questions, such as the following: What is your job or hobby? What do you do at work? Are you exposed to any chemicals at work or at home? Are there others at work with similar exposure and similar symptoms? How long have you been exposed to these chemicals? The clinical suspicion is the key to finding a potential toxic cause to a set of symptoms, and may lead to a wider public health problem.

INSPECTING THE PLACE OF WORK

The public health authority responsible for health at the place of work may be under the authority of a Ministry of Labor or under a public health authority. Site inspection provides a guide to management and workers for safety and health issues. Noncompliance with federal, state, or local standards should lead to regulatory action to correct deficiencies and should include, if necessary, punitive damages to management. Examination of the work site involves on-site observations as listed in Table 9.11. The inspection should be documented and made available to management, workers, and follow-up inspections.

RISK ASSESSMENT

Identifying and quantifying occupational and environmental risks is difficult, but clinical or public health observations, supplemented by epidemiologic analysis, can identify toxic or carcinogenic factors that can be reduced or eliminated by public health intervention. High levels of awareness by clinicians of potential health effects from environmental or occupational exposures can help identify index cases just as in infectious disease, leading to an investigation and removal of the cause. Similarly, epidemiologic small area analysis can identify populations at high risk for cancers or other toxic effects, giving a localization for further investigation.

Establishment of dose–response relationships requires well-conducted obser-

TABLE 9.11 Factors Observation during Walk-through Inspection of Work Sites

Marker	Observations of conditions, safety arrangements, and effects on workers
Sensory observation	Eye irritation, poor lighting, noise levels, metallic taste in air, smell, visible fumes, exhaust, temperature (heat/cold)
Safety devices	Use of hard helmets, welding masks, safety shoes and clothing, ear protectors, eye and face protectors, first aid facilities, respirators, scaffolding, monitoring procedures
Storage	Hazardous chemical substances closets; unlabeled bottles, containers
Toilets	Cleanliness, fixtures, soap, toilet paper, waste disposal bins
Worker hygiene	Changing place, showers, lockers, clothing change
Eating place	Separate tables, cleanliness, washup facilities
Workers' ages/condition	Children, teenagers, elderly, pregnancy
Workers' complaints	Headache, fatigue, dizziness, nausea, breathlessness, skin problems
Worker morale	Worker morale is reflected in turnover and absenteeism
Work site layout	Safety in movement of supplies, products, ventilation
Hazard control	Labeling, process recording, worker records, periodic screening
Cleanliness	Removal of waste products, oil or chemicals on floors, machines, tables
Vents, fans	Exhaust of fumes, odors, dust
Worker–management cooperation	Mechanisms for worker and management to consult and share responsibility to reduce hazards and improve performance

Source: Weeks J. L., Levy, S. S., Wagner, G. R. (eds.) 1991. *Preventing Occupational Disease and Injury.* Washington, DC: American Public Health Association.

vational studies. Some studies may be so insensitive as to dismiss risks which are at low levels of statistical significance, but still represent preventable risks that can be sufficient to warrant compensation. This occurred in the case of veterans in the United States who were exposed to Agent Orange in Vietnam in the 1960s and those suffering effects attributed to toxic exposures in the Gulf War in 1991.

Regulatory and compensation decisions must often be made in the face of inconclusive or contradictory evidence from epidemiologic studies. In the 1960s, the FDA used the Delaney Clause applied to food additives or coloring in which any degree of ill effect noted in animal studies was enough to disqualify a drug from acceptability, but this has not become an accepted legal standard. The topic remains one of controversy and contradiction, with cases providing precedents that affect future court and regulatory decisions. The contribution of epidemiology to resolving such issues also remains controversial.

PREVENTING DISASTERS IN THE WORKPLACE

A disaster in a workplace can affect the workers and the surrounding community. The major responsibility for prevention is with management, but the worker and society also have roles in the process. Prevention involves education of workers and management, and constant vigilance. Government has the overall responsibility to legislate and enforce standards, safe conditions of work, and control of toxic materials and ensure fair compensation for injury or disease. The simple qualitative observations listed in Table 9.12 can provide a useful picture of the disaster management capacity of a work site. These observations can be made by management, health professionals, and workers' representatives to monitor and promote improved worker health and safety.

The principle of "good worksite practice" is a parallel to good manufacturing practices required by food and drug authorities. It is based on the concept that current standards of acceptable safety involve standards of facilities, staffing, and operational criteria. The healthful and safe work site should be maintained and accredited on that basis.

OCCUPATION AND THE NEW PUBLIC HEALTH

Social class, often defined by occupation and education, is a key determinant of health status. A population of unskilled workers has much higher rates of coronary health disease, strokes, and cancer and their children have much higher rates of mortality and morbidity than higher skilled workers, or business and professional people. The evidence points to a feeling of having less control over one's own life as a major consideration. The worker who has little say in determining his or her own activities may be subject to higher stress on the job, such as on the production line, or in job security, advancement, and wages. Loss of work is a key factor in increasing the vulnerability of men especially to a variety of life threatening conditions, including suicide, alcoholism, violence, cardiovascular disease, and others. The phenomenon of downsizing, or reducing the work force, affects production workers disproportionately, but also reaches middle and upper management levels, so that the danger of losing a position at an age when finding new employment is unlikely may become a real health hazard. Awareness and respon-

TABLE 9.12 Markers and Indicators of Disaster Management Capability in an Industrial Setting

Markers	Indicators
Administrative	Occupational health disaster plan; access to first aid; frequent disaster drills; close supervision of subcontractors
Investigation	Thorough investigation of complaints, leaks, and spills
Monitoring workers	Monitor worker injuries, illnesses, toxic levels, use of safety measures
Technological	Fail-safe monitoring devices; real-time monitoring; minimal on-site storage; automatic alarm/shutdown devices; local incineration/neutralization
Transportation	Vehicle and container standards; driver training, fatigue, alcohol and drug abuse, traffic offenses
Information	Worker information; right-to-know of workers and community; community disaster plan

Source: Richter *et al.*, 1992. Recognition and use of sentinel markers in preventing industrial disaster. *Prehospital and Disaster Medicine,* 7:389–395.

siveness to a variety of risks associated with employment and occupation is part of health responsibility. Prevention may predominate in some situations, screening for case finding in others, and clinical management in others.

SUMMARY

Environmental and occupational health are increasingly prominent elements of the New Public Health along with concern for the ecology of the world, especially since the 1960s. The problems of this field have become more complex in the past several decades with numerous global ecological concerns emerging. These include global warming, hazards associated with nuclear accidents on the scale of Chernobyl, and chemical disasters occurring frequently in all parts of the world. Other massive environmental issues such as desertification, destruction of forests, and massive air pollution are health concerns, but also societal issues in general. Concern for the environment and the worker often clash with desire for economic growth, especially in the poorer countries as they try to cope with rapidly increasing populations and increasing expectations for a better life.

Important progress has been made in management of water, waste products, toxic wastes, and air quality standards especially since the 1970s. Worker's health and safety have improved dramatically over the past century in the industrial countries. Some of these gains are at the cost of moving hazardous materials and working conditions to newly industrializing or developing countries in the global economy. Even a vigilant health sector is, by itself, incapable of dealing with the problems of the environment and of occupational health. It requires many levels and agencies of government as well as the support of public opinion. The role of the public health community is to act in the professional and advocacy roles with intersectoral cooperation to address these complex and vital issues. Epidemiology provides tools to measure mortality, morbidity, or physiological change that may occur as a result of environmental damage, but these may not be sufficiently rapid nor sensitive. Both epidemiology and testing technology are improving steadily, providing hope for standards that one would expect to contribute to a cleaner, safer, and more aesthetic environment.

The New Public Health includes long-standing public health issues of the environment and occupational health, but widens the field to include clinical services, the community, and the individual. All need to be involved in healthy public policy, in case finding, and in documenting the results of workplace and environmental risks. For a society there are choices to be made in creating a less toxic and hazardous environment. Choices, for example, between private and public transport, between jobs in industries with toxic emissions, or between producing energy from fossil fuels or nuclear sources. The search for substitutes for toxic materials and raising the level of social consciousness are needed to reduce the gross pollution that was the price of industrialization over the twentieth century. Equally challenging is the need to prepare and deal with natural and man-made disasters that may involve conventional explosives or biological, chemical, and even nuclear methods of destruction. The price of unrestrained pollution and man-made destruction is too great to bear. Investment in a healthy environment is a health and quality of life issue in each community and for the entire planet.

ELECTRONIC MEDIA

Agency for Toxic Substances and Disease Registry, http://www.atsdr.cdc.gov/atsdrhome.html
  PCBs, http://www.atsdr.cdc.gov/HAC/PCB/b__pcb__cvr.html
  Full list of hazardous substances, http://www.astrdr1.astdr.cdc.gov:8080/97list.html
America Public Health Association, http://www.apha.org/science/
American Public Health Association, policy statements, http://www.apha.org/science.policy.html
Environmental Protection Agency (EPA), http://www.epa.gov/
EPA Office of Air and Radiation, http://www.epa.gov/oar/
  http://www.epa.gov/airsdata
National Institute of Environmental Health Sciences (NIEHS), http://www.niehs.gov/
Occupational Safety and Health Agency, http://www.osha.gov/
Unified Air Toxic Website, Office of Air Quality, Planning and Standards (EPA), http://www.epa.gov/ttn/uatw/basicfac.html

RECOMMENDED READINGS

Centers for Disease Control. 1995. *Vibrio cholerae* O1—Western hemisphere, 1991–1994, and *V. cholerae* O139—Asia, 1994. *Morbidity and Mortality Weekly Report,* 44:215–219.

Centers for Disease Control. 1999. Achievements in public health, 1900–1999. Improvements in workplace safety—United States. *Morbidity and Mortality Weekly Report,* 48:461–469.

Elling, R. H. 1998. Workers health and safety (WHS) in cross-national perspective. *American Journal of Public Health,* 78:769–771.

Froines, J. R., Dellenbaugh, C. A., Wegman, D. H. 1986. Occupational health surveillance: A means to identify work-related risks. *American Journal of Public Health,* 76:1089–1096.

Herz-Picciotto, I. 1995. Epidemiology and quantitative risk assessment: A bridge from science to policy. *American Journal of Public Health,* 85:484–491.

MacKenzie, W. R., Hoxie, N. J., Proctor, M. E., Gradus, M. S., Blair, K. A., Peterson, D. E., Kazmierczak, J. J., Addiss, D. G., Fox, K. R., Rose, J. B., Davis, J. P. 1994. A massive outbreak in Milwaukee of cryptosporidium infection transmitted through the public water supply. *The New England Journal of Medicine,* 331:161–167.

Rossner, D., Markowitz, G. 1985. A "Gift of God"?: The public health controversy over leaded gasoline in the 1920's. *American Journal of Public Health,* 75:344–352.

Wartenberg, D., Simon, R. 1995. Comment: Integrating epidemiologic data into risk assessment. *American Journal of Public Health,* 885:491–493.

World Health Organization. 1992. *Our Planet, Our Health: Report of the WHO Commission on Health and Environment.* Geneva: WHO.

BIBLIOGRAPHY—WATER QUALITY AND WATERBORNE DISEASE

Centers for Disease Control. 1993. Update: Cholera—Western hemisphere, 1992. *Morbidity and Mortality Weekly Report,* 42:89–91.

Centers for Disease Control. 1994. Assessment of inadequately filtered public drinking water—Washington, D.C., December 1993. *Morbidity and Mortality Weekly Report,* 43:661–663.

Centers for Disease Control. 1996. Surveillance for waterborne-disease outbreaks—United States, 1993–1994. *Morbidity and Mortality Weekly Report,* 45:SS-1:1–33.

Craun, G. F. (ed). 1986. *Waterborne Diseases in the United States.* Boca Raton, FL: CRC Press.

Esrey, S. A., Potash, J. B., Roberts, L., Shiff, C. 1991. Effects of improved water supply on ascariasis, diarrhoea, dracunculiasis, hookworm infection, schistosomiasis, and trachoma. *Bulletin of the World Health Organization,* 69:609–621.

Hurst, C. J. 1991. Presence of enteric viruses in freshwater and their removal by the conventional drinking water treatment process. *Bulletin of the World Health Organization,* 69:113–119.

Last, J. M. 1992. *Public Health and Preventive Medicine,* Thirteenth Edition.

Tulchinsky, T. H., Burla, E., Halperin, R., Bonn, J., Ostory, P. 1993. Water quality, waterborne disease and enteric disease in Israel, 1976–1992. *Israel Journal of Medical Sciences,* 29:783–790.

World Health Organization. 1993. *Guidelines for Drinking Water Quality,* Volume 1. *Recommendations.* Second Edition. Geneva: WHO.

World Health Organization. 1994. *Operation and Management of Urban Water Supply and Sanitation Systems: A Guide for Managers.* Geneva: WHO.

World Health Organization. 1996. *Guidelines for Drinking Water Quality,* Volume 2. *Health Criteria and other Supporting Information.* Second Edition. Geneva: WHO.

World Health Organization. 1997. *Guidelines for Drinking Water Quality,* Volume 3. *Surveillance and Control of Community Water Supplies.* Second Edition. Geneva: WHO.

World Health Organization. 1998. *Guidelines for Drinking Water Quality, Addendum to Volume 2. Health Criteria and other Supporting Information.* Geneva: WHO.

BIBLIOGRAPHY—OCCUPATIONAL AND ENVIRONMENTAL HEALTH

American Public Health Association. 1997. Policy Statement 9704. Responsibilities of the lead pigment industry and others to support efforts to address the national child lead poisoning problem, APHA, 1997. Available on APHA home page, http://www.apha.org/science.policy.html

American Public Health Association. 1998. Policy Statement 9806. Preventing adverse occupational and environmental consequences of methyl tertiary butyl ether (MTBE) in fuels, APHA, 1998. Available on APHA home page, http://www.apha.og/science.policy.html

Attfield, M. D., Castellan, R. M. 1992. Epidemiological data on US coal miners' pneumoconiosis, 1960 to 1988. *American Journal of Public Health,* 82:964–970.

Centers for Disease Control. 1993. Prevalence of work disability—United States, 1990. *Morbidity and Mortality Weekly Report,* 42:757–759.

Centers for Disease Control. 1994. Surveillance for emergency events involving hazardous substances—United States, 1990–1992. *Morbidity and Mortality Weekly Report,* 43 (SS-2):1–6.

Centers for Disease Control. 1994. Occupational injury deaths—United States, 1980–1989. *Morbidity and Mortality Weekly Report,* 43:262–264.

Centers for Disease Control. 1995. Agriculture auger-related injuries and fatalities—Minnesota, 1992–1994. *Morbidity and Mortality Weekly Report,* 44:660–663.

Centers for Disease Control. 1996. Mercury exposure among residents of a building formerly used for industrial purposes—New Jersey, 1995. *Morbidity and Mortality Weekly Report,* 45:422–424.

Centers for Disease Control. 1999. Adult blood lead epidemiology and surveillance—United States, second and third quarters, 1998, and annual 1994–1997. *Morbidity and Mortality Weekly Report,* 48:213–223.

Corn, J. K. 1992. *Response to Occupational Health Hazards: A Historical Perspective.* New York: Van Nostrand Rheinhold.

Deutsch, P. V., Adler, J., Richter, E. D. 1992. Sentinel markers for industrial disasters. *Israel Journal of Medical Sciences,* 28:526–533.

Doll, R. 1992. Health and the environment in the 1990s. *American Journal of Public Health,* 82:933–941.

Dwyer, J. H., Flesch-Janys, D. 1995. Agent Orange in Vietnam. *American Journal of Public Health,* 85:476–478.

Editorial. 1992. Environmental pollution: It kills trees but does it kill people? *Lancet,* 340:821–822.

Edling, C. (editorial). 1985. Radon exposure and lung cancer. *British Journal of Industrial Medicine,* 42:721–722.

Elliott, P., Cuzick, J., English, D., Stern, R. (eds). 1992. *Geographic and Environmental Epidemiolo-*

gy: Methods for Small Area Studies. World Health Organization Regional Office for Europe, Oxford University Press.

Environmental Protection Agency. 1989. Why Accidents occur: Insights from the Accidental Release Information Program. Chemical Accident Prevention Bulletin, Series 8 number 1, July. Washington, DC: US EPA.

Froines, J. R., Baron, S., Wegman, D. H., O'Rourke, S. 1990. Characterization of airborne concentrations of lead in U.S. industry. American Journal of Industrial Medicine, 18:1–17.

Ginsberg, G. M., Tulchinsky, T. H. 1992. Regional differences in cancer incidence and mortality in
Israel: Possible leads to occupational causes. Israel Journal of Medical Sciences, 28:534–543.

Gottlieb, R. (ed) 1995. Reducing Toxins: A New Approach to Policy and Industrial Decisionmaking.
Washington, DC: Island Press.

Harrington, J. M. 1999. 1998 and beyond—Legge's legacy to modern occupational health. Annals of
Occupational Health, 43:1–6.

Hunter. D. 1969. The Diseases of Occupations. Fourth Edition. London: The English Universities Press Ltd.

Kvale, G., Bjelke, E., Heuch, I. 1986. Occupational exposure and lung cancer risk. International Journal of Cancer, 37:185–193.

Landrigan, P. J. 1992. Environmental disease—a preventable epidemic. American Journal of Public
Health, 82:941–943.

McMichael, A. J. 1993. Global environmental change and human population health: A conceptual and
scientific challenge for epidemiology. International Journal of Epidemiology, 22:1–8.

Nadakavukaren, A. 1990. Man and Environment: A Health Perspective, Third Edition. Prospect
Heights, IL: Waveland press.

Nicholls, G. 1999. The ebb and flow of radon. American Journal of Public Health, 89:993–995

Office of Air Quality Planning and Standards. 1998. Latest Findings on National Air Quality: 1997
Status and Trends. Research Triangle Park, NC: Environmental Protection Agency. EPA-454/F-98-
009, December 1998.

Richter, E. D., Deutsch, P. V., Adler, J. 1992. Recognition and use of sentinel markers in preventing industrial disasters. Prehospital and Disaster Medicine, 7:389–395.

Rosenstock, L., Olenec, C., Wagner, R. R. 1998. The national occupation research agenda: a model of
broad stakeholder input into priority setting. American Journal of Public Health, 88:353–356.

Selikoff, I. J. 1986. Asbestos-associated diseases. In: Last, J. M. (ed.). Maxcy-Rosenau: Public Health
and Preventive Medicine. Twelfth Edition, pp. 523–525. Norwalk, CT: Appleton–Century–Crofts.

Sinclair, U. 1906. The Jungle. Classic Series. New York: Airmont Publishing Co. Reprint, 1965.

Stayner, L. 1999. Protecting public health in the face of uncertain risks: the example of diesel exhaust.
American Journal of Public Health, 89:991–993.

Steenland, K. (ed). 1993. Case Studies in Occupational Epidemiology. New York: Oxford University Press.

Stern, C., Young, O. R., Druckman, D. (eds). 1992. Global Environmental Change: Understanding the
Human Dimensions. Washington, DC: National Academy Press.

Sundin, D. S., Pederson, D. H., Frazier, T. M. (editorial). 1986. Occupational hazard and health surveillance. American Journal of Public Health, 76:1083–1084.

Tulchinsky, T. H., Ginsberg, G. M., Shihab, S., Goldberg, E., Laster, R. 1992. Mesothelioma mortality among
former asbestos-cement workers in Israel, 1953–90. Israel Journal of Medical Sciences, 28:543–547.

United States Congress, Office of Technology Assessment. 1985. Preventing illness and injury at the
Workplace. Washington DC: Office of Technology Assessment, OTA Publications #OTA-H256.

Weeks, J. L., Levy, B. S., Wagner, G. R. (eds). 1991. Preventing Occupational Disease and Injury.
Washington DC: American Public Health Association.

Wegman, D. H. 1992. The potential impact of epidemiology on the prevention of occupational disease.
American Journal of Public Health, 82:944–954.

World Health Organization. 1991. Drinking Water and Sanitation, 1981–1990: A Way to Health: A WHO
Contribution to the International Drinking Water Supply and Sanitation Decade. Geneva: WHO.

World Health Organization European Region. 1997. Health in Europe: Report of the Third Evaluation
of Progress Toward Health for All in the European Region of WHO (1996–1997). Copenhagen:
WHO European Region.

Zirm, K. L., Mayer, J. 1989. The Management of Hazardous Substances in the Environment. London:
Elsevier Applied Sciences.

10

ORGANIZATION OF PUBLIC HEALTH SYSTEMS

INTRODUCTION

This chapter examines the organization of public health and hospital services, mainly in the United States, not necessarily as the ideal, but to illustrate how existing separate systems of service co-exist and interact. Each system evolved in its own separate organizational and financing format, yet they are finding common ground as medical care and prevention become more mutually interdependent. Traditional public health systems will increasingly need to develop intersectoral cooperation with other components of the health sector, but also with government and nongovernment agencies in related fields, such as agriculture, business, social welfare, education, police, community-based organizations, and many others.

Governments have legislative, regulatory, and taxation powers set out in constitution and law for common action for the public good, including powers to restrict individual action, that may prejudice the health of others, and to promote health. City-states in ancient Greece provided sanitation for the entire community and medical care for the poor. The Elizabethan Poor Laws in Britain in the early seventeenth century established the responsibility of the local authority for health and welfare. Subsequent developments brought local, state, and national government into sanitation, disease control, and other aspects of public health and health planning, later extended to assuring provision of comprehensive health care on a social-equity basis for all or selected parts of a society.

Societies have learned to prevent disease by social action and have also learned that individual health depends on such action. Governments are involved in that process, whether the society is based on democratic and free market principles or is centrally managed with a command economy. Society has accepted some limitations on individual rights for the public good. This limits the individual from attacking and harming another person, or damaging goods whether private or public. A person is restricted from throwing garbage in the street, and an industry is prohibited from polluting the environment or endangering its workers.

Public health policy, legislation, and action involve common measures to protect the individual and the community. Such measures may take the form of mandatory reporting of an infectious disease, chlorinating or fluoridating a community water system, regulating food and drug industries, requiring children to be immunized prior to entry to school, or fining and imprisoning industry's managers whose negligence causes death and injury or whose industry pollutes the environment.

Achieving public health goals requires organization. Public health organization requires a formal structure for a defined population whose finance, management, scope and content is defined in law and regulations. It includes services to be delivered to people contributing to their health and health care, delivered in defined settings such as in homes, educational institutions, work places, public places, communities, hospitals, and clinics. It deals with the physical and psycho-social environment. A health system is organized at various levels, starting at the most peripheral, community or primary level. It includes district, regional, state, and national levels as well as international contact. International strategies for health, and national health systems, are focusing on investments that produce health gain as the central goal rather than merely the efficient management of existing medical care institutions and services.

Function and structure are interdependent; structure should evolve from the desired function, that is, to achieve national goals and objectives for health. This is fulfilled through legislative, regulatory, financing, and service functions which provide the underpinnings of health in any country. Some countries undertake to provide all health care through a governmental system. Others legislate financing of health care, while others focus on financing for subgroups of the population, placing greater emphasis on provision of facilities and research in health care.

This chapter describes public health organization primarily using examples from the United States, including federal, state, and local public health authorities. In contrast to most industrialized countries, the United States lacks a universal health plan. As a result, health care is provided through independent, private, and public agencies. This is often described as a "nonsystem;" however, it is in fact a complex network of interactive services. Yet, it lacks universality, leaving many out of the mainstream for health care. As a result, public health organizations in the United States plays a very important role in providing essential services for people or needs not otherwise met. Partly because of this fragmentation of health care in the United States, public health has played a leadership role, with a major role in advocacy, development, and achievement in promoting health.

GOVERNMENT AND HEALTH OF THE NATION

Public health involves a wide variety of issues which should be directly under governmental responsibility because they require legislation, enforcement, and taxing powers. These include, for example, environment, nutrition, food and drug control, sanitation, immunization, traffic laws, firearms control, and health education. Many of these functions are promoted by nongovernmental organizations (NGOs), with delegated governmental regulatory powers.

Financing and allocation of public funds for health are an important means of influencing health activities. This may mean allocation of public funds to support research, teaching facilities, and provision of service. National governments may actually provide services, but increasingly this is being decentralized to lower levels of government (regional, district, municipal) or to nongovernment health care providers. Academic, professional, and public advocacy organizations play important roles in the New Public Health, such as for manpower training, education, research, and professional standards setting. These functions can be diffused to a variety of professional, consumer, and academic institutions. This enables governments to act indirectly through setting standards and norms, as well as providing incentives, accountability, accreditation, and licensing activities, as well as by its direct regulatory standards or financing sanctions.

Federal and Unitary States

Public health requires a basis in law, public administration, and financing. The form of government may differ from country to country, some being federal, others unitary.

In a federal system, three levels of government, federal, state, and local, have separate but overlapping responsibilities for public health. Federal states have constitutions conceived and written when state rights were emphasized and health care was perceived mainly as a private activity between patient and doctor. Consequently, the primary responsibility for health was placed at the state or provincial level. However, because of greater resources at the national level, federal government roles have increased in the health field over the years. National governments, however, have responsibility to ensure equity of social policy. A growing federal role has been a historical process common to many countries. At a minimum, the federal level is responsible for national health policy, planning and setting national health targets. The United States, Canada, Russia, Argentina, and Nigeria are examples of countries with federal forms of government.

A unitary state is a form of government that has a central national and local levels of government, but no intermediary legislating level. This includes countries such as the United Kingdom and governments based on the French Napoleonic Code, including most Spanish-speaking countries. In these countries, the central government has great responsibility for health, but here, too, local government is still a major factor in sanitation and local public health. The powers of regional and local authorities are derived from the national structure. Public health grew at the local level with sanitation, business licensing, food control, and the like. In the United Kingdom, the national government promoted local public health organization, later organizing personal health services programs for the entire population in the centrally controlled National Health Service.

Diffusion of authority in health is common to all health systems although the degree is different, largely based on historical precedents. At the end of the twentieth century, national health authorities are largely responsible for overall policy, law, financing, standards, monitoring, research, and assurance of services to meet national health goals. Management of services is, however, generally decentralized, with responsibility at the state, regional, and local health authority level or at the institutional level. Diffusion or sharing of responsibility from each level of authority is common in current planning in order to cope with the wide range of activities and interests that make up the health sector of a society. Nongovernmental agencies often precede governmental authority in the field, and their presence and participation make up important elements of the health complex, whether as providers of services, as advocates, or as fund raisers for programs that a government cannot manage to include in its "basket of services."

Administration of services is often devolved to local authorities or to independent institutions or other agencies. Diffusion of responsibilities occurs in different degrees in administration of services, in education, training, and registry of health professionals, in research, in intersectoral cooperation between governmental agencies, professional and accreditation organizations, NGOs, and advocacy groups, and in academic and research facilities. Legislation may initiate and direct changes in health programs by regulatory and financing measures, but implementation also requires a broad spectrum of participation of individuals and organizations of consumers, providers, and other health interest groups. Health is not an isolated service, but a reflection of values and standards of social and economic development of a society, with a large degree of interdependence between health agencies and other governmental and nongovernmental elements of that society.

Checks and Balances in Health Authority

The balance between government intervention and private organization, between regulation and self-governance, is not easy to define nor to achieve in health. Historically, different elements of health care developed at different times and with different degrees of political, economic, and popular support. The accumulated experience of modern public health indicates that all elements of health need to be considered a part of a whole. Weakness in one area threatens the well-being of the totality. Poor levels of nutrition and sanitation breed disease, for which curative medicine is more expensive and less effective than prevention. At the same time low medical care standards due to inadequate training, motivation, resources, and supervision can lead to low standards of health among large segments of the population.

Organization for public health services, whether integrated into a total care system or separate from curative service systems, requires a combination of centralized and decentralized responsibilities. The overriding national responsibility can best be met by setting policy goals and standards, while assuring regional and social equity. Decentralization allows local authorities direct operational responsibility, with means and accountability. Diffusion of responsibility means that many agencies operate at different levels of the national entity. Each has its own sphere of interest, and these link together to form a working whole, with checks, balances, and cooperation among them.

A centralized health organization that controls policy, administration, financing, services, manpower training, research, and regulation may lack checks and balances needed to prevent authoritarian control. Formerly highly centralized health systems are seeking decentralization as a means of infusing additional funding, a sense of localism, pride, privacy, and quality in their health systems. Combining this with universal access and regional, ethnic, and social equity. Comprehensiveness and cost constraint are the challenges of organization of public health systems.

There are advantages to a federal structure in the division of responsibilities for health. The senior level of government serves as the overall policy level with financing and regulatory roles. By its very nature, the state level is close to the community and represents regional interests, while the local health government is closest to the community and, with state and federal backing, can serve to promote community health interests subject to state and federal guidelines and accountability.

The New Public Health seeks a balance and cooperation between government-operated health services and the diffused network of private often competing organizations, working together to use resources effectively to achieve common health targets to meet the needs of the individual and the population as a whole.

Government and the Individual

Conflicting ideas as to the overall role government should play affect public health in many ways. In 1869, John Stuart Mill, one of the founders of modern economics, wrote in the introduction to *On Liberty,* "the only purpose for which power can rightfully be exercised over any member of a civilized society, against his will, is to prevent harm to others. His own good, either physical or moral, is not a sufficient warrant."

The institutions of basic sanitation and community hygiene had to contend with such individualistic ideas. The issue of governmental interference in "private matters," such as in health, is not new and is actively debated in industrialized western societies, post-Soviet countries, and developing nations alike. Laissez-faire economists promote the idea of minimal governmental involvement in all economic affairs including social services such as health.

During the nineteenth and increasingly in the twentieth centuries, it became apparent and imperative for protection and promotion of health that the state intervene to set and enforce public health measures in all societies. At the other extreme, disillusionment occurred with experience of governments assuming total responsibility for health and total central management of health services. Most countries have their own balance between the two extremes. Paradoxically, the most decentralized and privatized of all national health systems, that of the United States, has emphasized development of national and professional standards, monitoring, national targets and regulation in health, and is in the process of profound change from individual care toward managed care systems.

FUNCTIONS OF PUBLIC HEALTH

The American Public Health Association (APHA), founded in 1872, periodically issues policy statements on the mission and essential services of public health

---

BOX 10.1 MISSION AND ESSENTIAL SERVICES OF PUBLIC HEALTH, AMERICAN PUBLIC HEALTH ASSOCIATION

Public health responsibilities or mission

1. Prevent epidemics and spread of disease.
2. Protect against environmental hazards.
3. Prevent injuries.
4. Promote and encourage healthy behaviors.
5. Respond to disasters and assist communities in recovery.
6. Assure quality and accessibility of health services.

Essential public health services

1. Monitor health status to identify community health problems.
2. Diagnose and investigate health problems and health hazards in the community.
3. Inform, educate, and empower people about health issues.
4. Mobilize community partnerships and action to solve health problems.
5. Develop policies and plans that support individual and community health efforts.
6. Enforce laws and regulations that protect health and ensure safety.
7. Link people to needed personal health services and assure provision of health care when otherwise unavailable.
8. Assure an expert public health work force.
9. Evaluate effectiveness, accessibility, and quality of health services.
10. Research for new insights and innovative solutions to health problems.

Source: Essential Public Health Services Work Group. 1997. In: *Morbidity and Mortality Weekly Report,* 46:150–152; see also http://www.apha.org/science/

---

organizations. These guidelines help government to provide or assure their provision through other agencies. The 1994 APHA statement of the overall vision and the mission of public health in America (Box 10.1) was endorsed by the Association of State and Territorial Health Officials, the National Association of County and City Health Officials, the Institute of Medicine, the Association of Schools of Public Health, the U.S. Public Health Service, and others. Periodic review and revision, with consensus among the many professional organizations concerned with public health, help maintain relevance for local and central public health organizations.

For many of the responsibilities legislated for public health agencies at the national, state, provincial, or local health authority levels, a combination of methods and approaches is needed. Table 10.1 represents a summary of activities of

TABLE 10.1 Responsibilities and Activities of a Public Health Authority,
National, State, or Local

Responsibilities	Programs/activities
Monitor health status	Maintain and monitor vital statistics, reportable disease, special disease registries, hospital utilization data, sentinel centers, growth patterns of children, school and work attendance, special surveys; health data acquisition and processing; evaluate effectiveness and quality of health services
Develop health policy and targets	Represent health interests in urban planning, environmental issues, social policies, health facility planning and operation, disaster planning; agency program planning; comprehensive state and regional health planning; disaster planning; health education of the public; health advocacy; involvement of nonagency health providers; research and development; community involvement; organization of the health agency; policy direction of the health agency; staffing; financing; financing; relationship with federal and state health authorities
Health promotion	Education of population in health-related issues; promotion of safety and and healthful community infrastructure and services
Community health services	Immunization, STD, TB, and communicable disease control; chronic disease control and medical rehabilitation; family health, including prenatal, well-child, crippled children, school health, and family planning; dental health; substance abuse; nutrition services and education
Regulate environment, food, drug, work, and other essential health related services	Regulation and surveillance; food protection; hazardous substances and product safety; water supply sanitation; liquid waste control; water polution control; swimming pool sanitation and safety; occupational health and safety; radiation control; air quality management; noise pollution control; vector control; solid waste management; institutional sanitation; recreational sanitation; housing conservation and rehabilitation; environmental injury protection
Financing	From tax and intergovernmental financing to promote health activities and services to meet health needs; provide incentive grants or funding to lower levels of government or to providers, researchers
Mental health services	Primary prevention of mental disorders; consultation to community resources; diagnosis and treatment services: outpatient, emergency, short-term hospitalization, day care and night care services, aftercare services, diagnostic and evaluation services for mentally retarded
Personal health services	Medical care for those without adequate private health insurance (the elderly, the poor, transient groups); health facilities operations; emergency medical services; employee health programs; medical care for inmates of prisons and institutions
Intersectoral cooperation	Coordination with many agencies in education, social services, urban planning, housing authorities; police, civil defense, military, prison services; coordination of NGOs, including voluntary organizations, community-based organizations, ethnic and religious organizations

public health authorities in the United States. Regulatory functions are those
based on the legal authority of a public health agency to set and enforce standards.
Setting health targets, policies and financing, and national or state standards are

important in promoting new program initiatives. Health promotion includes not only direct and formal teaching, but promotion of awareness of public health problems to the general public, health care providers, and other agencies. Services may be provided directly or may be funded and supervised by the public health agency. Direct service is the provision of services to the public, especially in areas where universal coverage of a standard program is needed (e.g., immunizations), or for high risk groups not able to access other services (e.g., prenatal care for the poor).

Intersectoral cooperation is the coordination with other agencies of government, NGOs, or service providers to work toward common objectives that will improve public health. This is an area where public health advocacy is important in that the public health authority tries to engage other agencies, as in development of water and sewage systems or policing of highways to reduce road crash deaths. NGOs, voluntary organizations, and advocacy groups have in the past and will in the future play a vital role in developing health programs.

Regulatory Functions of Public Health Agencies

The regulatory function in public health is based on a legal mandate to protect the public from nuisances and hazards that affect health and to assure certain standards for provision of care. Whatever degree of decentralizing occurs in the health service system, there are key central standards in public health that must be maintained at the federal level in essential areas such as nutrition, sanitation, food and drug control, and others over which the individual citizen or health provider has no direct control. The regulatory function covers a wide range of public health activities.

Regulated aspects of public health in the United States include

1. Birth and death certificates: local, state, and national authorities, such as the U.S. National Center for Health Statistics (NCHS);
2. Business licensing approval: local health authorities;
3. Building code compliance: local health authorities under state and federal codes;
4. Sanitation and environmental health: through municipal, state, and national agencies, such as the Environmental Protection Agency (EPA);
5. Regulation of health professionals: through state boards;
6. Licensing and certification of health facilities: local, state, and federal authorities;
7. Communicable disease control: local, state, and federal authorities with the Centers for Disease Control (CDC);
8. Food safety: local, state, and federal standards and inspections by the Food and Drug Administration (FDA);
9. Pharmaceutical standards: including safety, efficacy, labeling, and manufacturing standards by the FDA;
10. Occupational health and safety: local, state, and federal standards and in-

spections through the Occupational Safety and Health Administration (OSHA, for standards, regulation, and enforcement) and the National Institute for Occupational Safety and Health (NIOSH, for research).

Methods of Providing or Assuring Services—Direct or Indirect?

Whether a governmental agency provides or assures the provision of services varies from country to country. Canada's health insurance program is operated by the provinces with federal cost sharing. In the Nordic countries, the counties, which have many of the characteristics of provinces, operate most local health services. In centrally managed economies, such as in the former Soviet countries, health services have been operated with a high degree of central control. The present trend internationally is toward decentralization of management.

Many functions of public health can only be performed by government because they require legislative, taxing, and regulatory powers, or because they are directed at the total population. Central coordination is required for key public health functions such as epidemiology and disease control, monitoring of population health, and others discussed in the previous section.

In keeping with specific health targets as formulated by national or international public or professional bodies, local, state, or national health authorities may directly provide certain basic public health functions, such as specialized laboratories. In the United States, public health agencies provide services not available to high risk or otherwise underserved population groups. Many of these developed under special funding by higher levels of government to promote specific programs such as immunization, lead abatement, prenatal care, and HIV testing. They are generally services that are not adequately provided for by insured medical services of private practitioners or health care systems.

Immunization may be provided as a governmental service, as in Israel, or by private or managed care providers, as in the United Kingdom and the United States. Even in countries with well-developed primary care systems, additional service needs appear requiring special activities, such as screening for hypertension or congenital disease. Health education, a function of all levels of government and nongovernment health services, involves those activities centered on raising consciousness and knowledge in the health professions, the general public, or specific target groups, cutting across virtually all public health activities.

Financial incentives in the form of grants or other categorical funding may be directed to programs to promote a specific kind of public health service, research, or education. Financial incentives are used widely in seeking solutions to specific problems, such as incentive payments to physicians for achieving performance indicators or specific targets such as full immunization, or Pap smears and mammography for specific target groups in the United Kingdom. Proposals submitted are then reviewed as to their quality and relevance to the goals stated and their potential impact. Incentive or categorical funding is often a useful method to introduce a new set of activities, to strengthen a weak area of public health, or to promote a shift in emphasis in the health system.

NONGOVERNMENT ROLES IN HEALTH

Both the government and the private sector, including both not-for-profit and for-profit service systems, have vital roles to play in public health and health care. The private sector includes not only service providers, but professional organizations, universities, and consumer, voluntary, and advocacy groups. Because of their contribution to service delivery, professional standards, and education of health personnel, they can make a major contribution to any health system, no matter how it is financed or administered.

Nongovernmental organizations may be able to innovate through voluntary action and programming to meet very specific areas of need with which formal health systems may have difficulty. In the United States an outstanding example of a vol-

---

BOX 10.2 THE MARCH OF DIMES

Founded in 1938 with the participation of President Franklin Delano Roosevelt, himself a victim of polio, the March of Dimes played a major role in providing care for polio-stricken children and the search for a vaccine to prevent the disease. Thousands of volunteers helped to raise funds and to organize wide-scale clinical trials of the Salk vaccine. Following the eradication of polio in the United States, the March of Dimes shifted its focus to major health problems of children: birth defects, low birth weight, infant mortality, and lack of prenatal care.

The organization established the following targets for the year 2000:

1. Reduce birth defects by 10%, from 150,000 birth defects per year currently, a leading cause of infant mortality;

2. Reduce infant mortality to 7 per 1000; 33,000 infants die annually, and the United States is in twenty-fourth place for IMR internationally;

3. Reduce low birth weight to 5% or less; LBWs jeopardize the lives and well-being of 290,000 infants annually;

4. Increase the number of women receiving prenatal care in the first trimester to 90%, from the current rate of 79%;

5. Promote genetic research including gene therapy, testing, and counseling, and gene mapping;

6. Fund the Salk Institute for Biological Studies work in the Human Genome Project with genes relate to immune disorders, mental retardation, leukemias, improved blood test for newborn screening, and improved perinatal care for cerebral palsy and respiratory distress of the newborn;

7. Promote use of folic acid among women in the age of fertility to reduce risks of neural tube defects.

Source: March of Dimes Birth Defects Foundation. 1999. Website http://www.modimes.org/

unteer NGO and its contribution is the role of the March of Dimes in the development of the Salk polio vaccine and the care of people affected by polio and, more currently, in prevention of birth defects. There are many organizations raising funds for promotion of research and services for specific disease entities, ranging from diabetes to multiple sclerosis.

Voluntary organizations can often initiate services that the public sector cannot. Examples are numerous, but this may suffice. In Jerusalem, a man and his father established a voluntary organization in memory of their mother and wife (Yad Sarah) in 1976 to provide free, loaned medical devices and services from wheelchairs to home meals to day care centers and emergency call systems, with the mission of helping the elderly and handicapped to function in their own homes. Subsequently, branches were established in 70 cities all over Israel, other organizations established similar projects in over 25 cities of the former Soviet Union, and plans are in progress for a similar organization in New York City.

MEDICAL PRACTICE AND PUBLIC HEALTH

Public health and clinical services are both vital and interdependent to improve individual and population health. Ready access to high quality health care services is a right of the population and a requirement of good public health. This requires the availability of high quality providers of clinical and preventive care. The phenomenon of private payment to physicians working in public sector health systems is widespread, as is that of physicians in public service who practice privately after official hours. This is almost impossible to stop, but can be regulated.

In the United Kingdom, private practice by specialists employed by the hospitals is permitted and encouraged, allowing faster access to hospital care for private patients. This is often seen as a built-in injustice in the NHS. In Israeli teaching hospitals, a private medical service is organized using senior physicians on the hospital premises, with some of the generated funds going to the hospital.

Fee-for-service practice of medicine is still a common mode of practice in the United States and Canada, even though each has different methods of financing services. Canada's national health insurance program is based on private fee-for-service practice of medicine, but it bans extra billing by physicians, which could threaten equity of access for all the population, as part of federal criteria for support of provincial health plans.

The United States has a mixed situation of private health insurance, mainly through place of employment, Medicare for those over age 65, and Medicaid for the poor. Some 40 million persons lack health insurance and another 15 million have poor levels of coverage, with further difficulties for those who change jobs and lose their health insurance coverage. Growth of managed care plans is occurring as private medical practice is declining in the United States. Operated as for-profit or as not-for-profit programs, managed care plans provide lower cost and more comprehensive coverage than traditional insurance plans.

Reform of health care is going on in many countries. This requires incentives to promote ambulatory and community outreach services, through incentives and integration of hospital and long-term care. Managed care is important in the United States, and the model is relevant in other countries because of the linkage of restraint in unnecessary use of hospital and unreferred specialist services, placing emphasis on primary care and preventive care (see Chapters 11–13).

INCENTIVES AND REGULATION

Incentives and disincentives are important tools in health policy and management. Governments are responsible for assuring adequate supplies and quality of health facilities and personnel to meet the needs of the population and for assuring that financing of the system is adequate and efficient. This responsibility includes the use of public authority to ensure a balanced and high quality system of care equitably available to the people of all regions and social classes. Whether services are owned and administered through government, nonprofit agencies, or private auspices, the public authority is responsible and accountable for ensuring that the health needs of the population are met.

The appropriate balance between different elements of health systems serving the same regional or district population is an important public health planning issue. Health facilities such as hospitals and long-term and community care facilities are licensed and regulated by the appropriate public health authority. This regulatory power is necessary, but not sufficient without financing arrangements to combine incentives and disincentives (i.e., a "carrot and stick" approach). The combination provides mechanisms to encourage health facilities to develop, in keeping with national, state, or local needs. In developed countries, this may mean closure of excess hospital beds and reallocation of resources to community-based health services, as is under way in the United Kingdom, in Scandinavian and many European countries, and in Canada, the United States, and, to some extent, in Russia.

The ratios of hospital beds and medical personnel per thousand population are crucial determinants of health economics, so that national and state health authorities must use their regulatory powers to limit their supply and distribution. Excess supply of medical doctors is a problem in many mid-level developing countries, such as in Latin America. Regulatory or financial powers, as well as financial controls, can be used to reduce enrollment in medical faculties in order to reduce the oversupply of doctors, and to redirect doctors to underserved areas of a country.

A federal government authority can act to promote specific health programs by setting financial incentives and disincentives. The categorical fiscal grant approach provides grants specified for a certain purpose, or cost-sharing for a program that meets specified guidelines. The Canadian health insurance plan is based on provincial plans meeting federal guidelines to qualify for a specified share of the costs.

National health insurance was included in the proposed social security legislation during the Roosevelt administration but excluded from the Social Security Act

---

BOX 10.3 THE HILL-BURTON ACT

The Hill-Burton Act provided a federal–state–local partnership which channeled $16 billion to assist (Hospital Survey and Construction Act, 1946, Title VI of the Public Health Service Act) 3800 communities in 11,500 projects for hospitals, extended care, rehabilitation facilities, and public health centers. It brought national standards and financing to local hospitals. The program helped raise standards of medical care throughout the United States in the 1950s and 1960s. It led to an increase in numbers of hospitals in underserved areas and the renovation of obsolescent facilities. It promoted desegregation in the southern United States and provided a mechanism for treatment of the uninsured in the nation's hospitals.

The program also succeeded in limiting the buildup of an excess of hospital beds, setting standards at 4–4.5 acute care hospital beds per 1000 population (more for rural areas), without an increase in the total supply of beds. While it favored middle-class communities because it required local financial contributions, it also channeled federal monies to poor communities, thus raising standards of hospitals and equity in access to quality care. It required hospitals assisted by federal funding to provide a reasonable volume of free or reduced cost care to the poor (see Chapter 11) and emergency treatment of the uninsured. In setting upper limits on hospital beds, it limited hospital expansion and contributed to a continuing process of improving and shortening hospital stays.

The program had a number of basic failings, including the promotion of the hospital as the main center of health care, leaving community care out of the main flow of added funds. It led to an increase in the proportion of health expenditures going to hospital care. Expenditures for hospital care as a percentage of total health expenses increased from 34.5% in 1960 to a high of 41.5% in 1980, but has since declined to 35.4% in 1995. In the 1980s, the Hill-Burton Act was expanded to promote clinic and primary care facilities.

---

of 1935, because it placed the act in jeopardy politically. Following the end of World War II, in 1946, the proposed Wagner-Murray-Dingell Bill for national health insurance failed to reach the floor of Congress, dying in committee, under pressure of the American Medical Association and the health insurance industry. A portion of that proposal emerged, however, as the Hill-Burton Act to provide federal assistance to local agencies to build or upgrade hospitals. The Hill-Burton model is a relevant approach to problem solving in a federal state, using a categorical grant mechanism to promote what is seen as a health priority. Such an approach may be used to strengthen a weak health program such as immunization and maternity care in a developing country. It may be used to change the balance in supply of services and resources. A system of incentives or cost-sharing arrangements can provide capital funding, for example, to reduce total bed capacity and to promote integration of maternity, mental health, geriatric, and tuberculosis facilities into general hospitals. A "downsize and upgrade" conditional grant would provide for renovation and transition to an approved program of facilities to modernize hospital services. The federal grant would encourage the local authority to apply and match part of the funding, and meet federal criteria and guidelines for this process.

In countries where health systems were highly centralized, such as in Britain's National Health Service and in former Soviet health systems, decentralization and diffusion of power were promoted by financing mechanisms. The Canadian national health insurance plan was implemented with fiscal cost-sharing incentives by the federal government to induce the provinces to agree to federal conditions of universal coverage, comprehensiveness, portability, and public administration as criteria for the provincial plans. When the federal government moved from fixed percentages of expenditures to fixed or block grants, it lost some control over detailed management of provincial plans, but it retains a strong voice in requirements for equity, portability (transferability of insured benefits), public administration, and not allowing extra billing for covered benefits. As federal funding declined as a share of total provincial costs, the provinces were under pressure to reform mainly by reducing the hospital bed supply and promoting community wide health service organization.

Unregulated chronic care facilities operated by private interests resulted in proliferation of poor quality facilities and sometimes scandalous levels of care in many communities in the United States. Public health authorities were powerless to interfere except in cases of gross neglect or poor sanitary facilities. The introduction of Medicare for the elderly and Medicaid for the poor provided federal and state agencies with the power to set minimum standards for care facilities, by requiring all facilities serving Medicare patients to be accredited by a nongovernmental agency accepted by the federal health authorities. This has become a standard requirement throughout the United States, and the Canadian provincial health insurance plans also apply economic sanctions on unaccredited hospitals.

Another measure to increase regulation of health care facilities was the requirement for any hospital proposing expansion or renovation to seek state approval through a Certificate of Need (CON). The CON, as used in the United States under state health legislation, is contingent on demonstrating need and sources of funding that are in compliance with state regulations. This can be linked with incentive grants but can also be used as a simple regulatory mechanism. The CON approach implemented by state departments of health were only partially successful in limiting unbridled ambitious expansion of hospital facilities. In the 1980s and especially the 1990s, competition and changes in payment systems have resulted in hospital closures and downsizing in the United States.

Promotion of Research and Teaching

Medical research and education are the basis for future developments in health care. They foster new health science developments, such as vaccines or, through the Human Genome Project, treatment for genetic and chronic diseases. This contributes to the development of medical schools, but also safeguard and increase their quality, raising standards of care. Research in public health depends on the basic and clinical sciences, but equally on epidemiology and documented experience of field programs.

In the United States, the National Institutes of Health (NIH), starting with the National Cancer Institute in the 1930s, have done much to encourage high quality medical education and research. The NIH granting system has been a major factor in promoting standards of medical education by financing research and teaching faculties in medical schools throughout the United States. NIH funding has played a major role in moving the United States into the forefront of the biomedical sciences since World War II. There are 24 separate National Institutes of Health (including centers and divisions; Table 10.2).

A combination of collegial competition, the free publication and exchange of

TABLE 10.2   United States National Institutes of Health, Centers and Divisions, and Internet Addresses, January, 2000

Institutes (general home page   http://www.nih.gov)
    National Cancer Institute (NCI)   www.nci.nih.gov
    National Eye Institute (NEI)   www.nei.nih.gov
    National Heart, Lung, and Blood Institute (NHLBI)   www.nhlbi.nih.gov/nhlbi/nhlbi/htm
    National Human Genome Research Institute (NHGRI)   www.nhgri.nih.gov
    National Institure on Aging (NIA)   www.nih.gov/nia
    National Institute of Alcohol Abuse and Alcoholism (NIAAA)   www.niaaa.nih.gov
    National Institute Allergy and Infectious Diseases (NIAID)   www.niaid.nih.gov
    National Institute of Arthritis and Musculoskeletal and Skin Diseases (NIAMS)   www.nih.gov/niams
    National Institute of Child Health and Human Development (NICHD)   www.nih.gov/nichd
    National Institute of Deafness and Other Communication Disorders (NIDCD)   www.nih.gov/nidcd
    National Institute of Dental and Craniofacial Research (NIDCR)   www.nidr.nih.gov
    National Institute of Diabetes and Digestive and Kidney Diseases (NDDK)   www.nddk.nih.gov
    National Institute of Drug Abuse (NIDA)   www.nida.nih.gov
    National Institute of Environmental Health Services (NIEHS)   www.niehs.nih.gov
    National Institute of General Medical Sciences (NIGMS)   www.nih.gov/nigms
    National Institute of Mental Health (NIMH)   www.nimh.nih.gov
    National Institute of Neurological Disorders and Stroke (NINDS)   www.ninds.nih.gov
    National Institute of Nursing Research (NINR)   www.nih.gov/ninr
Centers and divisions
    Warren Grant Magnuson Clinical Center (CC)   www.cc.nih.gov
    Center for Information Technology (CIT)   www.
    National Library of Medicine (NLM) and MEDLARS[a]   www.nlm.nih.gov
    National Center for Research Resources (NCRR)   www.ncrr.nih.gov
    National Center for Complementary and Alternative Medicine (NCCAM)   www.nccam.nih.gov/
    Fogarty International Center (FIC)   www.nih.gov/fic
    Division of Computer Research and Technology (DCRT)   www.dcrt.nih.gov
    Center for Scientific Review (CSR)   www.drg.nih.gov/

Source: National Institutes of Health, Bethesda, Maryland 20892. Website Jan, 2000 http//www.nih.gov/
[a]MEDLARS = Medical Literature Analysis and Retrieval System.

scientific studies, and views in peer-reviewed journals and professional meetings in government agencies, promote scientific and applied progress in the medical sciences. The private sector manufacture of drugs and medical devices contribute to the continued development of medical and public health sciences. National centers of excellence in public health in other countries include the Pasteur Institute in France and Cambridge Laboratories in the United Kingdom. They receive national funding and have a critical mass of high quality researchers.

Federal funding of medical teaching centers supports academic standards of undergraduate education. Federal or external granting mechanisms can be used to promote schools of public health and health administration that are needed to prepare the next generation of health leaders and researchers. Research may be initiated by scientists in university or research institutes, or in the governmental or private sector, in response to requests for proposals. A competitive grant system can be useful to upgrade medical education and university academic standards by promoting research and graduate education.

Accreditation and Quality Regulation

While public health authorities have sufficient powers to regulate health facilities. In practice accreditation based on professional guidelines and systems outside the governmental structure (see Chapter 15) plays an important role in quality of health care provider organizations, as an important adjunct to the official regulatory approach of health departments.

Initiated in the United States in 1913, with Canada from 1951 to 1959 when the latter established its own accreditation system, the Joint Commission on Hospital Accreditation (JCHA), established by a consortium of the American College of Surgeons, the American Hospital Association, and other voluntary professional bodies, carries out voluntary peer review of hospitals throughout North America. The Commission established minimum standards in 1918, and has gone on to develop extensive guidelines based on physical, organizational, and professional criteria, to protect the safety and rights of the patient, standards of care, and efficient organization of services. The review was initially on the basis of voluntary request by the institution. Accreditation involves a process of external review of the facilities, organization, staffing, and many functions including staff qualifications, continuing education, medical records, quality assurance, and others (Chapter 15).

The JCHA review was initially conducted on the basis of a voluntary request by the institution, but accreditation has become virtually mandatory for the economic survival of a hospital in the United States and Canada. Since 1965, Medicare and Medicaid accept accreditation as compliance with federal standards for the purpose of payment, and refuse to pay for services in an unaccredited hospital. The renamed Joint Commission for Accreditation of Healthcare Organizations (JCAHO) has gone on to develop standards for accreditation of facilities for the mentally retarded (1969), psychiatric facilities (1970), long-term care facilities (1971), ambulatory facilities (1975), hospices (1983) with home care (1990), and managed care programs (1989) with ambulatory care in 1990. There is a growing emphasis on action plans for quality improvement for rural hospitals, health care networks, laboratories, and public health programs. The JCAHO has become active in promoting accreditation organizations in other countries such as the United Kingdom and Australia.

The New York State Department of Health has its own mandatory regulatory system for hospitals and long-term care facilities. While perhaps unpopular with providers, regulation is viewed by many state health officials as essential to the maintenance of quality standards and prevention of professional and human rights abuses. Israel established a national system of inspection of private long-term care facilities, which has improved standards of facilities and care during the 1990s. While opponents see this as excessive state interference, in principle they accept the need for state regulation. These models could be useful for raising standards in other health care systems.

NATIONAL GOVERNMENT PUBLIC HEALTH SERVICES

National governments can use their financial power to promote programs directly to the state, provincial, or local governmental level or indirectly through nongovernmental agencies. The latter include universities, voluntary teaching hospitals, and private nongovernmental organizations (NGOs). They can use direct or indirect funding to diffuse and promote national standards, such as in medical education and research. They can also ensure regional equality of services by cost-sharing or grants that favor poorer regions of the country. National governmental health agencies are responsible for external relations, including those with international bodies such as the United Nations, the World Health Organization, the Food and Agriculture Organization, the International Labour Organization (Chapter 16), as well as with parallel ministries of health in other countries, and other national agencies in the same country.

Prior to and following World War II, most western industrialized countries developed some form of national health program. In North America, health care was provided through private insurance, largely union-negotiated employment-based health plans. The United States, frustrated in its attempts to bring in a national health insurance plan in 1946, established many categorical programs for financing state and county public health services and research and teaching facilities through the National Institutes of Health (NIH). This promoted high levels of competitive, peer-reviewed programs throughout the country, but failed to assure universal access to health care (Chapter 13).

In all forms of government, the national responsibility for health has led to specialized public health services as well as supervisory and regulatory functions. These include provision of vital support services, such as public health reference laboratories, epidemiology and communicable disease control activities (e.g., national epidemiologic publications, airport, and port surveillance), national health statistics, supervision of drugs and biologicals, research and teaching facilities, and cooperation among federal, state, and local authorities. Standards bureaus and

---

BOX 10.4 FUNCTIONS OF FEDERAL OR NATIONAL MINISTRY OR DEPARTMENT OF HEALTH

National health planning
National health financing
National health insurance
Assurance of regional equity
Goals, objectives, targets
Standards and quality of care
Research promotion and quality
Professional standards/licensing
Environmental protection legislation, standards, monitoring
Food and drug legislation, standards, licensing
Epidemiology of acute and chronic disease
Health status monitoring
Medical/pharmaceutical industrial development
Health promotion
Nutrition and food policy
National reference laboratories
Social assistance
Social security

---

agencies provide the guidelines, monitoring, and/or supervision of health care at the lower levels of government and in the nongovernmental and private sectors.

The federal government entered the public health arena in areas where only a national jurisdiction could function. The Marine Hospital Service was established in 1798 to provide care for U.S. and foreign seamen. The federal Food and Drug Act of 1906 protects the consumer from adulterated foods and ineffective or dangerous medicines. The Social Security Act of 1935 provided pensions for the elderly and the handicapped. In 1965, the Social Security Act was extended to include Medicare as a federal program providing health insurance for the elderly. In the same year, Medicaid was also established, providing health care for the poor, set up as a cost-sharing program with state and local authorities. The history of development of public health in the United States reflects advancing scientific knowledge, societal demands for better health, and the evolution of interactive organization at federal, state, and local levels. In some respects, public health in the United States has provided professional leadership in the field internationally; in other respects the U.S. has lagged behind other industrialized countries.

The Department of Health, Education, and Welfare was established in 1953 under a cabinet level officer of the executive branch of the Eisenhower administra-

TABLE 10.3 United States Department of Health and Human Services (HHS)

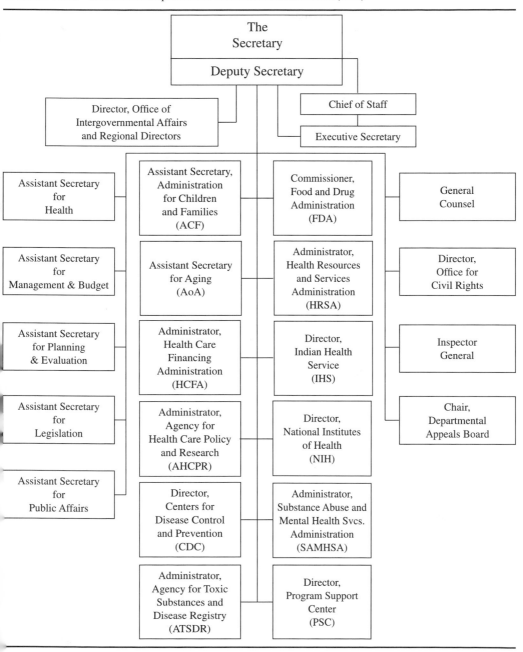

Note: The Assistant Secretary for Health is also the Surgeon General of the United States.

tion, bringing together a variety of federal agencies and programs. This and subsequent reorganization led to emergence of the Department of Health and Human Services (HHS) and its branches. The present organizational structure of HHS is seen in Table 10.3. The federal role in direct regulation and funding of projects deemed to be in the national interest helps to promote state and local health authority response to public health problems. The categorical grant system has been instrumental in advancing specific areas of activity, such as maternal and child health, which remain a major activity of local state and local public health departments. The initiatives of the Health Care Financing Administration (HCFA) in promoting changes in methods of paying for hospital care through diagnostic related groups (discussed in Chapters 12 and 13) helped reduce hospital lengths of stay, days of care, and the hospital bed to population ratio.

The Surgeon General of the Public Health Service provides important professional leadership to the public health movement in the United States. This was perhaps best exemplified by Dr. C. Everett Koop, Surgeon General for many years during the Reagan administration. The appointment of Dr. Koop, a pediatric cardiac surgeon, was initially opposed by the American Public Health Association. Within a short period of time he showed a degree of professional and moral leadership that served as a beacon greatly recognized by the public health profession.

The Centers for Disease Control and Disease Prevention (CDC) play a continuous role in dispersing epidemiologic data and evaluation throughout the country and the world (see Chapter 4). The training program of Epidemic Intelligence Officers for federal, state and local health departments continues to provide high quality medical epidemiologists capable of developing leadership in this field.

Other agencies of the federal government control health related programs, including the Departments of Agriculture, Defense, Environment, Interior, Labor, and Transportation. The Department of Agriculture operates a National School Lunch program and a food stamp program to supplement food purchasing power for the working poor. The Department of Labor operates the Occupational Safety and Health Administration. The Environmental Protection Agency is an independent federal agency responsible for air and water quality, pollution control, pesticide regulation, solid waste control, radiation and toxic substance hazard control, and noise abatement.

STATE GOVERNMENT PUBLIC HEALTH SERVICES

The state or provincial level of government has a leading role in health in most federal countries. State or provincial governments have responsibility to ensure adequacy in organization, set standards and targets, assist financially, and provide professional and technical support services to local health departments. State functions, such as financing and in some cases direct services and monitoring health status, include the following:

BOX 10.5 HEALTH RESPONSIBILITIES OF
A PROVINCIAL/STATE GOVERNMENT

Resource allocation
Health planning
Epidemiology
Vital statistics
Communicable disease control, immunization, investigation, reporting
Health promotion
Education of health professionals
Licensing and discipline of professionals
Licensing and supervision of health facilities
Laboratories operation, licensing, quality assurance
Health education
Quality promotion
Environmental health monitoring
Nutrition
Maternal and child health services
Mental health

---

1. Coordinate with other government departments: governmental planning and priorities, education, social welfare, labor, agriculture, mental health, and financing of universities;
2. Establish standards, develop, finance, advise, and supervise local health departments;
3. Legislate and regulate health-related matters: preparation, assistance, and enforcement;
4. Plan and set health priorities and targets;
5. Direct services to local health departments in epidemiology, laboratory services, biological surveys, program planning, and development;
6. Maintain and publish vital statistics, epidemiology, and health information systems;
7. Develop standards and monitor quantity, quality, and distribution of diagnostic and treatment services;
8. Ensure occupational health supervision;
9. Ensure environmental health supervision;
10. License and discipline health professionals and health care institutions;
11. Provide occupational and personal health services to state employees;
12. Coordinate with related state services, including social services, mental health, retardation, drug and rehabilitation, and prison services;
13. Coordinate with national and other state/provincial health authorities;
14. Monitor health status indicators of state/province and local authorities.

State or provincial departments of health are complex organizations with many responsibilities for financing, regulating, inspecting, and assuring health-related issues. In the United States, this includes administration of health insurance for the poor under Medicaid; in Canada, the provinces administer universal health insurance plans. States may initiate programs that are shared with local health authorities and with federal cost sharing, or respond to federal initiatives and seek their funds in a wide variety of programs through federal requests-for-proposals for maternal and child health or other categorical grants.

The New York State Department of Health (DOH) has a strong tradition of regulation in chronic care facilities, laboratories, and hospitals, and in environmental health. Certificates of need, regulation of methods of reimbursement for hospital care (see Chapter 13), health standards and surveillance systems, rural health systems, and many other regulatory functions of the department make it a powerful determinant of the operation of health care in the state. This state DOH is active in screening programs for congenital and infectious diseases of the newborn, laboratory certification, and quality assurance. An AIDS Institute is responsible for screening, prevention, and AIDS care programs. The Center for Community Health operates a wide range of public health programs, ranging from epidemiologic surveillance of infectious diseases, to prenatal and newborn care among the underserved, to the community health workers program, and to nutrition monitoring and many other intervention programs focused on high risk groups or topics. Environmental epidemiology and monitoring is also strong in the state which experienced the Love Canal incident (see Chapter 9). Table 10.4 is the 1996 configuration of the New York State DOH. This is not necessarily typical but does show the wide range of activities, including state, federal, and local initiatives.

The New York State Department of Health targeted the following issues during 1995–1997:

1. Reform of health care financing: converted hospital financing from a state regulated tariff to a competitive system in which the hospitals negotiate with the insurers to determine fees, with the object of introducing competition, greater efficiency, and lower health costs;

2. The Health Reform Act of 1996 created a fund of $2.6 billion from surcharges on specific medical services to finance charity medical care and expand health insurance for the working poor, support graduate education, develop primary care, and assist hospitals to adjust to the new economic climate;

3. Managed Care Initiative for 3 million Medicaid beneficiaries is negotiating with federal agencies (Health Care Financing Administration), including provisions for regulating managed care program recruiting practices;

4. Managed Care Bill of Rights: 6 million New Yorkers are enrolled in managed care programs, and this bill of rights assures that plans must advise members of their rights of potential treatment for their conditions, and assures a 2-day stay in hospital following births;

5. Physician discipline reforms through 1996 legislation enabled the Board of

TABLE 10.4 Table of Organization, New York State Department of Health, 1996

Commissioner

Health Facility Management
Office of Managed Care
Government Affairs and Strategic Planning
School of Public Health
Office of Minority Health

Office of Public Health (OPH)	Office of Health Systems Management (OHSM)
AIDS Institute	Health Care Financing
HIV care and community services	Elderly pharmaceutical coverage
Prevention	Financial management and institutional support
Center for Community Health	Health economics
Family and local health	Long-term care reimbursement
Nutrition	Primary and acute care reimbursement
Epidemiology	Health Care Standards and Surveillance
Chronic disease prevention and control	Home care
Center for Environmental Health	Hospital services
Environmental protection	Professional medical misconduct
Environmental health assessment	Emergency medical services
Occupational and environmental	Funeral directing
epidemiology	Controlled substances
District offices	Standards health facility planning
Laboratory Center	Miscellaneous
Molecular medicine	Architectural engineering
Infectious diseases	Health facility planning
Environmental disease prevention	Long-term care initiatives
Laboratory quality certification	Facility and service review
Laboratory operations	Financial analysis and review
Genetic disorders	Project management

Professional Medical Conduct to stop the practice of a doctor pending appeal for revocation of license; of 6000 investigations in 1995, 363 disciplinary actions were taken, including 15 immediate revocations of licenses;

6. Public Health Priorities for the Next Decade consultation was undertaken with hearings throughout the state;

7. Long-term Care Task Force appointed by the governor of the state recommended strategies for reform of long-term care financing, insurance, managed long-term care, and home care programs;

8. Newborn HIV screening was accepted by 90% of new mothers offered the service; this is important since antiviral therapy taken during pregnancy reduces risk of transmission to the newborn by 70%;

9. Emerging infectious diseases: control of *E. coli* O157 in undercooked hamburgers, cryptosporidiosis as a waterborne parasitic disease, Group A *Streptococ-*

cus, multiresistant strep pneumonia, Hantavirus related to exposure to rodent droppings, ehrlichiosis transmitted by ticks;

10. Response to health care crises by increasing quality of care oversight in hospitals, nursing homes, and other health care settings; a ban was imposed on sale of herbal stimulants carrying ephedrine which caused 15 deaths nationwide.

Other agencies with public health responsibilities in New York State include the Department of Education (school sanitation, health education, licensure of physicians and other health professionals), the Department of Labor (health and safety of workers, in-plant pollution and radiation control), Department of Environmental Conservation (control of pesticides, rabies control, air pollution, sewage and solid waste control), Department of Social Services (Medicaid), State University of New York (School of Public Health, student health services), Department of Mental Hygiene (mental institutions and community services), Narcotics Addiction Control Commission (treatment facilities, research, education), Department of Agriculture (licensure of meat dealers and slaughterhouses, inspection of restaurants, school-meal regulation of food additives), Department of Corrections (operation of prison hospitals and clinics, TB case finding), and the Department of Motor Vehicles (highway safety promotion). The New York State Department of Health is unique in that it is a cosponsor with the State University of New York (SUNY) of a School of Public Health at Albany, which involves departmental personnel as faculty and students in internships in branches of the DOH. While not necessarily representative of other states, this health department represents the broad scope of public health at the state level of government.

LOCAL HEALTH AUTHORITIES

Historically, the local health authority (LHA) is responsible for sanitation and the provision of direct care to the poor and other high risk groups. Boards of health were established in Philadelphia in 1794 and in New York City in 1796 for these purposes.

The city or county local public health department is the official public health agency closest to the population served. The LHA provides a range of direct supervisory sanitary functions to ensure compliance with local, state, and federal sanitary codes. The local public health department may also provide direct services, usually personal preventive services funded by the local government authority, or by higher levels of government. In the United States, the local public health department is the agency attempting to ensure services to persons inadequately served by voluntary or federal and state insurance plans. Programs may be funded by cost-sharing or may be based on categorical or block grants from state or federal governments.

Even though there has been massive growth in the involvement of higher levels of government in public health, the LHA remains the major force for public

---

BOX 10.6 HEALTH RESPONSIBILITIES OF A LOCAL HEALTH AUTHORITY

Registration and vital statistics
Epidemiology of infectious diseases
Health education, health promotion
Environmental protection, sanitation
Control of communicable diseases, STDs, HIV, TB
Preventive prenatal, infant, toddler care
Allocation of and search for resources from higher levels of government
Planning and management of services
Licensing and supervision of health facilities
Hospitals and home care
Care of handicapped
Rehabilitation and long-term care
Coordination of health services, public-private partnerships
Intersectoral cooperation
Mental health
Social assistance
Nutrition
Community participation

---

health at the community level. In the United States and Canada the LHA is organized in the form of city or county public health departments.

In new health initiatives, such as Health for All, district health systems, and Healthy Cities, the LHA is involved in a wider set of programs for health of its population. To meet the objectives of these programs, formerly highly centralized systems, such as those of the United Kingdom, the Scandinavian countries, developing nations, and republics of the former Soviet Union, are being decentralized to LHAs, with varying degrees of central funding, planning, and direction.

In 1940, the American Public Health Association adopted a recommended standard of six basic responsibilities of the LHA:

1. Vital statistics;
2. Communicable disease control: in the areas of childhood diseases, TB, STDs, and tropical diseases;
3. Environmental sanitation: water, food processing and marketing, sewage, garbage, sanitary condition of places of business, public eating places, and workplaces;
4. Laboratory services;
5. Maternal, child, and school health;
6. Health education.

BOX 10.7 ALBANY COUNTY HEALTH DEPARTMENT

Albany County, New York, had a population of 298,500 in 1993, a birth rate of 12.9 per 1000, an infant mortality rate of 5.7 per 1000, and a low birth weight rate of 7.6%, all similar to the rates for the rest of upstate New York (i.e., outside of New York City). Mortality rates for cerebrovascular disease and diseases of the heart were nearly 10% higher than those of upstate New York; motor vehicle accident mortality and homicide rates were lower. Infectious disease morbidity rates were similar to those upstate.

The Albany County Health Department receives income from program revenues (31%), county government (26%), state government (22%), and grants (21%). Grants are mainly from state government for primary care for infants and toddlers (1–5 years), prenatal care, and lead poisoning prevention. Services include programs for children with special needs, early intervention, preschool programs for children 3–5, and the physically handicapped program, accounting for the majority of expenditures ($12 million out of $18 million). The department is active in a wide range of programs including rabies control, environmental health, control of infectious and food-borne disease, health education (particularly HIV and STD related), smoking reduction, prevention of heart disease and stroke, TB control, travel immunizations, Lyme disease, Pap smear services, injury prevention, training of family practitioners, community health outreach workers, and promotion of fluoridation.

In 1995, the Department began to contract out for obstetrical and pediatric services to low income population groups to two Catholic-run hospital networks and a federally funded Community Health Center. Clinics operated by the county were turned over to these providers, which established services at the level of their other service programs. This "Healthy Partnerships" shifted care for over 1000 families from the county to these providers, using community health workers to follow up nonattenders and others in need, with improved access and satisfaction with care. This model of cooperation between public health and medical care systems represents a model and growing trend in other parts of the United States.

Source: Committee on Medicine and Public Health (Lazker, R. D., ed). 1997. *Medicine and Public Health: The Power of Collaboration*. New York: American Public Health Association and American Medical Association, New York Academy of Medicine, 1997; and Albany County Department of Health, 1998.

In 1950, the APHA adopted an expanded program of responsibilities for the LHA, which included the above plus the following:

1. Chronic disease control;
2. Housing and urban planning;

3. Accident prevention;
4. Coordination with other agencies;
5. Surveillance of total health status; births, deaths, chronic disease, morbidity data, surveys, reporting of morbidity, and evaluation of community needs;
6. Education of the public and professional community regarding health status and needs;
7. Supervisory and regulatory activities including health services providers;
8. Personal health services: direct provision and supportive services, varying from comprehensive service programs to services for those in need;
9. Planning of health facilities, urban planning and renewal;
10. Special diagnostic services: including STDs, TB, cancer, child development, and dental care.

Cooperation between the different levels of government is vital to define and achieve national health objectives. Each level of government has a unique role to play. Decentralized administration of public health, without national financing and policies, will not achieve the full potential of public health and will produce inequities between different regions of a country.

MONITORING HEALTH STATUS

As discussed in Chapter 3, public health depends on information, just as an army depends on intelligence in order to modify approaches in accordance with changing circumstances and need. Collection, collation, and analysis of this information is vital to informed health policy and must be available to all concerned with health for analysis and policy debate. All levels of government are engaged in health status monitoring, with geographic information systems (GIS), a multisource database related to health indicators for the population of a definable geographic entity, helping to identify specific localized problems for intervention.

The responsibility of gathering vital statistics lies largely at the local government level, as is the reporting of infectious diseases and other events. Initial collation of the data occurs at this level, and information is then passed on to state health authorities and subsequently to the national level. The gathering of information is a strongly developed tradition in the industrialized countries, and the United States has in many ways done this effectively. In the United States, the Centers for Disease Control (CDC) serves as a national leadership and reference center, not only for infectious diseases, but also for chronic diseases such as cardiovascular disease, nutrition, diabetes, perinatal epidemiology, and many other conditions.

Health statistics provide the ongoing data needed for monitoring the health status of a population. They provide routine diagnostic and population-based monitoring data that supply valuable epidemiologic information on congenital conditions, STDs, tuberculosis, and HIV infection. Centers of excellence of all kinds, funded or administered directly by federal or state government or by the NIH mechanism, provide tertiary level medical care and conduct biomedical and epidemiologic research, making important contributions to the information pool needed to promote quality analysis and health care.

The national health authority is responsible for the central collation and analysis of health information on the epidemiology of infectious and chronic diseases, vital statistics, utilization of services, and monitoring of national and regional variations in health. This information is only of value if gathered, processed, and published so that it is readily available to health administrators, planners, epidemiologists, care providers, and the public. Census data provide the population denominators for calculation of rates of death and disease incidence or prevalence.

Inexpensive technology of personal computers with modems, as well as telephones and facsimiles, enable local public health agencies to receive real-time information through Internet connections for continuous health profiles of their communities. Sources of data include the following:

1. Vital statistics and national centers for health statistics;
2. Epidemiologic reports of infectious and reportable diseases, including STDs;
3. State, national, and international reporting centers for disease control;
4. Census data;
5. Special disease registers (e.g., cancer);
6. Hospital discharge information systems;
7. Public health laboratories;
8. Poison control centers;
9. Central medical libraries with Medline;
10. Registries of medical, nursing, and dental professionals.

Geographic epidemiology has been important in the history of public health. Fragmentation of information systems has delayed applying modern information technology to multiphasic evaluation and integrating data from multiple sources.

NATIONAL HEALTH TARGETS

The U.S. Public Health Service has set national health targets since 1979. These are increasingly accepted at all levels of the national public health complex. They highlight areas of concern that require effort by all levels of government and the health care system. They also serve an educational role for health providers and the community.

Some of the progress made in reducing morbidity and mortality from epidemiologically important diseases is the result of that wider awareness and growing concept of "self-care." *Healthy People 2000 Review, 1994* issued by the National Center for Health Statistics and the CDC showed progress being made toward over 300 separate objectives designed to prevent disease and injury and to promote health in the United States. Some 8% of the goals had been met and progress made in another 41%. For 16% there was regression, and for 7% mixed or no change. The annual publication *Health, United States* provides continuous updating of a wide range of health statistics. Some examples of health target achievement include:

1. Infant mortality in 1994 was 7.9 per 1000, a decline of 5% from 1993; and further to per 1000 in 1996;
2. Heart disease and stroke mortality continue to decline, in both the white and black populations; however, the excess mortality for blacks as compared to whites has increased;
3. A number of risk factors for cardiovascular disease (elevated cholesterol, smoking, and high intake of fats) have declined; the proportion of the population overweight, has increased substantially, including teenagers;
4. The number of adults who exercise vigorously and regularly increased, as well as the number of work sites offering exercise programs;
5. Smoking among adults declined, with an increase in the number of states restricting or regulating smoking; a decline in lung cancer rates occurred between 1991 and 1993, after steady increases for 50 years;
6. Alcohol-related motor vehicle crashes declined, and licensing revocation laws were in effect in 37 states, and mandatory seat belt laws in 48 states; two-thirds of adults now use seat belts;
7. Suicide, a leading cause of death among teenagers (aged 15–19), is relatively stable, but suicide attempts increased;
8. Cancer mortality rates (e.g., colorectal and breast) have declined; increasing numbers of women are getting mammograms, and the percentage of women over 50 who have had clinical breast examinations increased from 25 to 55% from 1986 to 1993;
9. Occupational disease and injury mortality declined slightly;
10. Dental caries continue to decline; tooth loss is reduced in older adults.

The concepts of prevention and health promotion are integral to the setting and attaining of health targets. Increasingly the methods of public health are moving toward wider responsibilities in terms of health monitoring and organization to reach the stated goals and objectives. The New Public Health provides a conceptual basis for this process.

PUBLIC HEALTH ORGANIZATION AND THE NEW PUBLIC HEALTH

Because the United States lacks a national health insurance program it is commonly said that the United States has a "nonsystem." This is misleading; the United States has a very complex and unfinished health system. The pessimist says the cup is half-empty, and the optimist says the cup is half-full. The United States is a world leader in public health, not only in the development of new vaccines, but in implementation of important advances, such as fluoridation of community water supplies. The United States has the costliest health system, but it lags behind many other countries in important indicators of health status (see Chapter 13). Still, the United States is one of the few countries in the world with a universal school lunch program. Further, the U.S. health system is a complex interactive set of organizations, subject to system changes, that has pioneered many innovations in health sciences, health care administration, and public health.

Publicly administered universal access elements exist. The middle class are protected by employment-based health insurance, the elderly by Medicare, and the poor by federal–state–local Medicaid. The failure to adopt national health insurance providing equitable access to health care continues to be a major hindrance to improving health of the vulnerable poor and marginalized sectors of society. The public health services of all levels of government spend much of their energies and resources trying to cover the deficiencies resulting from inequities in access to services.

Increasingly the U.S. population is covered by managed care plans in which financial incentives are in play to promote decreasing use of hospital care and increasing use of ambulatory and preventive care. Collaboration between organized public health and medicine, long antagonists in the United States, took a new direction in the mid-1990s with development of a "new paradigm" of cooperation. The American Medical Association and the American Public Health Association agreed to work together to promote networking in the form of collaborative local programs to resolve unmet health needs of the community. This mutual awareness represents a recognition of the importance of both clinical medicine and public health. Intersectional dialogue helps identify potential for cooperation in the context of the dramatic changes taking place in the United States in health care organization.

In other countries, such as the United Kingdom and the Scandinavian countries, organization of health services moved to district health systems in which public health is a full partner with clinical services, and where prevention is integral to the economic and function of a population-based program. The managed care revolution in the United States in the 1990s may well promote a new level of cooperation between clinical medicine and public health. Integration of services financed by Medicare and Medicaid, with federal waivers of eligibility conditions for age and poverty, may allow a new approach based on residence in areas of need. Expanding Medicaid will occur largely through enabling enrollment into managed care programs of large numbers of eligible persons not currently enrolled.

Downsizing the hospital sector, constraint in health costs, increasing enrollment in managed care, focusing on health targets, and increasing coverage through managed care will constitute a national health program evolving toward some form of the New Public Health. The United States has been very innovative in financing systems to promote efficiency in use of services, and other countries have begun to apply those lessons in their national health insurance plans. The United States will benefit from examining the reforms going on in many countries, including Canada and European countries, as their health systems also evolve. The American public health community, including the schools of public health, has capacity and experience with professional leadership and advocacy, and it can make a great contribution toward adaptation of the New Public Health.

Expenditures on population-based public health services in 50 states is 3.0% of total health expenditures. Most expenditures by state health departments (66%) were for personal care services, mostly for people ineligible for health insurance or with benefits excluding preventive care. This reflects the predominant priority for hospital and ambulatory care services based on insured or personal outlay for services. While much ambulatory care involves preventive services, the relatively low expenditure for community-oriented public health functions reflects traditional values and underevaluation of the potential impact of community-oriented approaches to health promotion.

HOSPITALS IN THE NEW PUBLIC HEALTH

The hospital is an important element of the New Public Health. The individual institutional health facility, whether a general rehabilitation center, a nursing home, a mental, or other special hospital, is called a generic "hospital." Each has a defined role, administrative structure, budget, and modus operandi as a unique organization. Hospitals evolved under municipal, religious, voluntary, governmental, university, private, or other sponsorship. Hospitals have traditionally been separate administrative units from other health services, although often with a strong connection to medical and paramedical training programs. The organizational structure reflects the history of the organization but should also address the facility's mission, resources, and role as part of a larger community health system.

Hospitals remain among the largest employers in any country, with some three-quarters of all health personnel and, depending on the country and its traditions and reform processes, between 38 and 75% of total health expenditures. The magnitude of the hospital sector and the key role it plays in the health service system make it vital to rationalize its services, preventing duplication, bed surpluses, overemphasis on specialized services versus primary care, and depersonalization of patients.

The modern hospital is the most costly and visible element of a health system to the public; it employs the most personnel and it provides care for the seriously ill. Its management is therefore a important factor in managing the total health system. While health care is an organizational system, the component facilities such as hospitals are also living organizational entities that require structure, management, and planning.

Under managed care systems, the hospital will try to satisfy the patient and the economic constraints of the managed care system. The two clients may have different objectives and methods of assessment of the functioning of the institution and the community it serves. The insured patient, with the option to change health plans, will be able to exert some influence on the care he or she receives. Similarly, the managed care system can judge the quality of care rendered by a hospital and express dissatisfaction by choosing an alternative provider.

The mission of a hospital is to provide high quality of care and service to the patient within the limits of current standards of knowledge and resources. In addition, there are many other objectives of the hospital as an organization. They include professional and economic survival as an institution, teaching functions, research, and publication. The hospital provides an important contribution to its community, providing employment, financial stability and solvency, prestige, education research, and a system of access to health care.

To meet these diverse goals and objectives, hospitals have become complex organizations with an extensive division of labor (see Chapter 12). This involves many different professional areas, as well as the "hotel functions" of food, laundry, housekeeping, supplies, and the financial and personnel administrative functions. As a large organization of great complexity, a hospital must have a formal, quasi-bureaucratic structure with clear lines of authority and responsibility. Given that, however, the modern hospital cannot function under a traditionally authoritarian, paternalistic pattern of administration. Coordination of the many complex skills brought together in a hospital require lateral coordination between departments and staff at all levels or the machine simply will not function. As a result, the hospital is highly dependent on the motivation and integrity of its staff, and their ability to network with others in different departments or professional levels freely and without excessive bureaucratic constraints.

Nevertheless, basic acceptance of authority, discipline, and rigid requirements to maintain standards of care are still essential to hospital function and predictability of performance. A new demand on hospitals is efficiency so that waste, duplication of service, poor maintenance and function of facilities and equipment, corruption, negligence, or thievery cannot be tolerated by the organization. The modern hospital has formal bureaucratic lines of authority, and hundreds or perhaps thousands of examples of informal networks and sometimes formal organizations to carry out the daily work of patient care, while meeting the other needs of the hospital and "good standards of care" with efficiency in use of resources. There are many checks and balances in the structure with multiple lines of authority and responsibility, and sometimes even tension between administrative and professional elements.

Hospital Classification

Any bed that is set up and staffed for care of inpatients is counted as a bed in a facility. A bed census is usually taken at the end of a reporting period. The WHO defines a bed as one regularly maintained and staffed for the accommodation and full-time care of inpatients and situated in a part of the hospital where continuous medical care is provided. A bed is measured functionally by the number and quality of staff and support services that provide diagnostic and treatment care for the patient in that bed.

Hospitals are institutions whose primary function is to provide diagnostic and therapeutic medical, nursing, and other professional services for patients in need of care for medical conditions. Hospitals have at least six beds, an organized staff of physicians, and continuing nursing services under the direction of registered nurses. The WHO considers an establishment a hospital if it is permanently staffed by at least one physician, can offer in-patient accommodation, and can provide active medical and nursing care.

BOX 10.8 TYPES OF HOSPITALS

Short stay hospitals are those in which more than half of the patients are admitted to units in the facility with an average length of stay of less than 30 days. These include teaching, general, community, and district hospitals providing a broad range of services, as well as specialized hospitals that focus on special categories of patients by age, sex, or medical condition.

Long stay hospitals are those in which more than half of the patients are admitted to units in the facility with an average length of stay of more than 30 days. These may include special hospitals and may be jointly managed with short stay hospitals.

Nursing homes are establishments with three or more beds that provide nursing or personal care to the aged, infirm, or chronically ill. They employ one or more registered or practical nurses and provide nursing care to at least half of the residents. Skilled nursing homes provide more intensive nursing care, as defined by nursing care hours per patient day.

Hostels are residential facilities attached to a medical center for overnight stay of patients undergoing outpatient investigation or care.

Hospices are facilities related to a medical center especially organized to provide a humane, personalized, and family-oriented setting for care of dying patients.

Nonprofit hospitals are operated by a government, voluntary, religious, university, or other organization whose objectives do not include financial profit.

Proprietary hospitals are operated for profit by individuals, partnerships, or corporations.

General hospitals provide diagnoses and treatment for patients with a variety of medical conditions or for more than one category of medical discipline (e.g., general medicine, specialized medicine, general surgery, specialized surgery, and obstetrics). This excludes hospitals which provide a more limited range of care.

Community hospitals serve a town or city and are usually short stay (less than 30 days average length of stay) general hospitals.

District hospitals are general hospitals that serve a population of a defined geographic district and have, as a minimum, four basic services: general medicine, surgery, obstetrics and gynecology, and pediatrics.

Teaching hospitals are those operated by or affiliated with a medical faculty in a university or institute.

Special hospitals are single category in-patient care facilities such as a children's, maternity, psychiatric, tuberculosis, chronic disease, geriatric, rehabilitation, or alcohol and drug treatment center which provide a particular type of service to the majority of their patients.

Tertiary care hospitals are referral and teaching hospitals; a secondary level hospital is a community or district hospital providing a wide range of services; and a primary level hospital is a limited service community hospital in a rural area.

Source: *Health United States, 1996–1997,* and American Hospital Association website http//www.aha.org/

Hospitals include those operated not-for-profit and on a for-profit basis. Most are operated as not-for-profit facilities as public services by government, municipalities, religious organizations, or voluntary organizations. In the United Kingdom, hospitals formerly operated by the NHS are being turned over to public trusts to operate as not-for-profit public facilities. In the Scandinavian countries, county hospitals and other local health services are operated by the county health department. Private, for-profit hospitals, though increasing, are still a minority of general hospitals but include a large proportion of chronic care facilities.

In the United States, Canada, and Israel, long-term care for the elderly and infirm is dominated by private for-profit facilities. In these countries, private facilities arose because of inadequate public resources for direct provision of services. As payment systems evolved, private operators were encouraged to enter the field. Government supervision and regulation have diminished the abuses and exploitation that occurred in the 1960s, but the standards of care can be compromised by the profit motive. There are, however, good examples of large-scale operations of long-term care facilities run by private organizations that are efficient and provide good standards of care. As illustrated in Box 10.8, hospitals are also defined by the types of services provided, the population served, and average length of stay.

Supply of Hospital Beds

The supply of hospital beds is measured in terms of hospital beds per 1000 population. This varies widely between and within countries. Historically, hospital development was initiated by church or religious groups, municipalities or voluntary charitable societies, or by local, state, or national governments without national planning criteria. In all health systems, regardless of administration and financing methods, the supply of hospital beds and their utilization are fundamental to health economics and planning.

The hospital bed is often a political issue. In some countries, the hospital has been seen traditionally as a center of refuge from harsh conditions of life, climate, and social conditions. This is especially the case in rural areas with lesser access to health care. Pressures for more beds may come from physicians or from the public. Political figures tend to favor more hospitals because they provide jobs in a community, signify access to medical care, and create a public sense of well-being. The addition or closing of hospital beds is one of the difficult and controversial issues in health planning and health politics. However, if politicians are responsible for paying for hospital operational costs, they are able to resist. Pressures are difficult to resist, and political expediency to build more beds then faces the difficulty of financing the operation of those beds. It is even more difficult to close redundant or uneconomic hospital beds, because this means a loss of jobs in the community unless coupled with transfer of personnel to other services, itself a painful procedure.

The hospital bed has important economic implications for the health system. The cost per bed is measured by the total expenditure of the hospital divided by the number of beds. Building and operating costs, on average, are such that the cost of construction of a bed are usually equal to the cost of operating the bed over 2–3 years. The decision to build a bed obliges the health system to indefinitely fixed costs even if that bed is unused as a result of regulation or the reduced utilization from professional or economic incentives. Hospital planning is no longer left to the initiative of the facility itself even in the most competitive, market economy-oriented health system.

The tendency to build excess hospital beds and the resultant difficulty of maintaining them are common to developed and developing countries. Excess supply is associated with high utilization rates and long lengths of stay. Where there is no incentive for the hospital or the doctor to increase efficiency, patients tend to linger in the hospital. This results in higher overall costs of health care and is associated with medical mishaps, including falls in the hospital, errors in care, drug errors, anesthetic mishaps, and nosocomial infections. Excess bed capacity can be dealt with in a number of ways. Essentially it requires conversion of bed stock, and staff, to other purposes, or closure of obsolete facilities.

Especially since the 1980s, many countries are reducing excess hospital bed utilization by shortening the length of stay, increasing the efficiency in diagnostic procedures, decreasing unwarranted surgical procedures, and adopting less traumatic procedures (e.g., breast conserving surgery for breast cancer, and endoscopic surgery). Ambulatory services replace inpatient care for many types of surgery, including most eye, ear, nose, and throat surgery, and for medical care for oncology, hematology, mental health, and many internal medical problems (Chapter 11).

Development of alternatives to hospital care, such as organized home care, assist in earlier discharge of patients from acute care hospitals by providing services to the patient at home, such as nursing, physiotherapy, intravenous care, change of dressings, or removal of stitches following surgery. Rehabilitation facilities provide appropriate low-cost alternatives to lengthy recovery periods after surgery such as hip or knee replacements. Long-term care facilities for geriatric patients requiring extensive nursing care, but who do not benefit from lengthy stays in acute care hospitals, provide alternatives to hospital care. Closure or reduction of beds is important to assure that savings in one area of service are transferred to a common financing system to provide funding for those alternative services. This may require investment in these extended, community services before savings are realized from reduced hospital utilization.

The capitation system of payment provides incentives for district health or managed care systems to limit admissions and lengths of stay. Sweden succeeded in reducing the percentage of GNP spent on health care during the 1980s by reducing hospital bed supplies, with no slowing of improving health status indicators. Managed care systems and Diagnostic Related Groups (DRGs) are having the same effect in the United States. District health system capitation is leading to reduced hospital bed supplies in the United Kingdom. This is a complex and controversial issue, but managing the numbers of hospital beds is essential especially in view of aging populations with chronic diseases, and the highly intensive and expensive kinds of care needed by many patients (Chapter 11).

The Changing Role of the Hospital

Hospitals are increasingly technologically oriented and costly to operate. Under the influence of rising costs, incentives for alternative forms of care have led to the development of home care, ambulatory services, and linkages with long-term care. Forces acting on the hospital as an organization and economic unit place the hospital in a context where community-based care is an essential alternative that requires organizational and financial linkage to promote integration.

As a key element of any health system, the hospital will undergo changes as technology and health management sciences advance. Managing health systems with fewer hospital days requires reorganization within the hospital to provide the support services for ambulatory diagnostic and treatment services as well as home care. The interaction between the hospital-based and community-based services require changes in the management culture and community-oriented approaches. Involvement of all the staff in quality of the service has become part of this management culture.

Countries which operated hospitals as part of the Ministry of Health or National Health Service are tending to transfer hospital ownership and operation to not-for-profit agencies, or trusts as free-standing economic units, or integrated within service programs of district health authorities. Competition for patients and payment for services such as by a DRG system will increase competition and the need for excellence in hospital care and its management for the financial survival of the facility. There is a trend in the United Kingdom, Israel, and many countries in transition from the Soviet and post-colonial health systems toward less centralized management and greater competition in health care. The trend to include hospitals in district health authorities, such as in the Nordic countries, as part of geographic managed care programs is another important policy direction of health reform.

BOX 10.9 HOSPITAL MERGERS IN LOS ANGELES COUNTY—THE UNIVERSITY OF CALIFORNIA, LOS ANGELES (UCLA), HEALTH SCIENCES CENTER AND COMMUNITY OUTREACH

Los Angeles is a large, multiethnic, and rapidly growing metropolitan city of over 9 million population in southern California. It has built up a large number of hospitals operated by not-for-profit religious, ethnic, and other organizations, the LA County Department of Health Services, and university-owned facilities as well as for-profit facilities. The hospital bed-to-population ratio has been 3.5 per 1000 population during the 1980s and 1990s. Payment by DRGs in the 1980s, and growing membership in managed care, led to reduced hospital bed occupancy with 45% of beds occupied in 1996. In 1998, the vast majority of insured Angelinos belong to managed care programs. As a result, many hospitals are selling to for-profit hospital chains, or are under threat of closure, some being converted to long-term or ambulatory care facilities.

The UCLA Health Sciences Center is a teaching hospital owned and operated by the university. In the mid-1990s this center developed contracts to provide hospital care to many managed care programs including those covering UCLA employees and others serving many employers in Los Angeles County. In order to broaden its community service base, the center purchased several community hospitals and established affiliation agreements with medical group practices in adjacent areas of the city. This enabled the center to ensure its catchment population in a highly competitive market. The emphasis is increasingly on developing contractual arrangements with primary care medical services. The Health Sciences Center is replacing the hospital due to damage in the 1994 earthquake and will do so with substantially fewer number of beds. This is the survival strategy adopted to ensure its continuing role as a major teaching and community service hospital in the changing medical market and in the twenty-first century.

Source: UCLA Medical Center, *Health Sciences.* 1998.

In the United States, hospital networks are developing in the for-profit and not-for-profit sectors with integration of management and other cost savings in scale of purchasing and operation. Contracts with managed care organizations for hospital care has replaced the previous system where hospitalization of insured patients depended on where the attending doctor had privileges or worked on staff. The for-profit hospital corporations, along with similar managed care organizations, have brought health care to the stock market with profits larger than many other sectors of the private economy. Hospital mergers may then be seen in the context of any other business merger or corporate development.

Regulation of Hospitals

Governments have the responsibility and authority to assure standards of health for the population. Licensing of health facilities is one of the traditional methods used to ensure the public interest and prevent harmful practices in patient care facilities. The licensing procedure is the basis for regulation of standards and the content of the service, as well as in controlling health care expenditures (Chapter 11).

Governments have a number of methods of regulation of hospitals; one is control of the payment mechanism, which allows room for negotiation and influence on standards and content of care; the second is regulation of the number of hospital beds as the licensing and standards authority; the third is control of capital expenditures; a fourth method is to make payment for insured patients to accreditation of the hospital. The level of government responsible for regulation varies from country to country, usually depending on the constitutional division of responsibility between the different levels of government and the size of the country. Generally, the state and local authorities have the most influence because of their proximity. Where government agencies operate hospitals directly, there is a conflict of interests in the form of self regulation.

The combination of governmental roles in a highly centralized health system, of financing, operating, and regulating hospitals, may appear to have some advantages, but separation of these conflicting functions is important in promoting a high quality service. The separation of financing and regulation from operation of services is a widening trend in national health systems.

Governmental regulation makes nongovernmental accreditation virtually mandatory, with the accreditation agency's standards accepted by government. Mandating that hospitals and long-term care facilities in the United States be required to have full accreditation and using an existing NGO inspection system as a proxy for governmental standards save the government from the necessity of establishing large-scale regulatory and inspection systems.

THE UNINSURED AS A
PUBLIC HEALTH PROBLEM

The 40 million Americans uninsured and the large number with poor coverage constitute a serious challenge for public health in the United States. Much activity by county and municipal health departments focuses a great deal of attention on this population, who are largely poor and in need of health care. The coverage of the elderly and the very poor under Medicare and Medicaid has given a base of protection to these groups, but the near poor and the near elderly are still highly vulnerable, especially when job layoffs are a major part of the economic condition. While this problem is becoming more acute with a growth in the uninsured following the failure to enact national health insurance in 1994, there are increasing federal and state initiatives to widen coverage for Medicaid and especially target the approximately 20% of children who are uninsured. The United States, despite still being the only industrialized country lacking universal health insurance, has established other mechanisms that have had positive health effects, such as the school lunch program and many categorical health programs to promote prenatal care, immunization, lead screening, mammography, Pap smears, and other preventive services. No prediction can safely be made as to when or if the United States will adopt national health insurance. The extent of its continued delay will be a continuous burden on the full realization of America's national health potential, for its individual citizens and for the nation as a whole.

SUMMARY

Public health is organized at the local, state, and national levels to define and work toward health targets. A balanced health care system requires that resources be rationally allocated to the different preventive, curative, or environmental elements of health. Resources must be directed to all vulnerable groups in the population, recognizing that some groups have greater needs than others. At the same time, issues that affect all, such as nutrition, sanitation, housing, and socioeconomic conditions affect the poor and the elderly disproportionately. Sound public policy must also take into account the need to ensure adequate quality of care by health care providers and institutions, by developing and regulating standards, licensing procedures, and quality assurance mechanisms.

Impressive progress has been made in public health in the United States over the nearly two centuries since the establishment of the U.S. Marine Hospital Service, and the evolution of public health systems in the twentieth century. Despite lack of a national health system, the United States has been a leader in formulating administrative mechanisms to improve the efficiency of health care.

The revolution of health care in the United States in the 1990s toward managed care is causing a large-scale reorganization of hospitals with both vertical and lateral integration, that is, formation of networks of hospitals and linkage of hospitals with primary care. Adjustment to meet the health care organization environment of the twenty-first century requires changes for hospitals. These include downsizing, development of ambulatory and home care services, and linkages with primary care services to ensure a "catchment population." The competitive factors in which primary care providers and the community have a role in determining a hospital's utilization, occupancy, and ultimately its survival will help build a more community-oriented health system.

In the United States, public health has been separate and unequal in relation to medical services. The advent of managed care for a large portion of the population creates a professional and economic challenge for both sides. Organized public health in the United States needs to seek a closer liaison with managed care to promote a more comprehensive New Public Health approach. Managed care organizations need to develop health promotion while at the same time ensuring the interests of the patient to successfully promote its long term economic interests, and vice versa. If public health remains outside the issues of organization and financing of personal care services, the isolation of public health in the United States will deepen.

The New Public Health is a comprehensive approach to health care, stressing the interdependence of medical and hospital services with prevention and health promotion. Clinical medicine, management of health services, and community health approaches are interactive in many forms, in the United States and elsewhere. In northern Europe and the United Kingdom, district health systems are responsible for and are budgeted on a per capita basis to ensure community health and all levels of personal care services. In the United States, lack of universal health access and central payment systems for all has, paradoxically, promoted development of managed care systems linking all levels of health care. But public health remains detached from this process, being organized and financed separately.

The New Public Health approach seeks to link those activities of local, state, and national government that promote the health of a population, including the provision of personal care in hospitals, community, and long-term care or community settings. The New Public Health approach seeks to link those activities of local, state, and national governments with nongovernmental services that promote the health of a population, including the managed or individual provision of personal care in hospitals, community, and long-term care or community settings.

ELECTRONIC MEDIA

Agency for Health Care Policy Research, http://www.ahcpr.gov/
American Hospital Association, http://www.aha.org/
American Public Health Association, http://www.apha.org/
Bureau of Primary Care, DHHS, http://www.bphc.hrsa.gov/
Centers for Disease Control and Prevention, DHHS, http://www.cdc.gov/
Department of Health and Human Services (DHHS), http://www.hhs.gov/
Health Care Financing Administration, DHHS, http://www.hcfa.gov/
Health Care Indicators, United States, http://www.hcfa.gov/stats/indicatr/indicatr.htm and DNHS@hcfa (National Health Indicators)
Health, United States, 1998, http://www.cdc.gov/nchswww/products/pubs/pubd/hus/hus.htm
Hill-Burton Obligated Facilities, Health Resources and Services Administration, http://www.hrsa.dhhs.gov/osp/dfcr/obtain/hbstates.htm
Indian Health Services, DHHS, http://www.ihs.gov/
Institute of Medicine, National Academy of Science, http://www.nas.edu/IOM/IOMHome.nsf
Joint Commission on Accreditation of Healthcare Facilities, http://www.jcaho.org/index.htm
National Center for Health Statistics, http://www.cdc.gov/nchswww/
National Institutes of Health, http://www.nih.gov/

RECOMMENDED READINGS

Afifi, A., Breslow, L. 1994. The maturing paradigm of public health. *Annual Review of Public Health,* 15:223–235.
Alpert, J. J. 1998. Editorial: Serving the medically underserved. *American Journal of Public Health,* 88:347–348.
Centers for Disease Control. 1996. Historical perspectives of CDC. *Morbidity and Mortality Weekly Report,* 45:526–530.

Centers for Disease Control. 1997. Estimated expenditures for essential public health services—selected states, fiscal year 1995. *Morbidity and Mortality Weekly Report,* 46:150–152.

Centers for Disease Control. 1999. Ten great public health achievements—United States, 1900–1999. *Morbidity and Mortality Weekly Report,* 48:241–243.

Centers for Disease Control. 1999. Achievements in public health, 1900–1999: changes in the public health system. *Morbidity and Mortality Weekly Report,* 48:1141–1147.

Citrin, T. 1998. Topics of our time: Public health—Community or commodity: Reflections on healthy communities. *American Journal of Public Health,* 88:351–352.

Oberle, M. W., Baker, E. L., Magenheim, M. J. 1994. Healthy people 2000 and community health planning. *Annual Review of Public Health,* 15:223–235.

Rosenbaum, S., Hawkins, D. R., Rosenbaum, E., Blake, S. 1998. State funding of medical care service programs for medically underserved populations. *American Journal of Public Health,* 88:357–363.

BIBLIOGRAPHY

American Public Health Association. 1991. *Healthy Communities 2000: Model Standards,* Third Edition. Washington, DC: APHA.

Buttery, C. M. G. 1991. *Handbook for Health Directors.* New York: Oxford University Press.

Centers for Disease Control. 1991. *Profile of State and Territorial Public Health Systems: United States, 1990.* Atlanta: U.S. Department of Health and Human Services.

Centers for Disease Control. 1997. Estimated expenditures for essential public health services—Selected states, fiscal year 1995. *Morbidity and Mortality Weekly Report,* 46:150–152.

Green, L. W., Ottoson, J. M. 1994. *Community Health,* Seventh Edition. St. Louis: Mosby.

Institute of Medicine. 1988. *The Future of Public Health.* Washington, DC: National Academy Press.

Lasker, R. D. (ed.) and Committee on Medicine and Public Health. 1997. *Medicine and Public Health: The Power of Collaboration.* American Public Health Association and American Medical Association. New York: New York Academy of Medicine.

Mill, J. S. 1869. *On Liberty.* London: Bartleby Co. (www.bartleby.com/130).

Miller, C. A., Moos, M.-K. 1981. *Local Health Departments: Fifteen Case Studies.* Washington DC: American Public Health Association.

Patrick, D. L., Erickson, P. 1993. *Health Status and Health Policy: Quality of Life in Health Care Evaluation and Resource Allocation.* New York: Oxford University Press.

Pickett, G., Hanlon, J. J. 1990. *Public Health Administration and Practice,* Ninth Edition. St. Louis: Times Mirror/Mosby College Publishing.

Rohrer, J. E. 1996. *Planning for Community-Oriented Health Systems.* Washington, DC: American Public Health Association.

Turnock, B. J. 1997. *Public Health: What It Is and How it Works.* Gaithersburg, MD: Aspen.

United States Department of Health and Human Services. 1992. *Healthy People 2000: National Health Promotion and Disease Prevention Objectives.* DHHS Publication PHS 91-50212. Washington DC: U.S. Department of Health and Human Services.

United States Department of Health and Human Services. 1994. *Healthy People 2000 Review, 1994.* DHHS Publication No. (PHS) 94-1256-1. Washington, DC: U.S. Department of Health and Human Services.

United States Department of Health and Human Services. 1998. *Health, United States, 1998: with Socioeconomic Status and Health Chartbook.* DHHS Publication No. (PHS) 98-1232.

Williams, S. J., Torens, P. R. (eds). 1998. *Introduction to Health Services,* Fifth Edition.

11

MEASURING COSTS: THE ECONOMICS OF HEALTH

INTRODUCTION

Health economics is an important element of health policy, both at the strategic (macroeconomics) and tactical levels (microeconomics).[1] Macroeconomics in health deals with overall financing and allocation of health resources, while microeconomics compares alternative approaches to dealing with specific health issues. Monetary resources for health are limited, and difficult choices have to be made in their allocation. Management of health care requires an understanding of use of resources, priorities, and trade-offs in health. All professional health care providers and planners need a working knowledge of the fundamentals of health economics and how economic incentives and disincentives affect the supply, demand, and ultimately the cost of health services. This knowledge helps one to appreciate how health care, while giving benefits in terms of reduced morbidity and mortality, also has a cost in terms of resources used.

Health economic analysis provides a set of tools for management and decision making in the selection of priorities. It can add a measurable empirical element to policy formation as a necessary, though not sufficient, instrument for health policy decisions. Health economics sometimes conflicts with professional, ethical, and moral issues in solving the everyday problems of preventive and curative services. A balance between these issues is part of present-day health management and therefore of the New Public Health.

ECONOMIC ISSUES OF HEALTH SYSTEMS

Health care expenditures vary widely among different countries, ranging from under 4% to over 14% of GDP. The high and rising cost of health services is in-

[1]In standard economics macroeconomics is defined as the aggregate of economic activity; microeconomics is the theory of how individual firms (e.g., suppliers) and consumers behave.

creasingly coming under public scrutiny and economic analysis. The total of expenditures and how they are allocated are central issues in all health systems.

There are two basic economic problems in health affecting different countries in similar ways: underinvestment and overinvestment or, more accurately, misallocation of health resources. The World Bank's 1993 *World Development Report: Investing in Health* addresses problems besetting health care systems, mainly, but not exclusively, in the developing countries. This report stresses that health is essential for productivity and economic growth, and that allocation of limited resources to costly, relatively unproductive services, such as excessive expenditures for armed forces, prevents many countries from meeting their basic health needs.

Investing in Health

The *World Development Report, 1993* stresses the role of health in economic development, stating that a healthy population is not only a well-meaning social goal but, like an educated population is not only a well-meaning social goal but, like an educated population, is also essential to the development of a strong economy. Healthier populations are better workers and contributors to economic growth. Healthier children learn better in schools and, thus, also have prospects for contributing to economic development of their country.

This report addresses investment in health as follows:

1. Good health is a crucial part of well-being;
2. Spending on health can be justified on purely economic grounds;
3. Improved health contributes to economic growth:
    a. It reduces production loss by worker illness;

TABLE 11.1　Essential, Cost-effective Health Services for Developing Countries, World Bank, 1993

Public health interventions	Clinical services
1. Immunizations: DPT, polio, measles, hepatitis B, yellow fever, vitamin A, iodine supplements (EPI Plus)[a]	1. Short-term chemotherapy for tuberculosis
2. School-based health services: deworming, micronutrient supplementation, health education	2. Management of the sick child: ARI[a,b] diarrheal diseases, measles, malaria, malnutrition
3. Information on health, nutrition, family planning	3. Pregnancy care, prenatal and delivery
4. Tobacco and alcohol control programs	4. Family planning
5. Disease monitoring, surveillance, and vector control	5. Treatment of STDs[c]
6. AIDS prevention	6. Limited care for assessment, care of pain, trauma, infection, and more as resources permit

Source: World Bank. 1993. *World Development Report, 1993: Investing in Health: World Development Indicators.* New York: Oxford University Press, page 117.
[a]EPI = Expanded Programme of Immunization (Chapter 4).
[b]ARI = Acute respiratory infection.
[c]STPs = Sexually transmitted diseases (Chapter 4).

b. It permits use of natural resources that have been inaccessible because of disease;

c. It increases the enrollment of children in school and makes them better able to learn;

d. It frees for alternative uses resources that would otherwise have to be spent on treating illness;

4. Sound policy in financing and resource allocation is essential to achieve good health.

The World Bank report calls on governments to increase spending on health and also to foster an atmosphere in which individual families do the same. It calls for competition and diversity in health care, based on an economic rationale, with health development based on "a basket of essential public health and clinical services" (see Table 11.1).

This report is a landmark document in international health, which may rank in importance with such others as the Declaration of Alma-Ata and *Health for All 2000* (see Chapter 2). It establishes investment in health as a efficient tool for economic development, and thereby places health among the priorities for national and international financial investment.

BASIC CONCEPTS IN HEALTH ECONOMICS

All societies have limited resources and must, according to politically determined priorities, provide funds for health care in competition with funds for education, defense, agriculture, and others. The use of limited funds requires making choices. These choices reflect the overall political commitment to health and should, as far as possible, be based on an objective assessment of costs and benefits of available options.

The components of economic evaluation in health care are seen in Fig. 11.1. Expenditure of resources (in terms of financial and human resources), both direct and indirect, are targeted to a health program. This is expected to produce health benefits or utilities, which can also be both direct and indirect. Health benefits may be expressed in terms of a direct reduction in morbidity and mortality, or as improved productivity and quality of life.

Measurement of both input and output is an essential part of health management. Health inputs include expenditures on buildings, hospital or nursing home beds, equipment, personnel, home care, ambulatory care, and preventive programs. Other elements of health costs include patient's travel time, loss of work time for both the patient and caregivers, loss of full functioning years of life, and loss of quality of life. The "input–output" theory of health economics may sound simplistic, but it provides a useful measure when examining the benefits and costs of a specific health intervention. Alternatives can be examined and analysis of their cost-effectiveness made, in order for decision makers to select the most suitable alternatives.

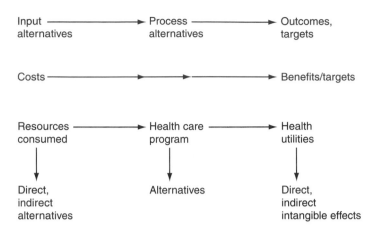

FIGURE 11.1   Economic models of health evaluation: resources–programs–benefits.

NEED, DEMAND, AND UTILIZATION
OF HEALTH SERVICES

Need and demand for medical service are not necessarily the same. Need in medical care exists when an individual has symptoms, illness, or disability for which there may be an effective or acceptable treatment or cure from which the patient can benefit. Need also refers to nonmedical conditions, such as for designated preventive services such as immunization. Demand for medical care exists when an individual considers that he or she has a need and is willing to spend resources of money, time, energy, loss of work, travel, and inconvenience to receive care. Utilization occurs when the individual actually acts on this demand or need and receives health services.

Normative Needs.   Normative needs are those services determined by experts to be essential for a specific need or for a specific population group. These include many examples of standard protocols for both clinical and preventive health care, such as prenatal care, immunization, child care for infants and toddlers, management of diabetes and hypertension, and screening for breast and prostate cancer. Consensus of expert opinion, including the published literature, may be considered objective, and is called "evidence-based medicine." There are very often legitimate differences of opinion on professional issues in public health which are based on alternative interpretations of the available information, or where the evidence is incomplete. Furthermore, as scientific knowledge advances, new information may not be absorbed into decision making processes as rapidly as needed. Professional value judgments may be biased by trends in medical opinion, or influenced by advances in clinical, technological, and epidemiologic evidence. Such normative needs should be under continuous review by qualified professionals and academics representing clinical and public health services, as well as managers and consumers of health care. Other disciplines such as health economics, sociology, health education, and planning add to the understanding of contributing factors to a disease and how to address it. Each field contributes to interpretation and decisions on standards for the health system.

The individual characteristics of people seeking care, or within the responsibility of a health system, including factors such as their age and sex, help determine the type and amount of health services needed. For example, a woman of 40 may not need a mammography as frequently as a woman over 50 years of age. An infant may need to be seen for preventive care assessment more often than a 3 year old. A male aged 45 needs his blood pressure checked more often than a 25 year old, and a teenager needs more attention paid to prevention of risk-taking behavior than a 35 year old.

Felt Need. Felt need is the subjective view of the patient or the community, which may or may not be based on actual physiological needs. Though subjective, felt need is a prerequisite to whether a person actually undertakes to seek care. There is a growing recognition of the importance of sharing health information with the population to increase the possibility that rational choices will be made (the health–belief model). Greater public knowledge is vital to acceptance of preventive programs such as immunization and compliance with treatment regimens for chronic diseases. Felt needs also affect health planning. A community or donor may, for example, feel that a community needs a new hospital, whereas the same resources might better be spent on developing primary care or health educational services that have a greater impact on health of the population. Even in an authoritarian society, public opinion may direct decision makers to make irrational choices, such as placing inordinate stress on hospital bed supply, or in the nutrition or smoking patterns of a community.

Expressed Need. Expressed need is a felt need that is acted on, such as in visiting a clinic or general practitioner. Felt needs may not be acted on because economic, geographic, social, or psychological barriers may inhibit a person from seeking or receiving care. Accessibility may be limited because the individual cannot afford to pay the fee. A service may be free, but not readily accessible due to such obstacles as distance, language or cultural barriers, difficulty in arranging an appointment, or a long waiting period. As a result, the person seeking care may not be able to receive it and may delay interfacing with the health system until a more urgent, and often more costly, problem arises. Distance, time, and cost of travel, inconvenience, and loss of wages affect the seeking of service, more so for preventive care than urgent surgical conditions, for example, even if the service is free of charge. Altering the supply, location and type of service availability can change these factors, thus improving equity of access. Elderly persons may sometimes avoid turning their felt needs into action as they may not feel comfortable with the fact that they are ill, or may not wish to become a burden. Religious, cultural, or even political factors may prevent a woman from practicing birth control even when further pregnancies may jeopardize her life. Lack of knowledge may also interfere with appropriate use of available clinical or preventive health care.

Comparative Need.   Comparative need is a term that relates needs of similar population groups, as in two adjacent regions with the same age/sex/ethnic mix and socioeconomic status. One region may have a certain service, such as fluoridation of the community water supply, while the comparison community does not. The population of the second community is objectively in need of that service or service according to the best present professional and scientific evidence. There are no definable absolutes in the extent of health care, but there are accepted basic standards that are part of world standards at a particular point in time for health care, for prevention or health promotion. These are derived from trial and error as much as from science and must be continuously reexamined in light of new information, as well as the measurable benefits and costs derived from them.

Demand.   Demand is based on the individual and community expectations (Fig. 11.2). Economists consider this to be a part of the economic demand theory of laissez-faire[2] in which the individual is seen as the best judge of his or her need. The individual may feel that he needs a service, but expert opinion may say that this is not a reasonable demand. A patient may ask a physician for an antibiotic to treat a viral infection which would not help and may even cause harm. A community hospital may wish to increase its bed supply or purchase a highly technical piece of medical equipment due to consumer demand. A patient may feel that denial of a referral by a doctor in a managed care plan is not justifiable, but there may be a legitimate and ethical reason for the refusal. Doctors may wish to have the prestige and convenience of certain equipment locally, but economic and planning assessments may say that this is not justified on economic or medical grounds. Such equations, however, are not immutable; costs of a procedure may change as technology or clinical experience accumulates, so what was once not justifiable may become so. Such conflicts are unavoidably part of health care planning.

Supply.   Demand may also be induced by the supply or provision of care. Making available more hospital beds may increase their use beyond justifiable need, or it may lead to an expectation or demand by patients or their families for an unnecessarily long stay in the hospital. Providing some services at no cost to patients may induce people to utilize those services more than they really objectively require for health reasons according to current best standards. An inappropriate or excessively frequent use of a service may be promoted, and used by the upper middle class, while other important services may be lacking to serve the poor due to inequitable allocation of resources. Sometimes the interest of the health care providers is such that they may act to promote the use of services because payments are received for each service rendered (fee for service). This occurs, for example, in situations where a greater supply of surgeons results in unnecessary surgery being performed.

---

[2]The term laissez-faire is used by economists to refer to minimal or noninterference by government in economic affairs, i.e., the free market of goods and services.

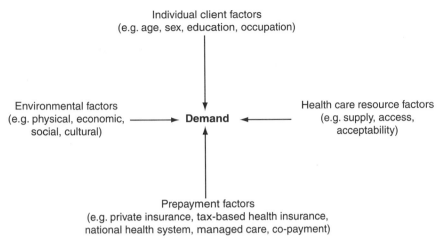

FIGURE 11.2   Factors in demand for health services.

Grossman's Demand Model.   A frequently used economic demand model is that described by Grossman. This method looks at health within the framework of a production function, that is, health status (output) is a result of health care activities (input) by the environment, the individual, and the health services system. Individual demand for health care is affected by many factors, such as socioeconomic, educational, and cultural barriers or incentives to health care, as well as age and health status.

In this model, everyone inherits a stock of health when born. Health depreciates over time, however, and investment is required to sustain health. As people age, there is an increase in the rates of illness and death and in utilization of health services. The rate at which a person's stock of health depreciates over time is represented in a health depreciation–time curve. The stock of health can be sustained by investment to maintain health, such as in use of health services and health-promoting activities (e.g., recreation and fitness facilities).

Change in health is thus a function of medical care received, as well as exercise, good housing, nutrition, lack of smoking, and societal factors that are difficult to quantify. Over their life cycle people will attempt to offset part of the increasing rate of depreciation in their health by increasing expenditures or use of medical care services. The production function depends on environmental variables, such as education, that alter the efficiency of the production process. Personal choices affect health, depending on resources allocated to its production, such as in allocation of free time for jogging, the choice of whether to eat fatty foods or to smoke cigarettes.

Health is also an investment good. Being unhealthy brings discomfort, a reduced sense of well-being, and a measurable loss of income from reduced work hours or performance. Health as a consumption good means that a health-related activity will improve the quality and enjoyment of life, prevent discomfort or illness, or improve appearance as in cosmetic surgery. Older people use more hospital and ambulatory medical and other services than younger people as their health declines, and they are subject to more disease. There are also factors within the health services system or in an insurance system that affect the way individuals act can influence the process including access to services and quality of care, such as if an insurance plan refuses to accept an enrollee on grounds of age, personal habits (e.g., smoking), or preexisting medical conditions, or does not cover preventive services.

In Grossman's model of health demand, rising income may adversely affect health because of an increase in unhealthy or risk-taking behavior. Excessive consumption of fatty foods, smoking, alcohol abuse, and motor vehicle crashes, which increased as per capita income rose with increasing death rates from cardiovascular diseases and trauma in the industrial countries in the 1940s and 1950s. This also occurred in the 1980s and 1990s in the growing middle class of developing countries. Very poor populations, whose basic health problems are inadequate food intake and infectious diseases, benefit from rising incomes in developing countries if family incomes rise to permit improved expenditures for food and nutrition, with positive effects on health status.

A consumption good has been interpreted by economists as use of resources in a manner to most benefit the individual and society, based on the best available current evidence. Free market economists, and institutions such as the World Bank, once considered health care as ineffective consumption and recommended that these resources be better invested in economically "productive" development per se. At the other end of the political spectrum, Marxist economics looked at health spending as nonproductive consumption, in comparison to investments in heavy industry or infrastructure. In contrast, democratic socialists, such as those in the Nordic countries, have long adopted the view that investment in improved health is justified on both social and economic grounds, with health, like education, being a social right and foundation of society. This concept gained international acceptance with the Health for All program for fundamental social justice and human rights. The World Bank's 1993 report, *Investing in Health,* takes a utilitarian view that spending money on health and education is an investment in economic growth, not a drain on the economy.

COMPETITION IN HEALTH CARE

Health services were historically established as private, charitable, religious, or governmental public services. In recent years, even where government finances the services, a market-oriented approach has evolved that gives the consumer choice to utilize competing health services. This is the basis of 1990s reforms in the United Kingdom's National Health Service, where the primary care provider hold the funds for the people who select him or her for their primary care. The fund-holding general practitioner selects the specialists and hospital service best able to meet the patient's needs. Similarly in the United States, managed care health plans receive insurance payments per capita and select hospitals, physicians, or other services on a competitive basis. This is a trend in other countries as well (e.g, Israel, Colombia, Phillipines, Jordan) where clients select health plans or sick funds that are then responsible for all care and costs, seeking competitive benefits from providers such as hospitals or medical practitioners.

Health care organizations, whether hospitals or primary care services, must provide quality service to the community in order to survive as institutions. The patient or consumer is both an end and a means for the institution, at the center of the process. Market analysis may be conducted for health care programs costs, contents, satisfaction of individuals and the community, as well as performance indicators.

Reforms in some countries with universal health entitlement include introduc-

---

BOX 11.1 OBLIGATIONS OF HEALTH SERVICE PROVIDERS

1. Availability: provides continuous service coverage, 24 hours every day and on holidays;
2. Accessibility: the client can readily reach the service within reasonable travel and waiting time;
3. Accountability: the system and provider explain programs, decisions, and actions to the client;
4. Affordability: services are provided at reasonable cost to consumer and insurer;
5. Acceptability: user-friendly in style and manner of staff to the patient and family;
6. Accredited: undergoes external evaluation and implements recommendations of accrediting agencies;
7. Equitable: provides fair and equivalent access to services needed regardless of age, sex, ethnic origin, religions, social, or political identification, place of residence, prior medical condition, ability to pay;
8. Efficient: in use of manpower, finances, and other resources;
9. Ethical: meets current professional, societal, ethical, and legal standards;
10. Consumer's rights: the patient and consumer are informed of their rights and alternatives of care, before medical or administrative decisions are made;
11. Financially sound: able to meet financial obligations;
12. Goals and objectives: defined, written, reviewed, and used as a basis for planning, monitoring;

13. Innovative: open to new methods in clinical and preventive approaches to health;

14. Quality: promotes high standards of facilities and services in accordance with current professional criteria and standards of leading providers;

15. Community acceptable: meeting community expectation with participation in promoting health standards;

16. Comprehensive: linked to a broad range of services provided or arranged to meet client needs.

tion of competition and market mechanisms. The concept of market mechanisms in health is ideologically difficult for those who believe that universal access and equity would be compromised, that the consumer may not be sufficiently informed to choose, and that they may make cost containment next to impossible. In fact, there are many "market" factors in health care, such as the supply of facilities and manpower, that affect demand, as well as method of payment for doctors and hospital services. All of these affect the economics of health. However, where consumers have no choice of service provider, there are dangers of exploitation by providers and consumers alike.

Managed care systems depend on reducing unnecessary hospital care and greater use of ambulatory and home care services. Services previously provided on an inpatient basis can be equally or more successfully provided on an outpatient basis. The manager or chief executive officer of a health facility needs to guide the transition from a passive receiver of the sick to an emphasis on operating an institution for meeting health needs in the community, while controlling costs in order to remain competitive and financially viable.

ELASTICITIES OF DEMAND

Demand may be affected by factors determined by the consumer, the provider, the supply, or the location of services. Elasticity of demand relates demand to the price of the goods or services. Cost to the consumer is a factor in choosing to purchase goods or to seek services. If the price goes up, then demand will decline, or vice versa. In other words, demand is not an absolute but will be fixed or affected by price and out-of-pocket payment required for the service.

In classic capitalist economic theory, the individual is seen as the best judge of his or her own needs and decides what to buy (consumer sovereignty). It assumes that consumers purchase services or a health plan based on factors such as cost and quality, as one would do when purchasing a refrigerator. Individual decisions are made on the basis of personal perception, information, and resources. Proponents of the market approach in health care suggest that it provides the consumer with more control and choice, and indirectly raises the competition for consumer demand and the quality of services while lowering costs.

Opponents to this view claim that this market approach fails in medical care because the issue is more complex than purchaser and provider. The market mechanism in health care is as much determined by supply, access, and method of payment as by consumer choice. Supply of services creates demand, as does prepayment. Consumers rely on their doctors to advise them; this is known as the agency relationship. Physicians make decisions for their patients on the basis of both patient needs and the supply of services. Payment for services by a third party, such as an insurance plan or the government, where the consumer may provide little or no direct payment, can lead to the provision of unneeded services, particularly if the doctor has incentives of the fee-for-service payment system.

Someone may wish to purchase additional or different services according to the expected benefits (marginal utility). The marginal cost is the cost of an extra unit of the commodity used. In classic economic theory, the consumer decides to purchase a service when the marginal utility is equal to or greater than the marginal cost, that is, the added benefit is worth the additional cost. This free market approach in health care may result in underprovision of vital preventive health services, especially to the population in greatest need. If medical care services operated in a free market, then consumers would not necessarily take into account benefits to persons other than themselves (externalities). For example, if vaccines are available purely on a free-market basis, many persons from their individual point of view would not purchase the vaccinations. This would increase the risk to the general public by reducing herd immunity. Externalities should be taken into account in public policy decisions, valuing the benefits to society as a whole.

Where there are many providers, they will compete with each other, in principle by offering services at lower prices in order to attract clients. However, this rarely occurs in medical services where fees or salaries are set by collective bargaining. In terms of hospital and insurance services, the existence of monopolies (i.e., only one provider) or oligopolies (i.e., a few providers) prevents price competition and often results in collusion to fix prices. A monopsony is a situation when there is only one purchaser, so that the provider may be subject to pressures to lower prices or meet additional demands of the buyer, such as a managed care organization purchasing hospital services.

In a highly privatized system of health care, such as that in the United States up to the 1990s, demand for care is rationed through fees, copayment by the consumer, or by limitations set by the indemnity insurance plan and lack of insurance benefits for many. Indemnity insurance plans require copayments and deductibles, so that the insured person has to take additional insurance to cover expenditures or pay part of the charges. Medicare beneficiaries in the United States must pay part of their health care expenditures, or they may join a managed care option which covers their service costs, but with a restricted list of physicians, so that the beneficiary must choose between extra payments versus limited choice of service provider.

In a public system of health care, demand is rationed through limitations on supply of services or by requiring the patient wishing to see a specialist to be referred by a general practitioner who plays a gatekeeping role, in which the consumer's choice is limited to that decided by the general practitioner. This has been adopted by many health systems, including managed care plans in the United States. This is a controversial limitation on the consumer, who may wish to consult other specialists, but may be essential to avoid frivolous "shopping" for care that drives up the cost of health care.

The market approach is based on choices being made by consumers on the basis of anticipated benefits in terms of improved health or health care or reassurance. The employer must offer the employee several options for health insurance. In a fee-for-service indemnity plan allowing the consumer greater choice of physician, the employee has to pay additional monthly premiums, as compared to a managed care option. This carries with it a measure of inequality because of differences within the population, both in health needs and in the ability to purchase or utilize needed services. Often those with the greatest need are those least able to get the desired or required services. Persons who have no source of income cannot be consumers or make decisions to purchase health services on the free market. Rather, they are dependent on free charitable services. This problem may be addressed by a number of economic alternatives, such as providing low income individuals with free health insurance under Medicaid or with vouchers to purchase services.

Decisions or actions of patients are influenced not only by cost and access, but also by knowledge and attitudes to care. A person unaware of modern birth control, or living in a society that discourages or limits its use on religious or political grounds, is not able to make an informed decision about its use. Market mechanisms in health work in different ways in different health systems. Even where care is a free service and is seen as a right for everyone, there are limitations in supply of services. By giving consumers choice of provider, and by the use of incentives to promote quality care or disincentives to reduce unnecessary or wasteful services, market mechanisms have a major role in the reform of many health systems.

MEASURING COSTS

Costs in health can be analyzed in various ways: direct costs to the patient; costs to the insurer or sick fund on behalf of the patient; costs to the hospital or other provider; and indirect costs of illness to the patient, his family, and society, including time off work due to illness or lowered productivity.

Opportunity costs refer to the resources used that could have been applied to other uses. Hospital land and building costs, for example, could be allocated for other purposes, such as primary health care facilities or facilities outside the health sector such as after school programs for children. Increasing the proportion of the GNP spent on health care may limit society's ability to spend money on education and other important social programs.

Social costs include indirect expenditures for health effects, such as the total value of lost production or costs of social support for a person whose health and work capacity has been impaired by illness. Private costs include out-of-pocket expenditures that an individual makes to purchase health care plus related expenses such as payments for health insurance, loss of wages, purchase of pharmaceuticals, and copayments for health services.

ECONOMIC MEASURES OF HEALTH STATUS

Economic analysis assesses not only input, as in costs and resources, but also output, as in morbidity, mortality, extension of years of life, and reduction of disability. Greater functional levels that improve the quality and quantity of life are output measures of health care. This should be part of an economic evaluation of the use of national or personal resources. Disability-adjusted life years and quality adjusted life years are measures of the total burden of disease as a guide to population health status (both of death and disability).

Disability-adjusted life years (DALYs) are calculated as the present value in years of disability-free life that might be lost as a result of premature death and disability occurring due to a disease in a particular year. Quality-adjusted life years (QALYs) measure life expectancy adjusted by changes in quality of life measured by assessing two or more aspects of health, such as pain, disability, mood, or capacity to perform self-care or socially useful activities such as paid employment or housework. DALYs and QALYs are constructed by using expert evaluation to estimate the degree of impairment (normal, impaired, or incapacitated) from specific diseases, These include impairments such as loss of ability to communicate, sleep disturbance, pain, depression, and sexual, eating, and mobility dysfunctions.

The value of the health status of an individual can then be assessed in terms of numerical values for comparisons. The values of the total scores are then added together and the overall score calculated out of a maximum value for comparison. This allows a measure of health status for purposes of comparison and may be used in comparing the effectiveness of alternative interventions. This is subjective depending on the perception of the assessor, and interobserver variability may be high. While such measures do not include all factors in determinants of disease, by pooling mortality and economic indicators they contribute to using the economic impact of disease as part of health planning.

DALYs and QALYs provide a common base for comparing mortality along with the dimensions of disability and quality of life as measures to compare different causes, settings, and change over time. They are used as proxy health status indicators to analyze different approaches to health policy and to justify specific interventions and determine priorities. The World Bank, WHO and other groups are examining alternative indicators to link health and its underlying determinants of the total burden of disease and disability, and to refine the process of establishing priorities for research and decision making for interventions.

Gains in life expectancy from preventive or curative medical interventions may be measured from published data sources. The gain in life expectancy calculated for a patient who survived cardiac arrest by placement of an implantable pacemaker is calculated as 36–46 months, and bone marrow transplantation for recurrence of non-Hodgkin's lymphoma is calculated at 72 months. For preventive measures, such gains if averaged over a total population appear smaller. Cervical cancer screening, for example, increases the life expectancy of all women by 3 months, but for a woman whose cancer of the cervix was detected early, the gain is an average of 25 years. This methodology relies on published studies, and can be of great value in comparison and analysis of alternative strategies and health care priorities.

COST–EFFECTIVENESS ANALYSIS

Cost–effectiveness analysis (CEA) in health care is the net gain in health or in reducing the burden of disease from a specific intervention in relation to its cost. It is used to determine the least expensive way of achieving the goal, by comparing alternative methods of intervention in order to make a choice. The most cost-effective method is the one that achieves the same goal using the least resources. A low cost per DALY gained indicates a high degree of cost-effectiveness, and therefore an intervention that should be of high priority, given limited resources.

Alternative methods of treatment may also be compared such as use of medication compared to surgery, day surgery as compared to inpatient care, or treatment in the community as compared to hospital inpatient care. As seen in Table 11.2, the most cost-effective services for developing countries are the expanded program of immunization, as vaccine-preventable diseases are major causes of DALYs lost. Other cost-effective programs include preventing iodine and vitamin A deficiency and treating intestinal worms, even though these are relatively lower causes of lost DALYs.

As part of the 1993 World Bank report, various interventions based on public health and clinical services in developing countries were compared various health interventions in terms of DALYs versus costs of the interventions (Table 11.2). Such analyses help to construct a basket of essential services on the basis of comparative cost-effectiveness. Highly cost-effective interventions include vitamin A supplementation, measles control, and directly observed chemotherapy for TB. A high cost but highly effective intervention presented as chemotherapy for leukemia in children under age 15. This intervention is justified as benefits are high. The same chemotherapy in a 75 year old would present a low DALY value. CEA studies examine issues such as day surgery versus inpatient surgery, operations versus medications (e.g., for peptic ulcers and coronary heart disease), public versus individual dental prevention (e.g., fluoride versus dental hygienist care), and community versus institutional care.

Comparison of life years gained for patients with end stage renal disease in the United States found that renal transplantation was less expensive ($3,600 per year of life gained) compared to home dialysis ($4,200 per life year gained) and hospital dialysis ($116,000 per life year gained). Moreover, transplantation provides

TABLE 11.2   Main Causes of Burden of Disease in Children in Developing Countries and
Cost-effective Interventions, 1990

Disease	DALYs lost (millions)	% Total DALYs lost	Appropriate Intervention	Cost-effectiveness ($ per DALY)
1. Vaccine-preventable childhood diseases	65	10	EPI[a]	12–30
2. Vitamin A deficiency	12	2	EPI plus[a]	12–30
3. Iodine deficiency	9	1	Iodine supplementation	19–37
4. Intestinal helminths	17	3	School health	20–34
5. Diarrheal diseases	92	14	Integrated management of sick child (MISC)	30–100
6. Protein-energy malnutrition	12	2	MISC	20–150
7. Perinatal morbidity and mortality	96	15	Family planning, prenatal and delivery care	30–100
8. Respiratory infections	98	15	MISC	30–100
9. Malaria	31	5	MISC	30–100
10. Congenital malformations	35	5	Surgery	High
11. Others	193	28	—	—
Total	660	100	—	—

Source: Bobadilla et al., 1994.

Note: Ranking is by intervention cost-effectiveness.

[a]EPI = expanded program of immunization; EPI plus = EPI with the addition of vitamins and iodine supplements (see Chapter 4); MISC = medically integrated services for children.

a higher quality of life. This was perhaps the first example of "cost–utility" analysis, where life years gained were weighted according to quality of life. This can be expressed as the cost-effectiveness per QALY. Policy decisions based on such findings are constrained by difficulty in obtaining sufficient donor kidneys and lack of manpower and facilities suitable to carry out transplantation effectively.

Surgical removal of the gallbladder now uses endoscopy instead of the traditional abdominal cholecystectomy. Endoscopy is less traumatic with faster recovery, so the patient is discharged from the hospital on the next day and returns to work within a day or two, while the patient with abdominal cholecystectomy requires much longer hospital stay and recuperation at home before returning to work. The newer procedure is easier on the patient and safer. Cost–benefit analysis must take into account not only the medical and hospital costs, but also the social costs of lost working time for the patient and caregivers. Computerized Tomography (CT) scanning, once considered costly and for special use, has become a valuable part of investigation of many conditions and is used frequently, often in place of costly, dangerous, and less effective procedures previously used to investigate many conditions.

Care of the infirm elderly, up to a certain level of disability, by means of care in a private home with outside help, including meal preparation and delivery, nursing, physiotherapy, and social work visits, is less costly than care of the same patient in an institution. Home care promotes earlier hospital discharge and recuperation at home. These assessments must take into account social costs and transfer to the patient's family of costs of services provided in an institution, such as food, laundry, heating, and electricity. Home settings promote improved recovery, avoidance of hospital infections, and a general feeling of well-being of the patient. For persons with severe illness or many disabilities, requiring a higher degree of nursing and/or medical care, institutions are more cost-effective than home care. Respite care sometimes provides support for a family in caretaking of a patient with multiple handicaps, delaying more costly institutional care.

Research comparing treatment of psychiatric patients in a large mental hospital, in the psychiatric ward of a general hospital, and in a day treatment center show day treatment center care to be least costly, but some measure of the severity of illness and care needs to be added to this assessment. Planning mental health services and facilities with reduced hospitalization requires adequate resources for mental health care in the community, to prevent chronic mental patients from becoming part of the homeless population as has happened in many large cities.

Sometimes the least costly method is the least effective. For example, a study showed that prevention of pregnancy by the withdrawal method is least costly but is far less effective than use of the birth control pill. Abortion as a method of birth control may be less costly than use of the pill, but, in addition to the ethical issues, it produces complications and contributes to excess morbidity and mortality in subsequent pregnancies, both for the mother and the newborn.

Cost–effectiveness analysis takes into account both the cost and effectiveness of interventions, as a measure of value for cost, but does not answer the question of whether or when the intervention should be done.

COST–BENEFIT ANALYSIS

Cost–benefit analysis (CBA) compares the expense of a specific program to its expected monetary yield or savings. Costs include direct expenditures as well as the indirect costs of loss of productivity and contribution to society. Direct benefits include reductions in morbidity and mortality and the associated savings in medical care costs, such as hospitalization, doctors' services and drugs used, and reduction in loss of life, with the attachment of economic value to this. Indirect benefits include savings to the patient's family in terms of expenditures to visit the patient (transportation costs) or time away from work to look after a sick child or parent. Other indirect benefits also accrue to society, in terms of savings in reduction of lost work time by the patient or his or her family during an illness.

The assessment of costs and benefits involves three stages: enumeration, measurement, and explicit valuation. Assessing a particular treatment, or enumeration, requires measurement of change in health status, in the cost of use of resources, as well as in change in the patients' productive output. Economic appraisal depends on medical determination of the factors that are needed in managing the problem and its expected outcomes. Explicit valuation, or estimation of the cost of a variable, is based on determination of the economic value of these factors. Often, many factors need to be taken into account and simplified; consequently, these are approximations rather than exact figures.

A CBA study of phenylketonuria (a congenital metabolic disorder) screening in the United States showed the cost of screening 660,000 newborns to be $1.39 million, including confirmation tests, special diet for those affected, and administration of the program. the benefits gained were $1.26 million for medical and other services, and $1.05 million for prevented loss of productivity, for a total of $2.31 million. The benefit:cost ratio was 2.31/1.39 = 1.66. For each dollar invested, the gain to society was $1.66. In CBA studies comparing addition of a second dose of measles or one dose of hepatitis B or *Haemophilus influenzae* B vaccine to an immunization schedule, the second dose of the measles vaccine was found to have a high benefit-to-cost ratio (CBA = 4.5/1) in both developed and developing countries. Hepatitis B vaccinations were found to have high benefit-to-cost ratios, even in countries with intermediate levels of endemicity (CBA = 4.5/1). For *Haemophilus influenzae* B vaccine, the social benefits were found to exceed the costs to society; however, many of the benefits fell outside the health sector, such as reduced need for special education for brain-damaged children. The benefit-to-cost ratio, if viewed solely from the point of view of the health sector, was lower. Such a result might prevent the health system from adopting a beneficial program if the benefit-to-cost study is too narrowly applied.

The decision to adopt a specific program may include a CBA but is often made on other grounds, including public and professional opinion as well as political factos. A CBA can give a prioritized ranking to alternative interventions, and thereby help in the decision making process. Ranking according to the relative costs and benefits can help a health ministry to choose among putting resources into a high-technology hospital, home care, expansion of an immunization program, or investment in home care and primary care services.

Both CBA and CEA include both initial and ongoing costs, but they must take into account that the value of money in the future will be less than at the present time, referred to as discounting. The costs as well as benefits to be derived from the project must be calculated as they accrue, so that a portion of the effect is observed next year, and a portion the following year. The cumulative discounted value is called the net present value (NPV).

BASIC ASSESSMENT SCHEME FOR INTERVENTION COSTS AND CONSEQUENCES

Assessment of effectiveness and costs of an intervention has become a basic part of policy making in health. An approach called BASICC (for basic assessment scheme for intervention costs and consequences) has been widely promoted by the Centers for Disease Control. A more complex approach looks at efficacy of the intervention, the cost, including direct outlays, productivity costs (i.e. loss of time for work or recreation) and intangible costs (i.e. pain and suffering).

Costs include fixed costs, or those that do not vary by the quantity of the service provided but also include a portion of rent, utilities, and equipment for their share in the program. The average costs are the total cost of a program divided by the total units of output produced. Variable costs are those that vary according to the level of service provided, such as the number of nursing visits required for a home care patient. Marginal costs are those additional costs to basic program costs such as expansion of staff or facilities to accommodate extra activities.

BASICC focuses on intervention costs and direct cost savings in terms of medical care. Net costs can be summarized as the cost of the intervention and its side effects for n persons minus the direct costs of the expected number of cases averted for the same n persons, calculated as follows:

$$\text{Net cost} = \text{cost of program} + \text{cost of side effects}$$
$$- \text{ cost of adverse health outcomes averted.}$$

The steps of BASICC include the following:

1. Describe the program, its objectives, target population, effectiveness of intervention, external constraints, resources required, management of the program, implementation strategy, scientific evidence of effectiveness;
2. Define the burden of the disease, its incidence and prevalence without the program;
3. Define outcomes anticipated in terms of improved quality of life, reduced incidence or severity of the disease and premature death;
4. Measure efficacy of the intervention, taking into account that interventions are rarely 100% successful in practice because of compliance and effectiveness of the intervention;
5. Measure intervention costs per unit;
6. Measure direct medical costs of outcome averted by the intervention;
7. Assess resources required for the intervention, which include fixed, variable, total as well as unit costs.

THE VALUE OF HUMAN LIFE

One anticipated benefit of health intervention is the saving of human life. Placing an economic value on life is useful in calculating the benefits of specific interventions or perhaps for compensation to the family of a person who loses his or her life as a result of, for example, a negligent doctor or plant manager.

The value of a human life in economic terms was first calculated by William Petty in 1699 while developing his idea of political arithmetic. William Farr in

1876 used life tables to calculate economic equivalents. More recently, economists have calculated the value of human capital, willingness to pay for services, and other methods of quantifying the value of human life.

Ethical and, indeed, political conflicts surround the issue of calculating the economic value of human life. A materialistic approach would evaluate human life based on the value of production that the individual might make to society. A humanistic approach would place virtually unlimited value on a human life according to the ethical value that saving of one life is as saving all human beings (sanctity of human life). By valuing human life as infinite, doctors may use precious resources to save one human, without considering that this may be at the expense of other lives. For example, the cost of a heart transplant, which may add quality and years to one person's life, could be alternatively applied to a preventive program that might save many more lives through the prevention of heart disease. Should international agencies spend hundreds of millions of dollars to eradicate polio, a much feared, crippling, but nonlethal disease, while measles, thought to be a common benign disease, kills over 1 million children annually? The valuing of human life is not meant to fuel ethical argument, but rather to provide a measurement tool for the planning of priorities and litigation needs of health economics.

In health economics some arbitrary measures are used in order to demonstrate alternative ways of using limited resources. The implicit social value of life (ISV) rates a program by the lives it saves and assumes that, in a democratic society, all lives have the same intrinsic value. Inconsistent valuations of ISV were seen in some decisions made by governments. A United Kingdom government decision not to introduce childproof drug containers implied a valuation of the worth of a life saved at less than $5,000, while the same government decided to change a building code that implied a valuation of $50 million per life saved. Estimated costs per person year of life saved may vary for specific public health interventions: annual

---

BOX 11.2 IMPLICIT SOCIAL VALUE OF LIFE—AN EXAMPLE

The implicit social value of life (ISV) is summarized in the following equation:

$$\text{ISV} = \frac{\text{sum of costs} - \text{sum of benefits}}{\text{sum of life years saved}}$$

As an example, in the United Kingdom in the 1960s, despite the fact that the social costs were $5,000 more than the social benefits, home dialysis was provided. In other words, society was willing to pay $5,000 to keep one member of society alive for 1 year. From this decision, we can infer an ISV of at least $5,000.

mammography for women aged 40–49, is estimated to cost $62,000 per life year saved, as compared to $2,700 for a program of mammography every 3 years for women aged 60–65; a smoking cessation advice program for men aged 50–54 yields a cost–effectiveness of $990 for a year of life saved (Brownson, Remington, and Davis, 1998).

Early economists valued life in terms of loss of net output to society, or the future loss of earnings minus the future loss in consumption resulting from the death of an individual. This human capital method is still widely used because of the simplicity of its calculations. However, it does not take into account the grief of the family. It places a negative value on the life of a pensioner who is no longer a "producer" in society, and gives no value to work done in the household, such as cooking, home maintenance, and rearing children. Nor does it give value to the intangible social and psychological benefits of the multigenerational family for all its members.

Another approach to valuation of life is based on court awards for compensation. It is a highly subjective method, often based on the court's interpretation of degree of contributory negligence, such as whether the injured person in a car accident was wearing a seat belt at the time.

A major method is the willingness-to-pay approach, where valuations of life are based on what individuals are willing to pay for reductions in their probability of dying. For example, how much would persons pay for new tires on their car, or how much extra would they pay in order to travel on an airline with a better safety record? How much will a patient be ready to pay above his or her insurance coverage to have a world-famous surgeon operate on him or her as opposed to accepting the service available within the health service. Such measurement is difficult and is often based on asking questions about hypothetical situations. Answers are also influenced by the income level of the respondent, by their attitude toward risk, and by the probability of death.

The issue is not only theoretical. If it costs $3,000 to prevent transmission of HIV to newborns, and the number of cases of HIV positive pregnant women who may transmit the virus is such that a very large part of a national budget for health in a developing country may go to this purpose, while there are insufficient funds for basic immunization, then choices need to be made, and they may be painful ones. All societies must make choices in priorities and in allocation of resources. Choosing to build large superhighways and neglecting public mass transit is a decision which assumes certain social values, but will cost lives and health because of downstream effects such as increased pollution, motor vehicle injuries, and deaths.

HEALTH FINANCING—THE MACROECONOMICS LEVEL

Financing health care has evolved from personal payment at the time of service to financing through health insurance (prepayment) by employer/employee at the place of work and governmental financing through social security or general tax-

TABLE 11.3   Sources of Financing Health Services

Public sources	Private sources	International coorperation
Federal, state, and local government general revunues, mainly from taxes; income, excise, resources, inheritance, value added, capital gains, property, special	Private health insurance	United Nations affiliates
	Personal expenditures	Foundations
	Private donations, wills	Religious organizations
	Private foundations	Other nongovernmental organizations
Social Security payroll tax	Voluntary community service	World Bank
Compulsory health insurance	User fees	Government bilateral aid
Lotteries		
Dedicated taxes: cigarettes, alcohol, gambling		

ation, supplemented by private and nongovernment organizations (Table 11.3). Ultimately every country faces the need for governmental funding of health care either for the total population or at least, as in the United States, for vulnerable groups, such as the elderly and the poor, as well as for services that insurance plans avoid or are inefficient in reaching, such as community-oriented services (see Chapter 13).

Health financing involves not only methods of raising money for health care, but also allocation of those funds. National health expenditures are derived from governmental and nongovernmental sources and are used to finance a wide array of programs and services. There is a competition for funds in any system, and how the money is allocated affects not only the way the services are provided but also priorities, as indicated in the "laws" of health economics in Box 11.3.

The economic consequences of decisions in resource allocations are major determinants of the economics of health care. Each country has to cope with similar

BOX 11.3 "LAWS" OF HEALTH RESOURCE ALLOCATION

1. Sutton's law: Willy Sutton was a bank robber and when asked why he robbed banks, he replied: "Well, that's where the money is." This expression is used to indicate that health services emphasize those aspects which are better financed. If more funds are available for treatment services, and preventive care is relatively underfunded, then treatment will have greater emphasis than prevention.

2. Capone's law: Al Capone, a well-known gangster, planning the division of Chicago among his colleagues, said: "You take the north side and I'll take the south side," i.e., let's divide things up according to our mutual interest. This expression in the health context is taken to mean that planning

may reflect interests of providers, as opposed to that of the general public. An alternative use of the concept is that macroeconomics planning may serve a general interest at the expense of the individual patient.

3. Roemer's law: "Hospital beds once built and insured, will be filled." The supply of hospital beds is a key determinant of utilization, especially where the public has insured benefits. This "law" has been modified by the experience of changing payment systems with incentives to reduce utilization. Following the introduction of the diagnosis related groups (DRG) method of payment in the United States in the 1980s, there has been a reduction in hospital bed occupancy and supply. Incentives to control both hospital bed supply and utilization are crucial elements of health planning in most industrialized countries.

issues in reforms to correct for changing health needs and the economic results of former decisions (see Chapter 13). A comparison of total national health expenditures is seen in Table 11.4. The United States has consistently been the highest spender on health care, but has succeeded in reducing the rate of cost increase in the 1990s. Canada also experienced high rates of cost increase in health during the 1970s and 1980s, but managed to reduce the rate of increase and moved from the second leading country in per capita health expenditures to fourth place after the United States, Germany, and France.

Health care expenditure involves money spent from all sources for the entire health sector, regardless of who operates or provides the services. The methods of financing health care include tax-supported, social security supported, employer–employee financed, or consumer payment at the time of service. The total of expenditures for health care and how those funds are spent are the most fundamental issues in health economics and planning. Allocation of resources requires a skillful planning process to balance spending on different subsectors of the system and to assure equity between regions and various socioeconomic groups in society.

What is the "right" amount of health care financing? This is a political decision that reflects the social and economic value placed on health by a nation. These attitudes affect such issues as how well medical and other health care staff are paid in comparison to other professions, and the supply of physical and human resources for health care in a given society. Virtually all developed countries have recognized the importance of national health and the role of financing systems to make health care universally available. The solutions vary from country to country, as discussed in Chapter 13, but is important to note that the system of financing greatly affects the services provided. The United Kingdom continues to operate its National Health Service at a relatively low percentage of GDP as do Denmark and Japan (Table 11.4).

There are vast differences in levels of expenditures on health between countries. In the established market economies 9.3% of GDP goes to health, while the former socialist economies expend 3.6%, and developing countries generally un-

TABLE 11.4   Per Capita Health Expenditures as Percentage of Gross Domestic Product for Selected Industrial Countries and Years, 1960–1997

Country	1960	1970	1980	1990	1995	1997
United States	5.1	7.1	8.9	12.2	13.6	13.5
Canada	5.5	7.1	7.3	9.2	9.7	9.0
France	4.2	5.8	7.6	8.9	9.9	9.6
Germany	4.3	5.7	8.1	8.2	10.4	10.4
Sweden	4.7	7.1	9.4	8.8	7.2	8.6
Japan	na	4.4	6.4	6.0	7.2	7.3
Denmark	3.6	6.1	6.8	6.5	6.4	7.4
United Kingdom	3.9	4.5	5.6	6.0	6.9	6.7

Source: *Health United States.* 1998, and OECD Health Data, 1998; Anderson, G. F., Poullier, J.-P. Health spending access, and outcomes: trends in industrialized countries. *Health Affairs,* op cit.

der 4.5%. Per capita health expenditures also vary widely. This does not reflect, however, the efficiency with which the resources are used. Unfortunately, many countries with low overall levels of health expenditures also allocate those meager resources inefficiently.

Regardless of how efficiently money is allocated, countries spending less than 4% of GNP on health will have poorly developed health care. Those spending between 4 and 5% of GNP may try to have universal coverage, but often do so by low staff salaries, inadequate equipment, and spreading limited resources too thinly. This is accentuated when a disproportionately large hospital system and excessive supply of physicians create a siphoning effect on health care spending, or when resources are concentrated in cities while most of the population is rural.

Developed countries that spend between 6 and 14% of the GNP on health care have made a value judgment placing health care among the vital priorities in their societies. In those countries with high health care expenditures, as in the United States, physicians' incomes are very high, even when compared with other highly paid professionals. Where financing is centralized in a single paying agency, administrative costs are less than in countries with multiple funding sources. Canada's provincial health insurance plans operate with administrative overheads of less than 5%, compared to some 30% in U.S. private health insurance.

The World Health Organization in 1981 defined a Global Strategy for Health Development which stressed efficiency in use of resources as a vital element of health development. The WHO recommends preferential allocation to primary and intermediate care services, especially for currently underserved rural populations. In most countries, some reallocation of resources will be required to strengthen primary care and to adopt new technology and health programs that are shown to be cost-effective in terms of costs as well as anticipated benefits.

Where there are multiple sources of health financing, it is difficult to develop effective national planning without regulation and supplemental funding by government to prevent inequity between socioeconomic groups and between urban and rural populations. When multiple agencies are involved in health insurance or direct government granting systems for specific services, there are gaps in services, usually to politically underrepresented sectors of the population, who may have the greatest needs. Very often, in such circumstances, public health services become oriented to provide basic services for persons excluded from health benefits because of lack of health insurance. This places a great burden on public health services, which are generally underfunded in comparison to clinical services. Further, such countries often bring in national health insurance for the disadvantaged groups (e.g., the elderly and the poor), setting them aside in insurance systems that pay less well than private insurance for the middle class and organized workers. This applies in the United States and in many mid-level developing countries (see Chapter 13).

Where financing of health care is centralized, the potential exists for rational allocation of resources. But this depends on adequacy of total financing and rational allocation policies to promote equitable in access to services, and a balance between one service sector and another. Allocation of monies within the total of health expenditures means selection from many alternatives. Misallocation of resources between sectors within the health field can lead to a wasteful and even counterproductive health system, such as excessive funding of tertiary care when primary care is lacking.

Where funds are allocated to regional or local health authorities, the potential for shifting resources to meet local needs should be greater. But this may be limited by lack of data or lack of analysis on a local or district basis to highlight priority areas of need. Where there is a highly decentralized management system, some centralized functions are essential to promote national health needs and equity between regions of the country. These include setting policy and standards, monitoring health status indicators, and determining health targets with funding to promote national priorities. The range of services or programs requiring funding for a population group are indicated in Table 11.5.

In the United States, 43.4 million persons or 16.1% of the population were uninsured in 1997. Medicaid, the largest public program providing health insurance for the poor, and long-term care for the aged, blind, and permanently disabled, in 1997 financed acute and long-term care services for 41.3 million beneficiaries, expending nearly $160 billion of federal, state, and local contributions. Medicare, which insures the elderly, the disabled, and patients with end-stage renal disease, in 1997 expended nearly $215 billion (per capita $4,083). Both are parts of the Social Security Act Amendments of 1965, with Medicare under federal administration and Medicaid shared between federal, state, and local administration.

COSTS OF ILLNESS

Direct expenditures for health care in the United States by type of illness are measured in periodic National Medical Expenditure Surveys of the civilian, non-

TABLE 11.5   Major Categories of Health Expenditures

1. *Institutional care*	6. *Categorical programs*
Teaching hospitals	Immunization
General hospitals	Maternal and child health
Mental and other special hospitals	Family planning
Long-term nursing care	Mental health
Residential care	STDs, HIV, tuberculosis
Hospice	Screening for cancer, diabetes, hypertension
2. *Pharmaceuticals and vaccines*	7. *Dental health*
3. *Ambulatory care*	8. *Community health activities*
Primary care, family practice, pediatric,	Healthy communities
prenatal, and medical	Health promotions, community, risk groups
Specialist medical, diagnostic, and treatment	Environmental and occupational health
Ambulatory and day hospital clinics, surgical,	Nutrition and food safety
medical, geriatric, dialysis, mental,	Safe water supplies
oncological, drug and alcohol treatment	Special groups
4. *Home care*	9. *Research*
5. *Elderly support activity/service centers*	10. *Professional education and training*

institutionalized population, covering thousands of persons and homes for self-reported expenditures. The largest items of health care expenditure were for cardiovascular disease, followed by injury, then neoplasms.

Since 1993, following especially high rates of increase in health expenditures during the 1980s in the U.S., measures were taken to restrain growth in health care costs, leading to a slower rate of increase. In part this was due to growth of managed care and incentives for lower hospital utilization, and by shifts in payment schedules for hospital and ambulatory care, while average Medicare payments increased by 36% for general physicians from 1991 to 1997, payments for medical specialists decreased by 15%, those for ophthalmic surgeons, for example, dropping by 18.4%, and for cardiac surgeons by 9.3%.

Costs and Variations in Medical Practice

Increasing costs of health care, waste, variations, and fraud in medical practice inevitably come under scrutiny whether prepayment is in the private or public sector. Variations due to different needs of various population groups may be justified. However, if through epidemiologic analysis there are no apparent reasons for the variations, then they become administrative problems which require other approaches. Comparing the quantity and quality of services between population groups is part of epidemiologic and administrative health practice. This approach, when supported by review of relevant current literature on methods of treatment provides a basis for what is termed evidence-based medical practice.

Analysis of medical practice by examination of medical and hospitalization data may show quite startling differences between different cities, regions, and countries. What has come to be called "small areas analysis" looks at patterns of practice and tries to determine what may be the cause of such differences. For example, there exists no evidence of benefit from higher rates of some types of surgery, such as hysterectomy, cholecystectomy, and tonsillectomy. Further, there is a cost attached to a surgical procedure that includes a certain mortality rate from anesthetic mishaps and other iatrogenic complications, that is, caused by medical care itself. For example, cholecystectomy rates in the early 1990s in Canada were 600 per 100,000 population, 370 in the United States, and 122 in the United Kingdom. These studies conclude that excess supply of surgeons and the payment by fee-for-service lead to unnecessary and potentially harmful surgical procedures. The cost implications for a health care system are high and can be calculated.

In the United States, health maintenance organizations (HMOs) have shown the capacity to provide comprehensive care over long periods to large population groups with relatively low hospital utilization. HMOs and for-profit managed care coverage increased dramatically in the 1990s. Hospital admissions, average length of hospital stay, and days of care for HMO members (non-Medicare) and for Medicare members as well, between 1993–1994, were well below comparable rates for fee-for-service based insurance plans.

Technological innovations of simpler, less costly, less invasive, and less risky procedures have led to important changes in health care standards. Continuous evaluation of criteria of good practice change with new knowledge and experience and consensus of leading opinion, meta-analysis, are essential to quality promotion measures in health care (see Chapter 15).

Cost Containment

High public and professional expectations from health care, along with increasing demands of an aging population, costly medical technology, and duplication of highly technological medical services, have all contributed to the health cost crisis in many countries. Cost containment became important as the costs in all health systems increased at rates well above economic growth during the 1970s and 1980s. Governments everywhere sought ways to restrain cost increases. Cost-effectiveness analysis and cost–benefit analysis have become a part of the planning and management review of ongoing or new interventions in health for both operational and capital expenditures as critical tools of health service planning for rational decision making to restrain health cost increases. Since hospitals are the major consumers of health care expenditures (between 40 and 60% in different countries), major focus in cost containment has been placed on reducing hospital utilization and developing alternative services or programs of ambulatory and community care.

Cost containment and high quality health care can coexist. Indeed, cost containment measures (Table 11.6) are associated with greater precision in care and more appropriate use of resources than previous patterns of care. Some measures relate to substitution of lower cost care for higher cost services while others relate to changes in medical care professional services, for example, surgery on an outpatient basis or shorter hospital length of stay following myocardial infarction.

Countries with public funding of health care systems are especially concerned with establishing cost containment in order to reduce the rate of increase in health costs. In Canada, governments have shifted their concern from assurance of ac-

TABLE 11.6   Example Health Service Programs Promoting Cost Containment

Program	Mode of operation
Home care	Reduces length of stay following medical or surgical hospital treatment; reduces the incidence of nosocomial (hospital-acquired) infections; helps elderly or chronically ill to remain at home rather than enter a long-term care facility
Long-term care facilities	For persons unable to be cared for in the family setting reduces lenght of hospital stay
Limitations	Limit supply of beds; limit medical services
Ambulatory or day care surgical, medical, mental	Reduces stay in hospital, with less secondary infections and iatrogenic complications
Prevention	Prevention (primary, secondary, and tertiary) reduces hospitalization for vaccine-preventable disease, cardiovascular disease, diabetes, and their complications
Environmental health	Chlorination of community water supplies prevents diarrheal diseases, hospitalizations; fluoridation reduces dental disease
Health promotion	Interventions to reduce trauma from road crashes; restriction of smoking leads to less lung cancer and coronary heart disease
Diagnostic related groups (DRGs)	Payment promotes reduced length of hospital stay
Health maintenance organizations and managed care organizations (HMOs and MCOs)	Promote alternatives to hospitalization and long stays, lowers hospital utilization; incentives for employers, employees, and governments to enroll beneficiaries in less costly managed care organizations; capitation provides incentive to prevent illness and institutional care, strengthen ambulatory and preventative care
District health system or regionalization	Promotes rationalization of services, elimination of excess facilities and duplication; promotes greater community orientation in service complex

cess to care to cost containment. Increase in health expenditure grew by 12.5% annually in the 1980s, well above growth of the economy. Canada's cost containment approaches include controls on fees, regionalization to reduce duplication, and excess hospital supply and utilization, as well as increased oversight in utilization of medical care. During the 1990s, these measures succeeded in slowing the rate of cost increase (Chapter 13).

MEDICAL AND HOSPITAL CARE— MICROECONOMICS

Policy in resource allocation, made at the national, regional, health insurance, or sick fund level, must address many specific factors affecting the way services are provided and paid for. Incentives and disincentives for efficient care include how doctors and hospitals are paid, and how services are organized. Payment for doctor's services includes fee-for-service, case payment, capitation, salary, or a combination of these methods. Each has its historical roots, its advantages and disadvantages, as well as proponents and opponents.

Payment for Doctor's Services

Fee-for-service is payment for each unit of service, such as a visit or surgical procedure. Payment for obstetrical care as a complete service including prenatal care and delivery, or other complete service over an illness or period of care, is called case-payment. Fee-for-service is historically the common method of paying for doctor's services and is still the norm in Canada, Germany, and other countries. In some places, payment may be according to a fixed fee schedule negotiated between the insurance mechanisms, whether public or private, and the doctors' representatives. Fee schedules are often weighted toward medical specialists who have greater prestige than primary care physicians.

Fee-for-service tends to promote an overabundance of the more expensive kinds of care, including surgery, often in excess of real need. This is especially so when the patient is fully covered by health insurance and is therefore better able to pay for the service than the person without insurance. Some insurance systems require participation of the user in the cost in what is called copayment or user fees or charges. This is often promoted by the idea that it provides the consumer with an incentive not to seek unnecessary care, as well as helping cover costs, while opponents justly reply that user fees affect the poorer sector of any population disproportionately and discourage preventive care.

Capitation for a doctor's services is payment by a fixed sum of money for the persons registered for care for a defined period of time. This can be for a comprehensive health service or for general practitioner services. Compared with salaried service, this method allows a greater degree of identification of the patient with the doctor. It has been in use in the United Kingdom since the introduction of national health insurance in 1911. Recent introduction of incentive fees for full immunization or screening programs have improved performance in these areas.

The United Kingdom's budget holder system pays a group of general practitioners (GPs) for their registered patients. They negotiate with hospitals, which act as contracting service providers. The GP fund pays the hospital on a diagnosis related group (DRG) basis. This system was initiated in the late 1980s and is gaining wide support as the National Health Service (NHS) is reducing hospital bed capacity and seeking to reallocate resources from institutional to preventive and ambulatory care.

Salary payments for doctors and other health workers is common in hospitals even where fee-for-service or capitation is the prominent method of payment. This has advantages for the physician in predictability of income, and it gives less incentive to promote unnecessary servicing. Salary payment may be combined with incentive payments for additional services.

The method of payment for doctors has an important impact on how medical services are used. Empirical evidence indicates that fee-for-service promotes excessive use of the system, including unnecessary surgical procedures, while salaried services are often criticized for diminished identification with patients and, perhaps, underservicing. Increasingly mixed systems of payment are emerging, with capitation as a predominant method.

Payment for Comprehensive Care

Per capita budgeting is a system of payment based on a defined population registered for care with a specific health service system providing a comprehensive range of services, as a district health system or a managed care organization. Capitation payment covers responsibility for total care, so that economies in hospital care can be applied to cost-effective alternatives such as strong ambulatory care, home care, and long-term care. The population may be enrolled on a voluntary basis, as in health maintenance organizations (HMOs) or prepaid group practice systems and managed care systems, or on a geographic basis as in regional or district health systems.

In some financing systems, the per capita payment takes into account the age and sex distribution of the region, locality, or the registered population. It applies national hospital utilization rates for different categories of age and sex. The capitation method provides an incentive against unnecessary admissions and decreases length of hospital stay; but it is not in a hospital's best interest to discharge a patient prematurely because of the potential for litigation and because the patient may later return in need of more care, adversely affecting hospital costs.

Capitation values may be adjusted by applying regional standard mortality rates (SMRs), as in the United Kingdom to account for age, sex, and morbidity differences. The British NHS is paying many of its general practitioners by a combination of capitation and DRG systems discussed later.

HEALTH MAINTENANCE AND MANAGED CARE ORGANIZATIONS

Health maintenance organizations (HMOs) are integrated health insurance and provider systems, responsible for hospital, ambulatory, and preventive care for an enrolled population. It is a system of prepaid health care in which the insured person joins a health plan that received a fixed per capita payment from the insurer to provide comprehensive health care for a defined time. This approach, which was developed in the United States, creates nonprofit organizations sponsored by industry, unions, and cooperative groups. Formerly called prepaid group practice, these plans were developed by Kaiser Permanente in California, during World War II and later in many other parts of the country.

Since the 1973 HMO Act, the HMO has become part of the accepted mainstream of health care in the United States. Some large HMOs operate their own hospitals, utilizing 1.5 beds per 1000 population, well below U.S. averages, even when adjusting for age and selection factors. They operate with 1.2 doctors per 1000 members, as compared to 4.5 per 1000 for fee-for-service health care sys-

BOX 11.4 HEALTH MAINTENANCE
ORGANIZATION MODELS

1. Staff and group: The traditional type of HMO is based on the prepaid group practice model in which the HMO employs or contracts with physician groups to provide comprehensive care to enrolled members, usually in health centers operated by the HMO and in hospitals owned or contracted with the HMO. Group HMOs may be partnerships that share in incentive payments.

2. Independent practice association (IPA): A medical organization of independently practicing physicians which contract to provide care at reduced fees or capitation for patients belonging to an HMO plan. The physicians may also provide care to private patients not belonging to the HMO or who belong to other HMOs.

3. Preferred provider organization (PPO): A formally organized entity, usually of physicians, hospitals, pharmacies, laboratories, or other providers which contract to provide care to HMO members on an agreed (discounted) fee schedule or capitation basis. Each provider works independently but agrees to contracted conditions, including utilization review. The beneficiary has choice of providers within the panel.

tems. Doctors working in HMOs may be paid on salary or capitation in a staff and group HMO, or on a fee-for-service basis in independent practice association (IPA), or Preferred Provider Organizations (PPO).

Health care in the United States has been influenced by the HMO experience and that of other health insurers using HMO-like cost control measures which limit unrestricted fee-for-service practice. The HMO or managed care approach to health care organization is less costly largely because of better management of patients in the community and lower hospital utilization patterns.

The major increase in managed care in the 1990s has been in for-profit managed care, which has been successful in taking a large part of the market share of health insurance because of its advantages of lesser cost and more comprehensive coverage than traditional fee-for-service health insurance. Managed care health plans undertake responsibility for the comprehensive care of its enrolled members. By 1995, more than 58 million Americans received care from HMOs and another 91 million were served by other types of managed care plans such as preferred provider organizations. In 1997, nearly 48% of Medicaid beneficiaries were enrolled in managed care plans, an increase from just under 10% in 1991. In 1997, 4.9 million out of a total of 38 million Medicare beneficiaries were in managed care plans (HCFA website). HCFA regulates HCOs and has instituted guidelines for reporting and quality assessment in an accreditation approach to quality assurance (see Chapter 15).

Managed care systems are being promoted by private employers, by insurance companies, by states for Medicaid beneficiaries, and by the federal government. Managed care has become the predominant mode of organization of health care. In 1997, one-quarter of the U.S. population were enrolled in HMOs, ranging from 18% in the south to 36% in the western part of the country. This was 67 million persons, double the HMO enrollment in 1991. Half of all HMO members in 1991 were in group plans, but this has declined to 17%, while membership in mixed plans increased from 10 to 43% in 1997, with the percentage enrolled in individual practice associations remaining steady at about 40%. As of May 1999, 181.4 million Americans, or 66.6% of the total population, were enrolled in managed care plans (Managed Care On-Line website).

Managed care, especially in the for-profit sector, is under criticism in medical and public health organizations and journal editorials, as well as in the media and state and federal legislatures. It is alleged that the system promotes denial of access to specialists and other needed care because of the economic incentives built into the capitation system, especially when administered by for-profit companies. The economic benefits are generally accepted. The controversy focuses on the incentives to underservice and on loss of choice by the consumer in for-profit managed care systems. The quality and ethical issues of managed care are discussed further in Chapter 15. Legislative efforts at state and federal levels to define patient's rights, grievance procedures, and minimum baskets of service have been under way, with a narrow (50–47) defeat in late 1998 of a Patient's Bill of Rights that was actively promoted by President Bill Clinton. This bill is likely to reappear in future Congressional sessions and broad political debates.

Opponents to the managed care approach argue that lower HMO hospital utilization may in part be achieved by selectivity of membership and that HMOs may underservice patients in order to reduce costs, or increase physician incomes or profits. Available evidence supports HMO experience as providing high quality medical care at lower cost than competing open-ended, fee-for-service insurance systems. The leveling off of expenditure for health in the United States from 1994 to 1999 is largely attributable to the move from fee-for-service care plans to managed care of a large percentage of the population. Managed care is also emerging in other countries, especially in Latin America (Argentina, Brazil, Mexico, Chile, Peru, and others), as well as in the Philippines, all seeking to restrain cost increases, while extending health care to more of their populations.

DISTRICT HEALTH SYSTEMS

In the United Kingdom and the Scandinavian countries a comprehensive service model has existed in the form of district health systems for many years. The residents of a district have their health benefits provided by or contracted out by the district. In principle, the geographic unit of service allows for efficiency in transfer of resources and patients from one service to another, based on need, and not on the financial interests of the insurance system or the provider.

The Scandinavian countries have a long tradition of management of health facilities at the county level with budgets derived from a combination of local taxation and national grants. In reforms since the 1980s, integration of various services into district health systems with reduced hospital bed supplies have resulted in a leveling off of cost increases for health.

In the United Kingdom, methods of budget allocation to health regions were the subject of a long and detailed study by the Regional Allocation Working Party (RAWP) in the 1980s. The decision was finally made that the optimal method of allocating funds to districts health authorities should be on the basis of per capita grants adjusted by the SMRs of the district. This adjustment takes into account age and morbidity differences.

New initiatives in health reform in Canada are combining a cutback in hospital beds to population ratios and regionalization of health services through integration of services in regional or district health systems. The Province of Saskatchewan in 1993 initiated 30 district health boards which amalgamated hospital, nursing home, and public health boards. Per capita funding allows transfer of funds from hospital care to other sector services such as home and community care. The province has managed to level off health expenditure increases to rates less than the growth of GDP. Rural health initiatives in New York State have also moved toward a district health system model, but rigidities in funding systems and lack of strong political leadership hinder the process.

Regionalization of hospital and other services is another approach to rationalization of health care and cost control. In communities with excessive hospital beds and competing services, regionalization provides a method of rationalization, with cooperative or mandatory elimination of wasteful, competing departments or investigative units such as *in vitro* fertilization, cardiac surgery units, advanced imaging devices (MRIs), or excess bed capacity. In the United States efforts to regionalize certain services, such as perinatal care systems in the 1980s, but did not lead to wider application of this approach. In the 1990s, hospital networks in both the for-profit and not-for-profit sectors have expanded aggressively to increase market share and vertical integration for service and management cost efficiency as part of the managed care dominance of the U.S. health insurance market.

PAYING FOR HOSPITAL CARE

Hospitals are the most costly component of a health service. Traditionally, hospitals were paid on a per diem or flat rate per patient-day. The per diem may be determined by actual costs or by national, state, or regional averages by dividing the daily operating costs by the number of beds, with perhaps adjustment for teaching or research functions. The per diem based on actual costs per patient in specific units in a hospital, such as intensive care, may be higher or lower than the amount received.

The per diem method of payment encourages long lengths of stay, rewards hospitals with low technology, and if based on national or regional averages may penalize hospitals with high levels of staffing and technology, such as teaching hospitals. When the service is insured, there is no financial incentive for shortening the patient's hospital stay. The per diem method is associated with inefficient use of facilities, such as admission to the hospital for diagnostic tests or prolonging a stay for additional testing or care that could be provided in alternative and less costly ways. The provider has an incentive to hospitalize and provide prolonged care for a relatively well patient, while the sicker patient is a financial liability, as are teaching and research functions, unless funded separately. This system lacks incentives to improve efficiency by developing alternative ambulatory or day care services, and it punishes more efficient hospitals which reduce length of stay or occupancy rates.

Fee-for-service payment for each service supplied in a hospital favors unnecessary marginal care, long lengths of stay, high admission rates, and the provision of duplicative or unnecessary services. This method was common in the United States with its multiple insurance systems but is increasingly being replaced by DRG payment (see below). There is no incentive to reduce costs or utilization in this model.

Historical budgeting is remuneration based on the previous year's budget, adjusted for inflation and the cost of new services. The budget may be reviewed line by line by the paying authority or be on a global or block budget basis, which frees the hospital to make internal reallocations within the overall allotment. Payment can include a capital fund for renovation. This method is often used when a hospital is directly operated by the ministry of health. As opposed to the per diem payment system, this method should theoretically provide some incentive to reduce length of stay and to search for efficiency in the use of hospital resources.

Payment by norms, as practiced in the Soviet health system, provided national incentives to maintain high bed to population ratios, low salaries, and low quality of care. Reform in post-Soviet countries requires cancellation of these historic norms, reducing excess hospital bed capacity and adoption of incentives for efficiency in health care (Chapter 13).

As a result of concern over high costs and utilization rates, alternative methods of payment were developed in the United States since the 1960s. The diagnostic related groups (DRG) system was adopted in 1983 by the U.S. federal Health Care Financing Administration (HCFA) as the basis for payment for hospitalization of Medicare patients. The DRG system is the basis for paying for hospital care in the United States in 1999, and it is increasingly being used in other industrialized countries, such as the United Kingdom and Israel, and some developing countries, such as the Philippines.

The DRG system is a prospective payment system for hospital care reimbursement, and the system pays the hospital according to 495 treatment classifications of diagnoses or procedures, each with a fixed hospital payment rate. This provides an incentive to reduce length of stay, more efficient use of diagnostic and treatment services, and reduced overall bed capacity. As a result, hospital outpatient services increased rapidly in the United States while bed occupancy rates and the hospital bed to population ratio declined steadily over the 1990s. The DRG system does not lead to fewer admissions and may encourage falsification of diagnostic criteria or increasing the diagnostic severity of case definition to increase revenues ("the DRG creep").

Different hospital budgeting methods have advantages and disadvantages. Payment by DRGs is most likely to promote rational use of hospital care. Regional budgets allocated on a per capita basis with hospital payment by DRGs may be the most effective way of combining ambulatory and hospital care, combining regional equity and incentives for efficient use of diagnostic and treatment services. Prospective payment systems must be associated with quality assurance mechanisms, a vital issue in health management (Chapter 15).

CAPITAL COSTS

The capital cost to build or renovate a health facility is based on long-term considerations but has important effects on current operating costs. The cost of operating a new health care facility may equal the capital cost in 2–3 years. Capital costs may be financed by public or private donations, risk-capital investment, or government-guaranteed loans. Government regulatory agencies may approve a capital project of construction or equipment of a hospital under a certificate of need procedure (CON) and then agree to a grant mechanism to provide funds to match local contributions or to budget or adjust rates to include repayment of long-term loans for capital costs. This occurred both in the United States under the Hill-Burton Act (see Chapter 10) and in Canada under the national health insurance system. Where hospitals are operated independently of government, they may borrow or raise money privately by long-term bonds or low interest loans. Repayment can be built into the operating costs and amortization of the loan over many years.

When governments finance capital costs, they have greater control over the direction, distribution, and supply of hospital facilities. Government norms may encourage an increased bed supply by encouraging hospital construction, or maintenance of high numbers of beds that may not be used or may be of poor quality. Norms may also be used to set upper limits or provide incentives to reduce bed supply. One of the common elements of cost-containment strategies in many industrialized countries in the 1990s is reduction in hospital bed supply, which is occurring without apparent harm to the quality of care. Hospital bed reduction is partly offset by transfer of long stay patients to home care programs or to nursing homes with a transfer of capital and operating costs. Overall, maintaining quality of care is not compatible with maintaining a large bed to population ratio (the converse of the aforementioned Roemer's law) because of the excessive resources required to maintain the beds at the expense of other needed services in the community.

HOSPITAL SUPPLY, UTILIZATION, AND COSTS

Acute care hospital bed to population ratios in the United States increased from the 1940s to the 1980s and declined thereafter. The supply and utilization of hospital beds is changing as economic incentives increase pressure to find less costly forms

of care, and as ambulatory and community-oriented care is perceived to be more effective in many instances. In the United States, hospital utilization, average length
of stay, and percentage occupancy show a decline mainly during the period 1980–
1995. Hospital staff per 100 patient days increased from 226 in 1960 to 583 in 1991,
reflecting increased support and technical services and greater severity of illness of
those hospitalized. Increased staffing, technological innovations, and expensive
medications increased the cost of patient care in hospitals. Table 11.7 shows the trend
in hospital bed supply and percentage occupancy in acute care, nonfederal general
hospitals in the United States from 1940 to 1996.

The trend in hospital bed supply and utilization in the United States have been
downward during the 1980s and early 1990s. Despite aging of the population, the
trend to lower overall hospital utilization has resulted from the following: changing morbidity patterns; substitution of ambulatory services in place of inpatient
care; adoption of the DRG system of payment, reducing length of stay; greater
stress on health economics and cost containment in medical considerations; more
efficient methods of care; greater health consciousness in the general population;
and improved self-care and prevention.

Mortality from coronary heart disease has fallen markedly during this period,
however, admission rates for heart disease overall have not declined, while total
days of care fell by 38%. This reflects changing patterns of care, with shorter length
of stay and a more aggressive rehabilitation approach to myocardial infarction, and
emphasis on ambulatory care. Medical treatment during the acute myocardial infarction stage is more effective than previous treatments, with technology such as
streptokinase, angioplasty, stents, and other interventions. All of this has been accompanied by a steady fall in mortality rates (Chapter 5).

Many western European countries began in the 1980s to reduce their hospital
beds, as seen in Table 11.8. Sweden and Finland reduced their hospital capacity by
53% and 36%, respectively, and western Europe as a whole by 26%. Countries of

TABLE 11.7   Acute Care Hospital Bed Supply and Utilization, United States, 1940–1996

Facilities	1940	1950	1960	1970	1980	1985	1990	1996
Beds per 1000 population	3.2	3.3	3.6	4.3	4.5	4.2	3.7	3.3
Discharges per 1000 population	na[a]	na	na	na	159	138	113	105
Average length of stay (ALS), days	na	na	na	na	7.3	6.5	6.7	6.5
Total days of care per 1000 population	na	na	na	na	1129	872	705	544
Percent occupancy	70	na	75	77	75	65	67	62

Source: *Health United States,* 1993 and 1998, and OECD, 1998.
Note: Does not include federal hospitals.
[a]na, Not available.

TABLE 11.8   Acute Care Hospital Beds and Total Health Expenditures as Percentage of GDP, Selected Countries, 1980–1997

Country	Beds per 1000 population			Total health expenditures as % of GDP		
	1980	1990	1996	1980	1990	1997
Italy	7.5	6.0	5.1	7.0	8.1	7.6
France	6.2	5.2	4.5	7.6	8.9	9.9
Denmark	5.6	4.6	3.9	8.7	8.2	7.7
Netherlands	5.2	4.0	3.4	7.9	8.3	8.5
Sweden	5.1	4.1	2.8	9.4	8.8	8.6
Israel	3.0	2.6	2.3	6.8	7.8	8.4
United Kingdom	2.9	2.3	2.0	5.6	6.0	6.7

Source: World Health Organization, Regional Office for Europe, Health for All Data Set, June 1999.

eastern Europe and the former Soviet Union remained with high hospital inventories but relatively low overall expenditures on health care per capita.

Reducing hospital utilization and bed supply creates a problem of staff and resource reallocation. Hospital facilities themselves can sometimes be converted to other purposes, as outlined in Box 11.5. Often, the most constructive use of obsolete hospital facilities is to transfer them out of the health sector, since the land may be of greater value than its continuing use for health care purposes.

BOX 11.5 DEALING WITH EXCESS HOSPITAL BED CAPACITY

1. Convert to long-term care (LTC) facility: extended care, rehabilitation, chronic care, or elderly persons housing;

2. Close maternity homes: replacement by opening maternity units in district general hospitals;

3. Develop acute psychiatric units in general hospitals: conversion of excess acute care beds, closure of long-term psychiatric beds, development of community services and group residential facilities;

4. Develop acute geriatric units in general hospitals: short-term care, with home and LTC facilities;

5. Develop tuberculosis units in general hospitals: short-term investigation and therapy, with closure of long-term beds in TB hospitals and strengthening community care systems;

6. Develop detoxification units for alcohol and drug abuse: general hospitals with community facilities support.

7. Convert to special needs shelters: homeless persons, abuse or rape crisis shelters;

8. Convert to ambulatory care facilities: use inpatient facilities and staff for outpatient, day hospital services;

9. Develop hospices: care of terminal patients;

10. Convert to other socially useful functions: community centers, schools, or vocational training;

11. Demolish and dispose of obsolescent facilities: land value may pay for part of new health programs.

MODIFIED MARKET FORCES

Classically, market forces are seen as a means of empowering the purchaser to seek the least expensive and/or best goods or services from competing providers. In health care, there are modifying factors that affect market forces. Understanding the modifiers of market forces summarized in Table 11.9 is part of the preparation of a manager, provider, and policy planner for a strategic role in health systems. Some of these modifiers are governmental regulatory factors, such as in supply of hospital beds. Others are related to access to services and the amount and method of payment and other factors that affect needs for care, as well as, the quality and efficiency of service. Market mechanisms are modified by regulations, incentives, and other factors used to promote a balance of preventive, curative, and rehabilitative services, including health promotion to improve health and help the individual seek and find the most appropriate care at any point in time (Table 11.11).

ECONOMICS AND THE NEW PUBLIC HEALTH

The New Public Health has a vital interest in methods of financing of health services, the economics of health care, and allocation of resources, whether in terms of money, manpower, or fixed assets. No other approach has taken the broad view of promotive, preventive, curative, and long-term care services. Their balance and interdependence are influenced by financing systems and choices among alternative ways of expending resources.

Economic analysis, like epidemiologic assessment, is a vital tool in health planning and management, particularly in evaluating allocation of resources within a health system. Failure to carry out such assessments results in inefficiency in health planning. Methods of economic analysis, described briefly in this chapter, are part of the armamentarium of the New Public Health. The new and veteran professional in health needs a basic understanding of the economic issues involved in setting priorities, service organization, utilization of services, and the complex of related ethical, political, social, and management issues.

TABLE 11.9   Market Forces and Modifying Factors in the Economics of Health

Determinants in health demand	Modifiers of utilization	Examples
Classic market factors	Supply	Providers combine to restrict supply
	Demand	Prepayment increases effective demand
	Competition in cost, quality	Managed care versus fee-for-service plans
	System macroefficiency	District health systems and HMOs
	Vertical integration	Multiservice systems for defined populations
	Lateral integration	Multihospital networks for efficiency
	System microefficiency	Quality improvement, computerization, staff attitudes, scheduling, and service hours
	Incentives and disincentives	Budget method of payment: block, per diem, DRG
	Reputation	Consumer, community, provider satisfaction
Regulatory factors	Regulate supply	Reduce hospital beds per 1000 and manpower
	Regulate demand	Gate-keeping functions
	Regulate price	User fees, fee control, income capping, salary, or capitation payment for doctors
	Regulate method of payment	DRGs, block budgets for hospitals
	Health promotion issues	Food enrichment, safety factors, seat belts
Health and societal factors	Differing population needs	Demographic and epidemiological transitions
	Social inequalities	Reduce social gaps, ensure universal access, target risk groups
	Improve infrastructure to reduce needs	Sewage, water treatment, road safety
	Socioeconomic improvements	National and family incomes
	Public social policies	Social security, pensions, compensation
	Health as a national and local priority	Health system reform
	Health promotion	Involve community and providers in prevention
	Improve KABP (knowledge, attitudes, beliefs, and practices)	Providers, beneficiaries, consumers' rights, needs, responsibilities
System determinants	Shift in resource allocation	Balance of institutional and community care
	Technological innovations	New vaccines, drugs, diagnostic equipment, oral rehydration, community health workers
	Substitution	Home care, generic drugs, nurse practitioners
	Total quality management	External accreditation, internal review systems, patient choice, continuous quality improvement

SUMMARY

The *World Development Report 1993* supports the basic view promoted by the WHO since the Alma-Ata Conference in 1978: that a balance between primary health care and hospital care is an essential element of the health system, and that primary care and health promotion are the most cost-effective interventions in improving the health status of the population. Where there has been an excessive emphasis on institutional care, there is real potential for transfer of resources and emphasis within the health system as part of the process of raising primary care and health promotion standards. This is the direction of health reform in many countries.

Innovations in health care financing and administration, such as HMOs, managed care, capitation payment, fund holding GPs, district health systems and DRGs, are all part of the search for more efficient ways of using resources and limiting cost increases. Improvements in care, such as endoscopic and outpatient surgery, and home care, are making an impact on health economics. As the population ages and technology advances, costs will inevitably increase unless resources are allocated to prevent or delay the onset of complications among chronically ill persons in the community.

Health must compete with other government programs for resource allocation. In the New Public Health, health care is an investment in human capital for the development of a country, as well as an ethical obligation of a society toward its individual members. The New Public Health is involved in management of health care in all its aspects so that an understanding of basic issues in health economics is as vital to its practice as is an understanding of communicable disease or any other element of the broad panorama of health.

ELECTRONIC MEDIA

American Public Health Association, Policy Statement 9615(PP), Supporting national standards of accountability for access and quality in managed health care, http://www.apha.org/science/policy.html

American Public Health Association, Policy Statement 9716(PP), The issue of profit in health care, http://www.apha.org/science/policy.html

American Public Health Association, Policy Statement 9802, Managed care and people with physical/mental disabilities, http://www.apha.org/science/policy.html

Health Care Financing Administration, http://www.hcfa.gov/

Managed Care, http://www.managedcare.hhs.gov

Managed Care On-Line, Claire@mcol1.mcareol.com, www.managedcaredigest.com

Organization of Economic Cooperation and Economic Development (OECP), http://www.oecd.org

World Bank Health Reform Online, http://www.worldbank.org/healthreform/

World Health Organization, European Region Health for All Data Set, http://www.who.org/dk

RECOMMENDED READINGS

Anderson, G. F., Ponllier, J.-P. 1999. Health spending, access and outcomes: trends in industrialized countries. *Health Affairs,* May/June:178–192.

Bobadilla, J.-L., Cowley, P., Musgrove, P., Saxenian, H. 1994. Design, content and financing of an essential national package of health services. *Bulletin of the World Health Organization,* 72:653–662.

Centers for Disease Control. 1995. Assessing the effectiveness of disease and injury prevention programs: costs and consequences. *Morbidity and Mortality Weekly Report,* 44:RR-10:1–10.

Centers for Disease Control. 1995. Economic costs of birth defects and cerebral palsy—United States, 1992. *Morbidity and Mortality Weekly Report,* 44:694–699.

Chang, W.-Y., Henry, B. M. 1999. Methodologic principles of cost analysis in the nursing, medical, and health services literature, 1990–1996. *Nursing Research,* 48:94–104.

Evans, R. G., Lomas, J., Barer, M. L., Labelle, R. J., Fooks, C., Stoddart, G. L., Anderson, G. M.,

Feeny, D., Gafni, A., Torrance, G. W., Tholl, W. G. 1989. Controlling health expenditures: The Canadian reality. *The New England Journal of medicine,* 320:571–577.

Murray, C. J. L., Govindaraj, R., Musgrove, P. 1994. National health expenditures: A global analysis. *Bulletin of the World Health Organization,* 72:623–637.

Wennberg, J. E. 1990. Outcomes research, cost containment, and the fear of health care rationing. *The New England Journal of Medicine,* 323:1202–1204.

Wickham, J. E. A. 1993. An introduction to minimally invasive therapy. *Health Policy,* 23:7–15.

World Bank. 1993. *World Development Report, 1993: Investing in Health. World Development Indicators.* New York: Oxford University Press.

Wright, J. C., Weinstein, M. C. 1998. Gains in life expectancy from medical interventions—Standardizing data on outcomes. *New England Journal of Medicine,* 339:380–386.

BIBLIOGRAPHY

Abel-Smith, B. 1991. Financing health for all. *World Health Forum,* 12:191–200.

Anderson, R. M., Rice, T. H., Kominski, G. F. 1996. *Changing the U.S. Health Care System: Key Issues in Health Services, Policy and Management.* San Francisco: Jossey-Bass Publishers.

Anderson, T. F., Mooney, G. (eds). 1990. *The Challenge of Medical Practice Variation: Economic Issues in Health Care.* Hong Kong: MacMillan.

Bodenheimer, T. 1999. The American health care system: Physicians and the changing medical marketplace. *The New England Journal of Medicine,* 340:584–588.

Brownson, R. C., Remington, P. L., Davis, J. R. 1998. *Chronic Disease Epidemiology and Control,* 2nd Edition. Washington, DC: American Public Health Association.

Bunker, J. P. 1970. Surgical manpower: A comparison of operations and surgeons in the United States and in England and Wales. *The New England Journal of Medicine,* 282:135–144.

Centers for Disease Control. 1992. A framework for assessing the effectiveness of disease and injury prevention. *Morbidity and Mortality Weekly Report,* 41:RR-3:1–12.

Centers for Disease Control. 1995. Economic costs of birth defects and cerebral palsy—United States, 1992. *Morbidity and Mortality Weekly Report,* 44:694–699.

Chassin, M. R., Kosecoff, J., Park, R. E., Winslow, C. M., Kahn, K. L., Merrick, N. J., Keesey, J., Fink, A., Solomon, D. H., Brook, R. H. 1987. Does inappropriate use explain geographic variations in the use of health services? *Journal of the American Medical Association,* 258:2533–2537.

Creese, A. L., Henderson, R. H. 1980. Cost–benefit analysis and immunization programmes in developing countries. *Bulletin of the World Health Organization,* 58:491–497.

Drumond, M. F. 1985. Survey of cost-effectiveness and cost–benefit analyses in industrialized countries. *World Health Statistics Quarterly,* 38:383–401.

Drumond, M. F., Stoddart, G. L. 1985. Principles of economic evaluation of health programmes. *World Health Statistics Quarterly,* 38:355–367.

Drumond, M. F., Stoddart, G., Labelle, R., Cushman, R. 1987. Health economics: An introduction for clinicians. *Annals of Internal Medicine,* 107:88–92.

Folland, S., Goodman, A. C., Stano, M. 1997. *The Economics of Health and Health Care,* Second Edition. Upper Saddle River, NJ: Prentice-Hall.

Friede, A., Taylor, W. R., Nadelman, L. 1993. On-line access to a cost–benefit/cost-effectiveness analysis bibliography via CDC WONDER. *Medical Care,* 31(Supplement):JS12–17.

Ginsberg, G. M., Tulchinsky, T. H., Abed, Y., Angeles, H. I., Akukwe, C., Bonn, J. 1990. Costs and benefits of a second measles inoculation of children in Israel, the West Bank, and Gaza. *Journal of Epidemiology and Community Health,* 44:274–280.

Ginsberg, G. M., Berger, S., Shouval, D. 1992. Cost–benefit analysis of a nationwide inoculation programme against viral hepatitis B in an area of intermediate endemicity. *Bulletin of the World Health Organization,* 70:757–767.

Ginsberg, G., Tulchinsky, T., Filon, D., Godfarb, A., Abramov, L., Rachmilevitz, E. A. 1998. Cost–

benefit analysis of a national thalassemia programme in Israel. *Journal of Medical Screening,* 5:120–126.

Inglehart, J. K. 1999. The American health care systems—Medicare. *New England Journal of Medicine,* 340:327–332.

Jacobs, P. 1991. *The Economics of Health and Medical Care,* Third Edition. Gaithersburg, MD: Aspen.

Kelley, J. E., Burrus, R. G., Burns, R. P., Graham, L. D., Chandler, K. E. 1993. Safety, efficacy, cost and morbidity of laparoscopic versus open cholecystectomy; a prospective analysis of 228 consecutive patients. *American Surgeon,* 59:23–27.

Kinnon, C. M., Velasquez, G., Flori, Y. A., World Health Organization Task Force on Health Economics. 1994. *Health Economics: A Guide to Selected WHO Literature.* Geneva: WHO.

Lieu, T. A., Cochi, S. L., Black, S. B. 1994. Cost-effectiveness of a routine varicella vaccination program for U.S. children. *Journal of the American Medical Association,* 271:375–381.

Lord, J., Thomason, M. J., Littlejohns, P., Chalmers, R. A., Bain, M. D., Addison, G. M., Wilcox, A. H., Seymour, C. A. 1999. Secondary analysis of economic data: A review of cost–benefit studies of neonatal screening for phenylketonuria. *Journal of Epidemiology and Public Health,* 53:179–186.

Mills, A. 1985. Economic evaluation of health programmes: Application of the principles in developing countries. *World Health Statistics Quarterly,* 38:368–382.

Mills, A. 1985. Survey and examples of economic evaluation of health programmes in developing countries. *World Health Statistics Quarterly,* 38:402–431.

Mitchell, J. B. 1985. Physician DRG's. *The New England Journal of Medicine,* 313:670–675.

Mooney, G. H., Drummond, M. F. 1982. Essentials of health economics: Part 1—What is economics? *British Medical Journal,* 285:949–950.

National Center for Health Statistics. 1998. *Health, United States, 1998, With Socioeconomic Status and Health Chartbook.* Hyattsville, Maryland.

OECD Health Data. 1998. A Comparative Analysis of 29 Countries. OECD.

Patrick, D. L., Erickson, P. 1993. *Health Status and Health Policy: Quality of Life in Health Care Evaluation and Resource Allocation.* New York: Oxford University Press.

Robinson, R. 1993. Cost–benefit analysis. *British Medical Journal,* 307:924–926.

Roemer, M. I. 1961. Bed supply and hospital utilization: A natural experiment. *Hospitals,* 1:35–42.

Smith, S., Freeland, M., Heffler, S., McKusick, D. and the Health Expenditures Projection Team. 1998. The next ten years of health spending: What does the future hold? *Health Affairs,* Jan/Feb:128–140.

Stocker, K., Waitzkin, H., Iriart, C. 1999. The exportation of managed care to Latin America. *New England Journal of Medicine,* 340:1131–1135.

Thompson, M. S. 1986. Willingness to pay and accept risks to cure chronic disease. *American Journal of Public Health,* 74:392–396.

Vayda, E. 1973. A comparison of surgical rates in Canada and in England and Wales. *The New England Journal of Medicine,* 289:1224–1229.

Weinstein, M. C., Stason, W. B. 1977. Foundations of cost-effectiveness analysis for health and medical practices. *The New England Journal of Medicine,* 296:716–721.

Wennberg, J. F., Gittelsohn, A. 1973. Small area variations in health care delivery. *Science,* 182:1102–1108.

White, C. C., Koplan, J. P., Orenstein, W. A. 1985. Benefits, risks and costs of immunization for measles, mumps and rubella. *American Journal of Public Health,* 75:739–744.

World Health Organization, Regional Office for Europe. 1996. *European Health Care Reforms: Analysis of Current Strategies, Summary.* Copenhagen: WHO.

12

PLANNING AND MANAGING
HEALTH SYSTEMS

INTRODUCTION

Health is a hugely expensive complex of services, provided by a wide range of professional and support service personnel making up one of the largest employers of manpower in any country. Services are increasingly provided by organized groups of providers in economic constraints requiring efficiency in use of resources. How organizations function is of great importance for their economic survival, but also, and equally important, for the well-being of clients and providers of care.

An organization is two or more persons working together to achieve a common goal. Management is the process of defining the goals and making effective use of an organization to attain those goals. Even small units of human organization require management. Management of human resources is vital to the success of an organization, whether in a production or service industry. Health systems may vary from a single structure to a network of many organizations. No matter how they are financed or operated, they require management.

Management in health has much to learn from management in other industries. Even if a health service is run as a nonprofit enterprise, theories and practice of management developed in profit-oriented business can be applied. Physicians, nurses, and other health professionals will very likely be involved in management of some part of the health care system, whether a hospital department, a managed care system, a clinic, or even a small health care team. At whatever the level, management means working with people, using resources, providing services, and working toward objectives.

Health providers require preparation in theory and practice of management. They require a management orientation in order to understand the wider implications of clinical decisions and their role in helping the health care system achieve goals and targets. Students and practitioners of public health need preparation in

order to recognize that the health care system is more complex than the direct provision of individual services. Similarly, policy and management personnel need to
be familiar with individual and population health needs and related care issues.

HEALTH POLICY AND PLANNING AS CONTEXT

Health has evolved from an individual one-on-one service to complex systems
organized within financing arrangements mostly under government auspices. As a
governmental priority, health may be influenced by political ideology, sometimes
reflecting societal attitudes of the party in power and sometimes apparently at odds
with its general social policy. Following Bismarck's 1881 introduction of national health insurance for workers and their families, paid for by both workers and
employers, most countries in the industrialized world introduced national health
plans. Usually, this has been at the initiative of socialist or liberal political leadership, but conservative parties have preserved national health programs once initiated. Despite the new conservatism of the 1990s with its preeminent ideology of
market forces, the growing roles for national, state, and local authorities in health
has led to a predominantly governmental role in financing and overall responsibility for health care.

Health policy is formulated to meet national, state, local, or institutional needs.
These are determined professionally and politically within constitutional, legal,
and financial constraints. Selection of which direction to take in organizing health
services is usually based on a mix of factors, including the political view of the
government, public opinion, and rational assessment of needs as defined through
epidemiologic data, cost–benefit analysis, and recommendations by experts. Lobbying on the part of professional or lay groups with particular interests they wish
to promote is part of the process of policy formulation and has an important role
in planning and management of health care systems. There are always competing
interests for limited resources, some in the health field itself and others outside of
the field. The political level is involved in health management, in determining the
place of health care as a percentage of total governmental budgetary expenditures,
and in allocation of funds in the competing priorities within the health sector. A
political commitment must be accompanied by resources adequate to the scope of
the task. Thus, health policy is in part determined by society and is not a preroga-
tive of government, health care providers, nor any institution alone.

As a result of long struggles by unions and political or advocacy groups, most
countries with well-developed market economies have come to accept health as a
national obligation and essential to a well-ordered society and have implemented
universal access systems. Once initiated, allocation of resources to health are high,
with relatively high salaries for health care professionals. In these countries, health
expenditures consume between 7 and 14% of GNP. Some industrialized countries,
notably those in the former Soviet bloc, lacking mechanisms for advocacy, including consumer and professional opinion, tended to view health as a political

objective of social benefits, but also as a "nonproductive" consumer of resources rather than a producer of new wealth. As a result, budget allocations and total expenditures as a percentage of GNP to health have been lower than in other industrialized countries. Salaries for health personnel are low compared to industrial workers in the "productive sectors." Furthermore, industrial policy did not promote modern health-related industries, as compared to the military or heavy industrial sectors. Typically, such countries allocate between 3 and 5% of their GNP to health. Most developing countries spend under 4% of GNP on health.

Financing of health care and resource allocation requires a balance among primary, secondary, and tertiary care. Economic assessment, monitoring and evaluating service, is part of determining the health needs of the population. Regulatory agencies are responsible to define goals, priorities, and objectives for resulting services. Targets and methods of achieving them help provide the basis for implementation and evaluation strategies. Planning requires written plans that include a statement of vision, mission objectives, target strategies, methods, and coordination during the implementation. Designation and evaluation of responsibilities, resources to be committed, and participants and partners in the process are part of the continuous process of management.

The dangers of taking a "wrong" direction can be severe, not only in terms of financial cost, but also in terms of excess morbidity and mortality. Health policy is often as imprecise a science as medicine itself. The difference is that inappropriate policy can affect the lives and well-being of very large numbers of people, as opposed to an individual being harmed by the mistake of one doctor. There may be no "right" answer, and there are numerous controversies along the path. Health policy remains more an "art" than the more quantitative and seemingly precise field of health economics.

Health policy, planning, and management are interrelated and interdependent. Any set goal should be accompanied by planning as to how to attain it. A policy should state the values on which it is based, as well as specify sources of funding, planning, and management arrangements for its implementation. Examination of the costs and benefits of alternative forms of health care helps in making decisions as to the structure and the content of health care services. This may involve both structures within one organization and linkages (intersectoral cooperation with other organizations). The methods chosen to attain the goal then become the applied health policy.

The World Health Organization's Health for All strategy was directed to the political level and intended to increase governmental awareness of health as a key component of overall development. Within health, primary care was stressed as the most effective investment to improve the health status of the population. In 1993 the World Bank's *World Development Report* adopted the Health for All strategy and promoted the view that health is an important investment sector for general economic and social development. However, economic policies promoting privatization and deregulation in the health sector may undermine this larger goal.

In the New Public Health, clinical and preventive care are part of public health because the well-being of the individual and the community requires a coordinated effort from all elements of the health spectrum. Defining and achieving national health goals requires planning, management, and coordination at all levels.

THE ELEMENTS OF ORGANIZATIONS

The study of organizations developed within sociology but has become a multidisciplinary activity involving many other fields, such as economics, anthropology, political science, and engineering. Organizations, whether in the public or private sector, exist within an external environment, and utilize their own structure, participants, and technology to achieve their goals. They must adapt to the physical, social, cultural, and economic environment.

Organizations participating in health care relate service providers and consumers, with the goal of better health for the individual and the community. The technology for this includes the legislation, regulation, professionalism, instrumentation, medications, vaccines, education, and other modalities of intervention for prevention and treatment. The social structure of an organization may be "formal" (or structured stability), "natural" (or groupings reflecting common interests, such as community based organization), or "open" (or loosely coupled, interacting systems adjusting themselves to achieve goals using materials, energy, and information).

Formal systems are deliberately structured for the purposes of the organization. Natural systems are less formal structures where participants work together collaboratively to achieve common goals defined by the organization. Open systems relate elements of the organization to coalitions of partners in the external environment to achieve mutually desirable goals. The social structure of an organization includes values, norms, and roles governing behavior of participants.

Government, business, or service organizations, including health systems, require organizational structures in order to function. An organizational structure needs to be tailored to the size and complexity of the entity and the goals it wishes to achieve. The structure of an organization is the way in which it divides its labor into distinct tasks and coordinates them. The major organizational models, which are not mutually exclusive, and may indeed by complementary, are the pyramidal (bureaucratic) and network structures. The bureaucratic model is based on a hierarchical chain of command with clearly defined roles. In contrast, the matrix or network organization brings together professional or technical persons to work on specific programs, projects, or tasks.

SCIENTIFIC MANAGEMENT

Some of the classic organization theory concepts help to set the base for modern management ideas as applied to the health sector. Scientific management was pioneered by Frederick Winslow Taylor (1856–1915).

His work was pragmatic and based on empirical engineering, developed in observational studies carried out for the purpose of increasing worker efficiency. Taylor's industrial engineering studies of *Scientific Management* were based on the concept that the best way to improve worker productivity was by designing improved techniques or methods used by workers. This theory saw workers as instruments to be manipulated by managers and assumed that efficient, rationally planned methods would produce better industrial results and industrial peace as the tasks of managers and workers would be better defined.

Time and motion studies analyzed work tasks to seek more efficient methods. Motivation of workers was seen to be related to payment by piece work and economic self-interest to maximize productivity. Taylor sought to improve the productive efficiency of each worker and to make management more scientific in order to increase earnings of employers and workers. He found that the worker was more efficient and productive if the worker was goal-oriented rather than task-oriented. This approach dominated organization theory for the first decades of the twentieth century.

Resistance to Taylor's ideas came from both management and labor, the former because it seemed to interfere with their managerial prerogatives and the latter because it expected the worker to function at top efficiency at all times. However, Taylor's work had a lasting influence on the theory of work and organizations.

BUREAUCRATIC PYRAMIDAL ORGANIZATIONS

The traditional pyramidal bureaucratic organization, existing since the dawn of human society, was analyzed by sociologist Max Weber between 1904 and 1924. This form of organization is classically seen in military and civil services, but also in large-scale industry, where discipline, obedience, and loyalty to the organization are demanded, and individuality minimized. Leadership is assigned by higher authority, and is presumed to have higher knowledge than members lower down in the organization. This form of organization is effective when the external and internal environments, the technology, and functions are relatively well defined, routine, and stable.

The pyramidal system (Fig. 12.1) has an apex of policy and executive functions, a middle level of management personnel and support staff, and a base of the people who produce the output of the organization. The flow of information is generally to the top level, where decisions are made for the detailed performance of duties at all levels. Lateralizing the information systems so that essential data can be shared to help staff at the field or middle levels of management is generally discouraged because this may promote decentralized rather than centralized management. Even these types of organizations have increasingly come to emphasize small group loyalty, leadership initiative, and self-reliance.

The bureaucratic organization has the following characteristics:

1. There is a fixed division of labor with clear jurisdiction and based on assignments, which are subject to change by the leader.

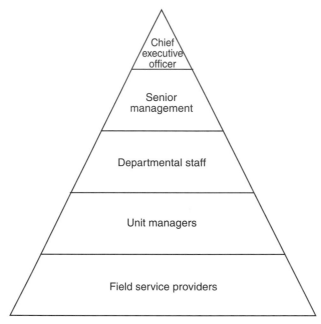

FIGURE 12.1   Pyramidal structure of organizations.

2. There is a hierarchy of offices, with each lower office controlled and supervised by a higher one.
3. A documented, stable set of rules governs decisions and actions.
4. Property and rights belong to the office, not the office holder.
5. Officials are selected on the basis of qualifications, and salaries and benefits are based on technical qualifications.
6. Employment is viewed as a tenured career for officials, after an initial trial period.

The bureaucratic system, based on formal rationality, structure, and discipline, is widely used in production, service, and governmental agencies.

ORGANIZATIONS AS ENERGY SYSTEMS

Health systems, like other organizations, are dynamic, requiring continuous management, adjustments, and systems control. Continuous monitoring and feedback, evaluation, and change help to meet individual and community needs. The input–process–output model (Fig. 12.2) depends on feedback systems to make the administrative or educational changes needed to keep moving toward the defined

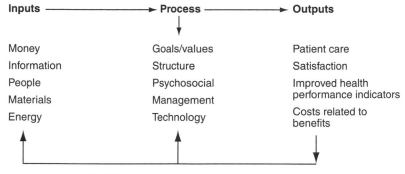

FIGURE 12.2   Organizations as energy systems.

targets. They use inputs that are processed to achieve outputs. The input is money, personnel, information, and supplies. Process is the summation of all activities taken to achieve the results intended. Output, or outcome, is the product, its marketing, and profit. In a service sector such as health, output or impact can be measured in terms of reduced morbidity or mortality, improved health, or number of successfully treated and satisfied patients. The management system provides the resources and organizes the process by which it hopes to achieve the established goals.

Program implementation requires systematic feedback for the process to work effectively. When targets are set and strategy defined, resources, whether new or existing, are placed at the service of the new program. Management is then responsible to use the resources to best achieve the intended targets. The results are the "output" measures which are evaluated and fed back to the input and process levels.

Health systems have many subsystems, each with organization, leadership, goals, targets, and internal information systems. Subsystems need to communicate within themselves, with peer organizations, and with the macro (health) system. Leadership style is central to this process. The surgeon is the leader of the team in the operating room, but he or she is dependent on the support and judgment of other crucial persons on the team, such as the anesthetist, the operating room nurse, and other support personnel, such as the pathologist, radiologist, and director of laboratory services, all of whom lead their own teams. The hospital director cannot function without a high degree of decentralized responsibility and a creative team approach to quality development of the facility.

Health systems management includes analysis of service policy, budget, decision-making in policy, as well as operation, regulation, supervision, provision, maintenance, ethical standards, and legislation. Policy formulation involves a set of decisions made in pursuit of a course of action for achieving defined health goals.

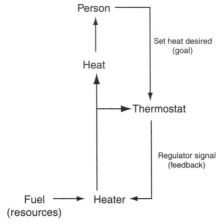

FIGURE 12.3 Cybernetic feedback control organization.

Cybernetics and Management

Cybernetics, developed by Norbert Wiener, refers to systems or organizations which are dependent on each other to function, and whose interdependence requires flexibility of response. Cybernetics gained wide credence in engineering in the early 1950s, and feedback systems became part of standard practice of all modern management systems. Its later transformations appeared in operating service systems, as information for management. Application of this concept is gradually entering health care. Rapid advances in computer technology, by which personal computers have access to Internet systems and large amounts of data, will enhance this process. In mechanistic systems, behavior is constrained and limited; in organic systems there is more interaction between parts of the system. The example used in Fig. 12.3 is the use of a thermostat to control the temperature and function of a heater according to conditions in the room. This is also described as a feedback system.

Cybernetics opens up new vistas on the use of health information for managing the operation of health systems. A database for each health district would allow assessment of current epidemiological patterns, with appropriate comparisons to neighboring districts or regional, state, or national patterns.

Data would need to be processed at state or national levels in comparable forms for a broad range of health status indicators. Furthermore, the data should be prepared for on-line availability to local districts in the form of current health profiles: Thus, data can be aggregated and desegregated to meet the management needs of the service, and can be used to generate real targets and measure progress toward meeting them. A geographic information system may demonstrate high rates of a disease in a region due to local population risk factors, and thus become the basis for an intervention program.

In the health field, the development of reporting systems based on specific diseases or categories has been handicapped by lack of integrating systems and a geographic reporting approach. The new technology of computers and the Internet should be used to process data systems in real-time and in a more user-friendly manner. These will enable local health authorities and providers to respond to actual health problems of their communities.

TARGET-ORIENTED MANAGEMENT

The management of resources to achieve productivity and measurable success has been accompanied by development of systems of organizing people to create solutions to problems or to innovate toward defined objectives.

Operations Research

Operations research is a concept developed by British scientists and military personnel to search for solutions to specific problems of warfare during World Wars I and II. The approach was based on the development of multidisciplinary teams of scientists and personnel. The development of the Anti-Submarine Detection Committee (ASDIC) for underwater detection of submarines during World War I characterized and pioneered this form of research. The common story of the camel being a horse designed by a committee is amusing, but such organization works. The famous Manhattan Project whereby the United States assembled a powerful research and development team which produced the atomic bomb is a prime example.

The team- and goal-oriented work produced dramatic effects in problem solving under the enormous pressure of wartime needs, and also influenced post-war approaches to developmental needs in terms of applied science in such areas as development in the aerospace and computer industries. The computer hardware and software industries are characterized by innovation conceived by informal working groups with a high level of individual competence, peer group dynamism, and commitment to problem solving. This is how the "nerds" of Macintosh and Microsoft beat the "suits" of IBM in innovation and introduction of the personal computer. Similar startup groups develop the Internet and much of the product of California's Silicon Valley and its imitators in many parts in the country and worldwide.

In the health field, innovation in organization developed prepaid group practice which became the health maintenance organization (HMO) and later the managed care organization, now a dominant factor in health care provision in the United States. Other examples may be found in multidisciplinary research teams working on vaccines or pharmaceutical research, and in the increasingly multidisciplinary function of hospital departments and especially highly interdependent intensive care or home care teams.

Management by Objective

The business concept of management by objective (MBO), pioneered in the 1960s, has become a common theme in health management. MBO is a process whereby managers of an enterprise jointly identify its goals, define each individual's areas of responsibility in terms of the results expected, and use these measures as guides for operating the unit and assessing the contributions of its members.

The common goals and then the individual unit goals must be established, as well as the organizational structure molded to help achieve these goals. The goals may be defined in terms of outcome variables, such as specified targets in infant or maternal mortality rates. They may also be defined in terms of intervening or process variables, such as achieving 95% immunization coverage, prenatal care attendance, or screening for breast examination and mammography. Achievements are measured in terms of relevancy, efficiency, impact, and effectiveness.

The MBO approach has been subject to criticism in the field of business management because of its stress on mechanical application of quantitative outcome measures and because it ignores the issue of quality. This approach has had great influence on the adoption by the World Health Organization of Health for All, and on the U.S. Department of Health and Human Services 1979 health targets for the year 2000. Targeting diseases for eradication may contribute to institution building by developing experience and technical competence to broaden the organizational capacity. On the other hand, categorical programs can distract the development of more comprehensive approaches. A suitable balance between comprehensive and categorical approaches requires very skilled management.

HUMAN RELATIONS MANAGEMENT

Management is the task of coordinating and integrating organizational resources, including people, money, material, time, and space. The purpose is achievement of objectives as effectively and efficiently as possible. Because workers are the key to achieving goals, whether in producing goods and profits or in delivering services effectively, management deals with human motivation and behavior. Knowledge and motivation of the individual client and the community are also essential for achieving good health, so management must take into account the knowledge, attitudes, beliefs, and practices of the consumer as much or more than of the people working within the system.

Management, like medicine, is both a science and an art. The application of scientific knowledge and technology in medicine involves both theory and practice. Similarly, management practice draws on organizational theory, which, in turn, draws on the behavioral and social sciences, and quantitative methodologies. Sociology, psychology, anthropology, political science, history, and ethics contribute to the understanding of psychosocial systems, motivation, status, group dynamics, influence, power, authority, and leadership. Quantitative methods including statistics, epidemiology, survey methods, and economics are also basic to development of systems concepts. Comparative institutional analysis helps to develop principles of organization and management, while philosophy, ethics, and law are part of understanding individual and group value systems.

Organizational theory is a relatively new discipline. Translation of organizational theory into management practice requires knowledge, planning, organization, assembly of resources, motivation, and control. Health organizations have become more complex and costly over time, especially in their mix of specializations in science, technology, and professional services. Organization and management are particularly crucial for successful application of the principles of the New Public Health, as it involves integration of traditionally separate health services. Delegation of responsibilities in health systems, such as intensive care units, is fundamental to their success in patient care with nurses taking increasing responsibility for management of the severely ill patient with multiple system failure.

The Hawthorne Effect

Elton Mayo of the Harvard School of Business carried out a series of observational studies at the Hawthorne, Illinois plant of the Western Electric Company between 1927 and 1932. Mayo and his industrial engineer and psychologist colleagues made a major contribution to the development of management theory. Mayo began with industrial engineering studies of the effect of increased lighting on production at an assembly line. This was followed by other improvements in working conditions, including reduced length of the working day, longer rest periods, better illumination, color schemes, background music, and other factors in the physical environment. These studies showed that production increased with each of these changes and improvements. However, the researchers discovered, to their surprise, that production continued to increase when the improvements were withdrawn. Further, in a control group where conditions remained the same, productivity also grew during the study period. This led Mayo to the conclusion that worker performance improved because of a sense that management was interested in them.

Traditionally, industrial management had looked on employees as mechanistic components of a production system. Previous theory was that productivity was a function of working conditions and monetary incentives. What came to be known as the Hawthorne effect showed the importance of social and psychological factors on productivity. Formal and informal social organizations among management and employees were recognized as key elements in productivity, now called industrial humanism. Research methods adapted from the behavioral sciences contributed to scientific studies in industrial management. Traditional theories of the bureaucratic model of organization and scientific management were modified by the behavioral sciences. This led to emergence of the systems approach, or scientific analysis to analyze complex structures or organizations, taking into account the mutually interdependent elements of activities, interactions, and interpersonal relationships between management and workers.

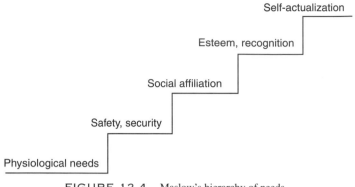

FIGURE 12.4   Maslow's hierarchy of needs.

Maslow's Hierarchy of Needs

Abraham Maslow's hierarchy of human needs was an important contribution to management theory. Maslow (1908–1970) defined a prioritization of needs, from those of basic human survival, including safety, shelter, food, and warmth, to those of social affiliation, esteem, and self-fulfillment. Others in the hierarchy include socialization and self-realization.

This concept is important in terms of management because it identifies human needs beyond those of physical and economic well-being, relating the social context of the work environment with needs of recognition, satisfaction, self-esteem, and self-fulfillment. It opened many positive areas of management research, not only in the motivation of workers in production and service industries, but also the motivation of consumers. Figure 12.4 represents Maslow's hierarchy of human needs.

Maslow's hierarchy of needs contributed to opening the discussion of management to consumers as well as employees. It played an important role in application of sociological theory to client behavior, just as the topic of personal lifestyle in health became a central part of public health and clinical management of many conditions, such as in risk factor reduction for cardiovascular diseases.

Theory X–Theory Y

Theory X–Theory Y, developed by clinical psychologist Douglas McGregor in the 1960s, examined two extremes in management assumptions about human nature that ultimately affect operations of organizations. Organizations with centralized decision making, a hierarchical pyramid, and externally controlled are based on certain concepts of human nature and motivation. McGregor's theory, drawing on Maslow's hierarchy of needs, set out an alternative set of assumptions that credit most people with the capacity for self-direction.

Management by direction and control (Theory X) assumes workers are lazy, unambitious, uncreative, and motivated by basic physiological needs or fear (Table 12.1). The Theory Y model provides a more optimistic view of human potential,

TABLE 12.1   Theory X–Theory Y

Theory X	Theory Y
1. Work is inherently distasteful to people.	1. Work is as natural as play in favorable conditions.
2. Most people are not ambitious, have little desire for responsibility, and prefer to be directed.	2. Self-control is indispensable in achieving goals.
3. Most people have little capacity for creatively solving organizational problems.	3. The capacity for creativity in solving organizational problems is widely distributed in the population.
4. Motivation occurs only at physiological and safety levels.	4. Motivation exists at the social, esteem, and self-actualization levels, as well as the physiological and security levels.
5. Most people must be closely controlled and often coerced to achieve organizational objectives.	5. People can be self-directed and creative at work if properly encouraged.

assuming that, if properly motivated, people can be self-directed and creative at work, and that the role of management is to unleash this potential in workers. Many other theories of motivation and management have been developed to explain human behavior and how to utilize inherent skills to produce a more creative work environment, reduce resistance to change, reduce unnecessary disputes, and ultimately create a more effective organization.

Variants of the human motivation approach in management carried the concept further by examining industrial organization to determine what effect management practices have on individual behavior and personal growth within the work environment. Theory X assumes that management produces immature responses on the part of the worker: passivity, dependence, erratically shallow interests, short-term perspective, subordination, and lack of self-awareness. In contrast, the other end of the immaturity–maturity spectrum was the mature worker, with an active approach, independent mind, capable of a broad range of responses, deeper and stronger interests, a long-term perspective, and a high level of awareness and self-control. This model has been tested in a variety of industrial settings, showing that giving workers the opportunity to grow and mature on the job helps satisfy more than basic physiological needs and allows them to use more of their potential in accomplishing organizational goals.

In *The Motivation to Work* (1959), U.S. clinical psychologist Frederick Herzberg wrote of his motivation–hygiene theory. He developed this theory after extensive studies of engineers and accountants, examining what he called hygiene factors (i.e., administrative, supervisory, monetary, security, and status issues in work settings). His motivating factors, which he proved had a greater positive effect on job satisfaction, included achievement, recognition of accomplishment, challenging work, and increased responsibility with personal and collective growth development.

These human resources theories of management helped to change industrial approaches to motivation from "job enrichment" to a more fundamental and deliberate upgrading of responsibility, scope, and challenge of work, by letting workers develop their own ways of achieving objectives. Even when the theories were applied to apparently unskilled work, such as plant janitors, the workers were changed from an apathetic, poorly performing group into a cohesive, productive team, taking pride in their work and appearance. This approach gave members of the team the opportunity to meet their human self-actualization needs by taking greater responsibility for problem solving, and it resulted in less absenteeism, higher morale, and greater productivity with improved quality and productivity.

Rensis Likert, who along with McDougal and Herzberg helped to pioneer the "Human Relations School" in the 1960s, applied human resource theory to management systems and styles. He classified his theory into four different systems as follows:

System 1. Management has no confidence or trust in subordinates, and avoids involving them in decisions and goal setting, which are made from the top down. Management is task-oriented, highly structured, and authoritarian. Fear, punishment, threats, and occasional rewards are the principal methods of motivation. Worker–management interaction is based on fear and mistrust. Informal organizations within the system often develop that lead to passive resistance of management and are destructive to the goals of the formal organization.

System 2. Management has a condescending relationship with subordinates, with some degree of trust and confidence. Most decisions are centralized, but some decentralization is permitted. Reward and punishments are used for motivation. Informal organizations become more important in the overall structure.

System 3. Management places a greater degree of trust and confidence in subordinates, who are given a greater degree of decision making powers. Broad policy remains a centralized function.

System 4. Management is seen as having complete confidence in subordinates. Decision making is dispersed, and communication flows both upward and downward. Economic rewards are associated with achieving goals and improving methods. Relationships between management and subordinates are frequent and friendly, with a sense of teamwork and a high degree of mutual respect.

Case studies showed that a shift from management system 1 toward system 4 radically changed the performance of production, cut manufacturing costs, reduced staff turnover, and increased staff morale. Furthermore, workers and managers both shared a concern for the quality of the product or service and the competitiveness and success of their industry. The health industry includes highly trained professionals and paraprofessional workers who function as a team with a high degree of cohesion, mutual dependence, and autonomy, like a surgical or emergency room team.

NETWORK ORGANIZATION

The network, or task-oriented working group, is basically an organization of relationships rather than authority, sometimes called an adhocracy. This is a more organic form of organization, best suited to be effective for adaptation when the environment is complex and dynamic, when the workforce is largely professional, and when the technology and systems functions are in flux. Complexities and technological change require information, expertise, flexibility, and innovation, strengths best promoted in free exchange of ideas in a mutually stimulating environment.

In a network organization, leadership may be formal or informal, assigned to a particular function, which may be temporary, medium-term, or permanent, to achieve a single defined task or develop an intersectoral program. The task force is usually for short-term specific assignment; a working group, often for a medium-term project, such as integrating services of a region; and a committee for permanent tasks such as monitoring an immunization program.

Significant advantages of this form of organization are the challenge and the sharing of information and responsibility, which give professionals challenges, creativity, and job satisfaction by providing the opportunity to demonstrate their creativity. Members of the task force may each report within their own pyramidal structure, but as a group they work to achieve the assigned objective. They may also be interdisciplinary or interagency working groups to coordinate activities, review past work, or plan common future activities.

An ongoing network organization may be a government cabinet committee to coordinate government policy and the work of the various government departments, or a joint chiefs of staff to coordinate the various armed services. This approach is commonly used for task groups where interdisciplinary teams of professionals meet to coordinate functions of a department in a hospital, or where a multidisciplinary group of experts is established with the specified task of a technical nature.

Network organizational activity is part of the regular functions of a health professional. Informal networking is a day-to-day activity of a physician in consultations with colleagues and also a part of more formalized network groups. The hospital department must to a large extent function as a network organization with different professionals working as a team more effectively than in a strictly authoritarian pyramidal model. A ministry of health may need to develop a joint working group with the ministry of transport, the police, and those responsible for standards of automobiles in order to find ways to reduce road crash deaths and injuries. If a measles eradication project is envisioned, a multidisciplinary and multiorganizational team, or a network, would need to be established to plan and carry out the complex of tasks needed to achieve the target (Fig. 12.5).

Organizational theory and practice has incorporated the network model into pyramidal systems as part of total quality management (TQM), or the World Health Organization's continuous quality improvement. IN TQM, the factory floor

Chairperson/facilitator/coordinator

A	B	C	D
E	F	G	H
I	J	K	L

FIGURE 12.5   Network organization structure. In a health service organization, a task group to determine how to eradicate measles locally might be chaired by the deputy chief medical officer. Members might include the chief district nurse, an administrator budget officer, pharmacist, the chief of the pediatric department of the district hospital, a primary school administrator, a health educator, a medical association representative, the diretor of laboratories, the director of the supply department, a representative of the department of education, a representative of a voluntary organization interested in the topic and others as appropriate.

production unit works as a team, with quality control as part of the production process. In a health care program, this may be developed as multidisciplinary teams for self-evaluations. Every hospital ward needs to have its own multidisciplinary meetings to discuss patient care and functioning of the department in order to maintain and develop its professional services. Every local health department needs to maintain regular staff meetings to develop the communication and trust that are vital to meet its program objectives.

Most organizations have both the pyramid and the network structures. It is often difficult for a rigid pyramidal structure to deal with parallel bodies in a structured way, so that the network approach is necessary to establish working relations with outside bodies to achieve common goals. A network is a democratic functional grouping of those professionals and organizations needed to achieve a specific target, sometimes involving persons from many different organizations. The application of this concept is increasingly central in health care organization as multilevel health systems evolve in the form of managed care or district health systems. They are vertically integrated management systems involving highly professional teams and units whose interdependence for patient care and financial responsibility are central elements of the New Public Health.

TOTAL QUALITY MANAGEMENT

In the United States during World War II, W. Edwards Deming, a physicist and statistician, developed a system of economic and statistical methods of quality control in production industries. Following the war, Deming was invited to teach in Japan and moved from the university to the level of industrial management. Japanese industrialists adopted his principles of management and introduced quality management into all industries, with astonishingly successful results within a decade. The concept, later called total quality management (TQM), has since been adopted in U.S. industry.

BOX 12.1 STANDARD MANAGEMENT THEORY

1. Quality is expensive;
2. Inspection is the key to quality and control, and experts and inspectors can assure this;
3. Systems are designed by outside experts—no input is needed from workers;
4. Work standards, quotas, and targets can help productivity;
5. People may be hired when needed and laid off when not needed;
6. Rewards and punishments will lead to greater productivity and creativity;
7. Buy at the lowest cost;
8. Change suppliers frequently, based on price alone;
9. Profits are based on keeping costs down and revenue high;
10. Profit is the most important indicator of a company.

In the Deming approach to management of companies, quality comes first and is the key responsibility of management, not of the workers. If management sets the tone and involves the workers, quality goes up, costs come down, and both customer satisfaction and loyalty increase. This means enhancing the pride of the worker, listening to his or her ideas, and avoiding a punitive inspection approach. Removal of fear and building a sense of mutual participation and common interest is the responsibility of leadership. Training is one of the most important investments of the company. The difference between management theory and TQM are shown in Boxes 12.1 and 12.2.

BOX 12.2 TOTAL QUALITY MANAGEMENT

1. Quality leads to lower costs.
2. Inspection is too late—worker involvement in quality services eliminates defects;
3. Quality is made by management;
4. Most defects are caused by the system, not the worker;
5. Eliminate all work standards and quotas in industry;
6. Fear leads to disaster;
7. Make workers feel secure in their job;
8. Judgment, punishment, reward for above or below average performance destroys teamwork essential for quality production;
9. Work with suppliers to improve quality and costs;
10. Profits are generated by loyal customers—running a company by profit alone is like driving a car by looking in the rear view mirror.

The TQM approach integrates the scientific management and human relations approaches by giving workers credit for intellectual capacity and expects them to use it to analyze and improve the tasks they perform. Even more, this approach expects workers at all levels to contribute to better quality in the process of design, manufacture, and even marketing of the product or the service.

These ideas were revolutionary and successful when applied in business management in production industries. The TQM concept is much in discussion in the service industries. The World Health Organization has adapted TQM to a model called continuous quality improvement, with the stress on mutual responsibilities throughout a health system to quality of care. The application of TQM and continuous quality improvement approaches is discussed in Chapter 15, including the external regulatory and self-development TQM approaches.

CHANGING HUMAN BEHAVIOR

Human behavior is individual but takes place in a social context. Changing behavior is needed to reduce risk factors for disease. Change is threatening; it requires alteration, substitution, transformation, or modification of purposes, procedures, methods, or style. Implementation of plans usually requires some change, which often leads to resistance. The resistance to change may be professional, technical, psychological, political, emotional, or a mix of all of these. The manager of a health facility or service has to cope with change and gather the support of those involved to participate in creating or implementing the change effectively.

The behavior of the worker in a production or service industry is vital to the success of the organization. Equally important is the behavior of the purchaser or consumer of the product or service. Diagnosing organizational problems is an important skill to bring to leadership in health systems. Even more important is the ability to identify and alter the variables that require change and adaptation to improve performance of the organization. High expectations are essential to produce high performance and improved standards of service or productivity. Conversely, low expectations not only lead to low performance, but produce a downward spiraling effect. This applies not only within the organization, but to the individuals and community served, whether in terms of purchase of goods produced or in terms of health-related behavior.

People often resist change because of fear of the unknown. Participation in the process of defining problems, formulating objectives, and identifying alternatives are needed to bring about changes. Change in organizational performance is complex, and this is the test of leadership. Similarly, change at the individual level is essential to achieve the goals of the group, whether this is in terms of the functioning of a health care service unit, such as a hospital, or whether it is an individual's decision to change from smoking to nonsmoking status. The health of an individual and a population both depend on individual health team members' motivation and performance.

The behavior of the individual is important to his own and community health. Even small steps in the direction of a desirable change in behavior should be rewarded as soon as possible (i.e., reinforcing positive performance in increments). Behavior modification is based on the concept that change in behavior starts with the feelings and attitudes within the individual. Change involves a number of elements to define "where you are at":

1. Knowledge: what is the level of adequate health information?
2. Attitudes: what is the person's perception of that information?
3. Behavior of the individual: what does the individual actually do?
4. Behavior of the group: what are the social norms and acts?
5. Behavior of the organization: what does the health system do to change these factors?

Change in behavior is vital in the health field. The health belief model (Chapter 2) is widely influential in psychology and health promotion. The belief–intervention approach involves programs meant to reduce risk factors or a public health problem. It may require change in organizational behavior, with involvement and feedback to the people who determine policy, those who manage services, and the community being served.

EMPOWERMENT

In the 1980s, major industries in the United States were unable to compete successfully with the Japanese in the consumer electronics and automobile industries. Management theory began to place greater emphasis on empowerment as a management tool. The TQM approach stresses teamwork and involvement of the worker in order to achieve better quality of production. Comparatively, empowerment went further to involve the worker in operation, quality assessment, and even planning of the design and production process. Results in production industries were remarkable, with increased efficiency, less absenteeism, and greater searching for ideas to improve quality and efficiency of production, with the worker as a participant in the management and production process.

Empowerment entered the service industries with the same rationale, namely, improvement in quality and efficiency of service requires the active physical and emotional participation of the worker. Participation in decision making is the key to empowerment. This requires management to adopt new methods that allow the worker, whether professional or manual, to feel like an active participant. Successful application of the empowerment principle in health care extends to the patient, the family, and the community, emphasizing patient's rights to informed participation in decisions affecting their medical care, and the protection of privacy and dignity.

Diffusion of powers occurs when management of services is decentralized. Delegation of powers to professional, NGO, and advocacy organizations is part of empowerment in health care organization. Governmental powers to govern or promote areas such as licensure, accreditation, training, research, and service can be devolved to local authorities or NGOs by transfer of authority or funds. Organizational change may involve decentralization. Local institutional changes by amalgamation of hospital, long-term care, ambulatory care, and public health services are needed to produce a more efficient use of resources. Integration of services under community leadership and management involves the power to transfer funds within a district health network from institutional care to community-based care. Such changes involve a test of leadership skills and behavior change and involves health workers in policy and management.

STRATEGIC MANAGEMENT OF HEALTH SYSTEMS

Strategic management emphasizes the importance of positioning the organization in its environment in reaction to its mission, resources, consumers, and competitors. It requires development of a plan of action or implementation of a defined strategy to achieve the mission or goal of the organization. Definition of the missions and goals of the organization must take into account the external and internal environment, resources, and operational needs to implement and evaluate the adequacy of the outcomes. Strategy of the organization matches its internal approach with external factors, such as consumer attitudes and competing organizations. It is a set of methods and skills for the health care manager to attain the objectives of the organization, including

1. Providing high quality care;
2. Innovating to avoid obsolescence;
3. Developing good internal and external professional relationships;
4. Utilizing human resources efficiently;
5. Ensuring accountability to the environment;
6. Promoting the service to improve market share;
7. Managing financial resources efficiently;
8. Promoting adequate and efficient use of physical resources.

Policy is the formulation of objectives and priorities. Strategy refers to long-range plans to achieve specified objectives, indicating the problems to be expected and how to deal with them. Strategy does not identify all actions to be taken, but it includes evaluation of progress made to a stated goal. While the term has traditionally been used in a military context, it has become an essential concept in management, whether of industry, business, or health care. Tactics are the methods used to fulfill the strategy. This strategic management by objectives is applicable to the health system, incorporating definition of goals and targets, and the methods to achieve them (Box 12.3).

Change in health organization may involve a substantial alteration in size of or

---

BOX 12.3 THE STRATEGIC MANAGEMENT PROCESS

1. Policy and planning
   a. Defining of mission, goals, objectives
   b. Surveillance
   c. Analysis of external environment
   d. Analysis of internal environment
   e. Assessment of capabilities
   f. Evaluating strategic choices, short range
   g. Developing strategic planning, long range
   h. Guiding the implementation process
   i. Communicating policy direction
2. Implementation
   a. Motivation: communicate clearly the goals and plans of the organization
   b. Differentiate between short- and long-term goals
   c. Ensure that staff know their responsibilities
   d. Ensure provision of adequate resources
   e. Promote sense of staff involvement
   f. Modify structure to meet needs
   g. Delegate authority, assign responsibility
   h. Promote interdepartmental coordination and interpersonal relations
   i. Promote capacity to deal with change
   j. Review policies in keeping with progress toward goals
   k. Promote understanding of change and resistance to change
3. Monitoring
   a. Evaluation of effectiveness
   b. Evaluation of outcome, lessons learned
   c. Revise strategic plan
   d. Redeploy resources in keeping with lessons learned

---

relationships between existing, well-established facilities and programs (Table 12.2). A strategic plan for health reform in response to the need for cost containment, redefined health targets, or dissatisfaction with the status quo requires a model or vision for the future. This requires a strategic plan and a well-managed program. Opposition to change may occur for psychological, social, and economic reasons, or because of fear of loss of jobs or changes in assignments, salary, authority, benefits, or status. Downsizing in the hospital sector, with buildup of

TABLE 12.2   Transformation of Health Care Paradigms

Old paradigm	New paradigm
Emphasis on in-patient care	Emphasis on continuum of care
Emphasis on treating illness	Emphsis on maintaining, promoting wellness
Responsibility for the individual patient	Accountable for defined population
Specialists rewarded more than generalists	Greater economic parity between providers
Surgery rewarded more than medical services	Prevention rewarded versus surgery
Goal to fill beds	Provision of care at appropriate level of care
Separate organization, funding of hospital other services	Integrate health delivery system
Managers run an organization, department	Managers promote market share
Managers coordinate services	Managers promote inter sectoral cooperation

Source: Modified from Shortel and Kaluzny, 1994.

community health services, is one of the major issues in health reforms in many countries. This can be accomplished over time by attrition from retirement, or retraining and reassignment. All of this requires skilled leadership.

HEALTH SYSTEM ORGANIZATION MODELS

The New Public Health is an integration or coordination of many participating health care facilities and health promoting programs. It is evolving in various forms in different places as networks with administrative and fiscal interaction between participating elements. Each organization provides its own specific services or groups of services. How they function internally and how they interact functionally and financially is an important aspect of the management and outcomes of health systems. The health system functions as a network with formal and informal relationships; it may be very broad and loosely connected as in a highly decentralized system, with many lines of communication, payment, regulation, standards setting, and levels of authority.

The relationship and interchange between different health care providers has functional and economic elements. This may be best shown by example. A pregnant woman who is healthy and receives comprehensive prenatal care is less likely than a woman whose health is neglected to develop complications and require prolonged hospital care as a result of childbirth. The cost of good prenatal care is a fraction of the economic cost of treating the potential complications and damage to her health or to the newborn. A health system is responsible to ensure that reproductive-age women receive folic acid before becoming pregnant, that she has had family planning services so that the pregnancy is a desired one, that the space between pregnancies was adequate for her health and that of her baby, and that she has adequate prenatal care. An obstetrics department should be involved in assuring or providing the prenatal care, especially for high risk cases, and the delivery should be in hygienic and professional supervised settings.

Similarly for care of children and the elderly, there are a wide range of public health and personal care services that make up an adequate and cost-effective set of services and programs. The economic burden of caring for the sick child falls on the hospital. When there is a per capita grant to a district, then the hospital and the primary care service have a mutual interest in reducing morbidity, and hence mortality. This is the principle of the health maintenance organizations and district health systems discussed elsewhere. It is also a principle of the New Public Health.

Health care organizations differ according to size, complexity, ownership, affiliations, types of services, and location. Traditionally health care organization provides a single type of service, such as an acute care hospital providing episodic inpatient care, or a home health care agency. In present-day health reforms, health care organizations, such as an HMO or district health system, provide a population-based, comprehensive service program. Each organization must have or develop a structure suited to meet its defined goals, both in the internal and external environments. The common elements that each organization must deal with include governance of policy, production, or service, maintenance, financing, relating to the external environment, and adapting to changing conditions.

Functional Model

A functional model of an organization perhaps best suited to the smaller hospital is the division of labor into specific functional departments, for example, medical, nursing, finance, pharmacy, and housekeeping, each reporting through a single chain of command to the CEO (Fig. 12.6). The governing agency, which may be a local nonprofit board or a national health system, has overall legal responsibility for the operation and financial status of the hospital, as well as raising capi-

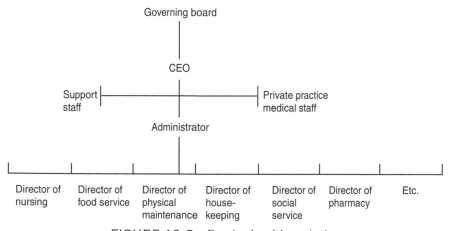

FIGURE 12.6   Functional model organization.

tal for improvements. The medical staff may be in private practice and work in the hospital with their own patients by application for this right as "attending physician," according to their qualifications, or the medical staff may be employed by the hospital much like any other staff. Salaried medical staff may include physicians in administration, pathology, anesthesia, and radiology, so that even in a private practice market system many medical staff are hospital employees.

This is the common arrangement in North American hospitals. The governing board of a "voluntary," nongovernmental, not-for-profit organization with municipal and community representatives may be appointed by a sponsoring religious, municipal, or fraternal organization.

Corporate Model

The corporate model in health care organization (Fig. 12.7) is often used in larger hospitals or where mergers with other hospitals or health facilities are taking place. This requires the CEO to delegate responsibility to other members of the senior management team who have operational responsibility for major sectors of the hospital's functioning.

A variation of the corporate model is the divisional model of a health care organization based on the individual service divisions allowing middle management a high degree of autonomy (Fig. 12.8). There is often departmental budgeting for each service, which operates as an economic unit, that is, balancing income and expenditures. Each division is responsible for its own performance, with powers of strategic and operational decision making authority. This model is used widely in private corporations, such as at General Motors in the United States. With increasing complexity of services, this model is also employed in corporate health systems in the United States, with regional divisions.

Matrix Model

The matrix model of a health organization is based on a combination of pyramidal and network organization (Fig. 12.9). This model is suited to a public health

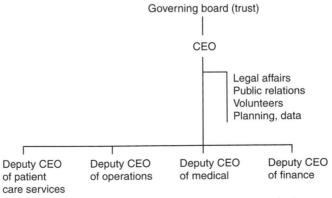

FIGURE 12.7   Corporate model of health care organization.

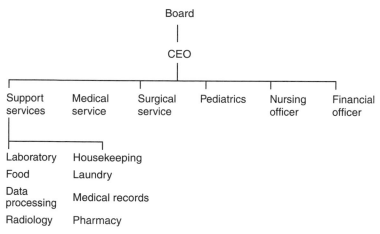

FIGURE 12.8 Divisional model of health care organization.

department in a state, county, or city. Individual staff persons report in the pyramidal chain of command, but also function in multidisciplinary teams to work on specific programs or projects. A nutritionist in the geriatric department is responsible to the chief of nutrition services but is functionally a member of the team on the geriatric unit. In a laterally integrated health maintenance organization or district health system, specialized staff may serve in both institutional (i.e., hospital) and community health roles.

The organizational structure appropriate to one set of circumstances may not be suitable for all. Even if the payment system is a regional or district health sys-

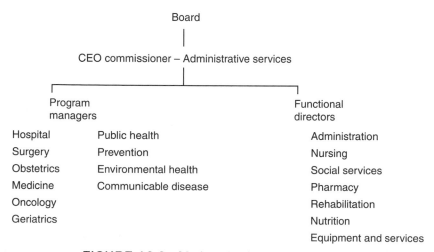

FIGURE 12.9 Matrix model of health care organization.

tem structure, the internal operation of a hospital will require a model of organization appropriate to it. Hospitals need to modify their organizational structure as they evolve, and as the economics of health care changes.

SKILLS FOR MANAGEMENT

Leadership in an organization requires an ability to define the goals or mission of the organization and to develop a strategy and define steps needed to achieve these goals. It requires an ability to motivate and engender enthusiasm for this vision by working with others to gain their ideas, their support, and their participation in the effort. In health care as in other organizations, it is easier to formulate plans than to implement them. Change requires not only the ability to formulate the concept of change, but also to modify the organizational structure, the budgeted resources, the operational policies, and, perhaps most importantly, the corporate culture of the organization.

Management involves skills that are not automatically part of a health professional's training. Skilled clinicians often move into positions requiring management skills in order to build an develop the health care infrastructure. In some countries, hospital managers must be physicians, often the senior surgeon. Clinical capacity does not transfer automatically into management skills to deal with manpower, budgets, and resources. Therefore training in management is vital for the health professional.

The manager needs the training for investigations and fact finding and the ability evaluate people, programs, and issues, and set priorities for dealing with the short- and long-term issues. Negotiating with staff and outside agencies are constant activities of the manager, ranging from the trivial to major decisions with wide implications. Perhaps the most crucial skill of the manager is communication: the ability to convey verbal, written, or unwritten messages that are received and understood and to assess the responses as an equal part of the exchange.

Interpersonal skills are a part of management practice. The capable manager can relate to personnel at all levels in an open and equal manner. This is essential to help foster a sense of pride and involvement of all personnel in working toward the same goals and objectives, and to show each member of the team that they are important to meeting the objectives of the organization. At the same time, the manager needs to communicate information, especially as to how the organization is doing in achieving its objectives.

THE CHIEF EXECUTIVE OFFICER
OF HEALTH ORGANIZATIONS

Hospital directors in the past were often senior physicians, often called the superintendent, without training in health management. The business manager CEO has become common in hospital management in the United States during the 1950s when the CEO was called administrator. He or she worked under the direction of a board of trustees who raised funds, set policies, and were often involved in internal administration.

Where the CEO was a nonphysician, the usual case in Northern American hospitals, often a conflict existed with the clinical staff of the hospital. In some settings, this led to appointment of a parallel structure with a full-time chief of medical staff. In European hospitals, the CEO is usually a physician, often by law, and the integration of the management function with the role of clinical chief is the prevalent model.

Over time the CEO role has changed to one of a coordinator as the cost and complexity of the health system has increased. The CEO is now more involved in external relations and less in the day-to-day operation of the facility. The CEO is a partner but *primum inter pares,* or first among equals, in a management team that shares information and works to define objectives and solve problems. This deemphasizes the authoritarian role and places stress on the integrative function.

The CEO is responsible for the financial management of operational and capital budgets of the facility, which is integral to the planning and future development of the facility. Budgets include four main factors: income, fixed or regular overhead, variable or unpredictable overhead, and capital costs, all essential to the survival and development of the organization.

The key role of top management is to develop a vision, goals, and targets for the institution, to maintain an atmosphere and systems to promote the quality of care, financial solidity, and to represent the institution to the public. The overall responsibility for the function and well-being of the program is with the CEO and the community board.

COMMUNITY PARTICIPATION

Community participation in management of health facilities has a long-standing and constructive tradition. The traditional hospital board has served as a mechanism for community participation and leadership in promoting health facility development and management at the community level. The role of hospital boards evolved from primarily a philanthropic and fund-raising one to a greater overall policy and planning function working closely with management and senior professional staff, as operational costs increased rapidly, as government insurance schemes, were implemented, and as court decisions defined the liability of hospitals and reinforced the broadened role of governing boards in malpractice cases and quality assurance. Centrally developed health systems such as the United Kingdom's National Health Services, have promoted district and county health systems with high degrees of community participation and management, both at the district level and for individual services or facilities (e.g., general practice and hospitals).

INTEGRATION—LATERAL AND VERTICAL

Rationalization of health facilities increasingly means organizational linkages between previously independent facilities. Mergers of health facilities are common events in many health systems. In the United States, mergers between hospitals, or between facilities linked to HMOs or managed care systems, are frequent events. Health reform in many countries is based on similar linkages.

Lateral integration is the term used for amalgamation among similar facilities. This is like a chain of hotels, and in health care involves two or more hospitals, usually meant to achieve cost-savings, improve financing and cost efficiency, and to reduce duplication of services. Urban hospitals often respond to market competition by purchasing other hospitals. This is often easier for hospital-oriented CEOs and staff to comprehend and manage, but it avoids the issues of downsizing and integration with community-based services.

Vertical integration describes organizational linkages between different kinds of health care facilities to form integrated, comprehensive health service networks. This permits a shift of emphasis and resources from inpatient care to long-term, home, and ambulatory care. This is the managed care or district health system model. Community interest is a factor in promoting change to integrate services, which can be a major change for the management culture, especially of the hospital.

The survival of a health care facility may depend on integration with appropriate changes in concepts of management. In the 1990s, a large majority of California residents moved to managed care programs because of the high cost of fee-for-service indemnity health insurance and because of federal waivers to promote managed care for Medicare and Medicaid beneficiaries. Independent community hospitals without a strong connection with managed care programs were in danger of losing their financial base. Occupancy rates in community hospitals in the United States fell from 75% in 1975 to 62% in 1996. Major medical centers responded with strategic plans to purchase community hospitals and develop affiliated medical groups and contract relationships with managed care organizations to strengthen their "market share" service population base for the future.

Vertical integration is important not only in urban areas, but can serve as a basis for developing rural health care in both developed and developing countries. The district hospital and primary care center operating as an integrated program can provide a high quality program. Hospital-centered health care common in industrialized countries channels most of the available funds for health to the hospital, preventing adequate funding to primary health care. To prevent this, developing countries can adopt the primary care approach, which is now the basis of managed care systems, including trying to limit hospital care by improving ambulatory and preventive care services.

NORMS AND PERFORMANCE INDICATORS

Norms are useful to promote efficient use of resources and promote high standards of care, if based on empirical standards proved by experience, trial and error, and scientific observation. Norms may be needed, even with insufficient evidence, but should be tested in the reality of observation, experience, and experiment. This requires data and trained observers free to examine, report, and publish findings for open discussion among colleagues and peers in proceedings open to the media and the general public.

Normative standards of planning are the determination of a number per unit of population that is deemed to be suitable for population needs, for example, the number of beds or doctors per 1000 population, or length of stay in hospital. Many organizations based on the bureaucratic model used norms of resources as the basis for planning and financing services (Chapter 11). This led to payment systems which encouraged greater use of that resource. If a factory is paid by the number of workers and not the number and quality of the cars produced, then management will have no incentive to introduce efficiency or quality improvement measures. If a district or a hospital is paid by the number of beds, or by days of care in the hospital, there is no incentive to introduce alternative services such as outpatient surgery and home care.

Performance indicators are measures of completion of specific functions of preventive care such as immunization, mammography, Pap smears, and diabetes and hypertension screening. They are indirectly measures of economy, efficiency, and effectiveness of a service and are being adopted as better methods of monitoring a service and paying for it, such as by paying a premium. General practitioners in the United Kingdom receive additional payments for full immunization coverage of the children registered in their practices. A block grant or per capita sum may be tied to indicators reflecting good standards of care or prevention, such as low infant, child, and maternal mortality. Incentive payments to hospitals can promote ambulatory services as alternatives to admissions and reduce lengths of stay. Limitations of financial resources in the industrialized countries and even more so in the developing countries makes the use of appropriate performance indicators of great importance in the management of resources.

NEW ORGANIZATIONAL MODELS

New models of health care organization are emerging and developing rapidly in many countries. Partly this is in a search for more economical methods of delivering health care and is also the result of the target-oriented approach to health planning that seeks the best way to achieve defined health objectives. The developed countries seek ways to restrain cost increases, and the developing countries seek effective ways to quickly and inexpensively raise health standards for their populations. The new organizational models that try to meet these objectives include district health systems and managed care systems, described in greater detail in Chapter 11.

MANAGEMENT AND THE NEW PUBLIC HEALTH

Management in health is for the purpose of improving health and not merely the maintenance of an institution. Separate management of a variety of health facilities serving a community has derived from different historical development and funding systems. The New Public Health looks at all services as part of a network of interdependent services each contributing to health needs, whether in hospital care or in enforcing public health law on food producers.

Separate management and budgeting of this complex results in a disproportion of funds, staff, and attention to traditional services such as hospitals and fails to redirect resources to more cost-effective and patient-sensitive kinds of services, such as home and preventive care. The effects of incentives and disincentives built into financing systems are central issues in determining how management approaches problem solving and program planning, and are therefore important considerations in promoting health.

The management approach to resolving this dilemma is professional vision and leadership to promote the broader New Public Health approach. The manager that persists in the outdated traditional emphasis, regardless of the larger picture, may find his hospital obsolescent and noncompetitive in a new climate where economic incentives promote downsizing institutions and upgrading health promotion. Defensive, internalized management will become obsolescent, and aggressively outward looking management will be the pioneers of the New Public Health.

SUMMARY

Health care is one of the largest and most important industries in any country, consuming anywhere from 3 to 14% of GNP. It is a service, not a production industry, and it is vital to the health and well-being of the individual, the population, and the economy. Because it employs large numbers of skilled professionals and many unskilled persons, it is often vital to the economic survival of small communities, as well as for a sense of community well-being.

Management includes planning, leading, controlling, organizing, motivating, and decision making. It is the application of resources and personnel toward achieving targets. Therefore, it involves the study of the use of resources, and the motivation and function of the people involved (the producer or provider and the customer, client, or patient). This cannot take place in a vacuum, but is based on the continuous monitoring of information and its communication to all parties involved. These functions are applicable at all levels of management, from policy to operational management of a production or a service system. Creative management of health systems is vital to the functioning of the system at the macro level, as well as in the individual department or service. This implies effective use of resources to achieve objectives, and community, provider, and consumer satisfaction. These are formidable challenges, not only when money is available in abundance, but even more so when resources are limited and difficult choices need to be made.

The scope of the New Public Health is broad. It includes the traditional public health programs, but equally must concern itself with managing and planning comprehensive service systems and measuring their function. Selection of targets and priorities is often determined by the feasible rather than the ideal. The health manager, either at the macro(health) level, or managing a clinic, needs to focus on real possibilities of achievements with current and appropriate methods. Good management means designing objectives based on a balance between the feasible and the desirable. This may be summarized in the saying, "the enemy of the good is the perfect." The New Public Health is not only a concept; it is a management approach to improve the health of individuals and the population.

RECOMMENDED READINGS

Kirsch, T. 1988. Local area monitoring (LAM). *World Health Statistics Quarterly,* 41:19–25.

McMahon, R., Barton, E., Piot, M., Gelina, N., Ross, F. 1992. *On Being in Charge: A Guide to Management in Primary Health Care,* Second Edition. Geneva: World Health Organization.

Scott, R. W. 1992. *Organizations: Rational, Natural and Open Systems,* Third Edition. Englewood Cliffs, NJ: Prentice-Hall.

Woodall, J. P. 1988. Epidemiological approaches to health planning, management and evaluation. *World Health Statistics Quarterly,* 41:2–10.

BIBLIOGRAPHY

Aday, L. A., Begley, C. E., Lairson, D. R., Slater, C. H. 1993. *Evaluating the Medical Care System: Effectiveness, Efficiency, and Equity.* Ann Arbor, MI: Health Administration Press.

Adler, P. S. 1993. Time-and-motion regained. *Harvard Business Review,* January–February: 97–108.

Aguayo, R. 1990. *Dr. Deming: The American Who Taught the Japanese About Quality.* New York: Simon & Schuster.

Cohen, J. 1990. Health policy, management and economics. *In* Lambo, T., Day, S. B. (eds). *Issues in Contemporary International Health.* New York: Plenum.

Darr, K., Rakich, J. S. 1989. *Hospital Organization and Management,* Fourth Edition. Owings Mills, MD: National Health Publishing.

Donabedian, A. 1976. *Aspects of Medical Care Administration: Specifying Requirements for Health Care.* Boston: Harvard University Press.

Duncan, W. J. P., Ginter, P. M., Swayne, L. E. 1995. *Strategic Management of Health Care Organizations.* Cambridge, Massachusetts: Blackwell Business.

Dutton, D. B. 1979. Patterns of ambulatory health care in five different delivery systems. *Medical Care,* 17:221–243.

Ellencweig, A. Y. 1992. *Analysing Health Systems: A Modular Approach.* Oxford: Oxford University Press.

Ezioni, A. 1964. *Modern Organizations.* Englewood Cliffs, NJ: Prentice Hall.

Glynn, J. J., Perkins, D. A. (eds). 1995. *Managing Health Care: Challenges for the 90s.* London: W. B. Saunders.

Griffith, J. R. 1993. *The Moral Challenges of Health Care Management.* Ann Arbor, MI: Health Administration Press.

Katzenbach, J. R., Smith, D. K. 1993. *The Wisdom of Teams: Creating the High-Performance Organization.* Cambridge, MA: Harvard Business School Press.

Maslow, A. 1954. *Motivation and Personality.* New York: Harper.

May, E. 1945. *The Social Problems of an Industrial Civilization.* Boston: Graduate School of Business Administration, Harvard University.

Mintzberg, H. 1994. The fall and rise of strategic planning. *Harvard Business Review,* January–February:107–114.

Mooney, G. 1987. What does equity in health mean? *World Health Statistics Quarterly,* 40:296–303.

Reinke, W. A. (ed). 1988. *Health Planning for Effective Management.* New York: Oxford University Press.

Robbins, S. P. 1990. *Organization Theory: Structure, Design and Applications,* Third Edition. Englewood Cliffs, NJ: Prentice-Hall.

Shortel, S. M., Kazluny, A. D. (eds.). 1994. *Health Care Management: Organization Design and Behavior,* Third Edition. Albany, NY: Delmar Publishers.

Sloan, M. D., Chmel, M. 1991. *The Quality Revolution and Health Care: A Primer for Purchasers and Providers.* Milwaukee, WI: ASQC Quality Press.

Tarimo, E. 1991. *Towards a Health District: Organizing and Managing District Health Systems Based on Primary Health Care.* Geneva: World Health Organization.

Taylor, C. E. 1992. Surveillance for equity in primary care: Policy implications from international experience. *International Journal of Epidemiology,* 21:1043–1049.

Taylor, F. W. 1947. Scientific Management. New York: Harper & Brothers, reprint.

Vaughn, J. P., Morrow, R. H. 1989. *Manual of Epidemiology for District Health Management.* Geneva: World Health Organization.

World Health Organization. 1994. *Information Support for New Public Health Action at the District Level.* Report of a WHO Expert Committee. Technical Support Series 845. Geneva: World Health Organization.

13

NATIONAL HEALTH
SYSTEMS

INTRODUCTION

Assuring access to health care for all is a basic principle of the New Public Health. Medical care is important, not only to the individual, but also to the population, as it has at its disposal effective methods of preventing many diseases and stopping the progress of others. Despite its value, medical care, by itself, is not sufficient to produce a high standard of health for the population. A person may have access to medical care, but personal or community risk factors produce ill health which medical care can only attempt to alleviate. Availability and access to care must be seen in the context of the activities of the individual and of society to prevent disease and to promote health.

Most industrialized countries have implemented national health programs as health insurance systems or national health services. Each was developed in the political context of the country and continues to evolve with its own and international experience. Developing countries are also struggling to achieve universal access to care and Health for All by expanding primary health care and social security plans providing benefits to workers. As they move up the scale of economic development, developing countries also address the problem of how to achieve equity in access to health care as well as now to expand the funding basis for health care through national health insurance.

Each national health system has its own characteristics and problems. Their management requires continuous evaluation based on well-developed information systems trained health management personnel, and involvement of society through professional organizations and advocacy groups. There is no single defined "gold standard" plan to provide universal access to health care that is suitable for all countries. Each country develops and modifies a program of national health appropriate to its own cultural needs and resources available. But there are evolving sets of patterns in health care, so that countries can and do learn from one another.

Barriers to care can be geographical, cultural, social, and psychological as well as financial. Removing financial barriers to care is necessary but not sufficient to optimal health. Equity in access to quality health care, with minimal interregional and sociodemographic differences, is vital to good public health standards. The services needed are not only those that are available on patient demand, but include those services that reach out to the entire population, especially to people at high risk who are often least able to seek appropriate care.

A program that provides equal access for all may not achieve its objective of better health for the population unless it is accompanied by governmental enforcement of environmental and occupational health laws, food, nutrition, and water standards, improved rural care, higher educational levels, and provision of health information to the public. Additional national programs are needed to promote health generally and to reduce specific risk factors for morbidity and mortality. Responsibility for health lies not only on medical and other health professionals, but also on society and its executive instruments, governmental and voluntary organizations, as well as on the individual, the family, and peer groups.

Individual access to an essential basket of services as a prepaid benefit (i.e., insurance) is integral to a successful national health program. Each country addresses this issue according to its means and traditions, but the benefits should be examined as to which are most effective at least cost to meet epidemiologic and demographic needs. Payments for heart transplantation may be beyond the means of a health system, but early and aggressive management of acute myocardial infarction is effective in saving lives at modest cost. Improved diets, smoking reduction, and physical fitness are even more effective and less costly. Prevention is cost-effective and should be integral to the development of service priorities within the basket of services.

In this chapter, national health systems are presented representing major models of organization, which may be influential in health care system formulation in both developing and developed countries, and in countries restructuring their health services. Health care systems and financing are under pressure everywhere, not only to assure access to health for all citizens, but also to keep up with advancing medical technology, and contain costs. Because a health system is judged by more than its costs and measures, included in this chapter are some indicators of health status of the population, including morbidity and mortality. This topic has developed a complex terminology of its own, and some of the key words are defined in the text and in the glossary.

HEALTH SYSTEMS IN DEVELOPED COUNTRIES

Evolution of Health Systems

The tradition of prepayment of health care goes back to ancient times, when municipal doctors, employed by local authorities provided care for the poor and the slaves. In the Middle Ages, the Church provided charitable care for the poor.

In the medieval and Renaissance periods, guilds provided prepaid health benefits to their members. These later evolved into the friendly societies (benevolent societies), mutual benefit programs that provided for burials, pensions, and health benefits for members. In the twentieth century, these evolved into union sponsored and private health insurance plans.

Social Insurance. Otto von Bismarck, Chancellor of Germany, introduced the first national health insurance plan for workers. It followed previous legislation in Germany establishing workmen's compensation on railroads (1838) and compulsory miners' benevolent societies (1854). Workmen's compensation and other benefits were extended in 1871 to many workers in other industries, such as those in domestic service, workers in mines, factories, and quarries, and seamen. Bismarck's 1883 legislation was intended to improve the health of workers and their families, and especially of potential army recruits, as well as to stave off the political advancement of the socialist parties. The program was based on the principle of social insurance, involving payroll deductions at the place of work, with contributions from the employer and employee, to cover medical care, unemployment benefits, and pensions for workers.

The Bismarckian model of national health insurance established sick funds (*Krankenkassen*) as insurers to provide payment to the physician, hospital, or other provider. In the years before World War I, many countries in central and eastern Europe implemented similar health plans. In the period between the World Wars, national health insurance programs were developed in many countries in the industrialized world. In Europe, most countries developed models based on the Bismarckian approach, with financing by worker and employer contributions to social security preserving private medical practice with fee-for-service payment through sick funds. The Bismarckian model is used widely in Europe and Israel. This model has also influenced post-Soviet health reforms in eastern Europe.

In 1911, the government of Great Britain, initiated by David Lloyd-George, Chancellor of the Exchequer, introduced national health insurance for workers and their families. General practitioners were paid on a capitation basis rather than on a salary, preserving their status as self-employed professionals. Initially this plan covered one-third of the population, but coverage increased to one-half by 1940. Administration was through approved mutual benefit societies (the friendly societies), and sometimes based on insurance companies and trade unions. European countries and Japan gradually developed compulsory health insurance following World War I, and completed universal coverage following World War II.

The Social Security model of health insurance for urban workers also became prominent in many countries in Latin America. Social security plans are financed by mandatory contributions of workers and employers, and they are administered by the state. The Social Security Act of 1935 in the United States was instituted in the depths of the Great Depression to provide benefits for widows, orphans, and the handicapped, as well as pensions for the elderly, and provided a base for future reform including health insurance. Since 1965, this legislation has provided the basis for U.S. medical and hospital coverage of the elderly under Medicare and the poor under Medicaid. The 1995 proposal for national health insurance in the United States was also based on a Social Security funding system.

National Health Service.   In some countries, the state directly assumed the responsibilities for both social security and health care. The welfare state took on measures such as unemployment and disability insurance, special disability benefits for the blind, widows, orphans, and the elderly through pensions, and child benefits to raise levels of child care and nutrition through general governmental revenues from taxation and other sources.

In 1918, following the Russian revolution, the new Soviet Union (U.S.S.R.) introduced its national health plan for universal coverage within a state-run system of health protection. The Soviet model, designed and implemented by Nikolai Semashko, provided free health care for all as a government financed and organized service. Its first objective was to provide free service for the total population, with a balance of primary and secondary care and universal and equitable access to care through district organization of services. It achieved control of epidemic and endemic infectious diseases and expanded services into the most remote areas of the country.

In the early days of World War II, the British government established a national Emergency Medical Service to operate hospitals in preparation for the large

---

BOX 13.1 OBJECTIVES OF NATIONAL HEALTH SYSTEMS

1. Universal access to a broad range of health services;
2. Promotion of national health goals;
3. Improvement in health status indicators;
4. Equity in regional and sociodemographic accessibility and quality of care;
5. Adequacy of financing with cost containment and efficient use of resources;
6. Consumer satisfaction and choice of primary care provider;
7. Provider satisfaction and choice of referral services;
8. Portability of benefits when changing employer or residence;
9. Public administration or regulation;
10. Promotion of high quality of service;
11. Comprehensive in primary, secondary, and tertiary levels of care;
12. Well-developed information and monitoring systems;
13. Continuing policy and management review;
14. Promotion of standards of professional education, training, research;
15. Governmental and private provision of services;
16. Decentralized management and community participation.

scale civilian casualties expected. The plan established national health planning and rescued many hospitals from near bankruptcy resulting from the Great Depression. During World War II, a postwar social reconstruction program was developed by William Beveridge, at the behest of the wartime government of Winston Churchill. The Beveridge Report of 1942, Social Insurance, and Allied Services, outlined the nature of the future welfare state including a national health service, placing medical care in the context of general social policy.

The wartime government coalition approved the principle of a national health service, which had wide public support, despite opposition from the medical association. In 1948, the Labour government of Clement Attlee implemented the National Health Service (NHS), a nationally financed, universal coverage system providing free care by general practitioners, specialists, hospitals, and public health services.

National Health Insurance. The Canadian system of tax based, national health insurance evolved from provincial initiatives led by Tommy Douglas, premier of the province of Saskatchewan. Initiated in 1946, provincial plans provided insured hospital services, later followed by medical and other services. Developed over the period 1946–1971, they were promoted by federal governmental cost-sharing, political support, and regulation. These plans were initially financed by taxation and premiums, but later by general tax revenues alone. The Canadian "Medicare" plans are publicly administered by the provinces with federal standards, cost-sharing, and comprehensive coverage. Care is provided by private medical practitioners on a fee-for-service basis under negotiated medical fee schedules. Hospitals are operated by nonprofit voluntary or municipal authorities, with payment by block budgets. This Medicare type of plan was later adopted in a number of countries including Australia.

THE UNITED STATES

In 1997, the population of the United States was 271.6 million, with a GNP per capita of $28,020 (1996) and a child mortality of 8 per 1000 live births (thirtieth place in world ranking). The United States has a federal system of government with 50 states each having its own elected government with legislative, judicial, enforcement, and taxing powers. The U.S. Constitution gives primary responsibility for health and welfare to the states, but direct federal services are provided to the armed forces, veterans, and Native Americans. However, the federal government has established a major leadership role in health by development of national standards, regulatory powers, and information systems. It also serves as a major financing agency for research, health services, and training programs.

Federal Health Initiatives

In 1798 the federal government undertook responsibility for health of seamen by establishing the marine hospitals (later the Public Health Service). In the late nineteenth and early twentieth centuries, the federal Department of Agriculture Extension Service promoted nutrition and hygiene education throughout the rural areas of the country. Later legislation provided federal grants to establish state, municipal, and county health departments. Health hazards in poor standards of food and drugs, nursing homes, lack of care for the elderly and the poor, dangerous automobiles, environmental pollution, and deficiencies in health services led to governmental intervention to protect the public interest. The Food and Drug Control Act of 1906 was promulgated to regulate and control commerce. In 1921, the Sheppard-Towner Act established the federal Children's Bureau that administered grants to assist states to operate maternal and child health programs, which were later incorporated into the Social Security Act.

In 1927, the Committee on the Costs of Medical Care, a commission funded by several private foundations, recommended that the United States implement a national health program in which medical care would be provided by group practices with voluntary prepayment to promote efficient health care for all. From the 1920s, labor unions won health insurance benefits through collective bargaining. However, these initiatives were slowed due to the Great Depression from 1929 to 1939. The Social Security Act of 1935 increased social support for millions with handicaps or occupational injuries, widows and orphans, and the elderly, thus alleviating some of the worst effects of the depression.

During World War II (1941–1945) millions of Americans in the armed forces and their dependents, previously with limited access to prepaid health care, were enrolled in a national plan for free health care (Emergency Maternity and Infant Care for the Wives and Children of Servicemen). At the same time, health benefits through voluntary insurance for workers were vastly expanded in place of wage increases, which were forbidden by federal wartime regulation. At the end of the war, millions of veterans were eligible for health care through the Veterans Administration (VA), which established a national network of federal hospitals for this purpose.

In 1946, an attempt to bring in national health insurance legislation (the Wagner-Murray-Dingell Bill) failed in the U.S. Congress. One section of the bill was approved, enabling the federal government to initiate a program of categorical grants to upgrade hospital facilities around the country under the Hill-Burton Act. Massive federal funding for health was also provided through the National Institutes of Health (NIH), established after World War II, to promote research and strengthen public and private medical schools, teaching hospitals, and research facilities. In the 1950s, the federal government also established the Centers for Disease Control (CDC) as well as increased public health grants providing assistance in state and local public health activities.

From the 1940s through the 1960s, voluntary health insurance became the major method of prepayment for health care needs, mostly through employment contracts. The private insurance industry developed rapidly, with minimal governmental regulation to ensure fair pricing and payment. During the 1970s and 1980s, employers became concerned about the costs of health insurance for their workers and pressed the government to act to restrain health care costs. Federal initiatives included public insurance for the elderly and poor, promoting efficiency in payment for hospital care, and later promotion of health maintenance organizations (HMOs) and managed care.

Medicare and Medicaid

In the mid-1960s, despite the growth of voluntary and employment-based health insurance, a large percentage of the elderly and poor American population lacked health insurance. In 1965, President Lyndon B. Johnson introduced Medicare for the aged (i.e., over age 65), disabled persons, and persons on renal dialysis as Title XVII of the 1935 Social Security Act. This brought some 10% of the population a limited form of national health insurance. Medicaid, Title XIX of the Social Security Act also enacted in 1965, provided federal sharing for acceptable state health plans for the poor, with local authority participation. These two plans together brought some 25% of Americans into public systems of health insurance. Limitations included variable definitions of poverty in each state, and copayments for Medicare beneficiaries.

Public funding for health care in the United States includes Medicare and Medicaid, research and medical education, and promotion of community health centers and services in impoverished or underserved areas (see Table 13.1). The percentage of public funding in the United States rose from under 25% in 1960 to over 46% of total health expenditures in 1995–1998. In 1996, 71% of the U.S. population under age 65 was covered under private insurance, mostly employment based insurance, 11% under Medicaid, and 2% other public coverage, while 16% were uninsured. The population enrolled in Medicare increased from 19 million in 1966 to 38.4 million in 1997, including 4.8 million disabled persons under age 65. The average number of persons served per 1000 enrollees increased from 570 in 1977 to 826 in 1995, with payments per person served increasing from $1,332 to $5,074 in those years. The Medicaid enrolled population increased from 28.2 million in 1991 to 32 million in 1997, with the proportion enrolled in managed care programs increasing from 9.5% in 1991 to 47.8% in 1997.

Medicare and Medicaid brought many previously noninsured persons under health insurance coverage. This contributed to increasing health expenditures in the public sector, a concern for both critics and supporters of public health care programs. Groups arguing against universal health coverage point to cost increases, and the 1997–1999 conservative majority in Congress talks of cutting back on Medicare benefits. Supporters of universal access call for cost reduction by restricting medical payments and excessive hospital utilization. All recognize that unrestrained cost increases will harm health care in the country. The rise in administrative costs also generates debate. With the large percentage of uninsured in the population (18%), and another 15% underinsured, the political pressure for some form of national health insurance will continue into the new century.

TABLE 13.1   Health Expenditures, Total, Public, and by Types of Service, United States, 1960–1998

Expenditures	1960	1970	1980	1985	1990	1995	1998
Total (billion $)	26.9	73.2	247.2	428.2	697.5	988.5	1,146.8
Per capita expenditures ($)	143	346	1,063	1,711	2,688	3,621	na
Expenditures as % of GDP	5.3	7.4	9.2	10.5	12.1	13.6	13.7
Public expenditures as % of total	24.5	37.2	42.0	41.3	42.2	46.2	46.2
Research and construction (%)	6.3	7.3	4.7	3.8	3.5	3.1	2.9
Total health services (%)	93.7	92.7	95.3	96.2	96.5	96.9	97.1
% Hospitals	34.5	38.2	41.5	39.3	36.8	35.4	33.4
% Nursing homes	3.2	5.8	7.1	7.2	7.3	7.9	7.6
% Home health care	0.2	0.3	1.0	1.3	1.9	2.9	2.9
% Subtotal	37.9	44.5	49.6	47.8	46.0	46.2	43.9
% Physicians	19.7	18.5	18.3	19.5	21.0	20.4	19.3
% Dentists	7.3	6.4	5.4	5.1	4.5	4.6	4.7
% Other professionals	2.3	1.9	2.6	3.9	5.0	5.3	5.8
% Drugs, medical supplies	15.8	12.0	8.7	8.7	8.6	8.4	9.3
% Others, vision, etc.	6.0	4.0	3.6	3.0	3.1	3.9	4.0
% Administration and health insurance net cost	4.3	3.7	4.8	5.6	5.5	4.9	6.5
% Government public health	1.4	1.8	2.7	2.7	2.8	3.2	3.6

Source: *Health, United States,* 1996–1997; 1998 data from Smith *et al.,* 1999.

The Changing Health Care Environment

From the 1960s through the 1990s, rapid cost increases were attributed to many factors: an increasing elderly population; high levels of morbidity in the poor population; the spread of AIDS; rapid innovation and costly medical technology; specialization, high laboratory costs; large scale public investment in medical education and research and health facility construction; high levels of preventable hospitalizations; an institutional orientation of the health system; high administrative costs due to multiple private billing agencies in the private insurance industry; high incomes for physicians, especially for specialists; and high medical malpractice insurance costs. The pressure for cost constraint came from government, industry, and the private insurance industry.

Most hospitals are owned and operated by nonprofit agencies, including federal, state, and local governments, voluntary organizations and religious organizations. Privately owned hospitals operating for profit increased from 7.8% of community, short-term hospital beds in 1975 to 12.7% in 1996. Private medical practice, with payment by fee-for-service, was the major form of medical care until the 1990s. HMOs and other forms of managed care have grown rapidly to become the predominant method of organizing health care in the United States.

Prepaid group practice (PGP) originated from company-provided contract medical care, especially in remote mining camps. The Community Hospital of Elk City, Oklahoma, established in 1929, is considered the first real medical cooperative or prepaid group practice. Later, many rural cooperatives were formed to provide prepaid medical care. Union-sponsored health services were developed to provide medical care in poor mining areas in the Appalachian Mountains, as well as in an urban cooperative in Washington, D.C., in 1937. In the 1940s, New York City sponsored the Health Insurance Plan of Greater New York to provide prepaid medical care for residents of urban renewal and low-income housing areas. This was later supported by organized union groups such as municipal employees and garment industry workers.

Prepaid group practice became best known in the Kaiser Permanente network developed for workers of Henry J. Kaiser Industries, at the Boulder Dam and Grand Coulee Dam construction sites in the 1930s. This experience was applied in Kaiser's rapidly growing industries in the San Francisco Bay area. Kaiser Permanente health plans now provide care for millions of Americans in many other states. Initially opposed by the organized medical profession and the private insurance industry, PGP gained acceptance by providing high quality, less costly health care. This became attractive to employers and unions alike, and later to governments seeking ways to constrain increases in health costs.

Since the 1970s, the more generic term health maintenance organization (HMO) has been used, especially by the federal government seeking to promote this concept. The HMO concept, linking health insurance and medical care in the same organization, was promoted through the HMO Act by President Richard Nixon in 1973. The HMO has become an accepted part of medical care in the United States and an important alternative to fee-for-service, private practice medicine (Table 13.2).

In order to encourage more efficient use of hospital care, the method of payment was changed during the 1980s. In 1983, a prospective payment system, called diagnosis related groups (DRGs), was adopted for Medicare, with payment categories of diagnosis (HCFA, 1998). This replaced the previous system of paying by the number of hospital days, or per diem. DRGs encourages hospital staff to treat

TABLE 13.2   Health Maintenance Organization Plans and Membership, United States, 1976–1997

HMOs	1976	1980	1985	1990	1995	1997
Number of plans	174	235	478	572	562	651
Enrollment (millions)	6.0	9.1	21.0	33.0	50.9	66.8

Source: *Health, United States, 1998.*

the patient effectively and expeditiously and to discharge the patient as quickly as possible. Payment for Medicare and Medicaid patients was shifted to this method. In many states this has also become standard for patients with private health insurance. Between 1980 and 1990, due to the DRG payment system and HMOs or managed care systems, which promote alternatives such as home and ambulatory care, hospital utilization was reduced in the United States. While total costs of health care increased in this period, without reduction of hospital utilization the increase would have been considerably higher.

During the late 1980s, the term managed care was introduced, expanding from HMOs of the Kaiser Permanente type to include other nonprofit and for-profit systems. Managed care plans of the HMO type operate their own clinics and staff (i.e., the staff model). Other managed plans operate on a not-for-profit or a for-profit basis. These are independent practice associations (IPAs), which operate with physicians in private practice, or preferred provider organizations (PPOs), which cover care with doctors and other providers associated with the plan providing services to the enrolled members or beneficiaries at negotiated prices (Chapter 12).

Following the failure of the Clinton national health insurance proposal in 1994 (see below), managed care experienced tremendous growth as employers sought to provide their employees with comprehensive coverage at reasonable costs. Managed care systems have been able to cut costs in health care in ways that the U.S. government could not. In 1996, 74% of insured American workers were enrolled in managed care plans, as compared to 55% in 1992. In California, with a long tradition of HMOs, by 1997 over 75% of insured persons, exclusive of Medicare and Medicaid beneficiaries, were enrolled in managed care. In the United States as a whole, in addition to the nearly 58 million enrolled in HMOs, another 91 million persons are enrolled in PPOs, with 25% of Medicaid and 10% of Medicare beneficiaries in managed care plans (Health Care Financing Administration website, 1998). In the United States, the search for cost containment has led to the development of a series of important innovations in health care delivery, payment, and information systems. HMOs have demonstrated that good care provided efficiently can operate with lower hospital admission rates than care provided on a fee-for-service basis.

The managed care revolution of the 1990s has brought about profound changes in health care organization in the United States. Hospitals and other specialty services are competing for contracts with managed care organizations and establishing community service systems of their own in order to compete for "market share" of insured clients. In many locales, excess hospital beds have become an economic burden, forcing many hospitals to downsize or sell out to larger hospital chains. Hospitals have responded by establishing contracts with managed care organizations and by reducing bed capacity; others have closed as they were unable to compete for sufficient patient flow.

Federal and state legislative initiatives are attempting to define patients' rights in managed care because of public complaints with limitations of managed care. At the same time, initiatives to increase Medicaid and Medicare beneficiary enrollments in managed care are underway. In response to widespread complaints regarding managed care restricting access to specialty services and shortened hospital stays. In 1998 the U.S. Congress passed a bipartisan-sponsored law that requires minimum 48-hour maternity stays. Many other pieces of legislation to protect consumers' rights and choice of doctor have been proposed in Congress and in state legislatures. The American Public Health Association in 1997 called for close monitoring and evaluation of managed care, especially when provided by for-profit organizations. The American Medical Association (AMA) promotes legislation restrictive to managed care plans. Federal and state legislative initiatives are unlikely to gain strong legislative action because of fear of initiating new inflationary pressures in health expenditures. However, restoration of the situation of fee-for-service as the predominant mode of health care provision in the United States is unlikely.

Health Information

The United States has developed extensive information systems of domestic and international importance. The CDC publishes the *Morbidity and Mortality Weekly Report (MMWR)*, which sets international standards in disease reporting and policy analysis. The U.S. National Center for Health Statistics (NCHS), the Health Care Financing Administration (HCFA), the U.S. Public Health Service (PHS), the Food and Drug Administration (FDA), the National Institutes of Health (NIH), and many nongovernmental organizations (NGOs) carry out data collection, publication, and health services research activities important for health status monitoring. National nutrition surveillance (NHANES, see Chapter 8) and other systems of health status monitoring are reported in the professional literature and in publications of the CDC. National monitoring of hospital discharge information facilitates the understanding of patterns of utilization and morbidity. These information systems are vital for epidemiologic surveillance and managing the health care system.

The 1965 U.S. Surgeon General's *Report on Smoking,* publications linking smoking and lung cancer had a major impact on public knowledge and behavior. The 1988 Surgeon General's Report on nutrition was another landmark document, setting international standards in public health. Media coverage of health events and topics is very extensive, and public levels of health knowledge grow steadily but vary widely by social class and educational levels.

Health Targets

Despite rapid increases in health care expenditures during 1970s and 1980s, improved health promotion activities, and rapidly developing medical technology, the health status of the American population has improved less rapidly than that in other western countries. Infant mortality in the United States in 1996 remained higher than that in 21 other countries; even the rate of the white population was higher than that of 16 countries which spent much less per person and a lesser percentage of GNP per capita on health care.

TABLE 13.3   Categories and Program Areas for Health Objectives, United States,
1979–1995

Category	Specific groups or activities with measurable targets
Health promotion	Improve physical activity and fitness
	Nutrition: reduce deficiency and excess dietary conditions
	Reduced tobacco use and exposure
	Reduce alcohol and drug abuse
	Promote family planning
	Promote improved care for mental illness, mental disorders
	Reduce violent and abusive behavior
	Promote educational and community-based programs
Health protection	Reduce unintentional injuries
	Improve occupational safety and helath
	Reduce environmental toxic exposure
	Improve food and drug safety
	Improve oral health
Preventice services	Reduce maternal and infant illness and risk factors
	Reduce risk factors for heart disease and stroke
	Reduce risk factors and morbidity from cancer
	Improve case finding and management of diabetes and other chronic disabling conditions
	Reduce incidence of HIV infection
	Reduce incidence of sexually transmitted diseases
	Improve immunization coverage to control infectious diseases
	Increase use of clinical preventive services
Surveillance and monitoring	Improve surveillance and data systems at federal, state and local levels
Age-related objectives	Improve services and utilization of servies for children, adolescents and young adults, adults, and the elderly
Special population	Improve access to and care of people with low income: African Americans, Hispanics, Asian and Pacific Islanders, American Indians, and other Native Americans

Source: Public Health Service, 1992, and Healthy People Website, 1999.

The 1979, U.S. Surgeon General Report *Healthy People* set forth a series of national health targets for a wide variety of public health issues. These defined 226 objectives in 15 program areas in the three categories of prevention, protection, and promotion. These goals and objectives were formulated based on research and consultation by 167 experts in different fields who participated in a conference by the U.S. Public Health Service. Consensus was based on position papers, studies, and conferences involving the national governmental health authority, the National Academy of Science's Institute of Medicine, and professional organizations, such as the American Academy of Pediatrics and the American College of Obstetrics and Gynecology. Many private individuals and organizations contributed to this effort, including state and local health agencies, representatives of consumer and provider groups, academic centers, and voluntary health associations.

These targets are periodically assessed as performance indicators of the U.S. health system and then updated (see Table 13.3). Progress made during the 1980s included major reductions in death rates for three of the leading causes of death: heart disease, stroke, and unintentional injuries. Infant mortality decreased, as did the incidence of vaccine-preventable infectious diseases.

Healthy People 2000, published in 1992 by the Surgeon General, details 332 specific health targets, in six groups, for the year 2000, in the areas of health promotion, health protection, preventive services, surveillance and data systems, and age-related and special population groups (see Chapter 11). It is estimated that nearly 1 million deaths occur annually in the United States from preventable causes. Reduction in risky behavior, health promotion at work, reducing regional inequities in health care provision, and education in health care self-management all have the potential to improve health and reduce the rate of increase in health expenditures.

Many states have adopted use of these targets as their own measures of health status and performance. Annual publications by the U.S. Public Health Service, in cooperation with the National Center for Health Statistics, make available a wide set of data for updating health status and process measures relating to these national health goals. The value of working toward health targets is widely accepted, and many states have adopted this approach. Development of *Healthy People 2010* has begun, with focus group sessions, public meetings, and a website (http://www.health.gov/healthypeople), enabling people across the country to make their voices heard. *Healthy People 2010* is expected to be released during 2000.

Review of progress in *Healthy People 2000* shows there has been continued progress in reducing infant mortality and mortality from heart disease and stroke, and fewer deaths from alcohol-related automobile accidents were recorded. Adults are using less tobacco, eating less fat, exercising more, and showing lower cholesterol levels and fewer stress-related problems. However, a 1998 panel of the U.S. National Institutes of Health established new guidelines which define some 55% of the U.S. population as overweight, a health problem associated with total costs of nearly $100 billion annually. Cancer incidence and mortality declined between 1991 and 1995 by some 8%. Measles and gonorrhea cases continue to decline. *Haemophilus influenzae* morbidity and mortality have largely disappeared. More people are living in communities with clean air, and blood lead levels in children are declining. AIDS is an enormous public health challenge, with death rates increasing rapidly from the 1980s to a peak in 1994, with a marked decline since, falling by 26% between 1995 and 1996 alone in the age group 25–44. Teenage pregnancy is very common, but teen birth rates have fallen steadily during the 1990s. Homicide rates have declined in most U.S. cities since 1994–1995, with improving economic conditions and low unemployment rates.

Health promotion has received wide public, governmental, and professional support in the United States over the 1980s and 1990s. In part, this reflects a long tradition of education on health matters in the rural agricultural sector and school health education. Nutrition and antismoking consciousness has grown in part because of wide media attention to many important epidemiological studies.

Advocacy

Consumer advocacy has been a potent factor for change in the United States in the twentieth century, and especially since the 1960s. It has contributed to strengthened governmental regulation in a wide area of public health-related fields (Chapter 2). These include automobile safety features and emission control, environmental standards, Mothers Against Drunk Driving (MADD), nutritional labeling and fortification of basic foods, and legal action against cigarette manufacturers. Food fortification, pioneered in the United States, is not mandatory as it is in Canada, but is nevertheless nearly universal, based on advocacy, informed public opinion, and an innovative, highly competitive food industry. Despite much public controversy, fluoridation of community water supplies covers 52% of the population, a higher coverage than most industrialized countries.

Advocacy groups can also promote regression in public health measures, as with groups currently fighting against immunization on the grounds of exaggerated concerns over reactions to vaccines. Some opposed to abortion have greatly affected public policy and promote sometimes violent activities against proponents and providers of abortions. Research and wide media coverage of health issues encourage a high level of individual and community consciousness of health-related issues and a climate receptive to health promotion.

Social Inequities

Lack of universal access and the empowerment it brings encourages an alienation or nonengagement with early health care, promoting inappropriate reliance on emergency room care and hospitalization in response to undertreated health needs. With large numbers of uninsured persons and many lacking adequate health insurance, access and utilization of preventive care are below the levels needed to achieve social equity in health, especially for maternal and child health and for chronic diseases such as diabetes, cancer, and heart disease. As measured by the infant mortality rate, the gap between the white and black populations of the United States remains high, even increasing between 1983 and 1995 (Chapter 6). Much of this social gap is due to inequities in health insurance coverage and socioeconomic factors.

Efforts to improve immunization coverage of U.S. infants to meet national health targets have been partially successful with efforts directed toward poor population groups. Still unsatisfactory levels of coverage in inner city and poor rural areas explained large outbreaks of measles in the period 1991–1992. Coverage rates improved from 83% for DPT, 72% for polio, and 83% for measles in the late 1980s to over 90% for most vaccines in the period 1996–1997. However, the percentage of those fully immunized according to the current recommended schedule was 78% in 1996–1997 surveys. Mandatory immunization as a condition of school entry has been established in all 50 states and has helped, but efforts to fully immunize the preschool population on time will require special attention of the public health system for years to come, especially as important new vaccines such as varicella are added to the routine immunization program.

School lunch programs and nutrition support for pregnant women- and children-in-need have reduced some of the ill effects of poverty in the United States, but lack of health insurance affects these groups severely.

Chronic disease and trauma are also excessively diseases of poverty with higher rates of morbidity and mortality in virtually all categories. Universal access to comprehensive care can reduce the social inequity even when income gaps are high. Conversely, increasing family disposable income for the poor is an effective way of reducing the social inequities in health. The two are complementary and equally important in social policy in the United States.

The Clinton Plan and Beyond

In the early 1990s, health insurance became a major political issue in national and state politics in the United States. A high percentage of the population was without any or adequate health insurance. Loss of health coverage with change of place of employment and the rapidly increasing cost of private health insurance generated widespread pressure for a national health program. The business community, too, had lost confidence in voluntary health insurance as costs of health insurance mounted rapidly as a cost of employment in an increasingly competitive international business climate. Former Arkansas Governor Bill Clinton promised a national health plan in his 1992 presidential campaign. In the fall of 1993, the Clinton health plan was presented to the American public, and the process of seeking Congressional approval began.

The Clinton health plan was based on a federally administered compulsory universal health insurance through the place of employment, with alternative plans available to choose from at different costs. A state could opt to form its own health insurance program and even designate its own department of health to fulfill this function. Physicians could contract with health insurance plans to provide care on a fixed fee schedule, or in HMOs, whether based on group or individual practice.

The Clinton health plan failed in Congress in late 1994. Apathy was widespread among the majority of the population who already had good insurance benefits under Medicare or employment-based health insurance. Their interest was in the status quo, and the insurance industry and organized medical community used this to defeat the bill. Federal legislation protects workers' benefits under collective bargaining, prevented states from mandating health insurance benefits. Federal assistance and waivers for state health insurance allow states to opt for managed care for Medicaid beneficiaries. Medicare and Medicaid waivers also allow states to include these beneficiaries in state health plans, but universal access to care will require enabling legislation in Congress. At the same time, conservative attacks on public programs such as Medicare keep the issue of national health insurance on the public agenda. State initiatives for universal health coverage are alternative approaches to achieving national universal access to health care.

The Managed Care Revolution

Many employers have switched to managed care coverage, offering indemnity plans to the employees at additional cost. The movement to managed care became an avalanche in the 1990s, with a high percentage of the population insured at their workplace becoming members of HMOs or other forms of managed care. The swing to managed care produced major effects in the health care system, not only for doctors increasingly pressed to join HMOs or PPOs, but also for hospitals and for the consumer who had to adjust to the rules of managed care. Restrictions on access to specialists and new procedures generated public and political criticisms but did not slow the economically driven changeover, and its profound effects on health care.

In 1996, many states introduced legislation to regulate HMOs, of which 56 laws were enacted in 35 states. Criticisms of for-profit HMOs are appearing frequently in the popular media, and there is a growing backlash of opinion against imposed limitations on specialist referrals, emergency room visits, hospitalization, and some therapeutic interventions (e.g., bone marrow transplants for terminal cancer cases). Some of these have also generated legal suits for malpractice with large settlements. A 1998 Commission on Health Quality appointed by President Clinton produced a bill of rights for patients that calls for additional information on health plans and for the right of appeal to an independent panel on health plan decisions regarding denials of coverage for emergency care or access to specialists.

The nonprofit prepaid group practice type of HMO uses over 90% of premiums for patient care, whereas the for-profit plans spend higher proportions of premiums for administration, including very high salaries for executive staff. The growth trend of managed care will certainly continue, but perhaps with greater regulation of for-profit HMOs in order to ensure access to services based on medical criteria in the patient's interest and quality assurance.

Summary

The United States has managed to achieve many of the targets set by the 1979 Surgeon-General's *Healthy People* report. At the same time, the annual average increases in health care expenditures in the United States slowed markedly from the 1986–1990 period with average annual increases of 10.7%, falling to 6.9% in the 1991–1994 period. This is partly due to lower general inflation rates (<3%), but also to cost containment measures being adopted by the health insurance industry, the growth of managed care, and rationalizing the hospital sector by downsizing and promoting lower cost alternative forms of care.

National health insurance was again delayed by Congressional rejection of the Clinton health plan, but the issue is likely to continue to receive attention at the federal level. However, state initiatives with federal waivers, and perhaps cost-sharing, could dominate health care reform in the early years of the twenty-first century. In the mid to late 1990s, employers are promoting managed care options for their workers, so that managed care is growing rapidly through market mechanisms. State governments are acting to regulate this by legislation, such as those requiring minimum hospital stays for obstetrics cases, by limiting managed care programs from certain kinds of contracts for services, and by establishing appeals mechanisms for managed care members. Increasing access to Medicaid by raising the income levels defining poverty will increase health insurance coverage, while managed care on a mandatory basis will help to sustain cost constraint.

The term nonsystem is often applied to health care in the United States, but the implied criticism is not fully justified. It is a health system, or more accurately, a diffused system. Social and regional inequities in health status are still present, but not necessarily greater than in countries with universal access to health care. Further, there are many parallel programs in the United States that have important positive public health benefits, such as universal school lunch programs, nutrition support for poor women, infants, and children (WIC program), food stamps for the working poor, fortification of basic foods, free care in emergency rooms and for urgent hospital care for the poor, Medicare for the elderly, and Medicaid for the poor. Nevertheless, equitable universal access is lacking, and the system is the costliest in the world, without the best levels of health as measured by process indicators, such as immunization and prenatal care coverage, nor in outcome measures such as infant and other mortality rates. And social inequities in these health status indicators are important indicators of failures of the health system in the United States to reach its full potential.

CANADA

Canada is a federal country with ten provinces and two northern territories, a population of 29.9 million, and a GNP per capita of $19,020 (U.S.) in 1996. Responsibility for health is defined constitutionally as being at the provincial level of government, except for the aboriginal Indian and Inuit populations, armed forces, and veterans. Despite many geographic, historic, cultural, and political similarities to the neighboring United States, Canada developed its own unique national health insurance program.

Starting in the 1930s, federal grants-in-aid were given to the provinces for categorical health programs, such as cancer and public health programs. Based on this precedent, Canada's national health program is a system of provincial health insurance with federal government financial support and standards. It developed in stages between 1946 and 1971, first with hospital and diagnostic services and subsequently with medical care insurance, now collectively known as Medicare. It brought all Canadians into a system of publicly financed health care, while retaining the private practice model of medical care. Hospital care is provided mostly through nonprofit, nongovernmental hospitals.

The Canadian health program differs markedly from those of the United Kingdom and the United States. Each of the three health systems reflects the political and cultural traditions of their country. Each is being modified to constrain the rate of cost increases and preserve, or develop, universal coverage. After decades of stress on developing national health insurance, Canada became a leading innovator in health promotion.

Evolution of National Health Insurance

Initiatives for national health insurance in Canada go back to the 1920s, but definitive action occurred only after World War II. The development of national health insurance was partly the result of the experience of the Great Depression of the 1930s, a strong agrarian cooperative movement, and the collective wish for a better society following the war. In 1946, the recently elected social democratic government of Saskatchewan, a large wheat growing province of 1 million persons in the western prairies, under the leadership of Tommy Douglas, the founder of Canada's Medicare program, established a hospital insurance plan. This plan provided free hospital care for all residents of the province on a prepaid basis under public administration. Within several years, other provinces developed similar plans, and in 1956, the federal government passed legislation (the Hospital Services and Diagnostic Services Act) to provide a cost-sharing plan for provinces, adopting universal, publicly administered hospital insurance plans. By 1961, all ten provinces and the two territories had implemented hospital insurance plans meeting federal criteria in a two-tiered national health insurance plan, that is, universal provincial health plans with federal standards and cost-sharing.

In 1961–1962, Douglas and the province of Saskatchewan again led the way by implementing a universal plan for medical services (Medicare). This was opposed by a bitter, 23 day doctors' strike which resulted in some compromises, but the universal plan came into effect, paying doctors' bills on a fee-for-service basis. Again, this was based on the principles of universal coverage, comprehensive benefits, and public administration.

Following the controversies over this plan, a federal Royal Commission on Health Services (the Hall Commission) recommended establishment of similar plans with federal cost-sharing. In 1966, the federal government introduced its Medicare Act, providing federal cost-sharing of approved provincial plans. Federal reimbursement to the provinces included 25% of national average medical care expenditures per capita and 25% of the actual expenditures by each specific province. This provided higher than national average rates of support to poorer provinces as well as portability between provinces. By 1971 all provinces had implemented such plans.

Reform Pressures and Initiatives

The Canadian health program established universal coverage for a comprehensive set of health benefits without changing the basic practice of medicine from individual medical practice on a fee-for-service basis. Poorer provinces were able to use the cost-sharing mechanism to raise standards of health services, and a high degree of health services equity was achieved across the country.

Rapid increases in health care costs led to a review of health policies in 1969 (the Federal–Provincial Committee on the Costs of Health Services). The resulting report stressed the need to reduce hospital beds and develop lower cost alternatives to hospital care, such as home care and long-term care. Federally led initiatives during this period extended coverage to home care and long-term nursing home care, while restricting federal participation in cost-sharing to the rate of increases in the GNP. Since then, many provincial and federal reports have examined the issues in health care and recommended changes in financing, cost-sharing, hospital services, and development of primary care.

In 1974, a new approach to health was outlined by the federal Minister of Health Marc LaLonde in a landmark public policy document, *New Perspectives on the Health of Canadians*. This report described the health field theory in which health was seen as a resultant of genetic, lifestyle, and environmental issues, as well as medical care itself. As a result, health promotion became a feature of Canadian public policy, with the objective of changing personal lifestyle habits to decrease risk factors such as smoking, obesity, and physical inactivity. The pioneering work in nutrition from the National Nutrition Survey published in 1971 led to the adoption of federal mandatory enrichment regulations for basic foods with essential vitamins and minerals (see Chapter 8). This and other initiatives in the 1980s led to the Ottawa Charter on Health Promotion (see Chapter 2).

In the mid-1980s, physicians' organizations pressed for the right to bill patients above the rates paid for by Medicare, but this was forbidden by national legislation (the Medical Care Act), passed unanimously by the federal Parliament. This act penalizes provincial governments which allow extra billing by physicians by withholding federal funding.

The Canadian health insurance program experienced high rates of cost increase, approximately 12% annually, during the 1980s, while GDP grew at 3% per year. National expenditures on health reached 10.1% of GDP in 1992, declining to 9.5% in 1996. During the late 1990s, the rate of increase in health care costs was reduced by painful measures of retrenchment, especially in hospitals. A decade earlier, when the rate of growth of health expenditures from 1975 to 1991 was averaging 11%, Canada was second only to the United States in percentage of GNP expended on health. In the period 1991–1996, health costs inflated by 2.5% yearly, so that Canada's health care costs as a percentage of GNP in 1998 was the fourth highest in the world (with 9.2% of GNP in 1996), after the United States, Germany, and France.

Of total health expenditures in 1996, hospitals accounted for 34%, physicians for 14%, and drugs 14%. Expenditures for other institutions, drugs, and public health grew as percentages of total expenditures, while hospital expenditures declined and costs of physician services remained stable, as compared to 1991 distribution of expenditures (39% for hospitals, 15% for physicians, 12% for drugs) (Statistics Canada, 1999).

Financing of total health expenditures in 1996 was 70% from public sector sources, including federal, provincial, and municipal governments and workmen's compensation, a decline from 74.6% in 1991. Federal governmental cost-sharing in health expenditures has gradually declined since the 1970s, so that provincial governments are facing difficulty with continued financing at current levels, and are pressured to control rates of increase. This led many provinces to reduce the hospital bed supply, from 6.9 beds per 1000 in 1979 to 4.7 in 1995; the most populous province, Ontario, for example, reduced acute care hospital bed supply to 3.9 per 1000 in 1995.

Other health reforms included a growing emphasis on health promotion and development of alternatives to acute hospital care. In the 1970s, the province of Manitoba instituted reforms to establish district health systems integrating hospitals, nursing homes, home care, preventive services, and medical practice, reaching many of the rural areas of the province over the next several decades. In the 1990s, Saskatchewan began development of similar integrated district health service and regionalized hospital systems. Other provinces, similarly under pressure on health costs, are reducing hospital bed supplies, increasing regionalization, and strengthening health promotion and community health services.

Health Status

Criticism of the Canadian health system focuses on waiting time (in comparison to the United States) for diagnostic and surgical procedures, lesser access to high-tech equipment and procedures, and reduction in hospital staff positions. Such comparisons, however, are not substantiated by objective analyses, nor in measurable health indicators. Since the implementation of Medicare, Canada's position in major health status indicators improved in comparison to other countries. Infant mortality rates were higher than those of the United States up to the 1960s (28 versus 22 per 1000), but were lower in the 1990s (6 versus 7 per 1000 in 1997). Canada's 1980–1997 maternal mortality rate was 5 per 100,000 compared to 12 for the United States and 10 for the United Kingdom. In 1997, Canada's life expectancy at birth was 79 compared to 77 years in the United States; the Canadian–U.S. gap increased from 2 years in 1988 to 2.7 years in 1993 for men, and from 2 to 2.9 years for women.

Immunization coverage for infants was reported in 1995–1996 as 89% for polio, 93% for DPT, and 98% for measles. Canada's immunization coverage of infants, while better than that of the United States, is below those of the United Kingdom and the Scandinavian countries. Canada ranks among the 10 countries with lowest mortality rates from overall mortality, cerebrovascular disease, and diseases of the heart, but is fifteenth or higher for neoplastic disease, diabetes, motor vehicle accidents, and homicide.

Summary

Canada's health system successfully established universal, tax-supported national health insurance in North America. The Canadian health program has important lessons for health care reform internationally. Universal health insurance was implemented without reform in the mix of services, changes in the way they are funded, and initially, inadequate attention to prevention. These are all issues that need to be considered in the Canadian experience. At the same time, Canada pioneered the idea of health promotion from the 1970s in such areas as Healthy Cities, fitness, and food enrichment. It is a model with drawbacks, partly because it has tended to freeze the private practice model, and was slow in producing reform measures until provinces were unable to cope with the burden of costs as the federal government reduced its share of health care financing.

The quality of care is high, and Medicare is one of the most popular public institutions in Canada of which most Canadians, including physicians, are proud. Current reforms appear to be succeeding in controlling the rate of increase in costs in some provinces. The health status of Canadians is among the highest in the world. Despite the financial burden and the need for economic analysis with priority selection, given the popularity of the program, the Canadian commitment to universal, publicly administered health care is likely to continue, and the Canadian model will be of importance in reform for other countries, particularly the neighboring United States.

THE UNITED KINGDOM

The United Kingdom's population in 1997 was 58.2 million and the 1996 GNP per capita was $19,600, ranking it economically well below Scandinavian, North American, and many European countries. The United Kingdom ranked twenty-second among the leading nations in child mortality rate in 1997, at 7 per 1000. Maternal mortality was 10 per 100,000 (1980–1997).

The United Kingdom is a unitary state with public health developing at the local authority level, with national health initiatives evolving slowly since the mid 19th century. National Health Service developed and maintained high professional and technical standards, despite modest levels of funding of the service. Immunization coverage in 1995–1996 was 94% for DPT, 96% for poliomyelitis, and 92% for measles (MMR), with a second dose of measles vaccine at school entry.

The British developed a unique and important model of health care as a tax financed public service that is widely influential in other national health systems. It is popular and was successful in achieving its initial goals, and it has undergone periodic reforms since its inception in 1948, surviving many changes of government and political philosophy.

The National Health Service

As described earlier, the United Kingdom developed its present National Health Service over many decades. Milestones included reform of the Poor Laws of the eighteenth to nineteenth centuries, the friendly societies, the National Health Insurance of 1911 for workers and their families, the National Emergency Medical Service of World War II and the Beveridge Report of 1942. In 1946, under the Labour government of Clement Atlee, parliament approved the National Health Service Act with implementation in 1948, under the leadership of Aneurin Bevan.

The National Health Service is financed through general tax revenues to provide comprehensive service to the entire population. The NHS was originally organized as three parallel services: the hospital service with salaried doctors; the general practitioner (and dental) services provided by independent practitioners with capitation payment; and the public health service with salaried staff. The hospital and general practice services were operated by separate public boards or councils; the public health service was administered by the local authorities.

Structural Reforms of the National Health Service

During several stages of reform in the 1970s and 1980s, the NHS was reorganized, reducing the number of administrative levels and trying to achieve integration and coordination between highly specialized and fragmented services. The 1974 reform established Regional Health Authorities, and integrated Area Health Authorities (AHAs) beneath them to replace the previous multiplicity of hospital management committees, boards of governors, and local health authority committees. The AHAs were nonelected lay bodies that absorbed public health and hospital management functions, consolidating many previously overlapping jurisdictions. Multidisciplinary management teams were introduced at the AHA and district levels, with decision making by consensus, stressing professional managerial competency.

A further reorganization in 1982 abolished the AHAs, placing the managerial responsibility at the District Health Authority level, with a stress on further decentralization of management authority to hospital and community service structures. Reviews of the NHS were conducted during the conservative government led by Margaret Thatcher, focusing on managerial efficiency, governmental and business viewpoints, the growth of the private sector, consumer group advocacy issues, and protection of consumer rights.

From 1977 to 1996, the number of acute care hospital beds fell from 3.0 to 2.0 per 1000 population. Despite the aging of the population, the average length of stay in acute care hospitals fell from 9.8 days in 1977 to 4.8 days in 1996. Community care services of all kinds increased during this period.

Reforms since 1990

In 1990, the National Health Service and Community Care Act attempted to further rationalize management of the NHS. Three types of statutory health authorities were redefined: Regional Health Authorities (RHAs), District Health Authorities (DHAs), and the Family Health Service Authorities (FHSAs). Regional and District Health Authorities (RHAs and DHAs) became the primary administrative levels. The FHSAs managed contracts with the general practitioners (GPs). The NHS operations in Wales, Scotland, and Northern Ireland operate under similar arrangements.

The 14 RHAs assess health needs, set strategic direction for service development, monitor quality of management and care, and allocate resources to promote cost-effective services. They also promote medical audits and specific program development (e.g., transplantation services) and provide assistance to health providers such as hospitals with management problems. RHAs do not provide services. The DHAs operate under the authority of boards similar to those of the RHA, and are the major purchasers of services from hospitals and other providers. They contract with hospitals for services based on assessed need and on satisfaction with hospital performance. They may also operate NHS hospitals or other services, such as ambulance services.

Reforms since 1990 included the introduction of competition between providers, the development of community health services, and further reduction of the supply of hospital beds. These were intended to introduce greater choice for the patient and the primary care provider (the GP), with incentives for efficiency and quality of care.

Family Health Service Authorities (FHSAs) under the 1990 Act are governed by boards similar to the RHAs and DHAs. The FHSA is responsible for contracting with GPs, general dental practitioners, optometrists, and community pharmacists. The role of FHSAs expanded to include formulation of policies, supervision of facilities and services, as well as remuneration of contracting providers. Patients register with GPs and are referred to hospitals and specialists in keeping with medical needs. GPs have traditionally been paid on a capitation basis for the patients registered with them, and the patient has the right to change GPs. Capitation is the allocation of funds per person registered as a service beneficiary for a specified period of time to cover care for a range of services. Weighted capitation is allocation per person, with adjustments made for factors such as age, sex, and regional standardized mortality rates, which reflect both need and demand for health services. Standardized mortality rates (SMRs) are used as a proxy for morbidity in capitation allocation (see Chapter 3). GPs are now paid extra premiums for performance indicators, for example, specific preventive services such as immunization, Pap smears, and mammogram screening.

A major new innovation has been to allow the FHSAs to administer budgets for fund holder GPs, in the form of per capita payments including both GP and hospital services. The GPs are increasingly working in health centers, along with district public health nurses. By 1995, some one-third of GPs worked as fund holders with per capita payment by the NHS for ambulatory and hospital care. This empowers GPs to negotiate with the hospitals, reduce waiting times, and improve other health care conditions for their patients, placing the hospital in the position of having to compete for the referral work of the GP. Experiments with financing of hospital care through the GPs were designed to raise the quality of care and promote cost containment. The GP fund holder movement seems to be a successful program, although not well evaluated.

Hospitals are encouraged to become National Health Service Trusts, which are nonprofit public corporations governed by boards of trustees appointed by the national government, usually representing the local authorities. Hospital Trusts must demonstrate management capacity and viability to operate as economic units. They must compete for referrals, striving for patient and GP satisfaction. Hospitals are no longer funded directly by the NHS, but derive their income from providing services to the Health Authorities, fund holding GPs, private insurance, and self-paying patients, paid for services by a DRG system. This permits them to operate as independent economic units, enabling them to charge for services, determine staff conditions, raise capital by borrowing money, and, within limits, to buy or sell land or facilities.

Financing of the NHS continues to be by governmental allocations from general tax revenues. Some revenues come from other sources, including user fees, such as for prescription drugs or dental services. Operating budgets are allocated to RHAs to cover costs of hospital, community health, and primary care services. The allocation is determined on the basis of population size and adjusted by SMRs, with some local weighting factors based on service utilization. The DHAs are in turn funded by the RHAs, using similar criteria. RHAs administer GP fund hold-ing units, whereas FHSAs are funded to pay for contracting primary care services. Capital allocations to replace and modernize facilities and equipment are based on long-term planning at the RHA level.

Market reforms in the United Kingdom are still being developed. Despite be-ing the subject of continuous scrutiny in the press and at political levels, the NHS has provided universal access and maintained high quality at reasonable costs. Health expenditures in the United Kingdom decreased from 4.5% of GDP in 1970, to 5.6% in 1980, 6.0% in 1990 and 6.8% in 1997. This was and continues to be lower than expenditures in most industrialized countries, generating criticism of under-funding of important areas, such as oncology services.

Social Inequities

Social inequities in the health status of the population of Britain, which were part of the justification for the establishment of the NHS in 1946, have persisted. The Black Report (Douglas Black, 1980) documented this problem, and subse-quent reports indicate the persistence and even worsening of social inequalities into the 1990s. Changes in definitions and distribution of the population in the dif-ferent social classes may explain some of the differences; however, there is a widening of the gap between the social classes, with continuous increase in the SMR of class V and continuous decline in SMRs for classes I and II (see Chapter 4). The ratio between class I and class V in SMR was 90/110 in 1931, 75/142 in 1961, and 65/168 in 1981. Higher cause-specific mortality in lower socioeconomic classes is seen especially in cardiovascular disease, trauma, and cancer.

This social gap is not easily explained on the grounds of the classic health risk factors alone. The health gap correlating to economic disparities may be due to poor diet, more smoking, less physical activity, and social and working conditions with less reward, personal satisfaction, and less control of life events than for the high-er social classes. There are also regional differences in SMRs in the United King-dom; the reasons for these are not always well understood but probably relate to a variety of social, economic, personal lifestyle, and environmental risk factors.

In 1998, Donald Acheson, a senior professor of public health in the United Kingdom, reported on his Blair-government initiated inquiry into social dispari-ties in health in Britain. His report confirmed the findings of the Black report and evaluated findings of the many studies of social gradients in health status since that report was issued. The Acheson report has been a factor in governmental policies in tax and welfare reform, preschool child care programs, and tobacco legislation, and in some aspects of NHS reforms.

Health Promotion

During the 1950s and 1960s, mortality from cardiovascular diseases increased in the United Kingdom, as in most industrialized countries. These rates began to decline in the 1970s in the United States, Canada, and other European countries, but remained high in the United Kingdom for another decade, with coronary heart disease mortality only declining substantially since 1985. This delay in reduction of cardiovascular disease mortality may be explained by then prevailing conservative attitudes toward treatment of acute myocardial infarction in the United Kingdom such as aggressive, intervensive methods of treatment and intensive care units. The NHS was also slow in response to changing approaches to health promotion and risk factor reduction. Mortality from ischemic heart disease continued to increase up to 1978. The United Kingdom continued to have higher mortality rates from cardiovascular diseases compared to many countries in Europe. These and other public health issues, including relatively low immunization coverage levels, led to formulation of health promotion strategies in the Department of Health.

In the late 1980s and early 1990s, a number of major initiatives sought to improve prevention and health promotion activities in the United Kingdom, including greater public awareness of healthy nutrition and the risks of smoking. From 1978 to 1997, mortality rates from ischemic heart disease declined by 41%. Incentive payments to GPs resulted in a sharp improvement in immunization rates. Local authorities are required to have specialized staff to promote motor vehicle safety, which contributed to a reduction in road crash mortality of more than one-third from 1980 to 1995. A new Water Act and the Environmental Pollution Act of 1990 increased the supervisory and regulatory role of the national government in these areas of public health.

The Health of the Nation report (Secretary of State for Health, 1991; see also Chapter 2) placed health promotion and national health targets as a major focus of a national health program. Declining mortality from the major causes of death (cardiovascular diseases, cancer, and trauma) may reflect an increasing effect of health promotion activities in the United Kingdom.

The New Labour government of Tony Blair, elected in 1997, is undertaking further reform in the NHS especially in methods of financing primary care and the market forces of GP fund holding. The government also committed itself to increase funding for the NHS by 4% above inflation over the period 1999–2003 to strengthen clinical service sectors which have suffered from excessive cutbacks in the previous decade.

In 1999, the Blair government initiated a new reform of the NHS, establishing primary care groups (PCGs) throughout the country with GP groups serving population groups of between 30,000 and 250,000 persons. The PCGs replace the purchasing of services previously performed by the fund holding GPs and the health authorities. They are able to retain annual surpluses to improve their service facilities, as an incentive for efficiency. The central management of the NHS gained new tools to improve monitoring with a Commission for Health Improvement (CHIMP) and a National Institute for Clinical Excellence (NICE) established in 1999. GPs gained on-line appointment systems, and patients a free 24 hour telephone nurse consulting service, in attempts to modernize the NHS and its front-line contact with patients.

Summary

Successive governments of the different political parties have supported the NHS, and despite criticism, it remains a popular institution with the British public, surviving many changes of political leadership over the past 50 years. Reform, on the average of every decade, has enabled the NHS to evolve with experience and to meet the changing economic and health needs of the country. Decentralization to DHAs has been achieved with payment to the DHA by the NHS on a per capita basis, with an adjustment factor based on age, sex, and morbidity. Cutbacks in clinical services have generated criticism that quality of care is jeopardized and national supervision of DHA performance needs strengthening.

Changing epidemiologic patterns have also led the Department of Health and the NHS to develop health promotion strategies. This has helped to reduce high rates of mortality from cardiovascular disease and trauma. This may help to reduce social and regional inequalities in health still present after nearly a half century of universal access. NHS reforms in the early 1990s were intended to promote clients' rights and GP satisfaction with secondary and tertiary care, with the GP role as gatekeeper. It is still too early to evaluate innovations launched in the early 1990s to promote cost-efficiency and cost awareness.

The NHS has guaranteed access to health care for all, but failed to alleviate social class inequalities in health status. This discrepancy may be due to slowness in adopting health promotion efforts as in other industrialized countries to reduce the burden of cardiovascular disease, cancer, and other diseases that affect the poor more than the well-to-do. The NHS has succeeded in its mission of providing universal access in a tax-financed and relatively economical service. The Beveridge model NHS has been influential in the Nordic countries since the 1950s and in countries of southern Europe (Greece, Italy, Portugal, Spain, and Turkey) in their various reform programs since the 1970s. The NHS continues to evolve, and is an important and successful international model of health care systems, and one of the most cost-effective.

THE NORDIC COUNTRIES

Each of the health care systems in the Nordic countries has its own characteristics with reforms in progress in each country. Denmark, Finland, Iceland, Norway, and Sweden, with social democratic governments both before and after World War II, in many ways pioneered the welfare state. They were later influenced by the United Kingdom's NHS, but with strong regional or local governmental organization and taxation have had more emphasis on a decentralized program of health services. Their achievements in social welfare and health care over many decades have been widely acclaimed successful models for social protection in prosperous industrial economies.

Commonly, between 50 and 70% of health system revenues are generated from personal income taxes levied at the regional (Sweden, Norway, Denmark) or municipal (Finland) levels of government. Most of the remainder comes from general revenues raised by the national government through value added taxes or excise and personal or corporate income taxes. The national funds are distributed as block grants to minimize interregional inequities, with additional grants for medical education. National sickness funds pay for ambulatory visits. Municipal governments pay for long-term care for the elderly. Patient copayments provide 2–3% of Sweden's county health expenditures and were introduced in Finland in 1993. The user fees are not a hardship because of the widespread prosperity and well-established social security systems of the Scandinavian countries.

The Nordic countries have traditionally emphasized maternal and child health and have achieved very low rates of infant mortality. They have high rates of mortality from cardiovascular diseases as compared to countries in southern Europe. This is thought to be related to traditional dietary patterns with high fat diets, along with smoking and heavy alcohol usage. These risk factors have been the subject of much successful effort at health promotion and are slowly declining.

Sweden

In 1997, Sweden's population was 8.8 million, and the 1996 GNP per capita was $25,710. The 1997 infant mortality rate was 4 per 1000 live births, and life expectancy at birth was 78 years. The 1997 crude birth rate was 12 per 1000 population, with 100% of births taking place in medical facilities, and the maternal mortality rate was 5 per 100,000 (1980–1997). Immunization coverage in infancy in 1995–1997 was 99% for polio and DPT, and 96% for measles.

Sweden's health insurance system evolved over many decades, and became compulsory in 1955, covering compensation for medical clinic and hospital services and private ambulatory care. Swedish health care is tax-financed, with funding mainly from employers and government, but patients are charged a copayment for services. Health expenditures as a percentage of GDP increased from 7.5% in 1972 to 9.6% in 1982, declining to 7.5% in 1993.

The county or municipality is the principal level of government responsible for management of health care. There are 23 county councils and three large municipalities with populations ranging from 60,000 to 1.5 million. In 1993, the counties or municipalities, which have an income tax base of financing, provided 75% of funding for health care, with 7% coming from the national government, 5% from national insurance, 2–3% from patient fees, and the remainder from miscellaneous sources. Current reforms include improved primary care coupled with reduction in the hospital bed supply. Primary care is provided in health centers staffed by salaried GPs, nurses, and other staff serving about 15,000 clients. In 1993, private practice accounted for about 20% of total doctor visits.

Sweden has traditionally had a very high ratio of hospital beds to population (Table 13.4). In the late 1960s this included 5 acute care, 5 long-term, and 6 mental hospital beds per 1000 population. Hospitalization was a common form of care, especially in areas with sparse population and long distances to hospitals and doctors. Reduction in hospital bed supplies has been a long-term strategy in Sweden

TABLE 13.4   Hospital Bed Supply in Sweden by Type of Bed, Selected Years, 1973–1996[a]

Beds	1975	1985	1990	1993	1996
Acute care (medical/surgical) beds/1000	5.4	4.4	3.8	3.1	2.8
Psychiatric beds/1000	3.9	2.5	1.7	1.2	0.8
Long term, including geriatric	4.6	6.2	5.4	1.2	0.5
Total beds/1000[b]	na	14.6	12.4	7.0	5.6

Source: Statistical Yearbook of Country Councils, Sweden, 1995 [in Swedish]. Data for 1996 from WHO European Region HFA data set.
[a]Reform in 1992 included transfer of 31,000 nursing home beds from health to social care agencies.
[b]na, Not available.

since the 1940s, and emphasized since the 1960s, with a steady reduction in medical, surgical, and community care beds, as well as psychiatric beds. Long-term social care for the elderly has been transferred to social service agencies. As a result of this policy, Sweden is one of the few countries in the world to actually reduce the percentage of GNP spent on health care, to 7.6% in 1994. This was accomplished while maintaining a high quality service and improving national health indicators, such as infant mortality rates and maternal mortality rates, which are among the lowest in the world. Figure 13.1 shows the decline in hospital bed supplies in Sweden, Finland, and Denmark from the 1970s to the late 1990s.

Recent reforms in Sweden allowed contracting out for public sector services. This strengthened the role of primary care providers, now able to select more efficient and user-friendly services. Hospitals operate as economic units, balancing revenues and expenditures, and must compete for patients in the new public market for health care. Public institutions must also compete with the private sector and, in some instances, purchase services from private providers. This has helped

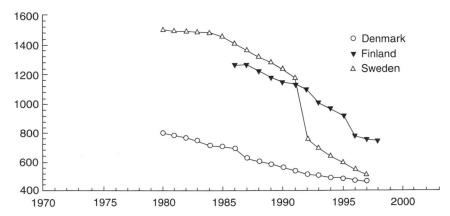

FIGURE 13.1   Number of hospital beds, per 100,000 population, in the Nordic countries. Source: WHO European Region, Health for All Data Set, 2000. Note: Includes acute care, psychiatric, and long-term care facility beds.

to reduce waiting times for operations and led to bankruptcy of inefficient or unacceptable hospitals.

Finland

Finland is a republic with a population of 5.1 million persons in 1997 and a 1996 GNP per capita of $23,240. Finland has achieved one of the lowest infant mortality rates in the world, declining from 22 per 1000 live births in 1960 to 4 per 1000 in 1997. Maternal mortality is 6 per 100,000 live births (1990–1997). Child care is provided free by the municipalities; this has resulted in vaccination coverage rates of 100% for polio, 100% for DPT, and 98% for MMR for 1-year-old children (1995–1997). Despite high immunization coverage for polio, Finland experienced an outbreak of polio due to use of an inadequately immunizing IPV vaccine in the 1980s. Longevity increased by 5.5 years for men and 5.1 years for women from 1971 to 1991, with a life expectancy of 77 years overall in 1997, an increase from 68 years in 1960.

Finland has three tiers of government. Strong municipal governments provide primary, secondary, and tertiary care services, as well as public health, education, and other social services. The states subsidize municipalities to provide these services, with management by locally elected officials. Taxes on income are shared between the municipal and national governments. Universal access to care is guaranteed to all.

Health policy is determined at the level of the national government, which regulates capital investment in health facilities and subsidizes municipalities, which are responsible for providing health and social services. State and municipal governments together collect approximately half of total taxation, high compared to other countries, reaching 46% of GDP. The economy was in recession during the early 1990s with a decline in GNP; as a result, the percentage of GNP spent on health care rose sharply, from 6.8% in 1982 to 9.4% in 1992, falling to 8.8% in 1993 and 7.5% in 1997.

High rates of mortality from cardiovascular diseases, injury, and suicide affect middle-aged men disproportionally. The widely known North Karelia project (see Chapter 5) to promote reduction in risk factors for heart disease stimulated national efforts and contributed to substantial reductions in mortality rates from these diseases. Cardiovascular mortality rates declined by 52% from 1970 to 1996, in part because of changes in diet with less meat and greater vegetable consumption. Smoking rates for men in the early 1970s reached 50% but have declined to one-third in 1993, with 20% of women smoking. Overall alcohol consumption is low, but binge drinking is common and relates to the high suicide and trauma rates.

Finland had high hospital bed ratios up until the 1980s when it changed health policy, recognizing the limitations of hospital care and placing greater emphasis on primary care, preventive and social services, and health promotion. Hospital bed supplies are still being reduced, with shorter lengths of stay and increasing ambulatory care and outpatient care. Mental hospital beds were decreased by 50% during the 1980s. The total hospital bed to population ratio declined from 15.6 in 1980 to

9.3 per 1000 population in 1995. Acute care hospital beds per 1000 decreased from 4.8 to 3.9 from 1975 to 1995 (see Fig. 13.1), and 2.4 in 1998 (HFA data set).

Reform in primary care services during the 1980s has reduced inefficiency, bureaucracy, and waiting times, and raised consumer satisfaction. A combination of capitation and fee-for-service payment is used. During 1993, reforms in health care financing converted national support for municipal health services to block grants based on capitation formulas to the municipalities, which now fund both the hospitals and primary care services. This allows the municipalities greater freedom in seeking a new balance of services and redirecting resources from the hospital to the primary care sectors. Local health centers provide most medical and health related services, including rehabilitative and addiction services.

Hospital-based physicians are permitted to do private practice. Over 90% of GPs work in publicly operated health facilities, but nearly one-third also conduct private practices in their spare time. GP satisfaction with the changes in the health system, with the combination of capitation and fee-for-service, is reportedly high.

The search for greater efficiency now includes a mix of planned and market economies in health. The strong tradition of publicly operated health services will continue despite introduction of market elements, but regional inequities may be an undesired result. Health reform in Finland will continue with decentralized service management and central planning and financial support.

WESTERN EUROPE

The countries of continental western and central Europe pioneered national health insurance through place of employment, with the national government regulating conditions of insurance, establishing fee schedules, and setting national health policies. the generic type is termed the Bismarckian national health insurance program, and is characteristics of Germany, France, Holland, Belgium, Luxembourg, Austria, and Switzerland, each with distinct characteristics and mixed features of social insurance with national service elements. These have been termed "sickness insurance," based on the solidarity principle of workers' benefits, including old-age pensions, disability benefits, and compensation for loss of working capacity. The funds have maintained a treatment-oriented approach, and only under exceptional circumstances have they undertaken disease prevention, much less health promotion.

Germany

Germany is a federal state with a century old tradition of social protection legislation. Most aspects of management are delegated to self-governing insurers and associations of providers. Germany's population in 1997 was 82.1 million following unification of western and eastern Germany, and its 1996 GNP per capita was $28,870. Infant mortality in 1997 was 5 per 1000 live births. Maternal mortality in 1990–1997 reported as 8 per 100,000. Life expectancy at birth in 1997 was 77 years. Immunization is given by private practitioners, and coverage for infants is relatively low: 45% for DPT, 80% for polio, and 75% for measles in 1995–1997 (UNICEF, 1999).

Bismarckian Health Insurance. Germany's system of national health insurance, based on Otto von Bismarck's plan introduced for low income workers, is financed through a social security system by employer and employee contributions. As discussed earlier, the Sickness Insurance Act of 1883 provided that all workers earning below a designated level be insured by a sick fund, with employer–employee contributions.

The sick funds (*Krankenkassen*) might be owned by unions or employer associations, which can operate their own health services to provide comprehensive medical and hospital services for enrolled members and their families. The side funds or mutual benefit societies may also provide cash benefits for accidental injuries, burial benefits, and widow's pensions. This plan was later extended to cover virtually the entire population and remains the foundation of Germany's health and social insurance up to the present time.

Premiums, or payments for the sick fund by the employee, are income-based, and not linked to medical usage. Associations of sick funds and physicians' associations negotiate fee schedules for services, and the sick funds are empowered to set employer/employee rates based on anticipated costs.

By 1914, 13% of the total population was insured. This increased to 32% in 1932, 85% by 1960, and 90% in 1986. In 1993, 85% of the population was covered by mandatory insurance through local or national sick funds, and 15% by private voluntary insurance schemes. At present, the statutory, compulsory health insurance is financed by employers (50%) and employees (50%). Health benefits are very broad, covering medical, dental, hospital care, and prescription drugs, as well as home care and rehabilitation at spas.

State governments have the authority to plan hospitals, of which half are operated by municipalities, one-third are run by nonprofit NGOs, and the remainder are for-profit operations. By 1972 and 1985 legislation, hospital capital costs were funded by state and local governments through a certificate of need. Operating costs were paid on a per diem basis by the sick funds by standard rates for all patients but differing among hospitals. There were no incentives for hospitals to reduce costs of utilization.

The professional associations and hospitals have had a strong role in determining the costs of health care by high salary levels and promoting an emphasis on high technology, high levels of surgery, and overlapping services. Patients have choice of physician but may be obliged to join one of the 1241 sick funds according to the choice of the employer or a person's professional grouping. The poor and unemployed must join the AOK sick fund (*Allegemeine Ortskrankenkasse*), which has special government support and is obliged to accept anyone.

Health Insurance Reform. In Germany, governments contribute 21% of total health expenditures, while employer/employee contributions make up 60%, out-of-pocket 11%, and private insurance 7%. Hospitals were paid on a per diem

basis, including salaried physician services. In 1986, global budgeting was introduced for hospitals, intended to promote cost-effective services, outpatient treatment, and hospital financing for greater ambulatory care and coordination of medical care. Germany has a high hospital bed supply and low occupancy rates. In
1988, and again in 1993, health reform laws were passed trying to restrain health
costs, the latter law limiting fee increases, the supply of physicians, and use of expensive technologies in ambulatory care. In 1986, Germany spent $1,014 per capi-
ta on health care (compared to $1,289 for Canada and $1,917 for the United
States), but these increased to $2,233 for Germany, $2,065 for Canada, and $3,898
for the United States in 1996. The percentage of GDP spent on health in Germany
increased from 9.2% in 1986 to 10.5% in 1996, the second highest in the world after the United States.

Germany's health care standards are among the highest in the world, but mortality rates from cancer, cerebrovascular disease, and heart disease place Germany between sixteenth and twenty-fourth place among the leading countries.
Health promotion approaches are not part of sick fund responsibilities, but are
under discussion in health reforms. Germany's health system has to cope with
the challenge of integrating the former East German health system and population. This is more than a matter of raising standards of medical care; poverty, unemployment, low standards of housing, and poor nutrition will be major factors
in this absorption process.

The Netherlands

In 1997, the Netherlands had a population of 15.7 million, with a 1996 GNP
per capita of $25,940 and life expectancy of 78 years, among the world's highest.
The infant mortality rate declined from 18 in 1980 to 5 per 1000 live births in 1997,
compared to U.S. rates of 26 and 7 per 1000, respectively. Maternal mortality was
10 per 100,000 live births in 1980–1997. Immunization coverage in infancy in
1995–1997 was 95–96% for DPT, polio, and measles. The Netherlands experienced two outbreaks of poliomyelitis among nonimmunized religious groups from
imported poliovirus in 1987 and 1992.

The health care system of the Netherlands is a combination of public and private financing, with private delivery of care. The system evolved from medieval
guilds and mutual benefit associations to health insurance through employer–
employee payments to nonprofit sick funds or private insurance plans. By 1933
health insurance offered by such groups covered 41% of the population. National
health insurance was introduced in 1941 (by the Germans!). Sick funds were established on a geographic basis covering a majority of the population. Employees
contribute 4.95% and employers 3.15% of wages. Physicians are paid on a fee-for-
service basis for insurance patients and by capitation for sick fund patients.

Patients must have a referral from their general practitioner before seeing a specialist (i.e., the GP as gatekeeper). This helps to prevent unnecessary referrals,
strengthening the role of the GP and helping to control health care costs. Most specialists are hospital-based and are paid on a fee-for-service basis. Hospitals are not-
for-profit and paid on a block budget negotiated with the sick funds and private in-

surers. The supply of hospital beds is closely regulated by the government as is technology investment, restraining cost increases for the hospital sector.

Reform of the health system in the Netherlands emphasizes competition and market-based approaches to give consumers greater choice of sick funds or private insurance. Health expenditures as a percentage of GDP increased from 6.7% in 1972 to 8.4% in 1982, remained relatively stable until 1993 when it rose to 8.9%, and subsequently fell to 8.6% in 1996. The acute care hospital bed to population ratio was reduced from 5.5 in 1970 to 3.9 beds per 1000 population in 1995.

Mortality patterns show the Netherlands population to be at relatively high risk for cancer, but at lower risk than most northern European countries for cardiovascular disease. The Dutch health system has been very successful in restraining cost increases as compared with the United States, Canada, and many other European countries, while providing universal coverage, preserving primary care medical services, and achieving health status measures among the best in the world.

JAPAN

Japan, a centralized industrialized democratic country with a 1997 population of 125.6 million and a per capita GDP of $40,940 in 1996, has one of the lowest infant mortality rates in the world, 4 per 1000 live births in 1997. Maternal mortality was reported at 8 per 100,000 in 1980–1997. Immunization coverage in 1995–1997 was 100%, 98%, and 94%, respectively, for DPT, polio, and measles vaccines. Longevity is among the world's highest, with a 1996 combined male and female life expectancy increase from 72 years in 1970 to 80 years in 1997.

Following World War II, the Japanese placed emphasis on maternal and child health, providing free maternal and child care services. Pregnant women receive maternity bonuses to encourage early prenatal care; child care services include an extensive immunization program, screening for diseases of the newborn, developmental testing, and special care for low birth weight or handicapped newborns.

Japan has very low rates of heart disease, diabetes, and malignant disease mortality, but relatively high rates of stroke and trauma (motor vehicle accidents and suicides). Coronary heart disease death rates in Japan are low, 25 per 100,000 for men as compared to 118–164 in Canada, the United States, Sweden, and the United Kingdom. However, cerebrovascular death rates are higher than in these and other countries. Stomach cancer rates are higher, but lung and breast cancer mortality lower. The Japanese diet is low in animal fat and cholesterol, which may relate to the low cardiovascular disease mortality rates, but high in smoked and salty foods, perhaps explaining the higher cerebrovascular disease and stomach cancer mortality rates.

National Health Insurance

The basic health insurance program was enacted in Japan in 1922 as an extension of the employment-related social insurance law of 1874. In 1935, health insurance was extended to all manual workers, then further expanded in 1938 to self-employed persons. By the mid-1960s virtually the entire population was covered by a health insurance plan, either through employers, local government, or trade associations. Government-managed health insurance covers employees of small businesses of less than 300 employees, which encompasses some 29% of the population. Large companies, or groups of companies, with more than 700 employees, as an alternative to the government health insurance plan, can set up independent insurance plans for their employees. This currently covers some 25% of the population. Mutual aid associations provide coverage for civil servants, educators, and others (approximately 10% of the population).

Two laws, one promulgated in 1972 and the other in 1992, provide coverage for the elderly and low income earners (32% of the population). Insurance for these groups is administered by local authorities or trade associations. There are also many health laws governing a wide range of issues including nutrition, tuberculosis prevention, communicable disease control, mental health, environmental sanitation, and health planning.

Financing and Services

Japan's health service is financed by a payroll tax with rates fixed by law at 3.6–4.55% for employees and 4.1–4.7% for employers. Government subsidies for health insurance cover 65% of health costs, with control of costs by national obligatory fee schedules for a basket of covered services. Copayments by patients include 10% for employees and 30% for their dependents for hospital care and outpatient care. Health plan benefits include medications, long-term care, dental care, and some preventive services, as well as medical and hospital services. Preventive care is provided free of charge through a nationwide network of health centers with costs shared by the central and local governments. Expenditures for health increased from 4.8% of GDP in 1972 to 6.8% in 1982, but have remained stable since, with expenditures of 7.1% of GDP in 1996, well below those of most industrialized countries. Expenditures per capita in Japan were $1,713 in 1996, compared to $3,926 for the United States, $2,112 for Canada, and $1,358 for the United Kingdom.

Japan has a very high hospital bed-to-population ratio, and few beds designated for long-term or nursing care. Hospital utilization rates are therefore distorted compared to other countries. Admission rates are lower than in western countries, with a greater proportion of care provided on an ambulatory basis, but average lengths of stay are 4–8 times as long. The total hospital bed ratio, 16.2 per 1000 in 1996, with an average length of stay of 49 days, is high compared to Organization of Economic Cooperation and Development (OECD) averages of 8.4 days and 14.4 beds per 1000, respectively. Hospitals, generally small with an average size of 166 beds, include both acute an chronic nursing care patients. On the other hand, Japan has relatively fewer physicians, with a ratio of 1.8 per 1000 in 1996 as compared to the OECD average of 2.7 per 1000.

Patients have free choice of doctors, two-thirds of whom work as private practitioners in both public and private hospitals. About one-third of physicians are solo general practitioners, paid on a fee-for-service basis, which favors primary care. National fee schedules promote primary care by financial incentives. Physicians also dispense medicines in their private clinics, so that the Japanese consume more medications than most industrialized populations. Physician contact rates are at least double those in western countries; 12.9 contacts per capita per year, compared to 2.8 in Sweden and 5–7 in Canada, the United States, and the United Kingdom.

Japan has had very low birth and fertility rates since the 1950s. This, coupled with low mortality rates, are contributing to an aging of the population, posing a problem for the health services in the years ahead. These include a need for geriatric facilities, nursing homes, home care, and support services for family care of the elderly. Proliferation of medical technology is a problem in the health system, and cost containment is now a major issue, with governmental regulation in health care likely to increase.

Summary

The Japanese health system is highly decentralized, but regulated by the national authorities. It has achieved success in lowering mortality rates for most ages and conditions to among the lowest in the world, while restraining health costs. Incentives for primary care seem to have been successful, despite the promotion of excess use of medication. Japan has a high total hospital bed ratio because it has a high percentage of elderly in its population and lacks alternative facilities for long-term care. The problem of caring for the elderly will be a challenge in the years ahead.

RUSSIA

The Russian Federation is the largest country in the world, stretching from Europe to the Pacific, with a highly urbanized (77%) and educated multiethnic population of 148 million persons (1997), a strong industrial base, and abundant natural resources. Despite major improvements in health status during the Soviet period, its health status lagged well behind other industrialized countries. Since the collapse of the Soviet regime, health conditions have worsened dramatically, with life expectancy declining from 69 years in 1970 to 65 years in 1997.

The Soviet health system, developed since 1918, brought health care to a vast, underdeveloped country. This system provides universal access, within a governmental operated system of service. It was a source of pride to the Soviet state, and was recognized internationally as an important model, because of its successes from the 1930s. With universal access of the population to preventive and curative care, control of infectious disease was achieved and the health status of the population improved. During the 1950s, the Soviet model health system was widely promoted and emulated in eastern Europe, and in newly independent countries in Africa, Asia, and the Middle East as well as in Latin America. It also influenced the development of the Alma-Ata approach of Health for All based on primary health care.

From the 1960s an epidemiologic transition occurred in varying degrees in the different republics and ethnic populations of the Soviet Union. This transition was characterized by declining mortality from infectious diseases and rising death rates from noninfectious diseases. Life expectancy remained static during the 1970s and 1980s. In the 1990s, life expectancy declined dramatically since 1990, especially for men, during the economic and social crisis following the breakup of the Soviet Union. In 1992, the President of the Russian Federation, Boris Yeltsin, issued a Report on the Health of the Russian Federation detailing the poor state of national health and the need for health reform.

After the 1991 breakup of the Soviet Union, the Russian Federation entered a period of political, economic, and social reform with important effects on the national health system and health of the population. In 1993 a compulsory national health insurance plan was adopted to augment funding and promote decentralized management of health care and movement toward a market economy in health. The health issues are, however, complex, and changing methods of financing medical care services alone may worsen the health situation by reducing access to care. For this reason, this section will deal not only with the system of care, but reform of content needs as well.

The Soviet Model

Prior to the 1917 revolution, Russia was a largely rural country with higher mortality rates than European countries. Public medical care and other social services for the rural poor majority were established in Czarist Russia in 1864 under the local district assemblies or *Zemstvos,* providing tax-financed services for medical and hospital care. Health insurance was established in 1912 based on the Bismarckian social security model, covering about 20% of industrial workers.

Following the tragic conditions of World War I, the 1917 October Revolution, and the Civil War, Russia was racked by mass epidemics and starvation. In 1918, reconstruction planning included the Soviet concept of health care formulated by Nikolai Semashko, based on the principles of government responsibility for health; universal access to free services; a preventive approach to the "social diseases"; quality professional care; a close relation between science and medical practice; continuity of care between health promotion, treatment of the sick, and rehabilitation; and community participation.

The state undertook to provide free medical services for all, through a governmental, unified health system. The "social diseases" referred to all diseases related to the poor living and working conditions of the workers, mainly infectious and occupational diseases, as well as maternal and child health problems. Epidemic control was implemented on an urgent basis, especially for tuberculosis, typhoid fever, typhus, malaria, and cholera. Community prevention approaches were enforced, often with use of draconian measures. Prophylactic measures such as quarantine were implemented, urban sanitation and hygiene improved, and malarial swamps drained. Medical prevention of social diseases focused on regular checkups for the population-at-risk. From the 1920s, emphasis was placed on prevention and control of infectious diseases. In order to meet the needs of the system of providing health care throughout the country, increases in the supply of hospitals, polyclinics, doctors, and nurses was a national priority. In 1937, all insurance and hospital-based sick funds were closed, and hospitals and other health facilities nationalized and organized under district health management. Virtually all health personnel became public employees. Parallel services were provided within industries and for special categories, especially party leadership, some ministries, defense and security personnel, miners, workers in heavy industries, and transport workers.

General government revenues provide financing of health services as part of national plans for social and economic development. The central administration directly employed staff, paid salaries, and provided supplies for all health care facilities and research and training institutes. Directors of health facilities therefore administered their allotted resources, supplies, and manpower with no opportunity for program management or financial accounting of service costs.

The health system was developed, financed, and managed under strong central government control. Mandatory norms for facilities and manpower were enacted by the Commissariat (later Ministry) of Health, under strict regulation of the central authorities of the Communist Party. These norms were revised periodically at Party Congresses, with expansion of services being the major policy orientation. The policy of continuing to increase the supply of hospital beds and medical personnel was reiterated in the mid-1980s, and continued into the 1990s.

During World War II the Soviet health system was mobilized for the war effort, providing care for huge numbers of military and civilian casualties. Despite harsh conditions for both military and civilian populations, no mass epidemics occurred. External observers including Garrison in the 1920s, Sigerist in the 1940s, Field in the 1960s, and more recently Roemer (1991, 1993), as well as Russian medical historians, Yeravinski, Smirnov, and others, have noted the remarkable achievements of reducing epidemic diseases, meeting wartime demands, and bringing health care to the whole country. Postwar stabilization allowed restoration of health services and makeup of trained personnel lost in the conflict.

In order to assure equal access, a district system evolved with each required to have sanitary–epidemiologic stations (SESs), hospitals, polyclinics, and specialized treatment facilities according to national norms based on population size. The SESs supervised water, sewage, air, and ground quality, as well as conducting epidemiologic investigations of infectious disease outbreaks and monitoring child health and nutrition status. *Medsanchest* clinics located in industrial plants provided on-site medical and occupational health services, and prophylactory health centers provided a variety of medical rehabilitation services, sanatoria, and vacation benefits. Originally, polyclinics in each district were linked as outreach facilities to the district hospital with staff rotation between them to promote continuity of care and improve professional education. However, this became impractical because of rapid expansion of the number of polyclinics. Prevention of disease continued to be based on routine screening checkups for workers and other specified groups.

TABLE 13.5 Hospital Beds and Doctors per 10,000 Population, Russian Federation, Selected
Years, 1913–1998

Resource	1913	1940	1950	1960	1970	1980	1990	1996
Hospital beds/10,000	15	43	59	82	113	130	131	111
Doctors/10,000[a]	1.5	7.4	14.5	20.9	29.0	40.3	40.7	42.1

Source: Adapted from Field, 1988, Ministry of Health of the Russian Federation, 1996, and Health
for All Data Set. WHO European Region.
[a]Includes all categories of doctors.

As a response to the increasing prevalence of chronic diseases in the mid-1960s,
the Communist Party Plenum in 1983 decided to implement annual *dispanser-
izatzia* or checkups as a uniform program for the general population, provided in
polyclinics, hospitals, and specialized clinics. The checkups and treatment involved clinical care, follow-up ambulatory or hospital care, sanatoria (i.e., rest
homes), and a change of work if necessary. The screening program increased demands for hospitalization because of limited ambulatory diagnostic resources.

Despite limited resources allotted to the health system, central government
health planners continued to place emphasis on increasing the supply of hospital
beds and manpower (Table 13.5). In the mid-1980s, the Ministry of Health enunciated the continued direction of health policy as concentrating on "development
of preventive medicine and improvement of health care facilities through a program for building general and specialized hospital establishments." With central
control of financing, the state set mandatory norms for manpower and hospital
beds, and controlled medical education to produce the manpower to operate the
system. The state monopoly on health produced, however, led to a conceptual stagnation which biased the system toward hospital care, without financial or epidemiological accountability for efficiency and effectiveness.

The government-operated health system was funded with health expenditures
of 3–3.5% of GNP, while other industrialized countries expend 7–13% of GDP.
The Russian per capita GDP declined from $3,220 in 1991 to $2,410 in 1996. The
1994 estimate of expenditures on health was 2.3% of GDP. Despite the declining
resources for health care, Russia maintains a very high hospital bed-to-population
ratio and continues to produce medical graduates at high rates.

Epidemiologic Transition

Life expectancy improved up to the 1960s, but since has lagged well behind
other countries. In 1960, life expectancy was 69 years, but declined in the early
1990s (61 for men and 73 for women) then increased to 67 years in 1998. Very
high mortality rates from cardiovascular diseases and trauma are primarily responsible for low and declining life expectancy. Cardiovascular disease mortality
is twice as high as in OECD countries. Trauma mortality in males is 2.5–3 times
that of western industrial countries.

TABLE 13.6 Age Standardized Mortality Rate per 100,000 Population for Selected Causes of Death in Russia and Other Countries, 1990–1991[a]

Country	Total mortality	Total cancer	Lung cancer	IHD	CVD	Respiratory disease	Infectious diseases	Trauma
Russia	1454	237	56	404	308	76	13	152
Lithuania	1246	218	47	546	154	59	10	129
Germany	1015	238	42	201	119	61	6	55
Denmark	997	258	57	241	90	72	8	67
United States	933	255	64	215	62	84	13	62
England/Wales	926	248	58	241	105	98	5	31
Israel	910	183	28	183	86	68	19	52
Sweden	845	188	25	226	89	62	6	27
Canada	826	227	60	189	62	72	6	51
Japan	720	185	32	43	101	94	11	33

Source: *World Health Statistics Annual 1993*, pp. 438–463.
[a]Data are standardized to the European population. Trauma includes accidents, poisonings, homicide, and suicide. IHD = ischemic heart disease; CVD = cerebrovascular disease.

Even before the impact of the collapse of the Soviet system was felt in 1991, mortality rates in Russia were much higher than those in other industrialized countries, as seen in Table 13.6. Standardized mortality rates in Russia were 1.5 times higher for total mortality and even higher in categories such as cerebrovascular disease, trauma, and infectious diseases. Control of infectious diseases has declined since 1991, with epidemics of diphtheria, primarily among adults, throughout Russia, with measles and pertussis increasing. There is also evidence of cholera in southern Russia.

The crude birth rate has declined in Russia from 17 per 1000 in 1987 to 9 per 1000 in 1998. The total fertility rate declined from 2.0 in 1989 to 1.2 in 1997, well below replacement levels. Infant mortality rates fell from 22 per 1000 in 1980 to 16 per 1000 in 1998, still three to four times those in western European countries. Abortion is the main method of birth control, and modern methods are not widely available or trusted. Maternal mortality in Russia declined from 65 per 100,000 live births in 1980 to 44 in 1998 (HFA). Overall high mortality and low birth rates create a situation of declining population. In 1993, deaths exceeded births by 800,000.

The decline in health status since 1990 cannot be blamed solely on the current economic crisis, nor entirely on the health care system. The worsening mortality pattern is due to a combination of factors: stress, alcohol, smoking, violence, lack of a balanced diet, lack of modern health care technology, environmental pollution, and a general mood of anxiety and depression related to the dramatic decline in economic and political stability since 1990. Males are not surviving to pension age at 60 years.

A combination of factors encouraged a medical bias toward care of individual patients and failure to apply the successful experience of the 1930s to the control of epidemics of noncommunicable disease. The concept of prevention took on a primarily medical orientation, stressing routine checkups. Health policy continued to promote increased supplies of doctors, polyclinics, and hospital beds. A vastly oversized hospital sector with a passive strategy of treatment and long hospital stays was unable to keep up with technological advances and consumed a large share of the very limited amount of funds allocated to health care. Poor medical care, inadequate diet, environmental pollution, and lack of public health efforts contribute to this public health crisis.

Post-Soviet Reform

The Russian Federation continues to provide basic social security and health care for all citizens. Until 1993, when compulsory health insurance was established, all social benefits were funded from the general budget of the government. The health insurance scheme was based on mandatory payment by employers to regional health insurance funds. Decentralization of management of state services, and financing of health care has increased regional and local health autonomy. Declining real expenditures and a severe shortage of foreign currency hamper supplies of drugs and other materials. The sudden and complete decentralization has hampered central management capacity to develop new policies for public health issues such as immunization, nutrition, and health promotion as well as health management issues such as national norms for hospital beds and manpower.

Epidemiologic, economic, and cost-effectiveness analysis is vital to reform in health care, especially in harsh economic conditions. Calculation of the cost of a service is fundamental. Regional or district authorities now have financial responsibility and power to reallocate funds and shift priorities from institutional treatment to prevention and ambulatory care, but lack of trained health management personnel to challenge old assumptions such as the norms for hospital beds and manpower, which are still accepted as guidelines.

Development of information systems, training of leadership personnel in modern management theory and practice, and reduction of the hospital sector with transfer of resources to primary care is needed to improve health care efficiency and quality. Reforms based, in part, on reallocation of existing resources, will require additional funding to meet the cost of the transition and raise the quality of care.

Health reforms are essential to preserve universal access and to raise population health status, medical care, and public health to international standards. Changes in financing of health care, adoption of international health targets, and changes in manpower development programs are needed. But these depend on a new set of priorities and new standards at the national, oblast (province/state in Russia), and local health authority levels of government. Decentralization and diffusion of the overly centralized system requires epidemiologic information and dialogue on health issues to raise health awareness and management practices to meet the health needs of the people.

The sanitary–epidemiologic stations have potential to lead health promotion activities at the community and district level, raising public awareness and knowledge of health issues. This will mean a change in attitude from being defenders of the old system to responders to community needs. That means redefining objectives, instituting training programs for new manpower, modernizing technology for laboratories, and redefining environmental quality and enforcement issues.

Training for epidemiologic data collection, analysis, and its distribution are essential to promote knowledge of risk factors and their control. Policy, provider, and community levels need to define cost-effective programs to meet local conditions. Wide distribution of relevant data to government and the general public is needed to help change knowledge, attitudes, beliefs, and practices related to risk factors.

The current system of health care organization and financing in Russia needs conversion to a modern managed care system by adoption of capitation payment as used in the Scandinavian countries, the United Kingdom, and the HMO model in the United States. Adaptation of the principles of HMOs has been started on a pilot basis in the cities of Tula and Kaluga. Experimentation and trial and error, the new health insurance plan, and decentralization of management of the health system is producing a potential for reform.

The Russian health system has important assets with potential for change. Universal access should be preserved. Decentralization is stimulating local initiatives in health care reform. Innovation in seeking cost-effective health measures and responsiveness to local and regional epidemiologic evidence are needed to respond to consumer needs and to advance medical standards. Health financing reform may help to raise the level of resources, but this is unlikely during the period of economic transition. Needed resources will have to be found by reordering priorities and reallocating expenditures.

Summary

Russia in the 1990s is in the midst of a health crisis with declining longevity, high and rising mortality rates, and the reappearance of epidemic diseases. This crisis relates not only to the period of economic transition, but goes deep into the former Soviet health system. This state-operated service provided free, universal health care with ample, indeed excessive resources in medical personnel, hospital beds, polyclinics, and other services, but with quantity compromising quality. The system operated as a state monopoly, with the central government controlling budgets, setting mandatory norms, and totally controlling manpower training and research. It lacked mechanisms for epidemiologic or economic analysis and accountability to the public. The epidemiologic transition from a predominance of infectious to noninfectious diseases was addressed by further increases in the quantity of services. Policy and funding favored hospitals over ambulatory care, and individual routine checkups over community-oriented preventive approaches.

Reform since 1991 has centered on national health insurance and decentralized management of services. However, the reform movement lacks a broad national health strategy to address the fundamental public health problems and especially the present enormous excess of preventable mortality. The obligation of finding a new direction rests primarily with the Russian federal and local government health authorities. Establishing adequate epidemiological data and defining and working toward health targets require political, moral, professional, and financial support, both at home and abroad.

ISRAEL

Israel in 1997 had a population of 5.8 million, with a GNP per capita of $15,870 (1996). Infant mortality in 1997 was 6 per 1000 live births, and maternal mortality was 5 per 1000,000 (1980–1997). Immunization rates are high, with 1995–1997 rates of 92% for DPT, 93% for polio, and 94% for measles. Despite high immunization rates, an epidemic of polio with 15 cases occurred in 1988, measles epidemics in 1991 and 1994, leading to the adoption of improved immunization policies.

Health indicators for Israel show an advanced state of health but with important ethnic, regional, and gender inequalities. In infant mortality, Israel ranked twenty-second in 1995 among the advanced countries, a rise in ranking from twenty-fifth place in 1990. For longevity, Israel is second among the leading countries for males and among the top 10 for female longevity.

Origins of the Israeli Health System

Israel's health system evolved gradually over the past century. Palestine under the Ottoman empire was a poor disease-ridden remote province rife with malaria, dysentery, and other infectious diseases. Immigration of Jews from eastern Europe and Arabs from surrounding countries since the 1880s led to initiation of charitable hospitals to provide care for the urban poor.

Jewish immigrants from eastern Europe formed labor brigades and mutual aid associations. Sick funds were initiated in 1912 based on mutual benefit principles derived from European models, associated with the union movement and later with other political organizations. The sick funds grew to provide medical care insurance and services to over 95% of the population. They provide services through neighborhood and specialized clinics, or affiliated doctors in their own clinics, purchasing hospital care from governmental or NGO-operated hospitals in areas where they lack their own.

Preventive care originated in 1911 by nurses from the United States sponsored by Hadassah, an international women's organization. Following the conquest of the area by British forces from the Turks in 1917, Hadassah sent the American Zionist Medical Unit from the United States to help establish a network of health facilities in Palestine. This consisted of 44 doctors, nurses, dentists, and other personnel with equipment and financial support from Hadassah and the Joint Distribution Committee. The Unit opened hospitals in many urban centers, and established nursing training and preventive care programs for immigrants and schoolchildren, as well as mother and child health stations (*Tipot Halav* or drop of milk stations). These were gradually located in towns, villages, and neighborhoods

throughout the country providing prenatal care and child care for infants and toddlers. They provide immunization, child development monitoring, and nutrition
counseling to almost all the infants in the country, and prenatal care for most
women in the country, others going to private doctors.

The British mandate from 1917 to 1948 brought successful colonial administrative experience and development of basic public health law and systems, licensing of medical professions, sanitation, food and drug laws, as well as public
health laboratories, malaria control, and many other features of public health of
the standard of the time.

From 1912 to 1948, the health system grew based on primary health care
through the *Tipot Halav* and the labor movement's sick fund clinics in towns and
villages throughout the country, providing ready access to primary care treatment
and referral services.

Following the establishment of the State of Israel in 1948, massive immigration from post-Holocaust Europe and the Middle East brought an enormous burden of health problems to the country. The new Ministry of Health established
regional hospitals throughout the country in abandoned British army camps,
providing acute care, rehabilitative services, mental health, and chronic care services. Other hospitals are owned by the major sick funds and by NGOs. Reliance
on ambulatory and primary care with regional medical and hospital centers was
the basis of the Israeli health system.

Health Resources and Expenditures

Health care expenditures as a percentage of GNP increased from 6.9% in 1980
to 7.9% in 1990 and 8.4% in 1997 (Table 13.7). Over the period 1985–1997, total bed capacity was reduced from 6.5 to 6.0 per 1000, but there was a shift in composition of the beds. There was an increase in geriatric and rehabilitation beds,
from 1.8 to 2.7 per 1000, fewer acute care beds, a decline from 2.8 to 2.3 per 1000,
and for psychiatric beds, from 1.9 to 1.1 beds per 1000.

Expenditures on hospital care increased from 34 to 41% of total health expenditures from 1975 to 1990, followed by a decrease to 39% in 1993 (see Table 13.8).

TABLE 13.7   Health Expenditures, Hospital Resources, and Utilization, Israel, 1960–1997

Resource/Utilization	1960	1970	1980	1985	1990	1997
Health expenditures as % GNP	5.5	5.4	6.9	6.8	7.9	8.4
Acute care beds/1000	3.1	3.2	3.0	2.8	2.6	2.3
Hospitalization (acute) days/1000 year	954	1148	997	911	834	785
Discharges (acute)/1000 year	115	129	158	163	158	181
Average length stay (acute care)	8.3	8.9	6.3	5.6	4.8	4.3
Mental health beds/1000	2.1	2.4	2.2	1.8	1.5	1.1
Mental diseases days/1000	615	631	721	623	496	379

Source: *Statistical Abstract of Israel,* 1975, and Ministry of Health, *Health in Israel,* 1998.

TABLE 13.8   Health Expenditures (%) by Type of Service, Israel, 1975–1995

Category	1975	1985	1990	1995
Hospitals and research	33.9	42.7	40.7	40.6
Public clinics/prevention	28.6	32.5	32.6	34.5
Dental care	11.5	11.0	12.0	11.1
Private physicians	3.1	3.4	4.9	4.5
Medicines, equipment purchased by individuals	4.5	4.7	4.8	3.5
Fixed capital formation	17.4	4.9	5.3	4.7
Government administration	1.0	0.8	0.8	1.1
Total	100	100	100	100
National expenditures as % of GNP	6.0	6.8	7.8	8.6

Source: *Statistical Abstract of Israel,* 1996, and *Health in Israel,* 1998.

There has been a steady decline in the hospital bed supply, days of care, and average length of stay. Ambulatory care and community health consume about one-third of total expenditures, an increase since 1975. Fixed capital formation has fallen steadily since the 1970s as a percentage of overall expenditures. Salaries in the health sector have been low compared to other sectors in the society, as has been capital investment. Cost restraint, improving the physical infrastructure, and keeping up with technological advances in medicine are a major challenge for the future.

Health Reforms

Israel implemented a national health insurance (NHI) plan on 1 January 1995 covering the total population through the universal National Insurance social security system after many years of debate and gradual reform of health services. The individual pays for this through a 3% deduction from his or her salary along with an equivalent employer's contribution to a mandatory National Insurance payment, which also covers old age and disability pensions, workmen's compensation, and other social benefits. Each family must select membership in a sick fund which functions as a health maintenance organization. The National Insurance Institute transfers funds to the sick funds according to a per capita formula, with a larger per capita payment for the elderly and for populations in development towns.

The Ministry of Health supervises the sick funds, which are required in the new law to provide a basic basket of services that is very comprehensive. The sick funds are obliged to provide all specified services and arrange for those services it cannot itself provide. They provide comprehensive care either through their own neighborhood clinics or through affiliated private physicians who are paid on a capitation basis. Additions to the obligatory basket of services, such as new medications, are funded by the Ministry of Health.

The sick funds are accountable for the services rendered, and the individual is entitled to change sick funds semiannually. Most hospital beds are operated directly by the Ministry of Health. Government hospitals are meant to be transferred to independent trusts, to operate as economic units able to allocate funds as they deem necessary and compete for clients, and paid on a DRG basis. Regionalization of services will be difficult to achieve in the present configuration of the NHI law because each sick fund has its own regional organization.

Health promotion is gaining strength in Israel, and health awareness has generally increased (Table 13.9). The compulsory seat belt law has met with compliance by a large majority of car drivers, and similar legislation requiring use of helmets for motorcycle drivers is also generally implemented. Similar requirements are proposed for bicycle users. Increases in permitted speed limits on major highways has been followed by a rise in motor vehicle deaths and case fatality rates.

Antismoking legislation banning smoking from public buildings and workplaces has helped to reduce smoking in the adult population, especially among males. Smoking is still practiced by some 30% of adults. Low fat foods are now commonly available in supermarkets. Private food manufacturers fortify their baby formulas and cereals with vitamins and minerals. Breakfast cereals are also enriched, but food fortification of bread, milk, and salt with essential minerals and vitamins is not practiced. Obesity is common especially in women over age 45, and while consciousness of the importance of physical fitness is increasing, it is still not at an acceptable level.

Regional, social, and ethnic disparities are still important in Israel's health status; the Arab population has higher rates of infant mortality than the Jewish population, 12 versus 6 per 1000 in 1996. Large scale immigration of Russians and Ethiopians in the 1990s brings together people with different risk factors. Traditional distribution of health resources favors the more concentrated population centers, while the more rural areas receive fewer resources per capita.

The health agenda has paid increasing attention to health promotion as well as to the structure of health services in the new health reform. The Ministry of Health concentrates on reform in health services and national health insurance, but needs a cohesive strategy with health targets and strengthening of health promotion strategies. Decentralized administration of services is through the sick funds now emphasizing the health maintenance function. Regional disparities in health resources and health status constitute a powerful argument for regional budgeting, which is now being discussed by the centrally managed sick funds.

Summary

Israel has achieved high standards of health care and health status indicators. The Israeli health system has been a quasi-national health service for many years. The system has helped the Israeli population achieve low rates of mortality from infectious and noninfectious diseases and life expectancies among the longest in the world. Medical and paramedical professional education, research, and medical and drug industries have reached high levels of excellence. The 1995 implementation of national health insurance provides greater equity in financing and reduces political manipulation of the health system. Primary care services still separate preventive and treatment community-based facilities, so that the sick funds are insufficiently

TABLE 13.9   Status of Health Promotion Initiatives in Israel and Their Effects, 1997

Topic	Action	Effects
Smoking	Antismoking legislation Restricted advertising Antismoking promotion by NGOs	Declined for men from 1973 to 1998 from 41 to 33%, and for women from 36 to 25%
Cancer prevention	Promoting mammography, reduced sun exposure Restricting smoking	Slowly improving public awareness; cancer increasing, but far lower than most European rates, except for breast and colon cancer
Nutrition	Health education Mediterranean diet; high vegetable, fruit intake; high but declining sugar consumption; low consumption of animal fats	Possible contribution to rapidly declining cardiovascular mortality and low rates of cancer
Motor vehicle accidents (MVA)	Mandatory seat belt law implemented Highway speed limits raised Inadequate highway patrols	MVA and fatality rates increasing Driver behavior unimpeded by fear of punishment; electronic monitoring demonstrated to be more effective
Water qualtiy	Mandatory chlorination 1988 No mandatory filtration, coagulation	Less waterborne diarrheal disease
Sewage treatment	Increasing treatment	Reuse of wastewater increasing
Enrichment of basic foods	Law permits but does not require Private manufacturers initiatives	Breakfast cereals enriched Infant formulas and cereals enriched Basic foods (bread, salt, milk) not enriched
Food quality	Food supervision strengthened Standards improved to inter- national levels	Public awareness increased Low fat foods now widely available
AIDS/STD prevention	School health education	Widespread information on use of condoms, and avoidance of trans- mission in drug use
Healthy cities	Active association of healthy cities	Several cities organizing health cities programs
Dental health	Fluoridation of municipal water supplies	Fluoridation reaches 47% of population: mandatory fluoridation for 2001
School health	Long delayed, improving in 1990s education	"Know your body" studies wide- spread in schools No school lunch program

Source: Adapted from Israel Center for Disease Control. Health Status in Israel, 1999.

prevention-oriented. Regionalization of financing and service management is needed to alleviate disparities, but the Israeli health system has contributed to high standards of health for the whole population.

HEALTH SYSTEMS IN
DEVELOPING COUNTRIES

In most developing countries, health services were inherited from colonial regimes and subsequently influenced by the Soviet model of health care in the 1950s and 1960s. The development of primary health care was neglected, underfunded, with excessive allocation of resources to teaching hospitals in the main population centers, leaving little for the rural majority. As a result, most developing countries are facing the need to reform their health systems.

During the 1980s emphasis slowly moved toward primary care under the influence of the Health for All initiatives sponsored by the World Health Organization. Achievements during the 1980s and 1990s included greatly improved immunization coverage, widescale use of oral rehydration therapy (ORT), and improving sanitation. There has been a decline in birth rates in most regions of the world, including sub-Saharan Africa, which had until the 1990s seemed totally resistant to birth control. National health programs emphasize primary care with immunization, ORT, promotion of breast-feeding, supplemental feeding for infants, and birth spacing.

Many countries in the developing world are raising productivity and per capita GNP so that a combination of universal primary education and improved economic status is furthering the potential for continuing improvement in health standards. In sub-Saharan Africa, low and declining levels of economic activity reduce the likelihood of increasing funds for health care. AIDS, malaria, tuberculosis, measles, other infectious diseases, poverty, malnutrition, and high birth rates with high child mortality aggravate a poverty–population–environment (PPE) problem, which impairs national growth potential. As western diets, lifestyles, and technology are absorbed in developing countries, they face a dramatic increase in chronic diseases such as hypertension, diabetes, stroke, coronary heart disease, and motor vehicle accidents. This, along with increasingly costly technology, places new burdens on health services.

Most developing countries spend less than 4% of their low national incomes on health and much of that on costly hospitals in the capital cities. Lack of adequate government funding raises interest in national health insurance, especially in the mid-level developing countries, with the purpose of bringing more of the population into the health care system and raising additional funds for health care beyond the little that is provided through government allocations.

In this section, following brief regional overviews, we give several examples of developing countries actively working to reform their health care systems. Health insurance is needed to increase funding for health care and provide for the growing urban employed and middle-class health needs, but at the same time, ministries of health must provide direct services to the rural poor majorities. As in developed countries there will be no uniform approach, but sharing of lessons learned will be helpful.

SUB-SAHARAN AFRICA

Sub-Saharan Africa includes 40 countries with a total population of 576 million persons in 1996 and a GNP per capita of $501. Annual births of 24.8 million in 1996 represent a decline in total fertility rates from 6.6 in 1960 to 5.9 in 1997. Life expectancy increased from 44 years in 1970 to 51 years in 1996. Mortality of children under age 5 declined from 256 per 1000 in 1960 to 170 per 1000 in 1997. Maternal mortality remains very high, with a rate of 597 per 100,000 in 1990, with only 37% of births attended by a trained attendant (1997). Immunization rates for the major childhood diseases hover around 52%, with tetanus immunization of pregnant women at 39%. Malnutrition (acute and chronic) further damages child health status. Tuberculosis, malaria, AIDS, measles, and other infectious diseases are major contributors to high rates of morbidity and mortality. In addition, chronic diseases and trauma are increasingly important contributors to the total burden of disease.

In the face of economic decline, political chaos in many countries, and the aforementioned health problems, resource allocations for health budgets have been jeopardized in many countries. Some countries in the region devote less than $2 per capita to health budgets. Despite these challenges, progress has been made in efforts to improve sanitation and expand primary care services to the underserved rural areas and urban slums, giving hope for effective public health in the twenty-first century.

Dramatic progress has been made in the eradication of poliomyelitis, dracunculiasis, and onchocerciasis. Less progress is seen in tuberculosis, schistosomiasis, and malaria control. The WHO recommends wide ranging new efforts in prevention of cancer by immunization against hepatitis B, screening transfusion blood for hepatitis C, schistosomiasis control, smoking and alcohol control, reducing risk factors for cardiovascular diseases and diabetes, mental health, and oral health. The effects of civil war, collapse of governments in some areas, and refugee situations have had dreadful effects on public health. Since the mid-1990s, signs of stabilization in the governments of some countries of Sub-Saharan Africa and economic progress offer new hope for the future of this potentially wealthy continent.

Nigeria

Nigeria is the most populous of the sub-Saharan countries of Africa, with 118 million population in 1997, approximately one-third urban and two-thirds rural, the latter living in 97,000 villages. The annual number of children born in one year is 5 million. Despite the vast oil wealth brought in during the 1980s, Nigeria, long ruled by repressive military regimes, had a GNP per capita in 1996 of $240. In basic health indicators this country is among the least developed countries.

Adult literacy rates for men increased from 47% in 1960 to 67% in 1995, and for women from 23 to 47%. The birth rate has declined among the educated urban population but remains high in the Moslem northern half of the country and the southern primarily Christian rural areas, with an overall total fertility rate of 6.0 children per woman in 1997. Only 37% of infants are delivered by trained health personnel.

Infant mortality declined from 122 in 1960 to 112 per 1000 live births in 1997. Child mortality in 1997 was reported at 187 per 1000, down from 207 in 1960. Maternal mortality was 1000 per 100,000 live births in 1990, with only 31% of births attended by trained health personnel in 1990–1996. An estimated 750,000 children and 42,000 women die in childbirth each year. Life expectancy at birth increased from 43 years in 1970 to 52 in 1997.

Infectious diseases are the major causes of death, predominantly malaria, measles, meningitis, pneumonia, yellow fever, dysentery, tuberculosis, and AIDS. Immunization rates increased during the early 1980s but declined in the latter part of the decade. Immunization of pregnant women against tetanus was 23% in 1995–1997; infant immunization for BCG was 29%, 21% for DPT, 25% for polio, and 38% for measles. Measles accounts for 12% of child deaths and AIDS is a major public health issue in Nigeria as in other sub-Saharan countries.

Health care insurance is through Social Security and state National Assistance together with special group coverage for members of the armed forces and organized urban groups such as those working in the transport sector. The public health services suffer from low salaries, lack of supplies, and inefficient administration. Private practice is common in the urban centers, serving mainly the middle class. Drugs are expensive and imported in an unrestricted fashion.

The health system inherited from the British colonial period included limited hospital care in the urban centers, and some medical training facilities. Following independence in 1960, the state-operated health system began to develop a widened network of primary care services, in parallel with state primary education. Health care expenditures in 1992 were $1.50 per capita, and health constituted 5% of the national budget. The present health system is seriously underfunded and covers only two-thirds of the population, with large parts of the rural population outside of the system. Access to health services in the period 1985–1992 was estimated by UNICEF at 85% for the urban population, 62% for the rural, and 66% for the total population. Curative services have received the major share of the fiscal resources in hospitals and primary care clinics. Proposed changes in allocation will allocate health resources in Nigeria as follows: 15% to federal government-operated specialty hospitals; 25% to state government-operated district hospitals; and 60% to local government-operated primary health care clinics, including maternal and child health, school health, and other aspects of primary health care.

Medical manpower is 1 per 10,000 population, but most physicians are located in the urban areas. Despite medical training being considered of a high standard, overall quality of care and efficiency in health management are low. The federal government is undertaking initiatives to broaden health insurance in order to raise revenues for health care and to increase equity of access to services. A National Health Care Fund is being established that will receive funds from federal government general revenues, rural cooperative health insurance premiums, and employed persons' health insurance. This is expected to expand coverage for basic health care to a large part of the rural and urban poor population.

Nigeria's health problems are severe even by developing country standards. Medical education has been given high priority with 12 medical schools, but there is no school of public health. This reflects priorities of curative services, while the basic health problems of the country require the application of well-known and cost-effective public health programs. Increasing death rates from noninfectious diseases and trauma require attention in planning preventive and curative services for the future.

Summary

The evolution from a colonial health service very limited in scope to a centrally managed service with serious underfunding, and then to a more universal system, reflects current trends in many countries. Facing a population explosion and contracting economies, African countries have gone through a very difficult transition in the 1980s. The primary care system needs strengthening to meet the challenge of preventable diseases, which has been exacerbated by a decline in immunization coverage in recent years. Decentralization of organization to increase the role of the state and local government authorities may improve community participation and efficiency of services. It may also increase revenues by providing a mechanism for local financial input.

LATIN AMERICA AND THE CARIBBEAN

The Latin American and Caribbean region includes 22 countries, with a population of 487 million persons. Average GNP per capita increased from $2,883 in 1994 to $3,681 in 1996. The crude birth rate fell from 42 to 23 per 1000 population between 1960 and 1997, and total fertility rates fell from 6.0 in 1960 to 2.7 in 1997. Crude mortality declined from 11 to 6 per 1000 population from 1970 to 1997, while child mortality fell from 158 to 41 per 1000 and infant mortality from 106 to 33 per 1000 live births from 1960 to 1997. These all indicate impressive economic and health care progress for the region. Maternal mortality in 1990 was still high, 190 per 100,000 live births, with 78% of women attended by trained personnel in childbirth. Life expectancy increased from 60 to 70 years from 1970 to 1997.

Continent-wide eradication of wild poliovirus and success in control of measles and other vaccine-preventable diseases have been achieved. Tuberculosis, malaria, Dengue fever, Chagas' disease, and cholera are still major public health problems. Communicable, maternal and perinatal causes account for 32% of all deaths, injuries 10%, while noncommunicable diseases account for the remaining 58% of all deaths. Cardiovascular diseases are increasing as the associated risk factors of smoking, fatty diet, inactivity, hypertension, and diabetes are more prevalent than in the past throughout the area.

Rapid urbanization has created sprawling urban slums, with serious sanitary and other health hazards, placing great strain on social and infrastructure services. Access of the rural population to health services remains one of the critical issues in improving population and individual health throughout the continent.

Colombia

Colombia is a mid-level developing nation of 37 million population in 1997, more than two-thirds urban, with a per capita GNP of $1,330 in 1992 increasing to $2,140 in 1996. The infant mortality rate decreased from 82 per 1000 in 1960 to 25 per 1000 live births in 1997. Child mortality declined from 130 in 1960 to 30 per 1000 in 1997 and life expectancy at birth increased from 57 in 1960 to 71 years in 1997. Maternal mortality remains high at 50 per 100,000 in 1980–1997. From 1970 to 1997, the crude mortality rate fell from 9 to 6 per 1000. Primary school enrollment is universal, and literacy rates are high (87% in 1990). The percentage of people living below the absolute poverty level in 1992 was 32% of the urban and 70% of the rural population.

The national territory is divided into 32 departments, further subdivided into 107 regions and local government authorities. The local governments are responsible for health centers, health posts, sanitary services, health promotion, and community health worker programs. Physicians are paid a salary, but often work several jobs or conduct private practice.

The first hospitals were established in the Spanish colony in 1513, with a leprosy hospital set up in 1564. In 1798 Spanish law required all provinces to operate a hospital under management of religious orders. Basic public health laws were enacted in 1601 and the first medical school was founded in 1636. Sanitation boards established in 1825, followed by the Department of Charity and Rewards in 1848, were in charge of hospitals for the poor and slaves. A central Board of Hygiene, established in 1887, later evolved into the national health authority and in 1946 became the Ministry of Health.

Health System.   The health system in Colombia consists of three major elements: the official sector, Social Security, and the private sector. Each sector has a separate constituency, and services, but with substantial crossover.

The largest sector is the National Health System enacted in 1975, meant to cover some 70% of the population. It is administered by the Ministry of Health, providing national policies and local implementation at the district level. The municipalities provide hospital and ambulatory care with extensive use of *promatoras de salud* or community health workers. The National Health System is funded by government allocations (36%), departments and local contributions and taxes (22%), sale of services to insurance plans (12%), and the rest from various other sources including endowments and properties.

Social Security, enacted in 1937, covers industrial and commercial employees, and the self-employed, for comprehensive medical services. Civil servants, police, and military personnel and families are insured under a parallel program. The Social

Security system covers some 24% of the population. The private sector includes private medical insurance for some 3% and for the armed forces (2.6% of the population).

The private sector with private hospitals and medical services covers a small percentage of the population. Overall, not more than half the population is covered with health insurance, but local health authorities, largely on their own budgets, provide basic child health care and, to a lesser extent, maternal care to their populations.

National Health Plans.   A 10 year national health plan (1968–1977) focused on reducing morbidity and mortality from preventable diseases by mass vaccination, programs in STD and malaria control, nutrition and feeding, occupational and dental health, as well as improving the organization and delivery of health services. In 1970, the main health problems were defined as diarrheal diseases, childhood infectious diseases, malnutrition, TB, and malaria. By 1986, a major epidemiologic transition had occurred, and the primary causes of death were homicide and violence, cardiovascular diseases, neoplasms, and malnutrition. Malaria, still a major public health problem in the coastal lowlands, is coming under control in areas served by specially trained community health workers able to diagnose and treat the disease.

Some 40% of the population lack access to health services, while there is serious overlapping of services for others. Regulation of hospitals and medical practice is weak. Hospital care is overutilized, while primary care is underattended. Health expenditures have been less than 4% of GNP. In 1993, Colombia had 105 physicians per 10,000 population, compared to 245 per 10,000 in the United States, but only 49 nurses and midwives compared to 878 per 10,000 in the United States. However, uncontrolled growth of private medical schools is creating a rapid increase in production of physicians. Increased training for family practice is expected to help reduce the trend to specialization and the oversupply, and sometimes unemployment, of physicians.

The official system managed at the municipal level is focused on providing primary health care using *promatoras* community health workers recruited from the population served and trained for 3 months to carry on intensive outreach and health educational functions. This has made a major contribution to the marked improvement in infant and child mortality, very high coverage of immunization (92–98% for various EPI vaccines), as well as successful national immunization days for polio eradication and measles elimination. Primary care has also contributed to family planning, with 72% of women using contraception and the crude birth rate declining from 37 per 1000 population in 1970 to 24 per 1000 in 1996.

During 1990–1993, a review of health led to formation of a new national strategy intended to promote efficiency and expand coverage of services. This reform started functioning in 1994, with the goal of integrating various national programs and instituting mandatory national health insurance based on employment, with employee and employer contributions, and tax support. The reform seeks to implement universal coverage, equity of access, free choice of provider, institutional autonomy, decentralized administration, and national regulatory mechanisms.

Mandatory health insurance under the Social Security plan covers salaried and self-employed workers. They contribute 12% of income (8% by the employer and 4% by the worker). Indigent people are included in this plan, paid for by the national government and local authorities. The plan coverage increased from 4.5 million persons in 1994 to 20 million in 1998 (i.e., over 50% of the population). The rest of the population receives care through public hospitals and primary care centers. There has been a shift in emphasis to hospital-centered care to ambulatory and primary health care both in the public and private sectors. This finances a national basket of services defined by a National Council of Social Security chaired by the Minister of Health, and includes consumer representatives.

Health reform is meant to introduce universal coverage and consumer choice of HMO-like organizations based on the concept of universal access to a market-oriented set of service alternatives. They will provide care paid on a capitation basis and be subject to accreditation and quality control with GPs as gatekeepers. Direct service development of primary health care continues as a responsibility of the Ministry of Health with some assistance by NGOs. *Promatoras,* are an important part of that strategy. This will provide an important experience in health care reorganization in a mid-level developing country, as well as a major step forward for Colombia's social security, but will not resolve the problem of providing care for the underserved rural population.

ASIA

UNICEF divides Asia into two groups: South Asia, and East Asia and the Pacific (Table 13.10). The former includes India and has a total population of 1.3 billion persons, while the latter, including China, has 1.8 billion persons. Japan is excluded, being linked to the industrialized countries. The countries of South Asia have progressed less rapidly than those of East Asia in terms of economic, demographic, and health status indicators. Each is a diverse group of nations, but many have common problems, including infectious diseases (e.g., AIDS, tuberculosis, and malaria), poor nutrition for the majority, problems related to rapid urbanization, and the growing problem of noninfectious diseases.

China

China had a population of 1.2 billion persons in 1997. The GNP per capita increased from $300 in 1988 to $370 in 1990 and $750 in 1996. China is in the process of rapid change and economic growth. Life expectancy increased from 61 years in 1970 to 70 years in 1997; infant mortality fell from 140 per 1000 in 1960 to 38 in 1997. Immunization coverage in 1996 was 95–97% for DPT, polio,

TABLE 13.10 Countries of South and East Asia, Demographic and Health Indicators

Indicator	South Asia	East Asia and the Pacific
Countries	India, Pakistan, Bangledesh, Nepal, Afghanistan, Sri Lanka, Bhutan	China, Indonesia, Malaysia, Korea (North and South), Phillipines, Thailand, Vietnam, Cambodia, Hong Kong, Singapore, and others
Population (000s) 1997	1,291,153	1,818,498
Annual births (000s) 1997	35,361	34,141
GNP per capita		
1996	380	1193
Total fertility rates		
1960	6.1	5.8
1990	4.0	2.5
1997	3.4	2.1
Under 5 mortality		
1960	239	201
1990	135	58
1997	116	52
Life expectancy at birth		
1960	44	49
1970	48	58
1997	61	68

Source: UNICEF, *State of the World's Children,* 1999.

measles, and tuberculosis. The crude birth rate fell from 33 per 1000 population in 1970 to 16 per 1000 in 1997. Maternal mortality in 1980–1997 was 60 per 100,000, a decline from 1500 per 100,000 in 1949.

There is a strong social value placed on health and education, with major achievement in development of a health care infrastructure during the twentieth century. Primary school education is universal. Adult literacy increased from 79 to 90% for males and from 53 to 73% for females between 1980 and 1995. Deliveries attended by trained health personnel in 1990–1996 were 84% of all births. As a result of falling birth rates and mortality patterns, the population pyramid is becoming similar to that of developed countries, with a rapidly aging population.

Ancient China had a rich tradition of medical care and vital statistics. The Confucian and Taoist streams of Chinese culture supported a "high order" medical system, emphasizing both preventive and curative services. Classical medical texts documented an empirical base of pharmacopoeias and therapeutic traditions. The yin–yang principle of resonant harmonies between alternative structures was in contrast to the single causation emphasis of western culture. Ancient Chinese medicine based itself on treatment with herbal medicines, and at the same time included a holistic, psychosomatic perspective. Preventive medicine included attention to diet, rudimentary sanitation, personal hygiene, destruction of rabid animals, inoculation against smallpox, and an orientation to the well-being of the individual as essential to health. However, this high order medicine was available only to the elite of a rigid feudal–bureaucratic society. The vast bulk of the rural population had to rely on folk medicine based on herbal and other traditional practices.

Western medicine was introduced to China with the advent of missionary activities in the nineteenth century. This was accepted as another eclectic element of medicine, and medical schools were opened in the early twentieth century to train medical personnel in western medicine. In the period 1911–1949, medicine and public health advanced with establishment of the national Ministry of Public Health (1927), 30 medical colleges, municipal public health departments, rural district hospitals, military medical services, a factory inspection service, and an array of public health professional departments including maternal and child health, and a large number of provincial medical centers. This brought vaccination, ophthalmic and other forms of surgery, western hospitals, clinics, and medical schools to the provinces and rural areas. The Japanese invasion and civil war that ravaged China from 1936 to 1948 halted this progress.

Since the establishment of the People's Republic of China under the leadership of Mao Tse Tung in 1949, China placed improvement of living and health conditions among the rural population as a high national priority. Between 1949 and 1965, China, acting with advisors from the Soviet Union, emphasized rapid expansion of training of mid-level health personnel—nurses, midwives, dispensers, and feldshers (Chapter 14)—as well as doctors, whose numbers increased from 13,000 in 1945 to 150,000 in 1966. Hospital bed supply was also expanded rapidly so that by 1965 every county had at least one modern hospital.

In 1966, as part of the Cultural Revolution, policy review placed emphasis on traditional medicine and self-sufficiency in health care at the community level. Western medical training was reduced in scope and duration. Auxiliary or barefoot doctors were trained briefly in a mixture of western and traditional Chinese medicine. The barefoot doctors brought health care to the rural population living in 27,000 communes, as well as to urban neighborhoods, focusing on sanitation, family planning information, immunization, and treatment of common illnesses. Since the 1960s, emphasis on family planning has resulted in a slowing of population growth. The policy of one child per family is enforced with many sanctions. This has led to widespread social problems including female infanticide and abandonment, and a high male to female population ratio in younger age groups.

The rural population of China constitutes some 80% of the total population. Rural health care was based on cooperative medical services (CMS) funded by the rural communes using barefoot doctor and referral services. This provided effective preventive and curative services to the vast bulk of the rural population of China during the 1960s and 1970s. However, the economic reform in agriculture as part of the transition to a market economy, in effect, abolished the rural communes. As a result, the CMS system had no organizational or financial basis, and was replaced by fee-for-service practice by the former barefoot doctors, who became private medical practitioners during the 1980s. By 1986 only 9.5% of the rural population was still covered by the CMS system, in comparison to 90% in 1978. This has resulted in greater use of emergency service and hospitalization, with less diligence in performance of preventive health services. In some areas, the CMS model is being restored as cooperative measures under local initiatives. The national Ministry of Health and provincial/regional or municipal departments of public health are responsible for health services in their jurisdictions, with a high degree of local autonomy.

From 1976 to the present, a new focus on modernization replaced the ideological zeal and violence of the Cultural Revolution, and has been associated with a period of rapid economic growth. Barefoot doctors were retrained and examined for licensing as village doctors. In 1993, China had 115 physicians per 10,000 population and 88 nurses/midwives per 10,000 population, compared to 177 and 641 per 10,000, respectively, in Japan.

Hospital bed ratios in China increased from 4.6 beds per 1000 population in 1985 to 6.1 in 1989 in urban areas and decreased slightly from 1.5 to 1.4 beds per 1000 in the rural areas during the same period. Similarly, health professionals increased in urban areas to 12.6 health professionals per 1000 urban residents compared to 2.3 per 1000 rural residents in 1989. Polyclinics and sanitary–epidemiological stations were established throughout the country; patients were charged fees for services to support the health system. The World Bank estimates expenditures on health at 3.3% of GNP. National or provincial governments covered 30% of total health expenditures, while employment-based insurance covered 31%, private expenditures 32%, and communes 7%.

In the 1990s a national campaign to eradicate poliomyelitis was conducted, showing good results in reduction of cases from 5,065 in 1990 to 1191 in 1992, by supplemental oral polio vaccine national immunization days (NIDs) for children up to age 4 years. the crude mortality rate fell from 25 per 1000 in 1949 to 9 in 1970 and 5 per 1000 in 1996. Maternal mortality fell from 1500 per 100,000 births in 1949 to 95 per 100,000 in 1990. Other health indicators are seen in Tables 13.11 and 13.12

Serious health problems in China include high rates of lung cancer in polluted industrial cities, with very high rates of smoking. The leading causes of death are similar to those in developed countries, but regional disparities are apparent, with

TABLE 13.11   Health Indicators, China, 1960 and 1997

Indicator	1960	1997
Infant mortality rate	140	38
Child mortality rate	209	47
Life expectancy	44	70
Crude birth rate	37	16

Source: UNICEF, *State of the World's Children,* 1999.

TABLE 13.12   Vital Statistics, People's Republic of China, 1949–1992

Indicator	1949	1952	1965	1970	1980	1990	1992
Crude birth rate/1000	36.0	34.0	37.8	33.4	18.2	21.1	18.2
Crude mortality rate/1000	25.0	10.9	9.5	7.6	6.3	6.7	6.6
Natural increase/1000	16.0	23.1	28.3	25.8	11.9	14.4	11.6

Source: Ministry of Public Health, Beijing, 1994.

rural populations having higher death rates in all categories. Urban health care has always been at an advantage in China.

China's achievements in control of infectious diseases have been matched by success in birth control and in arranging access to medical care for a population of over 1.2 billion people. China has achieved better outcomes in terms of infant and child mortality and life expectancy with low health expenditures (3.5% of GNP) than many other developing countries. This is probably the result of a long-term program of developing primary care for the rural and urban poor population. The transition to a market economy will leave many with no care. Aging of the population and the one child per family policy creates a situation where the tradition of family care of the elderly will be by a couple who will have sole responsibility for four parents. This will be compounded by the rapid movement of young people to the cities for economic opportunity, so that care of the elderly will be a major problem in the coming decades.

As the country rapidly expands its economic potential, national health insurance is in an advanced stage of preparation. The Chinese experience in health improvement for its huge population in a chaotic period is an enormous achievement considering the economic level of development in China. The country has successfully reduced fertility rates in an attempt to limit population growth and reduced infant, child, and general mortality rates, but it faces challenges not only in transforming the health system to a market economy but also from the effects of the profound demographic shift.

COMPARING NATIONAL HEALTH SYSTEMS

The major participants in national health insurance networks include governments, employers, insurers, consumers, providers, and the public. Governments have increasingly come to recognize the economic and social value of improving the health of the population. They carry this out through public health measures to ensure the basic health of the nation, as well as through legislation regarding the nature of health insurance, whether it is provided through private or public insurance mechanisms. In both the original United Kingdom's Beveridge and the So-

---

BOX 13.2 PARTICIPANTS IN NATIONAL HEALTH SYSTEMS

1. Government—national, state, and local health authorities;
2. Employers—through negotiated health benefits for employees;
3. Insurers—public, not-for-profit, and private for-profit;
4. Patients, clients, or consumers—as individuals or groups;
5. Risk groups—persons with special risk factors for disease, e.g., age, poverty;
6. Providers—hospitals, managed care plans, medical, dental, nursing, laboratories, others;
7. Not-for-profit provider institutions;
8. For-profit institutions, individual providers, and groups;
9. Teaching and research institutions—universities, hospitals, institutes;
10. Professional associations, societies, academies, colleges;
11. Social security systems—with employer and employee contributions;
12. The public, the community, public opinion;
13. Political parties, philosophies and social agendas;
14. Advocacy groups—age, disease, poverty, or public interest groups;
15. The media—advocacy and watch-dog roles;
16. Economies—national, regional, and local;
17. International health organizations and movements;
18. Pharmaceutical and medical technology industries.

---

viet Semashko models, the government directly finances and provides health care. Services in the United Kingdom are provided by independent contractors, general practitioners, and hospitals operated by freestanding hospital boards (now trusts). The Semashko model is a totally state financed and operated service, with decentralized management.

In the Bismarck model, health insurance is financed through social insurance, paid at the place of employment, with sick funds paying for services of private medical practice and nongovernment hospitals. The Canadian plan finances health services by provincial governments funded by general tax revenues with federal government financial support, but care is provided by private practitioners and not-for-profit community-based hospitals. In all variations of health insurance systems, the place of the government as provider and insurer is important to the care received by the consumer and the general state of public health.

There are many variations in methods of assuring national access to health care. Different approaches taken in the development and current structure of health systems in the United States, Canada, United Kingdom, European, and Nordic countries, Japan, Russia, Israel, and the developing countries are given as examples in this chapter. Improved health, as measured by outcome indicators such as reduced morbidity, mortality, or sociophysiologic dysfunction, is the major underlying objective of a national health system. This is sometimes forgotten in debates that may reflect interests of groups such as insurers, providers, institutions, governments, professional groups, or even political philosophies.

A typology of national health systems based on methods of financing and administration of health services provides a framework for their classification and for comparisons (Table 13.13). Mixed models have also developed as the dynamics of health system reform evolves in many countries.

TABLE 13.13   Typology of Financing and Administration of National Health Systems

Type	Financing source	Administration
Bismarckian health insurance through social security, e.g., Germany, Japan, France, Austria, Belgium, Switzerland, Israel	Compulsory employer–employee tax payment to sick funds or through social security	In Germany, governments regulate sick funds which pay private services; strong sick fund and doctor's syndicates; Israel's sick funds compete as HMOs with per capita payments for mandatory basket of services
Beveridge National Health Service, e.g., United Kingdom, Norway, Sweden, Denmark, Italy, Spain, Portugal, Greece	Government—taxes and revenues; U.K. national financing; Nordic countries combine national, regional, and local taxation	Central planning, decentralized management of hospitals, GP service and public health; integrated district or county health systems with capitation financing in the U.K. and Nordic countries
Semashko national health systems, e.g., former USSR	Government—taxes and revenues; post-Soviet national health insurance	Strong central government planning and control; financing by fixed norms per population; allocation of facilities and manpower promoted increase in hospital beds and medical staff; post-1990 reforms emphasize decentralization with capitation and compulsory health insurance, i.e., payroll taxation
Douglas national health insurance through government, e.g., Canada, Australia	Taxation—cost sharing between provincial and federal governments	Provincial government administration; federal government regulation; medical services paid by fee-for-service; hospitals on block budgets; reforms to regionalize and integrate services
Mixed private/public system, e.g., United States, Latin America (e.g., Colombia), Asia (e.g., Phillippines) and African countries (e.g., Nigeria)	Private insurance through employment and public insurance through social security for specific population groups	Strong government regulation (U.S.); mixed private medical services, public and private hospitals, state/county preventive services; DRG payment to hospitals, rapid increase in managed care; extension of Medicaid coverage

Economic Issues in National Health Systems

As discussed earlier, the costs of health care are a major issue in national health systems. This is in part due to the rising costs of technology in medicine and the increasing age of the population with associated increasing importance of chronic disease, but it is also due to the traditional emphasis on institutional care. Health expenditures for preventive care, health promotion, and environmental health are usually not well financed nor analyzed in routine economic data reporting. This makes economic analysis and comparison of interventions difficult, thereby handicapping the search for cost-effective interventions.

National expenditures on health care are usually expressed in terms of U.S. dollars as a percentage of gross national product (GNP) or gross domestic product (GDP). The two economic figures are expressions of the total goods and services in a country, but GDP excludes international transfer of funds. Health care costs are also expressed directly as expenditures per capita (per person, per year), and indirectly as resources such as the number of hospital beds or medical manpower per 1000 (or 10,000) population (see Table 13.14). The percentage of GNP spent on health care often is not necessarily directly related to health indicators, such as infant mortality or longevity, as funds may be allocated to or spent on less effective and more costly care. This said, countries with low GNP per capita that spend less than 4% on health have poorer health indicators because there are insufficient resources to provide a basic health level for all. Underfinancing and inappropriate allocation of funds have been severe problems in most Soviet health systems and in most developing countries.

The supply of health care services remains one of the difficult and controversial topics in health planning. Economic analysis usually focuses on methods of financing in health care, and on methods of reimbursement or payment for ser-

TABLE 13.14   Population, GNP Per Capita, Health Facilities and Utilization, and Health Indicators, Selected Countries and Years, 1994–1997

Country	Population (millions), 1997	GNP per capita ($), 1996	% GNP for health, 1997	Acute care beds/1000, 1995	Average length of stay (days), 1995	Discharges/ 1000, 1994– 1995	Infant mortality 1000, 1996	Life expectancy, 1997
United States	271.6	28,020	13.5	3.4	6.5	88	7.8	77
Canada	29.9	19,020	9.2	4.3	7.5	na	6.0	79
Sweden	8.8	25,710	8.6	3.0	5.2	163	4.0	78
Germany	82.2	28,870	10.4	6.9	12.1	92	5.0	77
Finland	5.1	23,240	7.2	4.0	5.5	209	3.9	77
Denmark	5.2	32,100	7.4	4.0	6.1	197	5.6	76
Israel	5.8	15,870	8.0	2.4	4.5	203	6.3	75
United Kingdom	58.2	19,600	6.7	2.0	4.8	117	6.1	77

Source: *Health United States, 1998;* UNICEF, *State of the World's Children,* 1999; OECD Health Data 1998.

vices, placing less emphasis on the supply and quality of services. The World Bank's 1993 *World Development Report,* discussed in previous chapters, places major emphasis on the economic benefits of prevention and cost-effective measures to reduce the burden of disease. Excessive hospital utilization is not cost-effective.

Roemer's law (see Chapter 11) states that hospital utilization under insurance varies with bed supplies. Despite its essential truth, evidence shows that payment systems for hospital care can be modified so that there are incentives to prevent unnecessary admissions and to shorten hospital stays. As health costs increased rapidly, the concept of providing health care with fewer hospitalizations and more emphasis on ambulatory service became one of the essentials of health policy in many countries since the 1970s.

Health resource indicators are quite variable among the developed market economy countries. Acute care bed ratios represent the number of general, short-term beds per 1000 population. Total beds per 1000 includes all institutional beds utilized for inpatient medical care, but not geriatric custodial care. This is no standardized set of international definitions, and some of the differences may be due to variability of definitions used, but acute care hospital beds per 1000 is a more precise and comparable definition (Table 13.14). Many countries are actively reducing hospital bed supplies (United Kingdom, the Nordic countries, most western European countries, the United States, and Israel), developing alternatives to hospital care, using incentive payments to promote ambulatory or day-hospital treatments.

The hospital bed ratio often reflects historical patterns, medical practice traditions and concepts, medical technology and organization, as well as existing financial incentives or disincentives for more efficient care. Since the elderly are greater consumers of health services than the young, another major factor that influences this ratio is the age distribution of the population. Investment in alternatives to hospital care and health promotion reduce morbidity, helping to control costs of health care. This requires investment in education, legal action, screening, nutrition education, group counseling and selective home support services, and many other elements of the broad concept of health promotion.

Important factors in determining costs of national health systems include salary or income of providers, levels of technology in the service, health planning criteria (norms), hospital bed supply and utilization, availability of home care and comprehensive community care services, use of integrated or regionalized models of health care delivery, methods of paying for hospital services, use of incentive payment systems to promote more efficient use of resources, and emphasis on prevention and health promotion. All of these are issues in the reform of national health systems. Table 13.14 shows a comparison of expenditure, resource, utilization, and outcome indicators for selected industrialized countries.

No analysis of a health system can be complete without addressing the importance of poverty as a major contributing factor to morbidity and mortality. Poverty is associated with high rates of mortality from stroke, coronary heart disease, trauma, asthma, and cancer. Poverty is also related to many specific risk factors for illness: low educational levels, poor housing conditions, poor nutrition, psychological depression, cigarette smoking, alcohol and drug abuse, teenage pregnancies, single parenthood, early bereavement or abandonment, lack of prenatal care, low birth weights, family and neighborhood violence, and others. Universal access to traditional medical care may alleviate some of these effects, but it fails to address the core issues. Social policy and health programs are interdependent, each contributing to improving the quality and length of life. Health planning, including economic indicators, must take this factor into account.

REFORMING NATIONAL HEALTH SYSTEMS

Health care systems are continuously evolving. Impetus for reform of a health system may derive from a need for cost restraint, universal coverage, or efficiency in use of resources, or an effort to improve satisfaction of consumers or providers (see Table 13.15). The objective of improving the health of the population is also a motive, but this is often expressed as improved access, equity, efficiency, quality of care, and outcomes.

Political and philosophical considerations for health reform often stress issues such as universal access and equity in resource distribution, manpower and hospital beds, but it is equally important to focus on targets for improving the health of the general population and special groups at risk. Philosophical and historical issues and arguments for national health insurance have included the need for social protection as a matter of national honor, but a system that fails to improve national health in terms of international outcome indicators does not meet this objective.

Debates and reforms in organizing health systems continue and increase in intensity as the political objective of Health for All meets the reality of rising costs. Efficiency in use of resources and satisfaction of the public and providers are major issues in all health systems. There is no single best means, despite claims by proponents of state-operated systems and equally ideological claims by market-force proponents. Direct importation of a total health system model is not feasible, because there are many factors contributing to the development of a health system relating to the political, social, and professional cultures of each country.

The assumption that market forces produce a better quality of health care is commonly expressed. This point of view has merit if taken in the sense that personal management of finances and choices in health care empower the individual to choose. This may be an advantage for a better educated urban population living near specialized services unavailable to others. Free choice for consumers and freedom of choice (autonomy) for providers are different aspects of the market force issue. Taken together, they provide a measure of protection of the rights of the consumer and provider to choose health systems. However, they diminish the responsibility and ability of the system to reach out and provide care and preven-

TABLE 13.15 Goals, Issues, Strategies, and Tactics for National Health Policies

Goals	Issues	Strategies/tactics
National political commitment to improved health for all	Health as a government responsibility Universal access Adoption of international standards Regional and social equity in access Free choice by consumers and providers Healthy lifestyle as national policy	Health promotion as policy Law/regulations Regulate consumers' rights in health Public information on health Advocacy groups—public, professional
Finacing within national means for social benefits	Adequate overall financing (>6% GNP) Shift from supply planning to cost per output Categorical grants to promote national objectives	Increase financing at national, state, and local government levels Health insurance as supplement Reduce acute care beds to <3.51/1000 District health authorities with capitation funding
Management for cost-effectiveness	Cost containment Cost-effective health initiatives Decentralized management National policy, monitoring, and standards Information systems/monitoring District health profiles	Increase primary care Increase home care, long-term beds Increase home care, nonadmission surgery, long-term care facilities Health information systems Managed care and DRGs
Defining national health targets	Define leading causes of morbidity, mortality, and YPLL, hospitalization with regional analysis Health promotion versus treatment philosophy Prioritization for use of available resources Use of relevant international standards	Social factor analysis in health Improve health KABP Community attitudes to health promotion Promote public health, nutrition, environment, immuization policies

tive services, so that important programs such as immunization, prenatal care, and care of the elderly may suffer as a result. This set of rights is also sometimes in conflict with the imperative of cost control and may also have the undesired effect of promoting excess services such as unnecessary surgery, which has costly and potentially harmful consequences. Market mechanisms that promote individual as well as health system responsibility can make important contributions in health.

Governmental responsibility to promote health requires initiatives that may limit individual rights. Adding chlorine and fluoride to water, or iodine to salt or folic acid to bread, requires people who may not directly benefit to accept this in the interests of the need of others in the community. A governmental authority may close a business that is hazardous to health, such as an unhygienic restaurant. Management of health care systems must address macroeconomic and microeconomic issues for efficiency. Communities and regions will often address health planning in terms of its impact on business, jobs, and prestige in the community, as opposed to national or regional plans and priorities.

Since the 1970s, there has been a growing stress on health promotion as a way of reducing the burden of chronic disease and the cost of health care for those diseases. This was stimulated and promoted by the health field concept (1974), the Alma-Ata Conference on primary health care (1978), and the WHO's Health for All concept (1978). Specific health targets in the United States (Healthy People 2000) and in the European region of WHO (1985) (see Chapter 2) placed emphasis on measurable objectives as the basis for health planning, affecting the planning process.

The 1990s was a decade of major reforms in national health systems. Industrialized countries attempted to restrain cost increases while retaining universal access. Sweden has brought down its health expenditures by reducing hospital bed supplies. The United States, building on its social security-based health insurance plan for the elderly and the poor, failed in its 1995 attempt for national health insurance, but is undergoing dramatic changes in the managed care revolution, propelled by the need to control the rate of cost increases. Canada is facing a crisis in sustaining its national health insurance system as federal withdrawal of funding leaves the provinces to finance a generous range of benefits and high levels of hospital bed supply. Israel has moved from voluntary sick funds to national health insurance, with the sick funds as managed care systems. The eastern European countries are in a state of transition away from the pre-1990 Soviet model, adopting national health insurance and decentralized administration of services. In developing countries there is concern that directly financing services through the government will hinder development of health services, so that there is a tendency to look toward national health insurance as a way to improve funding of services and bring more people into care. China has moved toward fee-for-service in its rural health care for some 70% of its population. All countries are struggling to develop adequate prevention models to reduce the burden of disease that can bankrupt a national health system.

Universal access to health care does not necessarily address social inequities in health. Removal of financial barriers by itself does not guarantee good health. Many social, cultural, and environmental health risk factors are not correctable or preventable by medical or hospital care. They may be of greater importance than the medical care provided (see Chapter 3). The models presented may serve as examples for other countries, and will continue to do so. It is therefore useful to understand how they evolved, their successes and failures, and how they are continuing to develop.

There are two basic directions for reform, which are sometimes in conflict. One is the primary health care approach. This is based on tackling the basic health problems of developing countries by promoting primary health care as a public service through decentralized delivery and administration. The alternative approach, based on the market economy theory, is to promote access to health care by national health insurance, funded through employer–employee contributions or through general taxation.

The fundamental differences in these two approaches presents a dilemma for the developing countries and in many ways for the developed countries as well as they struggle to control health care costs. A health insurance approach may increase funds available for health care, but it invites increases in expenditures for care, inequities in access to care, and an emphasis on curative as opposed to preventive service. This is decidedly a medical approach, promoting hospital and physician services, with public health inadequately addressed and left to the care of private medical practitioners.

The market approach assumes that promoting competition will increase the quality of care and attention to consumer needs, but it is often associated with overutilization of costly services and drives health costs to very high levels. It is a luxury available only to the very wealthiest countries and still not provide all citizens with equal access to services. Developing countries may not have adequate funds to provide health care for all. At the same time developed economies may not be able to fund health services on demand at levels that consumers and providers might consider ideal. This has led many countries to restrict access to specialist services and place other limitations on services and is the basis for the managed care approach in the United States.

The public service model often leaves a national program underfunded, leading to problems of quality and morale for the provider as well as the consumer. However, a national health policy is still essential for vulnerable population groups or areas, whether in a developing or developed country. Even the most developed countries have substantial population groups living in poverty, with poor health conditions. The United States has over 43 million persons, 16% of the population (in 1997), without health insurance and probably an equal number with very inadequate coverage. It also has rural areas ill-served by collapsed rural health services. Study of the international experience of health care systems helps to promote international standards and criteria.

The health sector is under great pressure to constrain costs. Employer–employee contribution systems are implementing changes to control costs since health costs are partly responsible for making their industry noncompetitive in the global market. At the same time, there are inflationary pressures of the aging of the population, medical technologic innovation, and high professional and public expectations. Health system reform includes downsizing the hospital sector and building up community health care.

HEALTH SYSTEMS AND THE NEW PUBLIC HEALTH

National health systems throughout the world are in process of change, seeking restraint in increasing costs, universal coverage, equity in access and quality, as well as efficiency and effectiveness in use of resources to achieve health targets. Many countries are looking for ways to provide universal and equitable care, while controlling costs and improving efficiency. There is no one answer to the search for a health system that works.

Social security and social welfare systems took up the task of assuring access to health services over the twentieth century. National health systems evolved to provide access to medical, hospital, preventive, and community health services. Financing of services through general taxation based on progressive income tax, resource taxes, and excise taxes may be the most equitable way of raising funds. Many countries use social security systems based on employer–employee contributions to pay for health services. Universal access is a means of assuring that the economic barrier is removed for the total population and may lead to increased access to medical and hospital services for those previously excluded. It does not, of itself, guarantee achievement of important health targets. Allocation of resources is an even more fundamental problem.

Beyond financing and resource allocation, there are many "nontariff" barriers to health. Even in highly developed national health systems, such as that of the United Kingdom, social class, place of residence, education, level, and ethnicity play important roles in morbidity and mortality rates. Factors other than medical or hospital care are vital, as classic risk factors for disease, such as diet, smoking, and physical fitness. Partly, however, social class differences in morbidity and mortality are the result of less well-defined aspects of poverty, such as depression, fear, insecurity, and lack of control over one's life. These are issues that are important to the achievement of national health goals and equity.

Health systems must be continuously evaluated. Traditional outcome indicators, such as infant and child, maternal, and disease-specific mortality rates, are important but not sufficient. Incidence of vaccine-preventable diseases, immunization rates for infants, anemia rates in infancy and pregnancy, and handicapping conditions are also necessary. Newer measures such as DALYs and QALYs (see Chapters 3 and 11) may help to change the emphasis from mortality to quality of life measures as part of the evaluation. National health systems require data systems that generate information needed for this continuous process of monitoring. High quality academic centers for epidemiologic, sociologic, and economic analysis are needed to train health leaders and managers and to carry out the studies and research vital for health progress.

Despite the structural diversity and underlying philosophical differences in national health systems, there are important common elements. They are large employers and among the largest industries in their respective countries. All face problems of financing, cost constraint, overcoming structural inefficiencies, and, at the same time, finding incentives for high quality and efficiency.

A national health system is a holistic entity with many parts. The quality of the community infrastructure (sewage, water, roads, communication), the quantity and quality of food, levels of education, and professional organization are all parts of this continuum. Narrow planning for health systems ignores this message at the risk of missing its targets of improved health indicators. National health systems are not only methods of financing and assuring access to services; they are part of the larger social and economic development of a society and a major contributor to improved quality of life.

ELECTRONIC MEDIA

Canadian Institute for Health Information, National Health Expenditure Trends, 1975–1998, website, http://www.cihi.ca/medrls/execsum.htm

Health Care Financing Administration website, http://www.hcfa.gov/stats/indicatr/tables/

Healthy People website, http://www.health.gov/healthypeople/

United States Department of Health and Human Services, *Health United States*, 1998, http://www.cdc.gov/nchsww/products/pubs/pubd/hus/hus.htm   (note: this requires Adobe Acrobat software to download; also available on CD-ROM including Lotus spreadsheet)

World Bank, Health reform.online website, http://www.worldbank.org/healthreform/index.htm

World Health Organization, European Region Health for All Data Set. Download from website: http://www.org/dk

RECOMMENDED READINGS

Bobadilla, J. L., Cowley, P., Musgrove, P., Saxenian, H. 1994. Design, content and financing of an essential national package of health services. *Bulletin of the World Health Organization,* 72:653–662.

Murray, C. J. L., Govindaraj, R., Musgrove, P. 1994. National health expenditures: A global analysis. *Bulletin of the World Health Organization,* 72:623–627.

Roemer, M. I. 1993. National health systems throughout the world. *Annual Review of Public Health,* 14:335–353.

Russel, L. B. 1993. The role of prevention in health reform. *The New England Journal of Medicine,* 329:352–354.

Schieber, G. J., Poullier, J. P. 1989. International health care expenditure trends: 1987. *Health Affairs,* 8:169–177.

Schieber, G. J., Poullier, J. P., Greenwald, L. M. 1991. Health care systems in twenty-four countries. *Health Affairs,* 10:22–38.

Young, Q. D. 1993. Health care reform: A new public health movement. *American Journal of Public Health,* 83:945–946.

BIBLIOGRAPHY

Abel-Smith, B. 1992. Cost containment and new priorities in the European community. *The Milbank Quarterly,* 70:393–416.

Aday, L. A., Begley, C. E., Lairson, D. R., Slater, C. H. 1993. Evaluating the Medical Care System: Effectiveness, Efficiency, and Equity. Ann Arbor, Michigan: Health Administration Press.

Avian, R., Abel-Smith, B., Tombari, G. 1990. Health Insurance in the Developing Countries—The Social Security Approach. Geneva: International Labor Organization.

Blendon, R. J., Leitman, R., Morrison, I., Donelan, K. 1990. Satisfaction with health systems in ten nations. *Health Affairs,* 9:185–192.

Chernikovsky, D. 1995. Health system reform in industrialized democracies: An emerging paradigm. *Milbank Memorial Fund Quarterly,* 73:339–372.

Dutton, D. B. 1979. Patterns of ambulatory health care in five different delivery systems. *Medical Care,* 17:221–243.

Ellencweig, A. Y. 1992. *Analysing Health Systems: A Modular Approach.* Oxford: Oxford University Press.

Kirsch, T. 1988. Local area monitoring (LAM). *World Health Statistics Quarterly,* 41:19–25.

Mills, A., Vaughan, J. P., Smith, D. L., Tabibzadeh, I. (eds). 1990. *Health System Decentralization: Concepts, Issues and Country Experience.* Geneva: World Health Organization.

Organization for Economic Cooperation and Development. 1990. *Health Care Systems in Transition: The Search for Efficiency.* Social Policy Studies No. 7.

Roemer, M. I. 1991 and 1993. *National Health Systems of the World,* Volumes 1 and 2. New York: Oxford University Press.

Saltman, R. B. (ed). 1988. *The International Handbook of Health Care Systems.* New York: Greenwood Press.

UNICEF. 1998. The State of The World's Children, 1998. New York: UNICEF.

World Health Organization. 1993. *Evaluation of Recent Changes in the Financing of Health Services.* Report of a WHO Study Group, Technical Report Series No. 829. Geneva: World Health Organization.

World Health Organization. 1997. *The World Health Report, 1997: Conquering Suffering, Enriching Humanity.* Geneva: World Health Organization.

World Health Organization. 1993. *World Health Statistics Annual 1993.* Geneva: WHO.

United States

Anderson, R. M., Rice, T. H. E., Kominski, G. F. (eds). 1997. *Changing the U.S. Health Care System: Key Issues in Health Services, Policy, and Management.* San Francisco: Jossey-Bass.

Bergner, L. 1993. Race, health, and health services. *American Journal of Public Health,* 83:939–941.

Blumenthal, D. (editorial). 1993. Administrative issues in health care reform. *The New England Journal of Medicine,* 329:428–429.

Brown, E. R. 1992. Health USA: A national health program for the United States. *Journal of the American Medical Association,* 267:552–558.

Fries, J. F., Koop, E., Beadle, C. E., Cooper, P. P., England, M. J., Greaves, R. F., Sokolov, J. J., Wright, D. 1993. Reducing health care costs by reducing the need and demand for medical services. The Health Project Consortium. *The New England Journal of Medicine,* 329:321–325.

Levit, K. R., Lazenby, H. C., Sivarajan, L. 1996. Health care spending in 1994: Slowest in a decade. *Health Affairs,* 15:130–144.

National Center for Health Statistics. 1997. *Health, United States 1996–1997* and *Injury Chartbook.* Hyattsville, MD: 1997.

National Center for Health Statistics. 1998. *Health, United States: With Socioeconomic Status and Health Chartbook, 1998.* DHHS Publication number (PHS) 98-1232.

Public Health Service. 1992. *Healthy People 2000: National Health Promotion and Disease Prevention Objectives.* U.S. Department of Health and Human Services. Boston: Jones and Bartlett.

Smith, S., Freeland, M., Heffler, S., McKusick, D., and the Health Expenditures Projection Team. 1999. The next ten years of health spending: What does the future hold? *Health Affairs,* 17:128–132.

Welch, W. P., Miller, M. E., Welch, H. G., Fisher, E. S., Wennberg, J. E. 1993. Geographic variation in expenditures for physicians' services in the United States. *The New England Journal of Medicine,* 328:621–627.

Canada

Canadian Public Health Association. 1996. *Focus on health: Public health in health services restructuring: An issue paper.* Ottawa: CPHA.

Deber, R. B., Hastings, J. E., Thompson, G. G. 1991. Health care in Canada: Current trends and issues. *Journal of Public Health Policy,* 12:72–82.

Deber, R. B., Thompson, G. G. (eds). 1993. *Restructuring Canada's Health Services System: How Do We Get There from Here?* Proceedings of the Fourth Canadian Conference on Health Economics. Toronto: University of Toronto Press.

Evans, R. G. 1989. Controlling health expenditures—The Canadian reality. *The New England Journal of Medicine,* 320:571–577.

LaLonde, M. 1974. *New Perspectives on the Health of Canadians.* Ottawa: Department of National Health and Welfare.

Naylor, C. D. 1999. Health care in Canada: Incrementalism under fiscal duress. *Health Affairs,* 18: 9–26.

Pineault, R., Lamarche, P. A., Champagne, F., Contandriopoulos, A. P., Denis, J. L. 1993. The reform of the Quebec health care system: Potential for innovation? *Journal of Public Health Policy,* 14:198–219.

Rachis, M., Kushner, C. 1994. *Strong Medicine: How to Save Canada's Health Care System.* Toronto: Harper-Collins.

Roos, N. P., Brownell, M., Shapiro, E., Roos, L. R. 1998. Good news about difficult decisions: The Canadian approach to hospital cost control. *Health Affairs,* 17(Sept/Oct):239–246.

Saskatchewan Health. 1992–1995. *A Saskatchewan Vision for Health: Introduction of Needs-Based Allocation of Resources to Saskatchewan District Health Boards for 1994–1995; and Planning Guide for Saskatchewan Health Districts.* Regina: Government of Saskatchewan.

United Kingdom

Acheson, D. 1998. *Independent Inquiry into Inequalities in Health.* London: Her Majesty's Stationery Office.

Black, D. 1980. *Inequalities in Health: Report of a Research Working Group.* London: Department of Health and Social Security.

Bone, M. R. 1992. International efforts to measure health expectancy. *Journal of Epidemiology and Community Health,* 46:555–558.

Coulter, A. 1995. Evaluating general practice fundholding in the United Kingdom. *European Journal of Epidemiology.* 5:233–239.

Day, P, Klein, R. 1991. Britain's health care experiment. *Health Affairs,* 10:39–59.

Dobson, F. 1999. Modernizing Britain's National Health Service. *Health Affairs,* 18:40–41.

Donaldson, R. J., Donaldson, L. J. 1993. *Essential Public Health Medicine.* Dordrecht: Kluwer Academic Publishers.

Glynn, J. J., Perkins, D. A. (eds). 1995. *Managing Health Care: Challenges for the 90s.* London: W. B. Saunders.

Illsey, R. 1999. Reducing health inequalities: Britain's latest attempt. *Health Affairs,* 18:45–46.

LeGrand, J. 1999. Competition, cooperation, or control? Tales from the British National Health Service. *Health Affairs,* 18:27–44.

Marmot, M. 1999. Perspective: Acting on the evidence to reduce inequalities in health. *Health Affairs,* 18:42–44.

McKee, M. 1995. What can we learn from the British fundholding experience. *European Journal of Public Health.* 5:231–232.

Pocock, S. J., Shaper, A. G., Cook, D. G., Phillips, A. N., Walker, M. 1987. Social class differences in ischemic heart disease in British men. *Lancet,* 11:197–201.

Secretary of State for Health. 1991, reprinted 1995. *The Health of the Nation.* London: Her Majesty's Stationery Office.

Smith, G. D., Bartley, M., Blane, D. 1990. The Black report on socioeconomic inequalities in health 10 years on. *British Medical Journal,* 301:373–377.

Europe

Brown, L. D., Amelung, V. E. 1999. "Manacled competition": Market reforms in German health care. *Health Affairs,* 18:76–91.

Elola, J., Daponte, A., Navarro, V. 1995. Health indicators and the organization of health care systems in western Europe. *American Journal of Public Health,* 85:1397–1401.

Hermanson, T., Aro, S., Bennett, C. L. 1994. Finland's health care system: Universal access to health care in a capitalist democracy. *Journal of the American Medical Association,* 271:1957–1962.

Hurst, J. W. 1991. Reforming health care in seven European nations. *Health Affairs,* 10:7–21.

Iglehart, J. K. 1991. Germany's health care system. *The New England Journal of Medicine,* 324:1750–1756.

Roberts, J. L. 1996. *Terminology for the WHO Conference on European Health Care Reforms: A Glossary of Technical Terms on the Economics and Finance of Health Services.* Copenhagen: World Health organization, Office for Europe.

Saltman, R. B. 1990. Competition and reform in the Swedish health system. *The Milbank Memorial Fund Quarterly,* 68:597–618.

Saltman, R. B., Figueras, J. 1997. *European Health Care Reform: Analysis of Current Strategies.* Copenhagen: World Health Organization, Regional Office for Europe.

World Health Organization Study Group. 1993. *Evaluation of Recent Changes in the Financing of Health Services.* Geneva: World Health Organization.

World Health Organization, Regional Office for Europe. 1997. *Health in Europe, 1997.* Copenhagen: World Health Organization.

World Health Organization, Regional Office for Europe. 1999. *Health 21: The Health for All Policy Framework for the WHO European Region.* Copenhagen: World Health Organization, Regional Office for Europe.

Japan

Ikegami, N. 1991. Japanese health care: Low cost through regulated fees. *Health Affairs,* 10:87–109.

Ikegami, N., Campbell, J. C. 1999. Health care reform in Japan: The virtues of muddling through. *Health Affairs,* 18:56–75.

Ingelhart, J. K. 1988. Health policy report: Japan's medical care system, parts 1 and 2. *New England Journal of Medicine.* 319:807–812 and 1166–1172.

Russia

Centers for Disease Control. 1992. Public health assessment—Russian Federation, 1992. *Morbidity and Mortality Weekly Report,* 41:89–91.

Centers for Disease Control. 1995. Diphtheria epidemics—New independent states of the former Soviet Union, 1990–1994. *Morbidity and Mortality Weekly Report,* 44:177–181.

Field, M. G. 1988. Union of Soviet Socialist Republics. In Saltman, R. B. (ed). *The International Handbook of Health Care Systems.* New York: Greenwood Press.

Golovoteev, V. V., Pustovoj, I. V. 1984. Public health finance and planning in the Soviet Union. *World Health Statistics Quarterly,* 37:364–374.

Tulchinsky, T. H. E., Varavikova, E. A. 1996. Addressing the epidemiologic transition in the former Soviet Union: Strategies for health system and public health reform in Russia. *American Journal of Public Health,* 86:313–320.

Willikens, F., Scherbov, S. 1992. Analysis of mortality data from the former USSR: Age–period–cohort analysis. *World Health Statistics Quarterly,* 45:29–49.

Israel

Central Bureau of Statistics. 1975 (and 1996). *Statistical Abstract of Israel, 1975 (and 1996).* Jerusalem: Herned Press.

Israel Center for Disease Control. 1999. *Health Status in Israel 1999.* Jerusalem: Ministry of Health.

Ministry of Health. 1998. *Health in Israel: Selected Data, 1998.* Jerusalem: Ministry of Health.

Penchas, S., Shani, M. 1995. Redesigning a national health-care system: The Israeli experience. *International Journal of Health Care Quality Assurance,* 8:9–18.

Tulchinsky, T. H. E. 1985. Israel's health system: Structure and content issues. *Journal of Public Health Policy,* 6:244–254.

World Health Organization, Regional Office for Europe. 1996. *Highlights on Health in Israel.* Copenhagen: World Health Organization, Regional Office for Europe.

Developing Countries

Barnum, H., Kutzin, J. 1993. *Public Hospitals in Developing Countries: Resource Use, Cost, Financing.* Baltimore, MD: The Johns Hopkins University Press.

Evans, J. R., Hall, K. L., Warford, J. 1981. Shattuk lecture: Health care in the developing world: Problems of scarcity and choice. *The New England Journal of Medicine,* 305:1117–1127.

Ron, A., Abel-Smith, B., Tamburri, G. 1990. *Health Insurance in Developing Countries: The Social Security Approach.* Geneva: International Labor Office.

Schieber, G., Maeda, A. 1999. Health care financing and delivery in developing countries. *Health Affairs,* 18:135–143.

Tarimo, E., Creese, A. (eds). 1990. *Achieving Health for All by the Year 2000: Midway Reports of Country Experiences.* Geneva: World Health Organization.

Colombia

Ministerio de Salud/Ministerio de Trabajo. 1993. *La Seguridad Social en Colombia, Ley 100 de 1993.* Bogota: Minsterio de Salud/Minsterio de Trabajo.

Pan American Health Organization/World Health Organization. 1986. *Evaluation of the Strategy for Health for All by the Year 2000.* Seventh Report on the World Health Situation, Volume 3. Region of the Americas. Washington, DC: PAHO.

Lujan, F. J. Y. 1988. Colombia. In Saltman, R. (ed). *The International Handbook of Health Care Systems,* pp. 57–72. New York: Greenwood Press.

China

Grogan, C. M. 1995. Urban economic reform and access to health care coverage in the People's Republic of China. *Social Science and Medicine,* 41:1073–1084.

Lawson, J. S., Lin, V. 1994. Health status differentials in the People's Republic of China. *American Journal of Public Health,* 84:737–741.

Zheng, X, Hillier, S. 1995. The reforms of the Chinese health care system: The Jiangxi study. *Social Science and Medicine,* 41:1057–1064.

Comparisons

Anderson, G. F., Poullier, J.-P. 1999. Health spending, access, and outcomes: Trends in industrialized countries. *Health Affairs,* 18:178–192.

Davis, K. 1999. International health policy: Common problems, alternative strategies. *Health Affairs,* 18:135–143.

Donelan, K., Blendon, R. J., Schoen, C., Davis, K., Binns, K. 1999. The cost of health system change: Public discontent in five nations. *Health Affairs,* 18:206–216.

Sochalski, J., Aiken, L. H. 1999. Accounting for variation in hospital outcomes: A cross-national study. *Health Affairs,* 18:256–259.

Tuohy, C. H. 1999. Dynamics of a changing health sphere: The United States, Britain, Canada. *Health Affairs,* 18:114–134.

14

HUMAN RESOURCES
FOR HEALTH CARE

INTRODUCTION

The New Public Health's concern with the total health system requires an understanding of issues related to the training, supply, distribution, and management of many kinds of human resources, including the balance between personnel working in institutions and in the community. Health systems require well-trained, up-to-date providers of care to promote health, prevent disease, treat illness, and rehabilitate in a compassionate, ethical, professional, and cost-effective manner. This means that health care providers must be educated not only for competence and humaneness in clinical functions, but also to be continuous learners and knowledgeable in the economic aspects of health care. They must be aware of and be able to adapt from related fields such as epidemiology, economics, and management as well as the social and behavioral sciences. The quality of the practitioner depends on the recruitment of socially motivated and talented people, on education, training, and professionalization as providers, as well as on the structure, content, and quality orientation of the health system in which they work.

Determining need and allocation of human resources are important health planning issues. An over- or undersupply of one or more heath professions creates a bias or imbalance in the health system and its economics. Mid-level practitioners and community health workers are being recognized as essential to ensure access to appropriate levels of service and to provide for unmet service needs in both developed and developing countries.

In seeking efficient ways of improving health, health systems have opened many new professional roles, to work in new organizational frameworks. As definitions of health service were widened to include health maintenance, new health professions were added to the total health service spectrum. Continuing education is vital to maintaining and upgrading quality in a health care system. Registration and data systems are important to provide basic information on all relevant aspects of health manpower.

The purpose of this chapter is to examine the elements essential in training of human resources and their importance in the New Public Health, in relation to the quantity, quality, and changing interaction between the health professions.

OVERVIEW OF HUMAN RESOURCES

The numbers, types, and distribution of personnel supply are major determinants of access, availability, appropriateness, and costs of health care. The training, quality, and performance of health personnel, and the technology they use, are all important health planning issues. Every health professional needs knowledge of the principles and current standards of public health in order to perform his/her functions, as all of health care now routinely involves prevention, teamwork, management, quality assurance, cost containment, and related ethical issues.

In many countries, the major focus of education of health personnel has been to prepare clinicians, without equal emphasis on preparation of public health policy analysts, health managers, and public health professionals. Yet the latter are especially important when health reforms are under way and when health promotion and prevention are needed to cope with changes in the health needs of a society.

The principal problems in human resources development vary from country to country but consistently include the following:

1. Imbalance in training of health professions: oversupply of doctors, shortage of nurses and other health professionals;
2. Excessive training of medical specialists, and insufficient training of primary care physicians, inflating health costs and compromising access to care;
3. Geographic maldistribution with poor supply in rural areas of different professional categories;
4. Underfinancing of public systems of health care in comparison to private practice, fostering poor work conditions, low remuneration, and indifferent career opportunities, with low staff morale, performance, and client satisfaction;
5. Insufficient standards and period of training of specialist physicians to produce well-qualified professional leaders;
6. Lack of orientation of all health providers to public health;
7. Lack of training of public health specialists in epidemiology, the health related social sciences, health system policy analysis, or health system management, compromising the ability of a health system to monitor its outcomes and its resource allocation, or to evaluate program effectiveness;
8. Licensing of health providers by the government, which may allow for compromises in quality to ensure adequate numbers of graduates; con-

versely, delegating licensing to professional syndicates may result in a protectionist approach, placing the interests of the profession above those of the public;

9. Compromising of the quality of human resources by inadequate recruitment and educational standards, inadequate continuing examination and recertification;

10. Conflicts of academic, professional, government, or insurer interests with public and individual patient interests in training policies.

From the 1950s to the 1970s, medical schools were opened and existing schools expanded to meet problems of access to care, perceived to be due to a shortage of doctors. It was thought that increased numbers of doctors would increase competition and lower doctors' incomes; however, medical incomes continued to rise, and problems of access to care were unresolved. With growing emphasis on health economics and health promotion and disease prevention, there was a realization that excess medical personnel would not contribute to the national health, and that in some countries the excess of medical personnel had become a liability. In both fee-for-service medicine and salaried health service, increases in physician supply generate increases in health expenditures. Supply and demand market forces do not adapt well to health care, because the consumer demand is to a large extent generated by provider decisions (e.g., for return visits, investigation, or hospitalization). Fees may be fixed arbitrarily or by negotiation with a public insurance mechanism; the service is paid by a third party, and the consumer is less knowledgeable than the provider.

Each country addresses the issue of how many and what kind of human resources to train in its own needs, related to the design and operation of the health systems. The province of Alberta, Canada, during the 1970s, had relatively stable expenditures for medical services and physician-to-population ratios. During the 1980s, the province experienced a marked economic downturn and zero population growth, but the supply of physicians and services per physician increased by some 20%. The reduced numbers of clients per physician led to an increase both in fees and in volume of services per capita so that physician incomes were sustained. As a result, total and per capita expenditures for health care increased sharply. In many countries during the 1980s, policies were reformulated to reduce the size of medical school training entry classes.

Excess supply of physicians can also be a serious problem for a health system, promoting a bias toward a medical orientation in health at the expense of other, more basic needs of public health, primary care, and fundamental support systems for vulnerable groups in society. An excessive medical orientation fosters misallocation of limited resources by creation of tertiary care and positions for doctors in the central cities, leaving rural and primary care underdeveloped. This is widely prevalent in developing countries such as Bangladesh, India, Pakistan, Mexico, Colombia, and other Latin American countries. In some developing countries the problem is often due to the inability of the health budget to employ needed numbers of physicians. Unemployment among young physicians is a substantial problem in some of these countries.

Globally, the shortage of nurses continues, although it is particularly pronounced in developing countries where nursing is a low prestige profession. Medicine, on the other hand, is highly popular as a profession and means of social advancement. The extent of geographical inequity is understated by regional comparisons, but even aggregate comparisons show sharp differences: for every 10,000 persons there were 56 nurses and midwives in the Americas, but only 3.3 in southeast Asia.

The achievement of the goal of Health for All through primary health care requires the effective and coordinated services of many types of health personnel within a national health system designed to reach this goal. Political policy may influence the preparation, composition, and work patterns of the health work force. National expenditures on health are dependent on the political priority given to health compared to other issues that may be equally or more pressing to the governing power. A strong national health policy can nevertheless be constructed even in a poor country by well-defined health programs. A community and rural health policy in China during the 1950s was based on a number of elements: development of a 3-year family doctor training program for rural service; upgrading of training of village doctors to assistant doctor level; and incentives to encourage work in the countryside and at a grassroots level. This program was successful in raising health standards in China beyond that which might be expected from its economic level. With reform in the economy during the 1980s and 1990s, this system is going through profound changes (Chapter 13).

HUMAN RESOURCES PLANNING

The health infrastructure of a country includes the resources available, and their organization. Human resources are essential to any health system. The supply of manpower and facilities, economic support of the system, management and policy, methods of payment of providers, and organization of the services are therefore a vital part of health planning.

Resources available to a health system include facilities, manpower, and financial resources for health care. The organizational and financial structure of a health system determines how these resources are allocated or expended, in the public as well as the private health care sectors. Both structure and methods of payment affect the way services are provided. health system requires economic support sufficient to include basic and continuing education of human resources of high quality, as well as for human resources management for their appropriate and optimum use.

Regulation of health manpower includes licensure and discipline of health personnel and is an important governmental function. Measures to control or limit the supply of medical practitioners, along with incentives to promote more efficient health care, are important issues in rationalizing health care systems.

**BOX 14.1 HUMAN RESOURCES AND THE
HEALTH SYSTEM INFRASTRUCTURE**

Management
and policy
$\updownarrow$

Supply of $\leftrightarrow$ Organizational $\leftrightarrow$ Method of
manpower,   structure   payments
facilities $\updownarrow$
                                               Economic
                                               support

Source: Adapted from *Reviewing Health Manpower Development,* WHO, 1987.

Fundamental to the process of determining manpower needs is a knowledge of the current manpower situation. Essential for this are data systems based on periodic registration or census taking of persons practicing a health profession. Practitioners may retire, die, move, emigrate, or leave the profession and so should be taken out of actively practicing registries. An accurate, up-to-date picture of actual manpower provides information on specialty, geographic distribution, age, sex, and current work activities. International comparisons of professional manpower help to place a national pattern in the context of other countries with similar socioeconomic and health standards. Human resources should be matched to the targets and resources of a country. Alternative approaches may be needed if the supply of workers is insufficient or inappropriate to meet heath care needs and targets.

Assessing current manpower supply and determining future needs are specific tasks of a government agency concerned with comprehensive national socioeconomic planning. They may be assigned to a planning agency, board, commission, or committee empowered by national or state authorities working with education authorities, consistent with general health planning. Academic training centers play an important role not only in training, but in implementation of national manpower policies, so they are part of the consultation process in determining policies.

Supply and Demand

A common form of quantitative human resources planning or nonplanning is a market-oriented approach, based on the needs of the training institution and demands of trainees. The demand for training as physicians may be high, and the schools have an interest in training more students for financial or prestige reasons. If unregulated, this may not take into account the needs or capacity of the country to absorb the graduates, leading to creation of more schools and training capacity, producing a surplus of practitioners.

During the 1950s and 1960s planners in many countries thought that universal access to medical care would solve most health problems and that more doctors would be needed to fulfill that dream. It was assumed, even in centrally planned health systems, that supply and demand would direct new graduates to underserved regions or professional specialties. However, increasing the supply of medical graduates is costly to society and resulted in increased supply especially in major urban centers, with increased utilization, and subspecialization of medical care.

Increasing the supply of physicians was expected to increase access to health care and to increase the number of doctors entering less popular fields of practice, such as primary care, and moving to less served geographic areas. This approach has been less accepted Since the 1980s. Even in free-market societies, as it inflates the costs of health care and fails to meet needs in underserved areas or specialties. Immigration and emigration of medical personnel, or departures from active practice, are also factors in supply and distribution of health personnel. In the 1980s, health care costs rose rapidly, and associated with increasing specialization, a growing realization that health needs depended more on prevention than on increasing supply of doctors. This led to a trend to reduce numbers of new students entering medical schools, decrease specialty training positions, and reduce immigration of doctors.

Medical and other health professional schools are costly to establish and operate, and they can generate high cost to a health system if they produce an excess number of graduates. Founding new schools and maintaining existing schools at present levels of enrollment requires careful consideration of the effects of the numbers of graduates on the health system. In either a regulated environment or a free-market situation, the supply of manpower can be a powerful engine driving up health care costs. A period of restraint in health expenditures calls into question the wisdom of continuous increases in manpower and unlimited service as free or insured benefits. Even in free-market settings such as the United States, governmental funding and regulatory powers are used to reduce the number of training positions in the specialties in favor of increased incentives and openings in primary care.

Table 14.1 shows trends in medical education in the United States between 1950 and 1996, with projections to the year 2000. The number of medical graduates increased threefold between 1950 and 1980 but has grown at a slower rate since

TABLE 14.1   Medical and Osteopathy Schools, Graduates, and Physician Supply, United States, 1950–1996, with Projections to the Year 2000

	1950	1960	1970	1980	1990	1996	2000
Medical, osteopathy schools	85	92	110	140	141	142	—
Graduates (000s)[a]	5.9	7.5	8.8	16.2	16.9	17.9	18.0
Doctors/10,000	14.1	14.0	15.5	19.6	23.2	24.7	26.8

Source: *Health United States,* 1996-1997, 1998.
[a]Includes all graduating and active doctors and osteopaths.

1990. Medical manpower per population increased by one-third from 1950 to 1980, and continued to increase by another third from 1980 to 1990, based on the cumulative pool of graduates over the past 30 years, and immigration of physicians. A leveling off in production of new doctors is now occurring.

A normative approach uses standards or norms derived in some systematic, arbitrary way. The standards may be based on empirical criteria of the number of physicians, nurses, or other health personnel required. This approach may be excessively rigid and unresponsive to changes in disease prevalence and technological changes in health service needs. Standards may also be adopted from ratios found in other countries or in other successful or "gold standard" areas of the same country.

Manpower planning may set certain goals intended to produce personnel in numbers maintaining or increasing the current supply to population ratio by a selected percentage, for example by 5 or 10% within a chosen time period. Thus, if a country has 50 nurses per 100,000 population, it might plan to increase this to 55 per 100,000 population in a 5-year period. This approach is less likely to lead to an oversupply but may maintain an arbitrarily high level of human resources despite changes in epidemiologic patterns or increased efficiency of the services. For example, as tuberculosis declined, fewer tuberculosis specialists were needed, but as the disease recurs, there is a demand to improve the training and numbers of

---

BOX 14.2 ISSUES IN HEALTH
PERSONNEL PLANNING

1. Current and projected demographic changes, i.e., population growth and aging of the population;
2. Current and projected supply of practitioners and their geographic distribution by specialty;
3. Technological advances requiring new professions;
4. Immigration and emigration effects on manpower supply;
5. Costs/benefits of increasing professional to population ratio versus prevention, health promotion measures;
6. Changing epidemiological patterns, such as the reduction of dental service needs by fluoridation of community water supplies, aging of population with increasing prevalence of chronic disease;
7. Health systems shift from institutional to ambulatory and preventive care;
8. Shift of tasks from higher level to other personnel specifically prepared for needed health services; increasing range of health personnel, such as optometrists, psychologists, social workers, midwives, dental nurses, nurse practitioners, and community health workers.

specialists in the field. Hospitals are becoming less the center of health care, and reduction in hospital beds has become part of restructuring of services. This should lead to a shift of personnel from institutional to community-based services, with provision for retraining and skilled system management.

Many countries require medical graduates to serve one or several years in rural areas. This exposes young graduates to the realities or primary care as part of their professional development and, it is hoped, infuses them with concern for the harsh realities of the living conditions of rural poverty. However, it places inexperienced young professionals in isolated locations without adequate collegial support or supervision, where they are unlikely to remain beyond a compulsory period of service. Efforts to require young graduates to work in rural areas are temporary solutions, and are generally frustrated by the desire of doctors and nurses to live in urban areas and practice in clinical subspecialties.

Partly in search of methods of constraining cost increases and in part searching for ways to improve access to care for high risk groups, training of mid-level health workers is increasingly accepted as part of human resources planning. Manpower planning should take into account the many different disciplines needed for both clinical care and public health, taking into account changing patterns of need, technology, and spread of health care responsibility among many professions.

The organization of care affects the numbers of different health workers required. Independent private practice and free choice of physician or specialist promote higher utilization patterns, create waiting lists, rapid cost increase, and an apparent shortage of manpower. Centrally controlled health systems such as the Soviet health system created inflated staff-to-population norms and low efficiency health services.

Together with quantitative planning, methods of qualitative planning are also necessary. Quality of training programs at the undergraduate and graduate levels, accreditation, licensure procedures, and ongoing quality assurance measures are important elements in the quality of national health systems.

BASIC MEDICAL EDUCATION

The education of medical doctors and training of specialists are, in principle, a state commitment. Governments have a responsibility to ensure an adequate number of well-trained health professionals to provide services. This is a combined function of health and education authorities, carried out by providing financial support and standards for the universities or medical training institutes where the education occurs. Funding support and accreditation of the educational institutions provides mechanisms for applying national or state policy in both quantity and quality of educational programs. National or provincial departments of education set guidelines and standards for funding through a university grants mechanism or commission, often based on enrollment. Standards may be set for curriculum, faculty, basic sciences, and clinical training, as part of approval for funding or through nongovernmental accreditation structures organized by the medical faculties themselves (see Chapter 15).

The long tradition of multifaculty, university-based medical education is widespread in the industrialized countries and in their former colonies, now independent states. Medical training gains from the environment that promotes research and service in an academic atmosphere with its associated standards. This tradition of linking research with education and service is important in promoting high standards. Having a research climate of peer reviewed work raises the aspirations of the institution and its faculty and sets a standard for students for their life's work. A university degree confers prestige to a profession, encourages the pursuit of peer recognition of excellence and academic criteria for student selection, curriculum, and faculty standards. This is widely the case for medical schools, and increasingly for schools of nursing and other health professions. However, a university degree is not required for all health professions. Community colleges may more appropriately provide a multifaculty educational environment and a broad education base for some health providers.

In the nineteenth century, medical training in the United States was primarily carried out by private, commercial schools of medicine with poor facilities, staffing, and standards. The Carnegie Foundation sponsored a study of medical education in the Untied States and Canada, carried out by Abraham Flexner, a nonphysician educator, who reported in 1910 on the poor quality of these commercial schools. This report promoted university-based medical schools modeled on the Johns Hopkins University, which itself was based on the successful, scientifically oriented German medical schools, combined with the strong clinical orientation of the British teaching hospital medical schools. Most of the 450 commercial schools in the United States closed down soon after this report and were replaced by the present 126 university based medical schools with high standards of medical education and academic research. Since the 1950s, U.S. medical schools were stimulated by large amounts of federal funds channeled into research and training through the National Institutes of Health (NIH) and nongovernmental sources, including private and foundation donations.

Medical schools are resources for their community as well as being centers of academic excellence. Their goal should be to provide a balanced education in an academic environment where teaching, research, and service interact to produce medical graduates competent and oriented to meet the needs of the population. This requires a balance among the biomedical, psychological, population-based, and sociological perspectives on health care. Teaching methods should be designed to promote the objectives of the program. Many medical schools teach primarily by lectures to very large classes, with limited supervised clinical experience. This reduces the chance for the student to develop patient-oriented and problem-solving skills. It promotes a didactic approach to medicine, and minimizes the opportunity for the student to work with multidisciplinary teams, or to

BOX 14.3 THE FLEXNER REPORT, 1910

"For twenty-five years there has been an enormous over-production of un-educated and ill-trained medical practitioners in absolute disregard and without serious thought to the interests of the public. Taking the United States, physicians are four or five times as numerous in proportion as in old-er countries like Germany. Over-production is due to the very large num-bers of commercial schools. Colleges and universities have failed to appre-ciate the great advance in medical education and the increased cost of teaching it along modern lines. A hospital under complete educational con-trol is as necessary to a medical school as is a laboratory of chemistry or pathology. Trustees of hospitals, public and private, should, therefore, go to the limit of their authority in opening hospital wards to teaching. Progress for the future would seem to require a very much smaller number of med-ical schools, better equipped and better conducted and the needs of the pub-lic would equally require fewer physicians graduated each year better edu-cated and better trained."

Source: Flexner A. Medical Education in the United States and Canada: A Report to the Carnegie Foundation for the Advancement of Teaching, 1910. Reprint, New York: Arno Press and the New York Times, 1972. Condensed from introduction by H. S. Pritchett.

see medical care as part of a complex team service. Working with students of oth-er sciences and professions in a collegial fashion helps the medical student under-stand the team role of health professionals in the health care system.

The purpose of training medical practitioners is to have skilled professionals to provide patient care and the professional leadership needed to develop and maintain a high quality health care system. In order to meet these goals, high standards are re-quired in selection of candidates. Medical schools in the United States are graduate schools requiring a prior university degree for candidates. In other countries such as Canada and the United Kingdom, medical education includes 2 years of premedical studies followed by 4 years of medical school. Medical education of quality requires curriculum development and review, as well as highly qualified teachers, library ac-cess, clinical training, and examination during and at the point of completion of train-ing. The nature of the undergraduate training will be a key factor in determining the lifelong practice habits of the providers, but equally important are the specialization period and ongoing education throughout their professional lives.

In most industrialized countries, enrollment of women and minority groups has increased dramatically in the 1980s and 1990s as part of social policy. While there are social and political reasons to assure access to professional schools for all el-ements of a population, academic standards of acceptance should not be allowed to adversely affect the quality of services provided to the patient or the population as a whole. Private medical schools are a highly lucrative business in some developing countries, which, if unregulated, may contribute to an over production of inadequately trained doctors and compromise national efforts to promote quality of training in public universities.

Where the language of instruction is not one used internationally for scientific literature, the local medical community may be limited in access to current textbooks and the peer reviewed professional literature. The language of instruction in most schools of medicine in the Arab world is English, and in many European schools English as a second language is required. Lack of a high level of English prevents or hinders access to the world literature, participation in international exchanges, and scientific progress.

Curriculum reform, as in the days of Flexner's recommendations, must be an ongoing process to meet the health needs of the population, in keeping with current international standards. This includes adequate attention to basic medical sciences, clinical experience and patient care, hospital and community-based training, and research. Access to libraries with an adequate supply of current international literature, textbooks, and computers with Internet and electronic mail services is essential to maintain standards.

Reform in medical education is focusing on producing practitioners for the twenty-first century, meeting the needs of both primary care and specialized medical services. In recent years, there has been a growing concern that there has been too much emphasis on science and specialization to the detriment of primary care in training U.S. physicians. All medical students should be exposed to patient contact earlier in their training than in the past, in different health care settings, including teaching hospitals, outpatient clinics, community-based clinics, as well as public health programs. They should also be familiar with community-based resources for the infirm, the handicapped, and the needy. Training should include multidisciplinary components so the student is familiar with the professional elements of other disciplines including those in public health, health-related economics, and social sciences.

Two major international conferences on medical education sponsored by the World Federation for Medical Education in 1988 and 1993 attempted to define a new direction for education of physicians to promote their role in promotion of health as well as treatment and prevention of illness (Table 14.2). These conferences, sponsored by the WHO, UNICEF, UNESCO, UNDP, and the World Bank,[1] provided a international forum for reevaluation of medical education for the twenty-first century in the context of changes in medical and public health technology, organization of health care, and needs of the population. Change in medical education is often made difficult because of competing concepts of what the medical student should know, and a lack of focus on what the practicing doctor should be.

[1]UNESCO is the United Nations Educational, Scientific and Cultural Organization.
UNDP is the United Nations Development Programme.
UNICEF is the United Nations Children's Fund.
World Bank is the International Bank for Reconstruction and Development.

TABLE 14.2   Medical Education Issues for the 1990s—The Edinburgh Declaration (1988) and the World Summit on Medical Education, 1993

1. Conducted in relevant educational settings—hospital, community, workplace, homes	13. Students involved in planning and evaluation of medical education
2. Curriculum based on national health needs	14. A multiscience-based medical graduate
3. Emphasis on disease prevention and health promotion	15. Ethical and moral basis of medical practice
4. Lifelong active learning	16. Curriculum options for dealing with information overload
5. Competency based learning	
6. Teachers trained as educators	17. Postgraduate education in relation to community needs
7. Integration of science with clinical practice	
8. Selection of entrants for social commitment, intellectual attributes	18. Health teams and multiprofessional education
9. Coordination of medical education with health care services	19. Community participation in medical education
10. Balanced production of types of doctors	20. Population based education—care for individual patients in context of needs for a defined population
11. Multiprofessional training	
12. Continuing medical education requirements	

Source: Adapted from World Federation for Medical Education. World Summit on Medical Education: The Changing Medical Profession, Edinburgh, August, 1993.

The costs of medical education are high and require public subsidies. University grants commissions are semiautonomous bodies with financial support from education departments of governments. They use both financial and regulatory powers to set criteria for accreditation of faculties of medicine. This represents an important diffusion of power and responsibility from direct control by government. Regulation by accreditation of schools is also strengthened by national organizations which promote national standards of medical education.

POSTGRADUATE MEDICAL TRAINING

Undergraduate medical training provides the educational basis, but is not adequate preparation for a medical practitioner. Training following graduation is essential in assuring the quality of health care services. Specialty training requirements should be regulated by a national or state authority, or a professional body (college or association) designated to have the legal right to licence practitioners. This includes designation of facilities accredited for training, academic and research areas within the curriculum, clinical experience, duration of training, and requirements for examination at several stages during the training period. National standards are needed to ensure equivalent quality and permit freedom of movement of professionals. However, this may put provincial areas at a disadvantage by promoting a "brain drain," or loss of professionals, usually from rural to urban areas, or from poor countries to wealthy ones. The rights of an individual practi-

---

BOX 14.4 STANDARDS FOR POSTGRADUATE MEDICAL TRAINING

1. Regulated by national board with professional, governmental, and public representation for quality, numbers, and clinical and community-based training;
2. Duration of training of 4–6 years, depending on speciality;
3. Supervised independent clinical experience;
4. Accreditation of training centers based on academic and service criteria of licensing body;
5. Supervised research period in basic science laboratory or epidemiologic study;
6. Familiarity with relevant international literature;
7. Rotation with part of training in a different medical center;
8. Demonstrated high levels of clinical ability, responsibility, knowledge, human relations, and ethical standards
9. Examinations in mid-training with written examinations based on international standards;
10. Examinations at end of training; clinical and written examinations based on international standards;
11. State board or professional college setting examinations and certification;
12. Recertification requirements.

---

tioner to select place and type of practice are limited by open positions in training centers or in practice settings.

In the United States, the Council on Graduate Medical Education in 1992 recommended limiting the number of positions (i.e., slots) for postgraduate training to 110% of the number of U.S. graduates, thus accommodating training for foreign graduates, some of whom return to their own country. The Council added a condition that 50% of these postgraduate physicians enter general practice at the end of their training. These recommendations were supported by other bodies such as the Association of American Medical Colleges and the American Medical Association. This represents a consensus that in the United States there is an excess supply of physicians and of these too many are specialists while general medical practitioners are in short supply.

Licensing of medical specialists is a state responsibility, but in some countries this is delegated to a professional association. In the United States postgraduate training is under the control of state and national boards, made up of state appointed officials and public and professional representatives. In Canada, licencing of physicians is delegated to the medical associations, while postgraduate examinations and certification are under the authority of a professional body of the Canadian Royal College of Physicians and Surgeons. In the United Kingdom, licencing of physicians is under a state appointed body, the General Medical Council, while specialty recognition is by a series of Royal Colleges.

Standards for specialty training must reflect the views of the specialty practitioners as well as the public interest. The public interest is best protected by a combination of state and professional supervisory systems with the force of law, including the regulatory and disciplinary measures needed to maintain professional and ethical standards demanded by the public interest.

The former Soviet countries provided specialization during basic medical training (i.e., internal medicine, surgery, pediatrics, or sanitary-epidemiology). They did not establish the equivalent of national boards to govern graduate training criteria including length of training and examinations before being given specialist status. This lack of standards compromised overall quality of care in their health systems.

The specialist-trainee requires supervised time and experience to mature as a professional. Supervised clinical experience, research, publication in peer reviewed journals, and continuing peer review are all essential in the training process to produce specialists motivated and capable of keeping up with rapidly evolving standards of modern medicine. Clinical specialization time requirements vary widely from country to country. Eligibility for specialty boards in the United States are generally 3–4 years of recognized training after graduation, with examination by member boards of the American Board of Medical Specialties (ABMS).

SPECIALIZATION AND FAMILY PRACTICE

Good medical care depends on access to primary care and referral for specialty care. Most systems utilize the primary care physician as the "gatekeeper" for referral for specialty care. Laissez-faire systems allowing unreferred access to specialty care face the difficulty of maintaining primary care medicine and continuing pressures on physicians to select specialty training as their career choice (Table 14.3).

TABLE 14.3   Patient Care Specialist and Generalist Physician Supply Ratios per 100,000 Population, United States, 1965-1995

Physicians	1965	1970	1975	1980	1985	1990	1995
Specialists	56	65	79	94	106	113	130
Generalists	59	50	51	55	60	63	68

Source: Council on Graduate Medical Education. Tenth Report: Physician Distribution and Health Care Challenges in Rural and Inner-City Areas, February 1998.

Maldistribution of medical practitioners is a widespread problem, with rural and urban poverty areas often suffering from lack of access to primary care physicians. Specialist physicians are less likely than generalists to live in rural areas. In 1995, large metropolitan areas in the United States had 304 physicians per 100,000 population; small metropolitan areas had 235 per 100,000 and rural areas between 53 and 168 per 100,000. Maldistribution by specialty is another problem in medical resource planning. Regulations to limit the number of training positions is now operational in the United States and common in many countries. Regulations governing medical teaching centers require training in primary care.

National health systems deal with these problems with regulations to require, and financial incentives to attract, physicians to underserved areas and understaffed specialties. In the United Kingdom, as in many other European countries, the National Health Service uses the general practitioner as the primary care provider for all beneficiaries, with specialty access through the GP. Managed care programs in the United States also stress and require beneficiaries to see primary care physicians. The changing economic environment of health care will be associated with changes in medical specialization more easily than the urban–rural inequities. These issues are leading to greater role delegation to nursing and new kinds of health workers.

TRAINING IN PREVENTIVE MEDICINE

Preventive medicine is recognized as a clinical specialty in the United States. Promoted since the 1970s, this speciality attempts to bring public health and clinical medicine closer together. Preventive medicine training is one of 24 accredited clinical specialties in the United States, with doctors becoming board certified in one or more subspecialties: General Preventive Medicine and Public Health, Occupational Medicine, and Aerospace Medicine. These programs are part of the postgraduate training program system of the American Medical Association, in conjunction with the American Board of Preventive Medicine. Master's or doctoral degrees are earned in more than 23 graduate programs situated in departments of community or preventive medicine within a medical faculty.

Preventive medicine is a specialized field of medical practice composed of distinct disciplines that utilize skills focusing on the health of defined populations to maintain and promote health and well-being and prevent disease, disability, and premature death. It involves training in biostatistics, epidemiology, administration, planning, organization, management, financing and evaluation of health programs, environmental and occupational health, social and behavioral factors in health and disease. It applies primary, secondary, and tertiary prevention measures within clinical medicine. Graduates in this field provide potential health planners, administrators, teachers of preventive medicine, researchers, and clinicians applying preventive medicine in health care settings of practice. They may also serve in governmental (local, state, national, and international) public health departments, educational institutions, organized medical care groups, in industry, other employment settings, and the community, voluntary health agencies, and professional and related health organizations. Requirements include a graduate year of clinical training and experience in the clinical field; a year of academic training in a fundamental field of preventive medicine; and a practicum or year of supervised field experience (e.g., occupational). Training of clinicians in health services research and clinical epidemiology also provides a potential career path for physicians entering one of the many fields of public health.

In the former Soviet countries, a limited version of public health training is offered in medical training institutes at the undergraduate level. Public health practitioners are trained as undergraduates within the medical institutes. This does not produce public health professionals with postgraduate training in a multidisciplinary environment. Research institutes in various fields of public health provide graduate level training up to the doctoral (Ph.D. and Candidate of Science) levels. As freestanding institutes, unaffiliated with universities, they do not provide an academic environment of teaching and service. Economic and policy analysis required to affect policy and priorities suffers from these deficits. Graduate schools have recently started in several eastern European countries, including Poland and Romania, and one in Moscow in 1997 at Moscow Medical Academy.

NURSING EDUCATION

Nursing is the backbone profession in hospital and community health care. The place of nursing in a health system reflects cultural values of the society and has an important effect on the health system. Whereas medicine is generally a high prestige profession, in many countries nursing is of low social status, with strong cultural biases against women entering nursing. Finland and Sweden have more than four nurses per physician, while developing countries such as India and Bangladesh have between one and two nurses per physician. This represents a widespread overemphasis on medical training and an underemphasis on training of nurses in developing countries. The health system thus suffers from a lack of personnel to develop and operate primary care services, with biases to high cost secondary and tertiary care services. Furthermore, lack of high level professional nursing personnel prevents full development of quality secondary and tertiary care services. Lack of nursing at the professional level may be one of the biggest factors in retarding development of health services in many countries.

The number of nursing graduates more than tripled from 1950 to 1996 in the United States (Table 14.4). The promotion of the academic aspects of nursing is seen in the growth of baccalaureate nursing education from 13% of all nursing graduates in 1960 to 34% in 1996. The decline in the number of nursing schools in the 1950s was due to closure or consolidation of individual hospital schools of nursing. North American schools for nursing education are now largely associated with university or college programs. University-based schools in the United States provide academic degree programs at the bachelor's, master's, and doctorate level. Nursing education at the master's and doctorate level provides the teaching, research, and management cadres needed for a progressive health care system.

The scope of activities that professional nurses are authorized by law and custom to carry out has gradually broadened to include procedures previously performed only by physicians in the United States. Nurse practitioners are trained to diagnose and to treat, usually under authorization from a supervising physician. In some developing countries, especially in rural areas, auxiliary nurses, as well as professional nurses, are expected to diagnose and treat common ailments, in addition to conducting health education and primary and secondary prevention.

Nursing specialization may be at a certificate or master's level. Certificate courses are in fields where the nursing role involves highly skilled practice crucial to patient outcomes such as in intensive care or emergency room nursing. Master's programs in areas such as pediatrics, geriatrics, or adult health produce a more broadly based and independent practitioner, researcher, or educator.

The first baccalaureate program in nursing was established at the University of Minnesota in 1909, and by 1962 there were 174 bachelor's programs for registered nurses. Decisions to upgrade educational standards for existing professions, such as nursing or midwifery, involves considerations of the costs and effects on personnel supply as well as the desirability of raising professional standards. The advent of degree programs in nursing raised the level of prestige, leadership, research, teaching, and service of the profession. The transition from hospital apprenticeship training to university-based education (i.e., "academization") was opposed by traditional interests such as hospital management and the medical profession, but this resistance subsided with the demonstration of greater capacity in the nursing profession to take responsibility and incorporate rapid scientific and technological advances.

TABLE 14.4 Nursing Schools and Graduates in the United States, 1950–1996, with Projections to the Year 2000

	1950	1960	1970	1980	1990	1996	2000
Nursing schools	1770	1137	1340	1385	1470	1508	na[a]
Total nursing graduates (000s)	25.8	30.1	43.1	75.5	66.1	94.8	79.7
Graduates with BA (000s)[b]	na	4.1	9.1	25.0	18.6	32.4	26.5
Registered nurses/10,000 population	na	na	35.6	56.0	69.0	79.8[a,c]	na

Source: Health, United States, 1998.
[a]na, Not available.
[b]Includes bachelor of sciences degrees
[c]Data for 1995.

BOX 14.5 NURSING SHORTAGE:
"NEED VERSUS DEMAND"

In the United States, the National League of Nursing estimated that 530,000 of a total of 850,000 nurses listed in 1970 were active in the field. By 1980, there were 1.3 million licensed nurses, most of whom were working at least part-time, increasing to over 2.0 million in 1994. The nurse to population ratio increased from 56 per 10,000 in 1980 to 69 in 1990 and 80 in 1996. Hospital bed supplies fell from 1.4 million in 1980 to 1.2 million in 1990 and 1.1 million in 1996, while the population of aged and hospitalized patients is, on average, sicker.

Enrollment in schools of nursing is expected to decline from 97,000 to 80,000 in the year 2000. The medical and nursing professions were in conflict over numbers and roles of nurses. The nursing profession struggled to establish greater autonomy and academic quality, while the medical association fought to maintain large numbers of nurses and a subordinate role for nursing. The conflict between "need and demand" is a matter of definition, viewpoint, and priority. Changing health care to a greater community orientation will affect employment places for nurses.

As health care needs to cope with both an aging and healthier total population, chronic diseases require care at the primary level with increased roles (and needed retraining) for physicians and for nurses in home care and other outreach programs of care for persons at high risk, e.g., secondary prevention for hypertension or diabetes.

National health authorities need to take processional views into account, but balance the vested interests of each profession with other factors such as changing hospital bed supply and utilization, alternative forms of care (which can be labor intensive), aging of the population, changing disease patterns, and technological changes in prevention and health care, especially increasing ambulatory and community health care.

Source: National League of Nursing website http://www.nln.org/ and Health, United States, 1998.

IN-SERVICE AND CONTINUING EDUCATION

Rapid change in all fields of medical science and practice makes in-service and continuing education a fundamental necessity of any health program or services to maintain professional standards. In-service education increases the sense of self-esteem of workers and motivates staff to better performance. It serves to reinforce knowledge, introduce new information, and is essential to facilitate change in an institution. It also provides opportunity for the supervisory staff to reinforce and raise standards of the service. Introduction of a new program or technology should be accompanied by staff orientation as part of an ongoing in-service education program.

Continuing education refers to ongoing professional education in the form of courses, conferences, workshops, and literature. Medical graduates who complete requirements for specialization need to continue to upgrade their training with periodic courses in specialty areas where rapid advances are continuous. In public health, staff may take summer courses in epidemiology at schools of public health or departments of clinical medicine. Many medical, nursing, and other professional organizations require proof of continuing education for continued licensure and for professional advancement.

Governments, professional associations, provider organizations, nongovernmental health agencies, and the general public all have strong interests in continuing education for the health professions. In-service and continuing education should be part of the working schedule of a health institution and included in budgetary planning for all levels of health personnel, from laundry room staff to hospital managers and from community health workers to medical officers of health.

ACCREDITATION OF MEDICAL EDUCATIONAL OR TRAINING FACILITIES

All facilities training health professionals should be accredited to do so by the national or provincial authority or by an agency recognized by them for this purpose. In Canada, accreditation is carried out by the Medical Council of Canada, which is also the examining body for graduates of all medical schools. Provincial licensing bodies accept the License of the Medical Council of Canada (LMCC) as the basic requirement for licensure. In the United States, the Association of American Medical Colleges provides guidelines and accreditation of existing schools and reviews applications for new schools wishing to be recognized. Medical schools are subject to the state educational boards governing higher education facilities. State boards are responsible for examination of graduates and their licensure.

Universities or colleges establishing schools for other health disciplines are subject to the requirements of the authorities governing postsecondary education. A university wishing to establish a medical, dental, nursing, pharmacy, or other professional school would need prior approval showing the need for the facility, financial resources, and a complete proposal including curriculum, staffing, facilities, affiliations, and objectives. Recruitment standards and policies, clinical affiliations, quality of library and basic sciences facilities, and budget would be scrutinized. Staff qualifications, tenure procedures and requirements, publications and research, access to international professional literature, availability of textbooks, and student ability to read them (i.e., in a foreign language) should be part of the accreditation process.

Some relatively new medical schools, such as McMaster University in Hamilton, Ontario, Canada, and Beersheva in Israel, were founded with a focus on preparing primary care physicians, but it is not clear to what extent they have succeeded in this objective. Curriculum review has become widespread in schools of medicine with concern that there may be an excessive emphasis on basic sciences and specialty clinical services so that the graduate has little orientation to family and community practice or public health. New professions such as the nurse practitioner are developed from graduates of degree programs and require a master's level of training in an accredited program.

THE RANGE OF HEALTH DISCIPLINES

Establishing or recognizing new health professional roles, such as nurse practitioners, optometrists, or community health workers, is dependent on and related to the needs of the health system. Development of curricula, criteria for enrollment, and site of the training program should be governed by the objectives of the program, but also should ensure wide acceptance of the new profession and potential for career advancement. Acceptance by the community is important, especially in programs intended to improve contact of persons in high risk groups with the health and social services systems. Traditional birth attendants and community health workers are categories of personnel providing health care where cultural adaptation is especially important.

Clinical medicine has evolved from primarily a medical and nursing service to involve a highly complex team of professionals. Similarly, in public health the range of professions is broad. Interdisciplinary training is important to the function of a department or service increasingly dependent on teamwork.

The complexity of modern public health and clinical services is shown in the number of different professions listed in the Table 14.5. This broad range of professions in public health requires graduate studies with an interdisciplinary approach to preparation of leaders, teachers, and researchers for the field. Public health professionals work within a variety of settings. They need a wide base of training in order to understand the broad professional aspects of public health that relate to a complex and rapidly changing field.

LICENSURE AND SUPERVISION

All countries have legal or regulatory systems by which newly trained health personnel are permitted to practice their profession. The requirements differ from country to country and for various types of personnel within a country. In some countries, health personnel have to take licensing examinations in addition to completing the prescribed training. In others, registration by the government is more or less automatic after the prescribed training, including the examinations, has been successfully completed. For some disciplines, such as medicine, dentistry, nursing, or pharmacy, the legal requirements for the license may be delegated to professional colleges or to state or national boards. Certification and relicensing of medical and other health care practitioners have become standard practice in the United States and some other jurisdictions to assure that the health care provider is up to accepted professional standards of the day and public expectations.

Examination of undergraduate students is generally by the teaching institution itself, but examination at completion of training for licensing to practice medicine should be by external examination, preferably at a national or even international level. National examinations, formulated and supervised by professional and governmental authorities, establish and maintain the standards of medical graduates. In the United States, state boards govern medical licensure and specialty certification.

TABLE 14.5 Major Types of Health Professional in Public Health and Clinical Services

Public health professions	Clinical health professions
Public health management	Health facility administrators—accounts, hotel func-
Health policy analyst	tions (dietary facilities, laundries, maintenance),
Medical officer of health	purchasing, planning, legal, public relations,
Epidemiologist	secretary, volunteers
Dental public health officer	Physician—general and specialty
Veterinary public health officer	Dentist
Industrial health physician	Nurses—administration, general, and specialty
School health officer	Midwife
Health economist	Pharmacist
Medical sociologist	Physical therapist
Medical anthropologist	Occupational therapist
Legal officer	Speech therapist
Information scientist	Respiratory therapist
Demographer	Social worker—medical, psychiatric
Statistician, biometrician	Laboratory personnel—biochemist, microbiology,
Health service researcher	genetics and others
Supervisor of midwives	Genetics counselor
Health educator—community, public,	Radiological technician
school, and industrial	Imaging technician
Nutritionist	Dental auxiliary
Computer personnel	Nutritionist
Public health nurse	Dietician
Sanitary engineer	Pharmacy assistant
Product safety engineer	Licensed nursing assistant
Food technologist	Health record clerk
Biochemist, microbiologist, food scientist,	Community health worker
and toxicologist	Vital records clerk
Veterinary scientist	Assistant midwife
Sanitarian	Others—kitchen, laundry, cleaning staff, supply, and
Community health worker	maintenance
Others	

Licensing of health professions in some countries, such as the United Kingdom, Canada, and others in the British model, allows the health professions self-government to set standards and govern the discipline within the profession as a form of peer review. National examinations and limitations of foreign graduates are spelled out in regulation or by decisions of the governing body of the profession. Foreign schools may be accepted for equivalent status or examination requirements established.

In Canada, the Medical Council of Canada, a consortium of provincial professional bodies, establishes and supervises medical graduation examinations, but licensure is by provincially licensed bodies. Other countries regulate medical licensure directly but delegate specialty training supervision to professional organizations. Many countries have developed national examinations for medical, dental, nursing, pharmacy, and other professional licensure to promote high level requirements and avoid the conflict of interests of a school examining its own graduates.

Graduates of Canadian schools of nursing are accepted for licensure by some states in the United States. Medical graduates are licensed by state boards in the United States. Graduates of U.S. medical schools in one state are accepted in other states for postgraduate training but not necessarily for medical practice, although some states have agreements of reciprocity. Canadian provinces in the past accepted graduates of British medical schools, but this was restricted in the 1970s to reduce the flow of immigrant doctors. In the United Kingdom, the General Medical Council is the legislated body empowered to license local graduates and immigrant physicians.

Licensing of physicians, nurses, midwives, psychologists, optometrists, nurse practitioners, or other professionals needs to be based on legislation or regulation under public health law to designate the scope of permissible functions in each profession, licensing and examination procedures, as well as a code of conduct. Diffusion of power in governance of medical practice has contributed to setting high standards of practice. Control of education and licensing by the same authority that operates the national service may tend to compromise standards. Development of multiple systems of accountability in a previously totally state controlled system, as in Russia, will require many changes in existing practices of medical education, examinations, licensure, specialty training, and examination and discipline, as well as development of independent professional organizations and accreditation bodies (see Chapter 16) and standards of care.

CONSTRAINTS ON THE
HEALTH CARE PROVIDER

Maintaining standards requires organized supervision of performance by public bodies, in written guidelines, based on accepted current standards of care. This is often based on a consensus of professional views, practices as well as recommended guidelines of professional bodies. Care should be taken to avoid penalizing legitimate innovations or differences of professional opinion, such as whether simple mastectomy is sufficient care for cancer of the breast as opposed to radical mastectomy. This is part of quality assurance discussed in Chapter 16.

Ideally, the constraints that impinge on the health care providers are the sum of training, licensing, practice, collegial relationships, and self-governance of ethical, humanitarian, and professional standards. These are under scrutiny and the potential disciplinary procedure from a variety of sources, including legal responsibility and standards of care expected by the employer or institution in which the provider functions. The provider is also under scrutiny in the eyes of the public or consumers who may select him or her for care or choose not to where consumer choice is part of the system. The total effect of peer review on a continuous basis, encouraging good standards of practice in the community, helps to assure basic standards but, at the same time, may promote conformity limiting medical innovation, especially in the areas of organization of health care. Recognition of new professions can lead to conflict of interests with self-governing professions as has happened in the case of optometrists and nurse practitioners. Similarly, professional groups may oppose changes in health care financing and organization, both for the public good and sometimes for professional self-interests.

Application of standard medical curricula, national standard examinations, national licensing and disciplinary boards, and standards of medical practice are essential for maintaining and improving health care. External peer review is consistent with current stress on total quality management (TQM) or continuous quality improvement. When entering medical practice, doctors seek access to hospitals where they apply for "hospital privileges" and are assessed by professional peers for experience and qualifications. Individuals are expected to be members of professional organizations in their specialty and participate in departmental staff meetings and programs of continuing education organized by professional associations, hospitals, and medical schools.

Health insurance plans monitor billing or practice patterns of physicians and investigate aberrant practice or potential fraud. This may be followed by administrative action against the offending physician or, rarely, by criminal procedure for fraud. Monitoring of surgical procedures may point out poor practice that could lead to disciplinary procedures. Criminal conviction means suspension of license to practice. Malpractice insurance is vital to protect any physician or other health care provider against litigation and may be very costly depending on specialty. No-fault insurance for injury resulting from medical care under legislation to protect the provider and the injured party, such as in vaccine use, may be a model for a more rational system than the litigation approach.

Professional accountability is specific in each country. In the United Kingdom, the General Medical Council is empowered by the state to issue medical licenses and discipline practitioners. A Patient's Charter sets out the rights, entitlements, and standards of service the citizen may expect in health care. This, coupled with the right to change general practitioner, empowers the patient to seek redress of grievances. Complaints regarding hospital care are investigated and can be pursued through stages of investigation. Consumer satisfaction is a factor in the recent innovation of GP fund holding in the United Kingdom (see Chapter 13), where the patient may, with the GP, select among hospitals or other support services.

Health care is complex and requires a skilled and integrated team functioning with mutual trust, based on a common set of professional and ethical goals and standards. This is clear in the drama of hospital settings as seen in popular television programs, but applies equally in the larger scale of health system organization and interaction between institutions, insurers, and pubic health networks. In addition to oversight by financial authorities and accreditation bodies, the scrutiny of the media, the political sector, the consumer, and the public at large is important. In short, the health provider and the health system are, and should be, under scrutiny.

NEW HEALTH PROFESSIONS

New professional roles emerge as the health needs of the community evolve. Public health and health management professionals and health care providers are all essential, tailored to the community they serve. The health system needs to assure that traditional health workers are available in the numbers sufficient to provide for community and national care needs. In addition, there has been a growing realization of the need for new levels of health workers.

Mid-level health worker experience in Russia with the feldsher was important to provision of primary care in rural, underserved areas. In many developing settings, experience with community health workers has been growing, and this has also been applied in some developed countries. The nurse practitioner and physician assistants have emerged as new health professional roles in the United States to provide health care in underserved areas, and in some cases to provide health care for targeted under-served population groups, such as the elderly, or diabetics.

Nurse Practitioners

In the 1960s health care providers and planners in the United States became aware of the growth of specialization in medicine and the decline of the general practitioner. The nursing profession promoted expansion of nursing roles to fill this gap. These factors and increasing costs of medical care fostered the wider role of nurses to provide medical services.

In 1965, the first program to train nurse practitioners was developed at the University of Colorado. Nurses were taught specific medical functions, not as a doctor, but as a logical extension of the nurse's traditional function of assisting patients to regain health and independence. In 1971, the federal government adopted the Nurse Training Act, which defined the role of nurse practitioners.

A nurse practitioner (NP) is a registered nurse (RN) who has advanced training and education in the medical field. Nurse practitioners perform detailed physical examinations, order laboratory tests for diagnostic purposes, assess the results of such tests, write prescriptions for medications and other medical devices, make referrals, and perform other delegated medical functions. A nurse practitioner can provide care in hospitals or the community as part of a health team.

A 1996 U.S. Department of Health and Human Services National Survey of RNs showed 71,000 NPs out of an estimated 2.5 million nurses currently licensed, or 2.8% of the nurse workforce. The NPs were 44% of all advanced practice nurses (APNs), who represent some 6.3% of the total RN population. The number of NPs, estimated as high as 90,000 in 1998 and 100,000 in 1999, is expected to grow rapidly as the number of academic institutions with NP track graduate programs grew from 101 in 1990 to 202 in 1995. The number of clinical tracks more than doubled from 210 in 1990 to 527 in 1995. The numbers of applicants and enrollees more than doubled in the United States between 1993 and 1995, and the number of graduates increased by a factor of 1.8. The number of NPs may reach 10% of all RNs in the first decade of the twenty-first century as primary care roles for NPs increase. These include family practice; women, adult, and school health; pediatrics; and gerontology. Specialty care tracks include neonatal and acute care, occupational therapy, psychiatric care, and others in institutional care settings.

In each state, the practice of nursing is established and regulated by nurse practice acts and common law. These acts establish educational and examination requirements, provide for licensing or regulation of individuals who have met these requirements, and define the functions of the professional nurse in general and specific terms. The criteria establish parameters within which each NP may practice. The Commission on Collegiate Nursing Education, an accreditation organization, is attempting to standardize NP training programs to ensure quality of the graduates.

Nurses constitute the largest single group of professionals among American health personnel. There is ample evidence that expansion of nursing roles has improved the quality of health care, especially in medically underserved areas. Equally important, nurse practitioners in underserved professional areas such as geriatrics and health promotion have innovated programs and provided a research and theoretical basis for new directions in health care. With a growing use of clinical guidelines, the roles of the nurse practitioner may be expected to increase in the twenty-first century.

Physician Assistants

In 1923, 89% of U.S. physicians were general practitioners. By the mid-1960s the figure had declined to about 25%. Concern with the shortage of primary care physicians in the country led Eugene Stead, an internist at Duke University in Durham, North Carolina, to develop the first physician assistant (PA) training program in October 1965. Physician assistant training programs were developed for three reasons: to help alleviate a perceived shortage of primary care physicians; to compensate for geographic and specialty maldistribution of physicians; and to help control escalating health care costs.

Physician assistant programs are not part of nursing, and this has engendered conflict with nurse practitioner programs and nursing authorities. There are currently 52 accredited physician assistant training programs in the United States. The disciplines include neonatology, pediatrics, emergency medicine, occupational medicine, and surgery. In addition, PAs are taught preventive health care, patient education, utilization of community health and social service agencies, and health maintenance. The main focus in basic physician assistant training is patient care in the primary care practice setting. The first 6 to 12 months of training is devoted to preclinical studies and clinical laboratory procedures, followed by 9 to 15 months of clinical training. The curricula are reviewed regularly and modifications made in keeping with changes in the health care setting.

On completion of a PA training program, an entry level competence examination is given by the National Commission of Certification of Physicians' Assistants. When passed, it allows the PA to append the title PA-C (physician assistant-certified) to their name. Then every 2 years PAs must register, documenting 100 hours of approved continuing medical education. To assure clinical competency, PAs must take a recertifying examination every 6 years.

Physician assistants perform tasks such as history taking, physical examination, simple diagnostic procedures, data gathering, synthesis of data for a physician, formulation of diagnoses, initiation of basic treatment, management of common acute and emergent conditions, management of stable chronic conditions, patient and family counseling, supportive functions, and in 18 states limited prescribing. Task delegation of PAs is determined by the State Board of Medical Examiners within each state. The PA is not a substitute for the physician, nor an independent provider like a nurse practitioner. The PA is not licensed for independent practice, and the physician must assume all responsibilities and bear all the professional and legal consequences of the PA's actions.

The American Academy of Physician Assistants lists nearly 35,000 persons eligible to practice. Some two-thirds of PAs had baccalaureate degrees prior to PA training. Some one-third of respondents were employed in hospitals, another one-third in solo or group practice offices, and 9% in community health centers. PAs work in over 60 specialty fields, but over 50% are involved in general/family practice. Employment opportunities are increasing for NPs in various specialties and practice settings. A PA may be able to handle one-half to three-quarters of the clinical services provided by the supervising physician, indicating that they are productive and cost-effective in their employment settings. PA salaries are one-fourth or one-third those of physicians, so the cost–benefits of using PAs are obvious. The literature on this topic reports that the quality of care offered by PAs is comparable to that given by a physician.

The PA, although initially controversial, is now a widely accepted role. The nursing profession has regarded this as a method of increasing doctors' incomes, an infringement of nursing roles, and a reaction on the part of physicians to the growing professionalization of nursing. The medical profession defends the PA as a way of extending the possibility of medical care to larger population groups and improving the economics of medical practice.

Feldshers

The feldsher is a unique Russian mid-level health worker. The role originated as military company-level surgeons introduced by Peter the Great in the seventeenth century. Retired army feldshers returned to rural areas unserved by physicians, becoming the sole providers of rural medical care. The feldsher was adapted to the Soviet health system to provide care in rural areas still underserved by doctors despite the increase in medical personnel.

The feldsher is trained in a course of 2–3 years following intermediate school graduation. Feldshers complement physicians in urban and rural practice, especially in small medical posts, in mobile emergency medical services and in industrial health stations. In the health reforms of the 1990s, the feldsher is a declining profession, which may result in serious difficulties in maintaining rural health care.

Community Health Workers

The concept of the community health worker (CHW) is not new but has found new expression in health programs in many parts of the world as part of the primary health care initiatives springing from Alma-Ata. It is an adaptation of traditional village practice of midwives and healers to modern, organized public health services. CHWs were first developed to provide care in rural areas in developing countries without access to health care. More recently there has been an interest in the CHW model for urban community health needs where access to health is limited by geographic or socioeconomic reasons.

Mid-level CHWs include generalists in independent practice, categorical or targeted health workers, and preventive community health workers. The exact functions of these personnel, the duration of training, and the framework within which they work have varied much more than their titles. The generalist village health worker programs in some areas lacked close supervision, and may seek fee-for-service practice. Volunteer community health workers, advisors, or *promators* are commonly used in Latin American countries, often as volunteers. CHWs are recommended for rural locations in developing countries without supervisory or organized contact with professional health services, providing a wide range of diagnostic and treatment services.

Community health workers may provide services on categorical target diseases. These include malaria control, tuberculosis directly observed therapy (DOT, providing medication under supervision to assure compliance), support services and counseling for multiproblem families in an inner-city poverty area, STD follow-up, and promotion of immunization. Prototypes of the task-oriented CHW include malaria control CHWs in Colombia and tuberculosis DOT and AIDS case workers in New York City. In Africa CHWs are crucial to programs for eradication of guinea worm disease and river blindness.

A CHW program with a focus on preventive services was developed since 1985 in Hebron, in the West Bank, first under Israeli and later under Palestinian jurisdiction. CHWs act as preventive care workers as on-site representatives of the government health service, providing on-site prenatal and child care, immunization,

---

BOX 14.6 COMMUNITY HEALTH WORKER
PROGRAM MODELS

Independent CHWs

1. Feldshers in Russia for rural health care;
2. "Barefoot doctors" in China for rural health care;
3. "Where There Is No Doctor" CHW providing all health care in remote villages in Latin America.

Categorical CHWs

1. Program specific CHWs, e.g., malaria control in Colombia, guinea worm disease and onchocerciasis in Africa, AIDS or TB care in New York City, immunization in Los Angeles;
2. Public health nurse extender CHWs, e.g., Albany County Health Department, New York State.

Preventive Oriented CHWs

1. Preventive village health workers providing on-site services as part of the public health system, visiting medical–nursing services and close supervision, e.g., Hebron, West Bank, Palestinian Authority;
2. "Urban villages," i.e., urban poverty areas in United States, with CHWs as part of county health department or community-based organization services.

---

nutritional counseling, pregnancy care with a medical-nurse support team, first aid, and emergency or nonurgent referrals to the district hospital or to nearby medical clinics. The CHW is trained and supervised to provide primary care and outreach services in a community as part of an organized health system. Visits by supervisory professional staff are vital to the function of the system of care. The emphasis may be on preventive and community services in small villages without on-site services, or as part of an outreach and support service in a large urban setting.

Community health workers should be recruited from the community to be served and trained in settings with both didactic and field experience. Training of village CHWs may be very different from those of schools for training conventional health personnel. Training usually takes place in rural health centers or hospitals, to which classrooms and student living quarters have been added. The training is nearly always sponsored by the ministry of health, sometimes in collaboration with foreign agencies or international organizations. Candidates for such training are typically young people from rural families, who have been selected by their communities. During training, the student should be salaried, and on completion return to work in the community of origin.

In the United States, public health nurses have been traditional providers of home visiting programs or outreach programs to provide health care and education for families in need. The emergence of community-oriented primary health care broadens this strategy, using selected and trained community residents as CHWs. The CHW concept was used to carry out many Great Society programs in the United States in the 1960s and 1970s. Volunteer and paid workers worked as lay home visitors or health guides in programs in selected areas or target populations, such as pregnant women or parenting families. Other CHW programs in the United States included Navaho communities, urban health centers, rural Texas, and Alaska.

Community health workers have been trained to provide outreach and case management in the complex environment of New York State, going into the community to assist high-risk families in undeserved areas of large cities and in AIDS, STD, and tuberculosis patient care, especially with DOT for TB. CHW experiences working with the homeless and mental health patients have shown positive results. More recently, CHW projects have been developed in urban poverty "villages" in Los Angeles, sites of low income, public housing projects, with backup services of on-site clinics and sponsoring medical centers.

Evaluation and cost–benefit justification of CHW programs are difficult in terms of establishing population denominators and control groups, and in determining changes in outcomes measures in mortality, morbidity, and physiological indicators, such as growth patterns. This limitation is shared with many health programs, not only in primary care but perhaps even more so in highly technological medicine. The village health worker concept has had criticism as well as advocacy in recent years. The CHW as an all-purpose health provider of both preventive and treatment services may be impossible to sustain and is undesirable conceptually. The more selective approach may be more feasible and manageable, with the CHW showing the promising potential to provide a new parameter in health care.

ALTERNATIVE MEDICINE

Use of alternative medical treatment has become a widespread phenomenon in industrial countries and remains a mainstay of health care in preindustrial societies. Alternative medical care is based on the belief that illness includes physical, mental, social, and spiritual factors. Alternative medicine views health as a positive state, rather than as the absence of disease, and believes in the natural healing capacity of the human body. Its interventions are noninvasive and less technological than conventional medical care.

The growth in popularity of alternative medicine in part relates to a widespread disillusion with the depersonalization and technological orientation of medical care. Other factors include medicine's failure to treat the patient as a whole person, bias toward single causes, and "magic bullet" treatments for disease. Disappointment with medical results is common, with failure of cures or complications of treatment itself as iatrogenic disease. Attenders of alternative medical care in the United States showed that the number of visits to providers of unconventional therapies is greater than the number of visits to primary care doctors, with high levels of out-of-pocket expenditures. Users of alternative medicine are largely in the 25–49 age group and are among better educated persons in upper income categories. Alternative medicine categories included chiropractic, massages, commercial weight loss programs, lifestyle diets, herbal medicine, megavitamin therapy, self-help groups, energy healing, biofeedback, hypnosis, homeopathy, acupuncture, and folk remedies.

Changing attitudes within conventional medical care is seen in a growing tendency for acceptance of some previously excluded professions, including optometry, chiropractic, and acupuncture, in insured benefits or even multidisciplinary health care systems as complementary or supplementary to conventional medical care. Conventional medicine is itself in a process of change with addition of many paramedical professions and awareness of limitations of the biomedical model as the sole basis for health care.

Consumer demand plays an important role in this change, as does willingness of medical practitioners to refer chronic health problems such as back and neck pain, stress and related problems, phobias and addictions, allergy and skin disorders, and hormonal and menstrual disorders. Conversely, alternative practitioners seem to be increasingly aware of limitations in treatment of conditions such as cancer, hypertension, and chronic and hereditary disorders, and the need for referral for treatment by conventional medical methods. The nursing profession in the United States has promoted use of touch therapy as an independent healing modality over the past several decades.

Absorption of "holistic methods" by medical practitioners has become widespread. Many major medical centers now include departments of alternative or holistic medicine, with acupuncture, hypnosis, and other nontraditional approaches in their roster of services. This increases legitimization, status, and incomes of practitioners, but is only achieved over long periods of conflict and opposition by orthodox medical practice. In 1992, the U.S. Congress mandated establishment, within the National Institutes of Health, an office of Alternative Medicine, in 1998 expanded to become the National Center for Complementary and Alternative Medicine. This shows the growing acceptance of a now major element of community health. European medicine and health agencies, with long traditions of healing baths and rest cures, has been more open to complementary and alternative medicine than their North American counterparts.

CHANGING THE BALANCE

As cost containment becomes more important and many countries cut back on health expenditures in order to control rates of increase, the stress of readjustment falls on health workers. If such measures are implemented on an emergency basis rather than over time as part of transition and realignment, then the process will generate hostility, defensiveness, and political opposition. If, however, longer term planning takes into account the human issues of manpower in a changing balance of services, then the burden of the individual or group of health workers who suffer from the downsizing can be minimized.

During the 1960s, Canada was embarking on its national health insurance program, and leading thinkers of the day called for rapid expansion of medical and nursing schools to meet future needs of the population. They assumed that national health insurance would bring a significant portion of the population who lacked access to care to the health services, overloading the medical and hospital services. In the mid-1990s, all provinces cut back on the hospital bed supply, creating unemployment for nurses and maintenance staff. Provinces are attempting to restrict the numbers of practicing doctors and to modify the payment systems away from fee-for-service. Inflation in health care utilization and costs have resulted in a crisis management approach, rather than a structural reform to promote a more integrated population-oriented health care organization. This can lead to a great deal of public and professional dissatisfaction within a health care system.

EDUCATION FOR PUBLIC HEALTH AND HEALTH MANAGEMENT

Successful implementation of the New Public Health requires that many health disciplines work together. The training milieu should have a capacity for interdisciplinary training in a comprehensive program, including fundamental and applied

---

BOX 14.7 CUTBACKS AND MANPOWER EFFECTS IN CANADA

"There are almost a quarter of a million RNs (registered nurses) in Canada but very few in primary care centers. Most nurses work in acute care hospitals, where they are at high-risk for lay-offs as hospital beds are cut and middle management jobs vanish. We also have 6,000 RNs working in physician's offices, 7,000 in home care, and almost 10,000 public health nurses. . . .

"Approximately 75% of hospital budgets are for personnel, meaning you can't make deep cuts all at once without laying off employees. Without a careful strategy to reallocate budgets and redeploy personnel to community services, nurses and hospital workers end up on the street. And that's exactly what has been happening. As of April 1993, over ten thousand nurses had been laid off and nearly twenty thousand other hospital workers were collecting unemployment insurance."

Source: Rachlis M., Kushner C. 1994. *Strong Medicine: How to Save Canada's Health Care System*, pp. 245, 291. Toronto: Harper-Collins.

research, as well as a relationship with service programs and community health assessments.

To facilitate teamwork all public health practitioners need a background of the medical sciences, epidemiology, economics, social sciences, environmental and occupational health, health systems analysis, and management theory. It is important that they be familiar with the terms and concepts of fields other than their own specialized discipline. This is more likely to be found or created in a university atmosphere and is difficult to foster in separate, categorical institutes.

University resources are essential to a school of public health to provide teachers and courses from other faculties such as schools of business administration and the social, physical, and biological sciences. Schools that are unaffiliated with a degree granting university lack the broad academic atmosphere and requirements, as well as the connection with parent disciplines, such as economics, sociology, microbiology, and business management faculties.

Preparation of personnel for public health and health management should be at the graduate school level followed by continuing education. The U.S. Institute of Medicine's 1988 report *The Future of Public Health* defined a need for schools of public health to teach not only professional and technical skills, but also an understanding of how a particular discipline relates to public health as a whole, and the value system that is part of public health's coherence.

Training of public health physicians began in the United Kingdom in 1871 in Dublin's Trinity College, which granted the Diploma in Public Health (DPH). The program was designed to provide for the training of medical officers of health to lead the work of the boards of health that were established under public health acts and were required to have qualified medical practitioners as medical officers of health. Other universities later offered DPH or equivalent training, supervised by the College of Physicians and Surgeons, as public health became a recognized medical specialty. In 1924, the London School of Hygiene and Tropical Medicine brought together several institutes and produced a major center of training and research in public health. Since the 1991 Acheson Report on schools of public health in the United Kingdom, there has been a rapid growth of schools of public health under various names in a number of universities. This has coincided with growth of interest in the NHS in working toward health targets as opposed to simply managing health services.

The tradition of schools of public health is especially strong in the United States, where the Johns Hopkins and Harvard Schools of Public Health were founded in 1913. In 1915, the Rockefeller Foundation sponsored a national program to promote public health education at the University of Michigan, Yale University, and the University of Pennsylvania, which established graduate training for public health professionals to meet the needs of crowded urban industrial cities of the United States. These schools saw their mission as the training of public health practitioners, and secondly academics, educators, and researchers. They attempted to develop and assimilate new knowledge into public health practice. The development of public health as a multidisciplinary field made it vital to be inde-

BOX 14.8 ELEMENTS OF PUBLIC HEALTH TRAINING

Objectives (ASPH)

    1. To prepare target-oriented practitioners, researchers, policy analysts, and managers;

    2. To provide continuing education for public health practitioners;

    3. To promote public health research and policy analysis;

    4. To advocate and promote health-related issues in public policy.

Requirements (ASPH)

    1. Postgraduate training in a multifaculty academic setting;

    2. Multidisciplinary approach;

    3. Problem-oriented skills training to identify targets and problem-solving management approaches;

    4. Link education, research, and service in public health.

Core curriculum (Roemer)

    1. Basic tools of social analysis: history of public health, demography, medical sociology and anthropology, biostatistics, population sampling and survey methods, political science of health systems, principles of program evaluation, and health economics;

    2. Health and disease in populations: vital statistics, major human diseases and zoonoses, epidemiology of diseases and risk factors, methods of clinical diagnosis and prevention, infectious and chronic diseases, nutrition, environment, special disease and risk groups, global ecology of disease, and risk factors;

    3. Promotion of health and prevention of disease: communicable disease control, chronic diseases and their prevention, environmental and occupational health, maternal, child, adolescent, adult, and elderly health needs, mental health, STD/AIDS control, nutritional and dental health, health education and promotion, rehabilitation, refugee, migrant, and prisoner health, military medicine, and disaster planning;

    4. Health care systems and their management: organization and operation of national health care systems, health insurance and social security, health services and workforce development, health facilities and their management, drugs and their logistics, health planning, principles of management and application to health programs budgeting, cost control, and financial management, record and information systems, health systems research, health legislation and ethics, technology assessment, accreditation and quality promotion in health care, information systems, monitoring and research methods for management, global health.

Source: American Association of Schools of Public Health website, http://www.asph.org; Roemer M. 1999. Genuine professional doctor of public health the world needs. *Journal of Nursing Scholarship,* 31:43–44.

pendent from but affiliated with a medical faculty. Schools of public health in the United States produced many generations of well-trained epidemiologists, social scientists, health educators, practitioners, and leaders who were crucial for development of the field.

In the 1960s, schools of public health were criticized because of their separateness from and lack of influence on clinical medical training. In some jurisdictions, this led to closure and replacement by departments of community medicine within faculties of medicine. This occurred in the United Kingdom and in British Commonwealth countries. Canada's two schools of public health were closed, replaced by departments of social and preventive medicine or community health, developed within medical schools. Departments of community health within a medical faculty serve as only one department among many clinical or basic science departments. Such departments generally may lack prestige in the hierarchy of medical schools, in an environment promoting a narrow, medically oriented approach to public health, with an insufficiently multidisciplinary program and faculty. This model provides training at the undergraduate, master of public health (MPH), and doctoral levels. The full academic potential of a graduate school of public health is most suited to be in an independent, multidisciplinary, university-based academic center for public health research and training.

Periodically, threats of absorption of schools of public health into other sectors of the university arise. In 1994, a plan to close the University of California, Los Angeles (UCLA), School of Public Health by transfer to the School of Public Policy was halted by university and nationwide protests. Health administration is sometimes considered as better integrated within schools of business, as happened at the University of Minnesota School of Public Health.

Since the 1980s, there has been a renaissance of schools of public health in the United States with an expanding market for graduates. Today, students in schools of public health come from many different backgrounds, including medicine, dentistry, nursing, engineering, economics, social sciences, statistics, mental health, veterinary sciences, and others. In 1995, 4145 master's degrees were awarded and 491 doctors of philosophy. The most popular specialties were international health and epidemiology. The career outlook for graduates of schools of public health now includes other traditional public sector positions, but also higher paying positions in the private sector, including consulting firms and managed care organizations. In the 1996–1997 academic year, 14,007 students were enrolled in schools of public health in the United States.[2]

In the United States, there are 28 graduate schools of public health accredited by the Council on Education for Public Health. They are based on a multidisciplinary approach to training for public health, including technical and administrative leadership in epidemiology of communicable and noninfectious diseases, biostatistics, management of personal health services, environmental health, maternal

---

[2]For further information, consult the Association of Schools of Public Health website at www.asph.org

and child health, health economics, health education, and other related fields. The interdisciplinary aspect of public health is emphasized by the wide range of backgrounds and experiences of the students and their many areas of specialization. Analysis of problems with the skills of epidemiology, sociology, and other related disciplines permits the graduate to enter practice with a problem solving approach. Another 26 university graduate programs in community health education, community health and preventive medicine are accredited extend public health training to most parts of the country.

Accreditation involves external review of facilities, faculty, curriculum, student selection criteria, internships or field experience, and academic standards. An accredited school is better able to generate research and scholarship funds and is more attractive to students for future career advancement in an expanding job market. These and many other nonaccredited schools or university departments provide operating public health and health care agencies with well-trained personnel with a wide range of undergraduate and professional experience, thus enriching the field, not only with practitioners, but also with researchers, administrators, and policy analysts of high quality.

Schools of public health should have close working relations with state and local health agencies, recruiting part-time faculty from service agencies and conducting research in real public health problems that confront health agencies. Schools of public health can provide important services to departments of health in research and consultation on public health issues and assessment.

Graduate schools of public health are ranked annually by surveys of deans, top administrators, and senior faculty conducted as part of continuous surveys of graduate schools in the United States by a weekly news magazine (*U.S. News and World Report*). These surveys are based on annual assessments, including reputation among deans and senior faculty (40%), research activity (30%), student selectivity (20%), and faculty resources (10%). In 1998, the top ranked schools of public health were Johns Hopkins, Harvard, the University of North Carolina (Chapel Hill), the University of Washington, the University of Michigan, the University of California–Berkeley, Columbia University (New York), the University of Minnesota, Emory University (Georgia), the University of Pittsburgh, and Yale University (Connecticut). While these surveys and rankings have no official status, they are important in that they are widely used as guides for student selection of graduate schools, possibly affecting research grants, fund-raising and faculty recruitment.

In continental Europe, training in public health was traditionally through a job-oriented, vocational approach carried out in government or independent institutes as courses for medical officers of health or hospital managers. Since the 1980s or early 1990s, European schools of public health of the broader model been established, in Germany (Bielefield), Holland, France, Spain, Poland, and Romania. The Nordic School of Public Health serving the Scandinavian countries was founded in the 1970s in Gothenburg. It added a sociological emphasis to health systems analysis, but is still lacking the broad epidemiologic, public health practice, and economic strengths of U.S. schools. An association of European schools of public health has developed and promotes the idea of standardization and reciprocity for master of public health degrees in the European Union.

Schools of public health are of special importance for developing countries because of the prime importance of public health approaches in meeting their health needs. Nigeria is a very populous country where the main health issues are those in the public health sector, yet there are over 20 medical schools and no school of public health. Departments of social or community medicine, primarily teaching the medical student, are common in most medical schools, but this fails to provide students at the graduate school level with an academic environment and multidisciplinary training and specialization that the field requires to provide the professional leadership needed to meet the health challenges of their societies.

Leadership positions in health systems at the local, state, and federal levels are now held by doctors trained in clinical medicine, often with added training in public health. Few, however, have training in management and many with training focused in specific subdivisions of public health. On the other hand, schools of public health at the doctoral level focus on preparation of scholars in research and teaching rather than health leadership and management. Some seek preparation in master of business administration (MBA) programs. A new approach in preparation of doctors of public health requires 5 years of postbaccalaureate training with broad areas of knowledge: tools of social analysis, health and disease in populations, promotion of health and prevention of disease, and health care systems and their management (see Box 14.8). Some schools of public health are moving in this direction by providing special part-time programs for health executives. This will be especially important in the preparation of leadership capable of coping with the complexities of managed care or district heath programs developing in many countries facing the organizational, economic, and ethical aspects of individual and population health.

HEALTH POLICY AND MANAGEMENT
OF HUMAN RESOURCES

Preparation for policy and management roles in public health and health systems has become widespread in schools of public health and in business schools in the United States. This trend will undoubtedly increase as managed care increases, and as intersectoral mergers or other functional arrangements become more common. Departments of management and policy or health services in schools of public health have the mission to study and seek methods of improving efficiency and effectiveness of personal and population-based health organizations. As academic fields they share a population perspective and include interdisciplinary faculty from economics, law, management, medicine, history, sociology, and policy analysis. They focus on societal, population, economic, and organizational perspectives.

Personnel costs are the largest single component of total health expenditures, so that management and utilization of this resource is of prime importance to a health system. Health personnel should be recruited, trained, and utilized in a manner appropriate to meet the health needs of the population. This means employing their skills under conditions that promote effective work. Human resource management includes determining which category of worker can best provide specific services, how many are required, and what organizational frameworks are required to provide needed care most effectively. This requires not only delegation of responsibility, but also the resources with accountability to carry out the tasks.

Human resources management includes determination of numbers and types of health personnel needed for the future health system. Delegation of responsibility from professional levels can be made to appropriately qualified paraprofessionals. This includes planning, management, financing constraints, licensing, and discipline procedures and quality control measures. Accountability for performance is essential for any system dedicated to provision of quality care and to meeting its goals of improved health outcomes.

Restricting numbers of professionals is sometimes done in the self-interest of a professional group to restrain competition. Oversupply can be costly and destructive to the public interest by misdirecting health resources in nonproductive or even harmful ways. An excess supply of surgeons generates higher rates of elective surgical procedures than necessary or safe, while shortages of primary care physicians prevent adequacy in basic health services. Poor supply or quality of nursing personnel compromises the quality of hospital care and primary care in the community. This requires long-term retraining and redeployment policies developed and implemented over time rather than spasmodic mass layoffs of nurses and other hospital workers.

Managed care in the United States and similar comprehensive service programs in other countries provide the opportunity to seek a new balance of services and introduce new roles in health care. Physicians, nurse practitioners, community health workers, and many other kinds of health care professionals and technicians will be part of the complex of health care provision when the economics of care necessitate cost-effectiveness, where prevention and treatment are part of the same complex, and where a health promotion approach is fundamental to the objectives of the organization.

SUMMARY

Education and training of medical and allied health personnel are important issues in health care systems development, and include issues of both quantity and quality. Regular reassessment is needed lest the numbers of practitioners produced be larger or fewer than the needs of the services, and lest standards decline.

Preparation of managers and planners skilled in data and program analysis and leadership in implementation are as important as training health care providers.

Training of health professionals should be accompanied by orientation to the broad sweep of the New Public Health, including its management and evaluation skills. New health professional roles will evolve based on individual patient and community health needs.

Management in the New Public Health is confronted with many difficult issues in manpower policies. These include not only the quantity and quality of training but flexibility in utilization, including redeployment of personnel from institutional care settings to community health and health promotion activities. Personnel constitute some 75% of costs of patient care, and any program for reallocation of resources toward community and preventive care must necessarily involve health workers, not only as an economic issue, but as a qualitative one. These professional and personal issues must be treated with care and sensitivity.

In the end, a health care system depends on the quality, ethics, pride, and professional skills of its team members. The training and retraining of such personnel is therefore a fundamental consideration of the New Public Health.

ELECTRONIC MEDIA

American Academy of Nurse Practitioners (AANP), http://www.aanp.org/ care@nurse.net and http://www.nurse.net/

American Academy of Pediatrics (AAP), http://www.aap.org/

American Board of Medical Specialties (ABMS), http://www.abms.org/

American Medical Association (AMA), http://www.ama.org/

American Public Health Association (APHA), http://www.apha.org/

Association of Schools of Public Health (ASPH), http://www.asph.org/

Association of Schools of Public Health in the European Region (ASPHER), http://www.ensp.fr/aspher/

Association of Teachers of Preventive Medicine (ATPM), http://www.atpm.org/

Council on Education for Public Health (CEPH), http://www.ceph.org/

Joint Council on Accreditation of Healthcare Organizations (JACHO), http://www.jcaho.org/

National League of Nursing (NLN), http://www.nln.org/info-history.html/

RECOMMENDED READINGS

Feil, E. G., Welch, H. G., Fisher, E. S. 1993. Why estimates of physician supply and requirements disagree. *Journal of the American Medical Association,* 269:2659–2663.

Handler, A., Schieve, L. A., Ippoliti, P., Gordon, A. K., Turnock, B. J. 1994. Building bridges between schools of public health and public health practice. *American Journal of Public Health,* 84:1077–1080.

Kahn, K., Tollman, S. M. 1992. Planning professional education at schools of public health. *American Journal of Public Health,* 82:1653–1657.

Legnini, M. W. 1994. Developing leaders vs. training administrators in the health services. *American Journal of Public Health,* 84:1569–1572.

Roemer, M. 1999. Genuine professional doctor of public health the world needs. *Journal of Nursing Scholarship,* 31:43–44.

Weiller, P., Hiatt, H., Newhouse, J., Brennan, T. A., Leape, L., Johnson, W. P. 1993. *A Measure of Malpractice.* Cambridge, MA: Harvard University Press.

BIBLIOGRAPHY

Bender, D. E., Pitkin, K. 1987. Bridging the gap: The village health worker as the cornerstone of the primary health care model. *Social Science and Medicine*, 24:515–528.

Berman, P. A., Watkin, D. R., Burger, S. E. 1987. Community-based health workers: Head start or false step toward Health for All? *Social Science and Medicine*, 25:443–459.

Council on Graduate Medical Education. 1994. *Fourth Report: Recommendations to Improve Access to Health Care through Physician Workforce Reform.* Rockville, MD: U.S. Department of Health and Human Services.

Council on Graduate Medical Education. 1998. *Tenth Report: Physician Distribution and Health Care Challenges in Rural and Inner-City Areas.* Rockville, MD: U.S. Department of Health and Human Services. February.

Deber, R. B., Thompson, G. G. (eds). 1992. *Restructuring Canada's Health Service System: How Do We Get There From Here?* Toronto: University of Toronto Press.

Eisenberg, D. M., Kesser, R. C., Foster, C., Norlock, F. E., Calkins, D. R., Delbanco, T. L. 1993. Unconventional medicine in the United States: Prevalence, costs and patterns of use. *New England Journal of Medicine*, 328:246–252.

Fee, I., Acheson, R. M. (eds). 1991. *A History of Education in Public Health.* Oxford: Oxford Medical Publications.

Fulop, T., Roemer, M. I. 1987. *Reviewing Human Resources Development: A Method of Improving National Health Systems.* Public Health Papers, 83. Geneva: World Health Organization.

General Medical Council. 1993. *Tomorrow's Doctors: Recommendations on Undergraduate Medical Education.* London: GMC, December.

Harper, D., Johnson, J. 1998. The new generation of nurse practitioners: Is more enough? *Health Affairs*, 17:158–164.

Institute of Medicine. 1988. *The Future of Public Health.* Washington DC: National Academy Press.

Kindig, D. A., Cultice, J. M., Mullan, F. 1993. The elusive generalist physician: Can we reach a 50% goal? *Journal of the American Medical Association*, 270:1069–1073.

Kohler, L. 1991. Public health renaissance and the role of schools of public health. *European Journal of Public Health*, 1:2–9.

McKee, M., Clarke, A., Kornitzer, M., Gheyssens, H., Krasnik, A., Brand, H., Levett, J., Bolumar, F., Chambaud, L., Herity, B., Auxilia, F., Castali, S., Lanheer, T., Lopes Dias, J., Ria, T. 1992. Public health medicine training in the European community: Is there scope for harmonization? *European Journal of Public Health*, 2:45–53.

Miullan, F., Rivo, M. L., Politzer, R. M. 1993. Doctors, dollars and determination: Making physician work force policy. *Health Affairs*, 12 (Supplemnent):138–151.

National Center for Health Statistics. 1998. *Health, United States, 1998, with Socioeconomic Status and Health Chartbook.* Hyattsville MD: DHHS (PHS) 98-1232.

Rachlis, M., Kushner, C. 1994. *Strong Medicine: How to Save Canada's Health Care System.* Toronto: Harper-Collins.

Richards, R., Fulop, T., Bannerman, J., Greenholm, G., Guilbert, J. J., Wunderlich, M. 1987. *Innovative Schools for Health Personnel: Report on Ten Schools Belonging to the Network of Community-Oriented Educational Institutions for Health Sciences.* Publication No. 102. Geneva: World Health Organization.

Roemer, M. I. 1987. Health manpower allocation in developing countries. *Asia–Pacific Journal of Public Health*, 1:20–24.

Schorr, T. M., Kennedy, M. S. 1999. *100 Years of American Nursing: Celebrating a Century of Caring.* Baltimore: Lippincott Williams and Wilkins.

Walt, G., Gilson, L. (eds). 1990. *Community Health Workers in National Programs: Just Another Pair of Hands?* Milton Keynes, PA: Open University Press.

Werner, D., Thuman, C., Maxwell, J. 1992. *Where There Is No Doctor: A Village Health Care Handbook.* Palo Alto, CA: The Hesperian Foundation.

Workshop on Alternative Medicine. 1992. *Expanding Horizons: A Report to the National Institutes of Health in Alternative Medical Systems and Practices in the United States.* Washington, DC: U.S. Government Printing Office.

World Federation fro Medical Education. 1993. *World Summit on Medical Education.* Edinburgh: WFME.

World Health Organization. 1987. *The Community Health Worker: Working Guide: Guidelines for Training: Guidelines for Adaptation,* Third Edition. Geneva: World Health Organization.

World Health Organization. 1989. *Specialized Medical Education in the European Region.* Euro Reports and Studies 112. Copenhagen: World Health Organization, Regional Office for Europe.

World Health Organization. 1993. *Reviewing and Reorienting the Basic Nursing Curriculum.* Copenhagen: World Health Organization, Regional Office for Europe.

World Health Organization. 1994. Nursing Beyond the Year 2000: Report of a Study Group. Geneva: World Health Organization.

15

TECHNOLOGY, QUALITY, LAW, AND ETHICS

INTRODUCTION

Management of a production or a service system requires attention to the quality of personnel, as much as to the system in which they work. Their motivation and sense of participation, the scientific and technological level of the program, as well as the legal and ethical standards of individual providers and of the system as a whole are all important to the quality of care provided.

Quality is the result of input and process. It is measured by outcome or performance indicators as well as perception of the service by the patients, the staff, and the community as a whole. Input refers to the financial, human, and institutional resources, supplies, and services available. Process refers to the use of those resources, and outcomes generally include measures of morbidity, mortality, and functional status. The definition and measurement of achievement of national health objectives and targets, the methods of financing of services, and the efficiency of organization help determine quality. Training, supply, and distribution of health personnel are all determinants of access to and quality of care. Continuous and adequate availability of essential preventive, diagnostic, or treatment services, as well as accountability and internal methods of promoting standards, are all elements of the quality of a health service.

The content and standards of service can be assessed by organized review by professional peers within an institution, and from outside. Peer review inside the institution and external evaluation by accreditation or governmental inspection contribute to accountability and improved quality. Team responsibility among staff includes anticipating problems, implementing changes needed, and determining whether action is taken and if it has the desired effects.

Technology is in a continuing state of change. Systematic review and absorption of new innovations promote renewal in health care methods. Public health serves in a regulatory role to assure high quality care to the individual and the community. New technology, whether in the form of diagnostic procedures, new drugs, devices, or vaccines, or new types of health manpower, requires evaluation for effectiveness and appropriateness to the system.

Legal and ethical standards in health reflect the mores of a society. Innovations in the technology of health care and the organization of its services raise new issues and challenges. Determining standards of "good practice" is a continuing process with the rapid development of new knowledge, technology, and experience.

INNOVATION, REGULATION, AND QUALITY CONTROL

Health care technology has advanced with an increasing stream of innovation since the seventeenth century definition of separateness of mind and body by Descartes and the discovery of smallpox vaccination by Jenner, to the dramatic innovations of the end of the twentieth century (Table 15.1). The pace of innovation is rapid, creating the need for regulation, quality control, and technology assessment.

National governments are responsible for assuring that pharmaceuticals, bio-

TABLE 15.1 Health Care Innovations from the Seventeenth to Twenty-first Centuries

Period	Examples of Scientific, technologic and organizational innovation
17th century	Biological basis of disease (Descartes), circulation of blood (Harvey), microscope (Leeuwenhoek)
18th century	Thermometer, lime juice supplements (Lind), vaccination (Jenner), surgical anatomy (Hunter), clinical sciences (Sydenham)
19th century	Stethoscope, anethesia, laryngoscope, ophthalmoscope, blood pressure cuff, sanitation, antisepsis, braille printing, hygeine in obstetrics, nursing, microscopic pathology, pathological cheimistry, microbiology, vaccines, X-ray, national health insurance, syringes, well child care
1900–1930	Biomedical education, salvarsan, insulin, blood groups, vitamins, conquest of yellow fever, vitamin B, cost-benefit analysis
1931–1945	Penicillin, randomized clinical trials, antimalarial and vector controls
1946–1960	Vaccines, antihypertensives, psychotropic drugs, cancer chemotherapy, prepaid group practice
1961–1980	DNA, oral rehydration therapy, vaccines, cost-effectiveness analysis, open heart surgery, pacemakers, organ transplantation, computerized tomography (CT), eradication of smallpox, HMOs, DRGs, district health systems
1981–2000	Magnetic resonance imaging (MRI), positron emission tomography (PET), endoscopic surgery, eradication of polio, *H. pylori* and peptic ulcers, managed care, improved treatment of myocardial infarction

logical products, drugs, food, and the environment are regulated to protect the public. In some countries these responsibilities are divided among ministries of trade, industry, commerce, health, and environment. In a federal system of government, there may be a division of responsibility between federal, state, and local government, but with the national government providing national standards and leadership in this area.

Regulation and control are meant to protect the public. The U.S. Food and Drug Administration (FDA) is part of the structure of the Department of Health and Human Services (DHHS). The FDA is responsible to enforce the Food, Drug and Cosmetic Act, the Fair Packaging and Labeling Act, sections of the Public Health Services Act relating to biological products for control of communicable diseases, and the Radiation Control for Health and Safety Act.

The FDA is meant to assure the consumer that foods are pure and wholesome, safe to eat, and produced under sanitary conditions; that drugs and medical devices are safe and effective for their intended uses; that cosmetics are safe and made from appropriate ingredients; and that labeling is truthful, informative, and not deceptive. FDA acts and regulations govern both domestic and imported products. FDA approval is widely accepted internationally as a gold standard, that is, meeting a high standard of safety and efficacy. Investigation of contents, manufacturing standards, potency, and toxicity are carried out by accepted laboratory procedures as published in the compendium *Official Methods of Analysis of the Association of Official Analytical Chemists*. When FDA investigators, known as consumer safety officers, observe conditions that may result in a violation of an act, they leave a written report with the manufacturers and recommendations for correcting the conditions. They may issue product recall or seizure orders for products in violation of FDA standards. Drugs and devices include all drugs, diagnostic products, blood and its derivatives, biologicals, veterinary medicines and medicated premixed animal products. All manufacturers and distributors are required by law to register these products with the FDA in order to be allowed to market or import them.

New drugs and biological products for human use are required to pass rigorous review before approval for marketing. Applications are submitted by the manufacturer or sponsor with acceptable scientific data including test results to evaluate its safety and effectiveness for the conditions for which it is being offered. All manufacturers of drugs are required to be registered with the FDA and to meet its requirements for each drug produced and marketed, including reporting of adverse reactions and labeling criteria. Manufacturers are required to operate in conformity with current good manufacturing practices, which include stringent control over manufacturing processes, personnel training, and computerized operations, and testing of finished products. The FDA publishes guidelines to help manufacturers familiarize themselves with current standards. The United States Pharmacopoeia and the National Formulary are the official listings of approved products.

Medical devices are also supervised by the FDA. Thousands of products for health care purposes require premarket approval, ranging from basic articles such as thermometers, tongue depressors, and infrauterine devices (IUDs) to more complex devices, such as cardiac monitors, pacemakers, breast implants, and kidney dialysis machines. These products are subject to controls of good manufacturing practices, labeling, registration of the manufacturer, and performance standards.

Monitoring for efficacy and potential hazards have been strengthened since the 1970s as a result of findings of long-term carcinogenic and mutagenic effects of estrogens, and toxic effects of chloramphenicol on bone marrow. The drug thalidomide, used widely in Europe, Canada, and Australia in the 1960s, was, however, not approved by the FDA. Thalomide caused large numbers of serious birth defects leading to it being banned in most countries. Controls of blood and blood products have been strengthened since the transmission of HIV, hepatitis B, and hepatitis C by contaminated blood products in the 1980s. The responsibility of this regulatory function is well illustrated by the 1995 criminal conviction of several senior health officials in France for failing to stop the use of blood products contaminated with HIV in the mid-1980s. Concern regarding possible carcinogenic effects of silicone breast implants led to legal action and greater controls of all implantable products. A balance between safety and well-regulated approval of new products requires a highly professional and motivated regulatory agency, well-developed procedures, and staff.

The concepts of standardization of good manufacturing practices for pharmaceutical products and written protocols for good medical practice or good public health practice are accepted norms based on best available evidence of current scientific knowledge and experience. Recommended immunization schedules, water quality or ambient air standards or food fortification, and screening programs for early stages of diabetes are examples of accepted practice that become recommended standards of public health practice, just as clinical care develops its qualitative measures.

APPROPRIATE HEALTH TECHNOLOGY

The topics discussed in the growing literature and meetings of the International Society of Technology Assessment in Health Care represent the dynamic field of technology assessment. The issues range from economic evaluation of pharmaceuticals, measures of quality of life, technology dissemination and impact, health care in developing countries, outcomes measurement, finance and health insurance, informatics, teleradiology, technologies for the disabled, modeling approaches, screening, and cost-effectiveness. A mix of clinical, epidemiologic, and economic approaches are necessary elements in the scrutiny of both high and low technology services. As health costs rise and populations age, medical innovation proceeds at a rapid rate, and both client and community expectations in health care are high.

Appropriate technology is defined by the World Health Organization as the level of medical technology needed to improve health conditions in keeping with the epidemiologic, demographic, and financial situation of each country. All countries have limited resources and so must select strategies of health care and appropriate technology to use those resources effectively to achieve health benefits.

In developing countries, training and supervision of traditional birth attendants for prenatal preparation and normal deliveries may be an important way to reduce maternal mortality in rural areas. Similarly, community health workers can provide care to underserved rural poor populations with a defined package of services that can be tailored to meet specific local needs, such as immunization, growth monitoring, nutrition counseling, and malaria and tuberculosis control.

A major example of appropriate technology has been the WHO initiatives to promote national drug formularies (NDFs) as a consensus list of essential drugs that are sufficient for the major health needs of a country, eliminating unnecessary duplication and combined products on the commercial market. The WHO called on all member states to ensure availability and rational use of drugs and vaccines and supported states wishing to select an essential list of drugs for economic procurement. Assistance with drug regulatory agencies, legislation, quality control, information, supply, and training was offered to help the member country. Standard reference laboratories, the International Pharmacopoeia, and the *WHO Drug Bulletin* promote international standards and provide guidance to states. The WHO Model List of Essential Drugs is a valuable tool to improve quality and cost management in national health systems.

The World Bank's *World Development Report* of 1993 defined cost-effective clusters of clinical and public health programs essential to improving health outcomes for low and middle income developing countries. The programs focus on those diseases which contribute heavily to the burden of disease and are amenable to relatively inexpensive interventions. The *World Development Report* defined interventions most able to reduce the burden of disease in low and middle income countries using clinical and public health interventions, as summarized in Table 15.2. The EPI Plus program, for example, prevents 6% of the total burden of disease in a low income country and costs $14.60 per immunized child, or $0.50 per capita.

The WHO promotes widespread use of basic radiologic units (BRUs) to increase access to low cost, effective, diagnostic X-rays, especially in rural areas in developing countries. BRUs are hardy, relatively inexpensive pieces of radiological examination equipment that can be used in harsh field conditions for simple diagnosis of fractures and respiratory infection. The WHO estimates that 80% of all diagnostic radiology can be performed adequately by simple, safe, and low cost equipment, supported by training of local people to operate and maintain the equipment. This is a consensus view of leading radiologists and clinicians helping the WHO to develop model equipment and training material.

In industrialized countries technological advances in the medical and public health fields have been major contributors to increasing health costs and have led to pressures for greater selectivity in adopting costly innovations without adequate assessment of benefits and costs. Many countries have adopted more cautious policies with regard to financing unlimited expansion of new technology in the field of medical equipment, procedures, or medications. The need for organized assessment of technology is now an essential feature of health management at the international, national, and local level of service delivery. The major responsibil-

TABLE 15.2   Priority Cost-effective Health Interventions in Low and Middle Income
Developing Countries, 1990[a]

| Minimum Package Service Type | Burden of disease averted (%) in countries | | | |
| | Low income | | Middle income | |
	% Disease Averted	Cost per capita	% Disease Averted	Cost per Captia
Public Health				
EPI Plus immunization (DPT, polio measles, BPG, hepatitis B, yellow fever + vitamin A)	6.0%	0.50	1.0	$0.80
School health program (including deworming)	0.1%	$0.30	0.4	0.60
Other public health programs (family planning, health, and nutrition education)	—[c]	1.40	—[c]	3.10
Tobacco and alcohol control	0.1	0.30	0.3	0.30
AIDS prevention program	2.0	1.70	2.3	2.0
Subtotal	8.2%	4.20	4.0	6.80
Clinical Care				
Tuberculosis DOTs[b]	1%	$0.60	1%	6.80
Integrated management of the sick child	14	1.60	4	$0.20
Prenatal and delivery care	4	3.80		8.80
Family planning	3	0.90	1	2.20
Treatment of STDs	1	0.20	1	0.30
Limited care: pain, trauma, infection plus as resources permit	1	0.70	1	2.10
Subtotal	24%	7.80	11%	14.70
Total	32%	12.00	15%	21.50

Source: World Bank, *World Development Report,* 1993.
[a]Low income = <$350 GNP per capita; middle income = >$2,500 GNP per capita.
[b]DOTs = directly observed therapy, short course.
[c]Benefits not quantified.

ity for technology assessment is at the national level, even with decentralization
of service management.

Appropriate technology in the health field is becoming increasingly complex,
laden with economic, legal, and ethical issues. Professional and public opinion
demands make this a highly sensitive area of health policy, but responsible management of resources requires decision making that includes consideration of the
effectiveness, costs, and alternatives of any new technology. Failure to adopt new
innovations can result in obsolescene, while excessive expenditures hospitals and
medical technology prevents a health system from developing more cost-effective

---

BOX 15.1 QUESTIONS FOR TECHNOLOGY
ASSESSMENT

Questions that form the basis of technology assessment for a medical innovation include the following:

1. Is it a new service, or does it replace a less efficient service?
2. Is there a need for it?
3. Where is it in the order of priorities of development of the facility?
4. Does it duplicate a service already available in the community?
5. Does it make medical sense, i.e., does it help in diagnosis and treatment for the patient's benefit?
6. What are the alternatives?
7. What are the resources needed in purchase, staffing, and upkeep?
8. Can the facility afford it?
9. What could otherwise be done with the resources it requires?

---

preventive approaches, such as improved ambulatory care, or supportive care for
the chronically ill.

Technology of unproven value can be a highly emotional and political issue, in
criticism of managed care or national regulatory agencies, but spending limited
national resources on ultrasound machines, magnetic resonance imaging equipment (MRIs), or unnecessarily long hospital stays denies resources needed for other aspects of health care. A society must be able and willing to pay for medical innovation or improving quality of life by medical and public health interventions.
Underfunding of a health system can deny these benefits just as misallocation of
resources does.

HEALTH TECHNOLOGY ASSESSMENT

Medical and health technology assessment (TA) is the process of determining
the contribution of any form of care to the health of the individual and community. It is a systematic analysis of the anticipated impact of a particular technology
in regard to its safety and efficacy as well as its social, political, economic, legal,
and ethical consequences. The technology may be a machine, a vaccine, an operation, or a form of organization and management of services. Analysis should include cost–benefit and cost–effectiveness studies (see Chapter 11) as well as clinical outcomes and other performance indicators.

Pressures from medical professionals, manufacturers of new medical equipment, and the public for adoption of new methods can be intense and continuous.
Care must be taken that the specialists involved in committees for assessment are

---

BOX 15.2 ENDOSCOPIC SURGERY

In the 1990s, endoscopic surgery has been an important advance in surgery with great benefit to patient care in reducing trauma, discomfort, and length of hospital stay, becoming the surgical method of choice for most cases. Since reports of the first 100 operations done in France in 1990, endoscopy spread rapidly to all parts of the world within a short time. It is now recognized by surgeons worldwide as a safer, less traumatic, and more effective approach to traditional invasive surgery. Although the operating time is longer, the patient is able to be discharged from the hospital within several days and return to work shortly thereafter. In contrast, following traditional abdominal surgery the patient was very sick for many days, requiring intensive care initially and a recovery period of many weeks. Endoscopic surgery for cholecystectomy and esophageal, colorectal, herniorrhaphy, renal, orthopedic, and other forms of surgery have become standard practice, extending the range of outpatient surgery. Fewer complications are seen and more cases are being performed, raising questions as to whether excessive operations may be encouraged by this less traumatic procedure. The economic implications are important.

---

not those who may directly or indirectly benefit from the exploitation of technology, who therefore may have a conflict of interests. Assessment must be multidisciplinary, involving policy analysts, physicians, economists, epidemiologists, sociologists, lawyers, and ethicists. The available information needs to include evidence from clinical trials, critical analysis of the literature, and the economic effect of adopting the technology on allocation of resources.

Medical technology varies in complexity and cost, not only to produce but to utilize. Medical technology that is inexpensive to supply and administer is low technology or lotech, while high technology or hitech refers to costly and complex diagnostic and treatment devices or procedures.

At the lotech end of the technology scale, ORT (oral rehydration therapy) was developed in the 1960s for oral replacement of fluids and electrolytes lost in diarrhoeal disease. It was described in *Lancet* as one of the greatest medical breakthroughs of the twentieth century. The introduction and wide scale use of oral rehydration therapy for prevention of dehydration from diarrheal diseases throughout the world has saved hundreds of thousands of lives that would otherwise have been lost. In polio eradication, Sabin oral polio vaccine has important advantages over its rival Salk inactivated vaccine, being lower in cost and easier to administer by less qualified personnel because of its oral mode of administration.

The dissonance between hitech and lotech procedures may lead to serious consequences in any health system. Choices require well-informed analysis of benefits, costs, alternatives, ethical considerations, and political consequences before distribution of limited health care resources between hospital based hitech medicine versus lotech primary care.

Hitech procedures are usually applied in hospital settings in costly care for seriously ill, often terminal patients. Computerized tomography (CT), invented in the 1960s, quickly proved itself to be an extremely valuable diagnostic tool. Advances in CT and subsequent imaging techniques have proven to be cost-effective and lifesaving, replacing less efficient and dangerous invasive procedures. The CT scan allows the clinician to come to a rapid diagnosis of many lesions before they could be detected by other diagnostic techniques, at stages where they are susceptible to earlier and more effective interventions.

Technology assessment also examines methods of preventing and managing medical conditions. Treatment protocols or clinical guidelines are based on decision-analysis of accumulated weight of evidence. Published clinical studies are assessed in meta-analysis, using statistical methods to combine the results of independent studies, where the studies selected meet predetermined criteria of quality. This provide an overview from pooling of data, but also imply an evaluation of the studies and data used. Clinical guidelines are part of raising standards of care, but also contribute to cost containment. Many countries form professional study groups to carry out meta-analyses on important health policy issues and new technologies.

Technology Assessment in Hospitals

Varying between countries, hospitals consume between 40 and 70% of total national health expenditures, with pressures for increased staffing and medical technology a continuing inflationary factor. Shorter stays and older patients have resulted in a drift toward intensive care, especially for internal medicine patients. Medical innovation is a continuing process with new diagnostic and treatment modalities reaching the market.

Hospitals no longer live in splendid isolation in the medical economy. A state government will have regulatory procedures to rationalize distribution of medical technology. The certificate of need (CON) is a form of TA used in the United States since the 1960s to assess and regulate the development of hospital services to prevent oversupply and costly duplication of services. It attempts to establish and implement use of rational criteria for diffusion of expensive new technology. Whether this has had a lasting impact on restraining the excesses of high technology medicine is arguable. This regulatory approach was limited to the hospital setting and failed to stop development of hitech medical services such as ambulatory for-profit CT and imaging centers.

Many countries have adopted national TA assessment systems to review topics as far ranging as guidelines for acute cardiac interventions, liver, heart, and lung transplantation, minimal access surgery, beam and isotope radiotherapy, diagnostic ultrasound, sleep apnea, molecular biology, prostate cancer, magnetic resonance imaging, and new medications for inclusion in national health systems' approved "basket of services."

Despite the limitations of this approach, where governments are no longer directly operating health care services, governmental regulation is necessary to prevent inequities in services by excessive development in some geographic areas at the expense of others, or by overexpansion of the institutional portion of health care at the expense of primary care. Regulatory mechanisms are essential in health care planning to restrain excessive and inappropriate use of hitech services, but need augmentation by fiscal incentives to promote other essential services.

Technology Assessment in Prevention and Health Promotion

Technology assessment of preventive care programs includes evaluation of the methodology itself along with the costs and measurable benefits, as in reduced burden of disease. The World Bank estimates that measles immunization at 80% coverage is more cost-effective per death prevented than 60% coverage, by preventing an additional 11,000 deaths in a province in Bangladesh. Ambulatory treatment for tuberculosis costs $1–3 per DALY gained in several African countries.

Two major epidemics of measles occurred in Canada in 1990–1991, despite high rates of immunization coverage. Following this, a 1993 Delphi conference of experts from 31 countries reached a consensus to recommend a two-dose measles immunization policy. As new vaccines enter the field, it is important to evaluate their effectiveness, costs, and the benefits to be derived. The cost of the hepatitis B vaccine initially was over $100 for an immunization schedule of three doses but has come down dramatically to less than $2.00 for bulk purchases (outside of the United States). The vaccine is now a cost-effective method to prevent liver cancer and long-term effects of chronic hepatitis.

Screening and education for thalassemia in high prevalence areas has nearly eradicated clinical disease but not the carrier status in Cyprus, southern Greece, and other countries. Screening and case management for phenylketonuria (PKU), hypothyroidism, Tay-Sachs disease, and some other genetic diseases have been shown to be far less expensive than dealing with the severely retarded and dependent children born with these diseases.

Annual screening with Papanicolaou smears for cancer of the cervix is indicated especially for high risk groups, and every two or three for other adult women. Mammography is cost-effective but only in women over age 35 who are tested on a biennial basis. The value of health promotion in reducing exposure to HIV and cigarette smoking has been shown to be very cost-effective despite its lotech or nontechnological methodology, involving primarily group or mass education. Hypertension screening and case management is lotech but highly effective in preventing strokes and blindness.

Lotech innovations have had important impact in reducing death and injury in the 1990s. These include mandatory use of car seat belts, children's car seats, air bags, and bicycle and motorcycle helmets. Iodizing of salt and vitamin A supplementation or food fortification prevent large numbers of clinical cases of severe retardation, death, and blindness at low cost per child protected. Education for reducing risk factors for the cardiovascular disorders is far less costly than the pre-

TABLE 15.3 Examples of Hitech and Lotech Health Problem Solving

Problem	Hitech	Lotech
Infectious disease	Treatment	Vaccinations
Breast cancer screening	Mammography	Breast self exam, nutrition
Colon cancer screening	Colonoscopy	Occult blood, nutrition
Acute myocardial infarction, primary, secondary prevetion	Coronary angioplasty	Streptokinase, aspirin, diet, exercise, β blockers
Gallstones	Lithotropter Abdominal cholecystectomy	Endoscopic surgery
Head injuries	Intensive care	Helmets for bicycle riders
Thalassemia	Transfusions, chelating agents Prenatal diagnosis, amniocentesis, Chorionic villus biopsy	Screening, education
Liver failure	Liver transplant	Hepatitis B vaccination Screening blood donations
Dehydration	Infusion	Oral rehydration
Neural tube defects	Surgery, abortions	Folic acid fortification of flour

mature deaths and high medical costs of stroke and congestive heart failure patients. Health education, condom and needle supply, and screening of blood donations are the most important effective community health measures against the spread of HIV. Table 15.3 compares hitech and lotech approaches to health problems, which often complement each other.

Technology assessments represent the current consensus derived from reviews of published studies and exchange of views of highly qualified clinicians, epidemiologists, and economists within a context of TA. They may change over time as new data or innovations are reported, and this possibility should be kept in mind in such discussions. Technology assessment mobilizes information and critically analyzes many aspects of medical technology in order to build a consensus that can influence policy decisions. Public opinion, political leadership, administrative practice, as well as the scientific merit of the case are all factors in developing a consensus.

Technology Assessment in National Health Systems

Technology assessment requires an organization within the framework of national regulatory agencies. The U.S. Food and Drug Administration serves this purpose as a statutory body within the U.S. Public Health Service. Sweden, Canada, Australia, the United Kingdom, the Netherlands, Spain, and other countries also have TA advisory or regulatory agencies established by national governments to monitor and examine new technologies as they appear. Sweden has a widely representative national Swedish Council for Technology Assessment in Health Care, with an advisory role to the national health authorities.

Traditional systems to regulate food and drugs for efficacy, safety, and cost are only recently being applied to new medical procedures. The unrestricted proliferation of new procedures presents serious dilemmas for national agencies concerned with financing health care and controlling cost increases. Nongovernmental health insurance shares this concern, as does industry which pays the cost of health insurance. Thus there is wide support for government leadership in TA and cost control.

In Canada, the Health Protection Branch of the Federal Department, Canada Health reviews medical devices, as well as drugs, and with the consent of the provincial governments now licenses new medical procedures. Mutual concern over cost implications of new procedures led to this practice, and since 1988, a network of government and professional bodies have formed a nonprofit agency for technology assessment. This strengthens the provincial administrations of health insurance in resisting professional or political pressures to add untested technology or procedures to the health system as covered benefits. Comparison of rates of procedure performance between provinces shows very high discrepancies as high as 2:1 in, for example, coronary artery bypass graft (CABG) procedures or prostatectomies.

Control of acquisition of hitech equipment by national or state authorities is essential to prevent expenditures on high cost equipment without adequate assessment. Table 15.4 gives standards adopted by the Israel Ministry of Health as a key to approval control. Some new health technologies, such as *in vitro* fertilization (IVF), have escaped ministry control, and the practice has become completely uncontrolled with anxious clients and willing providers. However, the ministry has established reasonable controls in most areas of major equipment purchases.

TABLE 15.4   Standards for Acquisition of Costly Medical Equipment, Israel, 1995

Technology	Conditions	Ministry permitted rate
Computerized tomography (CT)	Hospital > 300 beds	1:200,000 population
Cardiac catheterization unit (cath lab)	Hospitals > 300 beds with cardiac intensive care > 5 beds	1:200,000
Nuclear magnetic resonance imaging (MRI)	Hospitals > 400 beds with approved radiology department + CT	1:300,000
Gamma camera	Standard in hospitals > 300 beds	1:100,000
Linear accelerator	Only in ministry approved centers for radiation therapy with cobalt, CT scanner units	1–3 units in each of six regional; radiation therapy centers, or 1–3:1,000,000
Lithotropters (extracorporeal), fixed for mobile	Major hospital center	1:1,000,000
Positron emission tomography (PET)	Research purposes	1:5,500,000

Source: Adapted from Siebzehner and Shemer, 1995.

TABLE 15.5   Hitech medical Equipment and Procedures Rates (per Million Population) in Selected Countries and Years, 1986–1996[a]

Country	CTs		MRIs	
	1986	1993–1996	1986	1995–1996
United States	12.8	26.9	0.5	16.0
France	4.7	9.4	0.5	2.3
Germany	6.9	16.4	0.7	5.7
Netherlands	3.2	9.0	0.1	3.9
United Kingdom	2.7	6.3	0.3	3.4
Canada	—	7.9	—	1.3
Sweden	—	13.7	—	6.8
Australia	—	18.4	—	2.9
Japan	27.5	69.7	0.1	18.8

Source: Data for 1990 from Banta, 1995; data for 1993–1996 from Organization for Economic Cooperation and Development, 1998.

[a]CTs = computerized tomography scanners per million population; MRIs = magnetic resonance imaging units per million population.

DISSEMINATION OF TECHNOLOGY

The rapid spread of medical hitech equipment has played a substantial role in escalating health costs. Magnetic resonance imaging units increased in Australia from 0 in 1985 to 10 in 1990, 25 in 1992, and 42 in 1994, with growth mainly in private radiology units (55%). The spread of endoscopic surgery in the 1990s has been worldwide. Health professionals become almost instantly aware of new developments from the news media as well as professional diffusion of information at conferences, in exchange visits, and in published articles in scientific journals. Table 15.5 shows the increase in hitech medical instruments and procedures during the period 1985–1996 in selected countries. MRI availability in Japan increased from 18.6 per million in 1995 to 2,663 units or 24 per million population, in 1996 but the rates charged are only one-fifth and utilization is about half that in the United States in 1996.

National policy to foster introduction of appropriate new technology requires a careful program of regulatory and financial incentives and disincentives to encourage or discourage diffusion of new methods of prevention as well as of treatment and community health care. Limitation of new techniques or procedures to selected medical centers allows time to fully assess the merits and deficiencies of new technology before general diffusion into the health care system. Such limitation, however, is fraught with the danger of depriving the population of benefits of new medical technology, and the possibility of "restraint of trade," to the economic advantage of selected providers. New technology impacts on insurance and managed care systems are necessarily involved in decision making as to inclusion of new procedures in their plans.

Publication in professional literature is an accepted method of establishing the scientific merit of a treatment or intervention. Too rapid diffusion of a medical practice can lead to disillusionment and confusion as to the merits of a particular medical procedure, as happened during the 1960s and 1970s with anticoagulant therapy for acute myocardial infarction and gastric freezing for peptic ulcers. Reviews of the literature should be critical and should assess the scientific merits of published data, as well as the sources of their funding. Well-controlled large scale clinical trials are needed to establish the relative values of alternative therapeutic approaches.

Dissemination of information about new medical innovations in the popular media is almost immediate. Many major newspapers and television networks have well-informed medical writers who have access to electronic mail medical journals as quickly as do specialists in each field. News magazines may carry special articles on new innovations, creating instant demand for them as benefits in a health program. This has both benefits and dangers.

In the United States health insurers have led the way in developing TA and information synthesis, and in evaluating the costs and benefits of new procedures. The process is affected by public opinion, as well as by court decisions. A landmark decision against an HMO in 1993 awarded $29 million in damages to the family of a terminal breast cancer patient who died following refusal of the HMO to authorize a bone marrow transplant, which was at the time an experimental procedure. Denial of new technology may lead to increases in malpractice suits. In countries with limited financial resources selection of technological innovations in health care that can benefit patient care or the public health requires a careful balance in order to use limited resources well, and to gain from application of appropriate new health care technology.

Payment systems may be effective in control of technology diffusion. Block budgets for hospitals have been more effective in Canada in restraining proliferation of high technology equipment than in the United States, but not in reducing hospital bed ratios. But this has led to criticism of limited access of Canadians to beneficial medical technology, such as CT, MRI, (Table 15.5), and advanced cancer therapies. In the United States the DRG system had the effect of increasing ambulatory surgery very dramatically, from 16.4% of all surgery in 1980 to 53.6% of all surgical procedures in 1992 and 56.7% in 1995. Inpatient surgical procedure rates (per 1000 population) declined from 108.6 in 1980 to 87.0 population in 1995.

HMOs and managed care organizations are paid on a per capita basis and have a strong incentive for cost containment, and they have developed procedures and medical guidelines for investigation and intervention that seek to reduce unnecessary procedures. At the same time, HMOs are very active in promoting preventive care and nonhospital care insofar as this is compatible with good patient care.

Coronary bypass procedures have increased in frequency, but mainly for white males. In the United States this procedure is less frequently done in women and blacks, because of lesser access to health insurance for blacks and possibly because of biases in terms of case assessment criteria in women. Cardiac invasive procedures increased dramatically since the 1980s in most industrialized countries, but with wide variation in their use. The benefits of aggressive invasive management of cardiovascular diseases remain controversial, but many such procedures have proven beneficial in reducing mortality rates and improving quality of life.

Critical analysis of the need for surgery has resulted in lower tonsillectomy and radical mastectomy rates along with increased use of outpatient procedures. Tonsillectomy, a routine procedure until the 1960s, is now performed infrequently since it was found to be of little medical value. Cataract surgery is now largely done on an ambulatory basis. The technology of home care has come to play an important role in early discharge of patients from the hospital, as has the wide use of cancer chemotherapy and radiation therapy on an outpatient basis.

QUALITY ASSURANCE

Quality assurance (QA) is an integral part of public health function and involves ensuring the quality of both health practitioners and facilities. It is an approach that measures and evaluates proficiency or quality of services rendered. Hospital accreditation is a long-standing method of quality assurance, providing many generations of health providers in North America with firsthand experience with QA in community hospitals and long-term care facilities, as well as ambulatory and mental health services. Hospital accreditation has contributed to improvement in standards of facilities and patient care throughout Canada and the United States and provides a working model for replication or adaptation internationally.

Adverse Events and Negligence

Iatrogenic diseases are adverse events that occur as a result of medical management and result in measurable disability. Negligent adverse events are those events caused by a failure to meet standards of care reasonably expected of the average physician or other provider of care.

Iatrogenic disease is a major cause of morbidity, prolongation of hospitalization, and even death. Hospital acquired (nosocomial) infections are estimated to occur in 7–10% of hospital cases in Britain and the United States. Primarily these are caused by urinary and respiratory tract and wound infections, and they include organisms that are resistant to many antibiotics and therefore difficult to treat. Infection control in hospitals is therefore an essential part of hospital organization. Because hospitals are increasingly being paid by DRG, any secondary event prolonging hospital stays may have adverse financial effects on the hospital. There is, therefore, a strong economic as well as professional interest in reducing hospital acquired infections. Hospital acquired infections, anesthesia mishaps, falls, and drug errors are the commonest of iatrogenic events.

A study of 32,000 hospitalizations (out of 2.6 million hospital discharges) in New York State in 1984 carried out by a Harvard University team, showed that 3.7% of hospitalized patients suffered adverse events, or injuries, caused by medical mismanagement which resulted in measurable disability. Of these 28% were due to negligence, so that 1.03% of all hospitalizations involved medical negligence leading to measurable injury. Of the total of some 100,000 adverse events in the study group, 57% recovered within a month, and 7% had severe injury. Some 14% or 14,000 persons with adverse events died as a result; 51% of these deaths were due to negligence.

A 1999 report of the US National Institute of Medicine estimates between 44,000–98,000 persons die annually in the United States from medical errors occuring in hospitals. Higher rates are seen among the elderly and the poor. Rates were lower in teaching hospitals as compared to community hospitals. About 20% of the events were related to drug reactions or dosage errors. Less than 3% of those injured brought civil litigation for the negligence. The search for "bad apples," that is, unethical, criminal or incompetent health providers, is necessary, but not sufficient to stem the problems created by the health system itself. Prevention requires organized activity. Investigation of adverse events helps to identify methods of prevention and to protect the patient's rights. Active measures to reduce hospital infection requires epidemiologic analysis of recorded events in the search for common causes and preventable factors. Organized surveillance and control requires one infection control practioner per 250 acute care beds, a trained hospital epidemiologist, and routine reporting of wound infections to practicing surgeons (CDC, Hospital Infection Program).

In response to the high frequency and cost of medical litigation, many states in the United States have enacted legislation to restrict court awards for medical negligence. Proposals for alternatives to the tort system of medical malpractice compensation include arbitration, conciliation, an administrative system similar to that used for worker's compensation, and a no-fault system of compensation, such as exists in New Zealand, Sweden, and Finland. In a no-fault system the complainant need not prove negligence on the part of the provider, but only that he or she suffered an adverse event which is compensable. In the United States, federal legislation provides compensation for vaccine injuries, and three states have enacted restricted no-fault systems for birth-related neurological injuries.

In addition there is greater emphasis on adoption of "fail-safe" mechanisms, such as introducing warning systems in anesthesia machines to alert the anesthetist if oxygen flow in the patient's tubing falls below a safe point. This was tested in Boston hospitals and found to reduce adverse anesthetic events to zero cases over a 3-year period. Vitamin K injection was made mandatory for all newborns in New York State, as was already the case in some other states, when a study showed deaths from hemorrhagic disease of the newborn in cases when vitamin K was not given.

Inappropriate medical practice patterns are an equally or even larger problem for health systems. Comparisons of surgical rates within the United States for coronary bypass procedures, hysterectomies, and cesarean section show wide variation between different parts of the country. The costs of excess surgery is not only economic waste, but also involves risks for the patient from the surgery itself or anesthesia mishaps, infection, pain and discomfort with legal and ethical questions of unwarranted interventions not for the benefit of the patient. Health systems are increasingly required to evaluate and control excess surgical, investigative, or other medical procedures, not only for financial reasons but also for protection against litigation and infringement of patients' rights.

Licensure and Certification

The requirements that society establishes for allowing an individual to practice medicine, or any health profession, are vital to maintaining or improving the quality of care (Chapter 14). These standards require defining the training and experience needed by the individual, examination procedures, and recognition for continued education and maintenance of competence. This requires a statutory base and national bodies operating under a national authority, separate from the agency operating the health system. Separation of licensing from operation of the health service is essential in maintaining high professional standards. The licensing system should be accountable to the public and not dominated by the profession being licensed.

Health Facility Accreditation

Hospital accreditation in North America is by a voluntary grouping of professional associations, including the Canadian and American Colleges of Physicians and Surgeons, the hospital associations, and the College of Nurses. The Joint Commission, originally operating in both Canada and the United States, carries out regular inspections of hospitals. In Canada, other organizations including the federal Department of Health, provincial ministries of health, Canadian Diabetic Association, Public Health Association, and the Standards Council of Canada participate in the Joint Commisssion as observers. Initially focusing on acute care hospitals, accreditation has been gradually extended to cover special hospitals, long-term facilities, home care programs, public health departments, and ambulatory care services.

Health facility accreditation is a systematic, multidisciplinary inspection of the physical and organizational structure of the facility or program and the functioning of its component parts. Factors measured include staff qualification, facilities, organization, record keeping, and continuing education of staff.

The process of accreditation requires a request from the board of governors of the hospital for accreditation, implying acceptance of the standards of the Commission. The Commission is invited to conduct a survey, and resurvey as it sees fit. The hospital pays a fee and commits itself to provide all data requested and to cooperate with the site visit. The Commission issues a confidential report, giving the accreditation rating and interim statement of deficiencies, and requests for progress reports in correcting deficiencies. It is also empowered to carry out follow-up inspections and resurveys. Table 15.6 lists the sectors of a large community or teaching hospital included in the accreditation site visit.

The assessment survey examines the goals and objectives of the organization and its administration, the direction and staffing of the facility, policies, and procedures. Review includes medical staff organization, credentials and review procedures, clinical privileges, selection of department chairpersons and their re-

TABLE 15.6   Services Evaluated in Accreditation in Large Community and Teaching
Hospitals, Canada

1. Ambulatory care	20. Medical services
2. Child life services	21. Neonatal intensive care
3. Clinical record services	22. Nuclear medicine
4. Critical care unit (generic)	23. Nursing services
5. Diagnostic services	24. Nutrition and food services
6. Discharge planning services	25. Obstetrical services
7. Education services	26. Occupational therapy services
8. Emergency services	27. Operating suite, postanesthetic recovery unit
9. Governing body	28. Palliative care unit
10. Housekeeping services	29. Pastoral services
11. Human resource services	30. Pharmacy services
12. Intensive care, cardiac care, transplant units	31. Physical plant and maintenance services
13. Laboratory services	32. Physiotherapy services
14. Laundry and linen services	33. Psychiatric services
15. Library services	34. Psychology services
16. Long-term care/geriatric unit	35. Rehabilitation services
17. Management services—utilization review, risk management, infection control health and safety, disaster and emergency planning	36. Respiratory therapy services
	37. Social work services
	38. Speech language and audiology unit
18. Material management services	39. Standards for delivery of care by program
19. Medical equipment services	40. Volunteer services

Source: Canadian Council on Health Facilities Accreditation, 1992; now the Canadian Council on
Health Services Accreditation, http://www.chsa.ca.

sponsibilities, standing committees, schedule of meetings, bylaws, and the role of
the governing board of the hospital. The presence and nature of QA organization,
records review procedures, and continuing education are assessed. The quality of
clinical records is assessed by examination of charts for the completeness of histories and documentation of the course of the hospital stay including laboratory reports.

Each section of the program being accredited is assessed in the following categories:

1. Statement of purposes, goals, and objectives;
2. Organization and administration;
3. Human and physical resources;
4. Orientation, staff development, and continuing education;
5. Patient care;
6. Quality assurance.

Hospital accreditation was established in the United Kingdom and Australia in
the 1980s and is attracting interest in other countries seeking ways to maintain and
promote standards. The procedure for accreditation of hospitals is still voluntary
in Canada, but in effect has become universal for hospitals of medium and large
size (i.e., over 75 beds) and common for smaller hospitals. It is seen as advanta-

geous for the governing board and the community and also for the medical staff in terms of medicolegal protection. In the United States, hospital accreditation has become virtually universal since payment for federally funded health insurance (i.e., Medicare and Medicaid) beneficiaries is not allowed for nonaccredited hospitals, and many private insurers make this requirement as well. In some states accreditation is mandatory for all hospitals.

Licensing and regulation of health facilities are a government responsibility, but an independent accreditation authority has advantages. The national authority may fail to monitor its own facilities with the diligence or objectivity needed, and there may be conflict of interests. Where there is a national system of organization, distinct departmentalization of the operating and certification functions may provide a greater measure of objectivity. Assistance from countries experienced in voluntary accreditation can help to establish accreditation mechanisms and provide technical and professional support to countries wishing to establish such programs.

In the current period of transition from central to decentralized management of health services in many countries, health facilities are being transferred from government operation to independent operation as not-for-profit or even for-profit facilities. Present methods of regulation by national or state levels of government will require review as decentralization and privatization takes place. Regulation by governmental authorities and nongovernmental professional bodies are mutually complementary in promoting accountability, standards, and quality of services.

Peer Review

A large part of the work of clinical and departmental managers in hospitals or other care settings relates to quality assurance. A major method of improving quality in a health program is through peer review by which the staff organizes systematic review of cases and records, using statistics on performance indicators. In hospitals this includes review of deaths, maternal mortality and infant mortality cases, surgical rates, complications following surgery, and infection rates. Medical records and computer information systems permit record reviews by diagnosis. These can be utilized to assess other events in hospitals, such as time from admission to surgery, lengths of stay by diagnosis, response to abnormal laboratory findings, and many other indicators of the process of care. Obstetrical departments can review the frequency of and criteria for cesarean section deliveries. Surgical departments review their appendectomy rates to separate pathological findings from normal appendices. Organized peer review has also been called "medical audit" and essentially describes methods of self-policing and education to learn from mistakes and experience and to improve the quality of care.

In 1972, an amendment to the U.S. Social Security Act required hospitals and long-term care facilities to monitor the quality of care given to Medicare and Medicaid patients through professional standards review organizations (PSROs). These were medical audit committees with specified tasks to conduct utilization review, medical care evaluation, and profile analysis of physician or institutional performance compared to accepted standards of the medical community. In 1982, peer review organizations (PROs) were created by federal statutes to replace PSROs. PROs are nonprofit corporations, staffed by physicians and nurses, to review medical necessity, quality, and appropriate level of care under the Medicare program.

Hospitals have departmental clinical meetings, adverse incident or outcome committees, mortality rounds, and clinical–pathological conferences help to evaluate and learn from difficult cases. The presence of functioning peer review mechanisms indicates that quality is of concern to the professional and administrative network, raising the confidence of the consumer in the system.

Maternal mortality committees have been widely used to assess preventable factors in deaths related to maternity and to point out areas of needed improvement in services. Identification of high risk pregnancies emerged from this process and has become an important part of prenatal care. Infant mortality reviews by professional groups can similarly demonstrate areas of needed improvement in services. Death rounds are held to review cases of death following surgery, or closely following admission, or an "incident" such as inappropriate medication given in error.

The successive waves of peer review initiatives in the United States represent attempts by the federal government to establish mandatory quality of care review by professional peers for facilities providing care to Medicare and Medicaid patients. The concept of requiring standards of care review has probably contributed to greater awareness of accountability of hospital-based practice. Frequent litigation may have contributed more to the sense that the physician is accountable for services and outcomes of care. PROs are a form of quality regulation that represent a commitment by funding agencies to accountability in care systems and to identification of organizational and administrative weaknesses in health care generally and not only in hospitals. The generation of U.S. physicians and health systems managers trained since the 1970s accepts peer review as an integral part of health services. Other countries use this kind of mechanism to maintain and promote quality of care.

Tracer Conditions. Tracer conditions are common medical conditions for which diagnostic criteria are well established and clear, there are effective preventions or treatments, and a lack of treatment can cause significant harm to the patient. Examples of tracer conditions include otitis media, appendectomy, cesarean section, and hysterectomy. These conditions, if evaluated in terms of incidence and actual chart review, can provide useful insights into departmental medical standards. Incident reports by nursing staff and nosocomial infections are an example of the functioning of the tracer condition concept. Incident reports in hospitals are designed to determine the causes of errors and to help develop remedial action and prevention of similar events. Tracer condition studies have become such an accepted part of modern health management that absence of an organized review system could be considered a serious structural flaw in a health service.

Setting Standards.   Standards recommended by independent professional organizations or by advisory committees appointed by ministries of health can play important roles in defining standards of care for specified conditions. In addition, organized professional bodies can issue practice guidelines or help governments or health care agencies to develop standards or algorithms for management of specific topics and conditions.

Specifying standards for preventive care, such as for infants and adults, assists local health authorities in planning and evaluating their services. The American Academy of Pediatrics has an extensive professional committee structure that publishes periodic guidelines for pediatricians on a wide variety of infant and child topics including nutrition, immunization, prevention of anemia and lead toxicity, child safety, and school health.

The American Public Health Association (APHA) publishes the *Control of Communicable Diseases Manual,* now in its sixteenth edition (1995), which is also available on CD-ROM. It is the authoritative manual on this topic. The American Academy of Pediatrics "Red Book," on infectious diseases is used across North America by pediatricians in clinical practice. These organizations and their counterparts in obstetrics and many other clinical fields directly relevant to public health, continually update practitioners and policy personnel in "the state of the art" or "gold standard," discussed previously. This constitutes professional self-guidance system in standards. Managed care and other health provider systems also issue guidelines for member practitioners that serve to maintain standards of service.

The wide use of treatment protocols and scoring systems in hospital medicine helps define standards of care in a measurable way. The Apgar score for rating newborn status has become a standard in hospitals worldwide, helping to standardize infant assessment and care. The APACHE system (acute physiology and chronic health evaluation) is a scoring system used widely to assess the chances of survival of patients admitted to intensive care units and to compare outcomes, for example, in intensive care between teaching hospitals and community hospitals. It is also used in assessing patient outcomes with different modes of treatment. Scoring systems are also used in community health care, as in risk scoring for pregnancy care (see Chapter 6).

Algorithms and Clinical Guidelines

Algorithms are decision trees or a systematic series of decisions based on the outcomes of previous decisions, test, or findings. Derived from operations research, this approach applied to medicine identifies all available choices (e.g., exposed versus nonexposed) and follow-up decisions based on findings from each previous option substantiated by observation. It is often presented graphically like the branches of a tree, showing the alternatives and subsequent decisions to be made.

A clinical algorithm is a systematic process defining a sequence of alternative, logical steps depending on outcomes of previous ones, incorporating clinical, laboratory, and epidemiological information, applied to maximize benefits and minimize risks for the patient. It gives the provider a review of the relevant literature and recommended standards of practice on a particular topic for preventive care or case management. These guidelines are usually arrived at by consensus by multidisciplinary working groups taking into account published studies on the topic. The guidelines may suggest that some procedures not be carried out routinely.

Clinical guidelines are meant to establish accepted standards of care and may have important economic implications. *Medical Letter,* published by the Consumers Union, is a long-standing and useful publication that reviews therapeutic issues of everyday medical practice and the relevant studies. It represents a balanced, updated view of medical practice and summaries of current literature, reviewed by respected, experienced, and competent medical authorities.

Clinical guidelines are helpful in clinical practice and in preventive medicine. They are increasingly used in managed care environments to assure standards, quality of care, and cost-effectiveness as well as legal protection. Guidelines for preventive medicine and public health practice are also part of the process of promoting the quality of individual and community health, as discussed in Chapter 11. Annual revision of the infant immunization program, discussed in Chapter 4, is a prime example, as is the set of guidelines for preventive care for adult health maintenance in Table 15.7.

The Province of Saskatchewan Health Services Utilization and Research Commission publishes periodic reports presenting consensus positions of panels com-

TABLE 15.7   Adult Health Maintenance Checklist

Procedure	Age 19–39	Age 40–64	Age 65+
Checkup visit	Every 3 years	Every 2 years	Every year
Cholesterol	With checkups	With checkups	With checkups
Fecal occult blood	Age 40–49 if high risk	Yearly	Yearly
Clinical breast exam (CBE)	Every 1–3 years	Annually	Annually
Mammography	Baseline age 35	40–49 every 2 years 50–70 annually	Over 70, every 2 years
Pelvic exam	Every 1–3 years	Same	Annually
Pap smear	Annually if on birth control pill; others every 3 years	Same	Same
Sigmoidoscopy	No	After age 50, every 3–5 years	Same
Prostate	No	Annually	Annually
Immunization			
Tetanus–diphtheria	Every 10 years	Every 10 years	Every 10 years
Pneumococcal pneumonia	For high risk	For high risk	Every 6 years
Influenza	For high risk	For high risk	Annually

Source: Adapted from Guide to Preventive Services, Report of the U.S. Preventive Services Task Force, Second Edition, 1995. and website http://www.ahcpr.gov/clinic/usps.fact

posed of medical faculty, clinical specialists in pathology and physical medicine, and public health specialists in nutrition, community health, and epidemiology. Its reports are circulated widely and serve to update medical practitioners, reduce unnecessary testing, promote changes in use of laboratory procedures, and provide standards of care for individual patients and community services, such as long-term care facilities and home health agencies. At the national level, expert committee reports include the Canadian Consensus Conference on Cholesterol (1988), the Canadian Lipoprotein Conference (1990), the Toronto Working Group on Cholesterol Policy (1991), and the Canadian Task Force on the Periodic Health Examination (1993).

The Health Care Financing Administration (HCFA) and the National Institutes of Health (NIH) have consensus programs to develop guidelines that are widely disseminated and set standards of practice. In 1977, the NIH issued its first consensus paper on breast screening for cancer, and this has been followed by many other topics each year since. In 1997, evidence-bound consensus guidelines were issued on the following topics: breast cancer screening for women aged 40–49, interventions to prevent HIV risk behavior, management of hepatitis C, genetic testing for cystic fibrosis, acupuncture, and effective medical treatment for heroin addiction.

Clinical guidelines are increasingly being promoted by professional, governmental, and managed care organizations with the purpose of promoting rational use of health care resources and at the same time promoting standards of care to incorporate "good standards of clinical practice." Clinical practice guidelines are now common in the practice of psychiatry and other specialties. The American Medical Association listed 2200 practice guidelines in 1997. These come from managed care companies, pharmacy benefit managers, governmental bodies, and professional organizations such as the American Psychiatric Association and the American Academy of Child and Adolescent Psychiatry. Legal aspects of health care also increasingly recognize the importance of clinical guidelines where committees of appropriate medical professionals convene and set out average or minimum standards of care for defined clinical entities. Thus peer reviewed guidelines set an appropriate standard (silver if not a gold standard) for judging malpractice or adequate practice.

The American Academy of Pediatrics (AAP) policy statements, practice parameters, and model bills have a wide distribution and are published in the Academy's journal, *Pediatrics*. The AAP clinical practice guidelines recently issued include diagnosis and treatment of urinary tract infection in febrile infants and young children, long-term treatment of the child with simple febrile seizures, management of acute gastroenteritis in young children, management of otitis media with effusion in young children, and others. The policy statements of the AAP cover a wide range of topics from use of bicycle helmets, to 55 mile per hour maximum speed limits, folic acid for the prevention of neural tube defects, to ethics in the care of critically ill infants and children.

Empirically derived, peer reviewed, regularly updated guidelines become an appropriate standard for practice and for judging malpractice, as well as balancing quality and cost-effectiveness. Clinical guidelines may become restrictive, but they help to reduce practice by whim and unsubstantiated belief to improve the quality of care overall. In large health care organizations they provide a basis for continuing education for staff and advancement of standards of the organization.

ORGANIZATION OF CARE

Administrative and financing systems are essentially elements of quality assurance. They can be designed to promote standards of care and to reduce financial incentives to overservicing. The organization of financing of health care has important implications for quality, technology, and ethical issues in the New Public Health.

Diagnostic Related Groups

Diagnostic related groups (DRGs, discussed extensively in Chapter 11) were developed in the 1960s as an alternative way of paying for hospital care in order to encourage shortened lengths of stays. Experience of payment by days of care (per diem) showed that it promoted unnecessary, long, and potentially dangerous use of hospital care, an important factor in inflation of costs in the health system.

The provider hospital is paid by the insurer for a procedure or diagnosis rather than the number of days of stay in hospital. This has led to a large reduction in hospital days of care and a remarkable growth in the number of surgical procedures done on an outpatient basis. Outpatient surgical procedures grew from less than one-fifth of inpatient procedures to more than equal the inpatient cases. Outpatient surgery is safer for the patient and less costly to the insurer.

The DRG system is widely considered to promote quality of care as an active process focusing on quickly addressing the diagnosis and management of the patient with rapid mobilization and return home. Critics suggest that this may shorten stays in hospitals too much and deny patients the rest they need for full recovery, or may promote altering diagnoses to higher cost units of service. Others think that DRGs, by reducing length of stay, have turned hospitals into intensive care units with ultrasick patients. Despite these views, the trend toward short hospital stays and active treatment approaches seems to be compatible with better care and improved outcomes. The rapid decline in mortality rates from coronary heart disease is thought to be due in large part to the activist treatment approach, with lengths of stay of one week or less for acute myocardial infarction compared to six weeks on average up to the 1970s.

Managed Care

Managed care systems developed in the United States in response to rapid cost escalation for health care and as health maintenance organizations (HMOs) based on concepts of resource management, quality assurance with rationalized use of technology. The system developed over time with checks and balances to provide comprehensive care at lower cost than traditional fee-for-service systems by discouraging excessive utilization without compromising quality of service. Managed care systems include traditional HMOs and various other organizations which employ physicians or are made up of independent physicians working together who own or contract for hospital services.

District health systems in the United Kingdom, the Scandinavian countries, and the post-Soviet model of health care incorporate organizational and financial linkage between care systems and financing from tax sources. HMOs, sick funds, and district health systems provide both prepayment and health services. Even in traditional private health insurance systems, the insurer is increasingly taking on the role of regulating reimbursement for medical services in order to contain costs and curb abuses by providers. Clinical indications, utilization review, and organizational and professional standards are now becoming accepted parts of the health insurance milieu.

The competition between hospitals for referrals from managed care plans in the United States has created a market situation in which a high proportion of hospital beds are empty, and in which mergers or closures of hospitals are common. Closures or reductions in hospital bed supply are also occurring in the United Kingdom and in most industrialized countries of Europe.

PERFORMANCE INDICATORS

Performance indicators (PIs) are measures such as morbidity, mortality, or functional status, or immunization rates in a community, used to monitor the functioning of a health service. Routinely collected statistics are analyzed to compare performance against objectives, helping monitor efficiency and effectiveness, and to point out problem areas within the service. This is based on the use of the concept of management-by-objective in health administration to promote achievement of national health targets.

The United Kingdom has a strong tradition of mapping of diseases as a basis of epidemiologic analysis and has applied this strategy to mapping of PIs to assess health care performance. The U.K. financing system is based on capitation adjusted by standardized mortality rates on the premise that mortality rates standardized and compared to the national average serve as indicators of need. In this way, they help to promote equitable funding between wealthy and poorer regions of the country, and thereby improve services in areas of greater need.

Performance indicators were introduced into the NHS during reforms of the late 1980s, providing a series of outcome or performance measures that are used to adjust payments allocated on a per capita basis to district health authorities. DHAs can be penalized for low rates of immunization, whereas general practitioners receive incentive payments for full imunization coverage. The result of this has been a rapid improvement in immunization coverage of infants and children in the 1990s compared to rates in the previous decade. Incentive payments in many countries encourage women to go to hospitals for delivery or to attend prenatal care by making social maternity grants conditions on seeking care.

Use of PIs requires development of health information systems with district health profiles to provide ongoing monitoring of health indicators in a district, compared to regional and national rates and targets. Health profiles help to establish and monitor the prevalence of chronic disease and measure the impact of health services. This helps to study the performance of preventive and curative services, such as managing hypertension to reduce the incidence of strokes and related conditions.

CONSUMERISM AND QUALITY

With decentralization and the growth of managed care, health systems must increase their attention to the attitudes of the consumer. Quality is, in part, how the client perceives the system, and how the system meets client needs in an acceptable manner, where privacy, dignity, the right to know, and the right to a defined set of services are protected. However, the rights of the client are not unlimited. A public or private health plan have the duty to manage their basket of services responsibly, which includes limitations such as in access to specialist services.

Patients' rights and consumer protection in health care include the right to select and change a health care provider, the right to receive high quality care for a designated range of services. The consumer's formal protection includes the right to complain, to seek redress of grievance and compensation for injury suffered from neglect or incompetent care (Table 15.8).

The consumer needs to be informed and conscious of health care costs if efforts to restrain cost increases are to be effective. Public attitudes are vital in terms of self-care, demands on the health service, limitations to the potential of health care, and resources for health care. The public media and consumer organizations can play important roles in advocacy for health, in raising public consciousness of self-care, in advocacy for health issues, and as watch dogs on abuses.

Consumer acceptance is manifested through choice of health plan and practitioner, or by seeking alternative care privately when a service is unacceptable because of quality or style. Erosion of confidence in a public system of care can lead to a two-tier system with the public system serving the poor and a private parallel system serving the middle and wealthy classes. Such a division can seriously undermine a public system unless it is addressed by improving the quality and manner of the service and by establishing supervision and limitations on public and private practice.

Private practice is a chronic problem in the United Kingdom's NHS, in Israel's health system, and in many countries developing their health systems through parallel public and private care. The issue is also surfacing in the United States in the transition to managed care with its inherent limitations of choice for people insured through their place of work or covered under the Medicare and Medicaid programs. Extra-billing, banned in Canada's national health insurance plan, is a recurring issue with the medical profession.

Consumer knowledge, attitudes, beliefs, and practices are part of the health system, from health promotion to tertiary care. Informed and health conscious con-

TABLE 15.8 The Patient's Charter of Rights, United Kingdom's National Health Service, 1992

1. To receive health care on the basis of clinical need regardless of ability to pay
2. To be registered with a GP
3. To receive emergency medical care at any time through your GP or the emergency ambulance service, hospital accident, and emergency departmetns
4. To be referred to a consultant acceptable to you when your GP thinks necessary, and to be referred for a second opinion if you and your GP agree this is desirable
5. To be given clear explanations of any treatment proposed, including any risks and alternatives before you decide whether you will agree to the treatment
6. To have access to your health records and to know that those working for the NHS are under a legal duty to keep their contents confidential
7. To choose whether you wish to take part in medical student training
8. To be given information on local health services including quality standards and maximum waiting times
9. To be guaranteed admission for treatment by a specific date no later than 2 years from the day when your consultant places you on the waiting list
10. To have any complaint about the NHS investigated and to receive a full and prompt reply from the chief executive officer or general manager

Source: Websit www.doh.gov.uk/pcharter/patients.htm.

sumers are stronger partners in the health system in achieving improved health than an ill-informed and apathetic public.

THE PUBLIC INTEREST

Population-based interventions are often more effective ways to reduce morbidity and mortality than individual prevention or treatment services. A population-based preventive program may require behavior change by the individual, such as in mandatory seat belt and motorcycle helmet enforcement or banning smoking in public places. Fortification of flour, milk, and salt with essential micronutrients are well-established public health measures. There is an element of compulsion in this, with the social gain considered to be sufficiently important to outweigh individual rights. In these cases, community rights have precedence over individual rights with controversy in many areas of modern public health, from chlorination of community water supplies to managed care systems for health services. Each issue has to be examined on its merits, especially in terms of what is accepted as good public health practice, based on documented experience, clinical trials, and practice in other countries. The evidence of successful public health measures in improving individual and collective health status is powerful, yet must always be balanced within the context of individual rights and the public interest.

TOTAL QUALITY MANAGEMENT

Total quality management (TQM), discussed in Chapter 12, has also been adapted to health care in the 1990s and provides a basis for promoting continuous improvement in health care systems. TQM involves everyone in the system, from all levels of management to production or service personnel and support staff. It helps to raise staff morale because of the sense of involvement and participation. Health is provided through multidisciplinary groups which need to approach problems with open and shared scientific inquiry, hypothesis formation and testing, revision, to find operational solutions to problems.

Total quality management incorporates statistical thinking, comparing variations in patterns of service or use of resources. It employs epidemiologic methods to draw conclusions for policy needs. It looks for continuous improvement, encouraging cooperation, and motivation to achieve common goals of service and client satisfaction. Psychological theory helps to seek higher levels of motivation and conflict identification and resolution. Leadership is shared, and there is a basic need for cooperation. Cost and quality are interrelated, as poor quality leads to waste, inefficiency, and dissatisfaction of clients and staff. High quality, humane, and effective services are especially important in a competitive environment where clients have the right to choose and where costs and efficiency are factors in the well-being and indeed the survival of institutions.

Medical care is increasingly practiced in larger health care programs. To provide technically competent medicine is not by itself sufficient. The patient's rights and sense of personal worth are also of great importance. Financial incentives to redirect health care priorities, such as in reducing hospital lengths of stays and admissions, may result in the patient or the family feeling they are not receiving the best care. DRGs, HMOs, and other organizational and financial systems meant to increase efficiency of care may have the effect of alienating patients from a health care system. Staff attitudes toward patients are important for client satisfaction. The service must include ready access to supportive services, such as home care, and counseling so that the patient and family do not feel abandoned by the system.

A by-product of TQM is continuous quality improvement (CQI) by which institutions wishing to improve quality train and empower their staff to work in teams to assess their own performance and seek solutions to problems in their operational unit. People of different ranks and professions work in a network organization, as opposed to the traditional hierarchical organization. This community of practice is important for staff morale and a shared sense of responsibility for the patient and the insititution.

Continuous quality improvement involves multidisciplinary approaches, not only to review problems, but also to seek better ways of functioning and improving consumer satisfaction. The process includes all those involved in providing care, support services, and administration of a department, hospital, clinic, or community health program. This is not only professional self-policing, but a method to find better ways of meeting needs and using resources. The involvement of all providers improves motivation and promotes a sense of common purpose in the organization.

Applying these principles in a health care setting can take many forms. Selection of topics by TQM/CQI committees in a hospital may be based on surveys or interviews with staff, patients, or management. Satisfaction surveys among women following delivery in an obstetrics unit could point out remediable problems. The problems associated with an obstetrical department could include high or low volume of deliveries, staff training, equipment and supplies, communication among staff and between staff and patients and their families, cleanliness, sterile technique, staff satisfaction, client satisfaction, and many others. The team looking at such a problem should be multidisciplinary, and emphasis should be on client attitudes and satisfaction. Examination of the function of an emergency room (ER) in a hospital would similarly look at many functional and attitudinal aspects of the service including staff attitudes, training needs, waiting times, consultation services, and others. Addressing waiting times, for example, can lead to ways to reduce this substantially, improving client satisfaction and the efficient management of the ER. Any service is there to serve patients and the community. A service is not primarily for the benefit of the staff, but staff satisfaction is essential to successful service to a clientele.

The European Region of the World Health Organization and the European National Medical Associations in 1995 agreed that medical associations should take leading roles in programs of CQI to achieve better outcomes of health care in terms of functional ability, patient well-being, consumer satisfaction, and cost-effectiveness. This is in keeping with the European Regions Health for All targets: that there should be structures and processes in all member states to ensure continuous improvement in the quality of care and appropriate development and use of health technologies.

The 1990s introduction of general practitioner "fund holding" for hospital care for patients on the GPs roster in the United Kingdom encourages the hospital to maximize patient satisfaction with the care system. This promotes application of CQI to improving the quality and acceptability of care. Similarly, performance indicators provide regional and district health authorities in the United Kingdom

---

BOX 15.3 ORGANIZATIONS TO PROMOTE QUALITY IN HEALTH IN THE UNITED STATES AND CANADA

1. NCQA, National Council for Quality Assurance: This nonprofit organization founded in 1979 by the managed care industry conducts surveys among managed care plans to evaluate clinical standards, members' rights, and health service performance. It accredits over 50% of the 630 managed care plans in the United States (1997).

2. AHPCR, Agency for Health Care Policy Research: This is part of the U.S. Public Health Service. In 1995, it was mandated to develop an evidence-based practice program in 12 centers in the United States and Canada. It conducts systematic reviews of the literature and publishes analyses and findings of these reviews along with guidelines for care, quality improvement projects, and purchasing decisions for health plans. It funds re-

search on outcomes and cost–effectiveness studies, and disseminates new information and guidelines for medical practice and cost-effectiveness.

3. FACCT, Foundation for Accountability: FACCT is a nonprofit organization with the purpose of providing information to consumers, with a stress on quality of care for chronic diseases and measuring tools to develop standards and assess health plan performance in conditions such as diabetes, asthma, breast cancer, coronary heart disease, and alcoholism.

4. HCFA, Health Care Financing Administration: HCFA is the federal agency of the Department of Health and Human Services responsible for administering the Medicare and Medicaid health plans. In the 1990s, the HCFA established requirements for managed care organizations and quality improvement in health care. This requires health plans to provide evidence of improvement in the health care they provide, stressing a move from payment to quality assurance.

5. IHI, Institute for Healthcare Improvement: Founded in 1991, this nonprofit organization aims to improve health care in Canada and the United States by fostering collaboration among health care organizations. IHI examines office practices of physicians, educational reform, and promotes interdisciplinary teamwork in quality improvement.

6. NPSF, National Patient Safety Foundation: Sponsored by the American Medical Association as a response to findings of high rates of injury and death from iatrogenic disease in the United States, the NPSF promotes research into human error among health care providers, seeking ways to reduce the frequency and effects of medical error, such as misdiagnosis, medication errors, and mistakes during procedures.

7. JCAHO, Joint Commission on Accreditation of Healthcare Organizations: Established in 1951, The Joint Commission began accrediting hospitals in 1953, changing to a broader mandate in 1987. Accreditation is mandatory for Medicare and Medicaid payment. The JCAHO is changing its approach from standards-based assessment every 3 years to one of reviewing performance data quarterly as a continuous surveillance activity for risk-reduction.

Source: Medical news and perspectives, *Journal of the American Medical Association,* 1997, 278:155–156.

with tools for CQI approaches. The United States has a number of government and independent organizations dedicated to improving quality in health care systems.

PUBLIC HEALTH LAW

Law represents the consensus of society, as enacted by an elected legislature, put into effect by the executive branch of government, and interpreted by the courts. The legislative and executive branches are separate under the U.S. constitution, but united in the parliamentary system. The responsibility and power to provide for and protect the public health are a basic function of a sovereign government, which may be delegated to another level of government (higher or lower) or even a nongovernmental agency. The constitution of a sovereign state spells out or implies that responsibility, but accepted practice and court decisions (i.e., the common law) define the powers of the national, state, or local government to police and protect the health of its citizens.

In the United States, national legislation is enacted under two powers of the federal government, namely, the power to regulate interstate commerce and the power to tax and spend for the general welfare. State legislation is enacted under the basic power of the state to protect the health, welfare, and safety of its citizens. Under these federal and state powers, a wide range of health legislation and regulations are enacted affecting public health, financing of health services, agriculture, food, drugs, cosmetics, and medical devices, labor and occupational health and safety, environmental controls, and public welfare. Public health law relies on a wide range of constitutional, statutory, administrative, and judicial decisions in both civil and criminal actions.

A combination of the regulatory, persuasive, and financing approaches is widely used in public health in control of communicable and noncommunicable disease, in improving standards of facilities, and in providing health services. The regulatory, enforcement, policing, and punitive functions of public health are important in health promotion and assurance of health care. The taxing power of government is essential to public health to ensure that adequate facilities and access to care are shared by all members of the community and especially those in greater need and at greater risk for disease.

Protection of the public or an individual may require legal power to detain a person to prevent spread of a communicable disease, to protect a mentally ill patient, or to restrain a violent person. Such powers should be used as a last resort if persuasion and education fail and where the danger to the community or the individual is sufficient to convince a court of the public need to override the personal liberty of an individual. Public health has evolved to rely more on voluntary cooperation of a patient than on compulsion. Enabling legislation may permit a local authority to fluoridate its water supply, but enactment of local legislation and funding to implement it may also require a public referendum.

Appropriation of public funds to promote public health is a legislative act. Provision of public funds may be as categorical grants for specified services, such as immunization, prenatal care, school health, or special diseases such as tuberculosis control or AIDS education. Programs may be designed to promote certain types and quality of services, such as the Hill-Burton Act, which provided federal grants for hospital construction in the 1950s to 1970s, conditioning these grants on certain requirements concerning hospital licensure and hospital planning. Such legislation has both a "carrot and stick" effect of attracting lower levels of government to seek such funding but requiring them to accept the conditions and regulations that go with the grants. The Canadian federal government's cost sharing of provincial health insurance programs is based on federal criteria requiring public administration, portability between provinces, and comprehensiveness and banning extra billing by physicians (Chapter 13).

Public funds are also appropriated in the context of legislated programs in which people are entitled to the services defined in the appropriation legislation, such as in the amendments to the Social Security Act providing Medicare and Medicaid programs, or national health insurance legislation in many countries. These acts and their regulations spell out categories and specified benefits of entitlement.

Legislation and court decisions to protect the right of the individual are part of public health. Public health law covers individual and community life, including the need to protect the individual from potential abuse, in keeping with protection under law as in the U.S. Bill of Rights. Enforcement of public health law may infringe on individual rights by enforcing sanitation laws, and on civil rights by rarely used mandatory treatment of a person with dangerous contagious disease or mental illness. Freedom of religion may come into conflict with other laws in public health. Restrictive practices may deny use of publicly supported health facilities, as when a religious hospital refuses an abortion in a case of rape; Alternatively, religious practices may endanger others in the community, such as in refusal to immunize children when it is mandatory. Public health law forbids misleading or unethical advertising, limiting freedom of speech.

The 1973 U.S. Supreme Court decision of *Roe v. Wade* allows women to seek safe and legal abortion. This remains a highly controversial political issue in the United States and many other countries. The potential conflict between community and individual interests and rights is part of the dynamics of public health law and public health practice. The issues involved are complex and often involve ethical distinctions where "the greatest good for the greatest number" may harm legitimate rights of individuals and vice versa. The legal aspects of public health are vital to its operation and are increasingly complex as they overlap with ethical issues and public debate.

ETHICAL ISSUES IN PUBLIC HEALTH

Ethics in health are based on the basic concepts and values of a society. If the principle of saving a life is above all other considerations i.e., the sanctity of life or *Pikuah Nefesh* (see Chapter 1), then all measures available are to be used, irrespective of the condition of the patient or the cost. If sickness and death are seen as acts of God, possibly as punishment for sin, then prevention may be considered to be interfering with the divine will, and the ethical obligation is limited to relief of suffering. Humanism balances these two ethical imperatives: saving of life and relief of suffering. Materialism may see health care as primarily a function to preserve health for economic productivity. The LaLonde concept of individual behavior as a major health determinant (Chapter 2) places the onus of illness and its prevention on the individual. All these points of view are involved in the ethical issues of the New Public Health (see Table 15.9).

TABLE 15.9 Ethical/Legal Issues in Individual and Community Rights and Responsibility in Health

Ethical/legal Issues	Individual Rights and Responsibility	CommunityRights and Responsibility
Sanctity of human life	Right to health care; responsibility for self-care and risk reduction	Responsible for providing feasible basket of services, equitable access for all
Individual vs community rights	Immunization for individual protection	Immunization for herd immunity and community protection; education; community may mandate immunization
Right to health care	All are entitled to needed emergency preventive, curative care.	Community right to health care regardless of location, age, sex, ethnicity, medical condition, economic status
Personal responsibility	Individual responsible for health behavior	Community education to health promoting lifestyles; avoid "blame the victim"
Corporate responsibility	Producer and individual manager accountablitiy to criminal and civil action	Producer, purveyor of health hazard accountable for individual and community damage
Provider responsibility	Professional, ethical care and communication with patient	Access to well organized health care, accredited to accepted standards
Personal safety	Protection from individual and family violence	Public safety, law enforcement, protection of women, children and elderly Safety from war, terrorism, ethnic violence
Freedom of choice	Choice of health provider; limitations of gatekeeper function; right to second opinion; right of appeal	Community responsible to control costs while ensuring individual rights; limita- of self-referrals to specialist
Euthanasia	The individual's right to die; limitation by societal, ethical and legal standards	Assurance of individual and community interests; prevention of abuse by family or others with conflict of interests
Confidentiality	Right to privacy of medical information	Mandatory reporting of specified diseases for epidemiologic analysis; provider's right to know
Informed consent	Right to know risks vs benefits and agree or disagree to treatment or participation in experiment	Helsinki committee approval of research; regulate fair practice in right-to-know; Patient's Bill of Rights
Birth control	Right to information and access to birth control; safe abortion; women's control of her body	Political, religious promotion of fertility; alternatives to abortion; protection of women's right to choose
Resources for health	Universal access, prepayment; individual contribution through the workplace or taxes	Solidarity principle and adequate funding; right to cost containment, limitations on service benefits
Regulation and incentives to promote community health	Social security payments, e.g., for hospital delivery, attendance for prenatal care Care in the community and ambulatory care; supportive community care services	Incentive grants to assist communities for programs of national interest; limit instituional faciletes; transfer of resources to primary care
Global health	Human rights and aspiration; economic development, health, education and jobs	Danger of transfer of health risks; occupational hazards and environmental damage to poor or developing countries
Rights of minorities	Equality in universal access	Special support for high needs groups
Prisoners' health	Human rights; prevention of torture, executions, prison conditions	Security and human rights;
Allocation of resources	Lobbying, advocacy for equity and innovation	Equitable distribution of resources; targeting high risk groups; cost containment

Resources for health care are limited even in industrialized countries. Money spent on new technology with only marginal medical advantages is often at the expense of well-tried and proven lower cost techniques to prevent or treat disease. The potential benefits gained by the patient from more and more intervention are sometimes very limited in terms of length or quality of life. These are difficult issues when the physician's commitment to do all to preserve the life of his patient conflicts with the patient's quality of life and his right to terminate heroic measures of intervention. The suffering that a terminal patient may endure during radical treatment which may prolong life by only hours or days conflicts with the physician's ethical obligation to do no harm to the patient. The ethical value of sustaining the life of a patient suffering pain in a terminal stage of life has is an increasing medical dilemma. The issue is even more complex when economic values are included in the equation. There is a potential conflict of interests between the economic issues, the role of the physician in preserving life, the physician's obligation to do no harm, the felt needs of the patient and his or her family, and the needs of the community as a whole.

The state represents organized society and has a major responsibility to promote healthful conditions and to provide access to health care. The conflict between individual rights and community needs is a continuous issue in public health. Application of accepted public health measures for the benefit of some people in society may require applying an intervention to everyone in a community or a nation. The majority thus are subject to a public health activity to protect a minority, without designating which individual's life may be saved. A society may need to restrict individual liberties to achieve the goal of reducing disease or injury in the population. Raising taxes on alcohol and cigarettes, mandatory speed limits, driving regulations and seat belt usage laws restrict individuals but protect individual persons and the community at large.

Some forms of mass medication are accepted forms of public health practice to reduce the risk of disease in the population. Chlorination of community water supplies is a well-established, effective, and safe intervention to protect the public health. Fluoridation of drinking water to prevent tooth decay in children means that other persons are also drinking the same fluoridated water, which is of less direct benefit to them. Fortification of foods with vitamins and minerals is also a cost-effective community health measure with advocates and opponents. The addition of folic acid to food as the most effective way to prevent neural tube defects in newborns is an intervention mandated by the FDA since 1998.

Confidentiality to assure the right of the individual to privacy involves ethical issues in the use of health information systems. Birth, death, reportable infectious diseases, and hospitalization data are basic tools of epidemiology and health management. However, caution is needed in their use to avoid individual identification that could be used punitively, for example, in denial of access to health insurance for smokers, alcoholics, or AIDs patients because damage is self-inflicted. Case finding and follow-up are vital to good epidemiological management of transmissible diseases, including STDs.

The AIDS epidemic in the 1980s raised a host of public health and ethical issues. Management of the AIDS epidemic is in some respects in conflict with the long-established role of society in contacting and quarantining persons suffering from transmissible diseases. It is not acceptable or feasible in modern society to isolate HIV carriers, but failure of public health authorities even in the late 1980s to close down gay bathhouses in New York and other cities in the United States where exposure and transmission of the infection occurred could be interpreted as negligence. The politics of AIDS in the United States during the 1980s centered on concerns in the gay community that HIV testing would be used in discrimination against gay people, and thus AIDS was not treated as a public health problem, but as a civil liberties issue. Screening, reporting, and case contact follow-up were seen as invasive of privacy and counterproductive by increasing resistance to and avoidance of testing. The educational approach was adopted as most feasible and acceptable. The AIDS epidemic and public anxiety about catching AIDS through casual contact reinforced the need for public education on safe sex, an ethical issue because such education may be construed as condoning teenage and extramarital relations. The issue of HIV screening of pregnant women in general or in high risk groups took on a new significance with findings that treatment of the pregnant woman reduces the risk of HIV infection of the newborn, and that breastfeeding may be contraindicated.

A preeminent ethical issue in public health concerns assuring access to services and provision of services according to need. An important ethical, political, and social issue in the United States at the beginning of the twenty-first century is how to achieve universal access to health care. The solidarity principle in funding access to health care is based on equitable prepayment for health care for all by nationally regulated mechanisms through place of work or general revenues of government. A society may see universal access to health care as a positive value, but use incentives to promote use of services of benefit to the individual, such as hospital delivery, prenatal care, immunization, mammograms, and others. Some services may be excluded from health insurance, such as dental care, although this is to the detriment of children and a financial hardship for many. The United States, by defining eligibility for Medicaid at income levels of 185% of the poverty level, excludes a high percentage of the working poor, while efforts are made to include more children in the program.

Choices in health policy are often between one "good" and another. Limitations in resources may make this issue even more difficult in the future, with aging populations and rapid increases in technology and its costs. The United Kingdom's National Health Service at one point refused to provide dialysis to persons over age 65. When computerized tomography was first introduced, Medicare in the United States refused to insure its services as an untested medical technique. The Soviet health system, because of lack of facilities such as incubators and poor prospects for the survivors, considered newborns as living only if they weighed more than 1500 grams, so that others who would be considered living by international definitions would be placed in a freezer to die. At the opposite extreme, many western medical centers use extreme and costly measures to prolong life in terminally ill patients, preserving life temporarily but often with much suffering and great expense to the public system of financing health care.

In the United States, there is a lack of funds for immunization of poor children but virtually unlimited funding for procedures such as cardiac bypass procedures that are not equitably distributed according to need alone. Closure of rural hospitals involves ethical decisions and is a source of friction between central health authorities and local communities. Health reforms in many industrialized countries, such as reducing hospital bed supplies and managed care systems' promoting cost containment and reallocation of resources, raise ethical and political issues often based on vested interests such as private insurance systems, hospitals, and private medical practitioners.

Ethics in Public Health Research

The border between practice and research is not always easy to define. In public health, disease surveillance is mostly anonymous, but individually identifiable data are needed for infectious disease control. It may also be needed in monitoring the effects of chronic disease, for example, to ascertain repeat hospitalizations of patients with congestive heart failure to assess the long-term effects of treatment.

The general distinction between research and practice has to do with the intent of the activity. Clinical research uses experimental methods to establish efficacy and safety of new interventions or unproved interventions; many drugs, procedures in common use in the U.S. never subjected to randomized controlled trials. In practice, many methods are devised that are held to be effective and safe by expert opinion. Researchers comparing HIV or hepatitis B transmission rates among intravenous drug users not using needle exchange programs would be doing unethical research by giving needles to the experimental group and withholding them from the control group.

A 1996 U.S. Public Health Service study supported by the NIH and WHO compared a short course of Zidovudine (AZT) to a placebo given late in pregnancy to HIV positive women in Thailand, measuring the rate of HIV infection among the newborns. The experiment was terminated when a protest editorial appeared in a prominent medical journal. The study confirmed previous findings that AZT during late pregnancy and labor reduced transmission by half. The findings indicate that AZT should be used in developing countries, and the manufacturers agreed to make it available at reduced costs.

Ethics in Patient Care

Ethical issues between the individual patient and health care provider are important in the New Public Health. A doctor is expected to use diligence, care, knowledge, skill, discretion, and caution in keeping with practice standards accepted at the time by responsible medical opinion and to maintain the basic medical imperative to do no harm to the patient. The patient has the right to know his condition, available alternatives for his treatment, and the risks involved. He also has a right to seek alternative medical opinions, but this right is not unlimited, as any insurance or health service may place limitations on payment for further opinions and consultation without the agreement of a primary care provider.

Health care has a responsibility beyond that of payment of health service bills and individual care by a physician, in institutions, or through services in the community or the home. The contract for service is less and less between an individual physician and his patient, but more and more between a health system, its staff and the client. This places a new onus on the physician to ensure that the patients receive the care they require. Conversely, the U.S. provider often faces the dilemma of knowing that a patient may not access needed services because of a lack of adequate health insurance.

Human Experimentation

Human experimentation has been a subject of great concern since the Nazi and Imperial Japanese armed forces experiments on prisoners and concentration camp

TABLE 15.10 Ethical Issues of Medical Research Derived from the Nuremberg Trials and Declaration of Helsinki

Nuremberg Code, 1946	The voluntary consent of a human subject is absolutely essential, with the exercise of free power of choice without force, fraud, deceit, duress, or coercion. Experiments should be such as to be fruitful results, based on prior experimentation and the natural history of the problem under study. They should avoid unnecessary physical and mental suffering. The degree of risk should not exceed the humanitarian importance of the experiment. Persons conducting experiments are responsible for adequate preparations and resources for even the remote possibility of death or injury resulting. The human subject should be able to end his participation at any time, and the scientist in charge is responsible to terminate the experiment if continuation is likely to result in injury, disability, or death.
Declaration of Helsinki, 1964	Research must be in keeping with accepted scientific principles, and should be approved by specially appointed independent committees. Biomedical research should be carried out by scientically qualified person, only on topics where potential benefits outweigh the risks, with careful assessment of risks, where the privacy and integrity of the individual is protected, and where the hazards are predictable. Publication must preserve the accuracy of finding. Each human subject in an experiment shoud be adequately informed of the aims, methods, anticipated benefits, and hazards of the study. Informed consent should be obtained, and a statement of compliance with this code. Clinical research should allow the doctor to use new diagnostic or therapeutic measures if they offer benefit as compared to current methods. In any study, the patient and the control group should be assured of the best available methods. Refusal to participate should never interfere with the doctor–patient relationship. The well-being of the subject takes precedence ovet the interests of science or society.

Source: Summarized from the Nuremberg Trials (1948) and World medical Association, Declaration of Helsinki; from Basch, 1990. Website sources include:
http://www.wma.net/
http://www.unmc.edu/irb/source_documents/nuremberg.htm
http://www.health.gov.au/nhmrc/ethics/helsinki.htm

BOX 15.4 THE TUSKEGEE SYPHILIS
EXPERIMENT

The Tuskegee experiment carried out by the U.S. Public Health Service,
between in 1932 and 1972, was meant to follow the natural course of
syphilis in 399 African-American men in Alabama. The men were not told
that they were being used as research subjects and were not offered treatment with penicillin until 1972, when the study became public knowledge.
The experiment had been intended to show the need for additional services
for those infected with syphilis, but researchers did not offer treatment that
became available, and prevented the men from receiving penicillin when
drafted into the army in 1942.

The case is considered to reflect unethical research because the interests
of the individuals to receive available care was put aside in the interest of a
descriptive study which would not benefit these subjects. In 1997, President
Bill Clinton apologized to the survivors and families of the men involved in
the experiment on behalf of the U.S. government. The Tuskegee experiment
is the source of lingering widespread suspicion in the African-American
community to the present time.

Source: Website http://www.cdc.gov.nshstp/od/tuskegee.htm

victims during World War II. The Nuremberg trials set forth standards of professional responsibility to comply with internationally accepted medical behavior
(see Table 15.10).

First adopted by the World Medical Assembly in 1964, and amended in 1975,
1983, 1989, and 1996, the Helsinki Declaration sets out standards of medical experimentation requiring informed consent for subjects of medical research. These standards have become an international norm for experiments with national, state, and
hospital Helsinki committees regulating research proposals within their jurisdiction.
Funding agencies require standard approval by the appropriate Helsinki committee
before considering any proposal, with informed consent on any research project.

The Tuskegee experiment in the United States, conducted from 1932 to 1972,
was a grave and tragic violation of the Nuremberg Code. Black men infected with
syphilis were registered for a program of follow-up by officials of the U.S. Public
Health Service in order to monitor the course of the disease. The subjects were untreated even though effective treatment was available. This case, attributed to then-
common racist attitudes, provided an important indictment of unethical experimental behavior.

Sanctity of Life versus Euthanasia

The ethical imperative to save life has become an important ethical and practical issue in health care. Advocates of physician-assisted suicide (euthanasia) argue for the right of the patient to die with dignity when terminal and suffering. This

is not a medical decision alone, and is an agonizing issue for society to address. The Nazi euthanasia program and its human experiments provided the direst of warnings to societies of what may follow when the principle of the sanctity of the individual human life is breached. The issue, however, has returned to the public agenda in the 1980s and 1990s as advances in medical sciences have allowed the prolongation of human life beyond all hope of recovery. Legislation in the Netherlands, United States, and northern Australia have legally sanctioned euthanasia with various safeguards in a variety of circumstances, such as patients in long-term comas or with terminal illnesses.

Doctors, patients, relatives, and health care organizations need clear guidelines, orientation, procedures, legal protection, and limitations where failure to take utmost steps to "save" the patient by intubation, resuscitation, or transplantation may cause legal jeopardy. Even though a distinction can be theoretically be drawn between permitting and facilitating death, in practice, doctors in intensive care units face such decisions regularly where the line is often blurred. Hospital doctors routinely take extreme measures to prolong the life of hopeless cases. Such decisions should not be considered for economic reasons alone, but in practice, costs of care of the terminally ill will be a driving force in debate of the issue. Living wills allow a patient to refuse heroic measures such as resuscitation with "do not resuscitate" standing orders and assignment of power of attorney to family members to make such decisions. Family attitudes are important, but the social issue of redefining the right of a patient to opt for legal termination of life by medical means will be an increasingly important issue in the twenty-first century.

SUMMARY—TECHNOLOGY, QUALITY, LAW, AND ETHICS IN THE NEW PUBLIC HEALTH

In order to maintain and improve standards of care, health systems need quality assurance and technological assessment as part of their ongoing operation. Poor quality care is costly in terms of iatrogenic diseases and prolonged or repeated hospitalization, and failure to prevent disease or complications with currently available methods. If innovations such as endoscopic surgery are not introduced, then longer hospital stays are needed for the same operation, wasting the patient's time and productivity while utilizing expensive health care resources and incurring the risks associated with more invasive surgery.

Health care is provided by people, as well as by institutions and machinery. The quality of care is set by the people providing care more than the technological facilities. Health care is a knowlege based service. Nevertheless, progress on the technological side of medical care is vital to continuing development of the field. Modern medications, monitoring equipment, laboratory services, and imaging devices have made an enormous contribution to advances in medical care. Appropriate technology is a critical issue for international health, since the most advanced technology may be completely inappropriate in a setting that cannot afford to maintain it or lacks the trained personnel to operate it, or where it comes in place of more vital basic primary care services. Technology assessment needs to be seen in the context of the country and its resources for health care.

Ethical issues in public health are no less demanding than those related to personal medical care. The rights of the individual and those of the community are sometimes in conflict, generating controversy which may impede adoption of workable, cost-effective solutions used in liberal and progressive countries.

Technology, quality, the law, and ethics are closely interrelated in public health. Well-informed and sensitive analysis of all aspects of their development is a part of the New Public Health. The balance between individual and community rights is very sensitive and must be under continuous surveillance. The New Public Health is replete with technological and ethical questions, especially in a time of cost restraint, increasing technologic potential, the public expectation of universal access to health care, and the assumption that everyone will live a healthy and long life. Health status has always been linked with socioeconomic status, and, despite enormous gains, that remains true even in the most egalitarian countries. Expansions of market mechanisms, such as control of supply of hospital beds, and providers, access to referrals, competition, and incentives/disincentives in payment systems for hospital and managed care systems, contribute to a need for dynamic health management capacity. The New Public Health assumes a social responsibility for health for all, using community and personal care modalities as effectively as possible to achieve that overall goal.

ELECTRONIC MEDIA

American Academy of Pediatrics http://www.aap.org

Canadian Council of Health Services Accreditation http://www.cchsa.ca

Health Services/Technology Assessment (HSTAT) is a free electronic resource that provides access to a large collection of databases. HSTAT was developed by the National Library of Medicine to provide access to practical guidelines for quick reference for clinicians and public health policy persons. It is available at http://www.text.nlm.nih.gov/ and includes the following:

    1. Agency for Health Care Policy and Research (AHCPR);

    2. National Institutes of Health (NIH) consensus development conference and technology assessment protocols;

    3. HIV/AIDS Treatment Information Service (ATIS) resource documents;

    4. Substance Abuse Treatment (SAMSA/CSAT) treatment improvement protocols;

    5. U.S. Public Health Service (PHS) Preventive Services Task Force Guide to Clinical Preventive Services; (also www.ahcpr.gov.clinic/nsps.fact)

    6. *MMWR* selected special reports;

    7. Detoxification and Alcohol Reports;

    8. Others.

Hospital Infections Program (HIP), Centers for Disease Conrol http://www.cdc.gov./ucidod/hip

World Medical Association http://www.wma.net

Joint Commission on Accreditation of Healthcare Organizations, http://www.jcaho.org/

National Institutes of Health Consensus Program, Office on Medical Applications Research, http://opd.od.nih.gov/consensus/about/about.htm

RECOMMENDED READINGS

Al Assaf, A. F., Schmele, J. A. 1993. *The Textbook of Total Quality in Healthcare.* Delray Beach, FL: St. Lucie Press.

Anderson, G. F., Poullier, J.-P. 1999. Health spending, access, and outcomes: Trends in industrialized countries. *Health Affairs,* 18:178–192.

Centers for Disease Control. Monitoring hospital-acquired infections to promote patient safety United States, 1990–1999. 2000. Morbidity and Mortality Weekly Reports. 49:149–153.

Ellwood, P. M. 1988. Shattuck Lecture: Outcomes management. A technology of patent experience. *The New England Journal of Medicine,* 318:1549 1556.

Fineberg, H. V., Hiatt, H. H. 1979. Evaluation of medical practices: The case for technology assessment. *The New England Journal of Medicine,* 301:1086–1091.

Fuchs, V. R., Garber, A. M. 1990. The new technology assessment. *The New England Journal of Medicine,* 323:673–677.

Grodin, M. 1999. Is informed consent always necessary for randomized controlled trials? *New England Journal of Medicine,* 241:449–450.

Grodin, M., Annas, G. J. (editorial). 1996. Legacies of Nuremberg: Medical ethics and human rights. *Journal of American Medical Association,* 276:1682–1683.

Landau, C., Lange, R. A., Hillis, L. D. 1994. Percutaneous transluminal coronary angioplasty. *The New England Journal of Medicine,* 330:981–993.

McClellan, M., Kessler, D., for the TECH Investigators. 1999. A global analysis of technological change in health care: The case of heart attacks. *Health Affairs,* 18:250–255.

Mariner, W. K. 1997. Public confidence in public health research ethics. *Public Health Reports,* 112:33–36.

Organization for Economic Cooperation and Development. 1998. *OECD Health Data 98: A Comparative Analysis of Twenty-nine Countries.* Paris: Organization of Economic Cooperation and Development.

Van der Werf, F., Topol, E. J., Lee, K. L., Woodlief, L. H., Granger, C. B., Armstrong, P. W., Barbash, G. I., Hampton, J. R., Guerin, A., Simes, R. J., Califf, R. M. (for the GUSTO investigators). 1996. Variations in patient management and outcomes in acute myocardial infarction in the United States and other countries: Result from the GUSTO trial. *Journal of the American Medical Association,* 273:1586–1591.

BIBLIOGRAPHY

Adelman, A. G., Cohen, E. A., Kimball, B. P., Bonan, R., Ricci, D. R., Webb, J. G., Laramee, L., Barbeau, G., Traboulsi, M., Corbett, B. N., Schwartz, L., Logan, A. G. 1993. A comparison of directional atherectomy with balloon angioplasty for lesions of the left anterior descending coronary artery. *The New England Journal of Medicine,* 329:228–233.

Baker, L. C., Wheeler, S. K. 1998. Managed care and technology diffusion: The case of MRI. *Health Affairs,* 17:195–207.

Banta, D. 1995. The diffusion of medical technology in the Netherlands. In Shemer, J., Schersten, T. (eds). Technology Assessment in Health Care: From Theory to Practice. Jerusalem: Gefen.

Banta, H. D., Luce, B. 1993. *Health Care Technology and Its Assessment: An International Perspective.* Oxford: Oxford University Press.

Basch, P. E. 1990. *International Health.* New York: Oxford University Press.

Berwick, D. M. 1989. Continuous improvement as an ideal in health care. *The New England Journal of Medicine,* 320:53–56.

Berwick, D. M., Einthoven, A., Bunker, J. P. 1992. Quality improvement in the NHS: The doctor's role I and II. *British Medical Journal,* 304:235–240 and 304–308.

Bittl, J. A. (editorial). 1993. Directional coronary atherectomy versus balloon angioplasty. *The New England Journal of Medicine,* 329:273–274.

Botalden, P. B., Stoltz, P. K. 1993. A framework for the continual improvement of health care: Building and applying professional improvement knowledge to test changes in daily work. *Journal on Quality Improvement,* 19:424–452.

Canadian Council on Hospital Accreditation. 1985. *Guide to Accreditation of Long Term Care Facilities: Survey Questionnaire.* Ottawa: CCHA.

Canadian Council on Health Facilities Accreditation. 1992. *Acute Care: Large Community and Teaching Hospitals.* Ottawa: CCHFA.

Council on Ethical and Judicial Affairs. 1995. Ethical issues in managed care. *Journal of the American Medical Association,* 273:330–335.

DeVille, K. 1998. Medical malpractice in twentieth century United States: The interaction of technology, law and culture. *International Journal of Technology Assessment in Health Care,* 14:197–211.

Gil, A. V., Galarza, M. T., Guerrero, R., de Velez, G. P., Peterson, O. L., Bloom, B. L. 1983. Surgeons and operating rooms: Underutilized resources. *American Journal of Public Health,* 73:1361–1365.

Goldsmith, J. 1989. A radical prescription for hospitals. *Harvard Business Review,* 67:104–111.

Grad, F. P. 1990. *The Public Health Law Manual,* Second Edition. Washington, DC: American Public Health Association.

United States Preventive Services Task Force. 1996. *Guide to Preventive Services: Report of the U.S. Preventive Services Task Force.* Second Edition. Baltimore, MD: Williams & Wilkins.

Hailey, D. M. 1997. An assessment of the status of magnetic resonance imaging in health acre. *Journal of Quality in Clinical Practice,* 16:221–230.

Harvard Community Health Plan. 1992. *HCHP Screening and Prevention Guidelines, 1992.* Boston: HCHP.

Institute of Medicine. 1985. *Assessing Medical Technologies.* Washington, DC: National Academy Press.

Institute of Medicine. 1985. *Clinical Practice Guidelines.* Washington, DC: National Academy Press.

Joint Commission. 1990. *Primer on Indicator Development and Application: Measuring Quality in Health Care.* Oakbrook Terrace, IL: Joint Commission on Accreditation of Healthcare Organizations.

Joregensen, T., Carrlson, P. (eds). 1998. Early identification and assessment of emerging health technology: Special edition. *International Journal of Technology Assessment in Health Care,* 14:603–606.

Kearney, B. J. 1996. Health technology assessment. *Journal of Quality in Clinical Practice,* 16:131–143.

Kohn, L. T., Corrigan, J. M., Donaldson, M. (eds.). Institute of Medicine, Committee on Quality of Health Care in America. 1999. *To Err is Human: Building a Safer Health System.* Washington DD: National Academic Press.

Loeffel, G., Blumental, D. 1989. The case for using industrial quality management science in health care organizations. *Journal of the American Medical Association,* 262:2869–2873.

Manning, W. G., Leibowitz, A., Goldberg, G. A., Rogers, W. H., Newhouse, J. P. 1984. A controlled trial of the effect of a prepaid group practice on use of services. *The New England Journal of Medicine,* 310:1505–1510.

Shaffer N. Short course zidovudine for perinatal HIV-1 transmission in Bangkok Thailand: Transmission Study Group. 1999. Lancet. 354:156–158.

Siebzehner, M., Shemer, J. 1995. Regulating medical technology in Israel. In Shemer, J., Schersten, T. (eds). *Technology assessment in Health Care: From Theory to Practice.* Jerusalem: Gefen.

Sloan, M. D., Chmel, M. 1991. *The Quality Revolution and Health Care: A Primer for Purchasers and Providers.* Milwaukee, WI: ASQC Quality Press.

Sidel V. 1996. The social responsibilities of health professionals: lessons form their role in Nazi Germany. Journal of the American Medical Association. 276:1679–1681.

Snider, D. E., Stroup, D. F. 1997. Defining research when it comes to public health. *Public Health Reports,* 112:29–32.

Stocking, B. (ed). 1988. *Expensive Health Technologies.* Oxford: Oxford University Press.

Thomas, L. H., McColl, E., Cullum, N., Rousseau, N., Soutter, J., Steen, N. 1998. Effects of clinical guidelines in nursing, midwifery, and the therapies: A systematic review of evaluations. *Quality in Health Care,* 7:183–191.

Torogi, Y., Takahashi, M. 1997. Cost containment and diffusion of MRI: Oil and water? *European Radiology,* 7 (Supplement 5):256–258.

World Bank. *World Development Report 1993: Investing in Health.* 1993. New York: Oxford University Press.

RELEVANT JOURNALS

Health and Human Rights
International Journal of Technology Assessment in Health Care
Journal of Nursing Care Quality
Journal of Quality in Clinical Practice
Journal of Public Health Policy
Medical Letter
Quality and Participation
Quality Assurance in Health Laboratory Technology
Quality in Clinical Practice
Quality in Health Care: British Medical Association
Quality of Care and Technologies Newsletter: World Health Organization, Regional Office for Europe, Copenhagen.

RELEVANT ORGANIZATIONS

Agency for Health Care Policy and Research (AHCPR), U.S. federal agency founded in 1987
Australian Health Technology Advisory Committee
Canadian Coordinating Office for Health Technology
Institute of Medicine Council on Technology Assessment, Washington, D.C.
International Society of Technology Assessment in Health Care
Medical Technology Assessment and Policy Research Center, United States

16

GLOBALIZATION
OF HEALTH

WHY THE "GLOBALIZATION OF HEALTH"?

Events in any part of the world can affect the status of health of people in other parts of the globe. There is a dynamic of interaction and interdependence such that a global approach is essential to achieve health targets even locally. Without restating arguments elaborated throughout the text, future generations of public health professionals must be well aware of what is occurring outside of their communities and countries. This means not only learning from the news media of outbreaks of exotic diseases, but also recognizing that the political, social, and economic upheavals that define everyday existence, across the street or halfway around the world, affect us all. Even the most remote communities in the world are not immune to the global impacts of distant military coups, civil wars, natural disasters, economic crises, or epidemics.

Previous generations of public health advocates have made tremendous contributions to understanding disease, how it spreads, and how it affects all forms of life. But mistakes have been made; HIV infection was not detected early enough, or its impact realized, until it had already reached pandemic proportions. The twenty-first century must be characterized by "globalized" public health professionals. The health of all humans is globally linked; yesterday's Ebola or cholera outbreak 10,000 miles away may manifest itself today in the arrival hall of your local airport. With this very real possibility in mind, the future of public health will require advocacy of international economic, political, and social justice policies, with the necessity of organized common efforts to improve health around the world.

Transportation, colonization, and commerce have been responsible for the dissemination of disease throughout history. With rapid movement of large numbers of people by sailing ships, steamships, rail, and later air, the possibility of disease transmission by travelers has increasingly become a global public health problem. Noncommunicable diseases have also been transferred between populations by adoption of risk factors, such as smoking, the automobile, and western diets heavy in protein and fats, bringing rising waves of mortality from these causes to areas with previously low rates. The impact of economic, demographic, and epidemiologic changes are not uniform, neither within nor between countries. Domestically and internationally, poor health and poverty affect the stability of us all. A popular late twentieth century slogan calls to "think globally, act locally"; there is no better expression to convey the interdependent realities of public health.

Previous chapters address demographic and epidemiologic issues with examples from different countries, as well as regional and global trends. In this chapter, major trends and contemporary patterns of health and disease in the world are presented, along with policy issues for improving those patterns. Global trends can be analyzed by grouping countries by geographic regions, levels of economic development, and political, cultural, or ethnic characteristics. Global health requires international health organizations to stimulate and facilitate joint efforts to achieve common goals, such as preventing the transmission of communicable disease or, more generally, promoting Health for All.

THE GLOBAL HEALTH SITUATION

Global health status involves a wide diversity of social and economic standards, disease, disability, and mortality throughout the world. Environmental and socioeconomic factors and health interventions all play a role in health status. Differences between and among developed and developing countries in these factors are great, yet there are common concerns and shared interests in health development. Study of countries classified by geographic region, such as the World Health Organization (WHO) regions, or by economic status, such as the Organization for Economic Cooperation and Development (OECD) countries, Countries of Eastern Europe (CEE), and the former Soviet Union (Commonwealth of Independent States or CIS), helps to provide an overall picture of demographic transitions and epidemiologic shifts. Economic groupings of countries are usually measured by Gross National Product (GNP) per capita, a measure of national productivity, which in industrialized countries is more than 20 times greater than that of the developing countries.

Countries now considered developed had in the past disease patterns similar to developing countries today. For example, infant and maternal mortality rates in many developing countries today are similar to those in the United States in the 1920s. Further, within industrialized countries, there are social, ethnic, or immigrant groups whose current health status is characteristic of developing countries. In many developing countries the rising middle-class populations show epidemiological patterns similar to those in developed countries, such as rising rates of heart disease.

In 1999, the world's population passed the 6 billion mark. Some 77% of them live in developing and least developed countries. Selected data on demography and

TABLE 16.1   Health Status Indicators by Country Development Status, Selected Years, 1960–1997

	Least developed	Developing	CEE/CIS[a]	Industrialized
GNP per capita in US$ 1996	232	1,222	2,182	27,086
Life expectancy in years at age 0				
1970	43	53	66	72
1997	51	63	68	78
Crude birth rate/1000 population				
1970	48	38	20	17
1997	39	25	14	12
Total fertility rate/adult women				
1960	6.6	6.0	3.0	2.8
1997	5.3	3.1	1.8	1.7
% Female adult literacy				
1980	24	46	na	96
1995	38	62	95	na
% Births attended by	28	55	93	99
trained attendant 1990–1997				
Maternal mortality/100,000	1100	470	85	13
live births 1990				
Infant mortality rate/1000 live births				
1960	171	138	76	31
1997	108	65	29	6
Child mortality rate[c]				
1960	281	216	101	37
1997	168	96	35	7
% Low birth weight	21	18	7	6
1990–1997				
% Immunization 1995–1997				
DPT[d]	62	80	91	90
Polio	62	81	92	89
Measles	60	79	89	90

Source: UNICEF, *State of the World's Children,* 1998 and 1999.
[a]CEE = countries of eastern Europe; CIS = Commonwealth of Independent States (including Baltic States), i.e., republics of the former Soviet Union.
[b]na, Not available.
[c]Child mortality rate = deaths from birth to 5 years per 1000 live births.
[d]Diphtheria, pertusis, tetanus vaccine.

health status indicators for industrialized, developing, and least developed countries are shown in Table 16.1. The enormous differences in GNP per capita and in birth rates are reflected by differences in almost all health status indicators. Population growth in the developing countries, due to high fertility rates and declining child mortality, is a key factor in poverty and poor health status, and thus a major health problem.

Trends in demographic and health indicators for countries classified as industrialized, developing, and least developed all show positive changes in health status: fertility and birth rates have declined in developing countries, but also more recently in the least developed countries. Female literacy rates are increasing in the least developed and developing countries. Immunization coverage has improved globally, as have infant, child, and crude mortality rates, so that life expectancy is generally rising. However, there remains a large discrepancy between rich and poor in health status indicators; in the least developed countries, maternal mortality is more than 60 times higher and infant mortality more than 18 times that of the industrialized countries.

Despite the fact that health status indicators have improved for the least developed countries, on the basis of current trends, the United Nations Population Division estimates that the infant mortality rate in developing countries will still be close to 60 per 1000 live births in the year 2000. Nearly one-fifth of the world's population lacks regular access to local health services. In 1991, primary health care coverage was estimated at 69% for the least developed countries, 89% for developing countries, and 100% for developed countries.

While the gap between developed and developing countries in health is very wide, the adaptation and dissemination of health technology are having profound effects. Each can learn, for good or ill, from the other. A developing country may spend most of its health resources in central teaching hospitals, while primary care is neglected. Adoption of appropriate priorities, including new vaccines and other health technologies, can bring dramatic improvements in health in developing countries. Conversely, innovations in the health field from developing countries can also be applied in developed countries. For example, oral rehydration therapy and community health workers, which provide care in developing country conditions, can be applied to unmet needs in industrialized countries.

PRIORITIES IN GLOBAL HEALTH

Poverty–Illness–Population–Environment

The interactions among poverty, population growth, and environmental degradation combine to adversely affect many developing countries and hundreds of millions of people globally. In many developing countries, economic stagnation and political instability compound these issues, causing an inability to address basic human needs and condemning more generations to ill health and early death. Although the effects of low income, lack of basic sanitation, and crowding in rural poverty or megacities cannot be overcome by health measures alone, the potential for raising the quality of life and survival rates by public health measures is nonetheless very great.

Industrialized countries have overcome many of these problems but must still cope with substantial pockets of poverty, homelessness, violence, preventable disease, environmental degradation, and rising costs of health care. The Southern Hemisphere is largely made up of developing countries with massive economic and social needs. The north–south socioeconomic divide is one that will shape global health and politics in the twenty-first century.

TABLE 16.2   Global Selected Causes of Deaths and Morbidity, 1997

Cause of death	Deaths (000s)	Cause of morbidity	New cases annually (000s)
Ischemic heart disease	7,200	Diarrhea	4,000,000
Cancer (total)	6,235	Malaria	300,000–500,000
Cerebrovascular disease	4,600	Occupational	467,000
Pneumonia	3,745	Pneumonia	395,000
Tuberculosis	2,910	STDs[c]	385,000
COPD[a]	2,890	Mood disorders	122,865
Diarrhea	2,455	Alcohol dependence	75,000
Malaria	1,500–2,700	Hepatitis B	67,730
HIV/AIDS[b]	2,300	Pertussis	45,050
Prematurity	1,120	Measles	31,075
Hepatitis B	605	Diabetes	10,540
Measles	960	Cancer	9,240
Cancer of the lung	1,050	Tuberculosis	7,250
Suicide	835	HIV/AIDS	5,800
Total all causes	52,200	NA[d]	

Source: Adapted from WHO, *World Health Report,* 1997 and 1998.
[a]Chronic obstructive pulmonary disease.
[b]Human immunodeficiency virus/Acquired immunodeficiency syndrome.
[c]Sexually transmitted disease.
[d]Not applicable.

During the 1990s, a number of developing countries have entered a phase of rapid economic and industrial growth, combining the advantages of educated low-wage work forces with market economies. Some Asian countries moved ahead rapidly with economic development, creating strong rates of growth. The breakdown of traditional social patterns, after traditional rural ways of life and intergenerational family structure, with better education, upward mobility and small family units. The 1998 downturn in southeast Asian economies will compound these problems.

The recognition that poverty and ill health are interactive has led the industrialized nations (G-7 plus Russia) to take an important step in mid-1999 of relief of debt and aid-related loans of very poor countries by some $118 billion. This will help many countries in sub-Saharan Africa by reducing their debt repayment by one-third to one-half. Despite this important step, most poor countries pay more on debt service than they do on health and education for their people.

Child Health

In developing countries, one-third of all deaths are in children under 5 (about 13 million annually), compared with a total of 2.4% or (280,000) of all deaths in developed countries. A child under 5 years in the developing world experiences an average of three episodes of diarrhea per year. In 1995, diarrhea was associated with approximately 2.0 million deaths in children under 5 years, down from 4.0

million in 1983. Six percent of deaths due to diarrhea in children under 5 occur in association with measles. With widespread use of currently available interventions, mortality related to diarrhea can be halved within a short time.

About 2 million children die each year from vaccine-preventable diseases. In 1997, there were 960,000 deaths from measles (down from 2.5 million in 1983), 275,000 from neonatal tetanus (1.1 million in 1983), and 410,000 from pertussis. An estimated 35,000 cases of clinical poliomyelitis occurred in 1997, a reduction from 350,000 cases in 1983 (excluding China). Despite the fact that both are treatable and largely preventable conditions, malaria and tuberculosis (TB) continue to take a toll of life which exceeds mortality from AIDS.

The worldwide basic infant immunization coverage rate for BCG (tuberculosis vaccine), DPT, polio, and measles was only 20% in 1980. By 1992, an estimated 85% of children worldwide had been immunized with a third dose of poliomyelitis vaccine before reaching their first birthday. As seen in Table 16.1, coverage in 1995–1997 had dropped. Children in the least developed countries had an average coverage of 62%, in developing countries, 79%, in eastern Europe and former Soviet countries, 89% and in the industrialized countries 90%. Supplementary mass campaigns (national immunization days) all over the world have reduced polio toward the goal of eradication. Massive use of measles vaccine has arrested local circulation in North and South America, although outbreaks from inported cases still occurred in 1999.

Nonetheless, millions of children do not receive immunization against the main vaccine-preventable diseases. Use of the lifesaving salts of oral rehydration therapy (ORT) increased from 1988 to 1993: in East Asia and the South Pacific from 32 to 36% of all diarrheal episodes, in sub-Saharan Africa (SSA) from 28 to 49%, and in Latin America from 23 to 64%. The impact of this progress, if sustained, will greatly improve child survival in the years ahead, and the conquest or control of the major childhood diseases will be attained.

Maternal Health

It is estimated that about 500,000 women die each year from pregnancy-related causes, most of which are preventable. Virtually all (99%) of these deaths occur in developing countries, and the large majority (95%) in Asia and Africa. This is due to a combination of factors: lack of prenatal care and professional care at birth, poor nutritional status and iron deficiency anemia, lack of adequate standards of obstetrical services, and pregnancies that are too early, too late, or too close together. Maternal mortality rates range from 1100 per 100,000 live births in least developed countries to 13 per 100,000 live births in most industrialized countries (table 16.1).

The health of women in relation to fertility is fundamental to national health standards. Education for girls and women, access to modern birth control, spacing of pregnancies and adequate care in all stages of pregnancy are the means to achieve improvement. Traditional birth attendants (TBAs) provide care during most deliveries in developing countries. There are no adequate substitutes for good prenatal medical care, but the work of TBAs can be improved by a strict program of licensing, training, and supervision (Chapter 15). Simple, inexpensive measures can improve outcomes: routine iron and folic acid during pregnancy, prenatal care stations (MCH, maternal and child health), birth centers (in hospitals if possible), high risk identification and referral systems (see Chapter 5), and deployment of well-trained community heath workers for preventive health care (Chapter 15) can all make a difference.

Between 1985 and 1991, use of prenatal care globally increased from 58 women attended per 100 live births in 1985 to 67% in 1991. In 1991, coverage was 99 women attended per 100 live births in developed countries, 65 in developing countries, and 53 in the least developed countries. Coverage for childbirth attendance increased worldwide from 53 to 55 per 100 between 1985 and 1991. Deliveries by trained health personnel worldwide in 1990–1997 was 60%, ranging from 28% in less developed countries to 99% in the industrialized countries. Furthermore, in developing countries, only approximately 52% of pregnant women receive two or more doses of tetanus toxoid in 1995–97 (UNICEF, 1999).

Population Growth

Despite the fact that population growth is a religious and political controversy in many societies, falling birthrates are now seen in most parts of the world. Developing countries are increasingly recognizing that high fertility rates hinder economic development, perpetuating poverty, a fundamental cause of ill health. The politics of population has traditionally rested on the assumption that population increase is essential for economic growth and national power. At the micro level in traditional farming societies the assumption is that more children provide greater security for the family. In recent decades the growth of the technology of family planning has been accompanied by a gradual shift to the view that unrestrained population growth is a barrier to economic development.

In many poor countries, high rates of population growth perpetuate poverty and ill health for mothers and children. Improved child survival and reduced economic imperatives for more children to work farms or to contribute to family incomes have led most countries to lower overall birthrates. Higher education levels for women have increased knowledge and use of birth control. Religious injunctions against birth control no longer have the power to prevent use, so that birthrates have fallen worldwide, and in many countries to rates below replacement levels i.e. negative population growth.

Governments have a crucial role in family planning. Distribution of information and promotion of family planning as a national policy and priority must be part of a new emphasis on primary health care. In China in the 1950s, Chairman Mao Tse Tung called birth control a new form of genocide of the developed countries against the developing countries. The legacy of this tragic pronouncement was no less destructive to public health than the pronouncements of religious bodies which still equate birth control with mortal sin. Both had the effect of promoting high fertility rates in those populations that can least afford the health and eco-

TABLE 16.3   World Population, Actual and Projected, in Billions, by Economic Level
1965–2030

Level of development	1965	1973	1980	1991	2000	2030
Low and middle income countries	2.60	3.17	3.66	4.53	5.29	7.74
High income economies	0.67	0.73	0.77	0.82	0.86	0.92
Total world population	3.28	3.90	4.43	5.35	6.16	8.66

Source: World Bank. World Development Report, 1993.

nomic burden of raising large numbers of children. For the past several decades, birth control has been promoted in India and China, the two countries with the world's largest populations, but the momentum of population growth continues and is unlikely to level off in the next 20 years. In addition, China's one-child-per-family policy has reportedly led female infanticide, forced abortions, and sterilization in a primarily rural society valuing male children.

Despite the decline in fertility rates in most regions of the world, the world population has passed the 6 billion mark and is growing at an average annual rate of 1.73% (Table 16.3). Asia accounts for almost 60% of the world's population, while Europe less than 10%. In many Asian countries, the birthrate has declined precipitously to rates close to those in developed countries. Over the period 1960–1992, total fertility rates fell: in southeast Asia and the Pacific from 5.8 births per woman to 2.6; in Latin America and the Caribbean from 5.8 births per woman to 3.1; and in the Middle East and North African region from 7 births per woman to 5.1. Many countries in sub-Saharan Africa will see their population double within 20 years, although even here there has been what appears to be the beginning of a decline in total fertility rates.

A demographic transition occurs when the age makeup of the population shifts. As countries move from developing to a developed or industrialized status, population age patterns change. With greater life expectancy and declining birthrates, the ages of the population shift toward older age groups. Developed countries are experiencing a rapid growth of the old-old, more dependent population (i.e. over 75 and over 85 years). These trends are of vital importance to the future of individual countries as they try to sustain or improve economic and social conditions. All countries need a working age population sufficient to sustain the elderly and the young dependent groups.

High birthrates in developing countries still strain the potential for the proper care and nurturing of children. Food supplies have been expanded by improved agriculture, but this may not be able to sustain high rates of population growth. In addition, rising standards of living and aspirations place further demand on natural resources and the environment with great strees the Earth's fragile ecology.

Malnutrition

Food production has increased in most parts of the world, but has steadily declined per capita in sub-Saharan Africa, along with GNP per capita. Increased production in other parts of the developing world during the 1960s and 1970s slowed during the 1980s. In developing countries, the capacity to produce food faster than population growth is limited. The developed countries, with one-quarter of the world's population, produce over half of the world's food supply. They dominate food production but have low rates of population growth. Developing countries may purchase this surplus of food, but many lack the hard currency to do so. Gross national product alone cannot measure wealth; it must also be weighed in terms of the capacity to produce food.

Hunger, adaptation, and starvation are difficult to measure. Hunger is a subjective phenomenon; adaptation occurs when persons adjust to lower energy intake; and when energy output exceeds intake, starvation occurs. Starvation may be acute or chronic, Hunger and famine are associated with natural disasters and war, but they also occur chronically in settings where food production cannot keep up with population growth. Although hunger and famine affect all ages and sexes, the most vulnerable groups in the population are infants and children, pregnant women, women as a whole, and the elderly. Men are affected in terms of reduced capacity to work. The Chinese famine of 1959–1961, one of the most tragic disasters of the twentieth century, killed up to 36 million persons.

Estimating the number of persons lacking food is difficult because of adaptation. The nutritional status of the population, more specifically of children, is usually measured by birth weight, weight-for-age, and height-for-age. Low height for a given age, or stunting, is the most prevalent symptom of protein–energy malnutrition. Approximately 40% of all 2 year olds in developing countries are stunted. The prevalence of stunting may be as high as 65% in India, about 40% in China and sub-Saharan Africa, and more than 50% in the rest of Asia. According to WHO standards, some 780 million persons worldwide are estimated to be energy deficient or in a state of protein–energy malnutrition (PEM). This is not always manifested by hunger, but rather represents inadequate food intake, especially protein, for energy needs. Malnutrition may be so widespread among children that parents and health providers assume the children's lethargy and stunting to be normal.

Micronutrient deficiency conditions affect some 2 billion people worldwide, with serious sequelae including premature death, poor health, blindness, growth stunting, mental retardation, learning disabilities, and low work capacity (Institute of Medicine 1999). Iodine, iron, and vitamins A, B, C, and D are commonly deficient in diets in developing countries, adversely affecting the health of the whole population but especially vulnerable subgroups. Iron deficiency is the most common of these, affecting mainly women and children (see Table 16.4), but also men and the elderly. In developing countries, children and women are especially vulnerable because of frequent childbirth and poor diets.

Anemia is, as referred to in Chapter 8, the most common nutritional deficiency

TABLE 16.4 Prevalence of Iron Deficiency Anemia in Pregnant Women and Women Generally, by Region, 1992

Region	% of All women	% of Pregnant women
Industrial world	13	18
Latin America/Carribean	31	40
East Asia and Pacific	37	49
Middle East/North Africa	42	52
Sub-Saharan Africa	44	52
South Asia	60	75

Source: UNICEF, *State of the World's Children,* 1995.

in the world. It is primarily a micronutrient dietary deficiency but is often exacerbated by other deficiences (e.g., vitamin C), concomitant parasitic infection in children and multiple pregnancies in women. In industrial countries, anemia of pregnancy affects 18% of pregnant women, but 40% are affected in China and Latin America and 88% in India. Iron deficiency in Russian women exceeds 50%, and iron supplementation is not routinely practiced, that is, without hemoglobin testing. Iron deficiency in infancy causes reduced growth (in height) and potential learning capacity in school. In adults, it reduces work potential. Distribution of inexpensive iron (ferrous sulfate) to pregnant and lactating women could largely prevent this onerous health burden, which by 1997 was estimated by the WHO to affect 1.8 billion persons.

Iodine deficiency affects some 1.5 billion persons globally, with severity varying from subclinical deficiency to cretinism and severe retardation. Iodine is deficient in soil and water in many parts of the world, and deficiency conditions at subclinical and clinical levels are widespread. Routine iodizing of salt is widely used to prevent iodine deficiency disorders and has been recommended by the WHO and UNICEF; it was adopted as a major objective of the 1990 World Summit of Children, along with elimination of vitamin A deficiency, which affects over 650 million persons (Table 16.5). Vitamin A supplements reduce mortality from measles and prevent xerophthalmia and blindness in children. This has created a major change in public health nutrition in developing countries by demonstrating nutritional comorbidity and the vital importance of nutritional interventions to prevent morbidity and mortality in vulnerable population groups. The extent of iodine and vitamin A deficiency conditions is enormous and entirely within the scope of current technology to prevent at low cost. The WHO estimates the cost of eradication of iodine deficiency by iodination of salt at $0.05 per person per year. Elimination of vitamin A deficiency can be achieved by giving vitamin A capsules to children over 6 months of age three times per year at a cost of $0.02 per capsule, by dietary modification to promote Vitamin A rich foods, and/or fortification of basic foods (oil, margarine, milk, or sugar).

TABLE 16.5 Vitamin A Deficiency in Children and Iodine Deficiency Conditions in Total Population Worldwide

Vitamin A deficiency condition (VAD)	Population affected (millions)	Iodine deficiency disorders (IDD)	Population affected (millions)
Severe eye damage	0.5	Cretinism	5.7
Xerophthalmia	3.1	Brain damage	26
Night blindness	13.5	Goiter	655
Elevated risk of death from infectious disease	231	Total[b]	1600
Total[a]	562		

Source: UNICEF, *State of the World's Children,* 1995.
[a]Total number of children under 5 in developing world.
[b]Total population with IDD's.

In many countries in the African, southeast Asian, and eastern Mediterranean regions, infectious and parasitic diseases in association with malnutrition continue to be major public health problems. They cause much of the mortality among infants and children and shorten life expectancy (see Chapter 6).

Infectious Diseases

Globalization of the spread of disease is as old as the migration of humans, animals, or disease vectors. In the 1980s and 1990s, concern has reawakened with the experiences of HIV, reemergence of diseases thought to be under control, and emergence of newly identified diseases presenting public health threats. Tuberculosis, the world's largest killer among infectious diseases, increased by more than 25% in worldwide case notifications between the mid-1980s and the early 1990s. Cholera increased fourfold between 1990 and 1994. Diphtheria reappeared on a large scale, and *E. coli* and salmonella food-borne diseases are increasing in the 1990s.

The emergence of the AIDS pandemic has affected all countries of the world, regardless of their level of development. It is estimated that there are 27.9 million persons with HIV infection, and 5.8 million have died, including 1.4 million children. In many developing countries, AIDS is an additional burden to economic growth and health systems. The cost of care and the partial loss of the economically productive generation of young adults are enormous burdens on the troubled economies of sub-Saharan Africa and other developing countries. Estimates of the *World Development Report* 1993 are that developing countries have spent $340 million to care for AIDS patients, and this will increase to $1.1 billion by the year 2000. Lack of resources for treatment is complicated by lack of resources and priority for prevention.

As was discussed in Chapter 4, the emergence of "new" infectious diseases and the reemergence of well-known but still uncontrolled diseases pose great challenges for public health and clinical care. The problems of these diseases are compounded by the rise of resistant microbial strains. The basic priorities in control of infectious diseases remain the need for universal coverage with childhood immunization, high standards of food and water safety and sanitation, education to reduce the spread of HIV and STDs, improved primary care for diagnosis and management of TB and malaria, and rational use of antibiotic therapy. The achievements of the past several decades have been truly impressive, but there is little room for complacency as the potential global threats of infectious diseases remain.

Important technological advances are needed before TB, HIV, and malaria can be controlled globally, but much can be accomplished with existing technology. The early part of the twenty-first century will see new vaccines come into common use that will reduce the international burden of infectious disease morbidity and mortality. Important new vaccines for varicella, hepatitis A, and rotaviruses are already licensed and entering routine immunization programs. Breakthroughs in development of new therapies and vaccines for HIV and malaria may occur early in the twenty-first century. It is hoped that new antimicrobials will be found to address the multiple drug-resistant organisms. Improved food technology will be needed to prevent salmonella and *E. coli* infection from food sources. Medical care will need to improve its methods of control of infectious diseases to avoid emergence of resistant organisms through more restrained use of antibiotics. The achievements toward eradication of important infectious diseases such as smallpox, polio, measles, guinea worm, leprosy, onchocerciasis, generates optimism for new and perhaps equally dramatic breakthroughs in the new century, even though tempered by realistic appraisal of unsolved and new challenges of infectious diseases.

Chronic Disease

The epidemiologic transition from a predominance of infectious diseases to the chronic conditions that occurred in the industrialized countries by the mid twentieth century, is also occurring in the developing nations. The cardiovascular diseases, cancer, other degenerative conditions and mental disorders, and trauma are already the major causes of death in many developing countries (see Table 16.2). Trauma also constitutes a vast public health issue, with serious individual, social, and economic consequences. Worldwide, some 7.2 million deaths occur from ischemic heart disease and another 4.6 million from strokes. Some 2.7 million deaths from injury and poisoning are reported yearly, 2 million of which occur in developing countries, and they result in considerable loss of potentially productive years of life. Motor vehicle accidents (MVAs) rank first in causality, followed by domestic accidents, including falls, burns, poisonings, and drowning, all of which are particularly prevalent among young people and the elderly.

Diseases related to smoking, overeating, and lack of balance in the diet are increasing in developing countries among the middle-class and working class populations. Rising death rates from coronary heart disease and strokes in the former Soviet countries constitute an enormous contributor to premature death and a burden on underfinanced health systems. As infectious diseases are better controlled and as eating patterns shift in the urban middle and working classes to high meat and fat intake, patterns of cardiovascular disease seen in the industrialized countries occuring in developing nations. Public health practitioners need to prepare for this epidemiologic transition with interventions such as antismoking campaigns, nutrition education, and other health promotion. Similarly, the burden of mental, dental, and other health needs of societies are part of global health planning for developed countries in transition.

Mental health is gaining increasing recognition as a health issue of global proportions affecting hundreds of millions with moderate to severe disability, not only in the industrialized countries, but also in developing countries. As measures of the burden of disease include morbidity as well as mortality (such as in DALYs), then major depression (unipolar), alcohol dependence, bipolar disorders, and schizophrenia especially become high in the list of causes of disability, especially in young adults aged 15–44. These require attention in the health system and especially in primary care.

Disaster Management

Tragic events leading to large scale loss of property and life created by nature and by man require organized international response to limit the damage, to reduce suffering, and to restore normality. These situations may be natural disasters such as hurricanes, floods, droughts, earthquakes, or volcanic eruptions with horrific consequences. They may be larger scale events created by human initiatives, such as binational wars, civil wars, ethnic cleansing, and civil strife or repression. Such events can take on enormous proportions as displacement, murder, and other forms of violence disrupt human norms and civil society. The public health aspects of such events are within the context of restoration of safety, provision of safe water, shelter, food, and sustenance, and efforts to restore civil life. Such events are now brought to the immediate attention of the world's community in television coverage. International action is forthcoming, but often inadequately coordinated with overwhelmed local civil and security authorities. Preparation and organization for such disasters are an important element of global health.

Environment

The environment is a global health concern, not only because it affects every country but also because its maintenance requires joint action (Chapter 9). Air pollution caused by industry, power plants, and domestic use of coal are common in urban areas worldwide. The quality of air in the industrial countries has improved over the 1980s and 1990s, but in many developing countries and the former socialist countries air quality has deteriorated because of poor quality in power generation and urban congestion. Excessive production of carbon dioxide by use of fossil fuels is contributing to a global warming effect, and chemicals used in industrialized societies cause ecological damage, with potentially serious global consequences.

DEVELOPMENT AND HEALTH

Some 1.3 billion persons in developing countries lack access to clean water; nearly 2 billion lack adequate sanitation. Poverty, low educational and job skills, poor nutrition, an unsanitary environment, and poor housing conditions all contribute to the enormous burden of disease of all kinds in developing countries. Indoor pollution from use of cooking fuels with improper ventilation in developing countries contributes to high rates of acute respiratory disease and deaths in children, as well as to chronic lung disease in the elderly and damage to the fetus in pregnancy.

Health status and economic development are interdependent, and the prevailing social and political philosophies have a vital impact on health, not only in terms of the amount of funds allocated to health, but also in the form of the health care delivery system adopted. Rapid economic development also has its price. Environmental pollution and increases in occupational health hazards occur when new technology is transferred to developing countries. Further degradation also occurs with the tendency of the rural poor to move to cities, where basic sanitation and other infrastructure are often lacking.

Measurement of economic development by GNP alone is misleading. The distribution of wealth in a country is an important variable along with other measures such as school enrollment. The human development index (HDI) includes life expectancy, educational attainment, and measures of income (giving lower weight to income above the poverty level, since extra income for upper income groups is less important to survival). The human development index along with DALYs and QALYs (see Glossary and Chapters 3 and 11) add an element of the quality of life to the usual economic indices.

Equally important to the amount of money spent on health is how the money is used. Some countries have succeeded in achieving marked improvements in health, while they remain poor as measured by GNP per capita. Some countries have higher ratings in terms of HDI than their ranking by GNP. China, with a GNP per capita of $370, has succeeded in attaining infant mortality and life expectancy rates of mid-level developing countries by bringing primary care to the vast rural population. Sri Lanka, with a per capita GNP of $500, has an infant mortality rate of 15 per 1000, comparable to well-advanced developing countries. Kerala, India, is well above national standards in HDI, while being economically poorer than the national average. On the other hand, some countries with high per capita GNP have lower HDIs; for example, Kuwait and Saudi Arabia have large GNP per capita but fewer public health achievements than a much poorer countries such as Cuba and Costa Rica Jamaica. In some countries, this may be due to the large economic gap between the small, very wealthy ruling class and the large, poor population.

ORGANIZATION FOR INTERNATIONAL HEALTH

As seen in Chapter 1, from the decline of the Roman Empire in the fifth century AD, Europe passed into a millennium of scientific repression. Knowledge, including medical knowledge, passed into the hands of the Church, and the Greek and Roman writings that were preserved in the West survived in isolated monasteries of Ireland and Europe, and in Arab civilization. where during the next few hundred years Arabian, Byzantine, and Jewish scholars translated and preserved ancient medical knowledge in Europe. In the ninth century, a medical school was founded in Salerno near Naples, and medical schools spread to cities throughout Europe and the Arab world.

European colonial expansionism, beginning in 1415 with the Portuguese attack on Moslem settlements in nearby North Africa, had extremely important effects on international health. European ships brought smallpox and measles to the natives of the South Pacific and the Americas, decimating their populations. Syphilis is thought to have been introduced into Europe by sailors returning from the Americas. European adventurers and settlers were often cruelly treated by the many endemic diseases to which they had scant resistance. Additionally, the slave trade brought communicable diseases from Africa to favorable habitats in the Americas.

Colonialism led to near eradication of many of the world's native peoples, and changed the character of many populations, most dramatically in North America, Australia, New Zealand, South Africa, and parts of Latin America. Colonial governments introduced western medical organization and practice and influenced health with respect to concepts of causality and the treatment of diseases. Widespread education and medical training were important legacies in many, but not all, developing countries that gained their independence in the mid twentieth century.

The development of sanitation and later microbiology depended on the scientific and technological underpinning provided by the Industrial Revolution. In the latter half of the nineteenth century, repeated epidemics of cholera in Europe and continuing havoc from other communicable diseases were intense stimuli for researchers to identify the causal agent and means of transmission of almost ever major bacterial and parasitic disease. Asiatic cholera arrived in Europe in 1832 and spread throughout the continent in repeated epidemics during the nineteenth century. This led to a convening of the International Sanitation Conference in Paris in 1851, with follow-up meetings held in 1874, 1881, and 1885. These conferences were held more frequently between 1892 and 1903 regarding maritime quarantine and control of international transmission of cholera, yellow fever, and typhus. In the early 1880s, a pioneering step in international public health occurred when, at the request of the International Cholera Commission, Robert Koch led a team to investigate cholera epidemics in Egypt. This resulted in identification of the *Vibrio cholerae* organism and recommendation of preventive procedures.

The Health Organization of the League of Nations (1921–1946), established in Geneva, was an attempt to develop the idea of international collective security for health. As part of its function, a Health Office provided an Epidemic Intelligence Service. The Health Office organized many expert committees on infectious diseases and other public health problems, including the establishment of standards for biologicals, maternal and child health, nutrition, health insurance, and medical education. Malaria, leprosy, and rabies control activities were promoted, as were the establishment of cancer registries and preparation for an international classification of disease; pharmacopoeias were coordinated and standards for housing and nutrition developed. The scope of organized international work was broadened from prevention of international transmission of disease to disease control and improved health conditions for vulnerable groups in the population. The collapse of peace in the late 1930s led to the disbandment of the League of Nations.

During World War II, the United Nations Relief and Rehabilitation Agency was established by the allied powers to assist resettlement of the millions of displaced persons. This became part of the initiative to establish a new international health organization as part of an international consensus to build a better world after the war, in the context of a stronger, more coordinated United Nations.

THE WORLD HEALTH ORGANIZATION

The World Health Organization (WHO) was founded in 1948 as a United Nations agency in the spirit of cooperation and idealism following World War II. The WHO charter states that one of the fundamental rights of every human being is "the highest attainable standard of health," and the United Nations' Universal Declaration of Human Rights in 1948 stated, "Everyone has the right to a standard of living adequate for the health and well-being of himself and his family."

The World Health Organization has made an enormous contribution to global health. It fills the need for a single unified intergovernmental organization representing all countries, covering all fields of health. The WHO consists of member states working together and with other organizations toward achievement of the highest possible level of health. It replaced previous organizations, especially the Health Organization of the League of Nations and the Pan American Sanitary Bureau. A Technical Preparatory Commission developed the organization and, in the new optimism of the time, undertook the enormous task of dealing with global health problems. Its direction and coordinating functions are the primary assets of the organization, especially in definition of health goals and in initiating international cooperation to achieve them. Also effective are its technical services, epidemiology functions, statistics, standardized nomenclatures for disease and drugs, and their publications.

TABLE 16.6 Major Headquarters Programs, World Health Organization, Geneva, 1997

HIV and Sexually Transmitted Diseases (ASD)	Mental Health, Prevention of Substance Abuse (MSA)
Child Health and Development (CHD, formerly Diarrhoeal and Acute Respiratory Disease Control)	Non Communicable Disease (NCD)
	Prevention of Blindness and Deafness (PBD)
Control of Tropical Diseases (CTD)	Promotion of Chemical Safety (IPCS)
Essential Drugs (DAP)	Promotion of Environmental Health (PEH)
Drug Management and Policies (DMP)	Health Technology (PHT)
Emergency and Humanitarian Action (EHA)	Publishing, Language and Library Services (PLL)
Emerging and other Communicable Disease Surveillance and Control (EMC)	Development of Policy Programme, Evaluation (PPE)
	Substance Abuse (PSA)
Food and Nutrition (FNU)	Reproductive Health (RHT)
UNAIDS (former Global Programme on AIDS)	Resource Mobilization (RMB)
Global Programme on Vaccines, Immunization (GPV)	Strengthening Health Services (SHS)
Global Tuberculosis Programme (GTB)	Tropical Diseases (TDR)
Health Promotion, Education, Communication (HPR)	Women's Health (WHD)
Health Systems Development (HSD)	Budget and Finance (BFI)
Human Reproduction (HRP)	Conference and General Services (CGS)
Health Situation and Trend Assessment (HST)	Personnel (PER)
Statistical Information System (WHOSIS)	Social and Occupational Health (SOH)
Interagency Affairs (INA)	Special Offices—legal advisor (LEG), informatics (AOI),
Information System Management (ISM)	internal audit (IAO), research policy and strategy
Elimination of Leprosy (LEP)	coordination (RPS), world health reporting (WHR)

Source: WHO home page, http://www.who.ch/whosis/whosis.htm (February 25, 1997).

The organizational structure of the WHO includes headquarters in Geneva and regional offices for Europe (Copenhagen), the Middle East (Alexandria), Africa (Brazzaville), South-East Asia (Delhi), Western Pacific (Manila), and the western hemisphere [the Pan American Health Organization (PAHO) in Washington, D.C.]. The central headquarters in Geneva has many offices dealing with a diversity of topics (Table 16.6).

The WHO has led in the formulation of a worldwide consensus on a new direction in health policy. It formulated a strategy that incorporated the principles of government responsibility for the health of their peoples, the right of people to take part in developing and controlling their health care, and equality in health. It helped to formulate and promote the concept that cooperative activity between different parts of the public and private sectors (intersectoral cooperation) is necessary to advancing health causes. The concept of appropriate technology is also a WHO initiative (see Chapter 15).

The problems and limitations of the World Health Organization are important to assess. The organization is part of the United Nations system and became partially subject to political conflicts during the Cold War period and in regional conflicts such as that in the Middle East. This was to the detriment of its leadership and moral authority and has led to the withholding of membership fees by the United States for many years. This politicization limited contacts of the WHO with the

BOX 16.1 SUCCESSFUL AREAS OF
INTERNATIONAL HEALTH LEADERSHIP

Successes and important initiatives of the international health movement led by the WHO and UNICEF include the following:

1. Eradication of smallpox;
2. Massive increase in immunization coverage (EPI);
3. Control and possible eradication of poliomyelitis;
4. Reduced measles incidence (still 1 million deaths per year in late 1990s);
5. Massive reduction in incidence of tetanus, diphtheria, pertussis;
6. Improving control of diarrheal disease (CDD) and reduced death rates;
7. Improving control of acute respiratory illness (ARI);
8. Improved control in tropical diseases: onchocerciasis, leprosy, yaws, and potential eradication of dracunculiasis;
9. Leadership in principles of primary health care: influencing national health programs particularly in developing countries;
10. Raising public and political consciousness of health issues;
11. Health for All initiatives;
12. Health targets initiatives;
13. The "Healthy Cities" movement;
14. Health promotion, raising health in national priorities;
15. Increasing awareness of health information needs.
16. Intersectoral cooperation in vaccines & immunization.

highest quality of professional leadership, impairing its ability to relate to the forefront of medical science, epidemiology, and public health practice. It has also led to inadequate leadership in areas where the response of the WHO to important issues might offend national pride. At times, epidemiologic monitoring and information sharing has also suffered.

If the WHO did not exist, the world community would need to invent it. Although hampered by its political nature, it exists as an international body in health representing all countries and dealing with health in a broad definition. The WHO's leadership in the Declaration of Alma-Ata and Health for All 2000 represented an important step forward in international health with its major commitment to primary health care (see Chapter 2).

Tropical disease work on malaria, bilharzia, filariasis, tuberculosis, onchocerciasis, leishmania, schistosomiasis, helminthic diseases, and diarrheal disease control are of particular importance to the developing countries. WHO leadership in eradication of smallpox and virtual eradication of guinea worm disease, on-chocerciasis, and poliomyelitis have been outstanding contributions to improved global health. Its initiatives in reducing nutritional deficiency conditions, in chronic disease control, in defining health manpower needs, and in health services financing have also been important for both developing and the developed countries.

The World Health Organization develops programs of work that guide its activities and its regional offices as well as member states. The WHO's Eighth Programme of Work for the period 1990–1995 defined 15 objectives and a number of targets for each objective. This involved a global strategy for health, including promotion of food production and distribution, social progress in literacy, poverty reduction, and economic growth. Also included were the following: intersectoral cooperation; development of health care systems with a stress on primary care and improved management skills and efficiency; community involvement; improved levels of health resources, including financial support by governments and universities involved in training health manpower; research, technology, and cooperation between countries; and environmental sanitation. All were included as areas for action within this program. The Ninth Programme of Work for the period 1996–2001 establishes a new global health policy framework for member governments, international organizations, banks, NGOs (nongovernmental organizations), and other organizations related to health, economic, and social development.

Many other United Nations agencies and other organizations play important roles in international health. These include UNICEF (the United Nations Children's Fund), the United Nations High Commission for Refugees (UNHCR), the United Nations Development Programme (UNDP), the International Labour Organization (ILO), the Food and Agriculture Organization (FAO), and the International Atomic Energy Commission (IAEC).

The United Nations Children's Fund

Following World War II, the new U.N. General Assembly created the United Nations International Children's Emergency Fund (UNICEF, now the United Nations Children's Fund), principally to assist the children of war-torn Europe. The program gradually expanded to include other activities and other areas, particularly in developing countries.

This agency has spent large sums of money, especially on food and supplies, for the promotion of child and maternal health and welfare activities throughout the world. Beyond this, usually through partnership with the World Health Organization, UNICEF has been carrying out large and significant programs of BCG vaccination and yaws and malaria control. The promotion of family planning in developing countries is one of its major activities. UNICEF plays an important leadership role in fostering primary care and community preventive approaches worldwide.

NONGOVERNMENTAL ORGANIZATIONS

Nongovernmental organizations (NGOs) are numerous and carry out specialized activities worldwide. They vary widely in content, funding, ideology, and modus operandi. Many provide important support for developing countries, often succeeding where international agencies have failed, precisely because they work outside of the national political framework. This is particularly true in the case of emergencies and areas of conflict.

The earliest NGOs were those of the various church missions and sectarian organizations. Among the many that might be mentioned are the Unitarian Services Committee, the American Friends Services Committee, Catholic Relief Services, the American Jewish Joint Distribution Committee, the International Rotary Club, and the American Bureau for Medical Aid to China. The International Committee of the Red Cross (ICRC), Medicins Sans Frontieres (MSF), Terres des Hommes, and other European-based NGOs provide direct assistance in developing countries during crises. In 1999, Medecins Sans Frantieves was awarded the Nobel Prize for Peace in recognition of its worldwide health achievements.

Philanthropic foundations have and continue to make major contributions to international health. Private foundations such as the Ford, Soros, and Rockefeller Foundations carry out important international health work within their own exclusive structures. They are important sources of grants to promote pilot programs and research in health care systems. In addition, they contribute extragovernmental funding that can stimulate development of innovative programs later affecting general health services. In 1999, the Bill Gates Foundation donated $100 million to international Childrens Vaccination Program conducted by UNICEF, especially for hepatitis B vaccine.

Among the foundations, the Rockefeller Foundation is the best known in the field of international assistance in health. Since its inception in 1913, it has operated in almost every country worldwide. Its significant contributions are many and include support of control programs for malaria and yellow fever, the development of recognized centers of learning in medicine and public health, postgraduate fellowships, and the demonstration of sound methods of organization and operation of health programs.

Despite the many positive aspects of NGOs, they can be a source of distortion in health care services in both developed and developing countries. They tend to focus on one kind of service, are very proud of their independence of government, and can create pressure for services that will place a burden on the system of financing or provision of health care. NGOs or bilateral aid can promote hospital development in places where there is already an oversupply and limited primary care. They can provide a primary service but be unwilling to coordinate with the basic governmental program in immunization, so that no one agency is fully responsible.

Coordination of NGO services into a comprehensive population-based service program may be compromised by political and international sensibilities, which can create chaos in an emergency situation. The balance of services for a population requires inclusion of governmental, NGO, and private services as a coordinated if not integrated whole. This may be impossible with highly independent NGOs, but the state public health authorities are responsible for overseeing the functions of NGOs, no matter how charitable the cause and well meaning the intent.

THE WORLD BANK

The International Bank for Reconstruction and Development (IBRD), also known as the World Bank, is based in Washington, D.C. It was established by the industrialized countries following the Bretton Woods Conference toward the end of World War II. It was an important financial institution to facilitate the reconstruction of postwar Europe. It has since become a major source of financing for development projects throughout the world. Traditionally it focused on large scale infrastructure, industrial, and farming development projects. Its policies in health development focused on promoting market mechanisms and privatization of health care in countries lacking infrastructure. This fostered an inappropriate stress on medical and hospital care when a community health orientation was needed. The World Bank examined the health sector and its importance for economic development in its *World Development Report* of 1993.

The 1993 *World Development Report: Investing in Health,* which examined the interaction of health status, health policy, and economic development, stated that, contrary to the views held by many traditional economists, health is essential for economic growth, and not a burden on the economy. The report advocated a four-pronged approach by governments to improve health in developing and former Soviet countries:

1. Foster an economic environment that will enable households to improve their own health by promoting income gains for the poor and expanding social investment in raising standards of education, especially for girls;
2. Redirect government spending away from specialized care toward low cost and highly effective activities such as immunization, programs to combat micronutrient deficiencies, and the control and treatment of infectious diseases;
3. Encourage greater diversity and competition in the provision of health services by decentralizing government services, promoting competitive procurement practices; and
4. Foster greater involvement of nongovernment (NGO) and private organizations, and regulate insurance markets.

World Bank assistance to health-related projects is growing steadily. The World Bank and the WHO have worked together on projects such as a special Program for Research and Training in Tropical Diseases (TDR) and the Onchocerciasis

BOX 16.2 HEALTH AND ECONOMIC DEVELOPMENT

"Good health . . . is a crucial part of well-being, but spending on health can also be justified on purely economic grounds. Improved health contributes to economic growth in four ways: it reduces production losses caused by worker illness; it permits the use of natural resources that have been totally or nearly inaccessible because of disease; it increases enrollment of children in school and makes them better able to learn; and it frees for alternative uses resources that would otherwise have to be spent on treating illness."

Source: World Bank, *World Development Report,* 1993, p. 17.

Control Project in West Africa. The World Bank pays special attention to health costs and financing, hospitals, pharmaceuticals, and nutrition.

The World Bank is a lending bank with a potentially very important role in health development internationally because it can command large financial resources and because it sees its role as promoting development. Its view of health as a productive investment sector makes its contribution in developing countries, and formerly Soviet countries, vital for future progress in health.

TRENDS IN GLOBAL HEALTH

It is now widely understood that the socioeconomic environment is a basic determinant of the state of health of an individual or population, even though the precise nature of intervening variables may not be sufficiently elucidated. The Southern Hemisphere has witnessed, along with its demographic explosion, the persistence of chronic problems plaguing the education, food, and housing sectors. In addition, more acute situations have emerged over the past few decades in relation to conflict, employment, migration, trade, and degradation of the physical environment. The Northern Hemisphere has enjoyed a rising level of affluence, with negative aspects that have made a sizable impact on public health: overeating, overdrinking, smoking, pollution, illicit drugs, and motor vehicle accidents.

The concept of primary health care as the basis of health system development has been almost universally accepted, yet evidence of public commitment to its implementation is still lacking. Problems reported include poor distribution of resources and inadequate orientation of health workers to primary health care, with continuing emphasis primarily on curative services. The community is often insufficiently aware of the role it should play and is frequently willing to accept competing demands for expensive secondary and tertiary care. Lack of resources to develop preventive services and health promotion is likely to erode the confidence and commitment of health workers and the community to primary health care.

The formulation and analysis of health manpower policy have emerged as a growing concern in the world (Chapter 14). There is a consensus regarding the urgent need to ensure the relevance and quality of manpower to the requirements of the health system, to avoid imbalances in the production of professional health manpower, especially with regard to physicians, nurses, and dentists. In most developing countries health manpower development plans either do not exist or are in the process of being developed.

Global changes in the 1980s and 1990s constitute challenges to continued progress in health. Population growth, aging of the population and the increasing incidence of chronic diseases, high expectations of the public for health care, increasing costs and medical technology, economic recession, and limited resources for health have all contributed pressures for health system reforms to maintain universal coverage. During the 1970s and 1980s, many industrialized countries developed health reforms that include reduction in hospital bed supply, financial incentives to promote development of community-based services, and a combination of decentralized management and integration of services in those countries with national health service type programs (e.g., United Kingdom). For countries with national health service insurance, control of supply and utilization, particularly of hospital beds, is also a feature of reforms for cost containment. In the United States, rapidly rising costs led to the rapid expansion of managed care systems seeking cost-effective health care combined with health promotion to reduce disease prevalence and dependency on treatment services (Chapters 13 and 14).

The relationship between disease and society is such that many of the factors needed to reduce preventable diseases largely lie outside of the biomedical framework of genetics, medical care, public health, and health promotion, but are determined by social preconditions that are in the realm of human rights. This, however, does not absolve governments or the health community from the imperative of applying known measures of prevention and curative services for all as a basic human right.

EMERGING INFECTIOUS DISEASE THREATS

International health began as an activity to prevent the spread of epidemics and communicable diseases. This involved the collection and dissemination of information in a timely fashion, preventive measures such as appropriate immunization campaigns to control the spread of a disease, and subsequent follow-up. Success in eradication of smallpox and increasing control of the vaccine-preventable diseases led to enthusiastic assessments that epidemic diseases were under control. This optimism has been tempered by setbacks in malaria and TB control, along with a host of other emerging and reemerging disease issues.

BOX 16.3 EBOLA VIRUS

Ebola virus was first identified in 1976 in the Sudan and in Zaire as the cause of a deadly hemorrhagic fever, with high mortality. An outbreak of Ebola virus in Zaire in 1995 was of major international concern, because of its 77% mortality rate and a fear that it could spread rapidly. This outbreak in Kikwit was limited to 316 cases with 245 deaths, many of whom were hospital workers. The organism has been isolated in specific species of monkeys and can be transmitted to man via blood and secretions.

An international team mobilized by the WHO and CDC came to the site to assist the Zairian public health staff. Stress on case detection, laboratory confirmation, isolation, and staff protection helped to limit the disease spread. The WHO Collaborating Center on Arboviruses and the Viral Hemorrhagic Fevers reference laboratory at the CDC in Atlanta played an important part in management of this epidemic. Rapid international response and heightened surveillance are part of the global concern for newly emerging infectious diseases, partly resulting from the lessons learned from the slow response to the AIDS epidemic.

Source: WHO, *World Health Report,* 1996.

The spectrum of infectious disease in a community evolves rapidly with changing conditions of the environment and society. Population growth, crowding in urban slums, homeless populations, massive migration, and travel contribute to international transmission of once-localized diseases internationally. Resistance to available antimicrobial medications is creating a new dilemma for modern medicine and public health. The synergism of infectious diseases such as AIDS with TB or cryptosporidium causes deterioration in the patient and spread by secondary infection to other persons. The "postantibiotic era" is widely discussed as a serious threat to modern public health. Strategies to prevent the loss of some of the important gains of the twentieth century in communicable disease control will require new research and strategies. Among the lessons learned from AIDS are that improvements in early warning systems and attention to new threats are vital tasks of public health.

In the United States, a number of new or resurgent infectious diseases are of increasing public health concern. These include HIV/AIDS, *E. coli* O157:H7 disease, cryptosporidiosis, coccidiomycosis, multidrug-resistant pneumococcal disease, multidrug-resistant tuberculosis (MDRTB), vancoymcin-resistant enterococci, influenza A Beijing/32/92, hantavirus infections, leishmaniasis in veterans of the 1991 Gulf War, Legionnaire's disease, and Lyme disease.

Newly emerging and reemerging diseases of concern internationally include

HIV/AIDS, multidrug-resistant malaria, tuberculosis, and cholera, *Shigella dysenteriae,* diphtheria, and *E. coli* O157:H7. Tropical diseases, such as yellow fever and dengue, are reappearing in Asia and Latin America, and Rift Valley Fever in Egypt, Lassa fever in West Africa, Ebola virus in Congo (formerly Zaire), Marburg virus via imported monkeys, Oropouche arbovirus and Sabia in Brazil, Junin virus in Argentina, and Machupo virus in Bolivia all are health concerns as they may spread from their natural habitat to other countries before the carrier shows symptoms.

It can no longer be assumed that such diseases will remain in their natural habitats; they can be transmitted all over the world via human or animal carriers and, in appropriate conditions, become serious local or even general public health concerns. The 1995 outbreak of Ebola virus in the former Zaire raised international concern of the very real possibility of widespread transmission of this deadly dis-

TABLE 16.7   Strategies for Coping with Emerging Infectious Diseases

Goals/topics	Activities	Examples
I. Surveillance	Detect, promptly investigate, and monitor emerging pathogens, the diseases they cause, and the factors influencing their emergence	Monitoring in sentinel surveillance networks, e.g., blood banks, emergency rooms, laboratories, sentinel settings Population-based surveillance Increase field investigation of outbreaks Dissemination of epidemiologic data locally and internationally using electronic media; Internet, PROMED website, and others Rapid laboratory diagnosis Monitor vector-borne diseases
II. Applied research	Integrate laboratory science and epidemiology to optimize public health practices	Promote reporting by sentinel laboratories and clinical settings Improve laboratory diagnostic techniques, gene typing, subtyping, and DNA mapping ("fingerprinting"), e.g., *E.coli*, cholera, polio, measles, menigitis
III. Prevention and control	Enhance communication of public health information about emerging diseases and ensure rapid implementation of preventable strategies	Wide and immediate dissemination of health information on infectious diseases to health professionals, general public, groups at special risk Promote health education on prevention of spread of communicable diseases
IV. Infrastructure	Strengthen local, state, and federal public health infrastructures to support surveillance and implement prevention and control programs	Improve laboratories, reporting, and training

Source: Modified from Centers for Diseases Control, *Addressing Emerging Infectious Disease Threats.* Atlanta, Georgia: U.S. Public Health Service, 1994.

ease. The public has been made aware of this kind of situation in graphic detail in the news media, novels, and movies.

In July 1996, a large-scale epidemic of food poisoning in Japan involving *E. coli* O157:H7 (first described in 1982) spread through contaminated school lunches and caused over 3000 illnesses, hundreds of hospitalizations for bloody diarrhea, many with severe hemolytic uremic syndrome, and 7 deaths. The identification of the source proved to be very difficult. Milder epidemics have occurred in many other countries, including Australia, Canada, the United States, and various European countries, so that there is continued concern for recurrence of this potentially severe form of food poisoning.

International cooperation to identify new infectious or other health threats to prevent global epidemics is an urgent priority for both international agencies and national health systems throughout the world. Control of communicable diseases requires medical, laboratory and epidemiologic intelligence services of a high order, with rapid means of communication, publication, and coordination, backed up by skilled professional services (Table 16.7). Examples of international activity in control of infectious disease are numerous. The crowning achievement in this field was the eradication of smallpox. This great feat may soon be matched by the international eradication of poliomyelitis. Such achievements can only be made with international cooperation and commitment.

NEW AND RENEWED TARGETS AND GOALS

Since the formation of the World Health Organization there has been a continuous search for effective ways of promoting international health. The 1978 Alma-Ata conference sponsored by the World Health Organization and UNICEF articulated a new agenda for national health authorities with an emphasis on primary health care. This stimulated a strengthening of activities at the primary care level. It has been helpful in developing countries with a severe limitation of resources and serious health problems, but the lesson has also been important for developed countries that are showing the increasing strain of maintaining highly costly health systems which often do not meet all the needs of the population.

During the 1980s there were encouraging indicators of progress in health in the developed and developing countries, including those classified as least developed countries (LDCs). In 1990, the WHO-sponsored World Summit for Children again focused attention on the continuing need to revise health targets and to mobilize international efforts to attain those goals, with the benefits of success shared by all. This conference accepted a number of specific targets for the years 1995 and 2000 (box 16.4), recognizing that the health of a population of a region, a country, or of a subgroup within a country is dependent on the following:

1. Economic development levels and distribution of poverty;
2. Education (especially of women);
3. Access to primary health care;

**BOX 16.4 THE WORLD SUMMIT FOR
CHILDREN 1990—TARGETS FOR 1995**

1. Raise immunization coverage to at least 80%;
2. Eliminate neonatal tetanus;
3. Reduce measles deaths and cases;
4. Eradicate poliomyelitis in key areas;
5. Increase use of ORT to 80% of cases of diarrheal disease;
6. Make maternity hospitals "baby-friendly," supporting breast-feeding and avoiding formula feeding;
7. Make universal iodizing of salt;
8. Virtually eliminate vitamin A deficiency;
9. Virtually eradicate guinea worm disease;
10. Ratify the Convention of the Rights of the Child in every country.

---

4. Resources available for health care;
5. Organization and management of health care;
6. Standards of health manpower and facilities;
7. Environment and sanitation; and
8. Family planning and birthrate.

The WHO considers that a majority of the targets are being met by many developing nations. Malnutrition has been reduced; immunization levels are being maintained or increased; measles deaths are down 80% compared to preimmunization levels; large areas of the developing world, including all of the Western Hemisphere, have become free of polio; iodine deficiency disorders are being addressed; vitamin A deficiency is in retreat; the use of ORT is rising, preventing more than a million deaths a year; guinea worm disease has been reduced by some 90% and complete eradication is in sight; thousands of major hospitals are now actively supporting breast-feeding; progress in primary education is being resumed; and the Convention of the Rights of the Child has become the most widely and rapidly ratified convention in history.

The principles and targets of Alma-Ata and the World Summit of Children remain the international health agenda for the beginning of the twenty-first century. Each country needs to develop its own national plan of action to achieve these objectives. These plans require

1. Strategic targets;
2. Resources needed and their appropriate allocation;
3. Intersectoral cooperation;
4. Training of staff;
5. Education of the general public and target groups;
6. Community support.

The acceptance of such international targets helps individual countries allocate limited resources to priority areas. The challenge is enormous, but the potential for reducing the burden of disease and raising the quality of life is very real.

Despite progress and optimism, the tragic toll of death and mental or physical disability caused by preventable infectious or vitamin deficiency conditions numbers in the hundreds of millions. Other issues remain to be dealt with in global health, especially regarding women's health, including family planning, reduction of maternal mortality and morbidity, reduction of violence and abuse against women, and improved education and job opportunities. Child abuse, child labor, female child murder, and sexual exploitation remain large scale global health problems. Increasingly, noninfectious conditions affecting young adult and middle-aged males are becoming issues of global health, for example, the cardiovascular diseases and diabetes mellitus, trauma, and cancer. Care of special needs groups in the population, such as the mentally ill, the handicapped, and the elderly (see Chapters 6 and 7), are global health problems that require attention in each country and locality.

International, individual government, and community action is vital to deal with these issues, not only in the poorest countries, but also in developed countries. Known public health measures applied effectively can reduce the burden of these conditions within a very few years. This is a challenge of historical importance and necessity.

EXPANDING NATIONAL HEALTH CAPACITY

The idea of a *cordon sanitaire* to protect a nation's health from invading epidemics is a form of passive defense that has not been effective in major epidemics in the latter part of the twentieth century. A forward defense is now part of the New Public Health. Countries need to reach out to other countries to help improve international public health capacity as their own first line of defense. The tragedy of the late discovery of AIDS and inadequate early response was matched by the equally poor handling of the first phase of the 1991–1996 cholera epidemic in South America.

Building the first line of defense means strengthening the capacity of individual countries to detect, report, and request help in controlling potentially serious disease outbreaks. Help is available from the World Health Organization, the Centers for Disease Control (CDC) in Atlanta and newly strengthened counterparts in France and the United Kingdom, as well as international organizations such as the International Red Cross, Medicins sans Frontieres, and many others. Training in basic epidemiology, sterile techniques, and laboratory services can mean the difference between local containment and widespread infection of hemorrhagic fever viruses, with person-to-person transmission amplified in a hospital setting.

Even the industrialized countries are in need of strengthening of epidemiologic capacity. Few have adequate information systems to collect hospitalization data that can provide vital measures of morbidity and the economics of health services. Few have the training capacity for public health epidemiologists, economists, sociologists, sexologists, psychologists, or anthropologists, let alone entomologists, geneticists, and the many other professions making up the New Public Health team.

Many industrialized countries, satisfied with universal access to doctors and hospitals and a feeling that infectious diseases are going away under the power of sanitation, vaccination, and antibiotics, allowed their public health infrastructure to decline with poor pay, reward, recognition, and motivation, and lack of training capacity. The 1990s have brought a different reality of emerging and reemerging infectious diseases and other plagues such as violence and trauma, drugs, heart disease, cancer and stroke. Failure to prepare public health professionals and support systems is an invitation to disaster, both epidemiologically and economically. No country can afford such laxity. Training of public health professionals requires graduate schools of public health, which are more essential than the excess of medical schools that already exist in most countries.

GLOBAL HEALTH AND THE
NEW PUBLIC HEALTH

The New Public Health is concerned with globalization of health in several senses. First, it includes all health activities in any one country, and, second, what happens in the rest of the world is of direct interest to each country, no matter how wealthy, industrialized, or isolated. The lessons of the bubonic plague may not be convincing, but the lessons of HIV should surely be; John Donne's idea that "no man is an island unto himself" expresses the issue clearly. Global health means identifying and addressing the acute infectious and chronic diseases as early as possible before they spread or amplify by common risk factors.

At the end of the twentieth century, more developing countries are reaching an epidemiologic transition that took place in the industrialized world in the mid century. The resurgence of long known diseases and the emergence of new and sinister infectious disease threats are occuring worldwide. The industrialized countries are again facing serious infectious disease challenges, including those imported from developing countries.

In the 1950s and 1960s, control of infectious diseases looked extremely promising. Vaccines and antibiotics seemed to provide the answer to age old infectious diseases. But in the 1970s and 1980s new infectious organisms appeared, as well as a frightening increase in resistance of microorganisms to therapeutic agents. Diseases spread from country to country, as did HIV in the 1980s, and cholera in Peru and diphtheria in Russia in the 1990s, and the Plague outbreak in India in 1994.

Conversely, the chronic diseases associated with overnutrition and smoking are invading the nonindustrialized countries just when the public health field is gaining momentum in controlling infectious and childhood diseases. And all countries are facing the strains of health expenditures and the painful process of health reform. The legal, ethical, and technological challenges are increasingly important in managing health care systems.

All health systems are obliged to face these challenges through the sharing of information, and improved monitoring of the use of resources, as well as seeking effective ways of preventing diseases and managing them to promote early and complete return to function. All industrialized countries are facing serious problems financing health care in its traditional form, and reform is taking place amid aging of the population, increasing technology, and high expectations of health care. Reforms shifting emphasis and resources from hospital to ambulatory and primary care show a strong return to the idea of health promotion by regulation and education.

Some answers to unconquered infectious diseases have came from simple technology, such as the use of oral rehydration therapy to reduce morbidity and mortality from diarrheal diseases. The resurgence of TB and multidrug-resistant organisms has been successfully handled by another simple innovation of directly observed therapy by community health workers to ensure compliance and completion of treatment, especially in high risk groups. Malaria control, using specially trained community health workers, is another application of inexpensive and simple, appropriate technology.

Simpler technologies are also having a major impact on the chronic diseases. Cardiovascular mortality rates are falling in most industrialized countries as a result of healthier lifestyles and improved treatments. Trauma deaths are falling as a result of both mandatory improvements in car and road safety, occupational safety, and treatment of substance abuse. For the chronically ill, simpler technology of home care returns people to their home with less lengthy, costly, and risky hospitalizations.

The future of health care will see tremendous change and adoption of new modalities of preventing and managing disease: recombinant vaccines will reduce costs and introduce new vaccines brining more infectious diseases under control, including the viral hepatitides and respiratory and diarrheal diseases. Vaccine technology for cancer and genetic disorders is evolving. Congenital disorders will be controlled by education, screening, and appropriate interventions. Dietary change will help in the control of cancer, as will screening and reduced exposure to carcinogens. Recognition of infectious causes of chronic disease, such as *helicobacter* and peptic ulcer, and synergies between micronutrient deficencies with infectious and duonic disease, such as folic acid deficiency and birth defects, open many new vistas for research and applied public health.

Health for All means access to care for everyone. This requires sound management of finances and other resources to provide the needed services, efficiently and by reducing the waste and extravagances of unnecessary servicing. It also requires a social and physical environment that enables people to experience healthful, satisfying, and productive lives. To attain these lofty goals, broad partnerships or coalitions of health services and providers working with communities and an increasingly knowledgeable and participating general public must be achieved. This is especially important for compliance with immunization, healthful infant and child nutrition and care, self-care in pregnancy, and healthful adult nutrition. Paternalistic, traditional services of doctors dominating both the health systems and patients are not able to raise the level of patient and community participation needed.

The goal of better health requires a sharing of tasks and resources between the clinical and community levels, and between countries. Assisting countries in developing the staff and infrastructure of epidemiology in infectious and chronic disease is an investment in the frontline of public health protection and self-defense. This is the substance of work by international organizations and bilateral aid. In international partnerships in Europe, the industrialized countries help each other: this needs to be applied to help promote public health infrastructure in developing countries as well.

SUMMARY

Health for all sounded like a hopeless idealistic wish when first promulgated by the WHO in 1977. Yet the progress since in lowering mortality and birthrates, raising longevity rates, and improving quality of life has been dramatic. Globalization of health means that what happens anywhere is the concern of everyone everywhere, as the world learned with plague in the fourteenth century and AIDS in the late twentieth century. At the same time, globalization means all aspects of health for a population, because of the interaction of health care, economics and the political priority given to health.

Global action means that countries must be committed to health, at all levels, including state and local governments as well as voluntary, educational, and many other elements of a society. The potential gain is enormous, and this requires systematic organization, information, with well-defined targets, strategy, and tactics. Measuring disease, both infectious and chronic disease, family health, special groups in the population, nutrition, environmental and occupational health, organization of pubic health, management of health systems, comparing with other health national systems, human resources, and technology assessment, quality assurance, law, and ethics, all chapter topics in this book, are the substance of the New Public Health. All together, they are the subjects of day-to-day life in health systems.

Eradication of smallpox will be followed by other successes with great health and economic benefits. The skills of infectious disease control will also bring change in chronic disease control. New acute and chronic disease challenges will emerge; preparation will increase the chances of coping with them before they reach epidemic proportions. Recognizing the rising tide of cardiovascular disease mortality in the former Soviet Union and many developing countries is no less important than recognizing the resurgence of TB or multidrug-resistant infectious diseases.

The conceptual basis of the New Public Health provides an idealized yet practicable model for developing countries. This idea, however, has since grown with many influences, including health promotion and health targets. It has come to involve the actual management of health systems and integration of secondary and tertiary care services of hospitals, and the whole range of programs or services that relate to improving the health of the individual and the society.

The international health community has succeeded, in part, in changing the health agenda of many countries toward prevention, primary care, and health promotion. The development of goals and targets with international sanction helps each country to resist pressure to place most of its health care resources into curative/tertiary services. An international commitment to Health for All has taken on an important meaning in member countries. It has helped national and regional health leadership tackle the difficult task of changing priorities to an emphasis on primary health care and modern public health.

Coalitions of forces are needed to take up the challenges that the heath community cannot do alone. Isolation of health from other sectors, or of parts of the health spectrum from each other, lowers the capacity of all to reach common goals. Networks of international agencies, including the WHO, UNICEF, the World Bank, the Food and Agriculture Organization, UNDP, private donor organizations, the private sector, and many others, are needed to face the health challenges and tasks. Similarly, at the national, state, and local levels, globalized approaches and networks of organizations can help to define targets and mobilize the resources needed to achieve them.

It is appropriate to end this book with the International Declaration of Health Rights, adopted by schools of public health in the United States as a personal committment for public health professionals graduating in 1998:

> We as professionals, do hereby commit ourselves to advocacy and action to promote health rights of all human beings.
> The enjoyment of the highest attainable standard of health is one of the fundamental rights of every human being. It is not a privilege reserved for those with power, money or social standing.
> Health is more than the absence of disease, but includes prevention of illness, development of individual potential, a positive sense of physical, mental and social well-being.
> Health care should be based on dialogue and collaboration among citizens, professionals, communities and policy makers. Health services should emphasize equity, accessibility, community participation, prevention and sustainability.
> Health begins with healthy development of the child and a positive family environment. Health must be sustained by the active role of men and women in health and development. The role of women, and their rights, must be recognized, respected and promoted.
> Health care for the elderly should preserve dignity, respect and concern for quality of life and not merely extend life.
> Health requires a sustainable environment with balanced human population growth and preservation of cultural diversity.
> Health depends on more than access to health care. It depends on heathy living conditions and the availability to all people of basic essentials: food, water, housing, education, productive employment, protection form pollution and prevention of social alienation.
> Health depends on protection from exploitation and discrimination on account of race, religion, political belief, ethnic group, national origin, gender, sexual preference or economic or social status.
> Health requires peaceful and equitable development and collaboration of all people. (UCLA School of Public Health, 1998)

The New Public Health is a conceptual framework and methodology for implementation of these lofty goals. It requires the managing of health systems as well as health promotion and disease prevention so that changes in priorities can be implemented by suitable shifts in resources to meet the health needs of individuals and communities.

ELECTRONIC MEDIA

Food and Agriculture Organization, http://www.fao.org/

Global Health for All Indicators database, available via the WHO home page (http://www.who.int), may be downloaded to IBM compatible computers.

International infectious disease monitoring network, http://www.promedmail.org

Morbidity and Mortality Weekly Report morbidity and mortality tables are available on the World Wide Web, http://www2.cdc.gov/mmwr/distrnds.html

Organization for Economic Cooperation and Development, http://www.oecd.org/

United Nations High Commission for Refugees, http://www.unhcr.ch/

United Nations Children's Fund (UNICEF), http://www.unicef.org/

Weekly Epidemiologic Record (WHO), http://www.who.int/wer/

World Bank, http://www.worldbank.org/

World Health Organization mail and email addresses for headquarters and regional offices are as follows:

WHO Headquarters, Avenue Appia 20, 1211, Geneva, Switzerland (postmaster@who.int)

WHO Regional Office for Africa (AFRO), Parirenyatwa Hospital, PO Box BE 773, Harare, Zimbabwe (regafro@whoafr.org)

WHO Regional Office for the Americas (PAHO), 525 23rd St., N.W., Washington, D.C. 20037, USA (postmaster@paho.org)

WHO Regional Office for the Eastern Mediterranean (EMRO), PO Box 1517, Alexandria 21511, Egypt (postmaster@who.sci.eg)

WHO Regional Office for Europe (EURO), 8 Scherfigsvej, DK 2100, Copenhagen, Denmark (postmaster@who.dk)

WHO Regional Office for South-East Asia (SEARO), World Health House, Indaprastha Estate, Mahatma Ghandi Road, New Delhi, India 110002 (postmaster@whosea.org)

WHO Regional Office for the Western Pacific (WPRO), PO Box 2932, 1000 Manila, Philippines (postmaster@who.org.ph)

RECOMMENDED READINGS

Berlinguer, G. 1992. The interchange of disease and health between the old and new worlds. *American Journal of Public Health,* 82:1407–1413.

Doll, R. 1992. Health and the environment in the 1990s. *American Journal of Public Health,* 82:933–941.

Center for Disease Control, Addressing Emerging Infectious Disease Threats. Atlanta, Georgia: U.S. Public Health Service, 1994.

Eisenstein, B. I. 1990. New opportunistic infections—More opportunities. *The New England Journal of Medicine,* 323:1625–1627.

Krause, R. M. 1992. The origin of plagues: Old and new. *Science,* 257:1073–1078.

Morse, S. S. (editorial). 1992. Global microbial traffic and the interchange of disease. *American Journal of Public Health,* 82:1326–1327.

Sommer, A. 1997. Vitamin A deficiency, child health and survival. *Nutrition,* 13:484–485.

UNICEF. 1993 through 1998. *The State of the World's Children.* New York: Oxford University Press.

Yach, D., Bettcher, D. 1998. The globalization of public health I and II: Threats and opportunities; the convergence of self interest and altruism. *American Journal of Public Health,* 88:735–741.

Winkelstein, W. (editorial). 1992. Determinants of worldwide health. *American Journal of Public Health,* 82:931–932.

BIBLIOGRAPHY

Basch, P. F. 1990. *Textbook of International Health.* New York: Oxford University Press.

Center for Disease Control, Addressing Emerging Infectious Disease Threats. Atlanta, Georgia: U.S. Public Health Service, 1994.

Garrett, L. 1994. *The Coming Plague.* New York: Penguin.

Howson CP, Kennedy ET, Horowitz A [eds]. 1998. Prevention of Micronutrient Deficiences: Tools for Policymakers and Public Health Workers. Washington DC: Institute of Medicine, National Accademy Press.

Institute of Medicine. 1992. *Emerging Infections, Microbial Threats to Health in the United States.* Washington, DC: National Academy Press.

Porter, D. (ed). 1994. *The History of Public Health and the Modern State.* Amsterdam: Rodopi V.

Wilson, M. E., Levens, R., Spielman, A. 1994. *Disease Evolution.* New York: Academy of Sciences.

World Health Organization. 1996. *World Health Report 1996: Fighting Disease, Fostering Development.* Geneva: World Health Organization.

World Health Organization. 1997. *World Health Report 1997: Conquering Suffering, Enriching Humanity.* Geneva: World Health Organization.

World Health Organization. 1998. *World Health Report 1998: Life in the 21st Century, A vision for All.* Geneva: World Health Organization.

World Health Organization. 1999. *World Health Report 1999: Making a Difference.* Geneva: World Health Organization.

World Bank. 1993. World Development Report 1993: Investing in Health. New York: Oxford University Press.

PUBLICATIONS AND JOURNALS

Bulletin of the World Health Organization: World Health Organization, 1211 Geneva 27, Switzerland.

Bulletin of the Pan American Health Organization: Pan American Health Organization Division of Vaccines & Immunization 525 Twenty-third Street, N.W. Washington, D.C. 20037 USA

Emerging Infectious Diseases, Centers for Disease Control and Prevention. 1600 Clifton Rd Ne #A23, Atlanta, GA, USA.

Health for All Series: World Health Organization, 1211 Geneva 27, Switzerland, 1978–1982.

International Agency for Research on Cancer: IARC Press, International Agency for Research on Cancer, 150 cours Albert Thomas, F-69372 Lyons cedex 08, France.

International Digest of Health Legislation: World Health Organization, Avenue Appia 20, 1211 Geneva 27, Switzerland.

Population Reports: Population Information Program, The Johns Hopkins School of Public Health, 111 Market Place, Suite 310, Baltimore, MD 21202-4012, USA.

Technical Report Series: World Health Organization, Avenue Appia 20, 1211 Geneva 27, Switzerland.

Weekly Epidemiologic Record: World Health Organization, Avenue Appia 20, 1211 Geneva 27, Switzerland.

World Health Forum: World Health Organization, Avenue Appia 20, 1211 Geneva 27, Switzerland.

World Health Statistics Quarterly: World Health Organization, Avenue Appia 20, 1211 Geneva 27, Switzerland.

GLOSSARY

A glossary is provided in order that the reader may have an easy reference for understanding basic definitions in public health in all its major aspects. We have attempted to cover the primary concepts in clear and internationally accepted terminology.

ABBREVIATIONS

AIDS	Acquired immune deficiency syndrome
APHA	American Public Health Association
ARI	Acute respiratory infection
BCG	Bacillus Calmette-Guérin
BOD	Burden of disease
CBA	Cost–benefit analysis
CBR	Crude birthrate
CDC	Centers for Disease Control and Prevention
CDD	Control of Diarrhoeal Diseases—WHO
CHD	Coronary heart disease
CHW	Community health worker
CINDI	Countrywide integrated noncommunicable disease intervention
DOPC	Community-oriented primary health care
CVD	Cardiovascular disease
DALY	Disability-adjusted life year
DHHS	Department of Health and Human Services
DHS	District health system
DOT/DOTS	Directly observed therapy (short term)
DPT	Diphtheria, pertussis, tetanus vaccine

DRG	Diagnosis related group
EPA	Environmental Protection Agency
EPI	Expanded program of immunization, includes DPT, BCG, measles, polio
EPIplus	EPI, hepatitis B, yellow fever vaccines, plus vitamin A and iodine supplementation
FAO	Food and Agriculture Organization
FDA	Food and Drug Administration
GDP	Gross domestic product
GNP	Gross national product
GOBI	Growth monitoring, oral rehydration, breast-feeding, immunization
GOBI/FFF	GOBI plus family planning, food production, female education
HCFA	Health Care Financing Administration
HFA	Health for All
HIS	Health information system
HIV	Human immunodeficiency virus
Hib	*Haemophilus influenzae* B
HMD	Health Manpower Development—WHO
HMO	Health maintenance organization
IBRD	International Bank for Reconstruction and Development—World Bank
ICD-10	International classification of disease, 10th edition
IDA	Iron deficiency anemia
IDD	Iodine deficiency diseases
IMR	Infant mortality rate
ILO	International Labour Organization
KABP	Knowledge, attitudes, beliefs, and practices
LDC	Less developed country
LLDC	Least developed country
MAP	Malaria Action Program—WHO
MCH	Maternal and child health
MMR	Maternal mortality rate
MMR	Measles, mumps, rubella
MMWR	*Morbidity and Mortality Weekly Review Report*
MOH	Ministry of Health, Medical Officer of Health
MONICA	Multinational monitoring of trends and determinants in cardiovascular disease—WHO
NGO	Nongovernmental organization
NHANES	National Health and Nutrition Examination Surveys
NHS	National Health Service, United Kingdom
NID	National immunization days
NIH	National Institutes of Health
OECD	Organization for Economic Cooperation and Development
ORT	Oral rehydration therapy
PAHO	Pan American Health Organization
PHC	Primary health care
QALY	Quality-adjusted life year
STD	Sexually transmitted disease
TBA	Traditional birth attendant
UNICEF	United Nations Children's Fund

UNDP	United Nations Development Programme
UNFPA	United Nations Population Fund
UNHCR	United Nations High Commission for Refugees
USAID	United States Agency for International Development
VHW	Village (community) health worker
WHO	World Health Organization
YPLL	Years of potential life lost

GLOSSARY OF TERMS

Abatement, A process to reduce, remove, or discontinue a nuisance disease vector, or pollutant.

Absolute Poverty Level, The income level below which a minimum nutritionally adequate diet plus essential nonfood requirements are not affordable.

Accessibility, Potential to reach needed and appropriate health services by members of a population by local means of transport in no more than 1 hour; it is affected by distribution of services, but also by convenience, cost, insurance coverage, and transportation as well as by cultural and social factors.

Accountability, The degree of responsibility for activities carried out in the health care sector, for which the care provider, health care organization or public health system can be held administratively, professionally, morally, and legally responsible.

Accreditation, A systematic, periodic, voluntary, external evaluation of a health care facility and its organization carried out by an outside agency, with a formal assessment of deficiencies according to written standards and as judged in site visits by a professional team.

Accuracy, The degree to which a measurement or estimate based on a measurement represents the true value of that which is being measured, and can be confirmed by repeat testing.

Activities of Daily Living (ADL), Ability/disability of an individual to perform daily tasks of self-care, such as dressing, toileting, feeding, home making, without assistance; it is used to measure outcomes of interventions for various stages of aging, a disease process, or a chronic and disabling condition.

Acute Illness, A brief, intense, or short-term episode of illness, or exposure to a pathogen; sometimes used to indicate severity.

Adaptation, The process by which an organism or person makes a successful change to accommodate new circumstances.

Addiction, A physiological and/or psychological dependance on a substance or activity.

Admission, Hospital, Entry to a hospital inpatient facility for care for a 24-hour period or longer.

Adult Literacy Rate, The percentage of persons aged 15 and over who can read and write.

Adverse Event, Any disease or injury resulting from medical management that leads to premature death or unnecessary morbidity.

Adverse Reaction, An undesirable or unwanted side effect as a consequence of a preventive, diagnostic, or therapeutic procedure. Where these are reportable events, an adverse drug reaction registry provides a mechanism to detect medically or legally important side effects.

Age-Adjusted Rate, A rate of occurrence of a condition in a population, adjusted to take into account the age distribution of a specific population.

Agent, Infective/Toxic, A microorganism, chemical substance, or form of radiation that is essential but not necessarily sufficient to cause a disease or disorder. A disease may have a single agent as a cause or may occur as a result of the agent in company with other contributory factors.

Aging of the Population, A demographic term meaning an increase in the number and percentage of elderly persons in the population, associated with declining birth and death rates and increasing life expectancy.

Alcoholism, Alcohol use in terms of drinking consistently in amounts that are excessive in terms of community standards, associated with dependence, habituation, and frequent intoxication.

Allocation of Resources, Distribution of resources, including money, manpower, and supplies, to different categories of health service, either nationally, regionally, locally, or within an organization providing health care.

Alma-Ata, A city in Kazakhstan where the Joint WHO/UNICEF International Conference adopted a Declaration on Primary Health Care in 1978, providing the basis for the WHO program of Health for All by the Year 2000.

Ambient, Surrounding conditions of the environment, usually referring to air quality and pollutant levels.

Ambulatory Care, Medical and paramedical care services provided to a patient attending a care setting on a visiting basis, usually in outpatient or community or specialty clinics.

Ambulatory Surgery, Surgical procedures, with or without general anesthesia, provided to a patient who returns to his/her own home following the surgery without an overnight hospital stay.

Analytic Study, A study designed to examine associations, hypothesized to be causal relationships, to identify or measure risk or health effects of specific exposures.

Anemia, Hemoglobin levels below accepted normal values, depending on age and sex, due to loss of blood, rapid breakdown of red cells (hemolysis), or a dietary deficiency of iron or other essential elements of red cells.

Anopheles Mosquito, The mosquito responsible for transmission of malaria parasites from one human host to another.

Anorexia Nervosa, An eating disorder in which the patient suffers from an obsession to lose weight leading to deliberate self-starvation, sometimes resulting in death.

Antenatal (Prenatal) Care, Care of the pregnant woman throughout pregnancy, ideally beginning before pregnancy and certainly during the first trimester up through delivery, with follow-up afterward in postnatal care.

Antibody, A protein molecule produced in the body in response to a foreign substance (i.e., an antigen) or acquired by passive transfer. Antibodies bind to a specific antigen that elicits its production, causing its destruction, thereby providing an important mechanism for protection against infectious disease.

Antigen, A substance (protein, polysaccharide, glycolipic, tissue transplant, etc.) that is capable of inducing a specific immune response in a host, e.g., by invasion of infectious organisms, immunization, or ingestion. An immune reaction occurring to self antigens may cause an autoimmune disease.

Antigenicity, the ability of an external agent to produce a systemic or local immunologic reaction in a host.

Antigenic Drift, Evolutionary genetic changes which can occur in viruses, that are responsible for antigenic changes of, e.g., influenza, necessitating annual reformulation of the influenza vaccine.

Antigenic Shift, A sudden mutation in the molecular structure of an organism, creating completely new subtypes of the organism and producing epidemics since hosts previously exposed to other strains may have little or no immunity to the new strains.

Apgar Score, A standardized numerical method of assessing and documenting the score of vital signs and activity level of a newborn performed 1 and 5 minutes after birth, as a measure of the physiological status of the newborn (e.g., heart rate, skin color, response to stimuli).

Appropriate Technology, An intervention, medication, device, or service meant to reduce morbidity and mortality, to improve health status, and to promote efficient use of resources that is in keeping with the financial ability of a country.

Appropriation, Placement of funds in a budget category previously authorized as a step in a legislative process intended to lead to action in an administrative branch of government.

Arbovirus, A mixed group of viruses sharing a common epidemiologic mode of transmission between vertebrate hosts by arthropod vectors, such as mosquitoes, ticks, or sandflies.

Atherosclerosis, Accumulation in the inner lining of arteries of fatty deposits causing a narrowing of the lumen along with reduced elasticity, resulting in a decreasing flow of blood in the vessel.

Attributable Risk, The rate of a disease, or other outcome, that occurs in persons ex-posed to a particular risk factor that is attributed to the exposure, i.e., beyond the normal level of the disease in a similar population group not exposed.

Average Length of Stay (ALS), The total number of patient days accumulated at the time of discharge, counting the date of admission but not the date on which patients were discharged during a reporting period, divided by the number of patients released from an inpatient care facility.

Bacillus Calmette-Guérin (BCG), A weakened or attenuated strain of tubercle bacillus used widely for vaccination to prevent tuberculosis, especially in children in high incidence areas.

Bacteriology, The branch of science concerned with the study of bacterial microorganisms as part of the science of microbiology, which includes bacteria, viruses, parasites, and other organisms.

Bacteriostasis, Slowing or cessation of multiplication of bacteria, enabling the host's defenses to destroy the invading organism.

Basket of Services, A defined set of public health and clinical interventions included under a national or other health insurance or service program, usually including preventive care, medical and diagnostic services, hospital and home care, possibly prescribed medications, but often excluding mental health services and dental care.

Bed Occupancy Rate, The percentage of hospital beds occupied over a specified period, such as a week, month, or year, divided by the total of beds at the national, regional, or local, hospital or departmental level; it helps management in tracking utilization of facilities.

Behavioral Risk Factor, A characteristic, behavior, addiction, or fixation associated with increased probability of a specific disease or outcome; it does not necessarily imply a causal relationship.

Benefit-to-Cost Ratio, Total benefits of an intervention divided by total costs of the condition and its complications, when both values are discounted, i.e., at their worth today, or at "present value."

Bias, A tendency that produces results which deviate systematically from the true value.

Bimodal Distribution, A distribution of two regions of high frequency separated by a region of low frequency of observations; a twin peak distribution curve.

Bioassay, Assessment of chemical substances in human, animal, or plant tissue.

Biological Plausibility, An association thought to be causal is strengthened when explainable in terms of current biological or medical knowledge.

Birth Certificate, A state-mandated legal document recording details of a live birth.

Birth Interval, The time between completion of one pregnancy and initiation of the next one.

Birth Weight, The weight of an infant at the time of birth as recorded on the birth certificate. Those below 2500 grams (5 lbs, 8 oz) are low birth weight; very low birth weight includes those below 1500 grams (3 lbs, 5 oz), and ultralow birth weight is a birth weighing below 1000 grams (2 lbs, 3 oz).

Births Attended, %, Percentage of births attended by physicians, nurses, midwives, trained primary health care workers, or trained traditional birth attendants (TBAs).

Blind Study, A comparative study in which the observer and/or the subjects are unaware of in which group a subject is situated. When both observer and subject are unaware, the study is called "double-blind," and if the statistical analysts are also unaware, it may be called "triple-blind."

Block Budget, A method of budgeting a health facility, such as a hospital, which is not designated line-by-line in detail, nor by department or other internal breakdown, but is a global sum negotiated with the funding agency, allowing the management freedom to change internal allocations, staff realignment or reductions.

Block Grant, A grant from government or other agency to operate a health program not based on units of services or specific category of service, but directed toward more general program content.

Body Mass Index (BMI), An anthropometric measure calculated as follows: BMI = weight, kg/(height squared, m^2), or $704.5 \times$ weight, lbs/(height, in^2).

Burden of Disease (BOD), Quantification of the impact of a disease on a population, as measured in terms of numbers of deaths; the impact of premature death and disability on a population, is combined into single units of measurement (See DALYs and QALYs).

Capital Investment/Costs, Expenditures on land, buildings, and equipment with a life or more than 1 year.

Capitation, A method of paying for designated health services whereby funds are allocated per person registered as a service beneficiary for a specified period of time. Weighted capitation adjusts for factors such as age and sex specific mortality rates, representing differences in health care needs of the population.

Carcinogen, A cancer-causing substance.

Carcinogenesis, The process of initiation of changes in a cell that becomes malignant and leads to unrestrained cellular growth and can cause serious clinical effects in the spread of malignant cells and tumors.

Cardiovascular Disease, One or all of the diseases of the heart and circulatory system, which include coronary heart disease, cerebrovascular accidents, rheumatic heart disease, congenital heart disease, and some others.

Carrier, A person or animal that harbors a specific infectious agent without discernible clinical disease, who serves as a source of infection or contamination of food, water, or other materials. A carrier may have an inapparent (subclinical) infection, or may be in the incubation, clinical, or convalescent stage of the infection, but is still infectious to others, directly or indirectly.

Case–Control Study, An observation or study of persons with a particular disease or other outcome variable in comparison to a reference or control group, which is otherwise similar to the study group. The results are often given as odds ratio.

Case Fatality Rate (CFR), The proportion of cases of a specified condition which are fatal within a specified time period. CFR = number of deaths from a disease/ number of cases of the disease × 100.

Case-for-Action, The weight of evidence accumulated from all possible sources of clinical, epidemiological, societal, and economic aspects of a health or illness condition which justifies an intervention program.

Case Management, Organized care to help a client/patient receive the services he or she needs, setting priorities for these services, and helping to integrate services with monitoring of progress.

Categorical Service, A service program funded and developed specifically to address one category of health problem, service, or risk group.

Cause of Death, Causes of death include those diseases, conditions, or injuries that resulted in or contributed to death,

and the accident or violence that produced the injuries. The *underlying cause of death* is recorded on the death certificate as the disease or injury that initiated the train of events leading to death, or the circumstances of the accident or violence which produced the fatal injury. These are coded according to the International Classification of Disease (currently ICD-10).

Cause-Specific Rate, A rate that specifies a health event, such as death or illness, according to their cause by disease classification.

Census, A national enumeration of a population, recording all persons, every place of residence, with age or birth date, sex, occupation, national origin, marital status, income, relationship to head of household, and additional information, such as educational levels and health-related data, e.g, a permanent disability.

Cerebrovascular Disease, Atherosclerosis and other diseases of the cerebral vascular system that can result in cerebrovascular accidents or strokes.

Certificate of Need (CON), A formal procedure in which a health facility wishing to invest in capital equipment, renovation, or expansion must present a detailed plan and justification for the plan to the state or national regulatory authority for consideration, modification, and approval, even when the source of funding is nongovernmental.

Certification, The issuance of a permit to practice or function as a health care provider or facility with specified rights in providing health services. The state may issue a permit to operate, retaining the right to inspect and cancel the license or require changes in operation of facilities.

Child Mortality Rate, The probability of dying between birth and age 5, expressed per 1000 live births. The term *under-five mortality rate* is also used.

Chronic Disease, An adverse health condition lasting along time, i.e., 3 months or longer.

CINDI—Countrywide Integrated Noncommunicable Disease Intervention, A WHO program established to improve community health by prevention and control of risk factors for chronic disease and injury.

Circulating Antibodies, Antibodies measurable in the blood, in contrast to antibodies within the cells, such as polio antibodies in the cells of the intestinal tract.

Clinical Guidelines, Evidence-based practice standards based on meta-analysis by multidisciplinary professional bodies with comprehensive reviews and rigorous analysis of relevant scientific information and technical reviews. They provide the provider and manager with professionally acceptable criteria for preventive and curative management of conditions of clinical, epidemiological, and economic importance.

Cluster Sample, Sampling by selecting groups rather than individuals, e.g., homes, city blocks, school or clinic attenders.

Coagulation, A method of water treatment by adding material, e.g. alum, to which particles suspended in water are adsorbed and settle to the bottom.

Cohort Studies, Prospective studies over a period of years of large population groups initially free of a disease, but exposed to factors thought to cause that disease.

Cold Chain, The continuous maintenance of vaccines and biological materials at cold temperatures designated by the producers, from the point of manufacture to the point of use, including all stages of storage and transportation.

Coliform Test, A standard bacteriologic examination of drinking water safety, using *Escherichia coli* as an indicator of fecal contamination.

Communicable Disease, An illness due to a specific infectious agent or its toxic products which arises through transmission of that agent from an infected person, animal, or inanimate reservoir to a susceptible host, either directly or indirectly through an intermediate plant, animal host, vector, or the environment.

Community-Based Programs, Programs and services serving a population living in a localized area, or a definable population with a common set of norms, values, and organization. Health services in the community usually imply services or programs offered to people in their homes, or other community settings and through ambulatory care centers.

Community Health Workers (CHWs), Nonprofessional health care providers selected from a designated community and trained to provide general or specific primary care services to ensure access to care for underserved population groups.

Community Medicine, The study of disease and health in the population of a defined community or group in order to identify the population's needs, to determine the means by which these needs should be met, and to evaluate the extent to which health services effectively meet those needs.

Community-Oriented Primary Health Care (COPC), An integration of epidemiologic analysis of community health needs, or community diagnosis, with primary health care of individuals in the community.

Community Participation, The involvement of the community in health related issues, through organizational activities to promote health awareness and health programs.

Compliance, The degree to which an individual patient or target group implements health recommendations specifically designated as needed to treat or prevent illness.

Confidence Interval (CI), The calculated range of numbers in a data set which, with a specified degree of probability (e.g., 95%), includes the true values of variables such as the mean, proportion, or rate of that set of numbers. The upper and lower boundaries of the confidence interval are called the confidence limits.

Confounding Variable (Confounder), A factor or variable that can potentially cause or prevent the outcome or disease being studied (the *dependent variable*), and must be taken into account or it will prevent reaching a conclusion regarding the impact of the *independent variable,* or hypothesized cause of the outcome disease being investigated.

Contact, A person or animal that has been in association with an infected person or animal or contaminated inanimate object or environment that might provide an opportunity to transmit the infective agent. *Direct contact* is a mode of transmission between an infected host and a susceptible host. A *primary contact* is a person who has been in contact with a case of the infectious disease; a *secondary contact* has been in contact with a primary contact, but not an actual case of the disease. Primary and secondary are both types of *indirect contact.*

Containment, A set of coordinated efforts to establish local or regional control to prevent transmission and reintroduction of the organism of an infectious disease during an outbreak or epidemic.

Contamination, The presence of toxins or infective agents in food, water, soil, air, or inanimate objects, which can be transferred to susceptible hosts.

Contraception, Prevention of conception through one the following methods: hormones (the pill, injectables, and subdermal implants), intrauterine devices (IUDs),

barrier devices (condom, diaphragm), chemical (spermicides), and "natural" methods (rhythm and withdrawal).

Contraceptive Prevalence, The percentage of married women in the age group 15–49 currently using contraception.

Copayments, The proportion of a service fee that the consumer pays at the time of use, which is meant to deter wasteful and unnecessary utilization of services.

Cost, Expense in terms of monetary losses (e.g., health care expenditures, loss of wages) and nonmonetary losses (e.g., pain, grief, suffering, and disability) to an individual, health system, or society from a disease or disability.

Cost–Benefit Analysis (CBA), A form of economic evaluation where all the costs and consequences of an intervention to prevent or treat a condition are expressed in money terms in order to assess whether a particular objective is worth achieving, or to evaluate competing or alternative priorities.

Cost–Benefit Ratio (CBR), A CBA in which all costs and benefits are converted into monetary values, and expressed as dollars ($US) of benefit per dollar expended.

Cost Containment, Reducing the rate of increase in overall health care costs to the rate of increase of an economic indicator of growth, such as GNP per capita.

Cost–Effectiveness Analysis (CEA), A form of economic evaluation where all costs are expressed in monetary terms, but the consequences are expressed in nonmonetary terms, such as life years gained, cases detected, or cases prevented.

Cost Utility, An economic analysis assessed as a quality adjusted outcome per net cost expended.

Coverage, A measure of the extent to which services rendered cover the potential need for these services in a community.

It is expressed as a proportion in which the numerator is the number of services rendered, and the denominator is the number of instances in which the service should have been provided according to the accepted standard.

Cross-sectional Study, A study of the relationships between diseases (or other health-related characteristics) with other variables of interest in a defined population at a particular point in time. This establishes the disease prevalence and the presence of the characteristic being studied (e.g., smoking or toxic exposure) in the diseased compared to the nondiseased person in the group.

Crude Birthrate (CBR), The annual number of live births per 1000 population in a given year.

Crude Death Rate, The annual number of deaths from all causes per 1000 population.

DALY—Disability-Adjusted Life Year, A unit used for measurement of both the global burden of disease and the effectiveness of health interventions. It is calculated as the present value of the future years of disability-free life that are lost as a result of premature mortality or disability occurring in a particular year.

Days of Care, The total number of patient days accumulated by patients at the time of discharge from short-stay hospitals during a reporting period, including day of admission, but not the day of discharge; usually expressed as rates, i.e., days of care per 1000 population, or as age or diagnosis specific rates.

Death Certificate, A death certificate is required by law for each death, signed by a health provider designated and licensed for this purpose. It includes identification of the patient, cause of death, age, sex, birth date, date of death, and place of residence. The immediate cause of death is recorded on the first line, followed by conditions giving rise to the immediate cause, and the underlying cause is recorded last (see Cause of Death).

Decentralization of Health Services, Transfer of operational authority, budget and accountability for health services, including management of budget and personnel, from a central authority at the national or state level to a local health authority, or to independent not-for-profit corporations or trusts, or community or voluntary organizations.

Delayed Mortality, This refers to death occurring after the acute phase of a disease, as in increased mortality in children following an outbreak of measles.

Demand for Health Services, Desire, willingness and/or ability to seek, use, and pay for health services. *Expressed demand* is actual use, while *potential demand* is an expression of need.

Demographic Transition, Transition in the age distribution of a population with an increase in the percentage of the elderly (i.e., over age 65) and a reduction in the percentage of children as a result of changes in mortality and fertility patterns; sometimes referred to as "aging of the population."

Demography, The study of populations, especially with reference to size and density, fertility, mortality, growth, age distribution, migration, vital statistics, and the interaction of all these with social and economic conditions.

Dependency Ratio, The proportion of children (<15 years) and elderly in a population (65 years and over) in comparison to all others, i.e., the proportion of economically inactive compared to the economically active.

Dependent Variable, An outcome or manifestation we seek to account for by the influence of an independent variable(s) or intermediate factor(s) in a hypothesized relationship being studied.

Deregulation, An administrative strategy to reduce governmental procedures for supervision of an economic or service sector, while retaining overall responsibility for the standards of the outcome.

Developed and Developing Countries, Countries are defined as to their level of development by GNP per capita by the World Bank. Those with 1995 GNP per capita of $765 or less in 1998 are defined as *low-income economies* or least developed countries. Those between $765–3,035 and $3,035–9,386 are considered lower and uppper income developing countries respectively. Countries with $9,386 or more are defined as developed (*high-income economies*).

Development Indicators, Measures which reflect the health, educational, and social progress made in a country, for comparison with other countries.

Diagnosis Related Group (DRG), A defined disease entity, condition, or procedure that serves as a module of payment for a hospitalized patient or for an outpatient service.

Diarrheal Disease, Any infectious disease that produces loose or watery stools, caused by viruses, bacteria, or parasites.

Direct Costs, Costs directly associated with prevention and treatment activities of the health care system for a condition, as opposed to indirect costs, such as loss of productivity or wages from work during illness.

Disability, Any temporary or long-term reduction of a person's activity as a result of an acute or chronic condition, measured as disability days or the number of days that a person's regular level of activity has been reduced in activities of daily living or bed-disability, work-loss, and school-loss days.

Discharge, Hospital, The completion of any continuous period of stay or one night or more in hospital as an inpatient, excluding the well newborn.

Discount Rate, The annual rate at which the value of a future cost or consequence is reduced to find its present monetary value.

Discounting, A method of adjusting for the value of future costs and benefits, expressed as present value; it is based on the current value of money to be spent in the future, since a dollar today is worth more than it will be a year from now, because of factors such as financing costs.

Disease, Disease, illness, and sickness are sometimes used interchangeably. *Disease* is a physiological or psychological dysfunction. *Illness* is a subjective state of a person who feels aware of not being well. *Sickness* is a state of social dysfunction, or the role a person assumes when ill. Disease has a natural history from preclinical to clinical and recovery or resolution stages, including the effects of preventive or treatment interventions.

Disinfection/Disinfestation, Any physical or chemical process serving to destroy or remove dangerous or undesired small animal forms, particularly rodents, insects or other arthropods, or microorganisms in order to reduce the chance of disease or transmission.

Distributional Effects, The manner in which costs and benefits of a preventive strategy affect different groups of people, in terms of demographics, geographic location, and other descriptive factors.

District Health Authority (DHA), A term describing a health service operated by a local authority that involves administrative responsibility for a number of services and overall responsibility for the health of the population of the district, with accountability for these to a higher level of government.

DPT Vaccine, Diphtheria, pertussis (whooping cough), and tetanus vaccine.

Ecology, The study of living organisms, including humans, in their physical, social, economic, and behavioral environment as an interactive whole.

Effectiveness, The improvement in health outcome that a prevention strategy can produce in typical community-based settings.

Efficacy, The extent of benefit or improvement in a health-outcome effect produced by a specific intervention under optimal conditions, ideally as shown in randomized controlled trials.

Efficiency, Relates the benefit or output to the unit cost of resources employed in an intervention.

Elimination, Control of the local or domestic presence of a disease, which can still occur due to importation of cases or the causative organism.

Endemic, The continuous presence of a disease or infectious agent in a given geographic area. Endemicity also refers to the usual level of disease in an area.

Enriched Foods, Basic foods to which nutritionally important vitamins and minerals are added to enhance, or fortify the nutritional value of the food, or restored to replace those lost in food processing or preparation.

Enteric Disease, Gastroenteric diseases due to infection with specific enteric pathogenic organisms including bacteria, viruses, parasites, and helminths.

Environment, All factors external to the individual that may affect its health behavior or well-being. Includes physical, biological, social, economic, and other factors.

Environmental Protection Agency (EPA), A United States federal government agency with responsibility for standards or all environmental media, including air, water, and land quality.

Epidemic, The occurrence in a community or region of a substantial number of epidemiologically connected cases of an illness in excess of the usual or expected number of cases in a similar time period; "substantial" may vary from one case of poliomyelitis or three cases of measles to larger numbers of other diseases.

Epidemic Curve, A graphic plotting of the distribution of cases by the time of onset to help define epidemiologic characteristics of the episode.

Epidemiologic Intelligence Service (EIS), A training and service program of the U.S. Centers for Disease Control and Prevention in which clinicians are trained to carry out epidemiologic investigations as part of their development as public health professionals.

Epidemiologic Transition, A change in the patterns of predominant diseases in society, usually associated with a decline in the absolute and relative importance of infectious disease with increasing importance of chronic disease and injury. New epidemiologic shifts include newly emerging infectious diseases, antibiotic resistance, and infections causing chronic diseases.

Epidemiology, The study of the distribution and determinants of disease, injury, and associated risk factors and effects in population groups, and identification of methods of preventing or reducing their impact on individuals and society. This includes monitoring of disease, mortality, and health care patterns, as well as active investigations, experiments, and studies to test hypotheses, within ethical limitations and professional standards.

Eradication, The total elimination of a specific disease and its causative organism from the world so that control mechanisms are no longer necessary, e.g., eradication of smallpox.

Essential Drug List, The Model List of Essential Drugs developed by the WHO suggests a list of drugs considered essential for dealing with health problems in developing countries.

Etiology, The origins or causes of disease, health conditions, or risk factors.

Evaluation, A process that attempts to determine as systematically and objectively as possible the relevance, effectiveness, and impact of activities in the light of their objectives. Evaluation is multidimensional, incorporating description, comparison, and analysis of input (i.e., resources), process (e.g., utilization), and outcome (e.g., morbidity, mortality, and functional status) indicators.

Expanded Programme of Immunization (EPI), Health for All immunization program of the WHO and UNICEF, including diphtheria, pertussis, tetanus, poliomyelitis, measles, and tuberculosis vaccines.

Expanded Programme of Immunization Plus (EPIplus), EPI plus immunization against hepatitis B and yellow fever, as well as vitamin A and iodine supplementation.

Exposure, Quantity and duration of contact between a person exposed to a toxic, carcinogenic, microbiological, or other harmful agent; exposure may be continuous, periodic, or episodic.

Externalities, A spillover of benefits or losses from one individual to another, e.g., herd immunity.

Family, A group of persons united by blood, adoptive, marital, or equivalent ties, usually sharing the same dwelling unit. The extended family is multigenerational; the nuclear family, in contrast, is a single generation family, usually husband–wife–children, but is often headed by a single parent, and sometimes by partners of the same sex.

Family Physician, A doctor who provides general medical care to individuals and families in the community, with continuity through long-term familiarity with a family and its medical and functional profile. The family physician's activities include both curative and preventive services.

Family Violence, Physical, sexual, or emotional abuse among family members. This can include spousal, child, and elderly abuse.

Federal, A system of government of a country with elected national, state, or provincial governments, each with legislative and taxing powers as defined in a national constitution and interpreted by supreme court decisions.

Fee-for-Service, Payment of a charge per item of health care received, such as examination, consultation, or diagnostic test.

Feldsher, A mid-level health worker with 2 to 3 years of training who operates independently in rural outposts in Russia.

Fertility Rates, *Total fertility rate* is the average number of children that would be born alive to a woman during her lifetime if she were to go through her cildbearing years conforming to the age-specific fertility rates of a given year. *General fertility rate* is the number of live births per 1000 women aged 15–49 in a year.

Filtration, A stage in the treatment process for community water supplies, following preliminary treatment, for removal of particulate matter in source water by percolation through sand or other porous material prior to further treatment with disinfection.

Fluoridation, Addition of fluoride to drinking water to a level of 1 part per million (ppm), in order to reduce dental decay in a population of children.

Food Balance Sheet, A table presenting an overall picture of the pattern of a country's food supply showing types and quantities of food produced, imported, exported, and used, and per capita supplies available for human consumption, in terms of energy and specific nutrients.

Food Consumption, Food consumption is the actual amount of food consumed as determined in population surveys, in contrast to national food balances which relate to supplies of food.

Formerly Socialist Economies (FSE), The republics of the former Soviet Union and the formerly socialist countries of eastern and central Europe.

Formula, Baby, Substitutes for maternal milk, mostly based on modified cow's milk, recommended for use during the first year of life if breast-feeding is stopped.

Fortification, See Enriched Foods.

Frequency Distribution, A summary of the frequencies of the values or categories of measurement made on a group being observed. The distribution tells what proportion of the group had each value, and the range of values, out of all possible values of the quantitative measure used.

Friendly Societies, Mutual benefit associations, or sick funds, provided health care among other social benefits to members of trade associations, professional groups, unions, or community groups in seventeenth to nineteenth century Britain; they became approved societies for the National Health Insurance Act of 1911.

Fund Holding, Funding of general practitioners (GPs) on a per capita basis, not only for their services but also for hospital and specialist care for their registered patients, as an experimental program in Britain's National Health Service since 1991.

Gatekeeper, The role of the general practitioner or other primary care provider to control or regulate the referral of patients to specialist or hospital services, as in the United Kingdom's NHS or in managed care plans.

General Hospital, A hospital providing short-term diagnostic and treatment services for patients with a variety of medical departments, both surgical and nonsurgical.

General Practitioner (GP), A primary care physician trained to provide continuing care for the family or individual. Sometimes used as synonym for *family physician.*

Germ Theory, The theory that specific microorganisms cause characteristic infectious diseases. This is in contrast to the miasma theory which attributed disease to influences spread in the air as a result of decaying organic matter.

Global Burden of Disease (GBD), An indicator developed by the World Bank and World Health Organization that quantifies healthy life from disease, as measured in disability-adjusted life years (DALYs); used to measure and compare trends over time, and between countries, and as an outcome measure of the effectiveness of specific interventions.

Global Funding, A method of providing money for health services that is based on a block of funds, instead of a specific amount of money being paid for each program or other unit of service, such as clinic visits, hospital admissions, days of care, or procedures.

Good Manufacturing Practice (GMP), Standards of manufacturing of pharmaceuticals, vaccines, biologicals, medical devices, or food products that are mandated by government to promote high levels of product safety.

Gross Domestic Product (GDP), GDP is

the GNP plus an adjustment for the value of productive services performed by foreign nationals minus the value of productive services performed abroad by nationals; it represents the net value of the domestic economy, expressed in U.S. dollars.

Gross National Product (GNP), GNP is the total goods and services produced for sale by the nation plus the estimated value of certain goods and services that are neither bought nor sold. Annual GNP per capita values are expressed in current U.S. dollars.

Gross National Product per Capita (GNP per Capita), The total value of goods and services produced in the country divided by the number of the population, in U.S. dollars.

Growth Rate, Annual rate of increase of the population in a given year (total live births plus net migration minus total deaths), expressed as a percentage of the population in the base year.

Hawthorne Effect, Improved performance of a worker under study due to the fact of being studied.

Health, A complete state of mental, physical, social, and emotional well-being, not merely the absence of disease (WHO, 1946).

Health Education, Provision of information that educates and promotes understanding of body function, health principles, and methods to reduce risk factors and promote healthy lifestyles.

Health Expenditure, National health expenditures include those of governments at all levels, nongovernmental organizations, and private groups or persons. *Recurrent expenditures* or *current operating costs* include items that recur year after year, including staff, food, "hotel costs," equipment, and maintenance of buildings; *capital expenditures* include

land, building, and equipment to establish or extend health facilities.

Health Field Theory (LaLonde Report), The concept that health is a resultant of heredity, environment, and lifestyle, as well as of medical care services.

Health for All by the Year 2000 (HFA 2000), An initiative of the World Health Organization, based on the 1978 Alma-Ata Declaration on Primary Health Care. It places responsibility for planning and implementing progressive improvement in the health of all people, with specific targets, and an emphasis on primary health care.

Health Insurance, Prepayment for health services based on regular payments, whether privately, through place of employment, or through payroll deduction or taxation via governmental mechanisms.

Health Maintenance Organization (HMO), A prepaid health plan that provides comprehensive health care to enrolled members through designated providers. Members pay a fixed monthly payment for health care services. The HMO covers a comprehensive basket of health maintenance and treatment services through its own clinics or affiliated services.

Health Promotion, A concept, set of activities, or process aimed at increasing people's ability to control and improve their health, and to reduce specific diseases and associated risk factors that reduce the health, well-being, and productive capacity of the individual and the society.

Health Targets, An intermediate measurable health status indicator as a defined target that a program seeks to achieve.

Health Cities, Development projects instituted by coalitions of local government and community organizations in metropolitan areas around the world to address priority urban health and environmental recreation and other societal problems.

Healthy Worker Effect, Workers generally have lower death rates than the same age/sex groups in the general population because disabled or chronically ill persons are usually excluded from employment.

Herd Immunity, The immunity of a group or community to invasion or spread of an infection because a high proportion of individual members of the group have actively or passively acquired immunity to that disease.

Hierarchy of Needs, Humans need to satisfy basic survival needs as the first priority in terms of safety, shelter, warmth, food, clothing, and medical care before addressing higher needs of acceptance and self-fulfillment. This has been very influential in promoting attention to worker satisfaction in job performance and in promoting personal behavior contributing to good health.

Home-Based Services, Services that are brought into the home for those who need acute, rehabilitative, palliative, or supportive services, including nursing care, meals, personal care, and housekeeping as an alternative to prolongation of care in a hospital.

Hospital, An establishment which offers inpatient and outpatient accommodation with active medical and nursing care. It provides diagnostic and therapeutic patient services for medical conditions, with an organized physician staff and continual nursing services by registered nurses. Hospitals are classified by duration of stay (short-term or long-term), services provided (general or multiple service department facility, single specialty), ownership (private, voluntary, municipal, religious, or state), and purpose (for-profit or not-for-profit). Also see General Hospital.

Hospital Bed, Any bed that is set up and staffed for care of inpatients, regularly maintained and staffed for the accommodation and full-time care of inpatients at the end of a reporting period.

Hospital Bed Ratio, Number of hospital beds per 1000 population.

Hospital Day, A hospital inpatient day is the number of adult and pediatric days of care rendered during a reporting period, and excludes days of care for newborns.

Hospital Utilization, Use of hospital bed services are measured by admissions per 1000 population and days of care per 1000 population. Average lengths of stay are calculated by total days divided by the number of discharges. An occupancy rate is the average percentage of rated beds being filled by patients over a given period of time.

Host, A person or other living animal, including birds and arthropods, that provides a place for growth and sustenance to an infectious agent under natural, as opposed to experimental, conditions. A *transport host* is a carrier in which the organism remains alive but does not develop.

Human Capital, Increasing the skills and capabilities of a population by investing in education and health to improve both economic growth and quality of life.

Immunity, Individual resistance to infection resulting from normal body defense mechanisms and the presence of antibodies resulting from prior exposure to the pathogen, naturally or as a preventive measure. Resistance to disease in a host may be natural, passive, or acquired.

Immunization, Protection of susceptible individuals from communicable diseases through administration of a living modified agent, a suspension of killed organisms (e.g., pertussis), a non-infective portion of an infective agent (e.g., hepatitis B), or an inactivated toxin (e.g., tetanus toxoid).

Incidence Rate, Incidence is the number of new health events during a prescribed period. The rate is calculated with the ratio of the numerator, i.e., the number of new cases, and the denominator, i.e., the population at risk. Also see Prevalence.

Incubation Period, The interval between initial contract with an infectious, toxic, carcinogenic, or teratogenic agent and the appearance of the first sign or symptom of the disease in question. Infectious organisms have characteristic incubation periods which may range from several hours (e.g., *E. coli*) to some years (e.g., HIV).

Index Case, The initial case in an outbreak or epidemic whose identification is crucial for investigation and control measures.

Indicators, Variables that help to measure changes in health, directly or indirectly, and help to assess the extent to which the objectives and targets of a program have been met. Often used as Performance or Health Status Indicators.

Inequity, An imbalance in resource allocation or access to services between one population group and another, whether the difference is based on location of residence or ethnic or socioeconomic group as opposed to others in the population.

Infant Mortality Rate (IMR), The number of deaths of infants under 1 year of age per 1000 live births. More specifically, this is the probability of dying between birth and exactly 1 year of age.

Infection, Entry and development (of parasites) or multiplication of an infectious agent in the body of persons or animals, which may be inapparent or manifest as an infectious disease.

Inpatient Care, Health services that require a patients to stay overnight in a health care facility.

International Classification of Disease (ICD), A classification and coding system for causes of death and disease, which is revised at approximately 10-year intervals. ICD-10 was approved in 1993. The ICD is important to standardize vital statistics, disease registry, and hospital morbidity data for time trends, regional, national, and international comparisons.

Intersectoral Cooperation, Joint activity by different governmental and nongovernmental organizations to improve health.

Intervention, A specific activity meant to reduce disease risks, prevent or treat illness, or ameliorate the consequences of disease and disability.

Iodine Deficiency Disorders (IDD), Disorders or conditions due to inadequate dietary intake of iodine, which are the most common causes of preventable brain damage and mental retardation.

Knowledge, Attitudes, Beliefs, and Practices (KABP), Levels of awareness of information on a topic in health are important in how people feel about it and the beliefs and behaviors or practices associated with it. Attempts to change health status need to take these factors into account if there is to be a chance to succeed.

Latency Period, The time from first exposure to a cause of a chronic disease and the appearance of that disease (e.g., asbestos exposure and mesothelioma).

Levels of Care, This refers to *primary, secondary,* and *tertiary* health care which are defined, respectively, as ambulatory care in the community, hospital care at the basic level of service, and specialized referral care in a teaching hospital or referral center.

Life Expectancy at Birth, The number of years that newborn children would live if subject to mortality rates prevailing at the time of their birth.

Life Expectancy, Age Specific, The average number of years of life remaining to a person at a particular age if current mortality rates continue to apply.

Life Tables, A summarizing technique used to describe and compare patterns of mortality and survival in populations.

Live Birth, The complete expulsion or extraction from its mother of a product of conception, irrespective of the duration of the pregnancy, which after separation breathes or shows any other evidence of life such as a heartbeat, umbilical cord pulsation, or definite movement of voluntary muscles, whether or not the umbilical cord has been cut or the placenta is attached.

Local Health Authority, The municipal or county government traditionally responsible for sanitation, including safe water, sewage and garbage collection and disposal, town planning, zoning, control of nuisances, animals, pests, inspection of businesses, factories, and food products.

Macronutrients, Food components that the body needs in large quantities, including proteins, carbohydrates, and fats.

Malnutrition, A pathological state resulting from relative or absolute deficiency or excess in the diet of one or more essential nutrients, as clinically manifest conditions or those detectable only by nutritional histories and surveys, or biochemical or physiological tests.

Malpractice, Treatment of patients falls below currently accepted standards, or personal behavior in patient contact exploits patient vulnerability for financial or sexual favors, subject to professional disciplinary and/or criminal proceedings.

Managed Care Organizations, Health insurance plans that provide or contract for a comprehensive range of health services for an enrolled population.

Management by Objective (MBO), A management concept that identifies targets, planning, staffing, training, and budgetary needs to achieve those objectives and intermediate targets.

Maternal and Child Health (MCH), Services related to women and fertility, including pregnancy and childbirth, as well as care of infants, and children through adolescence.

Maternal Mortality Rate, The annual number of deaths of women from pregnancy-related causes during pregnancy and up to 42 days following termination of pregnancy per 100,000 live births.

Maternity Hospital or Home, Separate, free-standing "lying-in hospitals" for delivery, as opposed to maternity units in general hospitals.

Measles, Mumps, Rubella (MMR) Vaccine, A vaccine combining measles, mumps, and rubella given at 12–15 months of age and with a second dose recommended at age 6 years.

Median, A set of measurements is divided equally into those above and those below the central "median" value as a measure of central tendency.

Medicaid, The U.S. national health insurance plan for low income and needy people established in 1965 as Title XIX of the Social Security Act; it is administered and cost-shared by federal and state governments. Health Care Financing Administration (HCFA) administers the program federally, covering some 36 million persons in 1999.

Medical Audit, A health service evaluation procedure in which selected data, including chart review, are analyzed to ascertain medical performance as part of quality assurance, in comparison with other facilities.

Medical Technology, Applied medical science and technical developments and not simply the machines used in medicine. The term is used in reference to lotech, inexpensive interventions (e.g., ORT and medications), as well as hitech or costly complex procedures or equipment (e.g, MRI).

Medicare, The U.S. national health insurance plan for the elderly, disabled, and end-stage renal disease patients, introduced in 1965 as Title XVIII of the Social Security Act of 1935. Medicare is administered directly by the federal government Health Care Financing Administration (HCFA), covering 39 million persons in 1999.

Mental Disorder, A heterogeneous group of disorders ranging from exaggerated response to stressful events to altered mental state from specific neurologic or genetic abnormalities, as listed in the ICD-10.

Mental Health, The emotional and social well-being of the individual and his or her psychological resources and functional status.

Mental Health Organization, An administratively distinct public or private agency, institution, or department whose primary concern is the provision of direct mental health services to the mentally ill or emotionally disturbed.

Mental Retardation (Handicapped), Arrested or incomplete development of the mind.

Meta-analysis, A systematic quantitative method of combining information from multiple studies to derive the most meaningful answer to a specific question. Assessment of different methods or outcome measures can increase power and account for bias and other effects.

Miasma Theory, The concept that epidemic disease transmission is due to "bad air" from decaying organic matter. Although proved to be scientifically unsound, it led to sanitary reforms that resulted in enormous progress in public health. See Germ Theory.

Micronutrient, Nutritional trace elements (vitamins and minerals) needed for optimal growth, development and health.

Micronutrient Deficiency States, Subclinical or clinical deficiency states due to inadequate regular intake of essential minerals and vitamins.

Morbidity, Illness.

Morbidity and Mortality Weekly Report (MMWR), A weekly publication of the Centers for Disease Control and Prevention in Atlanta, in which epidemiologic data are reported, along with special reports on reportable infectious diseases and noncommunicable diseases of epidemiologic interest.

Mortality, Death.

National Health and Nutrition Examination Survey (NHANES), First conducted by the National Center for Health Statistics (NCHS) in 1971, it monitors indicators of health and nutrition status of the U.S. population through questionnaires, dietary intake data, biochemical tests (e.g., lead), physicals (e.g., blood pressure, obesity). Findings from NHANES III, conducted in two phases during 1988–1991 and 1991–1994, are available on CD-ROM and in articles published in peer reviewed journals.

National Health Expenditures, The amount of money spent on all health services, supplies, and health-related research and construction during a calendar year, compiled from many sources. This includes estimates of out-of-pocket payments, private health insurance, and government programs: expenditures are broken down by type of service, e.g., hospital care, nursing home care, physician services, dental care, home care, and drugs.

National Health Insurance (NHI), Mandatory health insurance administered or regulated by the state, with universal coverage, comprehensive services, and public administration.

National Health Service (NHS), A tax-supported national program of comprehensive service free to all at the time of service, including hospital, ambulatory care, and public health, provided as public services, as established in the United Kingdom in 1948, and since replicated in many countries.

Natural History of Disease, The course of a disease from onset to resolution. Many diseases have well-defined stages that characterize the natural history of that disease.

National Plan of Action, When a national health strategy has been defined, a national plan of action should be drawn up, including the steps and programs needed to meet objectives and specified targets to be attained.

Needs, Assessment of a problem by expert opinion (*normative*), by the individual (*felt*), in comparison with other individuals or groups (*comparative*), and acted upon (*expressed*).

Needs Assessment, An organized attempt to gather information in order to assess community and patient health needs, including both demands and needs for health care.

Neonatal Mortality Rate, The number of deaths under 28 days of age per 1000 live births.

Nongovernmental Organization (NGO), An organization in the health field that is operated by a private nonprofit organization.

Normal ("Gaussian") Distribution or "Bell Curve", A continuous, symmetrical pattern of graphed data observations appearing as a bell-shaped curve with both tails extending to infinity, and the arithmetic mean and mode being identical. The shaped of the bell is determined by the mean and standard deviation.

Norms, Use of expected standards or patterns of facilities, manpower, performance, or behavior as a method of health planning.

Nosocomial Infection, An infection occurring in a patient in a hospital or other inpatient care facility, which was contracted during the time in the facility, and which may appear during or after the stay in the hospital. It includes occupationally acquired infections among hospital staff and iatrogenic infections.

Notifiable (Reportable) Disease, A disease in humans or animals that must, by law, be reported by health care providers to state or local public health or veterinary officials; diseases are made reportable because of their contagiousness, severity, frequency, or other public health importance.

Null Hypothesis, A hypothesis that two or more populations do not differ from one another statistically, and that any observed differences are due to chance alone. The hypothesis is then tested statistically to show the degree to which the observed differences may be greater than could be explained by chance factors so that the differences are "statistically significant."

Objectives, The end result a program seeks to achieve.

Occupational Health, The medical specialty and branch of public health concerned with all factors of workplace circumstances and conditions, including exposures to injury, infective, toxic, carcinogenic, and teratogenic substances or hazardous conditions and noise in or resulting from the environment of the workplace, and their impact on health.

Occupational Safety, Prevention of injury and poisonings from exposure at the workplace.

Occupational Safety and Health Administration (OSHA), The U.S. federal government agency charged with responsibility for setting standards and monitoring occupational safety and health in the country.

Operating Costs, The costs of operating an enterprise or service, calculated annually, which vary with the volume of the service; fixed overhead costs are costs of operation which are not dependant on volume.

Operations Research, The study, by observation and experiment, of the workings of a system, such as a health service, with a view to finding ways by which it can be improved.

Oral Rehydration Salts/Therapy (ORS/ T), A combination of salts and sugar used to replace electrolyte loss from diarrhea. The ORT salt solution includes sodium chloride, potassium bicarbonate, and glucose.

Organization for Economic Cooperation and Development (OECD), A organization of many developed or industrialized countries to promote trade and socioeconomic cooperation between its members and internationally. It includes Australia, Austria, Belgium, Canada, the Czech Republic, Denmark, Finland, France, Germany, Greece, Hungary, Iceland, Ireland, Italy, Japan, Luxembourg, the Netherlands, New Zealand, Norway, Poland, Portugal, South Korea, Spain, Sweden, Switzerland, Turkey, the United Kingdom, and the United States.

Outbreak, A localized or small-scale epidemic, with more cases than usual for the area, but not at the level of a general epidemic, depending on the disease and the level of control achieved.

Outcome Measures, Morbidity, mortality, disability, and physiological indicators of health risk factors that are the target of prevention or treatment programs.

Out-of-Pocket Payments, Money spent by the consumer on health care for services not covered under his or her insurance plan or a national health service; or payments, legal or illegal, to health care providers for services rendered to which he or she is entitled.

Outpatient Services, Care and treatment services in a facility for patients who are not admitted overnight.

Output, The immediate result of professional or institutional activity, or productivity, usually expressed as units of service, such as patient days, outpatient visits, laboratory tests performed, or infants immunized.

Overnutrition, A state of malnutrition due to caloric intake in excess of body requirements for energy, growth, and maintenance.

Palliative Care, Symptomatic and supportive care for the dying patient and his/her family.

Pan American Health Organization (PAHO), Regional office of the World Health Organization for the Americas, located in Washington, D.C.

Parasite, An animal or vegetable organism that lives on or in another and derives its nourishment therefrom. Some, called *obligate* parasites, cannot lead an independent life, while others, *facultative* parasites, can also lead an independent nonparasitic existence.

Particulate Matter, Particles of solid or liquid matter in the air, including nontoxic materials (soot, dust, dirt), heavy metals (e.g., lead), and toxic materials (asbestos, suspended sulfates, nitrates).

Pasteurization, The process of partial sterilization of a substance by heating it to a certain temperature for a given period of time required to kill pathogenic organisms, without major chemical alteration of the substance.

Pathogen, An organism, toxin, or other agent capable of causing human, animal, or plant disease.

Pathogenicity, The capacity of an infectious, toxic, carcinogenic, or teratogenic agent to cause disease in an exposed and susceptible host.

Payer, An organization responsible for payment of health care costs specified in a prepaid health plan.

Payroll Tax, A deduction from the individual worker's pay or from the firm's overall payroll established by the national government for a designated program, such as for health services or health insurance.

Peer Review, Mechanisms for quality assurance and promotion involving inspection of facilities, qualifications, records, or performance of health facilities, departments, and individual health providers, performed by professional colleagues as a regular method of operation of the department or unit to improve quality of care and service.

Per Diem, A method of reimbursement for hospital expenditures, for the days of care received by a hospitalized patient, usually based on average daily costs and not reflective of actual costs of treating an individual case.

Perinatal Mortality Rate, The number of late fetal deaths *in utero,* with gestation of 28 weeks or more, plus infant deaths within seven days of birth per 1000 live births.

Physician, A medical graduate from a recognized and accredited medical school, who is licensed by the state to practice medicine and who is subject to professional standards and discipline by a government appointed or designated body

that supervises medical professional licensure.

Physician Contact, A medical visit or consultation in person or by telephone to a physician for the purpose of examination, diagnosis, treatment, or advice.

Physician-to-Population Ratio, The supply of physicians as measured in relation to the population for comparison purposes; often expressed as population per physician ratio, i.e., 400 persons per physician.

Planning, A continuous process of establishing priorities, diagnosing problems, defining objectives and strategies, and monitoring progress or failure to achieve the goals selected, alternative courses of action, and resource allocation.

Planning Program Budget System (PPBS), A budgeting system in which allocation begins with a *de novo* program for a new budget year, as opposed to an automatic continuation of the previous year's budget plan.

Point Source, A common exposure of a group of persons to a toxic or infectious agent.

Poison Control Center, A national or regional reference center for emergency information, treatment, diagnosis, and management of poisonings and acute toxicities of all kinds, for clinical emergencies, forensic medical needs, security services, as well as the courts.

Policy Analysis, A process of analysis and revision of policy based on changing population health needs and technology in order to maintain effective and efficient use of resources.

Pollutant, Any substance that renders the atmosphere or water foul or noxious or a health hazard.

Population at Risk, The specific population group with higher than average risk for a specific disease or for all diseases, or the population base for whom a specific service program is developed.

Population-Based Health Services, Services directed toward all members of a community such as in a district health program, or for a specific population subgroup, such as infant and child immunization.

Population Density, Population per square kilometer.

Population Pyramid, A graphic representation of the age and sex composition of a population. It is constructed by computing the percentage distribution by age and sex group.

Post-Neonatal Mortality Rate, The number of deaths that occur from 28 days to 365 days after birth per 1000 live births.

Preferred Provider Organization (PPO), Association or organization of physicians and hospitals that contract with managed care organizations, employers to provide care on a fee-for-service basis for enrolled beneficiaries.

Premature Mortality, Deaths from specific and all causes, and the years of life lost by such deaths occurring before the life expectancy of a population, defined as the average years of life expectancy in a low mortality population; used by the World Bank in combination with assessments of disability used to calculate the total burden of disease.

Premiums, Payments by the beneficiary or on his behalf for health insurance.

Prepaid Group Practice (PPGP), Group medical practice coupled with prepayment with subscriber selection based on voluntary choice; precursor of health maintenance organizations.

Prevalence (Rate), The total number of all individuals who have a disease, a health-related event (e.g., motor vehicle accident) or attribute (e.g., smoking), including new and previous cases, in a given population at a designated time (*point prevalence*) or time period (*peri-od prevalence*), expressed usually as a rate per 1000 persons during a year.

Prevention, Any activity or approach that avoids or reduces the likelihood of disease (*primary prevention*), prevents it from getting worse or having complications if it does occur (*secondary prevention*), or which promotes maximal functional recovery (*tertiary prevention*).

Primary Health Care (PHC), As primary medical care, includes essential care at the first level of contact with the health system. As defined by the WHO in the Declaration of Alma-Ata, PHC includes a wide range of health education, prevention and treatment of common diseases, proper nutrition, safe water and environmental health, maternal and child health, control of communicable diseases and immunization, and provision of essential drugs.

Private Health Insurance, Health insurance operated by nongovernmental organizations for individual and group subscribers, on a nonprofit or for-profit basis, subject to regulation under state insurance and health law.

Private Practice, A physician or other health care provider working independently or as part of a partnership providing services to patients who agree to pay for the service or on contract to a sick fund or insurance agency to pay for the services provided to a beneficiary.

Private Voluntary Organization (PVO), Nongovernmental organizations with the legal status of a not-for-profit corporation entitled to engage in providing health care, to hold assets and properties, and to hire staff or contract for services.

Procedure, A surgical or nonsurgical special treatment ordered by the attending physician and recorded in the medical record of the patient discharged from an inpatient short-stay hospital, including medical (e.g., chemotherapy), surgical, or other special procedures (e.g., radiation therapy for cancer).

Professional Standards Review Organization (PSROs), Peer review organizations to review, assure and improve standards of performance in medical and hospital practice and to restrain cost increases.

Prospective Pricing System, Methods of paying for health services in which the payment rate is determined in advance. It includes paying for hospital care by diagnosis related groups (DRGs) for medical care for competitive selection of managed care organizations. These are now standard practice in the United States and becoming widely used in other countries.

Provider, General term for health care facilities and professionals who give diagnostic, treatment, or other care to patients.

Public Administration, Administration of a health program by a governmental agency.

Public Sector, The part of the economy of a country that comes within the scope of federal, state, or local government, or public corporations.

Quality-Adjusted Life Years (QALYs), A unit of measurement reflecting both quantity and quality of life gained from health programs. It measures life expectancy adjusted according to changes in quality of life, measured by two or more aspects of health, such as pain, disability, mood, or capacity to perform self-care or socially useful activities such as housework or paid employment.

Quality Assurance (QA), All measures taken to ensure and improve the quality of care in a health care facility or system.

Quarantine, Restriction of activities of well persons or animals who have or have been exposed to a communicable disease during its period of communicability to prevent transmission during the period of infectivity.

Rapid Assessment Procedure, Application of common anthropological techniques in the study of health issues, based on interviews to provide data for decision making.

Rate, A measure of the frequency of a phenomenon in a defined population in a defined time period. This can be presented as a ratio (fraction).

Recurrent Costs, Costs that recur in running an enterprise, such as salaries and supplies, also known as *operating costs*.

Reform, The process of change in health systems organization, financing, total supply of resources (e.g., hospital beds, manpower), allocation of administrative responsibility, and financial resources.

Regional Hospitals, Hospitals located in urban regional centers that provide primary hospital care to residents of part or all of the city, and secondary care to people from the city and the surrounding communities and health service region.

Regulation, Government mandated standards, supervision and licensing of food, drugs, medical devices, health facilities, insurance systems, businesses, restaurants, recreation facilities, or anything that might endanger the public interest in health or safety.

Rehabilitation, Restoration to the fullest degree of physical, mental, social, vocational, and economic usefulness of which an individual is capable following disabling illness or injury.

Reimbursement, Methods of paying for doctor or other health services, including salary, fee-for-service, capitation, or a mix thereof.

Relative Risk (Risk or Odds Ratio), The ration of risk of disease or death among the exposed, as compared to that of an otherwise comparable nonexposed population group.

Replacement Rate, The fertility rate at which births exceed deaths in a defined population.

Reservoir, The natural habitat where an infectious agent lives and multiplies, from which it can be transmitted directly or indirectly to a human host.

Resources, The inputs used to produce and distribute goods and services; these include land and buildings, financing, labor (people), and capital (equipment for diagnosis and treatment; it also includes patient's time).

Respite Care, Temporary care of chronically ill or disabled persons who normally reside at home and are dependent on family members or friends for continuous or intermittent care.

Restructuring, Reform in a health service that changes the quantitative balance in resources personnel and program emphasis from one sector to others, e.g., from hospital inpatient care to ambulatory and home care.

Risk Approach, The planning of services and interventions based on the highest risk groups in the population, described as "something for all, but more for those in need—in proportion to that need".

Risk Factors, Those characteristics or behavioral patterns known to increase the risk of disease.

Risk-Taking Behavior, The phenomenon of people engaging in unhealthy or potentially injurious behaviors, despite knowledge of the risks involved.

Sampling, A sample is a subset of a population selected to be as representative as possible of the total population. A sample may be random or nonrandom, representative or nonrepresentative. The major categories include: *cluster sample,* a group of persons not individually selected, i.e., all persons in a city block; *grab sample,* a simple survey among people who happen by or show up at a service offered, such as a street fair, from which no general conclusions can be drawn; *probability (random) sample,* where all individuals have an equal or known chance of being selected, or if stratified, subgroups may be assigned greater weight in the design; *simple random sample,* all persons in the group are assigned a number and the selection of the sample is according to a random numbers table, until the needed sample size is achieved; *stratified random sample,* where the population is divided into subgroups, and each of these is sampled randomly; and *systematic sample,* where the sample is selected on the basis of a predetermined method, such as alphabetic order or birth dates.

Sanitation, The sum of activities related to improving the physical environment by setting and enforcing water standards as well as those related to sewage disposal and treatment, garbage collection and disposal, waste management to maximize recycling, reduction of ambient air pollution, and conservation of the environment.

Screening, The presumptive identification of unrecognized disease or defect by use of tests, examinations, or other procedures applied to separate apparently health persons who probably have some disease process from persons who do not. *Mass screening* involves a whole population. *Prescriptive screening* aims at early case finding in presumably healthy individuals in specific risk groups, e.g., newborns for birth disorders. *Multiphasic screening* involves a variety of tests carried out on the same occasion, e.g., in middle-aged women and men.

Secondary Care Facility, A hospital or other facility that offers services to a community or district that include medicine, general surgery, obstetrics and gynecology, and pediatric services as a minimum, supported by adequate diagnostic and support facilities to provide high quality care, and with possibilities for referral to tertiary care centers for more specialized services.

Selective Primary Care, Focusing programs of prevention or treatment of common conditions to those issues selected as being the most common and most health damaging, with a high benefit-to-cost ratio.

Self-care, The element of personal responsibility in health and the methods of improving one's own health outside of the professional network of services.

Sensitivity, The proportion of persons truly having a disease in a population screened who are identified as diseased by the screening test. A test with high sensitivity will be positive whenever the condition is present, i.e., 100% sensitive, whereas a test of low sensitivity will give some false negative results.

Sentinel Center, Convenient but generally representative points of service where monitoring of cases presenting, or the population served, provides an index of events of public health importance.

Sentinel Event, A health related condition or event indicating change in the health level of a population, indicated by reported morbidity, mortality, or health service utilization data.

Sexually Transmitted Disease (STD), Any disease transmitted through sexual contact.

Sick Funds, Health organizations that insure and provide medical care to their members.

Sin Taxes, Taxes on cigarettes, tobacco, and alcohol as activities that are antihealth and are increasingly defined as antisocial or risk factors for disease and injury for the user and for others.

Social Class Classification, A population classification system used to reflect socioeconomic stratification of a society, useful for epidemiologic purposes. The British Registrar-General's Classification of Occupations is used on death certificates and used to study health experi-ences of social classes even within a universal access health system.

Social Cost, The cost of an activity or disease to society and not merely the agency carrying out the activity.

Social Diagnosis/Policy, Assessment of factors influencing the quality of life, and promotion of changes to raise those standards.

Social Pathology, Failure of adjustment leading to antisocial acts that are destructive to others or to individual or community property. The term is also applied in a community setting when a segment of the community is alienated and may be subject to or cause physical and social tension or violence in the community, and suffer from excesses of acute and chronic disorders.

Social Security, Nationally legislated systems of providing social benefits such as health insurance, and compensation to the individual for injury or illness related to work, pensions, survivor's benefits, health insurance, and pensions for orphans, widows, and disabled persons; a national system of mandatory insurance, financed by payments of the employee and employer, in parallel to income tax.

Social Support System, At the level of an individual, the family, friendship, and work colleague support system that helps an individual cope with stress and adaptation needs in life. At the community or national level, this refers to legislative and other support systems providing welfare and pensions as well as other programs (e.g., WIC food support) to prevent people from falling below minimal standards of living.

Social Welfare, This involves programs of governments at all levels to assist persons in distress with sustenance of basic needs by financial assistance, entitlement to food purchasing, or special entitlement to health services where these are not a universal right or entitlement.

Standardization, Statistical techniques used to remove as far as possible the effects of differences in age, or another confounding variables, when comparing two or more populations.

Standardized Mortality (Morbidity) Rate (SMR), The ratio of the number of events (e.g., deaths or diseases) observed in the study group or population compared to the number that would be expected if the study population had the same specific rates as the standard population, multiplied by 100.

Standards of Practice, Practice standards are the professional level practiced by the average provider in the category of service being rendered and is that which is expected of a practitioner in that field. Professional associations, health insurance systems, and managed care plans provide leadership in evidence-based practice standards and education of members to reach and maintain those standards.

Staple Food, A food that is regularly consumed in a country or community from which a substantial portion of total dietary energy is obtained.

Statistical Significance, Methods to estimate the observed degree of association between independent and dependent variables in a study group and a comparison group. Statistical tests help to indicate the degree of probability that the observed differences occurred by chance alone, or by a possibly causal relationship between the variables.

Strategic Planning, Process by which an organization defines its long range goals and selects activities for achieving them.

Stress, Response to challenge ("fright–fight–flight") of changes in the status quo of an individual causing overt or hidden psychological pressure that may manifest themselves in overt psychological symptomatology or in physical illness.

Stroke, Cardiovascular disease caused by a thrombus (blocking arterial blood supply) or bleeding into brain tissue, causing paralysis and other brain damage such as loss of speech.

Stunting, Moderate and severe growth retardation in linear growth, below minus two standard deviations from median height-for-age of the standard international reference population.

Subclinical, Predisposition for disease that develops gradually in terms of risk factors or biological changes preceding the onset of clinical symptoms and signs; inapparent on medical examinations.

Substance Abuse, Impaired social or occupational functioning caused by pathological use of alcohol or drugs, continuously or periodically, over a period of 1 month or more.

Supplementation, Providing nutritional supplements of specific vitamins or minerals to prevent deficiency conditions in vulnerable population groups, and to promote optimal health status.

Supply Induced Demand, Availability of a service increases its utilization.

Surveillance of Disease, The continuous scrutiny of all aspects of occurrence and spread of disease, in order to determine methods of control, e.g., infectious diseases, cancer, birth defects.

Susceptible, A person with insufficient resistance or with associated risk factors to a particular pathogenic agent or process, so that there is real danger of this person contracting the specific disease if or when exposed to the agent.

Taxes, Sources of funds for a government; includes progressive personal income tax, excise tax, resources tax, property tax, "sin" tax, social security tax, value-added tax, and others.

Technology Assessment, Measuring the cost-effectiveness and feasibility of application of existing and new methods of providing preventive and curative services.

Technology Transfer, The process of promoting the wide application of scientific discoveries, methods, procedures, techniques, and equipment that may promote health and socioeconomic development.

Terms of Reference, The definition of objectives and the task of a working group or committee established to carry out an investigation, review, or development of a task assigned by an organization.

Tertiary Care Facility, A hospital or other facility that offers a specialized, highly technical level of health care, associated with teaching programs and serving the population of a large region.

Total Fertility Rate, The number of children that would be born to a woman if she were to reproduce to the end of her childbearing years and bear children at each age according to prevailing age specific fertility rates.

Total Quality Management (TQM), A management approach in a production or service industry promoting a sense that employees at all levels feel part of a common effort to produce the highest quality of goods or services with customer or consumer satisfaction as the measurable end point.

Tracer Conditions, Indicator conditions or health states that are easily diagnosed, reasonably frequent, whose outcomes are believed to be affected by health care, and that reflect a wide array of patients and health problems seen in medical practice; used for peer review and also called *sentinel events.*

Traditional Birth Attendant (TBA), A person without formal training who delivers babies in the community, and who learned by an apprenticeship or self-taught experience.

Transmission, Any mechanism by which an infectious agent is spread from any source or reservoir to a person. *Direct transmission* is by contact as in touching, biting, or sexual intercourse, or by droplet transmission at close quarters such as by sneezing, coughing, or spitting. *Indirect transmission* is in vehicle-borne contact with contaminated inanimate objects such as bedding handkerchiefs, instruments, water, food, milk, blood, or sputum. *Vector-borne transmission* is via an insect in which the organism may or may not multiply. *Airborne transmission* is by dissemination of microbial aerosols or dust particles suspended in air (more than 1 meter).

Under-Five Mortality Rate, The probability of dying between birth and 5 years of age per 1000 live births.

Undernutrition, A condition arising from inadequate intake of food and micronutrients. Inadequate energy intake results in reduced body weight as its principal manifestation, but in children it can result in stunting (low height-for-weight) and wasting (low weight-for-height). Inadequate micronutrients result in a variety of deficiency states with individual and community health importance, such as iodine deficiency diseases.

Underweight, *Moderate,* below minus two standard deviations from median weight-for-age of the reference population; severe, below minus three standard deviations from the median weight-for-age of the standard reference population.

Urban Population, Percentage of the population living in urban areas according to the national definition used in the most recent census.

User Fees, A charge for a medical or other health service at the time of service in addition to the coverage the patient has from health insurance.

Vaccine, Immunobiological substances used for active immunization by introducing into the body a live, attenuated, inactivated, or portion of an infectious organism or its toxin. The vaccine is capable of inducing an immune response in the host, and as a result, the host becomes resistant to the infection from that specific organism.

Vaccine-Preventable Diseases (VPDs), Diseases preventable by currently available vaccines.

Variables, A variable is an attribute, phenomenon, or event that can have different values. An *independent variable* is the factor suspected of the causal relationship with the *dependent variable* or the outcome effect being studied. A *confounding variable* is a parallel factor that may cause or hinder the outcome being studied and unless adjusted for in an epidemiologic study would obscure the potential for defining the relationship of the independent and dependent variables.

Variation, Observer, Individual, Variation or error due to a failure of the observer to measure or identify a phenomenon accurately. All observations are subject to variation, including between the same observer (*intraobserver variation*) and between different observers (*interobserver variation*).

Vector, An insect or any living carrier that transports an infectious agent from an infected animal or human to a potentially susceptible individual or its food or immediate surroundings. The organism so transported may or may not undergo development during transportation.

Ventilation, The movement transfer of outside air into a building and of inside air to the outside, and the control of temperature, humidity, and internal movement of air.

Virulence, The degree of pathogenicity of an infectious agent; the disease-invoking power of a microorganism in a given host; it may be indicated by case complication or fatality rates, and/or its ability to invade and damage tissues of the host.

Vital Records/Statistics, Certificates of births, deaths, marriages, divorces, or immigration/emigration required for legal and demographic purposes. The statistics derived from vital records provide tabulated information based on registration of these vital events.

Wasting, *Moderate wasting* and *severe wasting* denote two or three standard deviations below the median for weight-for-height of the reference population respectively (see Undernutrition).

Waterborne Disease, Disease transmitted by contamination of community or non-community water supplies.

Weaning Foods, A food contributing to the nutritional needs of the infant during the weaning period in which the infant or child is in transition from breast-feeding to other feeding.

Wellness, Improving quality of life and physical, mental, and social well-being.

Women of Reproductive Age, All women in the childbearing age group, i.e., 15–49 years.

Women's Health, Specific health issues of women, including but not exclusively those related to reproduction, risks and benefits involved with pregnancy and childbirth, abortion and contraception, management of menopausal problems, and screening for common female cancers. Violence against women and children, STDs, chronic disease manifestations in women and related field of nutrition, osteoporosis, and many other issues of public health concern come under this grouping.

World Bank, The International Bank for Reconstruction and Development, Washington, D.C. This bank was established following World War II by the industrialized countries and has become a key source of loans for development of most of the countries in the United Nations.

World Development Report (WDR), The annual publication of the World Bank relating to international development. The 1993 WDR, entitled *Investing in Health,* examined the relationship between human health, health policy, and economic development.

World Health Assembly, The annual meeting of the WHO is held in Geneva to discuss current international health issues and to stimulate member countries to increase their priority and resources allocated to health.

World Health Organization (WHO), An United Nations affiliated agency established in 1946 currently with representation from 189 countries, which serves as the international forum for cooperation in health. The WHO has six regional offices that carry out regional assessment and cooperative activities to promote health.

World Summit for Children, The World Summit for Children held in 1990 attended by 71 heads of state and signed by an additional 71 heads of state committed their countries to national plans of action to deal with primary health care, family planning, safe water, environmental sanitation, nutrition, and basic education.

Years of Potential Life Lost (YPLL), Calculated over the age range from birth to 65 years. The number of deaths for each age group is multiplied by the years of life lost (the difference between 65 and the midpoint of the age group). For example, the death of a person aged 15–24 counts as 45 years of lost life. Then the years of potential life lost are summed over all age groups.

Zoning, Restriction on types of buildings that can be constructed or operated in specific areas of a community, so as to reduce exposure to nuisances or health risks to a residential population.

Zoonoses, Diseases transferred to humans by vertebrate animals under natural conditions, in contrast to disease transferred by arthropods (insects, ticks, spiders) or other invertebrates; may be epizootic or periodic, or enzootic, i.e., endemic.

REFERENCES

Benenson A. S. (ed). 1995. *Control of Communicable Diseases,* Sixteenth Edition. Washington, DC: American Public Health Association.

Centers for Disease Control. 1992. A framework for assessing the effectiveness of disease and injury prevention. *Morbidity and Mortality Weekly Report,* 41(RR-3):1–12.

Donaldson R. J., Donaldson L. J. 1993. *Essential Public Health Medicine.* Dordrecht: Kluwer Academic Publishers.

Green, L. W. 1990. *Community Health,* Sixth Edition. St. Louis, MO: Times Mirror/Mosby College Publishing.

Last J. M. (ed). 1995. *A Dictionary of Epidemiology,* Third edition. New York: Oxford University Press.

UNICEF. 1999. *The State of the World's Children, 1999.* New York: Oxford University Press.

National Center for Health Statistics. Health, United States, 1996–97 and Injury Chartbook. Hyattsville, MD: U.S. Department of Health and Human Services.

National Center for Health Statistics. Health, United States, 1998 with Socioeconomic Status and Health Chartbook. 1998. Hyattsville, MD: US Department of Health and Human Services.

World Bank. 1993. *World Development Report 1993: Investing in Health.* New York: Oxford University Press.

World Health Organization. 1984. *Glossary of Terms Used in the Health for All Series.* Geneva: World Health Organization.

World Health Organization. 1999. *World Health Report 1998: Life in the 21st Century: A Vision for All.* Geneva: World Health Organization.

SUBJECT INDEX